pages 595 & 596
not printed

History of Tribology

History of Tribology

D. Dowson B.Sc., Ph.D., D.Sc., C.Eng.F.I.Mech.E.,
Fellow A.S.M.E., Honorary Member A.S.L.E.

Professor of Engineering Fluid Mechanics and Tribology
and Director of the Institute of Tribology, University of Leeds.

Longman
London and New York

To the memory of Stephen Paul Dowson (3.10.1956–10.10.1968)

Longman Group Limited London

Associated companies, branches and representatives throughout the world

Published in the United States of America by Longman Inc., New York

© Longman Group Limited 1979

First Published 1979

Library of Congress Cataloging in Publication Data

Dowson, D
   History of tribology.
   1. Tribology—History.   I. Title.
TJ1075.D65   621.8'9   77-13054
ISBN 0-582-44766-4

Contents

Appendix   Men of tribology   529

Preface

In 1969 I started to write a textbook on tribology. The first chapter was to be devoted to the history of the subject, not only because this seemed to be the logical and traditional way to start, but because it reflected my long-standing interest in the relationship between our existing state of knowledge and the achievements of the past. Furthermore, I had acquired through my teaching of lubrication theory and aspects of the mechanics of machines to undergraduates since 1955 and tribology in its broadest sense to M.Sc. course students since 1964, a conviction that engineering students also found it valuable to see the technical aspects of their subjects in a historical setting. Experience has taught me that engineering students, contrary to popular belief, are genuinely interested in the interactions of technical, scientific, social and political developments.

As I formulated the opening chapter of my book on tribology I became dissatisfied with the superficial and repetitive narratives which formed such introductions. I had already acquired and read many of the original papers on the subject, but I gradually became interested in the background factors which promoted and influenced the work, as well as in the technical merit of the content. Furthermore, I soon realized that fascinating and significant developments in the field took place long before recorded history, in archaeological rather than historical times. This personal awareness coincided with a common and most enjoyable growth of interest in archaeology in general in my family and it soon became clear that the original outline chapter was totally inadequate. At this stage the concept changed and the proposed chapter became a book! It is of course no longer a conventional textbook, but I was much encouraged in treading the new path by the enthusiasm of colleagues, students, friends and the publishers. In particular, I was impressed by the interest and encouragement of a large number of fellow teachers and research workers in tribology.

It has been a privilege and a source of inspiration to prepare this text as a member of a university with a tradition of research and teaching in friction, wear, lubrication and bearing design dating back to the last century. Most of the manuscript was prepared in Leeds, but some sections were written in other European countries, the United States, Canada and Australia.

Delving into historical and archaeological records in any subject can be fascinating yet frustrating, and is almost always a great consumer of time. The process has drawn me into several unfamiliar sections of libraries, while many pleasant hours have been spent in museums in the United Kingdom and overseas. Throughout the years I have endeavoured to ensure the accuracy of relevant historical and archaeological facts, since this must form the foundation of all such studies. It has sometimes taken weeks, months and even years to unearth a single detail, but the exhilaration of finding the original work or that missing piece of the jigsaw of progress, has always justified the effort.

If readers find errors of fact or additional information I shall be most pleased to hear from them. The research will no doubt continue for many years, and although there is always scope for different interpretations of evidence, it is as well to be sure of the extent and validity of the original information.

It is entirely accidental, but nevertheless noteworthy, that this book is being published about 100 years after the Institution of Mechanical Engineers created its Research Committee. That Committee, formed on 20 December 1878, gave birth to the 'Committee on Friction at High Velocities', which in turn appointed Mr Beauchamp Tower to undertake his now famous experimental work on journal bearing friction and lubrication. It is a good time to take stock and to review progress, or in some cases the lack of it, in the past century.

I have been most fortunate to experience the twin joys of research in both the subject and the history of my chosen field in an aspect of science and engineering which is attracting much interest at the present time. I hope that those who continue to unravel the mysteries of tribology in the future will find in the present text the incentive and encouragement to explore and appreciate past achievements.

Adel   Duncan Dowson
Leeds
Christmas 1977

Acknowledgements Throughout the long period of collecting and arranging material for this text I have received much assistance and encouragement. A number of specific acknowledgements are made throughout the text, but I would like to take this opportunity to thank everyone who has helped to trace relevant historical and archaeological information and those with whom I have held many valuable and stimulating discussions.

Librarians have been helpful, patient and encouraging and I would particularly like to express my appreciation of the service offered by: the Brotherton and South Libraries of the University of Leeds; the Lending and Reference Sections of the City of Leeds Library; the National Lending Library; the British Library; the Library of the Institution of Mechanical Engineers; the staff of the American Societies of Mechanical and Lubrication Engineers; the Library of the Victoria and Albert Museum; the Imperial College of Science and Technology Lyon Playfair Library; the staff of the Novosti Information Service; the Institut de France Académie des Sciences; the Conservatoire National des Arts et Métiers; the Biblioteca Reale, Turin; the Universitätsarchiv of the Technische Universität, Dresden; and the Royal Library, Windsor Castle.

The staffs of several museums and keepers of special collections have also been most generous with their time. In particular I would like to recognize the assistance of the British Museum and the Science Museum, London; the Bridewell Museum, Norwich; the Smithsonian Institution, Washington; the Egypt Exploration Society, London; the Royal Scottish Museum and the National Museum of Antiquities of Scotland, both in Edinburgh; the Prehistoric Society, London; the Nationalmuseet, Copenhagen; and the Antiquarian Horological Society, London.

On several occasions it was necessary to obtain translations of original and often ancient texts and I was fortunate to be able to turn to my colleagues Professor Derek Bradley, Mr Brian Jobbins, Mr John Schwarzenbach and Dr Richard Wakelin for this assistance. Many years ago Mrs Margaret Bradley kindly translated a good deal of Petrov's work for me, and this proved to be particularly useful during the present exercise. Professor Max White and Mrs Juliet White translated substantial sections of Italian and early French texts and Mr G. K. Orton of the Department of Scandinavian Studies in the University of Hull provided a translation of Danish work.

I have enjoyed contacts with a number of individuals in many countries throughout the preparation of this text. Their prompt response to correspondence and their willingness to share their knowledge has been of immense value. Professor Robert Courtel, Directeur de Recherches au C.N.R.S. and Professor Maurice Godet of the Laboratoire de Mécanique Des Contacts, Institut National Des Sciences Appliquées De Lyon, provided valuable background information on the substantial contributions to tribology from France. Generous contributions were also received from Professor Dr-Ing. Horst Czichos of the Bundesamstalt Für Materialprüfung, Dr George Graue of Gesellschaft Für Tribologie e.v., Professor Dr G. H. Göttner of the Abteilung Materialprüfung und Tribologie, Institut Für Erdölforschung, Hannover, and Professor Dr-Ing. Jörn Holland of the Institut für Reibungstechnik and Maschinenkinetik, der Technischen Universität Clausthal, Germany. The late Professor Dr-Ing. George Vogelpohl's personal interest in the subject of the text was greatly appreciated.

On several occasions Professor Harmen Blok of the Laboratory for Machine Elements and Tribology of the University of Technology, Delft, Holland, kindly offered his advice, unearthed facts which were unknown to me and prepared copious notes on specific topics of tribological significance. His kindness and assistance is gratefully acknowledged. The late Professor Ladislao Reti of Monza, Italy, whose authoritative writings on the work and life of Leonardo da Vinci will be known to many, was most helpful in providing material on the Renaissance period. I hope that Chapter 7 goes some way towards meeting his view that Leonardo deserves a leading place in the history of tribology.

I was fortunate to talk to Professor F. Hirano of Kyūshū University, Japan, about the contributions of the East and from across the North Sea, Professor Helge Christensen, of SINTEF, Trondheim, Norway, Professor Bengt Jakobsson of Chalmers University of Technology, Göteborg, Sweden, and Professor Bo Jacobson of the University of Luleå, Sweden, kindly provided their interpretations of a number of developments.

Mr Frank R. Archibald of Needham, Massachusetts most generously sent me his personal collection of historical notes on famous contributors to the science of fluid-film lubrication. Mr E. D. Brown, Jr, tribologist with the General Electric Company, Waterford, New York, was equally generous and enthusiastic, particularly in drawing my attention to the exploits of Count Marin Carburi of Cessalonia in the late eighteenth century. Professor Jack F. Booker and his wife Barbara kindly provided background notes on the contributions to tribology of earlier workers at Cornell University and Darle W. Dudley of the Solar Division of the International Harvester Company advised me on matters relating to the history of gearing. Professor C. Stewart Gillmor of the Department of History in the Wesleyan University, Middletown, Connecticut, was most helpful when I was studying the contributions of Charles Augustin Coulomb. I have enjoyed general encouragement in this venture from the late Mr P. M. Ku of the Southwest Research Institute, San Antonio, Texas, and I am particularly grateful for his enlightenment on the development of Chinese technology in the seventeenth century. Mayo D. Hersey's writings on the history of lubrication, and particularly his book, *Theory and Research in Lubrication*, have provided most valuable sources of information, which have been supplemented by personal communications. During the preparation of the manuscript Professor Fred F. Ling of Rensselaer Polytechnic Institute, Troy, New York, who first referred to the exercise as project H.O.T.!, Professor Jack F. Booker of the College of Engineering, Cornell University and Dr Bernard J. Hamrock of NASA, Cleveland, Ohio, each spent one year working in the Institute of Tribology in the University of Leeds. Their interest and advice is greatly appreciated.

Professor I. V. Kragelskii, Head of the Laboratory of Friction and Friction Materials in the Institute of Mechanical Engineering, Moscow, kindly provided me with a copy of his book, written with V. S. Shchedrov, *Development of the Science of Friction*, which forms a most valuable source book on work in this field. His kindness and help are appreciated. Mr George Kinner and Mr Vic. H. Brix kindly translated parts of this text for me.

In the United Kingdom I have talked and posed questions to a number of colleagues and fellow tribologists. Some have been approached many times and I would particularly like to thank: Professor Alastair Cameron of Imperial College; Professor Gordon R. Higginson of the University of Durham; Professor Kenneth L. Johnson of the Engineering Department, the University of Cambridge; Dr H. Peter Jost of Angel Lodge Laboratories; Dr Philip B. Neal of the University of Sheffield; Mr Michael J. Neale of Michael Neale & Associates, Farnham, Surrey; Professor David Tabor of the Cavendish Laboratory, Cambridge; and Dr G. H. West of the Department of Polymer and Fibre Science in the University of Manchester Institute of Science and Technology. Mr A. J. S. Baker of the Esso Research Laboratories, Abingdon has shown interest and provided welcome encouragement throughout the preparation of the text.

During the preparation of brief accounts of the development of the early plain and rolling-element bearing companies I received willing support and cooperation. In particular, I would like to thank Mr F. A. Martin of the Glacier Metal Company; Mr D. F. Green of Vandervell Products Ltd; Mr R. S. Gregory of Kingsbury, Inc.; Mr S. D. Advani of Vickers Limited Engineering Group (Michell Bearings); Mr T. J. Rosinski of the Fafnir Bearing Company; Dr T. E. Tallian, Dr I. Fernlund and Dr A. Palmgren of SKF; Dr S. Y. Poon of RHP; Mr D. E. Eagon, Jr of the Timken Company and Mr E. F. Instone of British Timken.

Three people read and criticized the draft manuscript and their helpful comments have shaped the final text. This was a lengthy and onerous task and my sincere appreciation is extended to: Mr Donald F. Hays of the General Motors Corporation, Michigan; Professor Harry Naylor of the Thornton (Shell) Research Centre; and Dr Christopher Malcolm Taylor of the Institute of Tribology in the University of Leeds.

The manuscript was typed and prepared in excellent style by Mrs

Sheila Moore. Her interest in the project was a great encouragement. Mr Ronald T. Harding and Mr Michael C. Wigglesworth helped me with some of the drawings and Mr Michael Reynolds demonstrated great skill in photographing faded original plates and pictures in ancient manuscripts.

I have enjoyed the fullest possible encouragement from my family during the preparation of this text. My wife Mabel and son David Guy not only showed interest, but actively participated in seeking material and visiting museums and libraries as the research and writing progressed. I offer my deepest appreciation of their support.

The publishers and editorial staff have been most helpful and I am pleased to acknowledge their encouragement and skill.

Finally, I would like to thank all those undergraduate and postgraduate students whose interest and response to the subject of the text over many years has been a constant inspiration. I can only hope that they have learned as much from me as I have learned from them.

The Publishers are grateful to The Controller of Her Majesty's Stationery Office for granting permission to reproduce patent material, and to the copyright owners for permission to reproduce the following illustrations: *Acta Archaelogia*, (Copenhagen) for Figs. 5.15 and 5.16 (from Klindt-Jensen, 1949); Agricultural Research Council for App. 17; The American Society of Mechanical Engineers for Figs 10.19, 10.30(a) and App. 9; F. R. Archibald for App. 18; Edward Arnold (Publishers) Ltd for App. 11, from *John William Strutt: Third Baron Rayleigh* by Lord Rayleigh (1924); Blackie and Sons (Publishers) Ltd for Fig. 10.20, from *Lubrication—Its Principle and Practice* by A. G. M. Michell (1950); Robert Bosch GmbH, Stuttgart, for App. 14; The British Library Board for Figs 4.3(a), 6.1, 6.8, 7.1(a) and (b), 7.3 (from the *Codex Atlanticus*), 7.14, 7.15 (from Besson, 1569), 8.1, 8.2 and 8.3 (from Zonca, 1607), Cambridge University Press for Fig. 5.17, from *Science and Civilisation in China* Volume 4 by J. Needham (1965); Constable & Co. (Publishers) Ltd for Fig. 4.4(a) and (c) (from Newberry, 1900); Conservatoire National des Arts et Metiers, Paris, for App. 5(b); C. A. Coulomb for App. 5(a); Det Kangelige Danske, Videnskabernes Selskab for Fig. 5.6 (from Drachmann, 1932); Dean and Chapter of Wells Cathedral for Fig. 6.3; J. M. Dent and Sons Ltd (Publishers) for Fig. 9.32(b), from *The Story of Sprowston Mill* by H. C. Harrison (1949); Dover Publications Inc. for Figs 7.9 and 7.10, from their 1950 edition of Agricola's *De Re Metallica* (1556); Egypt Exploration Society for Fig. 4.15; *The Engineer* for Figs 10.2 (1944a), 10.6, 10.7, 10.8 (1912), 10.39 and 10.40 (1878); *Engineering* for Apps 7 (1890) and 15 (1935); Fafnir Bearing Company, Connecticut, for Fig. 9.11; Gregg International (an imprint of Avebury Publishing Company) for Figs 7.16, 7.18, 7.19, 9.25 and 9.26, from their 1970 edn of Remelli's *Le Diverse et Artificiose Machine* (1588); D. F. Hays for App. 24; App. 2(b) is reproduced by gracious permission of Her Majesty Queen Elizabeth II; Institute of Civil Engineers for Fig. 10.41, from 'Recent researches in friction' J. Goodman, Proc. *ixxxv*, Session 1885–6, Pt III, 1–19; Institution of Mechanical Engineers for Figs 8.20, 10.3, 10.9, 10.10, 10.11, 10.12,

10.13, 10.18, 10.28, 10.29, 10.42, 10.43 and 10.44, Apps 12 and 19, and cover illustration; I.C.I. Ltd for Figs 3.1, 4.3(*b*), 4.5, 4.6, 5.3 and 7.13; Instituto Poligrafico Dello Stato, Liberia Dello Stato, Rome, for Figs 5.8, 5.9, 5.10, 5.11, 5.12, 5.13 and 5.14, from *Le Navi Di Nemi* by G. Ucelli (1950); H. P. Jost for Fig. 11.2; The President of Kingsbury Inc., Philadelphia, for Fig. 10.31 and App. 16; The University Library, Leeds, for Figs 8.4, 8.14, 8.15, 8.16, 8.17, 8.19, 9.16, 9.17, 9.18 and 9.19; G. Lerche for Figs 6.6, 6.9 and 6.10 (from Lerche, 1970); Professor Ling for App. 18; Macmillan, London and Basingstoke, for App. 13 from *Miscellaneous Papers by Henrich Hertz*, translated by D. E. Jones and G. A. Schott; Simon Engineering Laboratories, University of Manchester, for App. 10 kindly supplied by Professor J. Diamond; John Murray (Publishers) Ltd for Figs 4.13 and 4.14, from *Discoveries in the Ruins of Nineveh and Babylon* by A. H. Layard (1853); The National Library, Madrid, and the McGraw-Hill Book Company for Figs 7.2(*c*) and (*d*), 7.4, 7.6, 7.8 and 7.7; National Museum of Antiquities of Scotland for Figs 6.4 (from Corrie, 1914) and 6.5 (from Fenton, 1965); National Portrait Gallery for Fig. 8.22; Norfolk Museums Service, Bridewell Museum, Norwich for Figs 9.28 and 9.33(*b*); Novosti Press Agency for Fig. 9.12 and App. 8; the Oriental Institute, University of Chicago, for Fig. 4.10; C. W. Phillips for Fig. 6.7 (from the *Proceedings of the Prehistoric Society*); The Prehistoric Society for Figs 4.7, 4.9 and cover illustration (from Childe, 1951); Biblioteca Reale, Turin, for App. 2(*a*) and cover illustration; John Reynolds, and Hugh Evelyn Ltd for Fig. 9.29, from *Windmills and Watermills* by J. Reynolds (1970); P. Rahtz and E. M. Trent for Figs 9.23 and 9.24; RHP Ltd, Chelmsford, for Fig. 10.36(*a*) and (*b*) kindly supplied by S. Y. Poon; T. R. Robinson for Fig. 6.2; The Royal Society for Figs 8.18, 9.20(*a*) and (*b*), 9.34, 10.14, 10.15, 10.16, 10.17, 10.21, 10.22, 10.25 and 10.26; the publishers of *Schmiertechnik and Tribologie* for App. 22 (from Vol. 23, No. 2); The Director of the Science Museum for Figs 4.4(*b*), 7.11, 7.17, 8.5(*a*) and (*b*), 8.6, 8.7 (from Van Natrus et al., 1734), 9.1(*b*), 9.2(*a*) and (*b*), 9.8 and 9.33(*a*), and cover illustration; *Scientific American* for Figs 4.8, 4.11 and 4.12; University of Sheffield for App. 21 kindly supplied by P. B. Neal; Shell Photographic Service for Fig. 10.1 and cover illustration; S.K.F. Ltd for Fig. 10.37 and cover illustration; Smithsonian Institution, Washington, for Fig. 4.19; Springer-Verlag, Berlin, for App. 20 (from *Jahrbuch der Schiffbautechnischen Gesellschaft*, 1924); Professor D. Tabor for App. 23; Tate Gallery for Fig. 9.32(*a*); the Timken Company for Figs 10.33 and 10.34; Ulster Museum, Belfast, for Fig. 9.22 and cover illustration; The Director of the Victoria and Albert Museum for Fig. 7.12(*a*) and (*b*); G. H. West for Fig. 9.1(*a*); and R. Wailes for Figs 9.30 and 9.31(*a*) and (*b*).

Chapter 1

History is not exclusively chaos or chance: a degree of observable order and pattern, of partially predictable regularity, exists in human behaviour.
'The Social Sciences in Historical Study,'
New York (1954).

Introduction

1.1 Tribology

The subject of this history was first defined in a Department of Education and Science Report (1966) as: '. . . The science and technology of interacting surfaces in relative motion and the practices related thereto.'

The topics enveloped by the new word are in the main ancient and well known and include the study of *lubricants, lubrication, friction, wear* and *bearings*. The word *tribology* is based upon the Greek word *tribos*, meaning rubbing. But why should '. . . interacting surfaces in relative motion', which essentially means the rolling, sliding, normal approach or separation of surfaces, particularly in machines, have exceptionally attracted the attention of engineers, scientists, economists and even politicians in recent years? The eternal answer is that surface interactions dictate or control the functioning of practically every device developed by man to enhance the quality of life through his inventiveness and the utilization of the resources of the physical world.

To seek an explanation of the emergence of the activity in the 1960s is more difficult and, and I hope to demonstrate, probably misguided, since the basic challenge of the subject had been accepted long ago. If an answer is to be found to the reason for the quickening pace of events, it must be sought in the gradual yet inevitable increase in concern for *conservation, reliability* and *efficient utilization* of everything made and used by man in this world of strictly limited resources. It is only now, when we recognize with increasing clarity that one day the mineral oil and fossil fuels will run out, that avaricious man has started to show concern for the consequences. But the full reasons for the immediacy of the study of tribology are more complex, being linked with the wider aspects of history itself, scientific and technological progress, the nature of twentieth-century civilization and even the form of specialization in educational patterns in different countries.

Engineering developments throughout the ages have been hindered by the incapacitating effect of friction, yet our daily lives rely upon the very existence of the phenomenon. Everything that man makes wears out, usually, but not always, as a result of relative motion between interacting surfaces. To illustrate tribology by such statements is to suggest that the subject must have enjoyed a position of rare eminence in the long history of science and technology, yet this is not the case. It was known that lubricants could be used to influence both friction and wear, yet the physical and chemical processes by which this is achieved have been explored relatively recently and are still being revealed to us.

Individual and impressive studies of specific problems related to friction, lubrication and wear, coupled with evidence of great ingenuity in the engineering of bearings have been traced throughout history, yet a recognition of the underlying coherence of these various aspects of surface interactions is a relatively recent phenomenon. This recognition arose at a time of increasing specialization in science and engineering. The natural philosopher, who was in many cases as much an engineer as he was a scientist, had given way to the chemist, physicist, mathematician, materials scientist, the mechanical and other forms of engineers. Yet the subject of tribology calls for contributions from all these specialists. This *multi-disciplinary* nature of tribology is at once its strength and weakness. Students of the subject can gain immense satisfaction from tackling problems requiring the application of fundamental concepts whose roots are found in many of the traditional subjects of science and engineering, yet it is difficult within existing schemes of study for anyone to feel fully in command of a subject dealing with interacting surfaces.

Machines can fail by breakage or by wear; the former being spectacular and sudden, the latter inconspicuous yet insidious. Studies of the bulk properties of both solids and fluids have dominated science and engineering for many centuries and continuum mechanics has become a sophisticated and essential basis for the successful design and construction of machinery. Studies of surfaces are at least as important and generally more challenging than considerations of materials in bulk, while many of the physical, chemical, mathematical and engineering problems of most immediate concern are of great complexity and associated mainly with boundaries, surfaces or interfaces.

It might be argued with demonstrable justification that an engineer knows sufficient about the steady and dynamic properties of the traditional and most of the newer engineering materials to enable him to build structures and make machines which will support known loads without suffering the indignity of breaking. There are, of course, a few spectacular exceptions to such generalizations, and even though the winds sometimes blow down cooling towers, waves wash away oil-drilling rigs, tower-blocks of flats crumble, airframes fatigue and blades fly off turbine rotors, such events are fortunately rare. Compared with his sophistication in matters relating to the '*strength*' of structures and machines, the engineer is in many respects *inadequate* and *primitive* in his approach to problems of *reliability* and the provision of adequate *life* for machinery. An analysis of machine breakdowns, which must surely coincide with the everyday experience of most readers, shows that many of the stoppages and failures of domestic equipment, motor cars, trains, workshop machinery and process plant are associated with interacting, moving parts–the gears, bearings, couplings, seals, cams, clutches and brakes of modern equipment. In short, many of the problems are tribological.

The growing recognition of the need to combine strength and safety of structures and machines with adequate reliability and life is radically changing the approach to engineering philosophy, theory, practice and education. Many of the leading exponents of this shift in emphasis elicit economic evidence to support their views, and there is no doubt that this is valid and of importance in terms of efficiency and political debate. Others prefer to concentrate on the need for reliable operation of complex and expensive machinery in a world in which scarce resources have to be used to the greatest advantage. While not disagreeing with these views, my own fascination with the question arises primarily from the scientific and engineering aspects of the subject. There can be few truly multi-disciplinary subjects offering a greater intellectual challenge. If an understanding of the nature of surfaces calls for such sophisticated physical, chemical, mathematical, materials and engineering studies in both macro- and molecular terms, how much more challenging is the subject of '. . . interacting surfaces in relative motion'. This is the challenge awaiting the student of tribology.

Several books dealing with the science and technology of tribology are now being prepared and academic institutions are showing an increasing interest in the subject. The economic and technological importance of tribology has been demonstrated to the satisfaction of industrialists, economists and many politicians. Professional bodies concerned with tribology are springing up throughout the world and there is now a vast literature on the subject in the form of books, papers and reports.

It is against this background that I have chosen to write this history, in the belief that a recognition of the present position, and perhaps future developments in the subject, can benefit greatly from an appreciation of the achievements and frustrations of the past.

1.2 History

To any engineer or scientist having a normal degree of scientific curiosity and an interest in the influence of social, economic and historical factors upon the development of his subject, there can hardly be a more fascinating study than the history of tribology.

The main theme of the book is the story of the development of both concepts and devices which have had an impact upon tribology throughout the ages. It is a short history, the accounts of many developments being necessarily brief, but I hope that it will encourage the reader to study the works listed in the references, to transpose himself to the period in which the original work was carried out and to reflect upon the impact of the work of those classical contributors upon both technical and social changes in the era.

In teaching this and other subjects to engineering students in the university, I have found repeatedly that there is great interest in the opportunity to study the influence of the past on the present state of knowledge. It is part of the process of seeing science and technology as an integral part of society and culture and the very development of civilization. Contemporary and restricted glimpses of a scientific and engineering subject can be stark, unsatisfactory and perhaps even misleading if they are not viewed against the wider backcloth of history. This book has been written by an engineer, or tribologist, but I hope that any lack of professionalism in the archaeological and historical fields will be seen to be compensated by enthusiasm.

1.3 **Structure of the book**

The history of tribology is told in chronological form from prehistoric times to the present day. The decision to follow this course rather than to trace the development of individual topics like friction or rolling-element bearings throughout the ages in single chapters was not taken lightly. However, by concentrating on the swell of interest in particular topics in each era, it is possible to place developments in perspective and time sequence in a manner which at least provides the opportunity for a broad-brush approach to the subject, which readers often miss in detailed examinations of restricted topics. Each chapter is sub-divided into sections which reflect major progress in the period under review. A reader interested in a single aspect of tribology, such as lubricants, can thus trace the history of his special interest by scanning the sub-headings in each chapter.

There are eleven chapters, together with one substantial Appendix of biographical sketches. The latter was stimulated by a series of pen sketches, *Men of Lubrication*, written by Frank Archibald (Archibald, 1957), which developed into a thirst for knowledge about the wider aspects of the lives of outstanding contributors to the subject of tribology.

Each chapter opens with a quotation, which I hope is both relevant and interesting, notwithstanding the view of that highly independent American lady, Abigail Adams, expressed in a letter addressed to John Adams on 2 April 1777: '. . . I think the author of common sense somewhere says that no persons make use of quotations but those who are destitute of ideas of their own . . .'

Chapter 2 is devoted to a brief and most general chronology of tribology presented alongside major historical, evolutionary and even geological time scales. The summary shown in Table 2.1 has the merit of brevity and neatness, combined with the disadvantage of distortion created by logarithmic scales. Subsequently, each chapter contains its own more detailed chronology of tribology alongside a time scale of general scientific and technical developments and a more selective collection of social and historical events. If the non-tribological content of these chronologies helps to provide a framework of conditions against which progress in tribology can be judged, they will have served their purpose. If the general state of daily life is overlooked, the history of a scientific and technological subject can become little more than a list of events.

Early chapters in the book are based almost entirely upon archaeological evidence; excavation reports, papers in journals, articles and exhibits in museums. At a later stage historical records supplement and in due course overtake the archaeological evidence, particularly after the introduction of movable-type printing in the fifteenth century. From the time of formation of the learned scientific societies in the seventeenth century and the professional engineering institutions in the nineteenth century, the publication of research papers and articles dealing with practical experience has grown at an impressive rate. Accounts of recent progress are thus readily accessible, if bewildering in their number.

In Chapter 3 the sparse but illuminating early examples of prehistoric interest in tribology are noted. Evidence of general technological progress and suggestions of some quite remarkable developments in tribology associated with the early civilizations discussed in Chapter 4 rest firmly upon archaeological evidence. The discussion of classical Greek and Roman times in Chapter 5 also relies predominantly upon archaeological evidence, but this is supported by a few historical records of enormous value and interest. The demarcation between ancient and modern times adopted by many historians is associated with the period from about 1500 B.P. to 500 B.P. This period is reviewed in Chapter 6, where the main features are the support for general technological progress in the development of machinery, transportation, instruments, water- and windmills and agriculture by empiricism in tribology.

In Chapter 7 the impressive, but at times merely suggestive, evidence of Leonardo da Vinci's contributions to tribology has been drawn together and balanced by an account of the hard facts of progress evident in Renaissance industry; particularly in relation to ingenious bearings for machinery, the mining and metallurgical industries, instruments like astronomical clocks, and the potter's trade. The period leading up to the Industrial Revolution discussed in Chapter 8 was dominated, in tribological content, by remarkable advances in handmade bearings for machinery in various industries, coupled with extensive early scientific studies of friction in Germany, Russia and particularly in France and England.

During the Industrial Revolution, considered in Chapter 9, the development of mechanical power-generating equipment based upon water, wind and steam, together with a high degree of sophistication achieved in road transportation and the enthusiastic development of railways, did much to promote bearing development by practical engineers. At the same time further scientific studies of friction were combined with an interesting but somewhat isolated experimental study of wear, the empirical development of liquid- and grease-like lubricants of animal and vegetable origin and the formulation of concepts of viscosity and the foundations of slow viscous flow theory which would one day provide the basis for modern fluid-film lubrication theory.

The period from 1850 to 1925 was dominated by the emergence of mineral oil as the major lubricant and the careful exposure by scientists of the physical and chemical basis of the mechanism of lubrication. This forms the subject of Chapter 10 and special reference is made not only to the scientific studies of lubrication and friction, but also to the development of the vast petroleum industry and the formation of specialist plain and rolling-element bearing manufacturing companies.

One of the fascinating features which emerges from this study is the evidence of *historical coincidence* in the field of tribology. The near-simultaneous discovery of basic knowledge or the development of similar tribological devices in different laboratories and often in different lands at times when communications were poor or even non-existent is intriguing. Three examples will suffice to make the point. Early yet similar forms of rolling-element bearings developed apparently independently, in Greece (battering-rams); Rome (revolving platforms or capstans on ships), Celtic lands (wagon wheels) and in China almost 2,000 years ago. In 1883 Nikolai Petrov in Russia and Beauchamp Tower in England independently, and by different means, concluded that successful journal bearings operated by developing a coherent film of lubricant between the bearing and the journal. Michell in Australia and Kingsbury in the United States independently invented the tilting-pad bearing, one of the most remarkable machine elements ever produced, within the first decade of the twentieth century.

I chose to indulge my interest in the history of tribology in my Inaugural Lecture in the University of Leeds (1969) in February 1968, and I am grateful for the opportunity to develop this theme in the present text. I had intended to end with the year 1925, by which time much of the basic work on lubrication had been completed, working rules for allowances for friction were generally adequate and the more recent explosion of scientific studies of the basic elements of tribology had barely begun. I once heard that '. . . a wise historian usually stops 20–30 years before his time, because he cannot see the wood for the trees'; the so-called thirty-year rule. My decision to close with Chapter 10 would have represented but a small extension of these guidelines to a fifty-year rule. But the word *tribology*, with all its implications, came on the scene in 1966, and it seemed necessary to attempt to close the time gap. For this reason Chapter 11 is somewhat different to the others. It does not attempt to be comprehensive or strictly chronological. Topics have been selected which seem to me to be particularly relevant to the evolution of present-day understanding and thinking on tribology, but I am conscious of the difficulty of interpretation of the significance of recent events and the danger of offending those fellow scientists and engineers whose work may not be mentioned and yet who have contributed so much to progress in the subject.

I can merely crave the indulgence of those whose views on the recent history of tribology do not coincide with all that I have written in Chapter 11, hope sincerely that the student and general reader will find the linking nature of the chapter useful and interesting and confess that I have acquired a sound appreciation of the wisdom of the fifty-year rule!

Chapter 2

We should regard the present state of the universe as the effect of its antecedent state and the cause of its subsequent state.
Laplace, 1814

Chronology

2.1 Introduction

The importance of friction and resistance to motion has no doubt been recognized throughout the ages, but a full appreciation of the significance of tribology in a technological society is a recent phenomenon. In this chapter a chronology will be presented to engender a perspective for the reader wishing to develop an understanding of the present state of the science and technology of tribology.

Most of the book will be concerned with recorded history, a period of 5,000 to 6,000 years, but the achievements of prehistoric man cannot be ignored. It would clearly be difficult and probably futile to try to specify with any precision the date or period when important tribological developments occurred in prehistoric and early historic times. Indeed, it is not always easy to determine exact dates for developments in recent history. We can, however, occasionally recognize recorded events and archaeological evidence which demonstrate that ideas and devices had reached a certain stage of development. I have attempted to attribute a date, however approximate, to the significant advances in tribology mentioned throughout the text.

If any justification is required for starting this history of tribology with a chronology, it is to be found in the words of Laplace at the beginning of the chapter.

2.2 A chronology of tribology

A full understanding of the beginning and development of a subject can be obtained only if events are related to the history of human evolution. A comparative chronology of geology, evolution, history and tribology is presented in Table 2.1. The logarithmic scale allows the presentation of all events since the formation of the earth's crust on a single diagram.

Table 2.1(a)   Chronology (10^{10}-10^6 years B.P.)

Time years before present B.P.	Geology	Evolution	History	Tribology
— 1×10^{10}	Pre-Cambrian			
—— 9				
—— 8				
—— 7				
—— 6				
—— 5				
—— 4	Age of earth's crust			
—— 3				
—— 2				
— 1×10^9				
—— 9				
—— 8				
—— 7		Worms		
—— 6				
—— 5	Palaeozoic	Crustacea, spiders		
—— 4		Land plants, tribolites		
—— 3		Trees		
—— 2	Mesozoic	Reptiles		
		Ammonites, birds		
		Large dinosaurs		
— 1×10^8				**Formation of mineral oils**
—— 9				
—— 8				
—— 7				
—— 6				
—— 5				
—— 4				
—— 3		Lemurs, apes		
—— 2				
— 1×10^7	Cainozoic	Mammals becoming abundant	Prehistoric	
—— 9				
—— 8				
—— 7				
—— 6				
—— 5				
—— 4				
—— 3				
—— 2		MAN		
1×10^6				

Table 2.1(b)   Chronology (10^6-10^2 years B.P.)

Time years before present B.P.	Geology	Evolution	History			Tribology
—1 x 10^6		**MAN** Palaeolithic (old stone age)	Prehistoric			
9						
8						
7						
6						
5						
4		Heidelberg man				
3						
		Swanscombe man				
2						
		Fontéchavade man				
		Neanderthal man				
—1 x 10^5		Homo sapiens				
9						
8						
7						
6						
5						
4						
3			Cave drawings			
2						
—1 x 10^4	Cainozoic	Mesolithic (middle stone age)				
9						
8						
7						Door sockets – wood and stone
6						
5		Neolithic (new stone age)	Mesopotamia Egypt			Potter's wheel – stone bearings
						Bearings on wheeled vehicles
4						
		Bronze Age	Stonehenge			Lubricated sledges – Egypt, Assyria
3			Greece			Use of bitumen and seeping mineral oil
		Iron Age	Rome	Buddha		
				Alexander the Great		Early ball and roller bearings – Rome
2				Christ		
			Middle Ages	Mohammed		Anti-wear stone inserts in wooden implements – Denmark
—1 x 10^3			Vikings			Use of vegetable oils and animal fats
9						
8				Marco Polo		
7						
6			Renaissance	Columbus		Leonardo da Vinci's studies of friction, wear, bearings
5				Galileo		
4						
3				*Newton's Principia*		Amontons – friction
		Steam Age	Industrial Revolution			Desaguliers – cohesion
2						Coulomb – friction
			Rapid expansion of railways			Leslie – friction
			Inst. Mech. Engrs.			Babbitt – white metal
1 x 10^2						Drake's well – mineral oil

Table 2.1(c)   Chronology ($10^2 - 10^0$ years B.P.)

Time years before present B.P.	Geology	Evolution	History	Tribology
——1×10^2	Cainozoic	Steam age		Petrov (Russia); Tower (UK); Reynolds (UK)
——— 9			ASME founded	
——— 8		Oil age		Kingsbury (USA)
——— 7				Tilting pad bearing
——— 6			First World War League of Nations	ASME Research Committee on Lubrication (RCL)
——— 5				Boundary lubrication–Hardy (UK)
——— 4			Second World War	Inst. Mech. Engrs. General Discussion on Lubrication (1937)
		Nuclear age		ASLE formed (1944)
——— 3			Atomic bomb United Nations	ASME Lubrication Co-ordinating Committee (1948)
				Polymeric bearings and solid lubricants
——— 2			Sputnik	ASME Lubrication Division (1955) Inst. Mech. Engr. Lubrication and Wear Group (1955)
			Gargarin (USSR) first man to orbit earth	Studies of elastohydrodynamic lubrication
				Increasing use of gas bearings
				TRIBOLOGY—Jost Report (1966)
——1×10^1		Space age		Inst.Mech.Engrs.–Tribology Group (1967)
——— 9			Apollo moon landing (Neil Armstrong–USA)	
——— 8				
——— 7				
——— 6				
——— 5				First European Tribology Congress (1973)
——— 4				
——— 3				
——— 2				
1×10^0 (1977)				

Geology

Geological time is divided into four main eras which are in turn divided into a number of periods representative of major geological events and rock formations. The fourth and most recent geological era, the Cainozoic, includes the Quaternary period which spans the past million years.

Evolution

It can be seen from Table 2.1 that man, who appeared on earth at a late stage in the long history of evolution, has probably occupied less than 1 per cent of the age of recognizable life forms. Nevertheless, dating of skeletal remains from Africa and elsewhere has shown that man and his immediate ancestors have been in evidence for at least the million years ascribed to the Quaternary period.

Terms like Stone, Bronze, Iron, Steam, Oil and Nuclear Ages are used to denote significant periods of development of our civilization. They reflect the progress made by man in his struggle to utilize and control his environment. The first group of terms refers to the materials used for tools, weapons and machinery, while the second denotes the recent history of controlled power generation.

History

The logarithmic presentation of the history of the world shown in Table 2.1 restricts the period of recorded history and most of our knowledge of the development of tribology to the bottom right-hand corner. It is interesting to note how some of the great early civilizations arose so rapidly in relation to the evolutionary time scale. The pace of man's inventiveness immediately after the Stone Age period was truly remarkable.

It is generally believed that the first civilization arose in Mesopotamia in the district of Sumer, and the influence of this early development upon subsequent civilizations in Egypt, the Indus Valley, China, Central and South America is a subject of long-standing archaeological and historical interest. The impressive Greco-Roman period witnessed the further development of mechanical contrivances, substantial architectural and structural achievements and, perhaps most important in relation to our subject, the development of a scientific approach to physical problems.

The Middle Ages, when the Vikings were leaving their mark on history, were almost barren in scientific and technological terms; tribology being no exception. The great revival of classical art, architecture and science of the Renaissance (*c.* 14th–16th century A.D.) preceded the impressive contributions to science and engineering by Leonardo da Vinci (1452–1519), but we shall see that a full recognition of his studies of tribology is only now emerging. The feats of Leonardo's contemporary, Christopher Columbus, are more widely recorded in the history books.

The pace quickened in the seventeenth and eighteenth centuries, with impressive contributions to science being matched by technological progress at the beginning of the Industrial Revolution. The explosion of world trade, industry and communications in the nineteenth century promoted great progress in the field of tribology; the remarkable development of bearings, the wide-scale use of mineral oil lubricants and the growth of understanding of fluid-film lubrication forming a vintage period.

In the twentieth century the growth of large-scale industrial plants, the call for machinery to operate in hostile environments, the development of advanced air, sea and land transportation and the testing requirements of space exploration are all factors which have stimulated the emergence and recognition of tribology. Military requirements have promoted certain scientific and technological developments throughout history; an early example in our field of study being the wooden linear roller-bearing designed for a battering-ram in Greco-Roman times. The century which has known two world wars has exerted similar influences on tribological development.

Tribology

The history of tribology naturally spans a period similar to that of recorded history, but the form of a number of prehistoric artefacts makes it possible to comment on the feats of Stone Age man.

The events recorded in Table 2.1 are restricted to the major trends in tribology and it can be seen that they are encompassed by four cycles of time. The major events in tribology are summarized in chronological form at the end of each of the chapters concerned with specific periods.

Chapter 3

===

Prehistoric times – before *c*. 3500 B.C.

3.1 Introduction

The prehistoric Stone Age era is conveniently divided into Old (Palaeolithic), Middle (Mesolithic) and New (Neolithic) periods. Early man developed during the Palaeolithic period and although little can be said about tribology in this and the subsequent Mesolithic era, the development of skills formed a necessary preliminary to the more productive Neolithic period. Evidence will be presented to support the view that man made a conscious effort to develop simple bearings in Neolithic times.

3.2 Palaeolithic period   (*c*. 1,000,000–11,000 years ago)

A remarkable feature of the Old Stone Age period was the occurrence of the Pleistocene or Great Ice Age. It is interesting to speculate on the extent to which the severe climatic conditions influenced the development of man and his skill in making tools, weapons, fire and shelters. The marked fall in temperature caused the Arctic ice cap to spread over much of Europe, including northern Germany, Holland, Britain and Ireland, together with Canada and the north-eastern part of the United States.

There were four distinct ice ages in Palaeolithic times, each being separated by inter-glacial periods during which the weather was warmer than the present day. In these warmer periods the ice cap receded such that ancestors of animals now restricted to equatorial regions managed to survive in Europe.

The periods of glaciation in Palaeolithic times are dated about 600,000, 400,000, 200,000 and 50,000 years ago. As the ice sheet melted at the end of the last period of glaciation, lakes and rivers formed in northern Europe and the level of the oceans rose. It is a mere 12,000 years since the bed of the Mediterranean filled, Sweden was separated from Denmark and the North Sea and English Channel cut off Britain from the mainland of Europe.

In the early (lower) part of the Palaeolithic period, up to about 200,000 years ago, early man-like creatures were few in number and scattered over the habitable parts of the earth. Creatures like Pithecanthropus (Java), Sinanthropus (China), Atlanthropus (Algeria) and the later Heidelburg, Swanscombe (England), Fontéchevade (France) and Neanderthal (Germany) men were familiar with fire, chipped stone tools and weapons.[1]

There is little direct evidence of tribological skill from this period, but the conquest of fire and its controlled generation by means of the percussion of flint stones and the friction of wood on wood[2,3] must have drawn man's attention to the phenomenon of frictional heating. It has been suggested that these skills may have developed when early man saw the effect of dry boughs being rubbed together by the wind.

Human cultures evidenced by carvings and burials appeared in the middle Palaeolithic period in the time of Neanderthal man between 200,000 and 300,000 years ago. Cave carvings and paintings emerged in the later (upper) Palaeolithic period which ended about 11,000 years ago.

During the Palaeolithic period man was a food-gatherer and hunter, but it was during this period that *Homo sapiens*, man the thinker, evolved and gradually developed his tools and skills.

3.3 **Mesolithic period** (*c.* 11,000–5,500 years ago)

Transition between the end of the last Ice Age and the beginning of the era of human settlement is known as the Mesolithic or Middle Stone Age period. The dating of this period is determined by the recession of the ice cap from various regions, but for Europe it started over 11,000 years ago and lasted for some 6,000 years. The northern forests grew and the formation of deserts in the Near East led to a concentration of vegetation in the river valleys which played a large part in the siting of early civilizations.

There is little to report in the way of significant tribological activity in this period, but the slow improvement of tools and weapons suggests that it was an era of development of skills acquired earlier. The potter's art emerged at this time, but the potter's wheel, a device incorporating some of the earliest man-made bearings, belongs to a later period.

This transitional period brings us right up to an explosion in the development of human cultures. Its importance is to be found not in its own limited and largely unrecorded history, but in its delimitation of the history of man.

---

1. See Furon, R., 'The dawn of science: Prehistoric beginnings', in Taton, R. (1963).
2. Derry, T. K. and Williams, T. I. (1960).
3. See Forbes, R. J., 'The beginnings of technology and man', in Kranzberg, M. and Pursell, C. W. (1967).

3.4 **Neolithic period** (*c.* 5,500–3,500 years ago)

The New Stone Age marks the change from hunting to farming and the evolution of new processes for polishing tools and weapons. There is no uniqueness in the dates of this period in all parts of the world, but it started about 5,500 years ago and lasted for some 2,000 years in most of Europe and the Near East.

Archaeologists and anthropologists[4,5] use the terms 'savagery', 'barbarism' and 'civilization' to denote stages of cultural development. In this sense, and without pejorative implication, savagery describes the food-gathering Palaeolithic and Mesolithic periods, while barbarism, or food production, marks the beginning of the Neolithic age. The transition from barbarism to civilization took place later in Neolithic times.

Finding and hunting food was a near full-time occupation for early man, but in the Neolithic age his culture developed rapidly once the basic problem of food production had been solved by stock-keeping and tillage. He was able to develop crafts and trade and to solve the problem of constructing a more permanent home. Basketry, rope and mat-making, textiles, pottery, carving, building and agriculture must have introduced a recognition of the elements of tribology to preliterate man. An important common feature of many of these crafts was the development of hand tools for boring and drilling. The use of shell, stone, bone and antler tools spreads back to Palaeolithic times. The processes of boring and drilling are important to our subject because they represent one of man's first attempts to utilize rotary motion in manufacture. The tools were hand held and the motion discontinuous, but an appreciation of the advantages of rotary motion for drilling must have been an important stage in the development of the spindles and simple bearings for continuous rotary motion which emerged at the dawn of civilization. Some writers[6,7] suggest that simple hand-held antler, bone or stone bearings for the spindles of drills used for making fire or drilling holes were developed in the Stone Age period.

Firm evidence of man's utilization of continuous rotary motion in such devices as the potter's wheel and wheeled vehicles dates back to about the same time as the development of writing, and since the latter is usually used to mark the beginning of civilization and historic times, these notable developments are reserved for Chapter 4. There is, however, some very interesting evidence of the development of the earliest bearings in Neolithic times.

Prior to the use of metal hinges doors were swung on poles which rested in hollows in the wooden or stone thresholds, while the upper ends hung freely in loops of hide fixed to the door-frame. Wooden thresholds equipped with such sockets have been found in Neolithic

---

4. See Childe, V. G., 'Early forms of society' in Singer, C. et al. (1954).
5. Daniel, G. (1971).
6. Singer, C., et al. (1954).
7. Davison, C. St. C. B. (1957(*a*), (*b*)).

Fig. 3.1
Stone socket for lower pivot of door. The inscription records the restoration of a temple by Gudea, Prince of Lagosh, Sumer (*c.* 2500 B.C.).

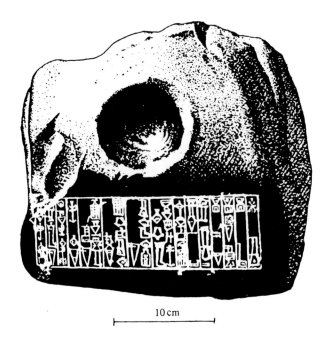

10 cm

houses preserved in Central Europe,[8] while stone sockets some 6,000–7,000 years old have been ascribed to Assyrian villages. A form of stone door-socket (*c.* 2500 B.C.) carrying an inscription is shown in Fig. 3.1. At a later stage the sockets were often lined with copper or bronze.

It is not clear whether prehistoric man deliberately made the door-socket by drilling a hole in the wooden or stone threshold, or whether he merely utilized convenient hollows in the material. In either event the oscillatory motion would cause the door-pole to produce further wear and, in many cases, the gradual development of a satisfactory footstep bearing. I wonder if our Stone Age ancestors used fats or oils to eliminate squeaks from the door?

3.5   Summary and chronology

During the period of about 1 million years covered by the Stone Age, man emerged to take an interest in devices requiring plain bearings. Activity in the field of tribology was prompted by the requirements of tools for various crafts, weapons and construction. Various forms of boring tools were produced and it is possible that the hand-held bearings on drill spindles developed in this period. The clearest evidence of plain bearing development comes from the reports of wooden and stone door-sockets.

The chronological chart shown in Table 3.1 looks a bit sparse, but it forms an important entry in the history of tribology.

8.   See Childe, V. G., 'Rotary motion' in Singer, C., et al. (1954).

Table 3.1

Chronology for the Stone Age

Years BC	The Stone Age		Tribology
10^6			
		Heidelberg man	
		Swanscombe man	
		Fontéchevade man	
		Neanderthal man	**Utilization of force of friction for the generation of fire by rubbing of wood-on-wood and the percussion of flint stones**
10^5	Palaeolithic period *(Old Stone Age)*	Homo sapiens	
		Cave drawings	
		Development of simple hand tools and weapons	
10^4			
	Mesolithic period *(Middle Stone Age)*	Early pottery	**Stone and wood footstep bearings for door-posts**
	Neolithic period *(New Stone Age)*	Change from hunting to farming	**Development of hand-held bearings of antler, bone and stone for drills**
10^3			

Chapter 4

The early civilizations (c. 3500 B.C.–A.D. 1500)

4.1 Introduction

The title of this chapter requires some elucidation before we proceed to the content. Indeed, we must start with the word 'civilization', since the specification of such an important milestone in the development of society or man's culture is necessarily somewhat arbitrary.

The *Oxford English Dictionary* describes civilization as an advanced stage in social development and it further notes that when we 'civilize' we 'bring out of a state of barbarism, instruct in the arts of life, enlighten and refine'. These definitions do little to assist us to date the dawn of civilization and it is for this reason that archaeologists and anthropologists frequently adopt more restrictive statements. One definition which has found some favour[1] was given by Professor Clyde Kluckholn when he said that a society, to be called civilized, must have two of the following:

- towns upward of 5,000 people;
- a written language;
- monumental ceremonial centres.

A less arbitrary definition of civilization has been given by Professor Piggott[2]

We should surely not be far from the mark if we thought of civilized societies as those which worked out a solution to the problem of living in a relatively permanent community, at a level of technological and societal development above that of

---

1. See Daniel, G. (1971), p. 31.
2. See the Preface to Professor Mallowan's *Early Mesopotamia and Iran* (Thames and Hudson, London) 1935, p. 7.

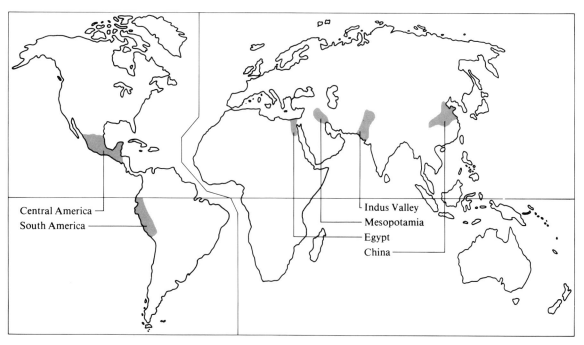

Fig. 4.1
Geographical location of the early civilizations.

the hunting band, the family farmstead, the rustic self-sufficient village or the pastoral tribe, and with a capacity for storing information in the form of written documents or their equivalent. Civilization, like all human culture at whatsoever level, is something artificial and man-made, the result of making tools (physical and conceptual) of increasing complexity in response to the enlarging concepts of community life developing in men's minds.

Archaeological studies have located the early civilizations in lower Mesopotamia (Sumer), Egypt, Pakistan and India (the Indus Valley), China, Central America (Guatemala, Honduras and Mexico) and South America (Peru). All these civilizations meet part or all of the Kluckholn and Piggott definitions. It is interesting to note that the first four are riverine cultures associated with the twin rivers of the Tigris and Euphrates, the Nile, the Indus and the Yellow rivers respectively. The geographical location of these early civilizations is shown in Fig. 4.1.

Much has been written by archaeologists and anthropologists[3] about the extent to which each of these early civilizations was influenced, and maybe initiated, by others. Certain similarities between the Mesopotamian, Egyptian and Indus Valley cultures support the concept of stimulus diffusion, but differences between the early American and Old World civilizations suggest that the two developed independently. There is more scope for controversy about the Chinese civilization which, although starting much later, does not appear to have been greatly influenced by the diffusion of ideas from Mesopotamia.

Although the ancient civilizations are at least 4,000 years old, the archaeological evidence of their existence and characteristics is recent.

3. See Daniel, G. (1971).

A common misconception is that knowledge of the distant past is itself ancient, but most of the evidence has emerged in the past two centuries. Much of the information from Mesopotamia followed the excavations started in the 1840s, while the discovery of both the Indus Valley and Chinese civilizations was announced as recently as the 1920s.

There appears to be little doubt that civilization began in southern Mesopotamia in the region of Sumer between 5,000 and 6,000 years ago. Since the archaeological evidence of tribological developments in both Mesopotamia and Egypt is rich and similar, it is convenient to consider the two civilizations together. The other civilizations will be mentioned only briefly, since, with the exception of China, records of tribological significance do not show sufficient novelty to warrant detailed description.

4.2 **Mesopotamia and Egypt** (*c*. 3500 B.C.–30 B.C.)

The first civilization recorded in the history of man developed in the fourth millennium B.C., probably about 3200 B.C., in a territory known as Sumer adjacent to the Persian Gulf at the southern end of Mesopotamia. The word 'Mesopotamia' means the land between the rivers (Tigris and Euphrates), while Sumeria was known as Babylonia after 2000 B.C. The Sumerian civilization lasted until about 500 B.C., but the somewhat later Egyptian civilization flourished in various forms until the Roman conquest about 30 B.C.

Five recorded features of the Sumerian and Egyptian civilizations are of great tribological significance:

1. *Drills* employing alternating rotary motion and simple bearings were developed for making fire and drilling holes.
2. The *potter's wheel* employing simple pivot bearings made from wood or stone was produced to facilitate the 'throwing' of clay at relatively high rotational speed.
3. The *wheeled vehicle* appeared.
4. *Lubricants* were used in a number of applications involving rotation and translation.
5. *Transportation of heavy stone statues and building blocks* on sledges.

Drills

Examples of the use of the simplest form of drill, in which a stick revolves between the palms, can be found in several of the earliest civilizations. Such drills were used for starting fires. More effective thong- and bow-drills were developed for making fire and drilling holes in the Sumerian and Egyptian eras and our present interest is centred on the hand-held bearings which were employed with these devices.

Rotation of the *thong-drill* is produced by the alternate pulling of ends of a cord or thong wrapped round the drill spindle in a simple loop. The drill is held upright in firm contact with the workpiece by a hand or

Fig. 4.2
Eskimo thong-drill in use in the
Hudson Bay area. (From an
illustration published in 1748.)

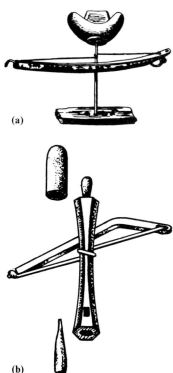

(a)

(b)

Fig. 4.3
Eskimo and Egyptian bow-drills.
(a) Eskimo fire-making bow-drill
with mouthpiece.
(b) Ancient Egyptian bow-drill
for making fire.

mouth-held bearing made of wood, bone, stone or coconut shell.[4] This type of drill is thought to be very old, but there is little direct evidence of its early use. It is used by the Eskimos and it is also found in northern Asia, India and Indonesia. An illustration of an Eskimo thong-drill in use in the Hudson Bay area over 200 years ago which shows an assistant guiding and applying pressure to the drill is shown in Fig. 4.2. The illustration appeared in the account of Ellis's (1759) voyage to the Hudson Bay.

The *bow-drill* represents an important improvement on the thong-drill since it can be operated comfortably by one man. The two ends of the thong or cord are attached to a bow which is moved to and fro by one hand while the guide bearing is held in the other. Typical Eskimo and Egyptian bow-drills for fire-making are shown in Fig. 4.3, while some of the excellent illustrations of the use of single and triple bow-drills by Egyptian carpenters and beadmakers found in Rekhmara's (*c.* 1471–48 B.C.) tomb at Thebes are shown in Fig. 4.4. These illustrations were found in perfect preservation on the upper half of the south-west wall of a passage leading to the shrine of the famous Rekhmara, Governor of Thebes and Prime Minister of his illustrious sovereign Thothmes III. Newberry[5] has given a full account of the scenes and in his description of the action shown in Fig. 4.4(*a*) he noted that the '. . . man with a bow-drill drills holes and prepares beads for threading by his three companions who are seated behind him. The beads are coloured blue and red (blue

---

4. See Harrison, H. S., 'Fire-making, fuel and lighting', in Singer, C., et al. (1954), p. 224.
5. Newberry, P. E. (1900).

Fig. 4.4
Egyptian carpenters and
beadmakers using single and
triple bow-drills.

(a) Egyptian beadmakers–a
triple bow-drill is being used
whilst the finished blue and red
beads are threaded by a
companion. A finished necklace is
illustrated between the two
workers.

(b) Egyptian carpenter using a
bow-drill.

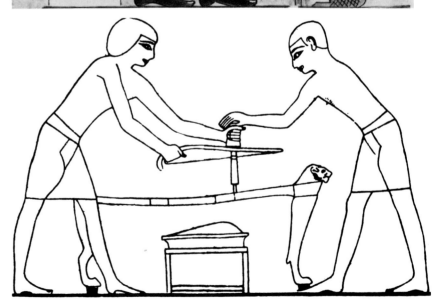

(c) Egyptian carpenters using a
bow-drill–the carpenter and an
assistant are shown drilling holes
in a chair or couch.

glaze or lapis lazuli and cornelian), and were threaded on fine twine without the need of a needle'.

Newberry also describes the scene showing the carpenters. In the first illustration a single carpenter is shown drilling a hole in a chair with a bow-drill. A Science Museum photograph of a drawing of this illustration made by Nina de G. Davies is shown in Fig. 4.4(*b*). Further to the right is the illustration shown in Fig. 4.4(*c*), of another carpenter making a chair or couch. In this case the carpenter is assisted by another worker who is seen to be pressing down on the hand-held bearing at the top of the bow-drill. It appears that the materials used for shrines and furniture were '. . . ivory, ebony, sesenezem wood, meru wood and new cedar wood from the hills'. The hand-held bearings were made from hollow horns, stone or ivory and the drill, which was recorded in Egypt from *c*. 2500 B.C. is still found in use in the Middle East.

The scenes depicted in this section of the tomb show Reckhmara, accompanied by forty attendants, '. . . inspecting all the handicrafts, and teaching each man his duties concerning the handicraft of all occupations'. An excellent example of the bow-drill was also found in Tutankhamun's famous tomb.

Some of these ancient drills had interchangeable wooden or flint bits, and there can be little doubt that early civilized man learnt a great deal about bearing materials from these simple devices some 4,000 to 5,000 years ago.

Potter's wheels

The oldest surviving portion of a potter's wheel was found by Woolley (1930) at Ur. It was dated at 3250 $\pm$ 250 B.C. and his description of the find reads '. . . All these vessels were wheel made, and at 9·90 m above sea-level there were found the fragments of a potter's wheel made of clay, a thick disc with a pivot hole smoothed with bitumen and small holes near the circumference at one point, doubtless for the stick which served to turn it; it was a wheel heavy enough to spin on its own momentum . . .'

This description contains a wealth of information; the pivot hole being a form of bearing some 5,000 years old, the reference to bitumen being an indication of the probable use of lubricants to reduce friction, and the ring of small holes showing how the wheel was started or speeded up by hand grip or the use of a stick.

It is quite possible that the early wheels were merely turntables used to orientate the clay in front of the potter, rather than spinning discs used to throw the clay. These simple wheels were probably developed in barbarian times, and if they were made from wood it is most unlikely that any will have survived.

The turntable and the true potter's wheel rotating at speeds in excess of 100 rpm or about 2 Hz in order to 'throw' the clay represent significant advances in the development of bearings and maybe in the recognition of the importance of lubricants. Childe[6] drew attention to the importance

---

6. See Childe, V. G., 'Rotary motion', in Singer, C., et al. (1954), p. 198.

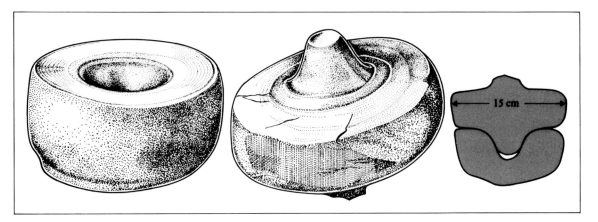

Fig. 4.5
Potter's wheel-bearing from Jericho (*c.* 2000 B.C.).

to be attached to tribological features of the potter's wheel when he noted that there was no evidence of either the pivot or socket being lined with metal. Archaeological evidence suggests that wood, stone and porcelain were used for the bearing components, while Childe wrote that lubricants in the form of fats, vegetable oils or bitumen would have been helpful to the functioning of these wheels.

It is clear that the early wheels were simple hand-operated devices and the frictional torque was probably too high to permit reasonably steady running at high speeds for any significant length of time. They were initially described as grinders; their link with the potter's wheel being established as late as 1939. At a later stage the potter was able to use both hands to form the clay when the foot-operated wheel was developed.

The drawing of a polished stone potter's wheel bearing from Jericho shown in Fig. 4.5 probably represents the bearing for a foot-wheel. The date is about 2000 B.C. The socketed member would have been embedded in the floor to provide a combined thrust and journal bearing for the spigoted stone. A large wooden foot-wheel would have been attached to the upper side of the rotating stone disc, with the work-table connected to the foot-wheel by a wooden axle. The work-table may have been cantilevered from the foot-wheel, but it seems more likely that the axle was steadied by a journal bearing in the form of a hole in a plank of wood or a simple loop of metal.

Archaeological evidence shows that the potter's wheel was used in Sumer and Iran before it appeared in Egypt, Europe or China. Tomb pictures showing the pivoted discs occur in Egypt after 2500 B.C. and Leach (1949) noted that the wheel was also known in China about 2000 B.C. Childe[7] gave a very good example of diffusion based upon archaeological finds in relation to wheel-made pottery which provided strong support for the view that the potter's wheel first appeared at the head of the Persian Gulf. The dates associated with finds in various places were:

Sumer	3250 ± 250 B.C.
Syria and Palestine	3000 B.C.
Egypt	2750 B.C.

7. See Childe, 'Rotary motion', in Singer, C., et al. (1954), p. 203.

Crete	2000 B.C.
Greece	1800 B.C.
Italy	750 B.C.
Danube and Rhine basins	400 B.C.
England	50 B.C.
Scotland	A.D. 400
The Americas	A.D. 1550

When, in addition, the date of *c.* 2000 B.C. is noted for China, it is clear that the centre of this diffusion was based upon Sumer.

Wheeled vehicles

More positive evidence of tribological development associated with continuous rotary motion is available in the archaeological records of transportation in the early civilizations. The first records are in the form of symbols in cave carvings in various parts of the world, together with the early pictographic script from Uruk, a Sumerian city in southern Mesopotamia.

Sledges had been used to transport heavy objects from about 7000 B.C.[8] while skis existed in Scandinavia from the late Neolithic period. A Stone Age rock carving from nothern Norway which depicts a man on skis is shown in Fig. 4.6.

The wheeled vehicle appeared at about the same time as the potter's wheel some 5,500 years ago; the delightful little picture from Uruk shown in Fig. 4.7 indicating that the transition from sledge to wheeled vehicle had been achieved by about 3500 B.C. A later Chinese pictograph (*c.* 1500 B.C.) for a chariot or cart, apparently with spoked wheels, is shown in Fig. 4.8.

The earliest remains of wheels of wheeled vehicles date back to the third millennium B.C. and excellent accounts of the early history have been presented by Childe (1951) and Piggott (1968*a*). Piggott (1968*b*) has noted that '. . . From the standpoint of the archaeologist wood is a miserable material; it is resistant to decay only in exceptional conditions of waterlogging or desiccation.' It is therefore fortunate that about fifty vehicles or wheels from the third millennium B.C. have been found in Europe and western Asia. The wheeled vehicle was rated so highly in the ancient civilizations that it was frequently buried with its owner in important tombs. Some of these buried vehicles have survived the normal processes of decay, or they have left recognizable impressions in the earth.

Fig. 4.6
Man on skis – Stone-Age rock-carving from Northern Norway.

Fig. 4.7
Uruk pictographs for sledge and wheeled vehicle (*c.* 3000 B.C.).

8. See Kranzberg, M. and Pursell, Jr, C. W. (1967), p. 23.

Fig. 4.8
Chinese pictograph for a chariot or cart with spoked wheels (*c.* 1500 B.C.). From 'The beginnings of wheeled transport' by S. Piggott. Copyright © 1968 by Scientific American, Inc.

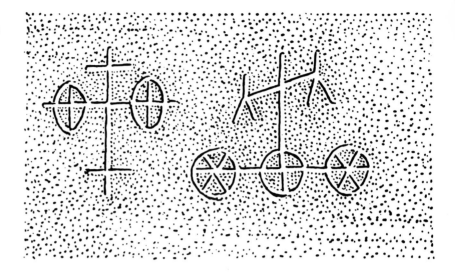

Remains of wheeled vehicles were found in the famous royal tombs built between 3000 and 2000 B.C. at Kish and Ur in Mesopotamia and at Susa in Elam. It appears that the early vehicles were built as two-wheeled carts, four-wheeled wagons or hearses. The military application of light vehicles was a later development associated with the emergence of the spoked wheel.

The early wheels were solid, as opposed to spoked, being carved from a single massive plank or fabricated from three or more planks dowelled and mortised together. Wooden battens, or copper clasps, were used to hold the several planks in place, and the rims of the discs were sometimes heavily studded with copper nails or bound with metal hoops. Childe[9] noted that '. . . The primary function of these nails was to protect the rims from wear.' This early attempt to defeat wear in a man-made contrivance was in evidence in wheels found in royal tombs at Kish (*c.* 2750 B.C.), Ur (*c.* 2500 B.C.) and Susa (*c.* 2500–2000 B.C.). Particularly fine examples of tripartite wheels complete with wooden felloes and rims packed with copper nails are shown in Fig. 4.9. The rims were sometimes bound with hoops of copper[9] and maybe leather. These hoops would no doubt have helped to hold the tripartite discs together, but they would also have served a tribological role in reducing wear of the rim.

Our interest in the development of the wheeled vehicle lies not only in the tribological features of the wheel rim but also in the form of the axle bearings. It is at this stage that we have to ask a most important question about the construction of these early wheeled vehicles. Did the wheels rotate on a fixed axle or was a solid axle-wheel arrangement allowed to rotate in bearings on the frame of the vehicle? Childe[10] thought that the archaeological and ethnographic data was inadequate

---

9. See Childe, V. G.,'Rotary motion', in Singer, C., et al. (1954), p. 208.
10. See Childe, op. cit. in Singer, C., et al. (1954), p. 207.

Fig. 4.9
Nail-studded tripartite wheels with wooden felloes from Susa Apadana, Tomb 280 (*c.* 2500 B.C.).

to decide this issue, but the *Encyclopedia Britannica*[11] tells us that '. . . the wheels were kept in position by linch pins on the stationary axle'. This view was supported by Kranzberg and Pursell (1967) who, together with Childe, doubted the popular view that the wheel developed from the use of logs as rollers under sledges. Childe presented a forceful argument when he pointed out that the tripartite form did not appear to fit in with the view that the wheel developed from the roller. Piggott has also confirmed[12] that in all the cases which he has seen, or which have been adequately published, the axles were fixed and the wheels turned on them, with provision for linchpins.

It would have been difficult for the early craftsmen to fashion long, round axles which could turn freely in bearings underneath the frame of the vehicle, whereas it would have been relatively easy to produce short lengths of round shaft at the ends of rectangular or square section axles. I am not aware of any strong archaeological evidence to show that a rotating shaft was normally carried in bearings formed in side planks or skirts round the early vehicles, and hence the view that the wheels rotated freely on axles fixed to the frame is the most acceptable. This arrangement has, however, some tribological penalty, since wear would take place between the axle and the wheel, the latter being by far the most complicated and expensive feature of the vehicle. Emmerton (1958) has discussed the development of both fixed- and rotating-axle vehicles in some detail and his broad conclusion was that the vast majority of road vehicles have always been of the fixed-axle variety. There are a number of recorded exceptions to this view, but the general case is clear. Emmerton drew attention to the interesting fact that vehicles using rails tend to adopt rotating-axle arrangements–the freely rotating wheel on a fixed axle being more prone to derailment on account of play resulting from wear and the facility for one wheel to advance either faster or slower than the other along the rail. The regular use of rotating axles is thus a recent trend associated with the development of vehicles guided by rails.

11.   Encyclopedia Britannica (1970) Vol. 23, p. 473.
12.   Private communication.

Fig. 4.10
Left side of a copper model of a
quadriga from the Sharma Temple
at Tell Agrab, Iraq (*c.* 2800 B.C.).

The early wheeled vehicles were pulled by oxen or asses. The spoked
wheel and lighter vehicles appeared much later, probably about 2000 B.C.,
while the horse-drawn chariot appeared in military applications soon
afterwards.

Some valuable additional archaeological evidence on the development
of the wheel comes from models made of clay or metal. Woolley (1930)
described an interesting clay zoömorphic vase in the form of an animal
on solid wheels which was excavated from the lowest levels of the cemetery
at Ur, while Piggott (1968*b*) described the copper model of a Mesopotamian chariot shown in Fig. 4.10 which illustrates the tripartite disc
wheels used earlier on carts.

Although archaeological evidence still favours Sumer as the location
of the origin of the wheel, Piggott (1968*a,b*) has given an account of
recent excavations in the USSR in the area between the Black and Caspian
seas just south of the Caucasus which suggests that the invention probably
occurred in a wide rather than a narrow area. Much of the recent evidence
comes from the region between the twin rivers Kura and Araxes, which
unite to flow eastward to the Caspian Sea. The Kura–Araxes culture has
been dated back to 3000 B.C. by carbon-14 determinations. It lasted for
at least 500 years and although the period is slightly later than the
Sumerian culture, the latter is located by historical computations which
are not necessarily consonant with the few radiocarbon determinations
available for that region and period.[13]

Piggott (1968*b*) has summarized the evidence for a wide occurrence
of early wheeled vehicles in the two most instructive maps shown in
Figs 4.11 and 4.12. There is, of course, the possibility of multiple invention

---

13.   See Piggott, S. (1968*a*), p. 309.

Fig. 4.11
Distribution of vehicles built before 2000 B.C. Note the two main concentrations in the Kura–Araxes and Tigris–Euphrates riverine systems.

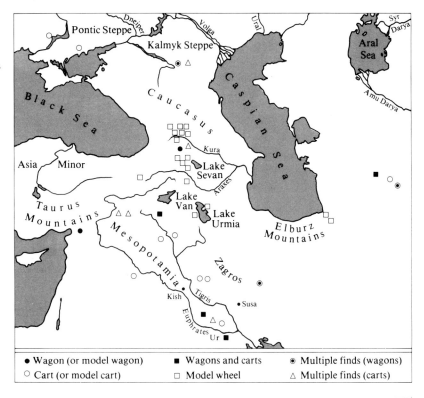

● Wagon (or model wagon)	■ Wagons and carts	⊙ Multiple finds (wagons)
○ Cart (or model cart)	□ Model wheel	△ Multiple finds (carts)

Fig. 4.12
Distribution of wagons in Europe before 2000 B.C.

Figures 4.11 and 4.12 from 'The beginnings of wheeled transport' by S. Piggott. Copyright © 1968 by Scientific American, Inc. All rights reserved.

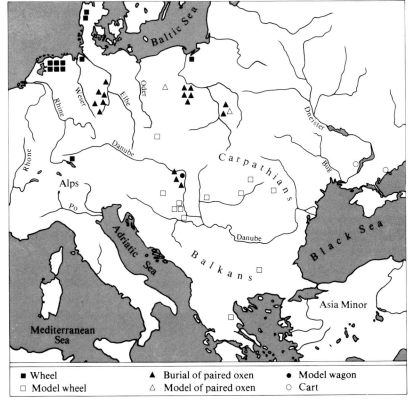

■ Wheel	▲ Burial of paired oxen	● Model wagon
□ Model wheel	△ Model of paired oxen	○ Cart

or rapid diffusion in the region between the Caucasus and the Persian Gulf and it may be a long time before a definitive answer can be given to the question of the origin of the wheeled vehicle.

Lubricants

There can be little doubt that the value of lubricants was appreciated at an early stage in the Sumerian and Egyptian civilizations. Attention has already been drawn to the traces of bitumen found in a potter's wheel bearing and it is known that there were a number of sites between the Nile and the Tigris where crude oil came to the surface. However, neither mineral oil nor natural gas appear to have found any favour in these ancient civilizations, and lubricants were mainly of vegetable or animal origin.

A chariot from about 1400 B.C. was found in the tomb of Yuaa and Thuiu, along with traces of the original lubricant on the axle. Parish[14] referred to an analysis of this lubricant by Lucas, official chemist of the museum at Cairo, which was reported by Quibell. The sample was small, weighing about 0·001 oz (0·038 g), sticky and slightly greasy. It contained road dirt such as quartz sand, compounds of aluminium, iron and lime. It had a melting point of 49·5°C (121·1°F), which suggested that it had been mutton or beef tallow, either of which would have proved suitable for axle lubrication in such a warm country.

At an even earlier stage lubricants had been used to reduce the friction between sledges used for the transportation of building blocks and the wooden logs or planks on which they moved. The tribological aspects of this problem will be discussed shortly, but it is worth noting that the use of a lubricant, possibly grease, oil or water, for this purpose, was recorded as early as 2400 B.C.

A most interesting story related to the early use of lubricants comes from the building of the Pyramids in the third millennium B.C. The blocks of stone were huge, often weighing tens of tons, and well dressed. Dry joints would have been perfectly adequate under these circumstances, but the traces or mortar between the layers of stones suggests that the builders used a 'squeeze-film' mode of lubrication to slide the blocks accurately into place. Hydrated calcium sulphate (gypsum) was used[15] to form the thin bed of viscid mortar, the efficiency of squeeze-film action being demonstrated by the precision of the final structure. Somers Clarke and Engelbach (1930) have considered this question in some detail. They point to the different role of mortar in modern and megalithic building. In the former, where relatively small stones or bricks are used, the mortar has to provide a strong bond to provide strength and stability, whereas in megalithic building the cohesive power of the mortar is of little importance owing to the high friction between the blocks. In both cases the mortar also acts as a lubricant to facilitate accurate 'setting' of the stones–a role which assumes great importance when the blocks are large.

14. See Parish, W. F. (1929), Vol. 14, p. 453.
15. See Lloyd, S., 'Building in brick and stone', in Singer, C., et al. (1954), p. 481.

It is not uncommon to come across accounts of horizontal joints in Egyptian buildings having a mean thickness of about 0·02 in (0·5 mm). Petrie (1885) has noted the remarkable ability of Egyptian masons to cut and dress stone to within 0·01 in (0·25 mm) over lengths of about 6 ft (1·8 m).

The mortar used in Egypt consisted of sand, gypsum and carbonate of lime and the mean of three samples tested by Lucas[16] showed proportions of 4·5 per cent, 83·5 per cent and 12 per cent, respectively. Lucas also noted that lime was not used in mortar in Egypt before Roman times and that the crude gypsum employed by the Egyptians would allow the mortar to set quite slowly.

The picture which emerges is that the large blocks of stone which formed the casings of the Pyramids were probably manoeuvred into position with levers and separated by temporary packing by an inch or so from the underlying block. The viscid mortar would then be inserted and when the packing was removed frenzied activity would ensue as the heavy block squeezed the 'lubricant' to give a final thickness of about 0·02 in (0·5 mm) as it sank into its most excellent location.

Aldred[17] has drawn attention to the possibility of a similar technique being used in the construction of columns. During the operations of the late Professor W. Emery at Buhem where a temple was being dismantled for re-erection at Khartoum, sand poured from between the joints as each drum was lifted. Emery pointed out to Aldred that the various courses had probably been bedded on to each other using wet sand as a lubricant and filler.

The dressing and transportation of large building blocks must have presented great problems to the Egyptians, but the ingenuity used to solve the equally difficult problem of placing the blocks in position by a tribological system calls for recognition and admiration.

Transportation of heavy stone statues and building blocks

For our last look at tribological innovation in the Mesopotamian and Egyptian civilizations we turn to the transportation of large stone blocks and carved figures used in the construction of the Pyramids and palaces. The haulage of heavy objects has evolved from simple dragging, through sledges, to the modern wheeled vehicle. For light loads sledges and skis can move freely over snow, grassland and even desert. In the Near East sledges were in use well before the wheel emerged *c.* 3500 B.C., as confirmed by the little picture from Uruk shown in Fig. 4.7. Indeed, it is believed that the sledge has been used by man for almost 9,000 years, while the delightful rock carving of a man on skis shown in Fig. 4.6 dates back to Neolithic times.

Layard (1849, 1853)–later Sir A. H. Layard–discovered the means by which large stone blocks and carvings were transported in Mesopotamia when he opened up the great mound of Kouyunjik at Nineveh.

---

16. See Clarke, S. and Engelbach, R. (1930), p. 79.
17. Aldred, C. Keeper, the Royal Scottish Museum, Edinburgh, Private communications.

Fig. 4.13
King superintending the transport
of a colossal bull at Kouyunjik
(*c*. 700 B.C.).

He made two expeditions between 1845 and 1851, the first being supported
by the East India Company and the second by the British Museum. His
detailed accounts, which run to two volumes for each expedition, make
fascinating reading.

Many bas-reliefs of great interest were discovered at Kouyunjik;
some showing the process by which large stone statues in the form of
human-headed bulls were transported to the palaces of which they formed
such remarkable features. It appears that the entrances to the great halls
of the palaces were flanked by these colossal statues. The stone was
transported from the quarries to a site near the palace by river, on a form
of boat pulled by nearly 300 men. The carving took place on the banks
of the river and the bull was then towed into position on a sledge, probably
with a tractive power of a group of men similar in number of those
employed in the river transportation. Some of the men appear to be
captives or malefactors since many were shown in chains.

Layard[18] noted that some of the largest sculptures were fully 20 ft
(6 m) square, thus indicating that they must have weighed between 40 and
50 tons (400–500 kN). The sledge was dragged by four teams of men and
impelled by leves from the rear. Each man was attached to one of the four
main towing cables by a lighter rope harness round his shoulders. All these
features of the process can be seen in Fig. 4.13, which shows the Assyrians
moving a large human-headed bull statue at Kouyunjik *c*. 700 B.C. The

18.   See Sir A. H. Layard (1853), Vol. I, p. 110.

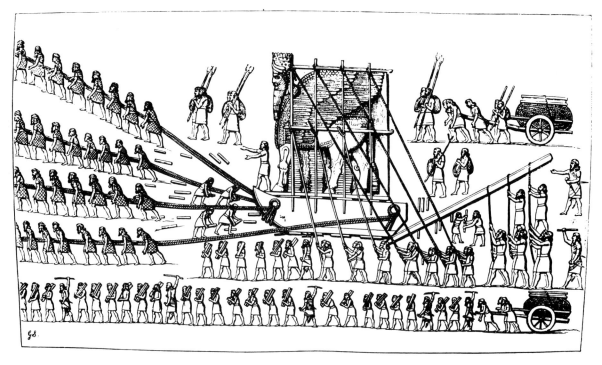

Fig. 4.14
Assyrians positioning a
human-headed bull – partly
restored from a bas-relief at
Kouyunjik (*c.* 700 B.C.).

bull is laid on the sledge on its side, with four officers riding on top
supervising the action, while the King overlooks everything from his
chariot. A second bas-relief shown in Fig. 4.14 shows the bull upright
on a sledge ready for final positioning. It is kept erect by beams and
cross-bars, with additional support from the long, forked poles and ropes
handled by pairs of men.

Perhaps the most interesting and controversial tribological feature of
these bas-reliefs is Sir A. H. Layard's claim that they show the use of
wooden rollers underneath the sledges. Both the reliefs depicted in Figs
4.13 and 4.14 show short wooden logs, or planks, being inserted between
the front of the sledge and the ground, while a number of men appear
to be engaged in the task of recovering the logs left behind the advancing
sledge. Layard wrote[18]

*. . . The sculpture moves over rollers, which, as soon as left behind by the
advancing sledge, are brought again to the front by parties of men, who are
also under the control of overseers armed with staves. Although these rollers
materially facilitated the motion, it would be almost impossible, when
passing over rough ground, or if the rollers were jammed, to give the first
impetus to so heavy a body by mere force applied to the cables. The Assyrians,
therefore, lifted, and consequently eased, the hinder part of the sledge with
huge levers of wood. . . .*

19. See Layard (1853), p. 106.

This quotation has been given in full since it is often referenced as evidence that logs were used in the form of linear roller-bearings in the ancient civilizations. However, Davison (1961) has disputed this view on the grounds that the *rollers* were shown with their axes parallel to the direction of motion of the sledge, and that they were clearly trimmed tree branches which were neither straight nor round. The merit of this argument is clear, but it does not with certainty dispose of Sir A. H. Layard's view that rollers were used to facilitate motion of the sledge. On another page[20] Layard wrote '. . . Although in these bas-reliefs, as in other Assyrian sculptures, no regard is paid to perspective, the proportions are very well kept.' It is undoubtedly true that some of the reliefs are lacking in perspective and it is difficult to know whether the orientation of the logs is a correct representation or an error in perspective.

Since the wheel was well known in Mesopotamia at the time of the action depicted in the bas-reliefs, it could be argued that the Assyrians would have been aware of the advantages of rolling motion over sliding. Furthermore, the Egyptians had moved statues of similar size without the aid of rollers or levers, but with the use of lubricants, some 1,000 years earlier at the beginning of the second millennium B.C. It would have been strange if the diffusion of ideas and the general development of transportation over 1,000 years had not led to a superior system in Assyria in the seventh century B.C. The question as to whether or not this evidence supports the view that the logs were used as rollers must remain unresolved, but with the weight of opinion against the suggestion.

The case for the use of rollers may also be thought to receive support from studies of the extraction and erection of obelisks. Engelbach (1923) has discussed these spectacular tapering shafts of stone with pyramidal tops in some detail. The known Egyptian obelisks vary in height from about 64 to 105 ft (19·5–32 m), with weights in the range 120 to about 500 tons (1·2–5 MN). The unfinished obelisk exposed in a quarry at Aswan in 1922 is a massive 1,168 ton (11·64 MN) block of granite some 137 ft (41·75 m) long. Engelbach[21] appears to favour, albeit cautiously, the view that the obelisks were transported over the land on sledges which themselves moved over wooden rollers. If the sledge moved over wet plants, as appears to have been the case with the El-Bersheh colossus to be described shortly, a reasonable estimate of the coefficient of friction leads to the view that well over 1,200 men would have been required to move a 500-ton obelisk over flat ground. The logistics of such an exercise is itself sufficient to raise doubts about reliance upon sliding motion. Furthermore, Engelbach reports that small rollers have been found, and in pictures of a model designed to illustrate a possible method for erecting obelisks he shows the sledge, apparently on rollers, moving up a large sloping embankment.

In a later book Clarke and Engelbach (1930) state that '. . . The evidence for rollers having been used in conjunction with sleds is comparatively slight, though their use is almost unquestionable.' A photo-

20.   See Layard (1853), p. 114.
21.   See Engelbach, R. (1923) pp. 58–60.

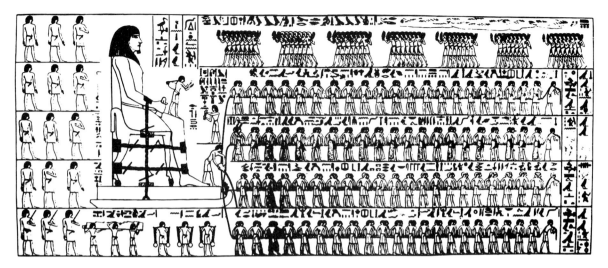

Fig. 4.15
Transporting an Egyptian
colossus – from the tomb of
Tehuti-Hetep, El-Bersheh
(*c.* 1880 B.C.).

graph is shown of such a roller found at Saqqara. It appears to be of acacia wood, to be slightly thicker in the middle than at the ends, which are rounded, about 3 in (76 mm) in diameter and some 18 in (457 mm) long. It is also reported that similar pieces of wood have been found in several foundation deposits.

The Egyptian method of moving large stone statues is depicted in the well-known painting from a grotto at El-Bersheh dated about 1880 B.C. shown in Fig. 4.15. The colossus is secured to a sledge similar to that employed at a later stage in Assyria, but there are no rollers or levers. A most interesting feature of this painting is that it shows an officer standing at the front of the pedestal pouring lubricant from a jar on to the ground immediately in front of the sledge. Sir A. H. Layard[22] stated that the lubricant was probably grease, but the scant archaeological evidence offers little support for this view and others have described the lubricant as water. Three men are each shown to be carrying two jars at the side of the sledge, presumably in support of the man on the pedestal, while a similar number carry a wooden plank on which the sledge was probably towed. Even if water had been used as the lubricant it could have led to a significant reduction in frictional resistance.

I am inclined to agree with the view that water was used as the lubricant, since the firmest evidence emerges from the translation of the inscriptions on the wall pictures from El-Bersheh. In Part I of the special publication of the Egypt Exploration Fund on the Archaeological Survey of the Tomb of Tehuti-Hetep at El-Bersheh, Newberry[23] recorded the following inscription relating to the three men with yokes and jars:

fat mu on per zet   (carrying water by (men of) the house of eternity)

Newberry also wrote '. . . Another figure standing on the base pours water from a jar in front of the sledge, perhaps only a ceremonial act,

22. See Layard, Sir A. H. (1853), Vol. 1, p. 115.
23. See Newberry, P. E. (1893), Pt I, pp. 9, 20.

since even in large quantities water poured upon the ground could not assist the dragging. . . .'

He thus misses the possible connection between the pouring of liquid from the jars and the inscription above the three men who '. . . carry on their shoulders a great block of wood with curiously jagged outline at the top . . .' which reads:

fat <u>kh</u>et ne seta an hemtiu　(carrying logs of conveyance by the workmen).

Indeed, in his description of the possible route followed by the colossus from the quarries via a stone road, the river Nile and finally cultivated land, Newberry[24] stated that '. . . free use would be made of great wooden beams laid longitudinally upon the ground . . .'. He also referred to Professor Petrie's suggestion that the piece of wood with a notched edge carried by four men was a beam, to be laid in the ground with the jagged side downwards, so as to grip the ground and prevent it from slipping with the movement of the colossus.

The manner in which frictional resistance was minimized in the transportation of the El-Bersheh colossus is of considerable tribological interest and I have devoted a good deal of time to the exposure and presentation of the facts. The inescapable conclusion is that the sledge was drawn over wooden planks lubricated by water.

In an earlier article[25] I described the officer on the pedestal as the first recorded *oiler, greaser, lubrication engineer or tribologist*, but I can now offer an earlier example. The transport of the statue of Ti (see Steindorff, 1913) was depicted in a tomb built at Saqqara, Egypt *c.* 2400 B.C.; some 600 years earlier than the El-Bersheh date. The general procedure for towing the sledge shown in Fig. 4.16 is similar to that at El-Bersheh (see Fig. 4.15). Since, in this case, lubricant was supplied from a jar by a gentleman standing in front of the sledge, it appears that the tribologist improved his status by being elevated to a riding position during the 600 years between El-Bersheh and Saqqara! The sledge appears to be moving over a firm, flat surface, probably made from wooden planks, while the tribologist applies lubricant as shown in detail in Fig. 4.17.

There is some merit in considering the tribological implications of the paintings and bas-reliefs of the movement of large stone statues on the assumption that they are accurate representations of the process. It is both interesting and instructive to attempt to quantify the coefficient of friction for the situation depicted in the records.

The weight of the Egyptian colossus (W) shown in Fig. 4.15 has been estimated at about 60 tons (600 kN). The painting shows 172 men pulling on all ropes, and Davison (1961) has estimated that each man can exert a pull of 120–240 lbf (534–1068 N). There is some evidence[26] that the higher figure could be achieved by present-day man, but it is doubtful if those toiling under the Egyptian sun nearly 4,000 years ago could have consistently achieved this figure. It will suffice to take the mean tractive effort

24.　See Newberry, P. E. (1893), Pt I, pp. 23, 24.
25.　See Dowson, D. (1969), p. 4.
26.　See ibid., p. 5.

Fig. 4.16
Transporting the statue of
Ti – from a tomb at Saqqara,
Egypt (*c.* 2400 B.C.). (Steindorff,
1913).

Fig. 4.17
The first recorded tribologist –
pouring lubricant (water) in front
of the sledge in the transport of
the statue of Ti (*c.* 2400 B.C.).
(Steindorff, 1913)

Fig. 4.18
Tribology in Ancient Egypt—
calculation of the coefficient of
friction in the transport of an
Egyptian colossus.

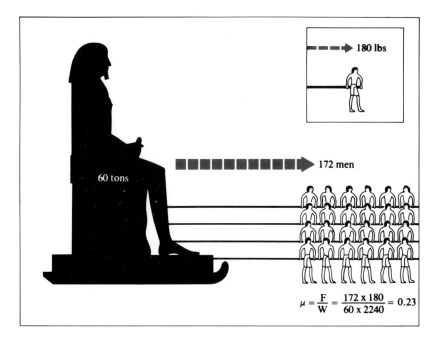

as 180 lbf (800 N) per man for the present calculation. On this basis the total effort, which must at least equal the friction force (*F*), becomes, 172 × 180 lbf (172 × 800 N).

Thus, the coefficient of friction (*μ*) is given by

$$\mu = \frac{F}{W} = \frac{172 \times 800}{600 \times 10^3} = 0\cdot23.$$

The basis of this calculation is summarized in Fig. 4.18. It is interesting to note that Bowden and Tabor (1950) quote coefficients of friction for wood-on-wood as follows:

wet   0·2   clean (and dry)   0·25–0·5

A comparison of the calculated coefficient of friction for the Egyptian sledge with the values quoted by Bowden and Tabor suggests that the sledge was indeed sliding over lubricated planks of wood.

This survey of the remarkable tribological achievements in connection with the drill, the potter's wheel, wheeled vehicles, the use of lubricants and the transport of heavy stone statues completes our study of the early civilizations in Mesopotamia and Egypt. Brief mention will now be made of a few similarities, and differences, in the developments in later civilizations in other parts of the world.

4.3   **The Indus Valley**   (*c.* 2500 B.C.–1600 B.C.)

It is usually quite a surprise to the new reader of archaeology when he learns that our knowledge of the oldest civilizations has emerged so very recently. In the case of the civilization in the Indus Valley most of our knowledge has come from excavations carried out since the early 1920s.

The area covered by this civilization was much larger than either the Sumerian or Egyptian cultures, being centred on two cities known as Mohenjo-daro and Harappa. The Indus Valley culture developed much later than Sumer and Egypt and it probably extended from the middle of the third to the middle of the second millennium B.C.

Although the archaeological evidence does not give precedence to the Indus Valley for any significant tribological development, most of the features attributed to Sumer and Egypt appeared in due course; occasionally with interesting variations. In the prehistoric Harappan civilization the door-socket was apparently made from kiln-fired brick rather than stone,[27] the recessed potter's wheel bearing was made from flint stone[28] while the wheeled vehicle appeared soon after 2500 B.C.[29]

Our knowledge of the Indus Valley civilization is growing steadily, with much of the recent understanding being attributable to the work and writings of Sir Mortimer Wheeler (1961, 1966), who was appointed Director-General of Archaeology in India towards the end of the Second World War.

4.4 China (*c.* 2500 B.C.–A.D. 100)

The first identification of a Neolithic community in China was reported by the Swede, J. G. Anderson[30] in 1921; the growth of knowledge on the Indus Valley and Chinese civilizations thus occurring at about the same time. There are a number of puzzling features about the development of Chinese culture, but the Yang Shao villages exposed in the 1920s seem to have arisen towards the end, or maybe in the middle, of the third millennium B.C. Most authorities seem to attribute a duration of 2,000 to 3,000 years to the Chinese civilization, with a starting date close to 2500 B.C. This makes the Chinese civilization much more recent than Sumer or Egypt; a view supported by the relatively late appearance of inscriptions in the fourteenth century B.C.

It has already been noted that the Chinese used hard porcelain cups for potter's wheel bearings, while the wheel appeared about 1500 B.C. The Chinese pictograph shown in Fig. 4.8, dating back to the middle of the second millennium B.C., confirms this view and at the same time draws attention to one of the puzzles relating to the development of Chinese culture. The wheels are spoked, but solid wheels always emerged first in the early civilizations of the Old World. Was there an earlier, undiscovered era of development involving solid wheels in China? Or did the diffusion of ideas from the Near East permit the Chinese to move straight to the more sophisticated design? The latter view on the effect of diffusion upon Chinese culture does not appear to enjoy much support.

Needham (1965) has discussed the origin of the wheel in his authoritative text on the development of mechanical engineering in China.

---

27. See Childe, V. G., 'Rotary motion', in Singer, C., et al. (1954), p. 194.
28. See Childe, op cit., in Singer, C., et al. (1954), p. 199.
29. See Childe, op cit., in Singer, C., et al. (1954), p. 211.
30. See Daniel, G. (1971), p. 114.

Fig. 4.19
South-pointing chariot–Chinese
(*c.* 255 B.C.).

He agrees with Davison's (1961) view that Sir A. H. Layard misinterpreted the reliefs of Sennacherib's palace at Nineveh when he suggested that logs were used as rollers. Attention is also drawn to the fact that the haulage of colossal statues by masses of men, whether of slave status or not, does not appear in any kind of ancient Chinese representation.

China had witnessed the development of elegantly constructed spoked wheels by the period of the Warring States (*c.* 4th century B.C.) and Needham wrote '. . . the hub was drilled through to form an empty space into which the tapering axle was fitted, and between the two a tapering bronze bearing was inserted, the exterior being covered by a leather cap to retain lubricant'.

This reference to metal bearings in Chou and Han China is interesting since it parallels developments in the Mediterranean area at this time which will form the subject of Chapter 5. The reference to leather seals and lubricants also confirms an interest in the tribological performance of wheels and a recognition of the need to retain lubricant in, and exclude dirt from, bearings. Needham has also expressed the view that China may have been deeply involved in the prehistory of ball-bearings.

I have not previously mentioned the development of the gear in this history of tribology, but it is appropriate to introduce the topic here since some claim that the earliest relics belong to China. In his book on the evolution of the gear art, Dudley (1969) attributed the earliest known relic of gearing from ancient times to the famous 'South-pointing Chariot', but the suggested date of *c.* 2600 B.C. is in error. An illustration of the miniature replica of the chariot displayed in the Smithsonian Institution in Washington DC is shown in Fig. 4.19, the original label carrying the above date. However, the label has now been changed[31] to read 'South Pointing Chariot of Ma Chün–Chinese, about A.D. 255'; the source of the new date being Needham (1965).

There may, however, have been an earlier model of this chariot, the purpose of which is well told in the following quotation.[32]

> *The south-pointing carriage was first constructed by the Duke of Chou [beginning of the first millennium B.C.] as a means of conducting homewards certain envoys who had arrived from a great distance beyond the frontiers. The country to be traversed was a boundless plain, in which people lost their bearings as to east and west, so [the Duke] caused this vehicle to be made in order that the ambassadors should be able to distinguish north and south.*

The south-pointing chariot shown in Fig. 4.19 was a differential gear train to ensure that the figure, once set, continued to point south regardless of which direction the chariot was going. The invention was no doubt most beneficial to the travellers in the Gobi Dessert. Dudley (1969) listed a number of sources of references to the use of gears in the 300 years B.C. He also concluded that the Egyptians and Babylonians must have been using gears as early as 1000 B.C. There is no mention of the use of gear lubricants at this early stage of development, but the beginnings of such a tribologically significant mechanism is of interest to our study.

4.5 Central and Southern America

(*c.* 500 B.C.–A.D. 1550)

The early civilizations in America appear to have developed in the absence of contact or diffusion of ideas from the earlier civilizations in the Old World, although Jairazbhoy[33] has claimed that the Egyptians anticipated Columbus by some 2670 years when they established the first civilization (Olmec) in America in the Mexican Gulf *c.* 1200 B.C. However, man probably reached the north-western part of the Americas 20,000 to 30,000 years ago and successive generations moved south until the tip of South America was reached about 7000 B.C. The American Indian was established in North America by 10,000 B.C.

---

31. Private communication, Smithsonian Institution (June 1971).
32. See Needham (1965) p. 286.
33. See Jairazbhoy, R. A., 'Clues to Egypt's conquest of Mexico', *The Times*, 26.8.1971.

The three great civilizations–Mayan in southern Mexico, Guatemala and Honduras, Aztec in Mexico and Inca in Peru–are all recent relative to Sumer, Egypt, the Indus Valley and China. All developed after the birth of Christ. The Mayan culture developed first in the early centuries A.D., and it is thought that simple forms of hand drill were developed at this stage. The Aztec culture developed in the twelfth century A.D., while the Inca in Peru emerged as late as the fifteenth century. All three civilizations decayed in the sixteenth century, the Mayan by internal disintegration and others as a result of Spanish conquest.

Tools and weapons made from wood and stone developed in the American civilizations, but there is no evidence of the wheel being used for making pottery or for transportation. Pottery was made in Mexico from the second millennium B.C., but the potter's wheel failed to emerge over a period of 3,000 years of development of the craft. The Incas in Peru moved and carefully positioned large dressed stones weighing up to 100 tons (1 MN), apparently without the use of mortar.

It can be seen that developments in the Old World civilizations had no recognizable impact upon the much more recent early American cultures. The first New World civilizations show very little evidence of tribological skills, certainly nothing to compare with the exciting progress in Sumer and Egypt some 3,000 years earlier.

4.6 Chronology

The six early civilizations span a time in excess of 5,000 years, but each developed and flourished at different times within this period. The approximate dates of each civilization are shown in Fig. 4.20.

The chronological chart of tribological developments in the early civilizations presented in Table 4.1 must be considered against the background provided by Fig. 4.20. Most of the significant developments are attributable to Mesopotamia and Egypt, with small contributions

Fig. 4.20
Approximate dates of the early civilizations.

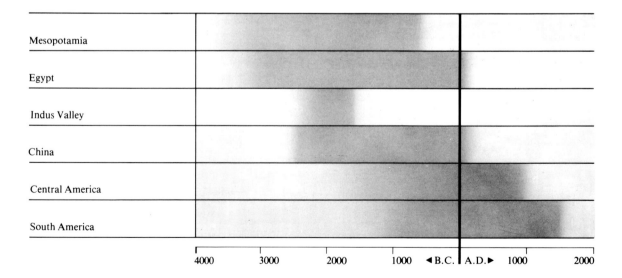

	4000	3000	2000	1000	◀ B.C. ┃ A.D. ▶	1000	2000
Mesopotamia							
Egypt							
Indus Valley							
China							
Central America							
South America							

Table 4.1
Chronology of the early
civilizations

Date	Political and social events	General technical developments	Tribology
		Simple sledges	
3500 B.C.	Development of earliest known civilization in Mesopotamia (between the Tigris and Euphrates)	Thong-drills	**Hand-held bearings (wood, stone, bone and shell)**
	Writing	Wheeled vehicles (solid or tripartite wheels)	
		Potter's wheel	**Stone bearings used in potters wheels – possible use of bitumen**
		Textiles and other crafts	
		Farming	
3200 B.C.	Early Egyptian civilization (Archaic period 3200 B.C.–2700 B.C.; Old Kingdom 2700 B.C.– 2160 B.C.)	Bow-drills	
		Construction of stone buildings; palaces and pyramids	**Possible use of mortar as a squeeze-film lubricant for large building blocks**
		Fine carved furniture	
		Use of sledges for transporting heavy stone statues and building blocks	**Use of copper nails to resist wear of rims of wheels (Kish, Ur, Susa)**
2500 B.C.	Evidence of civilizations in Indus Valley and China	Bronze (alloy of copper and tin) Age	**Door-sockets made of kiln-fired brick (Indus Valley).**
			Porcelain bearings for potters' wheels in China
			Use of lubricant (water or oil) to ease the passage of sledges over planks (statue of Ti, Saqqara, Egypt).
1880 B.C.			**Movement of statue on sledge over planks lubricated by water – El-Bersheh, Egypt**
1600 B.C.	End of early Indus Valley civilization		
1500 B.C.		Wheel appeared in China	
1400 B.C.		Iron Age	**Mutton or beef tallow used to lubricate chariot wheels in Egypt**
1000 B.C.		Use of wooden gears in Egypt and Mesopotamia	

Table 4.1 (*continued*)

Date	Political and social events	General technical developments	Tribology
700 B.C.			Movement of sledge carrying statue over logs; Kouyunjik, Assyria
c. 500 B.C.	Origins of early civilizations in Central and South America	Tools and weapons made from wood and stone in Central and South America	
	End of early Mesopotamian civilization		
c. 300 B.C.			Tapered, bronze bearing with leather seals on Chinese vehicles
c. 30 B.C.	Romans conquer Egypt		
c. 100 A.D.	Mayan culture in Central America		
	Approximate date of end of early Chinese civilization		
700 A.D.	End of 'Old Empire' in Mayan culture		
		Pictographic writing Deerskin books – painted	
1000 A.D.	'New Empire' period in Mayan culture		
1100 A.D.	Aztec culture in Mexico (Aztec Pyramids)	Pottery in Mexico (but no potter's wheel)	
1400 A.D.	Incas in Peru	Impressive building and movement of large blocks of carefully dressed stone	
		Use of soft metals (copper and gold for knives, tools, bells)	
c. 1500 A.D.	End of Central and South American civilizations following Spanish conquest		

coming from the Indus Valley and important, if more speculative, progress being allocated to China.

4.7 Summary

This is a period which, perhaps more than any other, adds substance to the thoughts so adequately conveyed by Leonardo da Vinci in the quotation which opens the chapter.

Well-established views on the order of development of the six civilizations are reflected in the chronological and geographical specification of early progress in tribology. Mesopotamia and Egypt provide a wealth of information on early forms of plain bearings, the use of lubricants and the development of systems for the reduction of friction and wear.

Attention has been drawn to the fact that most of our knowledge of the early civilizations has emerged within the last one and a half centuries, much of it within the past fifty years. There is, in consequence, a high probability that future archaeological studies will yield finds of significance to our subject. This chapter has been written against the background of present-day archaeological evidence, but many important pieces in the jigsaw of tribological history have yet to emerge.

It is clear that the subject of tribology received notable attention in these early civilizations. The second-generation Mediterranean civilizations discussed in Chapter 5 enjoyed a rich inheritance.

Chapter 5

The Greek and Roman period (*c.* 900 B.C.–A.D. 400)

5.1 Introduction

The great artistic merit of the Greek and Roman period has sometimes caused its technological achievements to be undervalued. It was a period of consolidation and development for many of the concepts which emerged in the early civilizations in the Near East; the era in which the icing was formed on a cake of technological progress equalled only by the events of recent times.

The period was marked by the application of scientific principles to the problems of the day; first to models and toys as an illustration of principles, then to the general development of machinery and finally to military engineering, construction and transport. It will be seen that the advantages of rotary motion were recognized in all these fields, that much of the tribological progress reflected this recognition and that the period witnessed the introduction of rolling-element bearings. The words of that flourishing Roman architect and engineer, Marcus Vitruvius Pollio, are indeed indicative of the engineering approach of the day.

5.2 The rise of Greece

As the technologically productive early civilizations in the Near East started to decline the classical Greek civilization developed at the north-eastern end of the Mediterranean. The reasons for the decline of scientific and technological activity in Mesopotamia and Egypt will not be pursued, but it would be wrong to envisage a disintegration of long-standing cultures in the absence of outside influences. Indeed, in the period from the third to the first millennium B.C. both regions suffered intrusions by conquering peoples from the north and east. In the second millennium B.C. the Hittites from the country now known as Turkey established a strong third civilization in which iron was used effectively for tools and weapons. The Hittite realm collapsed about 1200 B.C. and the Assyrians then assumed a dominant position until Nineveh fell to the Medes and Chaldeans in 612 B.C. Then came a great invasion from the east and the formation of the Persian Empire which overwhelmed Egypt and extended to the Danube in the sixth century B.C. It is clear that there was considerable turmoil in the Near East prior to the emergence of the Greek civilization, and it is hardly surprising that scientific studies and technological progress in non-military fields were in decline.

In the early part of the first millennium B.C. the Greeks were themselves divided, with frequent fighting between the city states. Unity came in the fourth century B.C. with conquest from the northern part of the Grecian peninsula by Philip, King of Macedon. Philip's son, Alexander the Great, went on to spread the influence of Greece over a large part of the Near East through his conquest of Egypt, Palestine, Persia and part of India.

Centres of learning flourished throughout the Near East under Greek influence, the period being renowned for its eminence in philosophy, politics, literature and science. It is sometimes argued that the same high position cannot be afforded to Greece in relation to technological achievements, and it is true that in many ways the period is devoid of stirring developments on the scale associated with the early civilizations of Mesopotamia and Egypt. It has also been suggested[1] that the classical attitude developed at this time which regarded the arts as superior to science and applied science had unfortunate consequences which can still be recognized in our present-day technological society. However, it may well be that important technological developments were merely overshadowed by the achievements in other fields, and a careful study of the evidence suggests that this is the case in relation to tribology. In this period of great intellectual activity there is little in the recorded literature which is directly concerned with our subject, yet it holds a few gems of tribological progress. It is necessary to comb the general scientific writings of the day and to put the evidence alongside subsequent archaeological finds in order to obtain a clear picture of the tribological significance of the period.

5.3 Philosophers, scientists and mechanicians

It is interesting to observe the transition from erudite pronouncements by philosophers and scientists to careful descriptions of ingenious machines and a concern for the principles of mechanics, pneumatics, architecture and construction throughout the Greek and Roman era. Pythagoras, whose philosophy was much concerned with immortality and transmigration of the soul, made his contributions to geometry, which still

---

1. See Burstall, A. F. (1963), p. 109.

provide excitement or despair for schoolboys, as early as the sixth century B.C. In the writings of the Greek historian Herodotus (484–425 B.C.) we find evidence of the ancient method of producing bitumen and a lighter oil from petroleum.[2] The main theme of his work was the enmity between Greece and Asia and he recorded his account of the treatment of petroleum after travelling in the Near East.

The great philosopher Socrates (469–399 B.C.) left no writings, but his approach was continued by students who included Plato (427–348 B.C.). One of Plato's famous pupils was Aristotle (384–322 B.C.) who, in his *Questiones Mechanicae*, recognized the force of friction and further observed that it was lowest for round objects.[3] This observation can hardly be taken as the beginning of the ball-bearing industry, but it does demonstrate that the Greek philosophers and scientists were starting to think about important aspects of our subject.

Archimedes (287–212 B.C.), a Greek mathematician of Syracuse, was perhaps the greatest exponent of mechanics and one of the outstanding mathematicians of all time. He is said to have uttered the exclamation, 'Eureka' upon discovering a means of determining the proportion of base metal in Hiero's golden crown by the principles of hydrostatics. He was the son of Pheidias, the astronomer, and although he developed great mathematical talent he succeeded in applying his skill to the solution of definite practical problems. He invented and made several interesting machines, one of which was the screw for raising water which bears his name. His approach was to have a great influence on later workers, particularly those at Alexandria.

The School at Alexandria was active in mechanics between the third and first century B.C. Its contributions were a mixture of theoretical and practical work, although the latter was often dissipated by its application to the construction of scientific toys rather than to the solution of technological problems facing society at that time. Euclid (*c.* 300 B.C.) completed his treatise on geometry at Alexandria and one of the notable mechanicians was Ctesibius (*c.* 300–230 B.C.), believed to have been the son of a barber, who himself started work in a barber's shop in Alexandria. He wrote the first work on the subject of pneumatics, but the manuscript was lost and it is now known to us only through quotations and extracts from later writers. Ctesibius was exceedingly talented, having invented the fire-pump, water-organ, catapults and a 'war engine'. He also improved the existing water-clocks.

Another notable contributor to the science of pneumatics was Philo of Byzantium–later called Constantinople. Very little is known about Philo, but he probably lived a generation or so after Ctesibius. He wrote a number of books on 'technics', one being devoted to pneumatics, but the extent to which his work was based upon the now missing writings of Ctesibius is difficult to determine. His writings on pneumatics are particularly interesting for their accounts of constant-head devices and

---

2. See Hersey, M. D. (1936), p. 1.
3. See Davison, C. St. C. B. (1957(*a*)), p. 158.

siphons. Philo's work is worthy of note in the present context since it had a considerable influence upon Hero of Alexandria, and we shall see that the latter made the most notable references to tribological topics of all the mechanicians of the School of Alexandria.

Some doubt exists about the dates to be attributed to Hero, but it seems to be fairly certain that he was alive in A.D. 62. He had shown how it was possible to determine the distance between Rome and Alexandria by observing the same lunar eclipse from both places, and in his commentary on Hero's *Dioptra* Neugebauer (1938) noted that the only eclipse that would fit the observations took place in the year A.D. 62. Drachmann (1948) has given the background to this piece of detective work in his fascinating account of the contributions made to the subject of pneumatics by Ctesibius, Philo and Hero. The originality of Hero's work was discussed in some detail and Drachmann concluded that, although Hero knew of Philo's work and sometimes copied his instruments, he generally took great pains to change them in one way or another. Hero himself wrote[4] '. . . You should also avoid the arrangements of your predecessors, so that your contrivance will look more fresh.' Drachmann (1963) has also presented a valuable chronology in his text dealing with the mechanical technology of Greek and Roman antiquity.

Hero wrote extensively on pneumatics, constant-level devices, siphons and toys based upon hydrostatic principles. He also used heated air and steam in a number of inventions. In the apparatus shown in Fig. 5.1 the opening and closing of the doors of a small temple was automatically controlled by the heat from a fire on a small altar. The altar contained an air chamber which was connected to a sphere containing water below floor level. Heat from the altar fire expanded the air which in turn displaced water from the sphere via a siphon into a bucket. As the bucket sank down it caused cords connected to balance weights to rotate the door-spindles, and since all the mechanism was hidden below floor level the effect upon spectators must have been pronounced. When the fire went out the water was siphoned back into the sphere and the doors closed again. Hero mentioned that some people used mercury because it was heavier than water and easily expanded by heat.

Tribological interest in the automatic temple arises from the pivot bearings for the door-spindles. These are clearly seen at the bottom of Fig. 5.1 and Hero must have paid some attention to the functioning of bearings in order to make his miniature machines operate in a satisfactory manner. In his treatise on pneumatics he wrote[5] '. . . Let the hinges of the doors be extended downwards and turn freely on pivots in the base. . . .'

Hero's aeolipile shown in Fig. 5.2 has been described as the first attempt to make a reaction turbine. A kettle is provided with a lid which in turn supports a sphere on two bent rods, one of which is hollow, while the other forms a pivot bearing. When the kettle is heated steam enters the sphere through the hollow vertical pipe and thence escapes to the

---

4. See Drachmann, A. G. (1948), p. 99.
5. See Cohen, M. R. and Drabkin, I. E. (1948), p. 328.

Fig. 5.1
Hero's model theatre with
automatically opening doors—note
pivot bearings. (After Cohen and
Drabkin, 1948)

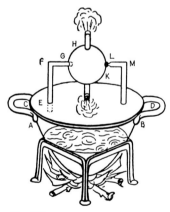

Fig. 5.2
Hero's Aeolipile—note trunnion
mountings. (After Cohen and
Drabkin, 1948)

atmosphere through two further pipes placed diagonally opposite each other and bent in such a way to make the sphere rotate. Although this ancient form of reaction turbine was a most interesting toy it was not a practical engine. The sphere merely turned without any attempt being made to drive any further mechanism. Once again the successful operation of the machine called for some skill in the manufacture of bearings; this time for the trunnion mountings of the sphere.

Hero was much concerned with the principle of moments, without addressing it as such, and one of the most interesting asides of tribological importance occurs in his description of the construction of the wheel and axle in his book on mechanics. He explained that there were five forms of machine which could be used to move a weight by a given force: the wheel and axle, lever, system of pulleys, wedge and the endless screw. In his *Mathematical Collection* Pappus, who lived in Alexandria at the end of the third and the beginning of the fourth century A.D., gave an abridged account of these five machines. Of the construction of the wheel and axle it was said that[6] '. . . One must take a strong log squared off like a beam, make its ends round by planing, and fit bronze end-pivots to the axle, so that when inserted in round openings in the immovable framework, they turn easily. For these openings also have a bronze lining upon which the end-pivots rest. . . .' It is clear from this account that the value of metal-clad bearings was recognized in the School at Alexandria and that bronze was the preferred metal of the day.

It may well be true that the Greek philosophers and mathematicians showed little concern for practical problems and that the mechanicians of the School of Alexandria were concerned with scientific and spectacular toys rather than real machines, but it is equally clear from the occasional references to friction, lubricants and bearings that some progress was being made in the field of tribology. We shall see shortly that this period witnessed more spectacular advances in tribology in connection with full-scale machinery and military machines.

5.4   The development of machinery

It is inconceivable that the general development of machinery in this period could have proceeded without a recognition of various facets of tribology. It is known that the lathe developed at this time and that it was probably driven in the same way as the thong-drill.[7] There are records of the use of such devices as the lever and the wheel, and the stamping of coins was a well-established process in Greco-Roman times. Mechanical systems which utilized rotary motion and which were developing rapidly included lathes, wheeled transport, pulleys, gears, cranes and mills of various kinds.

The lathe was used in the manufacture of furniture and it may have been used to fashion the spokes of chariot wheels. The solid wheel was still used for farm wagons and other heavy transport, but the spoked wheel played a large part in the development of light vehicles for rapid transport and military applications. On spoked wheels the metal tyre performed a tribological and structural role, since it helped both to reduce wear and stabilize the felloe. It is also worth noting that the horse-shoe, which emerged in Roman times, can be considered as a tribological device used to reduce wear and damage to the hoof! Human footwear was also prone to excessive wear and a particularly fine example from Fleet Ditch, London of the use of iron nails to reduce the wear of Roman *caligae* is shown in Fig. 5.3.

The origin of the pulley is not clear but the first evidence of its existence is found in the Assyrian relief of the eighth century B.C. shown in Fig. 5.4. The loads on the bearings would have been quite small and it seems most likely that wooden components were used. It is surprising that the earlier civilizations in Mesopotamia and Egypt, in which the wheel was well known, did not develop the pulley, but it appears that this useful device arrived at a relatively late stage in the Near East. The pulley was certainly

6.   See ibid., p. 225.
7.   See Burstall, A. F. (1963), pp. 68, 75.

Fig. 5.3
The use of iron nails to reduce the wear of Roman caligae–Fleet Ditch, London.

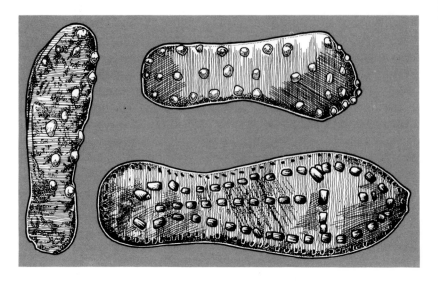

known to the Greeks and widely used in construction by the Romans. The structural arrangement of the simple crane was also well known and the use of pulley wheels to reduce friction and wear between the rope and frame must have represented a marked improvement in the performance of the machine.

The early development of gearing has been described by Dudley (1969). It appears that several well-known Greeks like Aristotle, Ctesibius, Archimedes and Hero made reference to various forms of gears made from wood and metal, but apart from recommendations for the general design, including the use of certain materials like iron, there is little evidence of the tribological performance of such devices.

The grinding of corn was responsible for an important mechanical development of tribological significance in this period. The early mills were hand operated and in the Egyptian saddle-quern a flat stone was rubbed to and fro on top of another. At a later stage the upper stone was fastened to a pole which was pivoted at one end so that a to-and-fro motion of the free end caused the stone to reciprocate along the arc of a circle. This *pushing mill* was superseded in Greco-Roman times by the true rotary quern shown in Fig. 5.5; a development which introduced the crank mechanism. The clearance between the stones was controlled by the height of the central pin and the efficiency of operation would owe much to the performance of the top pivot-bearing. Needham[8] has pointed out that the clearance between the stones could have been adjusted by allowing the central pin to pass through a hole in the lower stone on to a movable lever, and that this arrangement was common to all subsequent water- and windmills whether in China or the West. He has also drawn attention to the interesting evidence of parallel development in China and the Mediterranean. He concluded[9] that there was no evidence for rotary

Fig. 5.4
Assyrian relief showing pully–eighth century B.C.

8.   See Needham, J. (1965), p. 183.
9.   See ibid., p. 187

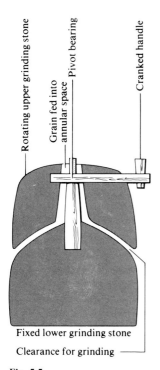

Fig. 5.5
Rotary hand-mill or quern.

mills of any kind in the West before Roman times and that the quern was in general use in China in the first half of the second century B.C.[10]

An interesting variant on the general line of development of milling equipment emerged in Greece in the fourth century B.C. in the form of an olive-crushing mill known as a *trapetum*. The essential features of the mill are shown in Fig. 5.6. It was designed to extract the stones from the fruit without crushing them; the oil being produced by pressing the pulp at a later stage.[11] The two hemispherical stones (*orbis*) were mounted on a wooden pole (*cupa*) which rotated about a vertical iron pivot (*columella*) fixed in a central pillar forming part of the base. In some cases the annular circular mill-trough (*mortarium*) was made of lava.

The tribological interest in the trapetum arises from the nature of the bearings in the orbis, which were apparently made from iron.[12] It is clear that wear between the cupa and the orbis would be excessive if the stones rubbed against the wood; this being an excellent early example of the recognition of the deleterious effect of wear upon machine performance and a tribological approach to the solution of the problem. The bearing formed between the columella and the cupa would not be highly loaded if roughly equal and opposite efforts were applied to the two ends of the cupa by animal power or manpower. The use of washers and collars on the mill is interesting and it is clear that successful operation of the machine called for accurate control of clearance, while adequate mechanical performance and life of the components called for satisfactory bearings.

The Romans later (*c.* A.D. 50) developed a simpler form of olive-mill consisting of two cylindrical wheels which ran on a plane surface. The stone wheels rotated about a horizontal axle which in turn rotated about a vertical shaft located in top and bottom bearings. The arrangement is essentially similar to the trapetum in mechanical and tribological terms, but it was probably easier to construct.

Fig. 5.6
The trapetum, an olive-crushing mill.

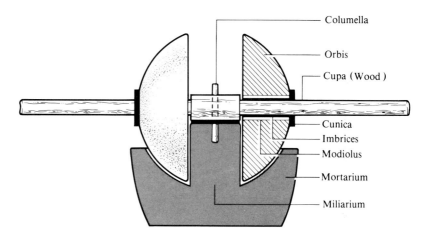

---

10. See Needham, J. (1965), p. 190.
11. See ibid., p. 202.
12. See Drachmann, A. G. (1932).

5.5   The Romans

In historical terms the great Roman civilization quickly followed and eventually engulfed the development based on Greece. Rome itself was founded in 743 B.C. as a small city state among many others in central Italy. The Romans fought and eventually conquered their neighbours, but instead of playing the traditional role of conquerors, they gradually permitted their enemies to adopt Roman citizenship, thus producing allies and an enlarged state.

By the middle of the third century B.C. the Romans had captured all the Greek cities in southern Italy and they had expelled the Carthaginians from Sicily. Carthage a city three times the size of Rome, was razed to the ground in 146 B.C. Eventually the Roman Empire extended from England to the Near East and the Danube to North Africa. Early expansion of the empire was guided by a republican form of government, but rivalries and quarrels between politicians and soldiers led to the breakdown of the Roman Republic and the creation of the Roman Empire in the first century B.C. The nephew and adopted son of Julius Caesar became the first Emperor and the title Augustus was conferred upon him in 27 B.C. The Roman Empire, which persisted until the latter part of the fifth century A.D., brought great stability to the Western world and an unrivalled influence upon subsequent generations. Nevertheless, there exists a measure of disagreement about the technological achievements of Rome. In the minds of most people the Romans were great engineers, but their contributions in architecture and construction were not matched in the mechanical sciences. Most of the forms of transportation employed by the Romans, including ships, wagons and chariots had been developed earlier, while the basic forms of machinery were known in Grecian times. Military effectiveness seems to have been the spur to many Roman developments, including roads and various forms of weapons.

Pliny the Elder (*c.* A.D. 23–79), Roman author of thirty-seven books on natural history (*Historiae Naturalis* 37) prepared a list of lubricants known to the 'ancients' of his time. Parish[13] took this and other evidence to mean that they had available and in use for various purposes the oils from about the same number and kind of seeds and plants as we have at present.

Some of the richest records of the state of tribology in Greco-Roman times are found in the writings of Marcus Vitruvius Pollio, the Roman architect and engineer who lived in the first century B.C. In the year 25 B.C. he dedicated his treatise *De Architectura* to Augustus, the first Roman Emperor, but his work was lost during the decline of the empire. Fortunately, a copy of this manuscript was located again in the fifteenth century and it is said to have had a considerable influence upon Renaissance architecture. The work comprises ten books on architecture and the student of Roman engineering is well advised to read the English translation by Morgan and Warren (1960). The opening paragraph of the preface to the first book, while a little ingratiating, tells a great deal about

---

13.   Parish, W. F. (1929), Vol. 14, edn. 14, p. 453.

Vitruvius, Augustus and the times in question and is repeated here in full.

While your divine intelligence and will, Imperator Caesar, were engaged in acquiring the right to command the world, and while your fellow citizens, when all their enemies had been laid low by your invincible valour, were glorying in your triumph and victory,—while all foreign nations were in subjection awaiting your beck and call, and the Roman people and senate, released from their alarm, were beginning to be guided by your most noble conceptions and policies, I hardly dared, in view of your serious employments, to publish my writings and long considered ideas on architecture, for fear of subjecting myself to your displeasure by an unreasonable interruption.

In his tenth book Vitruvius discusses machines which he defines as '. . . a combination of timbers fastened together, chiefly efficacious in moving great weights'. In his discussion of cranes[14] he digressed to explain how Chersiphron transported the shafts of the columns of the temple of Diana a distance of eight miles (12·9 km) from the stone quarries to the site at Ephesus. Carts could not be trusted '. . . lest their wheels should be engulfed on account of the great weights of the load and the softness of the roads in the plain'. The contractor found a satisfactory solution by fixing iron gudgeons shaped like dovetails into the ends of the cylindrical blocks of marble and then constructing a frame of 4 in (101 mm) timbers to contain the shaft like a garden roller. Oxen were used to pull the shafts over the soft earth and Vitruvius explained that the iron gudgeons were supported in 'rings' fixed into the wooden frame; a clear example of the use of metallic bearings in Roman times.

Chersiphron's son, Metagenes, used the same principle when the architraves were transported to the site. In this case the square-section architraves were built into 12 ft (3·7 m) diameter wheels to form axles, the projecting iron pivots again being supported in rings in a wooden frame. The wheels and axles rotated as one unit within the frame pulled by oxen, and according to Vitruvius the wheels soon reached the building. A sketch of the arrangements adopted by Chersiphron and his son is shown in Fig. 5.7.

Later, Paconius tried to move a block of stone some 12 ft (3·6 m) long, 8 ft (2·4 m) wide and 6 ft (1·8 m) high, which was to form a new pedestal for the colossal Apollo, in a slightly different way. He built the block into 15 ft (4.5 m) diameter wheels as Metagenes had done, but he then fastened 2-in (50·8 mm) crossbars from wheel to wheel round the stone with a maximum spacing of 1 ft (0·3 m). A rope was then coiled round the bars and yoked up to oxen. As the rope unwound the wheels rotated, but the structure swerved from side to side, making it necessary to draw back again. This drawing to and fro apparently made the exercise much more expensive than anticipated, for Paconius became insolvent.

Vitruvius referred to the economic aspects of engineering in various places in his books. In the introduction to his tenth book he drew attention to an ancient Greek law in which an architect became personally responsible for excess costs of a public building for which he was responsible if the charge exceeded the estimate by more than one-quarter.

---

14. See Morgan, M. H. and Warren, H. L. (1960), p. 288.

Fig. 5.7
Transporting carved stone from
the quarry to the site.
(a) Chersiphron's procedure for
conveying the shafts for the temple
of Diana at Ephesus.
(b) The method adopted by
Metagenes, son of Chersiphron, to
transport the architraves.

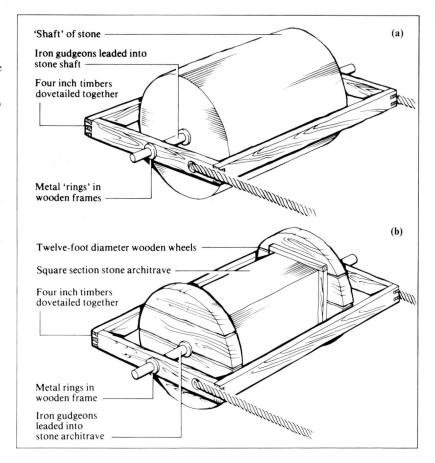

He cited this law to illustrate the need to complete construction in a satisfactory and timely fashion, and he utilized this theme to favour the inclusion of a section on machines in his treatise.

Chapters IV to VII in the tenth book by Vitruvius were devoted to hydraulic machines for raising water. In his account of the tympanum, a kind of treadmill for raising water from a stream, he referred to the use of lathes and once again to iron bearings. He wrote, '. . . An axle is fashioned on a lathe or with the compasses, its ends are shod with iron hoops, and it carries round its middle a tympanum made of boards joined together. It rests on posts which have pieces of iron on them under the ends of the axle.'

In this case it appears that the iron-shod axle merely rested upon iron supports and was not supported in full journal-bearings, but the more usual arrangement was mentioned in his account of the construction of an Archimedean screw where he wrote '. . . The ends of the shaft are covered with iron. To the right and left of the screw are beams, with crosspieces fastening them together at both ends. In these crosspieces are holes sheathed with iron, and into these pivots are introduced, and thus the screw is turned by the treading of men.'

A rare mention of the use of lubricant occurred in the section dealing with the bronze, twin-cylinder water-pump developed by Ctesibius when Vitruvius wrote '. . . Pistons smoothly turned, rubbed with oil, and inserted from above in the cylinders. . . .'

The Carthaginians discovered the merits of a battering-ram during their siege of Cadiz, although at that time the ram merely consisted of a beam driven repeatedly against the fortification by soldiers. According to Vitruvius '. . . a carpenter from Tyre, Bright by name and nature, was led by this invention into setting up a mast from which he hung another crosswise like a steelyard, and so, by swinging it vigorously to and fro, he threw down the wall of Cadiz'. At a later stage the swinging ram was mounted on a wheeled structure and covered with oxhide to protect the operators; the device then being called a 'tortoise of the battering-ram'. The history of the battering-ram is important to our subject since it led to the development of the first recorded roller-bearing by Diades, engineer to Alexander the Great. Vitruvius described the movable scaling towers, tortoise and borer devised by Diades and in his description of the machinery of the ram he wrote '. . . in which was placed a roller, turned on a lathe, and the ram, being set on top of this, produced its great effects when swung to-and-fro by means of ropes'.

Of the borer he wrote '. . . There are numerous rollers enclosed in the pipe itself under the beam, which made its movements quicker and stronger.' Several authors[15] have given priority to Diades (*c*. 330 B.C.) for the development of a roller-bearing. It appears that the rollers were made of wood, turned in a lathe, and that they were separated by a form of cage.

These references from the ten books of architecture show how the military requirements of the day promoted technological developments. Vitruvius gave an interesting account of the engineering scene and his digressions add great spice. He related the story of Archimedes and his bath with some relish and he expressed his amazement at a society which accorded greater honours to famous sportsmen than to authors like Pythagoras, Democritus, Plato and Aristotle '. . . whose boundless services are performed for all time and for all nations . . .'–a sentiment which has echoed through the ages!

We have seen that the Greek and Roman literature contains several references to the art of tribology, but it is to archaeology that we turn for the most exciting evidence of the development of rolling-element bearings.

There is in Italy, some eighteen miles (29 km) south-east of Rome, a small lake called Nemi among the Alban Hills. The lake occupies the crater of an extinct volcano. A small village of the same name resides in the north-east corner near to the Appian Way, while much of the lake is surrounded by thick woods.

The name 'Nemi' is associated with rites dating back to antiquity. Diana, the moon-goddess of virginity and hunting, was worshipped in the groves surrounding the lake. The celebrations involved human sacrifice

---

15. See Morgan, M. H. and Warren, H. E. (1960); Davison, C. St. C. B. (1957(*b*)); and Naylor, H. (1965).

with the priest holding office until an aspirant plucked a mistletoe bough from the sacred grove and slayed him in combat, a theme which formed the basis for Frazer's (1922) treatise on primitive superstition and religion.

It is strange how the virgin hunter and patroness of chastity twice enters our tribological history of Greco-Roman tribology. The construction of her temple at Ephesus caused Chersiphron and his son to introduce novel systems incorporating iron bearings for the transportation of carved stone columns and architraves across soft earth, while she now appears as part of the background to the story of Lake Nemi. If the subject of tribology ever required a more glamorous image it could well seek the patronage of the goddess Diana.

The fishermen had for a number of centuries passed on tales about ships on Lake Nemi which were said to have dated from ancient times, but it was not until the end of the nineteenth century that Eliseo Borghi began serious study of these stories. On 3 October 1895 one of the ships was located and divers recovered many objects including a fine metal ring associated with the rudder. On the indications of fishermen a second ship was found on 18 November of the same year and further material was recovered. These finds verified the existence of the two ships, but many of the most valuable objects were removed by divers.

The Italian dictator Benito Mussolini, who had a strong personal interest in exposing the earlier glories of Rome, made a passionate speech to the Reale Società Romana di Storia Patria on 9 April 1927 in which he announced an archaeological undertaking to recover the sunken ships in Lake Nemi.

It was decided that the level of the lake should be lowered to expose the two ships and the task of pumping and cutting the passages through the mountain progressed rapidly. A bronze from the first ship was exposed on 4 June 1929 and the second ship was recovered in 1930. Both are believed to have been constructed between A.D. 44 and 54 and an illustration of the scale of these ancient ships is shown in Fig. 5.8. Mussolini had approved a suggestion that all the material recovered from the lake should be preserved on the spot in a purpose-built museum, and a fine structure consisting of two main bays having a total plan area of $262 \times 255 \, \text{ft}^2$ ($80 \times 77 \cdot 6 \, \text{m}^2$) was completed in 1933.

The first edition of an excellent account of both the salvage operation and the finds was prepared by Ucelli in 1940. The most interesting mechanical device found on the ships was a platform revolving on trunnion-mounted balls. Borghi had salvaged a number of these bronze balls from the first ship in 1895 and sixteen are preserved in the Museo Nazionale Romano. A general display of part of the finds, including the bronze balls, is shown in Fig. 5.9.

The operation conducted in the late 1920s exposed more loose balls including two nailed with iron straps to a fragment of a wooden platform as shown in Fig. 5.10. A reconstruction of the thrust-bearing arrangement is shown in Fig. 5.11; the superficial similarity between this arrangement and modern thrust rings being quite remarkable. There is, however, one important difference: the balls are not free to rotate between the stationary and rotating tables since the trunnions are strapped to the

Fig. 5.8
Visitors viewing the first ship
salvaged from Lake Nemi.

turntable. Furthermore, the load is not transmitted across diametrically
opposite points on the balls, but rather between the lower point on the
balls and the flat base, on the one hand, and the upper generators of the
trunnions and the turntable, on the other. Ucelli shows that the balls
sit in relatively deep recesses in the turntable as illustrated in the upper
part of Fig. 5.11. In a remarkable passage he confirmed this observation
by noting that the machining marks left on the bronze balls when they
were turned in a lathe were visible everywhere except for a narrow band
corresponding to the contact between a rolling ball and the flat stationary
surface.

Fig. 5.9
Part of the material salvaged from
Lake Nemi by Borghi. Note the
wooden fragments and sixteen
trunnion-mounted bronze balls.
Displayed in the Museo
Nazionale, Romano.

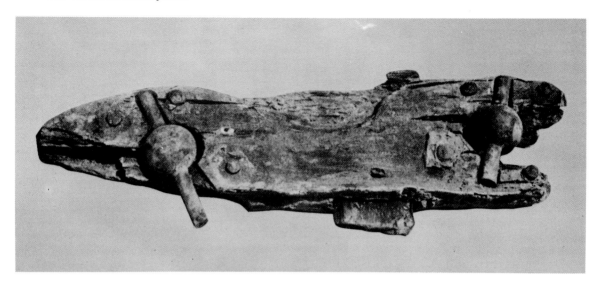

Fig. 5.10
Fragment of revolving wooden platform with bronze balls and straps from the ships in Lake Nemi.

Why were the balls trunnion mounted? It may be that the Roman engineers envisaged some unknown advantage for this arrangement of a ball-bearing, or that they used the trunnion mounting in an early attempt to perform the function of a cage in separating the rolling elements in a modern ball-bearing, but it seems far more likely that they were constrained to this form of rolling element by current manufacturing techniques. It is interesting to note that the balls were turned on a lathe, and since it would be extremely difficult to complete the turning of spheres, but relatively easy to leave reduced cylindrical sections as stub shafts, the latter view looks most reasonable.

Borghi had also salvaged ten little cylindrical rollers from the first ship in 1895, but all have been lost. The only record of the form of these interesting objects is contained in the scale drawing by Malfatti shown in Fig. 5.12. Nothing is known about the purpose of these rollers. They might have been used in a linear bearing rather like Diades' battering-ram, or they might have been used in a more complex machine. The trunnion mountings found on the balls is again a feature on the rollers, suggesting that the latter undoubtedly had a tribological application.

A third find in this extraordinary collection of early rolling-element bearings consisted of two fragments of another smaller circular platform, shown in Fig. 5.13, which appears to contain wooden rolling elements. These fragments, like the trunnion-mounted balls, were found in the bow of the ship. Ucelli suggested that the rollers were in fact conical and his reconstruction of a taper-roller thrust-bearing based on this view is shown in Fig. 5.14. It is difficult to state with certainty that the rollers, which show an unusual wear pattern, were in fact conical, but Ucelli seems to have been in no doubt.

The ships of Lake Nemi must surely present one of the outstanding features of tribological archaeology. We find at once three of the basic forms of rolling-element bearings: ball, cylindrical and taper rollers, albeit with trunnion mountings. The purpose of these rolling elements is not clear, but Ucelli wrote:

Fig. 5.11
Reconstruction of the revolving
wooden platform on
trunnion-mounted bronze
balls from the ships in Lake
Nemi.

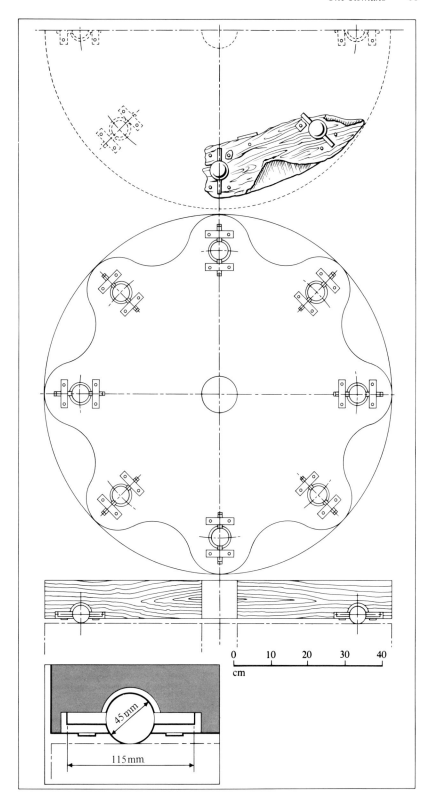

Fig. 5.12
Scale drawing of one of the ten
bronze cylindrical rollers
recovered by divers from the ships
in Lake Nemi in 1895.

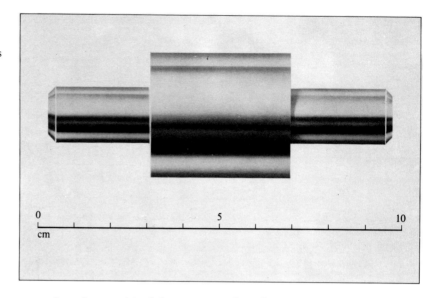

*. . . Some have explained them as items of naval rigging, capstans or windlasses;
others have recalled Cellini's description of the normal employment of revolving
bases for the support of statues; the engineer Jacono even thought of a stand for a
solar clock which could be oriented. But even if the function cannot be ascertained,
the evidence of the ingenious contrivance is undoubtedly of interest in the history of
technology.*

The story of this ancient Roman Navy as already told is remarkable
enough, but there is an almost unbelievable final twist. On 31 May 1944,
during the culmination of the Battle of Rome, the ships were burnt by
the German Army and most of the valuable remains were lost. A revised
version of Ucelli's (1950) book also deals with the tragic loss of the
ships and the reconstruction of the museum. This well-illustrated and
beautifully presented book forms a most valuable record of an amazing
aspect of tribological history. If Benito Mussolini had not displayed such
enthusiasm for the history of ancient Rome we would have had no

Fig. 5.13
Fragments of revolving wooden
platform and two wooden rollers
from the ships in Lake Nemi.

Fig. 5.14
Reconstruction of the revolving wooden platform on wooden taper rollers from the ships in Lake Nemi.

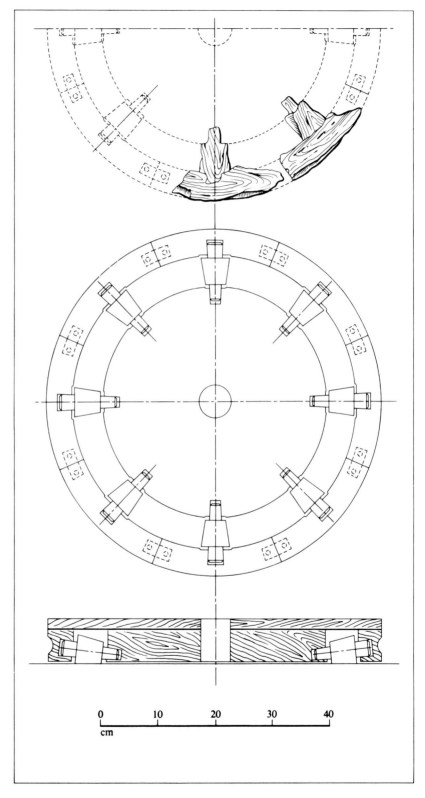

0   10   20   30   40
cm

knowledge of these impressive early rolling-element bearings but they would have been still preserved for posterity at the bottom of Lake Nemi.

5.6   Celtic vehicles and a Chinese puzzle

One of the features of Celtic armies admired by the Romans as they conquered north-western Europe was the lightweight two-wheeled chariot. This vehicle, which was used to carry the Celtic warriors to the scene of battle, but not as a fighting machine, was recognized as the product of a skilful race. These skills were not restricted to two-wheeled vehicles, and important finds in Danish bogs at the end of the nineteenth century have given rise to much exciting discussion and speculation about the role of the Celts in the development of rolling-element bearings.

The four-wheeled Dejbjerg carts, dating from the first century B.C., were found in 1881 and 1883 in a bog in the glebe land south-west of Dejbjerg Church in western Jutland. The carts had been taken to pieces and placed within an enclosure of stakes driven into the ground. The find was described in full in Danish by Petersen (1888) and later in English by Klindt-Jensen (1949). A view of one of the restored Dejbjerg carts as it appears in the National Museum in Copenhagen is shown in Fig. 5.15. Remains of eight wheels were unearthed, the hubs being of oak and the spokes of hornbeam. The centre of the hubs had the appearance of flattened spheres with cylindrical sections on each side encased in bronze collars of the form shown in Fig. 5.16. The axle bearing within these collars exhibited some essential features of an early roller-bearing,

Fig. 5.15
One of the Dejbjerg carts in the National Museum, Copenhagen.

Fig. 5.16
The bronze collars on the hubs of each Dejbjerg cart.

with some thirty-two circular section axial grooves as seen in the right-hand bronze in Fig. 5.16. Within these grooves were pieces of wood, described by Klindt-Jensen as cylindrical sticks, and it is easy to see how the association with rolling elements arose. This interpretation of the use of roller-bearings in Celtic carts in the first century B.C. has met with some acceptance by archaeologists, but Needham (1965) has questioned its validity on the grounds that Petersen (1888), in his original paper in Danish, described the pieces of wood which came out of the axial grooves in the hubs as flat strips and not rollers. The alternative view is that the strips may have been inserted to reduce wear, possibly in association with a lubricant, without actually rotating like true rollers. The true purpose of these bearing elements may never be known, but the complete arrangement of axial grooves, whether for true rolling elements or the application of lubricant, is a remarkable example of tribological progress in a part of Europe remote from the developments in Rome.

While expressing his doubts about the claims for wooden roller bearings in the Celtic carts from Dejbjerg, Needham[16] has drawn attention to the rival claims of China. Some remarkable annular bronze objects with internal grooves were found in Hsüeh-chia-yai village in Shansi. These objects, dated at least as early as the second century B.C., had four or eight internal compartments as illustrated in Fig. 5.17. The suggestion that the Chinese were actively involved in the early history of rolling-element bearings is strengthened by the fact that each of these internal compartments was filled with granular iron rust. Was a rotating shaft supported on balls or rollers constrained to rotate within these compartments?

---

16.   Needham, J. (1965), p. 94.

Fig. 5.17
Annular bronze objects
with internal
compartments–Hsüeh-chia-yai
village, China–second
century B.C.

Table 5.1
Chronology for the Greek and
Roman period

5.7 Chronology

The classical period of the Mediterranean civilizations considered in this chapter spans almost the millennium before to almost half the millennium beyond the birth of Christ. The Greek philosophers said little about tribological or even mechanical systems, but the mechanicians of the School at Alexandria showed clearly their concern for bearings in the models and machinery they devised.

During the period a number of threads of interest in rolling-element bearings emerged in various quarters and if they were not fully entwined into a firm thread of material the basis of a new fabric of tribology had been established by the time the Roman Empire collapsed in Western Europe. It is of some interest and importance to relate these events to historical, social and general technical developments of the period and it is to this end that Table 5.1 has been constructed.

5.8 Summary

There can be little doubt that many important tribological concepts emerged in the Greek and Roman period. The advantages of rolling over sliding were appreciated at an early stage while the mechanicians in Alexandria introduced effective plain metal bearings in numerous devices.

Date	Political and social events	General technical developments	Tribology
900 B.C.			
800 B.C.	800 B.C. Carthage built 776 B.C. First Olympiad 743 B.C. Rome founded	Pulley wheel used in Assyria	
700 B.C.	700–300 B.C. The Celts invaded and settled in England 612 B.C. Medes and Chaldeans take Nineveh		
600 B.C.	550 B.C. Buddha and Confucius lived about this time		
500 B.C.	484–425 B.C. Herodotus 469–399 B.C. Socrates 427–348 B.C. Plato		**Description of production of bitumen and a lighter oil from petroleum**
400 B.C.		Olive-crushing mill in Greece	**Use of iron bearings in olive-mill**
	384–322 B.C. Aristotle 323 B.C. Death of Alexander the Great		**Recognition of force of friction 330 B.C.** **Wooden cylindrical rollers, turned in a lathe, used in a linear roller-bearing on a battering-ram–Diades**

Table 5.1 (*continued*)

Date	Political and social events	General technical developments	Tribology
300 B.C.	300–230 B.C. Ctesibius; Philo	*c.* 300 B.C. Euclid's treatise on geometry	
	287–213 B.C. Archimedes	Pneumatics, constant-head devices, fire-pump, water-organ, catapults	**Simple gearing of wood and metal – Alexandria**
		Archimedean water screw or 'snail'	
		214 B.C. Great Wall of China begun	
200 B.C.		Quern in use in China	
	146 B.C. Carthage razed	Rotary quern in use in Mediterranean region	**Bronze rolling bearings (?) Shansi, China**
100 B.C.			
	55 B.C. Julius Caesar lands in England		
	27 B.C. First Emperor of Rome (Augustus) – Marcus Vitruvius Pollio	25 B.C. *De Architectura* Use of lathe well established	**Iron bearings used in moving stone columns, hydraulic machines and mills. Lubricant used in twin-cylinder bronze water-pump**
0 B.C.			**1st century B.C. Wooden rollers (?) in bronze hubs in Celtic wagons**
	A.D. 5–40 London originated as a city	Use of brass	
	A.D. 23–79 Pliny	Corn ground by water-mills	**List of lubricants presented Metal horseshoe used in Roman times and iron nails used in Roman footwear**
			A.D. 44–54 Bronze ball, bronze cylindrical roller and wooden taper roller-bearing elements (trunnion mounted) used – Lake Nemi
	A.D. 62 Hero	Pneumatics; hydraulics, constant-head devices, siphons, extensive use of wheels and pulleys	**Pivot bearings for model temple and aeolipile**
A.D. 100			
	A.D. 123 Hadrian's Wall built		
A.D. 200			
A.D. 300	Pappus A.D. 300–400 Saxons raid Roman Britain		**Description of bronze journal bearings for wheels and axles of machinery**
A.D. 400			
	A.D. 407 Roman Army withdraws from England		

Metal bearings were also used to improve the life and performance of olive-crushing and corn-grinding mills.

The outstanding tribological feature of the period was undoubtedly the progress towards effective rolling-element bearings. Alexander the Great must have felt well satisfied when his engineer devised the linear wooden roller-bearing in order to improve the performance of battering-rams. The amazing finds of metal balls and rollers, together with wooden taper rollers, in a single hoard in Lake Nemi ensures that the Roman Empire is credited with pioneer work in the field of rolling-element bearings *c.* A.D. 50.

If the Greek mechanicians laid the foundations for the great development of rolling bearings evident in Roman times, the remarkable Celtic and Chinese progress enjoyed no such background. These roughly simultaneous developments in such widely separated regions of the world are all the more interesting because they were apparently connected with land transportation by wheeled vehicles. While the archaeologists debate the purpose of the Celtic and Chinese finds there appears to be sufficient evidence to associate the objects with a major advance in the bearings of wheeled vehicles. If they were truly rolling-element bearings they represent a great tribological development which was to wait for almost 1,500 years for further exploitation.

The two-wheeled Roman chariots and carts were highly manoeuvrable, but their four-wheeled wagons did not enjoy this flexibility. In due course the swivelling front axle was to make a valuable contribution to the development of the wagon, but the Roman baggage vehicles, which were apparently based upon an earlier Greek design, did not possess this feature at the time of the invasion of Britain. All conquering armies rely upon the provision of supplies and the slow, cumbrous baggage vehicle was probably just as important to the Romans as the more glamorous chariot. One wonders if the long, straight roads which are afforded pride of place in many accounts of Roman achievements might not, after all, have been a consequence of the inability of the baggage vehicles to negotiate tight bends! The Celts were, of course, skilled vehicle builders and they had certainly incorporated the swivelling front axle in their designs of four-wheeled vehicles. If the Roman engineers had been able to utilize such manoeuvrable wagons in their long march across north-western Europe they might have introduced a more flexible and economic line for their highways.

The period under discussion undoubtedly witnessed a great awakening in the use of rolling-element bearings. The developing world seemed poised to exploit this major form of support for rotating shafts, but it was to be almost 1,500 years before the same interest and expertise flourished again in this field. It is just as important that we seek in Chapter 6 the reasons for this delay as it is to study and marvel at the tribological skills of the Greeks, Romans, Celts and Chinese.

Chapter 6

The Middle Ages
(*c*. A.D. 400–1450)

6.1 Introduction

The period covered by this chapter is generally described by historians as the demarcation between ancient and modern times. It is usually calculated from the fall of the Roman Empire in the West in the early part of the fifth century to the beginning of the Renaissance in the middle of the fifteenth century.

Historians appear to be divided on the question of the technological merit of the Middle Ages. The traditional view of cultural decline in Western Europe following the withdrawal of the Roman legions is well supported and the term 'Dark Ages' has been used to describe the period from the fifth to the ninth centuries. However, on a wider scale, and particularly in relation to science and technology, the term is excessively gloomy and misleading. By A.D. 328 Constantine the Great had already transferred the capital of the Roman Empire to Byzantium, which he renamed Constantinople and which is now known as Istanbul. When the empire was divided between the two sons of Theodosius in A.D. 395 a vigorous and thriving Byzantine culture was established which survived until Constantinople was captured by the Turks in 1453. Furthermore, Islam played a leading part in scientific work in the Middle Ages, particularly from the ninth to the twelfth centuries. It is thus clear that the Eastern division of the Roman Empire enjoyed a period of great activity and development while culture in the West was in decline.

The steady technological development in China continued during the millennium of the Middle Ages and Needham (1965) has presented firm evidence to support the view that many ideas and inventions originated in eastern Asia before being introduced into Europe. Some of these devices are of tribological significance and will be mentioned later in this chapter.

The view that technology, like the general level of post-Roman culture, was in decline in the West is itself open to debate. Compared with the rapid developments in the classical Greek and Roman periods the changes in the Middle Ages were certainly steady rather than spectacular. The disruptive influence upon culture of the Saxon, Viking and Norman invasions of Britain certainly influenced progress in the mechanical arts, although the changes sometimes promoted interesting developments. There was a general regression of bearings in the Middle Ages. The rolling-element bearings which emerged among the earlier Greek, Roman and Celtic cultures in the West and possibly in China in the East were not in evidence. The use of bronze and iron for bearings and even general machinery was in decline in the West with a consequent return to wood and even stone as tribological materials. One example of tribological inventiveness in relation to the use of stone which dates back to Viking times and is particularly noteworthy is the subject of Section 6.6.

Water- and wind-driven mills developed considerably in the Middle Ages and this trend is discussed in Section 6.5. The development of mechanical power was promoted by a reduced availability of human power; partly as a result of the spread of Christianity and the associated movement against slavery, but also by the disastrous ravages of the Black Death which reduced the population of the British Isles by almost 25 per cent in a mere two years in the middle of the fourteenth century. The population of England was then a mere 2 million.

The Church played an important part in the development of mechanical skills in the Middle Ages. White[1] has drawn attention to the involvement of monks in manual work – a most significant change in the attitude of intellectuals to physical effort. Burstall (1963) has argued that the orderly daily life of the church and the monastery provided the stimulus for the development of the mechanical clock – a device which forms the main subject of Section 6.4.

An important feature of the medieval scene which makes our studies somewhat easier than for earlier periods was the increasing use of writing to record events from everyday life. The Lindisfarne Gospel (700), the *Ecclesiastical History of the English Nation* completed by a Northumbrian monk the Venerable Bede (731) and the *English psalters* of the fourteenth century, to which we shall refer later for evidence of tribological development, provide examples of this further service of the Church. Other important publications of the period include that valuable economic census the Domesday Book prepared by order of William the Conqueror and published in 1086, the drawings of water-driven machinery prepared by the Cistercian monk, Villard de Honnecourt (*c.* 1250) and, towards the end of the era, the writings of Chaucer (*c.* 1340–1400). The ninth-century tapestry found with the Oseberg ship in southern Norway and the eleventh-century Bayeux Tapestry give valuable supplements to the written texts of the period.

---

1. See White, L., 'Technology in the Middle Ages', in Kranzberg, M. and Pursell, C. W. (1967), Vol. I, Ch. 5, p. 68.

The early universities, such as Bologna (1100), Paris (1150), Oxford (1200) and Cambridge (1229), provided not only an environment for learning and intellectual pursuits, but also a structure for information exchange on an international basis and repositories for vital knowledge of the current and earlier societies.

Parchment, prepared from the dermis of various animals, had enjoyed a growing popularity as a surface on which to write since Greek and Roman times. The traditional way of joining rectangular sheets of parchment inscribed in vertical columns into long *rolls* stored in cylinders gave way to the folded manuscript volume known as a *codex* after the second century A.D. The *codex* was a forerunner of the familiar printed book which emerged at the very end of the medieval period. In the twelfth century paper slowly began to replace parchment. It reached the West from China via Islam and Spain and its availability in Europe was essential to the development of printing in the mid-fifteenth century.

Perhaps the most interesting and remarkable tribological development in the Middle Ages was the use of hard stones embedded in agricultural machinery and carts to minimize wear of the wooden parts of the structures. I have chosen to devote a good proportion of the present chapter to this story, which appears to be unknown to tribologists, because it illustrates so well a recognition of the technological handicap and economic disaster which can be incurred by excessive wear of vital machine components. Our present-day concern with studies of wear and the economic aspects of tribology had an interesting curtain-raiser in Viking times.

Thus, while culture in Western Europe was in general decline in the Middle Ages a number of major technological innovations were taking place. It is a curious fact that while Western Europe was eclipsed by the vigour of Byzantium, the science of Islam and the influence of original concepts and devices from China, it nevertheless witnessed many of the major engineering developments of the Middle Ages.

6.2 The development of machinery

Most of the tools and machinery employed in the medieval period had origins in earlier times. There were, however, some significant engineering developments of previously known machines which introduced new but rarely excessive demands upon bearings.

Bearing materials

Wooden bearings were still widely used for lightly loaded machine elements although hard materials like stone were also used. There was a continuing utilization of metal and metal-clad bearings in the more severely loaded or more expensive (and complicated) machines, although Davison (1958) has suggested that most machine builders in the period between the fall of the Roman Empire and the fourteenth century were forced to employ hardwood owing to the expense and extra work involved in making metal bearings. The direct evidence from Western Europe does little to confound this view, but there is very little known about the detailed construction of many medieval machines and contemporary illustrations of machines usually fail to show the nature of the bearings. It would, however, be strange if some of the Celtic and Roman experience with metal-clad bearings had not been carried forward to the medieval road vehicles and the journal-bearings of some water-driven equipment. Furthermore, just beyond the period under discussion, accounts show that metal bearings were indeed in use in several forms of machinery, including the potter's wheel (see the reference to ceramics and Cipriano Piccolpasso (1524–79) in the following chapter).

Whatever the limitation in the use of metal bearings in the Middle Ages in Europe, Needham[2] has presented firm evidence to show that there was no such pause in development in China. Successful functioning of the mechanized armillary spheres between the second and eighth centuries depended upon the use of metal bearings, while reference is made to the use of steel bearings in an observational armillary about 720. Similar evidence of the use of iron-on-iron journal bearings comes from the Su Sung clock tower of 1088.

Lubricants

Medieval machines knew only vegetable oils and animal fats as lubricants, the widespread use of mineral oils being yet a long way off. Indeed, the main use of vegetable oil was not in the field of tribology at all, but in cooking, unguents, cosmetics and toilet preparations. The olive was widely grown in the Mediterranean area and the word 'oil' can be traced through *oleum* (Latin), *elaia* (Greek) and probably to the more ancient *ulu* (Semitic).[3]

Animal fats seem to have been preferred in north-western Europe where the climate was unsuited to the olive. In due course rape-seed and the poppy-seed were cultivated for their oils and came to compete with animal fats in various applications in the thirteenth century.

Herbert Maryon[4] has made an interesting comment to the effect that the craftsmen who made the silver bowls, each 9 in (229 mm) in diameter, found among the treasure in the burial ship at Sutton Hoo, Suffolk (*c.* 655), would have oiled the face of the silver in the lathe. These bowls were made by the then rare technique of spinning on a lathe.

Drills and lathes

Thong- and bow-drills were still widely used by carpenters throughout the medieval world, but the introduction of the crank mechanism enabled more rapid and effective drilling to take place as alternating movement was replaced by steady rotation. For light work the brace-and-bit became the most effective hand-drilling system which developed into the familiar modern form, while towards the end of the period lathes were operated by a foot-treadle and crank. Needham (1965) has given a full account of

2. See Needham, J. (1965), p. 93.
3. Forbes, R. J., 'Food and drink', in Singer, C., et al. (1956), Vol. II, Ch. 4.
4. Maryon, H., 'Fine metal work', in Singer, C., et al. (1956), Vol. II, Ch. 13, p. 449.

the history of the crank mechanism in both the East and the West, while Lilley (1965), in his neat account of the evolution of the crank and treadle, suggests that the application of this vital mechanism to other machinery was slow '. . . perhaps because of mechanical difficulties of making good bearings'.

There is distressingly little evidence of the construction of bearings or the use of lubricants in the evolving drills and lathes of this period. The requirements for the guide and thrust bearing on a medieval brace-and-bit would differ little from those of an ancient bow-drill, and a simple recess in a stone or metal plate would normally suffice. Likewise with the early lathes simple wooden, or occasionally metal, bearings were probably used.

It can be seen that the tribological demands of medieval machinery were modest and that existing forms of materials, lubricants and bearings were generally adequate. Indeed it has been argued that bearing development in the West was at a standstill for much of the period. However, the main tribological developments of the period were found outside the field of general machinery which had evolved from earlier years and these will form the subjects of subsequent sections.

6.3   Roads and land transportation

After the fall of the Roman Empire the roads in Europe were badly neglected and in some cases broken up for building purposes. They were, however, used, with little maintenance, by messengers and for the transportation of goods. The Roman roads were probably exceedingly costly to maintain even before the empire collapsed in the West and White[5] has noted that the wealthy Byzantine and Islamic empires decided that they were not worth the expense.

There were a few technical developments in land vehicles during the Dark Ages but the spring carriage emerged in Central Europe about the tenth century. The chariot rapidly lost much of its importance while transport vehicles assumed more significance.

In their efforts to reach the rich Eastern wing of the Roman Empire based on Byzantium, the Vikings found it convenient to sail across the Baltic and then as far as possible up the rivers of Russia. It is said[6] that they completed the land journey, before again launching their ships upon the rivers leading to the Black Sea, by drawing the vessels over wooden rollers in a manner similar to that propounded by Layard for the Assyrian statues.

The Vikings are undoubtedly best known for their seafaring, but the beautiful four-wheeled Oseberg cart shows that they were not unskilled in the art of constructing land vehicles.

---

5.   White, L., Jr., 'Technology in the Middle Ages', in Kranzberg, M. and Russell, C. W. (1967), Vol. I, pp. 66–79.
6.   I am grateful to Professor Bengt Jakobsson of Göteborg for drawing this to my attention.

Wheels

Some carts had metal hoops or tyres to bind the felloes and minimize wear, although they were sometimes banned from towns because of their effect upon the roads. It appears that this system for reducing wear and damage to the wheels created greater wear of the roads and the battle between transportation interests and urban amenities had begun!

It has already been noted (see Ch. 4) that metal nails were used to protect the rims of solid wheels some 4,700 years ago. In medieval times the nails were used once more when iron tyres were banned, but this frequently made matters worse. Protection for the wheel was then limited to flat nails. In some cases the dowels of the spokes were allowed to pass right through the felloes to take the wear in the same way as the metal studs. This arrangement would no doubt be equally effective in breaking up the roads.

Between 1300 and 1380 a number of *psalters* were produced in England in East Anglia. The manuscripts were copiously illustrated with the tinted drawings and paintings revealing much valuable information. In one such manuscript prepared for Sir Geoffrey Luttrell of Irnham in Lincolnshire between 1335 and 1340 the margins were adorned with fascinating country scenes. The Luttrell Psalter was purchased for the British Museum in the 1930s and in the chapter devoted to manuscripts in the *Treasures of the British Museum*, T. S. Pattie[7] has noted that the country scenes redeem an otherwise expensive but uninspired psalter.

The Luttrell Psalter provides the earliest documentary evidence for the use of wooden pegs rather than metal nails or iron tyres to prevent excessive wear of the rims of wheels. A number of vehicles are illustrated in the margins of this unlikely record of tribological history and one, a two-wheeled harvest cart with studded wheels, is reproduced as Fig. 6.1. There is, however, one feature of this illustration which must be noted. Although the description 'studded wheels' is used the studs or pegs appear to first pass through, or be an integral part of, a narrow band or rim on the wheel. The studs or pegs may therefore have been used not only to resist wear of the wheel but also to retain a rim or tyre on the felloe.

We shall have reason again to refer to the Luttrell Psalter in Section 6.6.

Footwear

Iron nails and spikes had been used to reduce the wear of *shoes* since the time of the *caligae* worn by the Roman foot-soldier.[8] Reference has already been made to the fine example of the use of iron nails for this purpose depicted in Fig. 5.3. The gradual replacement of oxen by horses as the motive force for both agricultural implements and general vehicles introduced a tribological problem which was also mentioned briefly in Section 5.4. Not only do horses' hooves tend to grow softer than those of oxen in wet climates, but they wear more rapidly when hauling heavy loads or moving quickly over paved surfaces. The Romans overcame this problem when they introduced the iron *horseshoe*, and this important

7. T. S. Pattie, in Francis, Sir Frank, ed. (1971), p. 166.
8. Waterer, J. W., 'Leather', in Singer, C., et al. (1956), Vol. II, Ch. 5, p. 168.

Fig. 6.1
Two-wheeled harvest cart with studded wheels. Luttrell Psalter (folio 173ᵛ) (*c.* 1338)

tribological device had found widespread application in northern Europe by the eighth and ninth centuries.

The wheelbarrow

This was developed in China from about the first century A.D. References to this single man-powered vehicle show that it was used for transporting supplies by General Chuko Liang as early as the ninth year of the Chien-Hsing reign (A.D. 231).[9] The wheelbarrow did not appear in Europe until the late twelfth and early thirteenth centuries, the earlier documentary evidence being provided by a mid-thirteenth-century manuscript.[10]

A most interesting difference between European and many Chinese wheelbarrows is that the former have the wheel well to the front while in the latter it is more central. The European arrangement permits greater manoeuvrability, but the Chinese vehicle, being balanced and independent of muscle power for load carrying, can transport much greater weights. This is another example of a Chinese engineering development which pre-dates considerably its medieval counterpart in Europe. Needham (1965) has given a full account of the Chinese development, together with a discussion of the similarities and differences with European forms.

The bearing loads on wheelbarrows tend to be relatively small, even in the Chinese arrangement, and there seems to be little doubt that simple wooden bearings would suffice in the early days. Even today when metal wheelbarrows have largely replaced their wooden forerunners the

9. See Needham, J. (1965), Vol. 4, p. 260.
10. Gille Bertrand, 'Machines', in Singer, C., et al. (1956), Vol. II, Ch. 18, p. 641.

bearings are extremely simple. Iron shafts running in simple bushes, sometimes of similar materials, appear to be adequate, and at an intermediate stage metal-clad wooden shafts probably ran in similar bushes.

6.4 Instruments

Two instruments will be discussed in order to illustrate the growing importance of tribology in medieval mechanical systems. For direction-finding on the surface of the earth the magnetic compass provided a system of great simplicity but one dependent upon low friction for its accuracy, while for time-keeping, mechanical clocks of considerable complexity and ingenuity were developed with metal bearings.

The magnetic compass

This originated in China and appears to have reached the West at the end of the twelfth century. The observation of directive properties in natural magnets or lodestones originated in the East and may have been associated with geomantic divination (see Needham, 1950, and an excellent account of the development of significant techniques and instruments in medieval times in Bernal, 1954). The water-compass in which a piece of magnetized iron was floated on a piece of wood in a bath of water was described in the eleventh century, but was probably known very much earlier. The use of hydrostatic principles to provide a low friction support for the *compass needle* through buoyancy is not without tribological interest.

The introduction of suspended and pivoted needles came much later and European centres probably played a role in the development of this early Chinese discovery. The magnetic compass was, of course, to play an important role in later exploration of the seas and remote areas of the earth.

The mechanical clock

This was probably the outstanding engineering achievement of the Middle Ages. It has been suggested by Mumford (1934) that the spur to this development in Europe was the need for regular communal life in the monasteries. A good account of the impact of horological developments upon medieval and Renaissance society has been presented by Hale (1971). Originally the hours were marked by the ringing of a bell (*cloche*) by a *watch*man, the intervals being determined by an hour-glass. The earlier water-clocks originated in the Bronze Age and were used extensively by the Greeks, Romans and, from about the sixth century, the Arabs. These devices sometimes achieved considerable complexity but they rarely enjoyed good accuracy. The mechanical system which replaced devices reliant upon the flow of sand in an hour-glass or water in the water-clock was thus most welcome.

An essential step in the development of the mechanical clock was the concept of an escapement. The idea seems to have originated in China in connection with the mechanism for slowly rotating astronomical

models.[11] The object of any escapement is to provide controlled movement in steps rather than very slow continuous motion. In the Su Sung
escapement of 1088 mentioned earlier the action was provided by a
trip lever and a number of scoops on the periphery of a wheel which
gradually filled with water. In the writings of the Cistercian monk,
Villard de Honnecourt (*c.* 1250) there is a description of a rope escapement
which many believe to have been the forerunner of the true mechanical
escapement. Towards the end of the thirteenth century the *foliot balance*
and *verge* escapement was invented. As the horizontal foliot beam
oscillated on its thread suspension two small pegs or pallets on the vertical
spindle or verge allowed a crown wheel to move forward with intermittent
motion. Duley[12] has noted that the efficiency of these pre-pendulum
escapement mechanisms would depend greatly on their friction, driving
weights and lubrication.

The foliot and verge provided the impetus for development of reliable
mechanical clocks in the thirteenth and fourteenth centuries. It became
an essential feature of clocks for the next 300 years and in many cases the
falling weight was replaced by a spring after 1450. The replacement of
hour-glasses and bells by mechanical striking clocks became a matter
of great civic pride. The early mechanical clocks were made by blacksmiths and millwrights, but in due course the clockmaker became recognized as a craftsman esteemed by the community. In Western Europe,
notably Germany, Switzerland and England, the spread of town clocks
and later portable timepieces was rapid, and one can imagine the enthusiastic atmosphere as local craftsmen vied with each other to construct
the mechanical device which would govern daily life in the community.

Not unnaturally, in the Middle Ages many churches were equipped
with mechanical clocks with foliot and verge escapements. The oldest
known clock of this type in England was installed in Salisbury Cathedral
about 1386 at the time when Ralph Erghum was the Bishop. It is also
thought to be one of the earliest complete mechanical escapement clocks in
working condition in the world. The clock was constructed entirely
of hand-wrought iron with no evidence of wooden components. This
example of medieval craftsmanship can still be seen in Salisbury Cathedral.
The clock, now nearly 600 years old, has struck the hours for about 500
years, although it has been moved and modified a number of times. The
foliot balance was replaced at some stage by a pendulum and the clock
was replaced by another timepiece in 1884. It was rediscovered and
recognized for its historic merit in 1929. In 1956 a complete repair and
restoration was effected and the pendulum was itself replaced by a new
foliot and verge. A photograph of the clock taken by Mr T. R. Robinson
in 1929 is shown in Fig. 6.2.

We now turn to the interesting question of the nature of the bearings
and lubricants in these ancient clocks. The not inconsiderable literature
on horology sheds little light on this matter, but since the form of escapement employed would work efficiently only if good bearings were

11. See Needham, J. (1965), Vol. 4.
12. Duley, A. J. (1972), Private communication.

Fig. 6.2
The medieval clock of Salisbury Cathedral – believed to be the oldest mechanical clock in England (*c.* 1386).

provided, it seems likely that the medieval craftsmen gave some thought to this tribological question. The photograph taken when the Salisbury Cathedral clock was rediscovered in 1929 clearly shows that brass bushes were fitted in the wrought-iron frame. But were these bushes present in the original clock – or did the iron shafts run in simple holes in the iron frame? Davison (1958), in his description of a similar but slightly later clock from Wells Cathedral which is now housed in the Science Museum, states that the axles of the early mechanical clocks ran in simple holes in the iron frame, and that brass bushes made from strip were introduced at a later stage. If true, this suggests that the performance of the early clocks was adversely affected by excessive wear or friction and that the sensible use of brass bearings was adopted to overcome the problem. There does not appear to be any direct evidence of the initial form of the bearings in the Salisbury clock, but Mr T. R. Robinson[13] appears to share the view that the early bearings were probably iron-on-iron and that the operating problems might well have given rise to the realization that bearings should be made of dissimilar metals.

Bishop Erghum moved to the see of Bath and Wells in 1388 and it is of some interest to find that four years later a similar clock to that at Salisbury was in use in Wells Cathedral. This is the clock described by Davison (1958) and the similarity between the Salisbury and Wells machines can be seen from a comparison of Figs 6.2 and 6.3.

Evidence of the use of lubricants in these early mechanical clocks has not been brought to my attention, but Mr Duley[14] has mentioned some interesting suggestions from a somewhat later period. Duley translated the Swiss publication *Die Uhrmacher von Winterthur und ihre Werke* and later wrote an article (1969) based upon the work. In the translation there is a church record dated 1603 which reads

> . . . *To Master Andreas Liechti for three times cleaning 8 Pfd and other things 3 Pfd, for correcting pivots and other things on the clock on the upper arch 3 Pfd, for a measure of beech wood for the watch maker for cleaning the clock 3 Pfd 4 Sch, for righting the items distorted in clearing the clock and for old brekages 8 Pfd 7 Sch 4 Hllr. . . .*

This account of the cleaning of a tower clock also suggests that there was no better way of removing dirty oil (probably animal fat) which had hardened than to burn the various parts in a wood fire. The cleaning took place in the open and was attended by dancing round the fire and other festivities.

This evidence suggests that lubricants were used in the bearings and that the solution to the problem of removal of degenerate material was not without hazard. Olive oil has been used extensively for many years for the lubrication of chronometers (see Aked, 1968) and it is not inconceivable that it was also used on the early mechanical clocks of medieval times.

6.5   **Mechanical power generation**

Very little is known about the bearings used in the *water-* and *wind*-driven machines which developed widely during the Middle Ages, but these machines formed such a vital aspect of the mechanical art of the period that they deserve a reference at this stage. The development of water-driven machines has been vividly described by Vitruvius (see Ch. 5), and we have already commented on the metal-clad bearings used in the first century B.C. The Vitruvian mills for the grinding of corn were invariably undershot vertical wheels with power transmitted through

---

13.   Robinson, T. R., Private communication, 1972.
14.   Duley, A. J., Private communication, 1972.

gears to the vertical shaft and the millstones. Water-mills were wide-spread in Europe by the eighth and ninth centuries. By about 1100 the traditional function of the corn-grinding water-mill was extended to include the fulling of cloth and the forging of metal.

The first evidence of Chinese water-mills is dated in the first century A.D., but a curious feature of these early records is their clear indication that the mills were used for blowing bellows used for metallurgical purposes. This suggests that there must have been a period of earlier development and application since European usage for this purpose was delayed until the twelfth century.

It is known that iron shafts were used in some medieval water-mills, but wood continued to be used on a wide scale. The nature of the bearings is largely a matter of conjecture, but they were probably made of wood or clad with metal as in the Vitruvian machines in the previous millennium. The early machines produced just 2 or 3 hp (1·5–2·2 kW), but later mills were capable of a 50 hp (37 kW) output. They became widespread, the Domesday Book of 1086 revealing the existence of no fewer than 5,624 water-mills south of a line drawn from the Trent to the Severn.

These mills played an important part in providing mechanical power in place of animal and human effort in the Middle Ages. They provided the basis for a technical expansion towards the end of the period and the foundation for the subsequent structure of technological development in Europe.

Needham (1965) and Forbes[15] have presented excellent accounts of the development of water-mills in both the East and the West.

The *windmill* is only a few centuries more recent than the *water-mill*, being known in China from about A.D. 400. It is believed that it originated from the wind-driven prayer wheels and it was certainly known in Islam in the middle of the seventh century and in Persia before the year 1000. It became an important prime mover for grinding corn and pumping water in China, India and Islam.

The horizontal-axle windmill developed in the lands surrounding the southern end of the North Sea in the 1180s.[16] It quickly spread through Europe and ultimately as far as Syria. All these early Western mills were of the *post* type in which the main driving shaft is orientated towards the wind by having the mill housing containing the grinding stones revolving round a central post or pivot fixed permanently in the ground. The *tower*-mill, built of brick or masonry, was a later development in which the shaft was mounted in a movable cap on the fixed structure. In this case the driving shaft drove the central vertical shaft through gearing at all orientations and the grinding stones were housed in the stationary tower. Once again we know little about the bearings in these wind-driven prime movers, but they clearly created a demand for substantial thrust and journal bearings. In later years the windmills of East Anglia were to play an important part in the development of ball thrust bearings.

15. See Forbes, P. S., 'Power' in Singer, C. et al. (1956), Vol. II, Ch. 17.
16. See White, L., 'Technology in the Middle Ages', in Kranzberg, M. and Pursell, C. W. (1967), Vol. I, p. 77.

6.6   Tribological stones in medieval ploughs and carts

The Society of Antiquaries of Scotland undertook extensive excavations at the Roman frontier post at Newstead between 1905 and 1910 and Mr John M. Corrie (1914) made a further systematic search for small objects on the site between 1911 and 1914. Among the finds reported by Corrie were twenty-three quartz-like pebbles, commonly worked or polished on one side, which he called burnishers or polishers. He was able to divide the pebbles into two classes, the larger group consisting of eighteen specimens which were roughly conical with a convex-ground end face showing distinct traces of striation, and five pebbles which presented a surface which had been worn flat and smooth. He was unable to ascertain the purpose of these pebbles although he conjectured that the worn, striated faces were consistent with their use with sand in the dressing of hides stretched on pegs preparatory to drying.

Mr Corrie was also able to report similar finds from the neighbourhood of Dryburgh in the parish of Mertoun, Berwickshire, and he was informed by Mr A. O. Curle that specimens had been found in both Wigtownshire and Aberdeenshire. The distribution of the objects thus appears to have been widespread. A single specimen had been found during the 1905 to 1910 excavations, but it was not reported by Mr James Curle in his volume on Newstead entitled *A Roman Frontier Post and its People*. Corrie notes that since the stones from Newstead had not been found outside the confines of the fort they were probably of Roman origin, although they had not been found in other Roman stations in Scotland.

An illustration showing six of Corrie's pebbles is shown in Fig. 6.4. An alternative explanation of the purpose of these finds emerged almost a quarter of a century later after archaeological studies on the other side of the North Sea.

Fig. 6.4
A selection of six quartz pebbles found by J. M. Corrie at the Roman frontier post at Newstead between 1911 and 1914.

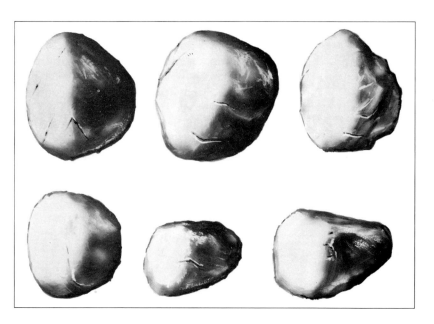

Agriculture has presented a number of tribological problems throughout the long history of civilization, ranging from the excessive wear of early wooden implements to the performance of modern mechanical and hydraulic equipment. It is the development of one particular implement in northern Europe which provides the next piece of evidence in our story of the fascinating use of stones in medieval times.

Neolithic man relied upon the wooden, bone or antler hoe and digging stick to till his land, but the plough had also been introduced as early as the Bronze Age. While the basic features of the plough have remained unchanged for thousands of years there have been numerous changes in design and modifications to meet local conditions. In dry climates, where the soils are light, the plough is required merely to loosen the surface while retaining the moisture in the soil beneath, but in wet climates, where the soil is heavy, good drainage requires a deeper furrow and turning of the earth.

The traction plough developed as early as the fourth millennium B.C. when oxen were used in Sumeria, and the wheeled plough (spoked) was known in Roman times. There is, however, little evidence of the widespread use of the wheeled plough in Europe until the eleventh century, and oxen were used more frequently than horses until the thirteenth century. A team of eight oxen was a common unit of assessment mentioned in the Domesday Book of 1086.

The importance of material properties for both structural and tribological aspects of plough performance was recognized and recorded from early times. Jope[17] has drawn attention to the early Greek sources, such as the poet Hesiod (*c.* 700 B.C.), in which reference is made to the value of different kinds of wood in the construction of the implement. The highly stressed beam was made from *holm*-oak while the stock, which is that part of the plough which breaks the ground, was made of *oak*. The pole was made of *elm* or *bay*. Theophrastus (*c.* 371–287 B.C.) noted that '... oak will withstand rot in the earth and elm in the air'. Some early Iron Age (*c.* 500–100 B.C.) ploughs were found to be made of a variety of woods, included *alder, birch, oak* and possibly *hazel*.

One of the major tribological problems was, and still is, the wear of the faces of the stock by abrasive rubbing against the sides of the furrow. The Egyptians shod their ploughs with flint stones and iron was used at a later stage throughout the Middle East and various European countries. Iron ploughshares were in use in Britain before the Roman invasion.

In a section dealing with Stone and Bronze Age cultivating implements in Scotland Fenton (1965) has drawn attention to a number of stone bars of the form shown in Fig. 6.5. The bars are 1–3 ft (0·3–0·9 m) long, circular or oval in section and all show a distinctive wear pattern on one side of the point extending some $2\frac{1}{2}$–4 in (64–102 mm) from the tip. Stevenson[18] was the first to suggest that they might have acted as shares for one of the simplest forms of plough, the *ard*. These stone bars, which

17. See Jope, E. M., 'Agricultural implements', in Singer, C., et al. (1956), Vol. II, p. 85.
18. See Stevenson, A. B. K., *Proc. Soc. Antiq. Scot.* 1955–56, **lxxxix**, 345–6.

Fig. 6.5
Stone or Bronze Age plough bar
shares from Shetland and Orkney.

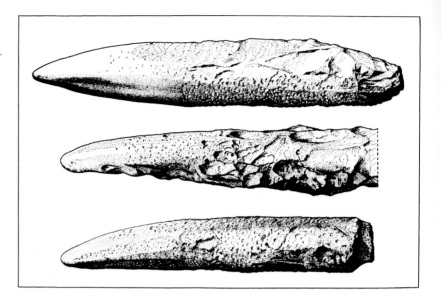

clearly took the brunt of the wear, would act as the ploughshare or perhaps protect an underlying arrow-shaped wooden share.

The long stone bars clearly played a tribological role in Stone and Bronze Age ploughs, but it is to the small pebbles described by Corrie that we now return.

Axel Steensberg (1936) published a most careful survey of evidence of the development of the plough in north-western Europe from prehistoric to medieval times. His evidence included remnants of early implements, illustrations, including rock carvings, traces of activity in the earth and literature. An ard found at Walle had been dated at 3500 B.C. by peat analysis from material found at the same depth and was thus attributed to prehistoric times. However, while not doubting the analysis, Steensberg had some reservations about the location of the depth of the find, and he felt that this ard might have belonged to a later age.

Steensberg gave a full account of part of a plough found in Tømmerby in the parish of Tem south of Silkeborg in Denmark in 1888. He thought that it might have been a remnant of a medieval wheel-plough, but was surprised when pollen analysis by Professor Jessen dated it to the pre-Roman Iron Age. Attention was again focused on the difficulties of dating by pollen analysis when radiocarbon dating later suggested that the true date was A.D. 1620 ± 100.[19] Within the oak remnants of the Tømmerby plough Steensberg found that '. . . the left side of the head is set with guard-stones, mainly quartz and quartzite, as was the custom with wheel-ploughs in Jutland right up to the beginning of the nineteenth century'. There were in fact ten pebbles still in position and some can be seen in Fig. 6.6(*e*).

This alternative tribological explanation of the role of the small pebbles found by Corrie stimulated Phillips (1938*a*) to suggest that

---

19. Lerche, G. (1970), p. 135.

although conditions in England rarely favoured the preservation of wooden ploughs a search for the pebble studs might well repay the effort. The call was quickly answered and in the same year Phillips (1938*b*) himself was able to report that a considerable number of the worn pebbles with characteristic striations had been located in some regions of vigorous Scandinavian occupation in England. Some were found in the fields of north Lincolnshire, others in large numbers in the neighbourhood of Atwick and Skipsea in the East Riding of Yorkshire and a few to the west of Sheffield. A selection of these stones is shown for comparison with Corrie's finds in Fig. 6.7.

The use of hard stones in wooden ploughs for tribological reasons appears to have been a well-established practice from medieval times until the nineteenth century. A selection of illustrations accompanied by brief details of some of the finds reported by Lerche (1970) in her most interesting and informative article is shown in Fig. 6.6. Since iron ploughshares had been used in Roman and even pre-Roman times it is necessary to ask why stone was favoured as an anti-wear material in the late Middle Ages. The reasons appear to have been mainly economic. Iron was expensive for the farmers of the day who naturally preferred the more readily available and easily worked wood. Pebbles were there to be found and Steensberg (1963) has written a delightful account of the way they were collected in the field, on the beach and in the gravel pits. Some farmers used 'will-o'-the-wisps' or fossilized belemnites, which not only had a good conical shape, were easy to handle and to drive home into holes bored into the wood, but were thought to have good protective powers to ward off evil. Perhaps the association of *wear* and *evil* represents a recognition of the significance of tribology in agricultural communities dating back to medieval times. In the nineteenth century[20] bags of belemnites were prized possessions which could be seen hanging up in peasant cottages. They were continually 'added to and passed on from father to son.

Protective stones for ploughs were also known and recorded in France many centuries later. Two primitive ploughs from Auvergne, resident in the Musée des Arts et Traditions Populaires, Trocadero, in Paris have been described by Haudricourt and Jean-Brunhes (1955). It appears that[21] the sole of one of them is 'shod' with *stone, bits of iron*, and *wooden pegs*, inserted in bored holes in the horizontal plough-head's sides and underneath, while the other is shod partly with the aid of *iron nails* with large heads and partly with *horseshoes*, which are similarly fastened with nails with large heads. Haudricourt and Jean-Brunhes also refer to a thesis by A. Dauzat dated 1934 which relates stories from the older people of that period indicating that the protective stones were sold commercially.

The earliest picture of a wooden plough protected by hard stone pebbles is possibly to be found in the illustrated margins of the Luttrell Psalter.

---

20. See Steensberg, A. (1963), p. 72, and his reference to J. C. la Cour, the writer on agricultural matters.
21. See ibid., p. 75.

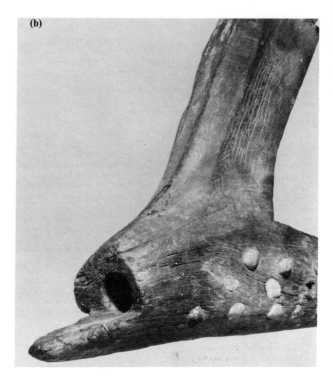

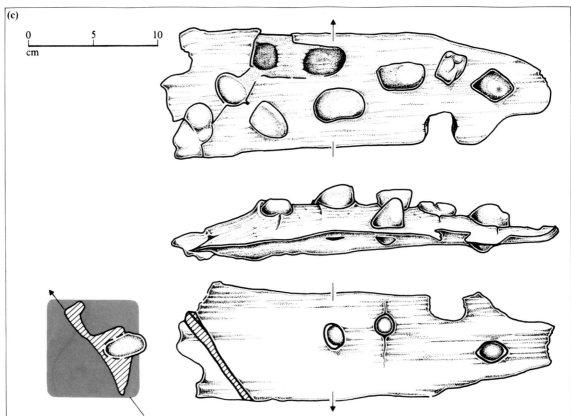

Fig. 6.6
Stone-studded wooden ploughs of
Medieval Denmark.
(a) The Linaa Plough. The oak
sole seen from the pebble-studded
land side. Found in 1959 in Linaa
parish in the Gjern district of the
county of Randers, Jutland.
There are five bored holes 2–5 cm
in depth and diameter with one
flint pebble remaining. Age:
1510 ± 100 (by radio-carbon
dating).
(b) The Andbjerg Plough. The
beech sole showing the hole for
the share. Found in 1945 in Dover
parish south of Silkeborg. The
land-side is studded with pebbles,
mostly of granite. Age: 1520 ± 100
(by radio-carbon dating).
(c) The Onsild Plough. Scale
drawing of beech-wood fragments.
Found in 1945 near the river
Onsild, in Øls parish in the
district of Hindsted, Ålborg
county, Jutland. There are six
inset pebbles with traces of seven
or eight more. Age: 1590 ± 100
(by radio-carbon dating).
(d) The Onsild Plough.
Photograph showing the worn
faces of the pebbles. The
horizontal wear-marks show the
direction of ploughing and the
worn faces of the pebbles have a
sharp edge at the rear.
(e) The Tømmerby Plough. The
oak sole showing the hole for the
tang of the share. Found before
1888 in the parish of Tem, south
of Silkeborg. The land-side of this
sole is protected by inset pebbles
of quartzite and granite of which
ten remain in position. Age:
originally thought to be
pre-Roman Iron Age by
pollen-analysis but now
dated at 1620 ± 100 by
radio-carbon analysis.
(f) The Boeslum Plough. The
sole of a nineteenth century
Danish wheel plough. A
share-shaped piece of iron plate
is now inserted in the tang-hole
instead of the share. The inset
pebbles show little sign of wear.
Date: 1820–30.

Fig. 6.7
A selection of eight pebbles showing heavy wear and characteristic striations found on the fields of Willoughton, Lincolnshire in 1938.

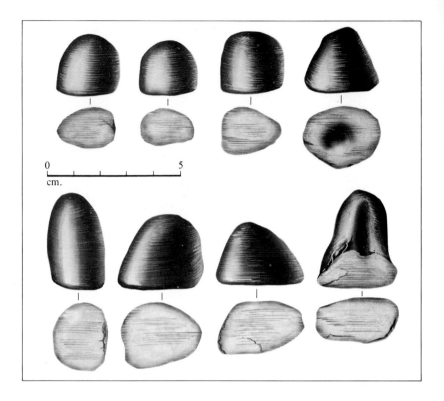

The mould-board of a swing-plough pulled by four oxen and shown in Fig. 6.8 is provided with twelve holes in a row. Steensberg[21] has suggested that these holes probably contained protective stones or wooden pegs. The date of this pictorial evidence (1335–40) is, of course, even earlier than the revised date for the earliest archaeological evidence from Jutland (1510 ± 100).

In summary, the evidence for this interesting development of a system of protective stones for wooden ploughs comes in the form of documentation (Luttrell Psalter), archaeological finds of wooden parts with embedded stones from the bogs of Denmark, loose pebbles in the areas of Viking occupation in England and Scotland and folklore in Denmark and France.

In our earlier discussion of the methods used to protect the felloes of carts from excessive wear (see Sect. 6.3) reference was made to the use of wooden pegs and metal nails. Steensberg[22] has also cited further evidence to show that in more recent times the felloes were protected by implanted stones.

Our final look at the development of wear-resisting hard stone inserts shows that the system was not restricted to the plough and the felloes of cartwheels. The wooden journals of wheeled ploughs and carts were also protected on one side by stones. Steensberg's (1963) accounts again

22.　See Steensberg, A. (1963), pp. 70–1.
23.　See Lerche, G. (1970), pp. 144–5.

Fig. 6.8
Illustration from the Luttrell Psalter showing the mouldboard of a swing plough provided with twelve holes for pegs or stones (*c.* 1335–40).

provide the evidence on this point, and they further show that *tar* was used to lubricate the wheel bearings on carts in which the wheels rotated on a stationary axle. Part of an oak forecarriage axle from a plough showing inserted pebbles was found during peat digging at the castle moat at Torp in the parish of Røer and brought to the Danish National Museum in 1877.[23] Mrs Grith Lerche has given an excellent account of this find and Fig. 6.9 shows her drawings of the remnants of the axle and her representation of the carriage.

There were twenty-four pebbles still in place and empty holes for thirteen more. Mrs Lerche has noted that the shape and wear patterns of the stones makes it possible to distinguish between plough and cart pebbles and also to determine the direction of the rotation of the shaft. Photographs of the pebble-studded Torp axle and the wear patterns on the stones are shown in Fig. 6.10.

6.7 Chronology

The millennium under review saw steady progress in technical matters in the East while the West was beset by various difficulties. The successive Saxon, Viking then Norman enterprises in Western Europe which followed the collapse of Rome were undoubtedly responsible for the slow progress, and in some cases regression, in the mechanical arts, although they also moulded the European character and set the scene for later developments.

Fig. 6.9
Illustration of the oak forecarriage axle of a wheel-plough from Torp showing the pebble-studded underside.

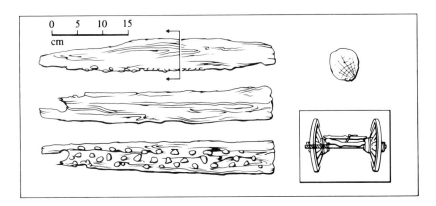

(a)

Fig. 6.10
Photographs of the
pebble-studded Torp axle.
(a) The complete pebble-studded
remnant of the axle.
(b) Close-up showing the wear on
the pebbles indicating rubbing in a
circumferential direction.

(b)

Medieval lubricants continued to rely upon vegetable and animal origins, while bearings showed little or no improvement. The return to wooden bearings proved to be a limitation on machine development only at the end of the period. The neglect of metals in the West prompted the interesting applications of stone as a tribological material in ploughs and in the wheels and axles of carts towards the end of the Middle Ages. It appears that the rolling-element bearings which emerged among the Greeks, Romans, Celts and Chinese in earlier times were not employed in the medieval period.

The limited but interesting tribological developments of the period are noted against a background of wider historical events in Table 6.1.

6.8 Summary

The Middle Ages is typified by a paucity of firm archaeological or historical evidence of tribological developments. We do, however, see an increase in general documentation towards the end of the period, and

Table 6.1
Chronology for medieval times
(A.D. 400–1450)

Date (A.D.)	Political and social events	General technical developments	Tribology
400	Fall of Roman Empire in the West Saxon settlement in Britain	Wheelbarrow in China Windmills in China	**(Regression in the use of metal bearings in the West in medieval times)**
500	English converted to Christianity	Water-clocks in use in Greece, Rome and Middle East	**(Vegetable oils and animal fats used as lubricants throughout the Medieval period)**
600	Muslims capture Jerusalem	Water-mills developed in China 1st century A.D. continued to be used for various purposes	**Possible use of oil in metal spinning in a lathe (Sutton Hoo)**
700	Venerable Bede First Viking raids on England		**Metal bearings in use in China**
800	Danes land in East Anglia		
900	End of Danelaw		
1000	Normans conquer England Domesday Book Jerusalem falls to Crusaders	Windmills in Persia Compass in China Water-mills widespread in Europe (wooden and iron shafts)	**Iron-on-iron bearings in China (Su Sung clock tower)**
1100	University of Bologna	Wheelbarrow in Europe Compass in Europe Windmills in North Sea area	
1200	University of Oxford Roger Bacon Magna Carta University of Cambridge Mongol dynasty Villard de Honnecourt	Rope escapement Foliot and verge escapement	
1300	Luttrell Psalter Hundred Years War Black Death Ming dynasty Chaucer	Crank mechanism Salisbury Cathedral–mechanical clock Wells Cathedral–mechanical clock	**Wooden pegs used to reduce wear of wheel felloes and plough mould-boards** **Brass bearings in mechanical clocks**
1400		Spring-driven mechanical clock	**Stones used to reduce wear of wooden ploughs, wheel felloes and axles (Denmark). Tar used as lubricant on carts (Denmark)**

some of it contains suggestions of the current form of bearings, lubricants and wear-resisting devices.

It is necessary to develop an appreciation of the general state of society and engineering progress in medieval times if the stirring events of the Renaissance are to be fully appreciated. It is for this reason that much of the present chapter is concerned with general aspects of machine development, but points of tribological significance are also included whenever the evidence is firm and relevant. Perhaps the most fascinating aspect of the period is the recognition of the importance of combating wear and the use of hard pebbles for this purpose in carts and ploughs in late medieval times.

The increasing use of mechanical power in place of human and animal effort and the ability to utilize natural resources provided late medieval society with a new spirit of optimism in relation to scientific and technical progress—a spirit so admirably reflected by the words of that thirteenth-century English Franciscan monk and scientist, Roger Bacon. It is true that few of the technical advances of the long medieval period presented excessive demands on the tribological skills of the day, but the subject was ripe for the new, or revised, approach of the Renaissance.

Chapter 7

The Renaissance
(c. A.D. 1450–1600)

7.1 Introduction

The information on tribological developments recorded in earlier chapters is based essentially upon archaeological evidence supported by accounts from the rare and precious manuscripts of the day, but from the end of the Middle Ages printed matter provided a rapidly expanding and most valuable source of historical material. Indeed, in the present chapter literature of the period will for the first time form the complete basis of our story. This does not imply a rejection of archaeological evidence, since artefacts will again be referred to in connection with later periods, but the value of contemporary Renaissance literature is enhanced by the relative paucity of remains of significant tribological devices of the period.

Printing developed first in China in the eleventh century and later in Europe in the middle of the fifteenth century. The importance of movable-type printing developed *c.* 1450 lies in the creation of numerous copies of relatively cheap books with the resulting dissemination of ideas and accounts of inventions on a hitherto impossible scale. Vitruvius' *De Architectura* referred to in Chapter 5 was printed in the latter half of the fifteenth century. It has been estimated[1] that about 8 million books were printed in the latter half of the fifteenth century. Davison (1958) has noted that the first printed book on engineering, which was concerned with various types of military machines and hydraulic equipment, appeared in Verona in 1472, while Bernal (1954) has stressed the significance of printing for the great scientific and technical changes in the sixteenth century.

---

1. See Clapham, M. 'Printing', in Singer, C., et al. (1957), Vol. III.

It became essential for the craftsmen of the day to become literate and this in turn produced a link between practical and learned men which was important in the development of the mechanical sciences.

The Renaissance was much more than the rebirth of interest in the classical Greek and Roman forms of art, architecture, scholarship and civilization which is so frequently romanticized. It was a period of intellectual revolution in which the scientific method kindled during the Middle Ages was applied increasingly to the problems of the day. The increasing use in the Middle Ages of mechanical power, based initially upon water-wheels and supplemented from the fifteenth century onwards by windmills, the development of trade and transportation by land and sea and the general interest in manufacture and distribution of goods, was both born in, and yet found to be incompatible with, the feudal economic system. Great merchant houses developed from the increasingly important trading companies while some of the merchants became bankers and industrial financiers to feed the system with capital for its further expansion. Capitalism thus replaced the feudal economic system throughout Europe in a rapid but far from painless process, and the intervention of merchant bankers was a significant factor in the industrial development of the Renaissance period.

The end of the fifteenth and the early sixteenth century was the great era of exploration by sea. In 1492 the Genoese navigator Christopher Colombus set out on his historic voyages to the west under the patronage of Ferdinand and Isabella of Spain. He landed near the mouth of the Orinoco in South America on his third voyage in 1498 and 12 October is now officially recognized as Columbus Day in many parts of the United States of America, the West Indies and South America. A summary of the state of development of tribology prior to the discovery of the New World has been presented elsewhere by the author (Dowson, 1973).

The southern route to the Indian Ocean had been found in 1488 by the Portuguese navigator Vasco da Gama, and in 1497 he again rounded the Cape of Good Hope to reach India. In the same year John Cabot, a Venetian navigator in the service of England, discovered Labrador, while the first circumnavigation of the earth (1519–22) was initiated but not completed by Fernao de Magalhaes (Magellan). These stiring voyages add lustre to our image of the times and are at once indicative of the new interest in discovery and trade, together with scientific and technological ability, to meet the needs of the day.

The need for navigational guides provided added importance to studies of navigation and astronomy. In a short time Copernicus (1473–1543) was disrupting the Aristotelian view of the universe by transferring the centre of rotation of our immediate celestial neighbours from the earth to the sun, while that Professor of Physics and Military Engineering at Padua, Galileo Galilei (1564–1642), was soon to invent the telescope (1600) which played an important part in confirming the Copernican system.

But what was the impact of this era upon tribology? In the first place the subject benefited from the inspiration of a man whose genius is almost beyond comprehension. Leonardo da Vinci's direct contribution to the Renaissance movement undoubtedly lies in the arts, although no more than twelve paintings can with certainty be attributed to him. His role as an engineer and scientist was not widely recognized until Bonaparte issued a decree on 19 May 1796 for the transfer to France of worthy collections of art. The manuscripts were then described by J. B. Venturi some 278 years after Leonardo's death. Hence, although Leonardo's works of art have been available to the public for over 500 years his notes on scientific matters have been lodged in private collections and libraries until relatively recent times. Valuable notebooks containing records in the form of sketches and notes of studies of bearings, friction and wear were found in Madrid as recently as 1967. A more complete account of the history of Leonardo's notebooks is presented in the Appendix (Sect. A.2).

An assessment of Leonardo's contributions to the mechanical sciences and tribology in particular cannot be based upon long-standing evidence alone. It must take account of current work and recent finds while paying cognizance to the possibility that further manuscripts might yet be found.

Some see Leonardo as a supreme artist with a real but secondary interest in science and engineering, while others see him as an engineer and scientist whose vision outpaced the technological capabilities of the day and who found it necessary to work as an artist in order to make a living. The latter school point to the overwhelming proportion of Leonardo's notes, drawings and sketches devoted to scientific and engineering topics, while the former point out that he was trained as a painter, but it seems clear that his contemporaries recognized him as both an engineer and an artist. The French burial certificate described him as:

Lionard de Vincy, noble millanois, premier peinctre et ingenieur et architecte du Roy, mescanichien d'Estat, et ancien directeur du peincture du Duc de Milan.

It might be argued that the limited availability of Leonardo's notes on engineering matters during the Renaissance should prohibit a full consideration of his work within this chapter. It is indeed a fascinating exercise to consider what impact the publication of a book by Leonardo might have had upon the development of Renaissance engineering. However, these speculations warrant delay in neither a consideration of Leonardo's work on tribology nor the allocation of precedence to many of his findings. The tragedy of the times was not so much that many of Leonardo's discoveries and findings had to be repeated, often hundreds of years later, but that his contemporaries were unable to benefit from his genius.

The scientific study by Leonardo of various aspects of tribology thus forms the subject of the early part of this chapter, while subsequent sections will be devoted to a discussion of the development of bearings based upon evidence from diverse forms of literature of the day. The period is richly endowed with early books on various trades and industries which provide a wealth of knowledge on bearing development. Many of the texts are copiously illustrated and much can be gleaned about tribological practice of the times from a careful study of the diagrams.

It will be seen that important progress was made in the development of both plain and rolling-element bearings in the period between the Middle Ages and the seventeenth century.

7.2   The tribological studies of Leonardo da Vinci

Whatever the view of Leonardo's relative contributions to science, philosophy and art, the arguments fade into insignificance when the range and merit of his talents are recognized. His manuscripts contain over 5,000 pages in the form of notes and sketches and it might be thought that the achievements in any one of his various fields of endeavour would have brought him long-lasting recognition. The present section should be read in association with the account of Leonardo's life given in the Appendix (Sect. A.2).

Many are unfamiliar with Leonardo's work in the field of tribology, but he applied himself to scientific studies of friction, the development of bearing materials, studies of wear and ingenious schemes for rolling-element bearings. His contributions to this subject alone make it necessary to adopt a measure of selectivity, but the material is of such inherent importance and the timing and the nature of the work so remarkable in historical terms that further justification of the present section would seem to be unnecessary.

Most of Leonardo's writings and sketches on tribological matters are found in the *Codex Atlanticus* published at the end of the nineteenth century and the manuscripts found in Madrid in 1967 (*Codex Madrid* I and II see Reti, 1967*a*,*b*) with smaller yet significant entries in the volume now in the British Museum known as the Arundel MSS.263. The fact that the original manuscripts were held in various private hands and later in museums before being printed, or even discovered, in relatively recent times has no doubt added to the stature of the man. His work was effectively lost and unknown to the world for many centuries and the impact of late recognition upon the scientific community was enormous. In recent years a reassessment of Leonardo's contributions and a discussion of the extent to which his notebooks reflected the general state of knowledge of the day rather than his own original work has been presented by various writers.[2] The discussion will no doubt continue for some time to come, but our assessment of Leonardo's contribution to tribology is placed on firmer ground by the consensus that his genius is best illustrated by his considerations of the application of mechanical principles to practical problems. Full details and histories of the Leonardo manuscripts have been presented by Richter (1883), MacCurdy (1938) and Richter (1952), while a most readable and beautifully illustrated book on Leonardo has been published by Lord Ritchie-Calder (1970).

---

2. See, for example, the foreword by E. A. Moody to the second edition of I. B. Hart (1963) and the careful studies by P. M. M. Duhem (1906) of the early history of mechanics.

Friction

Leonardo's studies of friction provide an excellent demonstration of the impact of the scientific revolution upon the Renaissance scene. It is not only the results of the studies which are of great interest, but also the scientific method adopted for their discovery.

What promoted Leonardo to study the phenomenon of friction? His ability to analyse the essential functions of machines and machine elements, together with his equal interest in synthesis for the promotion of machines to undertake a wide variety of tasks, would undoubtedly introduce a recognition of the restrictive nature of friction. He was, in fact, much concerned about the role of friction in the performance of screw-jacks and gears. It was entirely in keeping with his attitude that he initiated, for the first time, a study of friction before proceeding too far with detailed machine design. He constantly laid emphasis upon the importance of constructing a sound 'theory' supported and maybe promoted by careful experiment. His studies of friction conformed with this novel though now entirely familiar scientific approach.

While writing on precepts of the painter he wrote: 'Those who are enamoured by practice without science are like a pilot who goes into a ship without rudder or compass and never has any certainty where he is going.

Practice should always be based upon a sound knowledge of theory. . . .'[3]

Elsewhere he introduced a caveat: 'The supreme misfortune is when theory outstrips performance.'

His general philosophy on the importance of scientific method is epitomized by the words which open this chapter.

In assessing the significance of Leonardo's studies of friction it is important to realize that although the force of friction had been recognized by Aristotle some 2,000 years earlier, and numerous attempts had been made to minimize its effects in many of the early civilizations, his work represented the first recorded quantitative study of the subject. His experimental approach was essentially the same as that employed by Charles-Augustin de Coulomb some three centuries later and it still forms the basis of many honest though limited laboratory exercises.

Sketches from the *Codex Atlanticus* and the Codex Arundel shown in Figs 7.1 demonstrate that Leonardo measured the force of friction between objects on both horizontal and inclined surfaces. Cords attached to the object to be moved were allowed to pass over fixed rollers or pulleys to weights which gave a measure of the resisting force. The torque on a roller placed in a semicircular section trough or half-bearing was measured in the same way. The illustration in Fig. 7.1(*b*) is most interesting since it clearly relates to Leonardo's studies of the influence of apparent contact area upon frictional resistance.

A selection of Leonardo's notes will demonstrate the insight which resulted from these experiments. He soon recognized the difference between rolling and sliding friction and the beneficial effect of lubricants.

3.   From manuscript G of the Library of the Institut de France.

Fig. 7.1
Leonardo da Vinci's studies of
friction. Sketches from the *Codex
Atlanticus* and the *Codex Arundel*
showing experiments to determine:
(a) the force of friction between
horizontal and inclined planes;
(b) the influence of apparent
contact area upon the force of
friction; **(c)** the force of friction
on a horizontal plane by means of
a pulley; **(d)** the friction torque
on a roller and half bearing.

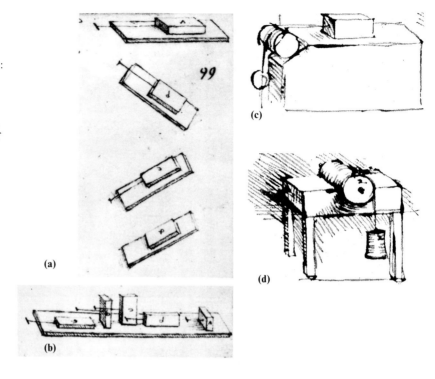

(a)

(b)

(c)

(d)

*The action of friction is divided into parts of which one is simple and all the others
are compound. Simple is when the object is dragged along a plain smooth surface
without anything intervening; this alone is the form that creates fire when it is
powerful, that is it produces fire, as is seen with water-wheels when the water
between the sharpened iron and this wheel is taken away.*

*The others are compound and are divided into two parts; and the first is when any
greasiness of any thin substance is interposed between the bodies which rub together;
and the second is when other friction is interposed between this as would be the
friction of the poles of the wheels. The first of these is also divided into two parts,
namely the greasiness which is interposed in the aforesaid second form of friction
and the balls and things like these.*

(Forster Bequest Ms. II, 131v[4])

It appears that Leonardo might have thought that the reduced friction
observed in lubricated sliding resulted from a rolling action for he wrote:

*All things and everything whatsoever however thin it be which is interposed in the
middle between objects that rub together lighten the difficulty of this friction.*

*Observe the friction of great weights, which make rubbing movements, how I
have shown in the fourth of the seventh that the greater the wheel that is interposed
the easier this movement becomes; and so also conversely the less easy in proportion
as the intervening thing is thinner as would be any thin greasy substance; and so
increasing tiny grains such as millet make it better and easier, and even more the
balls of wood or rollers, that is wheels shaped like cylinders, and as these rollers
become greater so the movements become easier.*

(Forster Bequest Ms. II, 132[5])

---

4.   See E. MacCurdy (1938), pp. 614–15.
5.   From manuscript G of the Library of the Institut de France.

Of the effect of apparent area of contact and the normal force upon friction, Leonardo wrote: 'The friction made by the same weight will be of equal resistance at the beginning of its movement although the contact may be of different breadths and lengths.' (Forster Bequest ms. II, 133r and 132v[6].) And–'Friction produces double the amount of effort if the weight be doubled.' (Forster Bequest ms. III, 72r[7].)

Note that nowhere does Leonardo refer to the 'force' of friction. The phenomenon of force, though evident in everyday life, was not adequately appreciated and defined by Renaissance mathematicians and engineers, and it was a further 200 years before Isaac Newton resolved the situation. Leonardo discussed the nature of force in some detail but his use of the term '*forza*' appears to be related to a concept of that which is used up in maintaining a constant velocity of an object against some resistance.[8] It is thus more akin to the modern view of energy or work than force.

The above observations are entirely in accord with statements of the first two laws of friction, namely;

1. The force of friction is directly proportional to the applied load.
2. The force of friction is independent of the apparent area of contact.

Although it is customary to refer to these general statements as 'Amontons' laws of friction' (see Ch. 8) it would seem to be entirely equitable to attribute them also to Leonardo. Leonardo's statements were more restrictive but they were, after all, the result of the first recorded scientific studies of friction and the first statement in scientific history of laws governing the phenomenon of friction. Perhaps justice would be done if we henceforth referred to the 'Amontons–da Vinci' laws of friction.

Leonardo introduced for the first time, and with impressive insight, the concept of the coefficient of friction as the ratio of the force of friction to normal load $\mu = F/P$. He observed that the frictional resistance depended upon the nature of the surfaces in contact, bodies with smoother surfaces having smaller friction. For 'polished and smooth' surfaces he concluded that 'every frictional body has a resistance of friction equal to one-quarter of its weight'.

Although the latter result is incorrect it is nevertheless interesting for two main reasons. The conclusion that the coefficient of friction was constant for all materials was reached again and quite independently by Guillaume Amontons in 1699. Secondly, the value of μ of 0·25 is quite realistic for the materials which were commonly used in bearings at the end of the Middle Ages and in the early part of the Renaissance. We have noted earlier that most bearings would consist of either wood-on-wood or wood-on-iron combinations, and in each case Bowden and Tabor (1964) quote coefficients of friction in the wet condition of 0·2. For unlubricated conditions Bowden and Tabor quote ranges of values

---

6. See E. MacCurdy (1938), Vol. 1, pp. 615, 616.
7. See ibid, Vol. 1, p. 621.
8. See p. viii of the Foreword to I. B. Hart (1963) by E. A. Moody.

of 0·25–0·50 and 0·20–0·60, respectively. With these figures in mind it seems that Leonardo's experiments were accurate and his finding of a constant coefficient of friction is entirely plausible.

We find that Leonardo's studies of friction were founded on the application of scientific method to contemporary problems. He appears to have conceived his experiments, reached conclusions and recorded findings which have, on the whole, stood the test of time. Whether this work resulted from or prompted his thoughts on perpetual motion machines we do not know, but he certainly developed clear views on the latter. On the first page of the Codex Madrid I he wrote,[9]

> *Among the superfluous and impossible delusions of man there is the search for continuous motion, called by some perpetual wheel. For many centuries almost all those who worked on hydraulic machines, war engines and similar matters, dedicated to this problem long research and experiments, incurring great expense. But always the same happened to them as to the alchemists; for a little detail everything was lost. My little work is going to benefit those investigators so that they won't need any more to run away because of the impossible things they promised to sovereigns and heads of state. I remember that many people, from different countries . . . went to Venice, with great expectations of gain to make mills in still water, and after much expense and effort, unable to set the machine in motion, they were obliged to escape.*

While the Forster Bequest manuscripts contain a shorter but more final commentary: 'O speculators about perpetual motion, how many vain chimeras have you created in the like quest? Go and take your place with the seekers after gold.'

Wear

Professor Reti (1971) has drawn attention to the evidence in Codex Madrid I which shows that Leonardo also studied experimentally the phenomenon of wear in bearings. A number of pertinent observations on the nature of this wear were recorded in the manuscript along with sketches of the form shown in Fig. 7.2.

It appears that Leonardo recognized that the amount of wear in a bearing depends upon the magnitude of the applied load, and that the wear groove formed by a horizontal shaft is not necessarily vertical but in the direction of the resultant steady load vector. The former result is now well established for many wear mechanisms and is embodied in Archard's (1953) wear law.

The third sketch shown in Fig. 7.2(c) illustrates so effectively Leonardo's observational skill. It shows the progress of the wear groove in a bearing. Since both the shaft and the bearing suffer the effects of wear, the diameter of the shaft will be progressively reduced as the process proceeds. The wear groove will thus adopt the tapered form shown in this delightful small sketch. 'There is no result in nature without a cause; understand the cause and you will have no need of the experiment.' (*Codex Atlanticus.*)

---

9. See Ritchie-Calder, Baron P. R. (1970) p. 134, and Reti, L. (1971) p. 103.

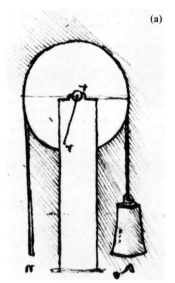

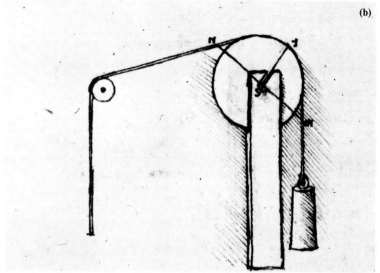

Fig. 7.2
Leonardo da Vinci's studies of wear. Sketches from *Codex Madrid I*:
(a),(b) showing that the wear groove is formed in the line of the steady load vector.
(c) showing the progressive reduction in shaft and hence wear groove diameter as wear proceeds.

Reti (1971) has also drawn attention to an early indication of interest in the economic significance of tribology in the records of a complex machine used in a Spanish waterworks in the sixteenth century. The evidence comes from the administrative documents of the Simancas Archives where reference is made to '... the continuous wear of the machine and of the needs for repair...'.[10]

Two items of heavy expense for maintenance are recorded; '... *tallow for lubrication* and charcoal for the forges that had to be operated constantly to repair the machine elements that broke down *or wore out*'.[11] It thus appears that the cost of the animal fats used for lubrication was not negligible and that there was concern about the cost of excessive wear in sixteenth-century machinery.

These references also draw attention to the little-known manuscripts of Juanelo Turriano now in the Biblioteca Nacional of Madrid. Juanelo was a man of considerable enterprise and skill, being variously described as an architect, mathematician, mechanician, silversmith, ironsmith and woodcarver. Born in Cremona, Italy in 1500 under the name Giovanni Torriani he entered the imperial household of Charles V about 1530 as a clockmaker. He served Charles and later his son Philip II for forty years in Italy and Spain, where he became established as Juanelo Turriano some time after 1545.

Although highly skilled as a clockmaker it is in the field of hydraulic engineering that he made his greatest contributions. He was responsible for the construction of the ... *artificio* ... of Toledo (1569); this being the waterworks mentioned above. A number of valuable accounts of Juanelo's life and his manuscripts have been presented by Reti (1967*c*, 1967*d*, 1968).

---

10. Reti, L., Private communication, 1972.
11. Reti, L. (1971), Vol. 224, No. 2, p. 104.

Bearing materials

There is in the Codex Madrid I a brief but most significant indication that Leonardo gave consideration to the development of a low-friction bearing alloy consisting of 'three parts of copper and seven of tin melted together'. He made reference to this plain bearing material (Lord Ritchie-Calder, 1970, and Reti, 1971) in a section dealing with his progressive design of a split-bush journal bearing.

The specification of a smooth 'mirror metal' or 'mother' as Leonardo called it anticipated by several centuries the pronouncements of Hooke (1684) and the invention of Isaac Babbit in 1839. However, the question of allocating precedence to Leonardo for this invention is not quite so straightforward. He may simply have recorded an accepted or new contemporary industrial practice, but since other and later writers of the Renaissance period do not appear to refer to the material the secret might well have been his and effectively lost for many centuries. The important point is that the concept of low-friction bearing materials should be placed firmly in early Renaissance times.

Bearings

Leonardo's basic studies of friction and wear and his specification of a low-friction bearing material provided for the first time the desirable background for a fundamental study of bearings. Just how effectively Leonardo combined his understanding of the theory and his practical genius in this field will soon become evident, particularly when his proposals are placed against a backcloth of late Middle Ages or early Renaissance technology. The majority of bearings of the day consisted of hardwood rubbing on hardwood, with a few examples of iron-on-hardwood and an even smaller number of steel-on-iron or iron-on-bronze combinations.

The advantages of rolling over sliding arrangements in the provision of low-friction supports for machinery was recognized in early Renaissance industry and particular examples will be presented in Section 7.3. One idea adopted for lightly loaded situations was to mount the shaft upon two wooden discs as shown in Fig. 7.3(a). The idea was for a long time attributed to Leonardo, but with the discovery of Codex Madrid I in 1967 it became clear that he was aware of the use of the 'roller-disc' arrangement in Germany through a German mechanic named Giulio who acted as his assistant. However, according to Reti (1971) Leonardo was responsible for the three-disc support shown in Fig. 7.3(b). A number of Leonardo's sketches of roller-disc bearings are shown in Figs. 7.3(c) and 7.4 and it seems clear that he was fascinated by them. Furthermore, he extended the concept to arrangements considerably in advance of contemporary practice.

In cases where the shaft merely oscillated through a small angle it was unnecessary to employ complete discs. Leonardo sketched various arrangements of pivoted arms representing sector-shaped supports as shown in Fig. 7.4. The arrangement is in many respects a good deal more complicated than complete discs, but it appears to have been widely adopted in machine construction. A good example is the application to a

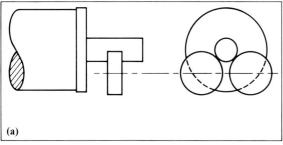 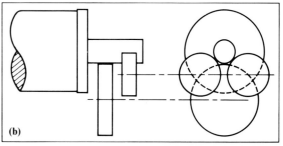

Fig. 7.3
Roller-disc bearings.
(a) Two-disc support.
(b) Three-disc support.
(c) Leonardo da Vinci's sketches
of roller-disc bearings.

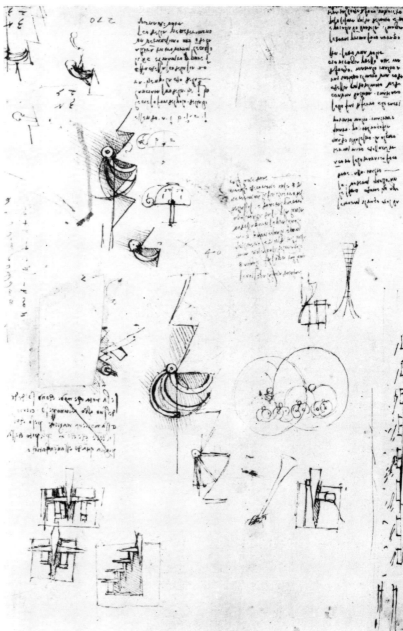

(c)

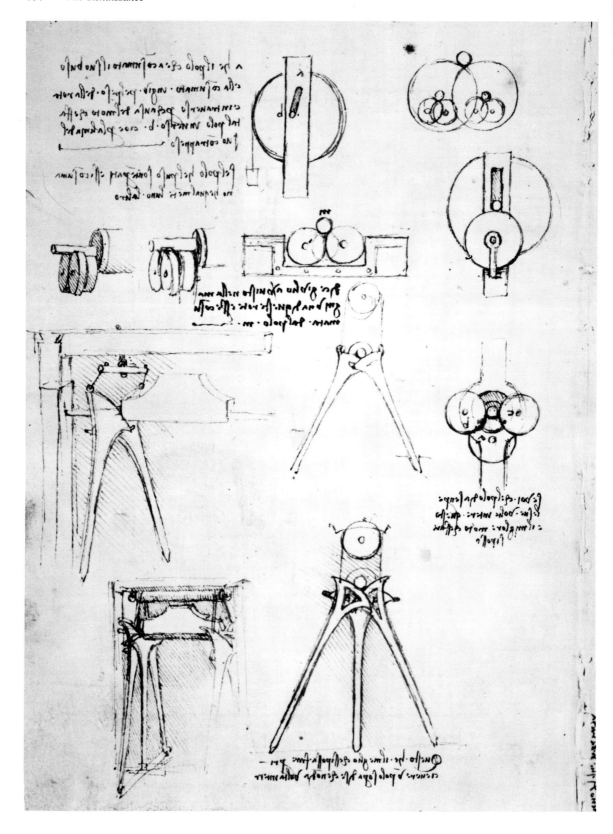

Fig. 7.4
Leonardo da Vinci's sketches in
Codex Madrid I of roller-disc
bearings for continuous and
oscillatory motion.

bell support, and the version built more than 200 years later and shown
in Fig. 7.5 demonstrates the extent of the trouble taken to provide low-
friction rolling supports.

The potential of true rolling motion for low-friction supports was
fully recognized by Leonardo for he wrote in the Codex Madrid I,

> *I affirm, that if a weight of flat surface moves on a similar plane their movement will
> be facilitated by inter-posing between them balls or rollers; and I do not see any
> difference between balls and rollers save the fact that balls have universal motion
> while rollers can move in one direction alone. But if balls or rollers touch each other
> in their motion, they will make the movement more difficult than if there were no
> contact between them, because their touching is by contrary motions and this
> friction causes contrariwise movements.*
>
> *But if the balls or the rollers are kept at a distance from each other, they will
> touch at one point only between the load and its resistance . . . and consequently it
> will be easy to generate this movement.*

This clearly takes the concept of rolling-element bearings well beyond
the early Greek, Roman, Celtic and Chinese arrangements discussed
in Chapter 5. It marks the big leap forward in rolling-bearing concepts,

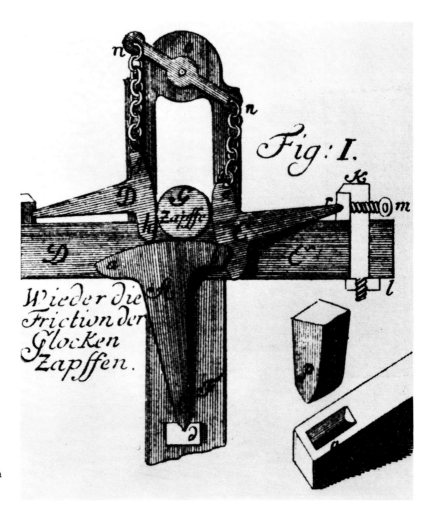

Fig. 7.5
Bell support described in Jacob
Leupold's 1724 Treatise on
Machinery and based upon the
concept of pivoted sector-shaped
supports sketched by Leonardo da
Vinci in the 1490s.

Fig. 7.6
Early form of 'cage' proposed by
Leonardo da Vinci in *Codex
Madrid I* for a ball-bearing to
prevent contact between the balls.

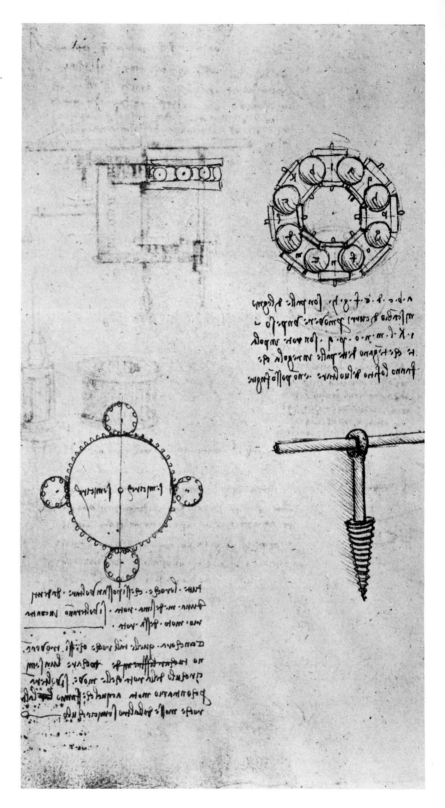

and although there is a lack of archaeological or historical evidence showing that any of the rolling bearings sketched by Leonardo were ever constructed, it might also mark the start of a new technology.

The insight signified by this recognition of the need to separate the rolling elements is remarkable. It is a point that can be made with merit in current introductions to the basic features of rolling-bearing performance. It leads, of course, to the concept of a bearing 'cage' or 'retainer' and Reti (1971) has noted that even this feature of modern bearings was anticipated by Leonardo in the Codex Madrid I. His design of a caged bearing is shown in Fig. 7.6, while several forms of remarkable rolling-element pivot bearings are shown in Fig. 7.7. The designs show a truly outstanding advance on the bearing arrangements of the day.

Two examples quoted by Reti (1967a) clearly indicate Leonardo's appreciation of the basic mechanics of rolling-element bearings. Commenting on the ball thrust bearing in the Codex Madrid I, Leonardo wrote '. . . three balls under the spindle are better than four, because three balls are by necessity certainly always touched, while using four there would be a danger that one of them is left untouched.' Leonardo appeared to favour a support based upon three conical rollers of similar proportions to the conical end of the shaft for he wrote '. . . we shall have three equal cones that are identical to the cone of the spindle, and by each turn of the spindle, the supporting cones will have each made a complete revolution.'

The wide range of rolling bearings discussed by Leonardo must not cause us to overlook his proposals for improvements in plain bearings. It has already been noted that he specified a low-friction bearing material consisting of a copper–tin alloy, and Reti (1971) has pointed out that in the Codex Madrid I he also considered two different lubrication systems for bearings.

His major contribution to the basic design of plain bearings is found in proposals for a split adjustable bush or block designed to support a shaft and to prevent it from jumping out of the bearing 'in spite of any strain'. The split bush could be tightened round the shaft to accommodate progressive wear and yet retain a real measure of location by either wedges or screw-operated levers as shown in Fig. 7.8. The advantages of split bushes were not recognized again for about 200 years. It was, incidentally, in connection with this invention that Leonardo's specification of 'mirror metal' is associated.

Leonardo's contributions in the fields of friction, wear, lubrication, bearing materials and bearing design were remarkable in both nature and timing. He was, in short, a great tribologist. His work in this branch of mechanics may not previously have received the recognition it deserves, representing as it does a most eminent early example of the merits of a scientific approach to basic engineering problems. There is sufficient evidence of Leonardo's personal contribution to these studies to render insignificant any anxieties about the originality of his work. Even when information exists that tribological systems similar to those considered by Leonardo were in use in the machinery of the day there is often additional evidence that he had considered the further development of the system.

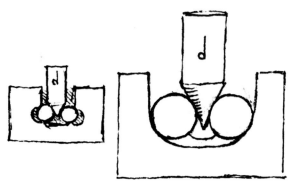

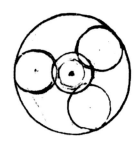

Fig. 7.7
Leonardo da Vinci's sketches in
Codex Matrid I of ball, cone and
roller pivot bearings.

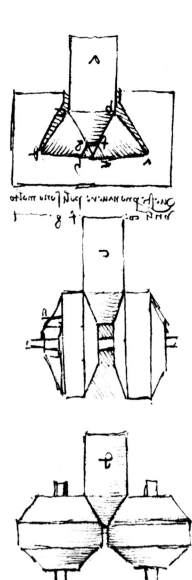

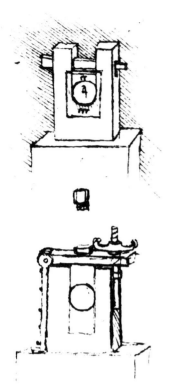

Fig. 7.8
Leonardo da Vinci's sketches in
Codex Madrid I of split,
adjustable bearing blocks.

In passing judgement on Leonardo's tribological work it is perhaps sufficient to observe that neither the problems nor the solutions discussed in his notebooks have changed very much in the past 500 years.

7.3   Tribological developments in Renaissance industry

The regression in bearing development in the long millennium of the Middle Ages has already been noted. Wooden bearings again dominated the scene and the opportunity to apply and develop the rolling-element bearings which had formed the subjects of trials by the Greeks, Romans, Celts and possibly the Chinese was ignored. Towards the end of the Middle Ages iron was being used increasingly as a bearing material, perhaps as a result of the steady growth in utilization of mechanical power in the form of water- and windmills, and the stage was set for a period of bearing development consistent with the technological demands of the Renaissance.

The notebooks of Leonardo da Vinci provide an exhilarating view of the first application of a scientific approach to tribological problems, but we now look to further evidence, primarily from other writers of the Renaissance period, regarding the general form of bearings in industrial situations. The selected evidence is found, in chronological order, in the works of Cellini (*c.* 1534), Biringuccio (1540), Piccolpasso (*c.* 1550), Agricola (1556), Baldewin, (1561), Besson (1572) and Ramelli (1588).

Globes of iron and wood

Benvenuto Cellini (A.D. 1534)   Early in the sixteenth century (*c.* 1520) cast-iron balls started to replace stone and Allan (1945) has drawn attention to the fact that they were employed in the positioning of gun carriages. Another reference to the use of spherical rolling supports was recorded shortly afterwards in the autobiography of the Florentine goldsmith Benvenuto Cellini.[12]

A commission from the King of France caused Cellini to provide in 1534 a silver statue of Jupiter about $4\frac{1}{2}$ ft (1·4 m) high. The translation of Cellini's description reads:

Having with the utmost diligence finished the beautiful statue of Jupiter, I placed it upon a wooden socle, and within that socle I fixed four little globes of wood which were more than half hidden in their sockets and so admirably contrived that a little child could with the utmost ease move this statue backwards and forwards, and turn it about.

This account is remarkably reminiscent of the finds from the Roman ships in Lake Nemi described in Chapter 5, although the balls employed by Cellini were apparently free-rolling and made of wood.

---

12.   See Roscoe, T. (1922), *Memoirs of Benvenuto Cellini*, with the notes and observations of G. P. Carpani, translated by Thomas Roscoe, 2 vols, H. Colban and Co., London.

Mining and metallurgy

Vanoccio Biringuccio (A.D. 1480–1539) and *Georgius Agricola* (A.D. 1490–1555)   The mining industry witnessed the introduction of many ingenious machines which were called upon to satisfy the rapid expansion of demand for metals in Renaissance times. The simultaneous development of printing fortunately provided us with accounts of the industry which included excellent descriptions and illustrations of the bearings of the day. The motive power was provided chiefly by humans and animals, but water-wheels and, to a much smaller extent, windmills were also used.

The first comprehensive book on metallurgy was probably that published by the Italian Vanoccio Biringuccio (1540) in Venice under the title *Pirotechnia*. French translations of the work appeared in 1556 and 1627 while a fifth Italian edition was published as late as 1678. Although heavily engaged in metallurgical work Biringuccio is described in the preface of the first edition of his book[13] as a mathematician.

There is in *Pirotechnia* one of the early expressions of interest in bearing materials when the author refers to the use of glass or gunmetal grooves as an alternative to iron-on-steel bearings for pendent bells.[14]

The classical Renaissance text on all aspects of mining, *De Re Metallica* by Georgius Agricola (1556) was published posthumously. The work consists of twelve books which were started about 1533, completed about 1550 and sent to the press about 1553. This mammoth literary task thus extended over twenty years, while the final publication was long delayed by the preparation of numerous woodcuts.

Agricola, whose real name was Georg Bauer (peasant), was born in Glauchen in Saxony on 24 March 1494, two years after Columbus had discovered the New World. He entered the University of Leipzig at the age of twenty and after graduating three and a half years later he taught Greek and Latin at the Municipal School at Zwickau. He became Principal of the School and also started his literary career with the publication of a small book of Latin grammar in 1520. He lectured at the University of Leipzig for two years before travelling to Italy to study philosophy, medicine and natural science in 1524. He returned briefly to Zwickau in 1526 and in 1527 he was appointed town physician at a thriving centre of the Bohemian mining region known as Joachimsthal. His interest in all aspects of mining must have developed rapidly during this period in Joachimsthal for he resigned about 1530 to spend two or three years travelling and studying mines. He again returned to medicine in the town of Chemnitz in Saxony in 1533. He was elected a Burgher and later Burgomaster of Chemnitz in 1546 when he was fifty-two years old and he died in the same city on 21 November 1555 aged sixty-two.

Although the tribological interest in Agricola's writings is found almost exclusively in Book VI, the remaining books make rewarding reading. The principles of underground mining and the art of surveying are dealt with in Book V and the opening paragraph reflects the wide-

GEORGII AGRICOLAE
DE RE METALLICA LIBRI XII· QVI-
bus Officia, Inſtrumenta, Machinæ, ac omnia deniçaad Metalli-
cam ſpectantia, non modo luculentiſſimé deſcribuntur, ſed & per
effigies, ſuis locis inſertas, adiunctis Latinis, Germaniciſçç appel-
lationibus ita ob oculos ponuntur, ut clarius tradi non poſſint.

E I V S D E M

DE ANIMANTIBVS SVBTERRANEIS Liber, ab Autore re-
cognitus: cum Indicibus diuerſis, quicquid in opere tractatum eſt,
pulchré demonſtrantibus.

FRO [ ] BEN

BASILEAE   M· D· LVI·

Cum Priuilegio Imperatoris in annos v.
& Galliarum Regis ad Sexennium.

---

13.   See Hoover, H. C. and Hoover, L. H. (1950), Appendix B, p. 614.
14.   See Smith, C. S. and Forbes, R. J., 'Metallurgy and assaying', in Singer, C., et al. (1957), Vol. III, Ch. 2, p. 41.

spread interest of Renaissance engineers in pumping equipment with the words: 'then I will speak . . . of the machines with which water is drawn from the shafts and air is forced into deep shafts and long tunnels, for digging is impeded by the inrush of the former or the failure of the latter.'

In the same chapter illustrations show windlasses mounted over mine shafts and although the sketches are small it appears that the drums were carried on simple shafts mounted in forks in wooden posts.

Book VI contains most excellent illustrations and detailed accounts of the Renaissance miner's tools and machines. The principles upon which the winding, pumping and ventilating machines operated had been known since the early Greek and Roman eras, but there is little recorded evidence of the development of the early mining machines. The depth of mines was largely limited by the ability to remove water, and it was no doubt for this reason that Agricola devoted considerable space to the description of majestic hydraulic machines driven by men, animals, water-wheels or windmills. The problem of water in mines was ultimately to provide the spur for Newcomen's invention of the steam engine in 1705.

The bearings on wheelbarrows were described as simple circular holes in wooden side-planks, and although it is not clearly recorded the axles were probably made of iron. The wooden wheels on open trucks certainly rotated on iron shafts having a square central section as shown in Fig. 7.9.

Descriptions of hauling machines range from the simple windlass to geared systems driven by human beings on treadmills or by animals.

Fig. 7.9
Agricola's sketch of an open mining truck showing wooden rollers (wheels) on iron axles.

A—Rectangular iron bands on truck.   B—Its iron straps.   C—Iron axle.
D—Wooden rollers.   E—Small iron keys.   F—Large blunt iron pin.
G—Same truck upside down.

There is clear evidence of the use of iron and steel bearings in these more highly loaded systems. Hoover and Hoover's (1950) translation of the original 1556 version of Agricola's work includes the following references.

In the description of the treadmill driven windlass ... (Book VI, p. 162) 'It consists of an upright axle with iron journals at its extremities, which turn in two iron sockets. ...' (Book VI, p. 163) '... A winding-rope is wound round this latter axle, which turns in iron bearings set in the beams. ...'

In the description of an animal-powered hoist there is mention of (Book VI, p. 164) '... a steel socket in which the pivot of the axle turns. ...'.

Elsewhere in the same section there is one of the most informative accounts of the form of contemporary bearings (Book VI, p. 164):

> *Every axle used in mining, to speak of them once for all, has two iron journals, rounded off on all sides, one fixed with keys in the centre of each end. That part of this journal which is fixed to the end of the axle is as broad as the end itself and a digit thick; that which projects beyond the axle is round and a palm thick, or thicker if necessity requires; the ends of each miner's axle are encircled and bound by an iron band to hold the journal more securely.* [One digitus is 0·726 in (18·4 mm) and four digiti make one palm.]

The section of Agricola's book dealing with water-raising machines is particularly interesting. There are detailed accounts of machines employing endless chains which raise water in buckets or dippers, piston pumps of numerous kinds and of rag-and-chain pumps. Consideration is given to various driving mechanisms ranging from human-operated cranks and treadmills to water-wheels.

The increasing use of iron for constructional purposes is evident from the description of a hand-operated chain-of-dippers machine.[15]

The gear-box for this machine was made entirely of metal, with iron axles revolving in bearings or wide pillows of steel. The teeth and the rundles of the wheels and pinions were also made of steel, while the individual teeth were screwed into the wheel to facilitate replacement if breakage occurred.

A most important feature of the account of this machine is a mention of the use of a roller-disc bearing of the type which caught Leonardo's imagination. According to Agricola (Book VI, p. 172), '... this axle proceeds through a frame made of beams which stands around the shaft, into an iron fork set in a stout oak timber, and turns on a roller made of pure steel'.

Here we have direct evidence of the industrial use of rolling bearings in the Renaissance period. We are even more fortunate in having an illustration, reproduced here as Fig. 7.10, to support the text.

In the accounts of piston-pumps there is reference to the use of metal discs and *leather washers* and the use of *intervening leathers* to prevent friction on pump rods.

To conclude reference to this most valuable record of both industrial and tribological practice in Renaissance times, attention is drawn to the

---

15. See Hoover, H. C. and Hoover, L. H. (1950), Bk VI, p. 172.

Fig. 7.10
Hand-driven chain-of-dippers water-raising machine showing the use of a steel roller-disc bearing (Agricola).

A—IRON FRAME. B—LOWEST AXLE. C—FLY-WHEEL. D—SMALLER DRUM MADE OF RUNDLES. E—SECOND AXLE. F—SMALLER TOOTHED WHEEL. G—LARGER DRUM MADE OF RUNDLES. H—UPPER AXLE. I—LARGER TOOTHED WHEEL. K—BEARINGS. L—PILLOW. M—FRAMEWORK. N—OAK TIMBER. O—SUPPORT OF IRON BEARING. P—ROLLER. Q—UPPER DRUM. R—CLAMPS. S—CHAIN. T—LINKS. V—DIPPERS. X—CRANK. Y—LOWER DRUM OR BALANCE WEIGHT.

rag-and-chain pump shown in Fig. 7.11. Two men operate the treadmill, thus causing water to be discharged each time the balls emerge from the pipe. The use of a simple wheel-and-pinion drive and the method of supporting iron journals in iron pillows is clearly shown.

Although greatly interested in the art of mining Agricola did not promote industry at any cost. He concluded Book VI by speaking of the ailments and accidents of miners . . . (Book VI, p. 214), 'for we should always devote more care to maintaining our health, that we may freely perform our bodily functions, than to make profits'.

A grim reminder of the price paid for industrial progress is given by Agricola's comment that . . . (Book VI, p. 214), 'in the mines of the Carpathian mountains women are found who have married seven husbands, all of whom this terrible consumption has carried off to a premature death'.

Fig. 7.11
Plain bearing on
sixteenth-century
rag-and-chain pump
(Agricola).

Fig. 7.12
Mid-sixteenth-century potter's
wheel–after Piccolpasso (*c.* 1550).
Crown Copyright, Victoria and
Albert Museum.

(a) Italian potters operating
kick-wheels.

(b) Bearing arrangement showing
'hard steel pivot' and 'upper
support' bearing.

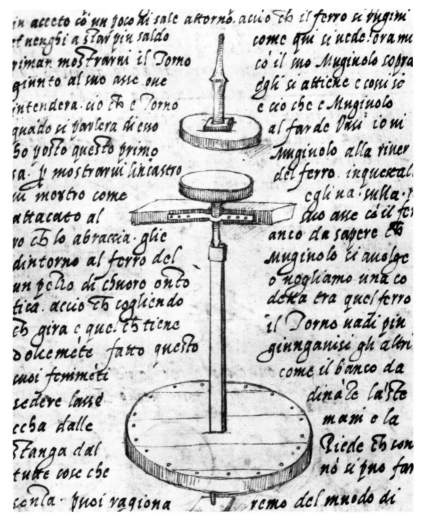

The potter's wheel

Cipriano Piccolpasso (A.D. 1524–1579)   A most interesting record of tribological development in the field of ceramics is found in Piccolpasso's[16] account of 'all the secrets of the whole Potter's art' as practised about 1550 in northern Italy. This area was a centre for Renaissance pottery and since Piccolpasso's brother was a potter the account is particularly valuable.

Piccolpasso lived in Castel Durante (now Urbania) and in his *Three Books of the Potter's art* he makes reference not only to the form of the wheel bearings but also to the use of a lubricant. He further provides us with the sketches shown in Fig. 7.12 and in the text he describes the wheel pivot bearing as a hard 'steel' point running on a 'flint stone' or a 'very hard steel plate'. The main pivot bearing of the potter's wheel had developed considerably from the heavy stone-on-stone arrangement described in Chapter 4, and the use of hard steel pivot points on hard supports in the Renaissance would undoubtedly have enhanced the production of wheel-thrown pottery.

The bearing towards the top of the axle was said to be 'cased with oiled leather'; a rare reference to the use of seals and lubricants in Renaissance times. In this instance sealing might well have been the main purpose of the supple oiled leather, since abrasive wear of the upper bearing would lead to irregular running of the wheel.

Roller bearings in astronomical clocks

Eberhardt Baldewin (A.D. 1561)   A further indication of Renaissance interest in rolling-element bearings is found in the history of horology. An astronomical clock of great complexity was constructed in 1561 by Eberhardt Baldewin, clockmaker to William IV of Hesse. That part of the gear train operating the dial of mercury depicted in Fig. 7.13 clearly shows that Baldewin used roller-bearings.

Renaissance machinery

Jacques Besson (c. A.D. 1540–1573) and *Agostino Ramelli* (A.D. 1531–c. 1604)   A number of early books on machinery produced in Renaissance times are of great interest as a result of both their timing and their nature. Those by Besson and Ramelli are particularly valuable. Specialist books relating to mining and ceramics have already been referred to, but the more general books on machinery concentrate upon devices like pumps and cranes which have a direct relevance to architecture, at the expense of accounts of vehicles and other machines employed in major industries of the day. This slightly curious imbalance is no doubt a reflection of the high respect for architecture established in classical times and revived in the Renaissance, but it leaves some serious gaps in our knowledge of mechanical devices of the period.

---

16.   See the facsimile edition with translation and introduction by B. Rockham and A. van de Put (1934), *Three Books of the Potter's Art*, by Cavaliere Cipriano Piccolpasso, The Victoria and Albert Museum, London.

Fig. 7.13
Roller-bearings in the gear chain operating the dial of mercury in Eberhardt Baldewin's astronomical clock of 1561.

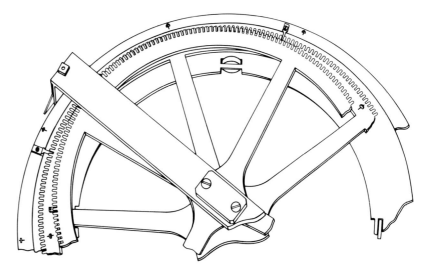

Descriptions of machinery did not appear for at least a century after the introduction of printing in the West and Keller (1964) has ascribed this delay to the protective attitude of the inventors and mechanicians of the day. It has to be remembered that there was little protection of ideas through copyright and patents and even Leonardo felt it necessary to take precautions by employing mirror-writing in his notes. Fortunately these difficulties were gradually surmounted and in due course important records of machinery started to emerge. An important and attractive feature of these books is their fine illustrations and we can frequently learn as much from the drawings as from the text.

Jacques Besson was probably the first person to publish his mechanical inventions and although we do not have a full account of his career the early references suggest that he came from Colombières, near Briançon in the alpine region of south-eastern France. He lived in Geneva for a few years from 1559 and published his first book in Switzerland on practical distilling. It appears that he later spent much time teaching mathematics in France where he published accounts of mathematical instruments which he had invented. After spending some time in Rouen, Paris and Orleans he presented himself to King Charles IX and returned to Paris as *master* of the King's engines. Besson's (1569) well-known book, *Theatre des instruments et machines* was certainly written by 1569, but the date of publication is a more open question. Gille (1966) suggests 1569 while Keller (1964) puts the date a little later, probably 1571–72, and others suggest that it was published posthumously. Besson was probably a Protestant for he died in London in 1573 after fleeing there in the previous year.

Besson's book shows a number of measuring instruments, the use of windlasses and pulley systems for moving and lifting heavy objects, several military machines and boats. The descriptive material is sadly brief, but the illustrations are rich in detail as shown by Figs 7.14 and 7.15. The first shows a pulley being used to lift and position a gun barrel. It is selected also to show that the use of studs to prevent the excessive wear

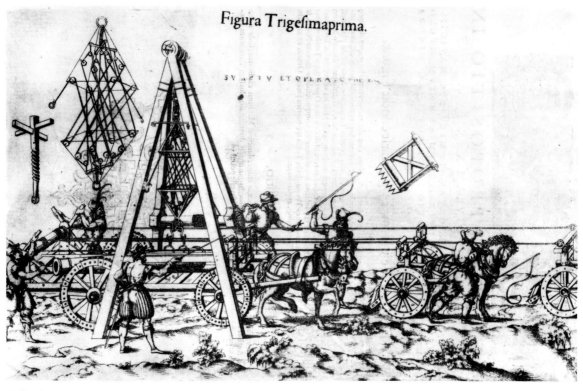

Fig. 7.14
Illustration of system for lifting and positioning gun barrels showing wheeled vehicles with studs in rims of wheels.

of the rims of wheels had continued from the time of the Luttrell Psalter (1338)[17] until the second half of the sixteenth century.

The second illustration from Besson's text shown in Fig. 7.15 shows the use of rollers and a windlass in the movement of an ornate carved pedestal.

Agostino Ramelli was born in Ponte Tresa on Lake Lugano in 1531. He was a Catholic and a military engineer who came to France in the service of Henri, Duke of Anjou, later to be King Henri III. While surveying the fortifications of La Rochelle for the Italian troops he was captured and spent some months in a Protestant dungeon. His famous book *Le Diverse Et Artificiose Machine* was published in Italian and French in 1588. It is clear that the Capitano had secured royal recognition from Henri for he is described on the frontispiece as 'Engineer of the most Christian King of France'. The last reference to our author appeared in 1604.

Ramelli's (1588) book is a most comprehensive account of sixteenth-century machines. About a third of the book is devoted to various forms of machinery for war, thus reflecting the author's military career. Hydraulic equipment is extensively discussed and elaborately illustrated, but the work also covers corn-grinding, including well-known illustrations of windmills, amphibious vehicles, screw-jacks and building hoists.

---

17.   See Ch. 6.

Fig. 7.15
The use of rollers in the movement of an ornate carved pedestal.

The illustrations of interest to tribologists are too numerous for the present text, but the following selection will give an insight into industrial practice of the latter half of the sixteenth century while also conveying a measure of the artistry of the records.

A remarkable four-cylinder reciprocating pump driven by a water-wheel through gearing and a fantastic swash-plate arrangement is shown in Fig. 7.16. Rollers were used to guide the pump rods and, as shown at H and N, for the successful operation of the swash-plate system. The arrangement is a remarkable commentary on the rapid development of rolling-element bearings in the latter half of the sixteenth century. The interest in both rolling-element bearings and gearing is further illustrated in the well-known picture shown in Fig. 7.17. A thrust ring, in which the axes of rotation of the rollers were apparently fixed, formed an essential feature of the support for the hand-driven gear-wheel on this simple well-hoist.

A fine example of the use of the roller-disc bearing featured by Leonardo in his notebooks, sketched earlier in the century by Agricola and shown in various places in Ramelli's text, forms the subject of Fig. 7.18. In this case the treadmill-operated chain-of-dippers water-raising machine has its two main shafts supported on roller-disc bearings at H and I. Several pages of the book are devoted to devices designed to utilize the potential of geared systems to give a large mechanical advantage in order to expand protective bars. Our final example in Fig. 7.19 shows such machines in which a toothed rack is clearly seen to be supported on rollers.

Fig. 7.16
The extensive use of rolling
elements in a water-wheel-driven
four-cylinder reciprocating pump
with swash-plate action (after
Ramelli, 1588).

It might be thought from this emphasis upon the evidence for the use
of rolling-element bearings that Ramelli cared little for plain bearings,
but this is not the case. Almost every illustration shows plain bearings,
but apart from the occasional use of elaborate decoration they do not
exhibit sufficient novelty to justify inclusion in this section.

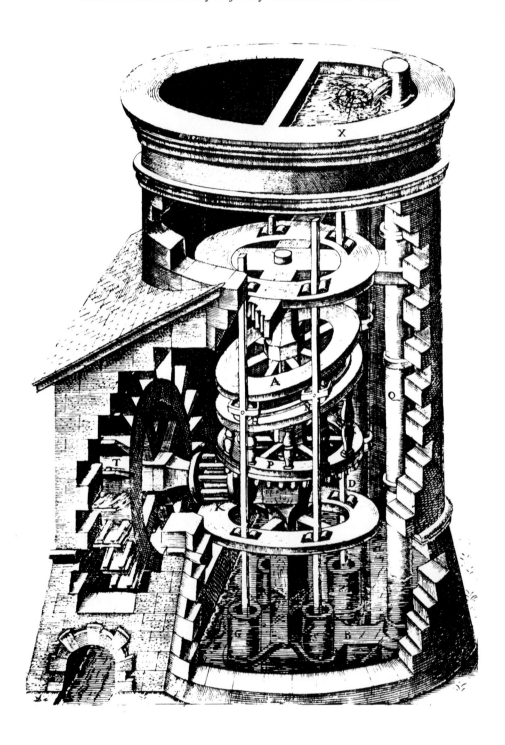

Fig. 7.17
Hand-driven well-hoist showing
roller-bearing support ring (after
Ramelli, 1588).

7.4 Chronology

Although the total time span considered here is a mere 15 per cent of the period covered in Chapter 6, there is much to report on the awakening interest in tribology. The story falls into two parts. In the first half of the Renaissance period from 1450 to about 1530, Leonardo's genius spilled over into the fields of friction, wear, bearings, gears and lubrication systems as summarized in Table 7.1. Leonardo probably started to write the Codex Madrid I which has yielded such valuable records of his work

Fig. 7.18
Roller-disc bearings supporting
the main shafts (at H and I)
on a treadmill-operated
chain-of-dippers (after
Ramelli, 1588).

on tribology a year or two after Colombus set out on his first voyage to
the west. The manuscript was completed four years later, in the year
before Columbus discovered South America on his third great voyage of
exploration.

The second half of the period from about 1530 to 1600 is dominated
by the publication of texts which confirm particular stages of bearing
development. It should be emphasized that none of the books mentioned
in the text or in Table 7.1 were devoted entirely or even mainly to tribology.
They were general accounts of particular industries or general forms
of machinery which, by their descriptive and illustrative content, showed

Fig. 7.19
The use of rollers to support 'racks' in hand-operated worm gear 'expanders' (after Ramelli, 1588).

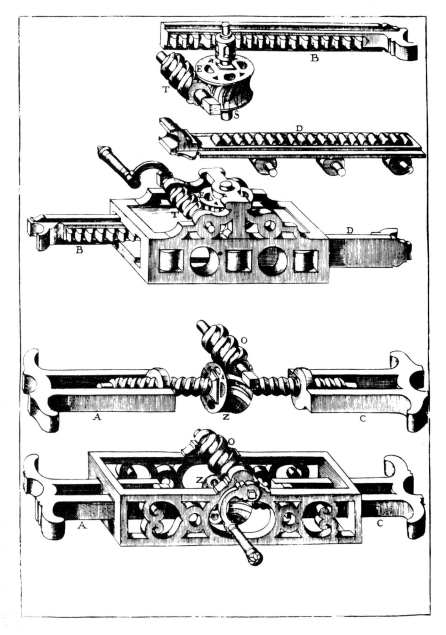

clear evidence of current tribological skills. Agricola published his great book on the mining industry two years before the crowning of Queen Elizabeth I, while Ramelli presented his beautifully illustrated book on contemporary machinery in the year of the defeat of the Spanish Armada.

Perhaps the most interesting general aspect of the Renaissance period is that early scientific studies of tribology were initiated alongside the impressive practical developments of the day. The interchange between research and the needs of industry has been an ever-present situation from the Renaissance to the present day.

It can also be seen from Table 7.1 that while European countries were

Table 7.1
Chronology for the Renaissance
period (A.D. 1450–1600)

Date (A.D.)	Political and social events	General technical developments	Tribology
1450	1451–1516 Amerigo Vespucci 1452–1519 Leonardo da Vinci 1453 End of Hundred Years War 1453 Fall of Constantinople 1455–85 Wars of the Roses	1450 Movable-type printing	
1460	1461–83 Reign of Edward IV 1462–1515 Louis XII of France		
1470	1470–1524 Emperor Maximilian 1473–1543 Copernicus 1473 Foundation of University of Pisa	Patents introduced at Venice 1476 William Caxton's printing press	**Leonardo da Vinci's–scientific studies of friction –observations on wear**
1480	1479 Wars with Naples and Rome 1483–85 Reign of Richard III 1485–1509 Reign of Henry VII		**–specification of low-friction bearing material (mirror metal) –design of plain bearings**
1490	1492 Columbus reaches San Salvador 1497 Vasco da Gama reaches India 1497 John Cabot discovers Labrador	1493–97 Codex Madrid I written	**–split bearing block –rolling bearing sketches –roller-disc bearing sketches**
1500	1498 Columbus reaches S. America 1509–47 Reign of Henry VIII		**–bearing 'retainer' proposed –lubrication systems considered –tribological aspects of gears and screw-jacks**
1510	1519–1522 Circumnavigation of the world (Begun by Magellan)		
1520	1525 Mogul Empire founded		
1530	1530 Charles V crowned by the Pope Henry VIII began his quarrel with Rome		**1534 Ball thrust bearing–Cellini**
1540	1546 Death of Martin Luther	1540 *Pirotechnica* published Casting of iron cannon	**1540 Bearing materials–glass and gunmetal**
1550	1550 Copernicus–solar system 1558–1603 Elizabeth I Potato introduced into Europe	1550 *Three Books of the Potter's Art*–Piccolpasso. Brass manufacture in England 1556 *De Re Metallica* published	**Hard steel pivot bearings and use of oiled leather in potter's wheel Plain bearings of wood, iron and steel Roller-disc bearing in industrial use**

Table 7.1 (*Continued*)

Date (A.D.)	Political and social events	General technical developments	Tribology
1560			**Gears of wood and steel** **Seals–leather** **Roller-bearing used in** **astronomical clock**
	1564–1642 Galileo Galilei	1561 Baldewin's astronomical clock Juanelo Turriano manuscripts	
1570		Mercator's map projection 1569 *Theatre des Instruments et Machines*	
1580	1588 Defeat of Spanish Armada	1588 *Le Diverse et Artificiose Machine* published 1589 Knitting machine invented (Lee)	**Plain and rolling bearings used in machinery**
1590	1591 *Revenge* sunk by Spaniards		
1600		1600 Galileo's telescope	

involved in various conflicts the period was also one of impressive exploration on a world scale.

7.5 Summary

The latter part of the fifteenth and the early days of the sixteenth century were dominated, as far as tribology is concerned, by the genius of Leonardo da Vinci. There is presently a tendency to play down his contributions to the physical sciences, mainly on the grounds that there is little evidence that any of his machines were built, that many of the devices which he sketched were impractical and that his notebooks reflected contemporary practice rather than his own original ideas. His work on tribology, which has not previously enjoyed the full recognition it deserves, appears to run contrary to the current criticisms.

Leonardo studied friction, wear, bearing materials, plain bearings, lubrication systems, gears, screw-jacks and, most importantly, rolling-element bearings. He conducted the first recorded scientific studies of these subjects, developed an understanding of the laws of friction and the nature of wear and devised bearings which bear a remarkable resemblance to their modern counterparts. It is indeed fortunate that his talents were applied to tribology, and full recognition of the merits of his work is long overdue.

The development of movable-type printing was probably one of the most significant innovations of the period. It yielded the first printed books in the West on the mechanical arts, and industrial techniques in the second part of the period are depicted in the fascinating texts of the day. The opportunity has been taken to provide a full background to the lives and work of the sixteenth-century mechanicians whose work is mentioned in this chapter. It is particularly interesting to note that many claim skills in several fields and the dichotomy between science and engineering was not yet in evidence.

The great sixteenth-century books on machinery contain a wealth of descriptions and sketches of bearings. Plain bearings increasingly employed metals, notably iron and steel, and there is also some evidence of a growing interest in low-friction bearing materials. In parallel with plain bearing progress there was much activity in the field of rolling bearings. It is known that the roller-disc bearing sketched by Leonardo was in use in machinery in 1494 and the texts of both Agricola and Ramelli confirm its wide acceptance. More conventional rolling bearings were also used quite widely; particularly in hydraulic machinery.

A recognition of the importance of tribology emerged strongly during the Renaissance. This is clear not only from the notebooks of Leonardo da Vinci but also from the sketches and accounts of those concerned with a wide range of machinery in the sixteenth century. The stage was now set for the steady development of bearings and the increasing use of scientific method in tribology prior to the Industrial Revolution.

Chapter 8

Towards the Industrial Revolution – early scientific studies of friction (A.D. 1600–1750)

8.1 Introduction

In the period leading up to the Industrial Revolution progress in tribology was profoundly influenced by the emergence of modern science. The convictions of the early promoters of the scientific revolution like Francis Bacon (1561–1626) and Galileo Galilei (1564–1642) inspired later workers to apply the scientific method and philosophy to both fundamental issues in natural philosophy and problems associated with the mechanical crafts of the day. The emergence of modern science and its application to technological problems was undoubtedly the outstanding development during the period.

The centre of activity in our subject in the period under review was located in those parts of Europe which nurtured the new scientific approach, but the translation by E-Tu Zen Sun and Shiou-Chuan Sun (1966) of Sung Ying-Hsing's (1637) interesting account of Chinese technology in the seventeenth century shows that there were impressive general developments of skills and machinery in the East to equal many of those in the West. It is recorded[1] that water-powered mills saved 90 per cent of the human effort required in pounding rice and in a rare mention of tribological matters reference is made to the deleterious effect of heat caused by the friction of rough millstones upon the grinding of wheat flour.[2] Elsewhere Sung Ying-Hsing included in a reference to the varied and numerous uses of oil a note to the effect that '. . . One drop of

---

1. See p. 93 of E-Tu Zen Sun and Shiou-Chuan Sun's (1966) translation of Sung Ying-Hsing (1637).
2. Op. cit. p. 95.

oil (in the axle) enables a cart to roll and one tan[3] of oil used in caulking a ship makes it ready for the voyage.'

France and England played major roles in the important tribological developments recorded in seventeenth- and early eighteenth-century Western Europe. The period opened quietly as a continuation of the Renaissance approach as far as bearings were concerned. Engineering progress can hardly claim a dominant position in the history of a period in which the works of William Shakespeare reflected the state of contemporary literature. Music, poetry and theatre flourished in England towards the end of the Tudor period, particularly in domestic circles.

The formation of great scientific Institutions like the Royal Society (1660) in London and the Académie Royale des Sciences (1666) in Paris reflected the strength of the scientific revolution in England and France. They provided a forum for the discussion of scientific matters and we shall see that the science of tribology was one of the many beneficiaries. Hooke's 1685 discourse[4] before the Royal Society on Stevin's sailing chariot is an example of the way in which seventeenth-century thoughts on bearings and other tribological topics were exposed and recorded. The seventeenth century also saw the publication of one of the greatest scientific works of all time, Newton's (1687) *Principia*. Understanding of the mechanism of lubrication remained at a primitive level compared with the enhanced knowledge of the physical basis of friction, but Newton's *Principia* contained the foundation of future understanding of fluid-film lubrication.

The beginning of the eighteenth century witnessed an explosion in studies of tribology. There are numerous accounts of interesting, successful and often spectacular applications of roller-disc bearings in the late seventeenth- and early eighteenth-century literature. Furthermore, discussions of a wide range of topics including plain bearing materials, rolling bearings, seals, lubricants and wear are recorded. However, pride of place must be reserved for the outstanding early scientific studies of friction reported between 1699 and 1750.

If the period had to be divided in terms of tribological progress it should undoubtedly be severed into sections covering 1600–98 and 1699–1750. It would, however, be unfortunate if such a diversion were to be introduced, since the first ninety-nine years laid the firm foundations which carried the great studies of friction through at least the next fifty-one years.

Some of the general industrial developments which provide evidence of the state of tribology in seventeenth-century Europe will provide the backcloth to our detailed discussion of subsequent scientific work. Special mention will then be made of progress in the development and application of bearings, particularly in the eighteenth century. Transportation problems feature largely in this story, particularly in relation to wheeled road vehicles, but it is also worth noting that horse-drawn

---

3. Op. cit., p. 215. (It appears that in the Ming Dynasty, 1368–1644, 1 tan was the equivalent to about 107 litre.)
4. See Gunther, R. T. (1930), Vol. VII, Pt II.

vehicles running on rails were in general use in harbours, coal-fields and heavy industries of the day.

No apology is offered for the detailed account of work on friction in this period, for it was in this subject area that the scientific revolution first impinged with some force upon tribology. In previous chapters the unfolding story has relied essentially upon archaeological evidence or historical accounts of the development of devices for combating friction and wear, but a new concept enters our history at this stage. For the first time the fundamental processes governing friction were studied theoretically and experimentally, with the findings being recorded in books and publications of the scientific societies for all to read.

This was truly a vintage era in the history of tribology.

8.2   Seventeenth-century industry

There were ever-increasing demands for mechanical power in the industries of seventeenth-century Europe. The mining industry required reliable and effective pumping and ventilation equipment to support its widespread and deepening activities. The metallurgical industry called for more powerful equipment to crush the ore and to operate the extraction plant, while the textile industry exercised demands for mechanically powered machines to improve both the quality and the quantity of its products. The growing population and the development of towns required a plentiful supply of water for both domestic and public purposes. The farming industry sought support from sources of mechanical power for drainage, irrigation and corn grinding.

Animal and human power sources were still widely used, but economic and technological factors promoted an interest in the widespread utilization of water- and wind-driven machines. Urbanization fostered an increasing degree of specialization in occupations while the very scale of developing industrial processes called for efforts beyond the capabilities of teams of men or animals. Water-wheels and windmills grew in both size and quantity throughout the period and the development of bearings in windmills will form the subject of a special section in Chapter 9. Indeed, these and other mechanical devices like piston-pumps achieved a degree of refinement and a physical size which approached the limits conceivable with the technology of the day. The scene was set for the introduction of better materials, design and production methods and, most importantly, the application of new principles of power generation based upon steam. The period is thus important, not only for its general progress in technology and a considerable development of the art and science of tribology in particular, but for the creation of conditions which promoted the Industrial Revolution.

Vittorio Zonca

A fine example of the general form of industrial machinery at the beginning of the seventeenth century is found in the writings of Vittorio Zonca (Zonca, 1607). The frontispiece of his book, which was published posthumously by the Paduan printer Francesco Bertelli, tells us that Signor

Vittorio Zonca was 'architect to the Magnificent Community of Padau'. We know little else about him, apart from the fact that he died in 1603 at the age of thirty-five, but his book is regarded as one of the classical representations of mechanical devices of the sixteenth and early seventeenth centuries. Keller[5] has described Zonca's *Teatro Nuovo di Machine et Edificii* as a more sober and realistic book of mechanical inventions than Besson's *Livre des Instruments Mathematiques et Mechaniques* or Ramelli's *Diverse et Artificiose Machine*. The wide-ranging work includes excellent illustrations of machinery found in various kinds of mills, particularly in the textile industry, printing presses and mechanical pulverizers. The advantages of various innovations are noted and the writings convey a sense of enthusiasm for the development of ingenious mechanical devices. It was an enthusiasm unbounded by a full appreciation of the physical laws governing the performance of mechanical devices, for his closing illustration shows a perpetual-motion machine to rival any of the period!

Zonca's work is important to our commentary on the state of early seventeenth-century tribology both from the point of view of his observations on wear, which included a comment on the need to use dissimilar metals in sliding pairs, and his sketches of various forms of bearings. He appears to have been the first person to record the view that,[6] *when running against steel, any sort of metal other than brass is consumed.* This comment is highly significant since it represents one of the few references to wear and the consequences of sliding steel-on-steel in the long interval between Leonardo da Vinci's writings and Robert Hooke's report to the Royal Society on Stevin's sailing chariot in 1685. It appears to be a clear recognition of the desirability of using dissimilar bearing metals and an indication that steel and brass formed a common combination for the bearings which supported the machinery of the day.

Each of the plates in Zonca's book is worthy of study, but some show more clearly than others details of tribological significance. The first section is concerned with doors, hoists, lock gates and machinery found in various kinds of mills. In the fulling of woven cloth the material was cleansed and thickened by beating in the presence of water. The original process, which continued well into the Middle Ages, consisted of the treading of cloth in vats. The water-driven fulling mill was probably developed as early as the twelfth century, but Zonca (1607) provides us with the earliest known pictorial representation of the process. It can be seen from Fig. 8.1 that a horizontal camshaft was driven by a large waterwheel. The cams lifted and then released large wooden hammers which struck the cloth in the vats. When not being used to improve the woven cloth the fulling mill became a giant washing machine for the community. The water-driven fulling mill not only saved labour, it also improved the quality of cloth delivered at the market. It provided an important finishing process for woollen material.

5. See Keller, A. G. (1964).
6. See Smith, C. S. and Forbes, R. J., 'Metallurgy and assaying', in Singer, C., et al. (1957), Vol. III, Ch. 2.

Fig. 8.1
Sketch of an early seventeenth-century fulling mill showing plain bearings on the wooden camshaft.

It can be seen from Fig. 8.1 that the plain bearings sketched by Zonca had the general characteristics of the fifteenth- and sixteenth-century bearings described in Chapter 7. The stub shafts rested on half-bearings consisting of a metal plate, probably iron, mounted on a wooden frame. There is a strong suggestion in the sketch that the radius of the bearing was considerably greater than the shaft, perhaps the result of excessive wear as the camshaft rotated under dynamic loading.

Elsewhere Zonca shows a pivot bearing (Fig. 8.2) supporting a hand-operated endless screw which is used to open heavy metal gates with the aid of a geared driving shaft. The heavy metal gear-wheels with their

Fig. 8.2
Sketch of a pivot-bearing
supporting a hand-driven endless
screw.

central square holes demonstrate an impressive state of machine construction in the early seventeenth century. The general history of gearing
cannot be considered in any detail in the present text, but the reader
should take advantage of the excellent account prepared by Dudley (1969)
and published by the American Gear Manufacturers Association.

Printing presses were in widespread use at the dawn of the seventeenth
century and Fig. 8.3 shows the plate used to illustrate Zonca's own account
of the contemporary process for producing such plates for books. The
engraving shows in some detail the materials and processes employed
and it is instructive to read the text which accompanies the plate. The

Fig. 8.3
Sketch of an early seventeenth-century printing press showing wooden rollers (A) and split-bearing blocks (B) of boxwood or pear.

construction of the press, the composition and preparation of the printing ink, the engraving and etching of the copper plates and their useful life are all discussed. Reference is made to the use of *common oil* on the plates to prevent corrosion and it appears that engraved plates would yield 1,000 impressions or even, with retouching, 2,000 prints.

The rollers (A) were apparently made of boxwood or pear and turned in a lathe. They were mounted in grooved pieces of *the same wood* as shown at (B), thus demonstrating that the split-bearing blocks proposed by Leonardo da Vinci were in regular use in machinery at the beginning of the seventeenth century.

Water raising

This became a subject of considerable importance in the period under review, primarily as a result of demands from the growing urban population. Significant mention of tribological aspects of water-raising equipment in earlier times has already been noted in connection with Vitruvius (Ch. 5); Agricola, Juanelo Turriano and Ramelli (Ch. 7). Various forms of pumps, storage towers and channels were constructed in Germany, England and France in the sixteenth and seventeenth centuries, with piston-pumps playing a significant role in meeting the demands of the day. It will suffice to draw upon evidence from the City of London to provide a representative example of contemporary developments in hydraulic machinery.

The first power-driven pumps installed in London were built by the German engineer Peter Maurice[7] in 1582. Water-wheels in the Thames were used to drive the first suction and pressure pumps which were set up near London Bridge. The pumps, which delivered water to the loftiest buildings in the highest part of the City, were so greatly admired that the Lord Mayor and Common Council granted permission for them to be installed in one of the arches of London Bridge. By 1594 Bevis Bulmer had established horse-driven chain pumps at Broken Warf and in due course further arches of London Bridge were occupied by the pumping engines of Maurice's successors. A valuable account of one of these engines near the north end of London Bridge was presented by John Bate (1654).

According to Bate this London Bridge engine '. . . by the Ebbing and Flowing of the Thames, doth mount the said water unto the top of a Turret, and by that means it is conveyed above two miles in compass, for the use and service of that City'.

It appears that John Bate saw the Waterworks shortly after the fire on London Bridge of 1633 and when he came home he '. . . drew a modell thereof, and have here presented it unto thy veiu'.

His sketch of the machine is shown in Fig. 8.4. The water-wheel XX driven by the tidal flow of the Thames caused a heavy wheel P to oscillate by means of a connecting-rod RR and a spoke attached to the lower side of P. The motion caused water to be pumped from the strong brass or iron cylinders WW through delivery pipes NN to the top of a nearby turrent where it was strained through a close wire grate before entering the main wooden pipe laid along the streets. Several smaller lead pipes '. . . serving each of them to the use and service of particular persons. . . .' were grafted into the main wooden pipe.

It appears that the wooden water-wheel was supported in brass bearings and it seems clear from this and several other references that brass was widely used as a bearing material in the seventeenth century. In this case Bate wrote '. . . the Gudgins of this wheele must be set to turn in strong brasse Sockets, firmly set in the two middle-beames of the Frame'.

THE
MYSTERIES
OF
NATVRE and ART

In four severall Parts.

The first of
VVater-VVorks.

The second of
Fier-VVorks.

The third of
Drawing, Colouring, Limming,
Paynting, Graving, and Etching.

The fourth of
Experiments.

By JOHN BATE.

LONDON.
Printed by R: Bishop for Andrew Crook, at the Green
Dragon in Pauls Churchyard 1654

---

7. See Wolf (1935), p. 532.

Fig. 8.4
John Bate's sketch of the London Bridge Waterworks (1654). (The water-wheel was supported in brass bearings.)

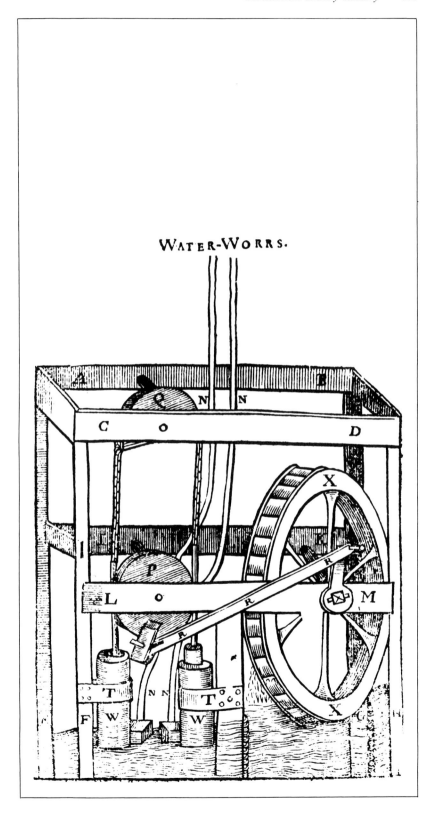

He further mentions that pump cylinders or barrels made of iron had 'more durance' than those made of brass.

Pumps mounted on wheels or fire-extinguishing engines developed rapidly in the seventeenth century and our author John Bate devoted part of his book to this topic. It appears that a large number had been constructed by 1666 and the reason that they played such a small part in controlling the Great Fire of London should be sought more in relation to poor organization and lack of mobility in crowded streets than in any lack of technical efficiency.

Further projects followed the London Bridge scheme in an attempt to keep pace with the demands of the growing city. With the exception of Sir Edward Ford's pump very few technical details are available of these schemes, but this one at least provides us with information of both historical and tribological interest.

Sir Edward Ford was born in 1605 in Harting, Sussex. He attended Trinity College, Oxford without taking a degree and took out his first patent at the age of thirty-five. In due course he was awarded one of about twelve letters patent issued by Oliver Cromwell for an invention relating to a pumping installation on the Thames near Somerset House. This fact is all the more surprising when it is found that Sir Edward had chosen to support the King during the Civil War! He became a Colonel in the King's Army, was later knighted and was captured by Sir William Waller after surrendering Chichester in 1642. He was soon released but after little more than a year he was recaptured again by Waller at Arundel Castle and sent to the Tower of London. This might have been the end of our story, but Ford escaped to the Continent in 1644, returned again to England and was once more imprisoned. King Charles I was executed in 1649 and six years later came the remarkable announcement that Ford had received letters patent from Oliver Cromwell. The fact that Sir Edward had married the sister of Henry Ireton, Cromwell's son-in-law, may have had some influence on the case.

The patent cover lasted fourteen years and it conferred the right to erect and lay engines, cisterns, pipes and other apparatus for supplying water to London, Westminster and surrounding parts. The document, written eleven years before the Great Fire, included the prophetic words '. . . and forasmuch as many parts in and about London and Westminster are in great want of water and thereby in the more danger of fire and other mischieff. . .'.

The French traveller Monsieur Balthasar de Monconys[8] who visited London in 1663 described the pump eventually built by Ford and his sketch is reproduced in Fig. 8.5. A series of four suction pumps worked by levers from a vertical rod delivered about 1·7 gal (7·7 l) of water through a head of 120 ft (36·6 m) at each stroke. The vertical rod rested on a horizontal face-wheel turned by horses and was activated by twelve cams. The cams were made of wood, faced with iron, and a bronze friction-roller acted as a follower at the end of the vertical rod. The intermittent action of

8. See De Monconys, B. (1666), Vol. II, p. 29.

Fig. 8.5
(a) Monsieur Balthasar de
Moncony's sketch of Sir Edward
Ford's horse-driven pump (1666).
(Note wooden cams faced with
iron on the horizontal face-wheel
and the bronze friction roller on
the vertical rod.)

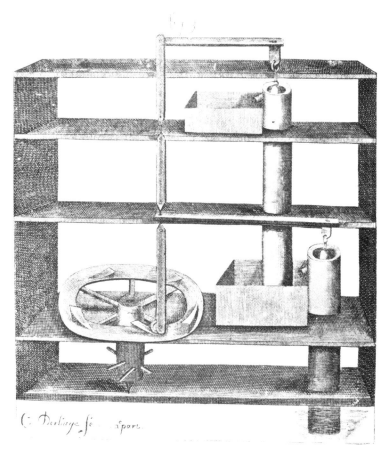

(b) Hollar's plan of western
London showing Ford's
'Waterhouse' near the Strand
Bridge.

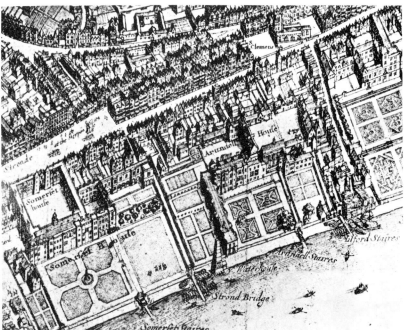

the cams would no doubt lead to severe loading on the friction-rollers and it has further been noted[9] that it was trying to the horses. It has been estimated[10] that the maximum efficiency of the pump was about 30 per cent.

The original site of Ford's Pump and Waterhouse can be seen on the reproduction of Hollar's plan shown in Fig. 8.5(*b*). De Monconys commented on the fine view to be had from the top of the tower, but the fact that it overlooked Somerset or Denmark House was to prove its undoing. Following the Restoration of the Monarchy King Charles II had installed his mother Queen Henrietta Maria in Denmark House and on 31 July 1664 the King himself ordered the removal of the tower within three months because of its nuisance value.

Several large towns and capital cities in Western Europe participated in this upsurge of interest in mechanical water-raising devices in the early seventeenth century. The prospect of a plentiful and reliable supply of water for domestic purposes not only provided the justification and the spur for innovation, it also enhanced the status of hydraulic engineers.

In Paris in 1608 a Flemish engineer by the name of Lintlaer constructed beneath the Pont-Neuf a piston-pump driven by the flowing Seine which supplied water to the Louvre and the Tuileries. A similar waterworks project was completed at Notre-Dame in 1669. A much larger and quite remarkable project was undertaken by the Dutch engineer Rannequin when he was called upon to design a system which would deliver adequate amounts of water for the palace and gardens at Versailles through a distance of about three-quarters of a mile (1·2 km) against a head of 533 ft (162·5 m). The elaborate but highly inefficient system involved two intermediate reservoirs and no less than 253 pumps driven by fourteen under-shot water-wheels in the river. Some of the pumps were located at the intermediate reservoirs and the mechanical transmission which took the power from the water-wheels to the elevated sites must have provided quite a spectacle. It has been reported[11] that about 90 per cent of the available power was lost in the drive system. The Versailles scheme was probably the most powerful pumping system in the world at the time it was completed, with a useful power rating of about 100 HP (74·5 kW) and a flow rate close to 1 million gal (4,546,000 l) per day. In due course the system was abandoned and replaced by a steam engine–a fate suffered by many water- and animal-powered hydraulic systems developed in the seventeenth century.

To conclude this section on water-raising equipment it is essential to make reference to the emergence of steam power, although consideration of the tribological aspects of reciprocating steam engines will be reserved for a later chapter. The reader will find articles by Stowers (1963) and Clarke (1963) useful summaries of the history of water-raising machinery and the steam engine, respectively. Interest in the power of steam dates back to Hero of Alexandria, but the full potential for raising

---

9.   See Wolf, A. (1935), p. 534.
10.   See the discussion by Mr Bruce Ball of Jenkins, R. (1930), pp. 43–51.
11.   See Op. cit. p. 536.

water by means of steam pressure or by a partial vacuum created by condensing steam was first recognized early in the seventeenth century. The Italian nobleman Porta (1606) described a machine for raising water based upon the condensation of steam and De Caus (1615) explained how a fountain could be built to take advantage of the expansive power of steam in a heated closed vessel partially filled with water. In a book written in 1655 and published in 1663 Edward Somerset, second Marquis of Worcester, described an engine for raising water which was essentially an improved form of the de Caus machine. Such an engine, capable of raising water to a height of 40 ft (12·2 m), was built at Vauxhall.

Towards the end of the seventeenth century the mining industry in England was faced with a serious problem of water accumulation as the workings reached to greater depths. This problem provided the incentive for the impressive attempts to utilize steam power in pumping operations which ultimately produced the reciprocating steam engine.

Captain Thomas Savery (*c.* 1650–1716) received a patent in 1698 for '. . . an engine for raising water by the impellant force of fire'. He demonstrated a model of the engine before the Royal Society in June 1699 and in due course a full-scale version built for the raising of water from Cornish mines was described in a well-known pamphlet published under the title 'The miner's friend' in 1702. Steam could be directed alternately from a boiler into two copper vessels. Entry of steam under pressure caused the water in the vessel to be discharged through a lift of about 30 ft (9·1 m) and subsequent condensation of the steam effected by cold water applied to the surface of the vessel created a partial vacuum which recharged the vessel with water from the suction or inlet pipe. Savery's machine, sometimes described as a pulsometer, was the first successful application of steam to water-raising problems for domestic and industrial situations.

The next and last entry in steam-engine history which enters the period under review was most important. Thomas Newcomen built the first atmospheric beam engine near Dudley Castle in Staffordshire in 1712. Steam admitted to a vertical cylinder fitted with metallic piston and rod caused one end of a trunnion-mounted beam to move upwards before condensation achieved by the injection of cold water effected the return stroke under atmospheric pressure. The other end of the beam carried a chain attached to a conventional pump. Separation of the pumping cylinder from the steam cylinder was a most important development. The early Newcomen beam engine had a very low efficiency, but it was soon adopted in many European Countries in piston sizes from 10 to 72 in (254–1,829 mm). Current machining standards were unable to provide close-fitting pistons and cylinders and the Newcomen atmospheric beam engine utilized a leather sealing disc to preserve the vacuum following condensation of the steam.

Steam power for water raising was well established by 1750, but the water-, wind- and even animal-driven machines still played an important and even a major role in power generation. In due course the vast increase in power for fixed machinery and transportation resulting from the emergence of steam was to introduce an excessive loading on bearings which in turn promoted the great nineteenth-century studies of bearings and lubrication.

8.3 Bearings in the seventeenth and early eighteenth centuries

An interesting and important feature of plain bearing development in the seventeenth and eighteenth centuries is the firm indication of interest in bearing materials. Evidence will be produced from the literature on lathes and carriage bearings to illustrate the use of bearing metals softer than the shafts which they supported. Plain bearing arrangements in machines were starting to reflect a move towards replaceable bearing components and this is again illustrated by accounts of early eighteenth-century lathes.

A more adventurous use of rolling-element bearings is demonstrated by accounts of early eighteenth-century windmills in Holland and wheeled vehicle construction in France and England.

Most of the firm information on both plain and rolling-element bearing development during this period comes from the early part of the eighteenth century and it is convenient to discuss the evidence in relation to the lathe, chronometers, the Dutch windmill and the horse-drawn carriage.

The lathe

The demand for wooden and iron machine components of greater accuracy than could be produced by the seventeenth-century carpenters and blacksmiths provided the spur for a move towards more sophisticated tools. The lathe was still treadle operated, but the advanced form of plain bearing arrangement was beautifully illustrated by Plumier (1701) in his account of the art of turning with perfection.

The bearings were made adjustable for wear by having split bushes and removable iron caps as shown in Fig. 8.6. The bearing halves were allowed to slide freely in vertical grooves in the wooden supporting blocks, but iron retaining caps bolted to the body carried a central wing-nut and screw for locating the upper half-bearing. Location of the shaft within the bearing could thus be adjusted continuously as wear of the softer bearing materials developed and in due course new bearing halves could be fitted with ease. The bearings were made of pewter, that soft grey alloy of tin and lead (Davison, 1958), or zinc (Davison, 1957b).

This illustration of the use of soft metals in plain bearings and the sophisticated nature of plain bearing design in the early eighteenth century is undoubtedly enhanced by the numerous fine sketches of composite and exploded views of lathe construction in Plumier's book.[12]

Chronometers and navigation

One of the prize-winners in an open competition instituted by the Board of Longitude, London, in 1714 for the determination of longitude at sea to within 0.5–$1°$ was an English watchmaker by the name of Henry

---

12. See, for example, Davison, C. St. C. B. (1958), Figs 3 and 4, and Naylor, H. (1965), Fig. 5.

Fig. 8.6
Sketch of early eighteenth-century lathe bearings from Charles Plumier (1701). (Note slit pewter bearing halves with locating screw in removable iron cap providing adjustment for wear.)

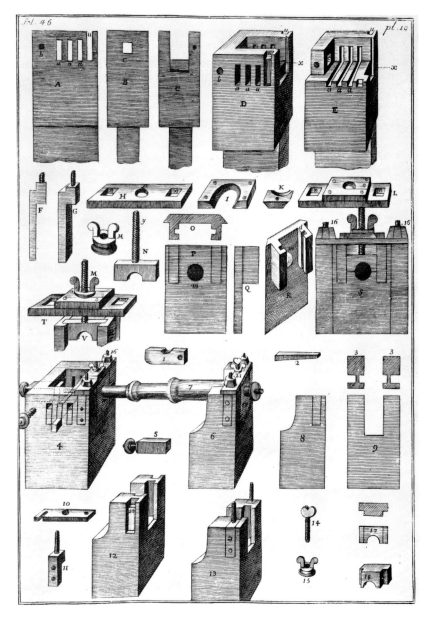

Sully. Mr Sully, who resided in Paris for many years and who wrote in the French language, enters our history of early eighteenth-century tribology because he fitted a form of roller-bearing in his highly accurate chronometer in 1716.

Rolling-element bearings have thus with certainty been used to improve the accuracy of precision instruments since shortly after the turn of the eighteenth century,[13] and apart from Baldewin's astronomical clock of 1561 Mr Sully's chronometer undoubtedly stands among the first recorded developments in this field.

---

13. See Ch. 7.

The Dutch windmill

Our main account of the history of windmill bearings will be reserved for Chapter 9, but there are two references and one important illustration of the introduction of rolling bearings into Dutch and German mills which call for special mention at this stage.

In a 1724 German text on engineering largely devoted to windmills and water-mills, Leupold makes reference to the use of roller-bearings for the hollow shaft of a windmill. At a later stage Leupold (1735) also made a recommendation for the use of tallow or vegetable oil for the lubrication of rough surfaces. The current concept of the physical mechanism of lubrication was nicely illustrated by his observation that: '... The roughness of the surface is made less by lubrication, and the bearing surface is rolled over as if it had small balls on it. . . .'

In the early windmills constructed in countries adjacent to the southern part of the North Sea the cap housing the near-horizontal windshaft was integral with the main part of the windmill structure, which in turn contained the machinery for grinding, pumping or sawing. The complete structure had thus to be rotated, usually by hand, to face the prevailing wind. At a later stage ingenious millwrights replaced the 'post' mill with a 'tower' mill in which the machinery remained in a stationary structure while the cap alone rotated to face the wind.

Some interesting bearing arrangements arose from these various cap support requirements and it is fortunate that many of the developments were recorded in such texts as Ramelli's *Le Diverse et Artificiose Machine* and the famous Dutch mill books published from 1725 onwards. It appears that the caps of Dutch windmills were initially supported and guided on hardwood blocks, but that by the turn of the eighteenth century 'roll-rings' in the form of trunnion-mounted roller thrust bearings were in use. This use of large roller thrust bearings was established by 1700[14] and a most excellent illustration of the arrangement employed in a *Dutch saw-mill* or *paltrok* is presented in the account of early eighteenth-century mills by van Natrus et al. (1734, 1736). The bearing, clearly shown in Fig. 8.7, consisted of an independently mounted cage-and-roller ring between tracks attached to the fixed brick support structure and the movable wooden cap. The rollers were kept in place by trunnions or pins through their centres,[15] but there is less certainty about the materials employed in these bearings. Wailes (1957; 1963) refers to the use of large square-section hardwood rollers about 7 in × 7 in (178 × 178 mm) mounted in iron cages in these roll-mill bearings, whereas Davison (1957*b*) attributes the first use of iron roller-bearings to this application: Wailes (1957) also notes that smaller cast-iron rollers were used in England at a later date. Van Natrus et al. (1734) appear to say little about the roller materials although the tracks were certainly made of wood. Stokhuyzen (1962) writes that the sill or lower track was made of oak, but he also states that the entire mill was constructed of wood and it thus appears

---

14. See Wailes, R. (1963), Vol. 1.
15. See Allan, R. K. (1945), p. 9.

Fig. 8.7
Early eighteenth-century Dutch
saw-mill or paltrok mounted on a
large roller thrust ring. (The
trunnion-mounted wooden rollers
were about 7 in (177 mm) long
with a diameter of about 7 in.)

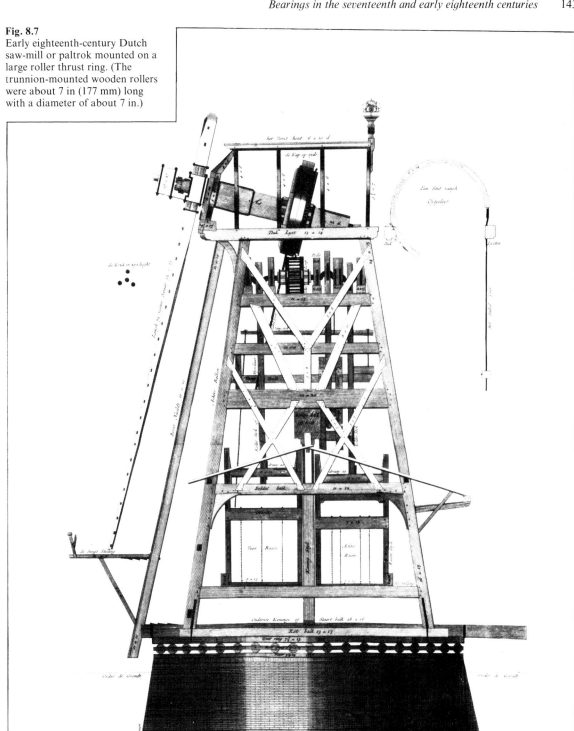

that this was most probably the material for the early rollers in these fascinating rolling-element thrust bearings.[16]

Carriage wheels and bearings

The simple journal bearing of wood, iron or occasionally brass continued to service the vast majority of seventeenth- and early eighteenth-century machines. Apart from the special and somewhat rare examples of sophistication in plain and rolling bearings mentioned earlier, the horse-drawn carriage undoubtedly provided the greatest stimulus to bearing development in this period. Increasing loads and speeds and economic pressures called for more efficient and reliable bearings in this major form of land communication and transportation. The seventeenth century witnessed the beginning of an era of carriage-bearing improvement which continued until the competition from railways accelerated in the nineteenth century. We shall find occasion to mention numerous important studies, designs and patents to illustrate this point in the present and subsequent chapters, but for the present we shall be concerned with three remarkable contributions by Hooke, de Mondran and Rowe.

Robert Hooke (1635–1703)

Hooke was born at Freshwater on the Isle of Wight on Saturday, 18 July 1635. He was a weak and sickly child and for at least seven years his parents had very little hope for his life.[17] His father, minister of the parish, was himself infirm during Robert's boyhood and this, coupled with Robert's frequent headaches which hindered learning, caused his early and further education to be wholly neglected. He learnt his grammar by heart, although apparently with little understanding, but he inevitably spent long periods alone. It is significant that he entertained himself during these periods of solitude in making small mechanical toys.

In due course he went to Westminster School where he lived in the home of the master, Dr Busby. He entered the University of Oxford and became a student of Christ Church while acting as servitor to a Mr Goodman of Oriel College. He took the degree of M.A. about 1662 or 1663.

Hooke had started to attend scientific meetings in Oxford from about 1655. His skill in the mechanical arts was recognized at an early age by Robert Boyle, his senior by about eight years, and it is generally believed that Hooke made all Boyle's apparatus and carried out most of his experiments on the vacuum and gases. He was Boyle's assistant and lifelong friend and in 1662 he became Curator of Experiments to the recently formed Royal Society.

Hooke is probably best known for his discovery of the shortest law in physics, . . . *ut tensio sic vis* . . . (extension is proportional to force) and his account of the microscopic world in the *Micrographia*. His

16. I am grateful to Mr R. J. Law, Assistant Keeper, Department of Mechanical and Civil Engineers, Science Museum, London for his advice on this point and confirmation from Mr Rex Wailes (private communication, 1973) that the rollers in these Dutch mills were indeed made of wood.
17. See Hooke, R. (1705).

excursion into tribology is less well known but highly significant. His penetrating observations on the problems of reducing friction and wear in carriage-wheel bearings arose in his discourse on carriages at the Royal Society on 25 February 1685.[18] Hooke's general discourse was concerned with '. . . the various ways of conveyance . . .', but he devoted a good deal of time to an account of the remarkably efficient sailing chariot made for the Prince of Orange by the Belgium civil and military engineer Simon Stevin (1548–1620).[19] Stevin, often referred to as Stevinus, was one of the great figures of post-Leonardo da Vinci times. He introduced decimals in 1585, made important contributions in the fields of statics and established the fundamental law of hydrostatics that the pressure depends only on the weight of the column of liquid above it.

Stevin's chariot was like a ship on wheels, with masts, sails and other convenient rigging. Hooke mentions that when running upon the sea-shore sands the vehicle travelled some 42 miles (68 km) in two hours while carrying twenty-eight men. Further references to the essential design features show that the chariot was steered by means of a pivoting axle-tree and that the wheels were widely spaced for stability. There were two chariots, the larger having two sails and steerable rear wheels and a smaller single-sailed vehicle with front-wheel steering. The rims of the wheels, were between 18 in (457 mm) and 24 in (610 mm) wide to avoid excessive sinkage in the sand.

Hooke concluded that Stevin's chariot was '. . . the swiftest carriage yet known, for so great a burthen, and so long a way. . .'. He argued that the resistance to motion would be small, '. . . save only some small matter in the rubbing of the ends of the axes in the naves of the wheels, which, being well oiled, will be very little. . .'. He conceived the application of this wind-driven vehicle to transportation on the plains and downs of southern England and he advanced suggestions for improvements which included the use of three wheels instead of four and the mounting of the wheels '. . . upon small steel pevots or gudgeons, in bell-metal sockets, well oiled, instead of being moved upon the large end of an axle-tree'.

This account of Stevin's sailing chariot and the general consideration of wheeled vehicles has been discussed in some detail since it provides the essential context for Hooke's illuminating remarks on rolling friction and plain bearings. His comments on rolling friction arose from a considera-tion of the impediment to motion of a wheel. He recognized two com-ponents of rolling friction: '. . . The first and chiefest, is the yielding, or opening of the floor, by the weight of the wheel so rolling and pressing; and the second, is the sticking and adhering of the parts of it to the wheel; . . .'.

The two now familiar aspects of frictional resistance associated with material deformation and adhesion thus entered Hooke's appraisal of the problem.

In relation to friction arising from deformation he further noted that the overall resistance was small, even for undulating surfaces; if the

---

18. See Gunther, R. T. (1930), Vol. VII, Pt II, pp. 666–79.
19. See Tokaty, G. A. (1971) pp. 49–52.

wheel and ground was hard, '. . . yet is there little or no loss, or considerable impediment to be accounted for; for whatever force is lost, in raising or making a wheel pass over a rub, is gain'd again by the wheel's descending from the rub, . . .'.

He further distinguished between the role of recoverable and non-recoverable deformation in the important passage:

Nor is the yielding of the floor any impediment, if it returns and rises against the wheel, for the same reason; but the yielding, or sinking of the floor, and its not returning again, is the great impediment from the floor; for so much of motion is lost thereby, as there is force requisite to sink such a rut into the said floor by any other means; whether by weight, pressure or thrusting directly down, or any ways obliquely.

On the question of sticking and adhesion he wrote:

. . . The second impediment it receives from the floor, or way, is the sticking and adhering of the parts of the way to it; for by that means, there is a new force requisite to pull it off, or raise the hinder part of the wheel from the floor, or way, to which it sticks, which is most considerably in moist clayie ways, and in a broad rimm'd wheel. . . .

A diagrammatic representation of Hooke's observations on rolling friction is shown in Fig. 8.8.

It is interesting to note that Hooke then addressed himself to the question discussed under the heading 'Wheeled vehicles' in Section 4.2 concerning the relative merits of wheels rotating on fixed axles or a fixed axle-wheel assembly rotating in sockets in the carriage. He comes down in favour of the latter arrangement in vehicles designed for speed, largely because of the structural advantages when the wheel is fixed firmly to the axle.

Finally, in the section of his discourse devoted to aspects of wheel design for vehicles constructed for celerity, Hooke presents a valuable commentary on advanced seventeenth-century concepts on bearing design, materials, lubrication and seals. The passage encompassing all these topics is quoted directly from Gunther (1930) for completeness.

. . . the less rubbing there be of the axle, the better it is for this effect; upon which account, steel axes, and bell-metal sockets, are much better than wood, clamped, or shod with iron; and gudgeons of hardened steel, running in bell-metal sockets, yet much better, if there be provision made to keep out dust and dirt, and constantly to supply and feed them with oil, to keep them from eating one another; but the best way of all is, to make the gudgeons run on large truckles, which wholly prevents gnawing, rubbing, and fretting. . . .

Here was a clear recommendation for the use of steel shafts or gudgeons running in softer metal bearings. Bell-metal, that alloy of copper and tin in which the proportion of tin is higher than in the majority of bronzes, would clearly be a good choice. The need for effective seals and an adequate supply of lubricant for effective bearing operation are also noted. The final reference to the great advantage of mounting the gudgeons on *truckles* is no doubt an interesting reference to the roller-disc bearing concept initiated in Renaissance times and developed for static machinery in the sixteenth and seventeenth centuries. The application by Jacob

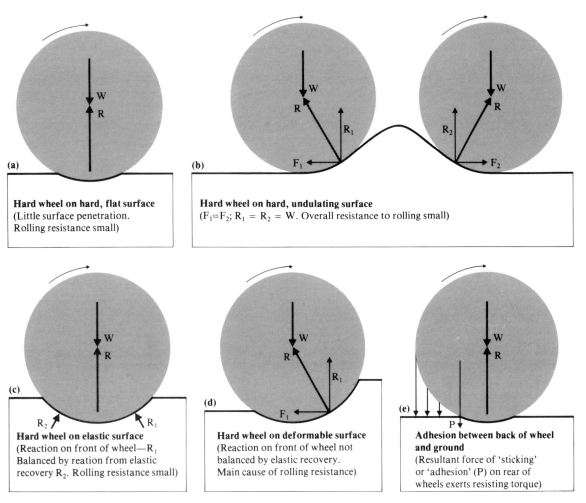

Fig. 8.8
Representation of Robert Hooke's
(1685) view of rolling friction.

Within the figure:

(a)
Hard wheel on hard, flat surface
(Little surface penetration.
Rolling resistance small)

(b)
Hard wheel on hard, undulating surface
($F_1 \approx F_2$; $R_1 = R_2 = W$. Overall resistance to rolling small)

(c)
Hard wheel on elastic surface
(Reaction on front of wheel—R_1
Balanced by reation from elastic
recovery R_2. Rolling resistance small)

(d)
Hard wheel on deformable surface
(Reaction on front of wheel not
balanced by elastic recovery.
Main cause of rolling resistance)

(e)
**Adhesion between back of wheel
and ground**
(Resultant force of 'sticking'
or 'adhesion' (P) on rear of
wheels exerts resisting torque)

Rowe of this form of bearing to road vehicles in the eighteenth century described later in this section shows that Robert Hooke had anticipated an important rolling-bearing development.

de Mondran (1710)

In 1710 the Academy of Science in Paris approved a design for a carriage submitted by a gentleman by the name of de Mondran. The journals were supported on *discs* or *rollers* and apart from Robert Hooke's reference to the advantages of mounting the gudgeons of carriages on truckles, this is probably the earliest reference to the use of roller-disc bearings on vehicles. The use of separately pivoted disc supports for rotating shafts had increased steadily in fixed machinery since Renaissance times and the merits of the device in wheeled vehicles is illustrated in relation to the design submitted by de Mondran by the claim that '. . . because of the rolling contact friction, one horse could easily do the work which could hardly be accomplished by two'.

Allan (1945) also mentions that another early eighteenth-century document refers to a gun carriage invented by Fahy which had a bearing arrangement similar to de Mondran's.

Jacob Rowe (1734)

Rowe provides our last and perhaps most colourful entry on the development of rolling bearing supports for wheeled vehicles in the period under review. He was awarded Patent Specification No. 543 of 1734 by His Majesty King George II and from this time onward the British system of patents provides a valuable source of reference on bearing development.

The frontispiece of Rowe's treatise makes it plain that his '. . . . wheels caused to turn under the axis of the heavy wheels, or against the point or place where the said axis must bear. . .', could be applied not only to wagons, carts and coaches but also to water-, wind- and horse-mills. He refers to the practice of using rollers in the weighing of anchors at sea and further claims the great advantages of his wheel supports in terms of charges and hard labour. It is interesting to note that Rowe, like de Mondran, claimed that the use of roller-disc bearings in carriages enabled a single horse to do work which previously required two animals.

Rowe wrote his treatise some seven years before it was published and in 1727 he sent it to a number of friends for comment. His friends approved the essential principles of the device and in 1732 he constructed demonstration models. He wrote: '. . . I then performed the Experiments before several Persons of Quality, and had their countenance and approbation so far, that they advised me to take out a Patent for my Inventions, part of which I have here described according to the best of my judgment.' The outcome was Patent Specification No. 543 of 1734.

Rowe's understanding of friction failed to match his ingenuity in the use of friction-wheels. He envisaged friction to be either natural or accidental: the former obeying Amontons' first law (see Sect. 8.4), and the latter representing resistance associated with the roughness of the rubbing bodies. He assumed that the resistance in the *pevets* was equal to the applied load and he then proceeded to demonstrate that: '. . . As the acting distance from the Center of the Axis is to the Semidiameter of the Pevets, so is the whole weight resting thereon to the resistance made by the Friction. . .'.

His development of a practical arrangement of friction wheels is illustrated in Fig. 8.9. The term 'friction-wheels' is introduced in the opening paragraph of Chapter II of the treatise which also summarizes the concept with clarity.

> . . . *As the Friction is caused by the Resistance of a dead and fixed Surface on which the Pevets of any Wheel, Etc. do rest, so can there be no other way of cancelling the said impediment but by causing the said Pevets or the end of the Axis to turn on the Vertex or Top of other wheels; which Wheels serving only for this purpose may be properly called pevet or friction Wheels: And by encreasing the number of these friction Wheels (one under the other) so as their Pevets may also rest and turn round on the Vertex or Top of each other (except the Pevets of the last which must turn on a dead Surface as common) the Friction of any great and heavy Wheel may be reduced to the lowest degree of Perception.*

The use of multiple friction-wheels each having a diameter twenty times greater than its axle is clearly shown in Fig. 8.10 in an arrangement which requires a turning force at z only $\frac{1}{8000}$ that required with dead pivots.

ALL

SORTS

OF

Wheel-Carriage,

IMPROVED.

Wherein it is plainly made appear, that a much less than the usual Draught of Horses, &c. will be requir'd, in Waggons, Carts, Coaches, and all other Wheel Vehicles, as likewise all Water-Mills, Wind-Mills and Horse-Mills.

This Method being found good in Practice, by the trial of a Coach and Cart already made, shews of what great Advantage it may be to all Farmers, Carriers, Masons, Miners, &c. and to the Publick in general, by saving them one half of the Expences they are now at in the Draught of these Vehicles, according to the common Method.

The whole illustrated with Copper Plates.

And an Explanation of the Structure of a Coach and Cart, according to this Method,

By *JACOB ROWE*, Esq;

LONDON:

Printed for ALEXANDER LYON under *Tom's* Coffee-house in *Russel-Street Covent Garden.* M DCCXXXIV.

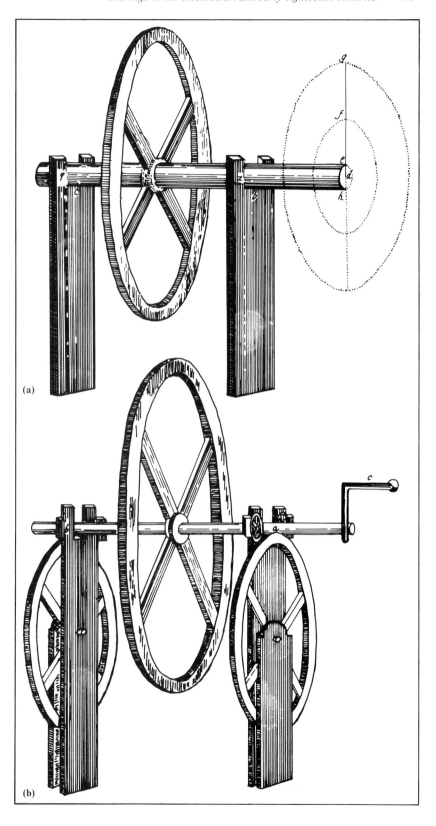

Fig. 8.9
Jacob Rowe's development of a practical arrangement of friction-wheels (Rowe, 1734). (Redrawn by author from originals.)

(a)

(b)

Fig. 8.10
Jacob Rowe's illustration of the
use of multiple-friction-wheel
supports (Rowe, 1734). (Redrawn
by author from originals.)

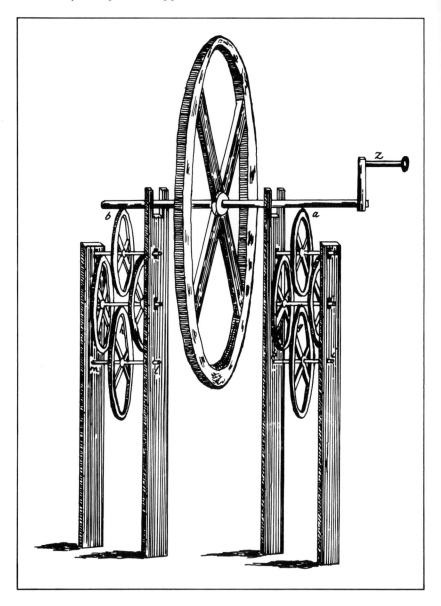

A section of the treatise devoted to the application of friction wheels
to '. . . Pullies and all sorts of Blocks used at Sea. . .'. '. . . Balances' and
'Pumps' is noteworthy, both as an extension of the application of rolling
friction supports and a reference to the use of various combinations of
materials including wood, cast iron, brass, wrought iron and steel for
the reduction of wear.

The most interesting technical part of the treatise is concerned with
the use of friction-wheels on wheeled vehicles. Various applications are
considered including a *Country Dung-Cart, Chariot and Carriages for any
kind of Coach and heavy Cannon.* The friction-wheels, commonly less than
2 ft (0·6 m) in diameter, were mounted on iron axles not exceeding 1 in
(25·4 mm) in diameter. In the fine illustrations of friction-wheels fitted to a

Fig. 8.11
Jacob Rowe's illustration of a coach fitted with friction-wheels (A), (B) (Rowe, 1734). The diameters of the friction-wheels (A), (B) were about 24 in (609 mm) and 18 in (457 mm), respectively. (Redrawn for the author from originals.)

coach and wagon shown in Figs 8.11 and 8.12 the rear wheels (A) had a diameter of about 24 in (610 mm) while the front wheels (B) were somewhat smaller with a diameter of about 18 in (457 mm). In a further account of these wheels which is reminiscent of Hooke's commentary on bearing materials and the importance of seals, Rowe wrote:

> . . . *both of which may be made of Wood, and hoop'd with Iron, with Iron Axis's, and their Pevets to turn in Bell-Metal Nuts or Coggs; but the Horizontal Wheels being but small are proposed to be of cast Iron, Brass, or Bell-Metal, and to prevent their clogging with the Dirt of the Roads, they may be covered with Leather, or be cased with thin Wood.*

Fig. 8.12
Jacob Rowe's illustration of a wagon fitted with friction-wheels (A), (B) (Rowe, 1734). The diameters of the friction-wheels (A), (B) were about 24 in (609 mm) and 18 in (457 mm), respectively. (Redrawn for the author from originals.)

The account occurs in a section of the treatise headed 'Advertisment' in which Rowe remarked that:

> *. . . On trial of a Coach full of People, having the Friction of the Wheels taken off, as before explained (which was performed by two ordinary Horses travelling full thirty six miles in a bad road, within the space of seven hours) I found where the road was even that the Braces were scarce ever seen to be drawn tight; but on the Coach's leaning considerably on one or the other side, the Draught of the Horses did appear to be considerably greater, the which was occasioned purely by the rubbing of the Shoulders of the Axis on the Cheeks, their being then a dead Bearing or Surface; and therefore I contrived a rememdy for the said Obstruction in the manner as follows.*

The remedy consisted of the use of horizontal friction-wheels which resisted axial loading as shown in the two alternative arrangements in Fig. 8.13. These are the 'Horizontal Wheels' mentioned in the description of the construction of the friction-wheels.

The last section of Rowe's treatise probably represents the first detailed attempt to quantify the savings resulting from tribological innovation. The economic aspects of tribology have been considered on several occasions since 1734, culminating in the Jost Report of 1966, and the subject is given special consideration in Chapter 11. Jacob Rowe's approach to the problem is full of interest, but for the present we will submit only the outline of his argument.

The starting-point is the observation that wagons and carriages fitted with the patent friction-wheels could be drawn by half the number of horses required for conventional vehicles. The work done by the 40,000 horses in the kingdom in 1734 could thus be done in 20,000, and since the labour of a horse was valued at 1*s*. 6*d*. per day this represented a direct saving of £1,500 per day or £547,500 per annum. Furthermore, the cost of keeping a horse was estimated to be £10 per annum and the savings on this account thus equalled £200,000 per annum.

Fig. 8.13
Jacob Rowe's alternative designs for friction-wheel supports to carry combined axial and radial loads on carriage- or cart-wheels (Rowe, 1734). (Redrawn by author from originals.)

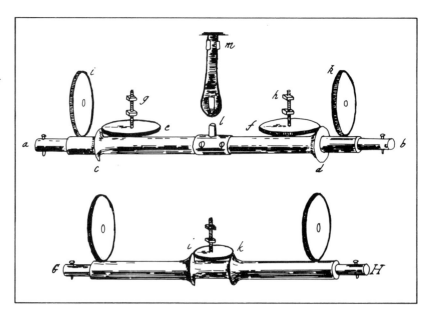

Rowe's estimate of the total potential savings on the operation of wheeled vehicles was thus £747,500 per annum. However, Rowe did not envisage a reduction in the number of horses, but rather an expansion of activity and production in the mines with the employment of more horses for speed rather than draught. He ended by stressing the possibilities of carrying '. . . Dung and all sorts of Dressings for Land . . .' so much more economically such that '. . . great quantities of barren Land will now be made fertile.'

The above calculation refers only to wheeled vehicles and Rowe further estimated that about one-third of the *Power* required to operate *Forcible Engines* employing *Wheels* could be saved by the introduction of friction-wheels. It is interesting to note that Rowe considered the cost of operating contemporary machinery in *Mines*, *Coal-Pits* and *Mill-work*, *etc.* to be £600,000 per annum. This was a mere 40 per cent of the total cost of the 40,000 horses and a good indication of the current state of industrialization.

The savings on power losses in machinery thus came to £200,000, giving a grand total of £947,500 – a sizeable sum in the account of the early eighteenth-century exchequer. Jacob Rowe's treatise is undoubtedly a fascinating document.

8.4   Early scientific studies of friction

The centre of gravity of seventeenth-century interest in mechanical devices was to be found in Western Europe, notably in France and England, and it is in these countries that we find evidence of the impressive early scientific studies of friction. Perhaps the best commentary on this formative period is that many of the ideas developed over 250 years ago have stood the test of time to form a central feature of our modern understanding of the friction process.

The seeds of this scientific activity had been sown in the early part of the seventeenth century when Francis Bacon (1561–1626) gave a practical basis to scientific learning. His insistence on the merits of an experimental approach to research attracted much attention, and in due course the posthumous publication of his *New Atlantis* provided the spur for informal meetings in London of a group of eminent men whose numbers included Robert Boyle (1627–91), Robert Hooke (1635–1703) and Christopher Wren (1632–1723). These meetings were held in the turbulent times of the English Civil Wars and from about 1649 there were similar societies in both London and Oxford. Such meetings were the precursors of the Royal Society, formed in the year of the Restoration of the Monarchy, 1660 and issued with a Charter of Incorporation, passed by the Great Seal in the reign of King Charles II on 15 July 1662. In France the Académie Royale des Sciences had been established in 1666 and it received its full statutes and further favours in a Royal Decree issued by King Louis XIV in 1699.

A forum had thus been established in both England and France for the free discussion of scientific investigations. The two features of seventeenth-century Europe which provided essential ingredients for scientific studies of friction were undoubtedly the formation of these learned scientific societies and the growing interest in machinery. The historical import of these tribological studies will emerge as we follow the details of some of the most significant investigations. The reader will find Bowden and Tabor's book (1964, Pt. II, Ch. XXIV), *The Friction and Lubrication of Solids*, and Kragelskii and Shchedrov's (1956) work, *Development of the Science of Friction* particularly valuable sources of information on this period.

The laws of friction and the role of asperities – the French School of Amontons, de la Hire and Parent

The work on friction was initiated in France at the end of the seventeenth century and it is to Guillaume Amontons[20] that we must afford pride of place in terms of both chronology and stature. His experiments and interpretations of his results were discussed in a classical paper presented to the Académie Royale on 19 December 1699.

In the introduction to his paper Amontons (1699) justifies the need to study friction with telling references to the current importance of machines and the effect of friction upon machine performance. 'The great use which all the arts are obliged to make of machines is a convincing proof of their absolute necessity. . . .' 'Indeed among all those who have written on the subject of moving forces, there is probably not a single one who has given sufficient attention to the effect of friction in Machines. . . .'

It is further argued that it is of little value to know that by classical mechanics the force required to move the beam AB (Fig. 8.14) up an inclined plane CD is equal to the weight multiplied by the sine of the angle of inclination when '. . . the resistance incurred in the rubbing of this beam against the earth may not simply be equal to this force, but may even exceed it by a considerable number of times'.

The apparatus used by Amontons in his 'Experiment concerning the rubbing of various materials one against the other' is shown in the simple yet informative sketch reproduced in Fig. 8.15. Test specimens like AA, BB were loaded together by various springs depicted by CCC, while the force required to overcome friction and initiate sliding was measured on the spring balance D.

The specimens tested were of copper, iron, lead and wood in various combinations and it is interesting to note that in each experiment the surfaces were coated with old pork fat. The laws enunciated by Amontons are frequently, but inaccurately, described by present-day writers as the laws of 'dry' friction and it is a salutary lesson to find that the seventeenth-century manuscript makes it clear that Amontons was in fact studying the frictional characteristics of greased surfaces under conditions which would now be described as boundary lubrication.

Amontons' main findings were:

1. That the resistance caused by rubbing only increases or diminishes in proportion to greater or lesser pressure (load) and not according to the greater or lesser extent of the surfaces.

---

20.   See the Appendix, Sect. A.3 for biographical details.

Fig. 8.14
Amonton's illustration of a beam
(AB) being drawn up a frictionless
plane (CD) (Amontons, 1699).

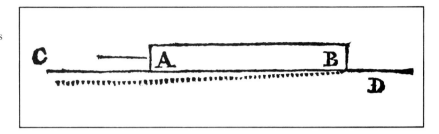

Fig. 8.15
Amontons' sketch of his apparatus
for friction experiments
(Amontons, 1699).
Test materials: A–A, B–B. Spring
providing normal loading:
C–C–C. Spring balance with scale
for friction measurements: D.

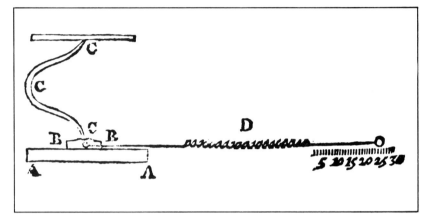

Fig. 8.16 .
Amontons' representation of
friction between multiple surfaces
(Amontons, 1699).

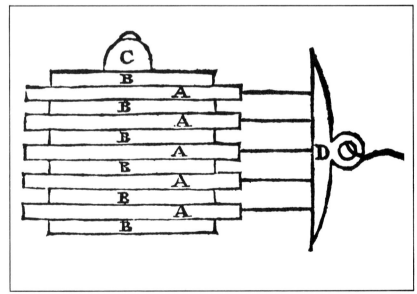

Fig. 8.17
Amontons' representation of
elastic asperities by springs (A)
(Amontons, 1699).

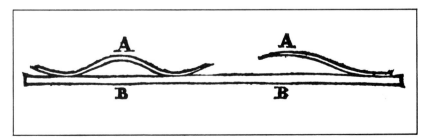

2. That the resistance caused by rubbing is more or less the same for iron, lead, copper and wood in any combination if the surfaces are coated with pork fat.
3. That this resistance is more or less equal to one-third of the pressure (load).

The first observation embodies the first and second laws of friction, namely:

1. The force of friction is directly proportional to the applied load.
2. The force of friction is independent of the apparent area of contact.

Amontons anticipated that these statements, particularly that regarding the null effect upon friction of the size of the rubbing bodies, might not meet with ready acceptance. This was perhaps fortunate since it prompted him to '. . . consider carefully the nature of friction . . .' before presenting his paper. He clearly thought of the fundamental cause of friction in terms of surface roughness and the force required to lift interlocking asperities over each other in sliding motion. He wrote:

> . . . *It is impossible that these irregularities shall not be partly convex and partly concave, and when the former enter upon the latter they shall produce a certain resistance when there is an attempt to move them, since in order to do this they will have to raise that which presses them against each other, and the action of these unevennesses or else the effect which these can produce is the same as that of inclined planes which are used in raising loads, it follows that the greater the pressure* (load) *the greater the resistance to movement; furthermore as in the case under consideration one must assume that the pressure is equally distributed over the whole area of the surfaces: it follows again that where you have several surfaces of differing areas loaded with equal weights each part forming the larger one carries less weight than each part of the same areas which form the smaller ones, this following the proportion these surfaces have between themselves.* . . .
> . . . *It follows further that the resistance caused by the rubbing of surfaces of differing areas is always the same when they are loaded with equal weights* . . .

He even disposes of an apparent paradox by explaining that the force required to move simultaneously a number of weightless blocks like A in Fig. 8.16 loaded with an arbitrary weight, is in fact dependent upon the number of sliding faces. It appears at first sight that this contravenes the second law of friction which says that the resistance is independent of the apparent area of contact, but Amontons points out that the resistance at each sliding surface is proportional to the weight C and that the total resistance is equal to this force multiplied by the number of sliding surfaces.

An interesting sequel to this suggestion of the role of surface roughness in the friction process was that Amontons recognized that the argument was valid for both 'rigid' and 'elastic' surface irregularities or asperities. Having presented the case in terms of rigid surfaces he proceeded to the elastic asperity situation and in relation to Fig. 8.17 wrote: '. . . one can suppose them to be capable of springing, since the force which will overcome the stiffness of a spring, and which will make it move, for example from A to B is in no way different from that which will raise to a comparable height a weight equal to the strength of that spring'.

In the one case the frictional resistance is seen to arise from the force required to pull rigid asperities up the sloping faces of opposing protuberances and in the other case the resistance arises from the force required to deform elastic asperities. Although the concepts have had to be modified in the light of more recent research on friction, the laws are still acceptable and of the greatest utility in many situations.

Evidence that mechanical problems of the day prompted Amontons' work on friction can be seen in the latter part of his extensive paper where he gives practical guidance on the calculation of losses of effective motive force attributable to friction in machines. He writes: '... Now, having sufficiently established what friction is, its nature and its laws, it only remains to say something about the rules by which it may be reduced to calculation in order to know how much friction there is in the most complicated machines.'

It is also interesting to note that both Leonardo da Vinci and Amontons concluded that the materials they investigated exhibited a constant coefficient of friction, the former finding a value of $\frac{1}{4}$ and the latter $\frac{1}{3}$.

Amontons considered the calculation of friction in wheel bearings and devoted much space to the influence of friction upon the tightness of ropes or cords wrapped round shafts and pulleys. He found that the resistance attributable to friction between cords and pulleys was proportional to the tension and the thickness of the cord. He also found that it increased with increasing pulley diameter, although not in direct proportion. It was left to Desaguliers (1734) in England to comment further on these observations.

In the *Histoire de l'Académie Royale* for 1699[21] it is reported that in his discourse upon his 'Moulin à Feu' Amontons mentioned, just in passing, that it was a mistake to think, as was quite commonly done, that the resisting force depended upon the size of the rubbing bodies. It is further reported that '... this new idea produced some astonishment in the Academy'. As a result that painter, architect, physicist and prolific writer of the day on scientific matters, Philippe de la Hire (1640–1718), was at once motivated to check the findings. He soon confirmed the result on the basis of experiments which involved the sliding of wood-on-wood and marble-on-marble under conditions of equal load and a range of different apparent contact areas.

Once the facts had been confirmed de la Hire proceeded to seek explanations on the basis of known physical principles. He had no doubt that the resistance arose from the texture of the surfaces which, if flexible, had to bend and lie flat or, if hard, had to disengage themselves and maybe become detached. His ideas were on the whole in accord with Amontons' views on the nature of friction. In the case of elastic asperities the deflection was seen to be inversely proportional to their number, while the total reaction remained constant, giving a resistance independent of area and hence the number of contacts. In the case of rigid asperities he again adopted the view that the resistance arose from the force required to lift

---

21. See 'Sur les frottements des machines', *Histoire de l'Académie Royal*, p. 128.

the surface rugosities over each other, thus yielding a direct relationship with load but no dependence on the apparent contact area. The French mathematician Antoine Parent (1666–1716) drew attention to the fact that in drawing a body up a slope of inclination θ, the ratio of the tangential to the normal force was $\tan \theta$. He discussed the impact of the new understanding of friction upon statics and equilibrium in two *mémoires* (Parent, 1704) to the Académie Royale.

Philippe de la Hire's suggestion that in some cases it may be necessary to detach asperities from the surfaces in order to initiate sliding is perhaps the major extension to Amontons' views on friction. It clearly introduces the concept of permanent surface deformation and shearing and, as he pointed out, the possibility that resistance would depend upon the number of asperities and hence the size of the surfaces. In addition, de la Hire pointed out that in most machines oil or lard was placed between sliding surfaces and that in such cases air would be excluded from the contact areas. The sliding surfaces would then carry all the weight of the atmosphere giving loads and hence frictional resistances proportional to the surface areas. The mention of atmospheric loading is misleading but de la Hire came close to a correct account of the resistance to sliding of lubricated surface for the wrong reasons. Philippe de la Hire also completed a full mathematical analysis of gears and recommended the use of involute teeth. He recognized the role of friction in gear performance for he stated that: '. . . tooth surfaces should be designed to roll on each other and so avoid friction; . . .'.

In the short space of five years the basic laws of friction had been enunciated by Amontons, confirmed by de la Hire and introduced into mechanics by Parent; a truly remarkable start to the eighteenth century.

Germany and Leibnitz

Germany was represented by no less a person than Gottfried Wilhelm von Leibnitz (1646–1716) in these early studies of friction. After spending some years in Paris, Leibnitz moved to Hanover to take charge of the Ducal Library in 1676. It was while he was in this post that he published (Leibnitz, 1706) his contribution to the study of friction. Although the work does not greatly extend the state of knowledge beyond that of the earlier French School, it is of interest in that a distinction is made between sliding and rolling friction.

François Joseph de Camus (1724)

De Camus was motivated by such practical considerations as the arts of carriage construction and clockmaking and the use of sledges in towns to undertake small-scale friction experiments. He investigated greased, oiled and wetted surfaces and found that the force of friction was directly proportional to load and independent of area. He also noted that when the radius of application of the force of friction on a pivoted body increased the resisting torque became greater. This led him by erroneous reasoning to believe that the force of friction increased with velocity.

The concept of cohesion—Desaguliers' work in England

Towards the end of the period under review a significant step was taken towards the generation of an alternative view of the friction process in England. The French were, however, also connected with this development through the parentage of the remarkable John Theophilus Desaguliers (1683–1744),[22] the son of a Protestant priest who fled to England after the Edict de Nantes. Desaguliers taught in Oxford before moving to London where his lectures on natural philosophy attracted considerable attention. He became chaplain to the Prince of Wales, a friend of Isaac Newton and a Fellow of the Royal Society.

The main exposition of his views on friction is contained in his remarkable book entitled *A Course of Experimental Philosophy* published in 1734, but he had previously presented an account of studies of the cohesion of lead to the Royal Society in 1724. He successfully introduced the notion of cohesive forces and he clearly thought that the action contributed to the overall frictional resistance experienced by sliding bodies. His views thus introduced a factor additional to the idea of interlocking asperities favoured in France. The account of his paper to the Royal Society[23] informs us that the Rev. J. T. Desaguliers (1725) took two leaden spheres weighing one and two pounds respectively and. . .

> . . . *having from each of them cut off a segment of about $\frac{1}{4}$ inch in diameter, he pressed them together by his hand, with a little twist, to bring the flat parts to touch closer. The balls stuck so fast, that when the hand H, by means of a string, sustained the upper ball A [Fig. 8.18(a)], the lower one B, was sustained by the contact of C, though loaded with the scale S, and weights E, which amounted to 16 lb. A little more weight added separated them, and, on viewing the touching surfaces, it appeared that they did not exceed a circle of $\frac{1}{10}$ inch diameter; but this surface can hardly be measured exactly, on account of its irregularity. The experiment was repeated several times, and the cohesion of the balls was different every time.*

On the basis of the figures quoted by Desaguliers the cohesion between the lead balls produced mean normal cohesive stresses in excess of 2,000 lbf/in² (13·8 MN/m²). In a further experiment involving identical lead balls of 2 in (50·8 mm) diameter, Desaguliers employed the apparatus shown in Fig. 8.18(b). In this case the separating force was applied though a steelyard and in one isolated test a cohesive force in excess of 47 lbf (209 N) was recorded.

On moving from Oxford to Channel Row, Westminster in 1713 Desaguliers started the presentation of a series of lectures from his house. His book of the course became *A Course of Experimental Philosophy* (Desaguliers, 1734) published in two volumes. In the first volume, dedicated to His Royal Highness, Frederick, Prince of Wales, he introduced the concept of the force of cohesion in a simple and persuasive manner;

> *As it is easier to raise most bodies from the ground than to break them in pieces; that force by which the parts cohere, is stronger than its gravity. That force, whatever be its cause, we shall call the Attraction of Cohesion. This attraction is strongest,*

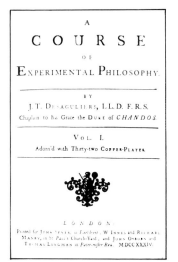

A
C O U R S E
OF
EXPERIMENTAL PHILOSOPHY.

BY
J. T. DESAGULIERS, LL.D. F.R.S.
Chaplain to his Grace the DUKE of CHANDOS.

VOL. I.
Adorn'd with Thirty-two COPPER-PLATES.

LONDON
Printed for JOHN SENEX, in Fleetstreet, W. INNYS and RICHARD MANBY, on St. Paul's Church-Yard; and JOHN OSBORN and THOMAS LONGMAN in Pater-noster Row. MDCCXXXIV.

---

22.   See the Appendix, Sect. A-4 for biographical details.
23.   *Phil. Trans. R. Soc. Lond.*, abridged, **vii**, 1724–34, 100.

Fig. 8.18
The Rev. J. T. Desaguliers'
demonstration before the Royal
Society of the cohesion of lead
(Desaguliers, 1725).

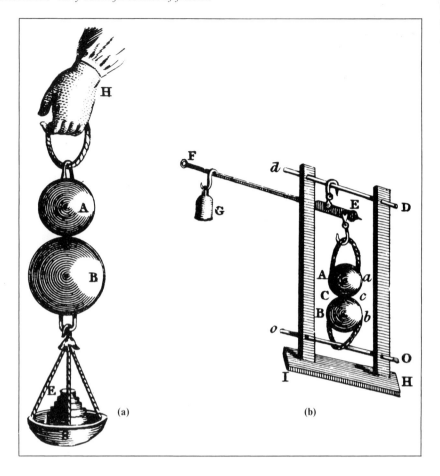

*when the parts of the bodies touch one another; but decreases much faster than
gravity, when the parts that were before in contact, cease to touch; and when they
come to be at any sensible distance, this attraction of cohesion becomes almost
insensible.*

In Lecture I of his course, from which the above quotation is taken,
Desaguliers describes numerous experiments which demonstrate cohesion.
His philosophical pragmatism is further illustrated in the same lecture
by the words:

*. . . that those properties of bodies, such as gravity, attractions, and repulsions, by
which we shall hereafter explain several phenomena, are not occult qualities or
supposed virtues, but do really exist, and are by experiments and observations
made the objects of our senses. These properties produce effects, according to
settled laws, always acting in the same manner under the same circumstances: And,
tho' the causes of those causes are not known, since we do not reason about those
hidden causes; it is plain that we reject occult qualities, instead of admitting them in
our philosophy, as the Cartesians always object to us.*

The second lecture is concerned with basic concepts in mechanics and
the third with '. . . Simple Machines or Organs, call'd by some Mechanical
Faculties, or Mechanical Powers'. He notes that '. . . All engines (however
compounded) for the uses of life, are made up of various combinations of

the simple machines . . .' of which he lists seven '. . . the Balance, the Leaver, the Pulley, the Axis in Peritrochio (or Axis in the Wheel), the Inclined Plane, the Wedge and the Screw'.

In the same lecture he emphasizes the need to consider the effect of friction upon machine performance in a passage which amply illustrates his concern to apply scientific principles to engineering systems.

> . . . *We are to have regard to the Imperfections of Engines and Materials, and the quantity of Stickage or Friction; which differ according to the number of combination of Parts and Nature of the Materials, of which the several engines consist: And having made use of the best Methods we can to discover the Imperfections above mention'd in each particular machine; we are to take care to allow enough to be deduced from the Calculation made concerning an Engine suppos'd Mathematically true.*

It is the fourth lecture which is of major import to our subject, since it is here that '. . . Several Methods of Finding the Quantity of Friction in Engines . . .' is considered. Desaguliers clearly intended to provide guidance for the designers and manufacturers of machinery of the day, although his theory of friction was no more than a confirmation of Amontons' laws. Those experimenters who have attempted measurements of the force of friction will appreciate the words of caution contained in the second paragraph of Lecture IV. 'Tho' there are so many Circumstances in the Friction of Bodies, that the same Experiment does not always succeed with the same Bodies, so that a Mathematical Theory cannot be easily settled; yet we may deduce a Theory sufficient to direct us in our Practice from a great Number of Experiments, always taking a Medium between Extremes.'

It is, however, noted that a single rule can be applied to the major engineering materials of the day such as wood, iron, brass, copper and lead, since all have nearly the same friction when greased or oiled, as is done in engines. The influence of cohesion is introduced in a passage in which the role of surface finish is considered. The major contribution to friction is clearly seen to be related to surface roughness and Desaguliers notes that when the sliding surfaces are highly polished the force of friction may actually rise due to the fact that '. . . the Attraction of Cohesion becomes sensible when we bring the Surfaces of Bodies nearer and nearer to Contact'.

The findings of Amontons are outlined and clearly form the basis of Desaguliers' theory and guide to designers on the allowances to be made for friction. The rule is succintly stated as: '. . . the Friction is equal to about one 3rd of the Weight, and arises from the Weight that presses the Parts together, and not from the Number of Parts that touch.'

Because deviations from the one-third rule were known to arise with variations in materials, surface quality and the type of lubricant, Desaguliers presented a tabular display of measured values of friction. The results were quoted from the work of Monsieur Camus, a gentleman of Lorraine, who prepared and published a book entitled *Traite des Forces mouvantes, pour la Pratique des Arts & Métiers, E.*, but Desaguliers appears to have confirmed the values in independent tests.

A sledge and pulley arrangement which anticipated in many ways the form of apparatus employed by Charles Augustin Coulomb in 1781 was used to evaluate the resistance to motion. Sledges made of wood, iron and wood shod with iron, lead and possibly brass and copper were caused to slide upon wooden boards or metal plates in the presence of water, grease or oil or simply under dry conditions. A strong silk thread passed over a pulley fitted with fine pivots to a silken purse which could receive a number of small lead balls to initiate and maintain motion. It was found to be necessary to incline the 2 ft (0·6 m) long baseboard by raising the pulley end a height of 1 in (25·4 mm) to achieve steady motion.

The lead balls were prepared such that twenty were equal in weight to the sledge and its load. The tabulated results recorded in Fig. 8.19 can thus be converted into values of coefficients of friction by dividing the number of balls by twenty. The values of the coefficient of friction range from 0·15 to 0·90 and it is interesting to note that the presence of a lubricant invariably made things worse. The author warned against the conclusion that lubricants should not be used in engines '. . . because we know them to be of use in great machines'. He saw the role of lubricant as a material to fill up the holes or surface imperfections in the sliding pairs, to facilitate movement by acting as rollers and to minimize wear. He noted that oil and grease generally produced lower friction than water, and that even in small machines like pocket-watches where oil caused slower running it produced a more uniform motion and better timekeeping.

Desaguliers did not leave the reader with the bare statement of his views on the nature of friction. He devoted many words to an account of the functioning of vehicles and machines, together with guidance on the procedures to be followed in making full allowance for friction in the evaluation of machine performance. The importance of wheeled vehicles in eighteenth-century society was emphasized by Desaguliers when he wrote: '. . . yet as Coaches, Waggons and Carts, and other wheel-carriages are so necessary for the Uses of Life, that only the Disuse of them for

Fig. 8.19
Desaguliers' (1737) table of frictions.

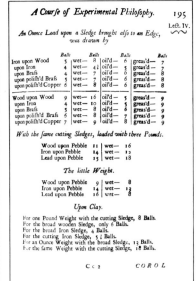

one Month would be enough to put a whole Nation in Confusion'. His views on the mechanics of wheeled vehicles were clearly in accord with those of the said Monsieur Camus for he went on to quote from the latter's writings on this topic.

In a later section he indulged in ergonomics and gave some guidance on the relative efforts which could be expected from horses and men. Once again he drew upon the experimental findings of others, and in a passage which can hardly be considered appropriate to the *Entente Cordiale* he wrote '. . . That five English Labourers are equal to an Horse, and only seven French Men or as many Dutch Men.'

However, he also noted that '. . . in Turkey the Porters will carry twice more than the strongest English Porters. . .'.

In one important respect Desaguliers disagreed with Amontons. The latter found that although the difficulty of bending a rope round a pulley or roller decreased as the diameter increased, the effect was somewhat smaller than that of inverse proportionality. Desaguliers found the effect to vary directly with the inverse of the diameter and he attributed the difference to the experimental procedure. In his own experiments he avoided friction between the coils of rope, whereas he believed that in Amontons' experiments the parts of the rope rubbed against each other.

How should we evaluate Desaguliers early eighteenth-century contribution to tribology? It could be argued that he introduced very little that was not already known to Amontons and that his concept of the role of cohesion was not built into his theory of friction or even evaluated by experiment. However, it is important that our appraisal should encompass wider issues.

Desaguliers' writings on friction found an honourable place in his wide-ranging lectures on natural philosophy. This is important since it brought the attention of English scientists and engineers and–through his general lectures–the public at large, to the significance of the subject. He provided the seed of the idea that cohesion might contribute significantly to sliding friction which was to germinate two centuries later in the care of Bowden and Tabor[24] in Cambridge. Perhaps the greatest merit of his work was that it provided clear guidance for more realistic estimates of the influence of friction upon machine performance. He drew together existing data, mainly from France, carried out his own experiments and showed by examples how a simple theory of friction could be incorporated into mechanics. He adjured all who were concerned with machinery and the utilization of animal power to take full cognizance of friction. This was the main impact of his work.

ARCHITECTURE
HYDRAULIQUE,
O U
L'ART DE CONDUIRE,
D'ELEVER, ET DE MENAGER
LES EAUX
POUR LES DIFFERENS BESOINS DE LA VIE.
PREMIERE PARTIE,
TOME PREMIER.

Par M. BELIDOR, Commissaire Provincial d'Artillerie, Professeur Royal des Mathematiques, aux Ecoles du même Corps, Membre des Academies Royales des Sciences d'Angleterre & de Prusse, Correspondant de celle de Paris.

A PARIS, RUE S. JACQUES,
Chez CHARLES-ANTOINE-JOMBERT, Libraire de l'Artillerie,
& du Génie, à l'Image Notre-Dame.

M. DCC. XXXVII.
AVEC APPROBATION ET PRIVILEGE DU ROY.

Surface modelling with rigid spherical asperities by Bernard Forrest de Bélidor (1697–1761)

Three years after the publication of Desaguliers' lectures in London, Bernard Forrest de Bélidor introduced an interesting representation of rough surfaces into the analysis of friction in his two-volume work on hydraulic architecture (Bélidor, 1737). Bélidor was a professor of

---

24. See, for example F. P. Bowden and D. Tabor (1950), Part I; (1964), Part II.

mathematics at the School of Artillery at la Frère who later became the Inspector of Arsenals and Mines in Paris. His work on friction was important more from the point of view of his representation of rough surfaces by arrays of spherical asperities than for the advancement of physical understanding of the friction process. He reverted to the view of interlocking rigid asperities (Fig. 8.20) and proceeded to calculate the force required to pull one set of spheres over the other. The force was found to be independent of the number of asperities and hence consistent with the laws of friction established by Amontons some thirty-eight years earlier. His analysis predicted a ratio of friction force to normal load of $1/(2\sqrt{2}) = 0.35$. The modelling of rough surfaces by means of spherical asperities has remained popular with tribologists concerned with theoretical studies of both friction and wear ever since 1737.

Friction and applied mathematics – Leonhard Euler (1707–1783)

To conclude our review of this most remarkable era of research into friction we call to mind that outstanding eighteenth-century mathematician Leonhard Euler. Euler was born in Basle, Switzerland but after studying theology, mathematics, physics and physiology he settled in St Petersburg in Russia at the age of twenty. After being a student of Jean Bernoulli he became a colleague of Jean's sons Daniel and Nicolas when he accepted an invitation from Catherine I to become an assistant in the Mathematics Department of the Academy of Sciences in St Petersburg. He became a professor of physics in 1730 and he succeeded Daniel Bernoulli to the Chair of Mathematics when the latter returned to Switzerland in 1733. After a further seven years in St Petersburg Euler was obliged to leave Russia during unsettled times. He moved to Berlin where in 1741 on the invitation of Frederick the Great he became a member of the Academy of Sciences. During his twenty-five years in Berlin he produced an amazing stream of publications, but when Frederick became less hospitable he decided once again to return to St Petersburg in 1766 on the invitation of Catherine the Great. He spent the rest of his seventeen years in Russia and in the last fourteen years or so of his life he produced 308 important papers on mathematical and physical topics.

Euler's skill and prodigious output are legendary. In an excellent account of his life and work Kragelskii and Shchedrov (1956) have noted that he produced some 750 original scientific contributions to knowledge. This incredible record assumes even greater proportions when it is recalled that he lost the sight of an eye after overtaxing himself with the compilation of navigational tables for Russian ships in 1735 and that he spent the latter part of his life after returning to St Petersburg in 1766 in total blindness.

In 1748, while working in Berlin, Euler (1750*a,b*) submitted two papers on friction to the Academy of Sciences. In the first paper the role of friction in determining the equilibrium, uniform acceleration or retardation of a solid body in contact with a plane was considered. Euler then showed in the manner of Parent that the horizontal force (F) required to move a solid of weight (P) up a slope of inclination (α) to the

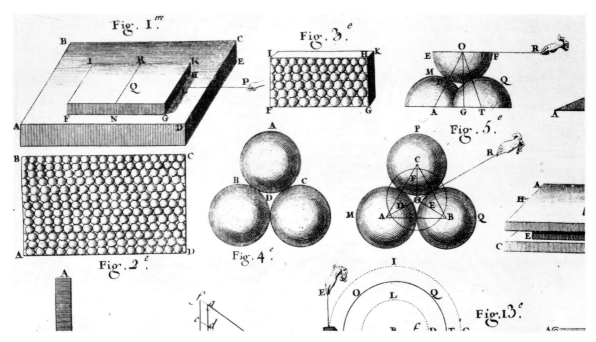

Fig. 8.20
Bélidor's representation of rough
surfaces with ideal spherical
asperities (Bélidor, 1737).

Fig. 8.21
Euler's (1750*a,b*) studies of
friction
(a) Representation of surface
roughness by triangular section
asperities. Having sides inclined at
angle (α) to the horizontal.
(b) Analysis of kinetic friction and
the motion of a block (P) down an
inclined plane.

horizontal was given by the expression

$$F > P \tan \alpha.$$

He then adopted the model of rigid interlocking asperities as the cause of frictional resistance and pointed out that on the basis of Amontons' hypothesis the slope of the asperities must be 19°29′ (sin $\alpha = \frac{1}{3}$), while for Mr Bilfinger's hypothesis, which apparently agreed with Leonardo's observations (sin $\alpha = \frac{1}{4}$), the slope would be 14°28′. He then suggested that the surfaces of solids were covered by small triangular-shaped asperities having sides of slope (α) as shown in Fig. 8.21(*a*). A simple extension of the above arguments led to the conclusion that if friction was the force required to lift one body over the asperities of the other, and if '. . . the friction is to the pressure as μ is to 1 . . .'

$$\mu = \tan \alpha.$$

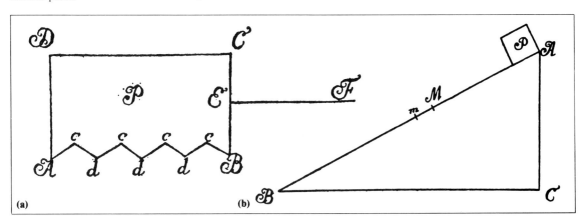

(a)   **(b)**

It was through Euler that the symbols e, π and i came into common use in mathematics and we see from the above quotation that he also introduced into the science of tribology the symbol μ for the coefficient of friction, although the latter term was not in use at this time.

Euler further pointed out that it was not necessary for all the slopes to be equal for the above result to be obtained, provided none exceeded α.

Perhaps the major import of his work is the clear distinction which he draws between static and kinetic friction. His analysis of sliding motion down an inclined plane with friction between the block and the plane can be summarized in relation to Fig. 8.21(*b*) as follows.

As the block of weight (P) moves a distance (s) from A to M on an incline of slope (α) to the horizontal the total loss of potential energy is

$Ps \sin \alpha.$

If the block acquires a sliding velocity (v) on reaching M the gain in kinetic energy is

$$\frac{P}{g} \frac{v^2}{2},$$

while the energy dissipated by the frictional resistance is

$\mu s P \cos \alpha.$

Thus,

$$s \sin \alpha = \frac{v^2}{2g} + \mu s \cos \alpha$$

or,

$$\mu = \tan \alpha - \frac{v^2}{2gs \cos \alpha},$$

and since, for uniform acceleration, the velocity after time (t) from a position of repose is given by

$$v = \frac{2s}{t}$$

we have

$$\mu = \tan \alpha - \frac{2s}{gt^2 \cos \alpha}.$$

Euler argued that if the plane was inclined at the limiting slope for static equilibrium and the angle was then increased by the smallest amount the body would move with an exceedingly small velocity. However, he found by experiment that this was not the case and that the body moved quite quickly once equilibrium was disturbed. He concluded that the kinetic friction must be smaller than static friction.

Euler's (1750*b*) second paper was concerned with the reduction of frictional resistance in the case of shafts and in a later (Euler, 1762) paper he analysed the role of friction in the equilibrium of ropes wrapped around shafts. His work is important mainly on three counts. It developed a clear analytical approach to friction, it introduced the well-known symbol μ for 'coefficient of friction' and it marked the transition from studies of static to kinetic friction.

8.5   Viscosity and viscous flow

Scientific studies of friction, most pertinent observations on wear and impressive developments of both plain and rolling-element bearings, so dominate the seventeenth and early eighteenth-century history of tribology that it may appear to be somewhat impertinent to introduce the subject of lubrication. There is nothing directly related to the scientific study of lubrication to compete with contemporary work on friction, although it seems clear[25] that the importance of adequate lubrication was recognized. Our intrusion into the present chapter is nevertheless well justified since the work of Sir Isaac Newton recorded here laid the foundation of fluid-film lubrication theory developed almost 200 years later by Osborne Reynolds.

Sir Isaac Newton, whose portrait is shown in Fig. 8.22, was born on Christmas Day 1642 in the hamlet of Woolsthorpe, Lincolnshire.[26] Galileo died in the same year and Leibnitz was born four years later. Newton's father, described by his stepfather the Rev. Barnabas Smith as a '... wild, extravagant and weak man. . .', died before Isaac was born and his mother remarried two years later.

Newton first attended two village schools where he learned to write, to read and to do a little arithmetic, but he moved to King's School in the nearby town of Grantham when he was twelve. His stepfather died shortly afterwards and when he was nearly sixteen his mother called him back to Woolsthorpe to manage the estate. He was not particularly successful as a farmer and he soon returned to school to prepare for university. He entered Trinity College, Cambridge, as a subsizar in 1661 where, in due course, he became acquainted with the Lucasian Professor of Mathematics, Isaac Barrow.

Newton, who did not excel in formal examinations, took the degree of B.A., apparently without distinction, in the year of the Great Plague, 1665. This was the worst and last of the epidemics of bubonic plague which had swept the country at intervals since 1348. The University was closed and Newton returned to the farm at Woolsthorpe where he lived for about two years. During this period of seclusion he invented the calculus, discovered the composition of white light and conceived the idea of universal gravitation.

Newton returned to Cambridge in 1667 and was made a Fellow of his college within five months. Two years later Barrow resigned his professorship to enable Newton to become the Lucasian Professor of Mathematics at the age of twenty-six. He was elected to the Royal Society in 1672 and

25. See, for example, Hooke, R. (1685).
26. See Magie, W. F. (1963), pp. 30–1 and North, J. D. (1967).

Fig. 8.22
Sir Isaac Newton. (Painted by Sir
Godfrey Kneller.)

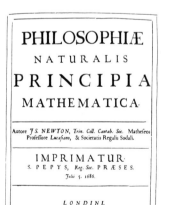

continued his scientific work at Cambridge for a further twenty years.
During this time he experienced the hostility of Hooke and his initially
friendly contact with Leibnitz descended into a bitter debate as to who
should have priority for the invention of the calculus. He became Warden
and later Master of the Mint before his election to the presidency of the
Royal Society in 1703. He was knighted by Queen Anne in 1702. Sir Isaac
Newton died in Kensington on 20 March 1727 and was buried at
Westminster Abbey.

The small part of Newton's work with which we are concerned was
included in his *Principia Mathematica Philosophiae Naturalis* (Newton,
1687). The *Principia*, one of the greatest scientific works of all time, was
translated from the Latin into English by Motte in 1729, revised and
supplied with a historical and explanatory appendix by Cajori in 1934
and published in paperback edition in 1966.

Section IX of Book II of the *Principia*, which is concerned with the
'Circular Motion of Fluids', opens with a 'Hypothesis' which embodies

a statement now described as Newton's law of viscous flow. The 'Hypothesis' reads:[27] '... The resistance arising from the want of lubricity in the parts of a fluid, is, other things being equal, proportional to the velocity with which the parts of the fluid are separated from one another. ...' The term '... want of lubricity. ..' or '*defectus lubricitatis* ...' would now be described as *internal friction* or *viscosity*, but it is important to realize that Newton did not use the latter terminology. It is interesting that the word *lubricity* has been the subject of debate and argument in recent times, apparently due to the difficulty in allocating a precise definition in accord with twentieth-century understanding of the physical and chemical processes which control fluid and boundary friction.

Viscosity is a word with an interesting background worthy of mention in the present section. Mistletoe berries contain a very sticky substance known as viscin which forms the main constituent of bird-lime. The latter was used traditionally as a sticky substance spread upon twigs to catch birds. The word *viscosity* has its roots in the Greek word for mistletoe ὀιξός and it comes to us via the Old French *viscosite* or medieval Latin *viscositas*.

The first recorded use of the adjective *viscous* dates back to A.D. 1400 (Lanfranc's *Cirurg.*) while the noun *viscosity* was recorded in 1425 (Arderne's *Treatise on Fistula*). The early use of the word thus appears to be related to descriptions in the medical literature of rather obnoxious sticky body fluids. The use of the word in physics in relation to fluid flow was not in evidence before the eighteenth century, perhaps as early as 1707, but certainly by 1786 in the writings of the military and hydraulics engineer Du Buat (1734–1809), who noted the influence of temperature upon the resistance to fluid flow (viscosity). However, 150 years were to separate Newton's famous hypothesis and Navier's (1785–1836) introduction of the coefficient of viscosity into the equations of motion.

To conclude, we examine the reason for Newton's concern with the circular motion of fluids and his important hypothesis. He was in fact seeking an explanation of the motion of celestial bodies and his work can be seen as a link between his general studies of the *Motion of Bodies* (in resisting mediums) and his mathematical statement of the *System of the World*. His studies showed that a fluid obeying his hypothesis would form a vortex around a rotating sphere with a periodic time proportional to the square of the distance from the centre of motion, whereas the planets moving round Jupiter and those revolving round the sun had periodic times proportional to the radius to a power $\frac{3}{2}$. He therefore demonstrated that vortex motion of a viscous fluid pervading the universe could not fully explain the observed motions of planets.

Fluids which obey the hypothesis are known as Newtonian fluids and fluid-film lubrication theory was built upon this foundation. There is no evidence that Newton considered this important application of viscous flow theory; his studies arose from the much wider field of celestial motion.

---

27. See Sir Isaac Newton (1687), p. 385 as translated by Motte, revised by Cajori in 1934.

8.6 Chronology

The time span of 150 years considered in the present chapter is identical with that of the Renaissance period covered in Chapter 7, yet there are some interesting differences between the respective contributions to tribology. The Renaissance developments were essentially of two forms: the technical and scientific works of Leonardo da Vinci and the development of bearings as evidenced by contemporary accounts of industrial machinery. Table 8.1 shows that in the period from A.D. 1600 to 1750 there were more numerous reports of scientific investigations and technical developments specifically related to those aspects of tribology which were of major concern prior to the Industrial Revolution. The formation of national scientific societies like the Royal Society of London and the Académie Royale des Sciences after 1660 provided a forum for the discussion of scientific studies of the major problems of the day, together with an avenue for the publication of such work.

Table 8.1
Chronology for the
pre-Industrial Revolution
period (A.D. 1600–1750)

Date	Political and social events	General technical developments	Tribology
1582		Peter Maurice's water-wheel-driven pumps set up near London Bridge	
		(1594) Bevis Bulmer's horse-driven chain pump at Broken Warf	
1600			
1603	Death of Queen Elizabeth I of England. End of Tudor period (1485–1603)		
1605	Gunpowder Plot–London		
1606		Baptista Porta's condensing steam pump	
1607	First successful settlement in Virginia	Vittorio Zonca's *Teatro Nuovo di Machine et Edificii* published posthumously	**Zonca's references to wear and a recommendation for brass bearings and steel shafts**
			Sketches of half-journal bearings; pivot bearings and split-bearing blocks (Italy)
1608		Lintlaer's water-pump built beneath the Pont-Neuf, Paris	
1615		Saloman de Caus' steam-driven water fountain	

Table 8.1 (*Continued*)

Date	Political and social events	General technical developments	Tribology
1616	Death of William Shakespeare (1564–1616)	(General development of coal-mining, agriculture, metal-working industry throughout the 17th century)	
1618	Start of Thirty Years War (1618–48) – Germany		
1620	Pilgrims settled at Plymouth, Mass.		
1625	Death of James I of England (James VI of Scotland)	Water-wheels and windmills developed to provide the power for machinery	
1626	Death of Francis Bacon (1561–1626).		
1632	Birth of Christopher Wren (1632–1723)		
1633	First settlement in Connecticut		
1633	Fire on London Bridge		
1634	Maryland settled		
1635	Robert Hooke born (1635–1703)		
1637		*T'ien-kung K'ai-wu* published	**Evidence of use of oil as lubricant in cart axles in China. Also concern for frictional heating between grinding stones**
1642	Start of First Civil War (1642–46) in England		
1642	Isaac Newton born		
1643	Louis XIV of France (1643–1715)		
1644	Battle of Marston Moor		
1644	End of Ming Dynasty (1368–1644)		
1644	Start of Ch'ing (Manchu) Dynasty (1644–1911)		
1645	Battle of Naseby		
1646	Birth of Leibnitz		
1647		Pascal reported on his experiments with siphons, syringes and bellows. Contributions to the laws of hydrostatics	
1648	Second Civil War in England		
1648	Battle of Preston		

Table 8.1 (*Continued*)

Date (A.D.)	Political and social events	General technical developments	Tribology
1654		John Bate's account of water-pumps and fire-engines in London	**Reference to brass bearings in water-wheel (England)**
1655		Patent issued by Oliver Cromwell to Sir Edward Ford for cam-driven water-pump in London	**Reference in De Moncony's (1666) account of Ford's pump to wooden cams faced with iron and a bronze friction-roller follower (England)**
1660	Monarchy restored in England. Charles II (1660–85)		
1660	Royal Society founded in England		
1662	Robert Hooke appointed Curator of Experiments to the Royal Society		
1663		Marquis of Worcester's account of a steam-pressured water-pump	
1666	French Royal Academy of Sciences established		
1666	Fire destroyed much of the City of London		
1670		Anchor escapement for mechanical clock takes over from verge and foliot	
1674		Hooke's joint invented and described in Robert Hooke's Cutlerian lectures	
1682	Pennsylvania and The Delaware territories settled		
1683	Birth of J. T. Desaguliers		
1685	Death of Charles II of England James II on throne of England (1685–89)	Robert Hooke's discourse on carriages to the Royal Society including an account of Stevin's sailing chariot	**(1685) Hooke's discussion of the nature of rolling friction; optimum wheel-on-axle arrangements; bearing materials (steel-on bell metal); the need for adequate lubrication and the advantages of 'truckles' (England)**
1687		Publication of Newton's *Principia*	**(1687) Newton's hypothesis on viscous flow which now forms the basis of fluid-film lubrication theory (England)**
1689	William III and Mary on throne of England (1689–94), William III reigned until 1702		

Table 8.1 (*Continued*)

Date (A.D.)	Political and social events	General technical developments	Tribology
1693	Beginning of the National Debt		
1694	Bank of England founded	Philippe de la Hire studied cycloidal form of gear teeth and recommended use of involute profiles	
1698		Savery's pulsometer pump	
1699			Amontons' experiment on friction of various materials lubricated by 'pork fat'. Two laws of friction enunciated (1) $F \propto P$ (2) F independent of apparent contact area
			$F = \frac{1}{3}P$ for all materials. Force of friction arises from the work done in sliding rigid or elastic asperities over each other (France)
			Philippe de la Hire confirmed Amontons' findings Reference to the practice of using lard or oil in most machines (France)
1700		Use of 'roll-ring' in Dutch windmills	
1701		Publication of Charles Plumier's account of the lathe and the art of turning	(1701) Plumier's reference to the use of soft bearing metals (like pewter) and split bushes with adjustment for wear in bearings for machine tools (France)
1702	Anne Queen of England (1702–14)	Savery's pulsometer described in 'The Miner's Friend'	
1703	Death of Robert Hooke		
1704			Publication of Parent's *memoires* on the role of friction in statics and equilibrium (France)
1706			Publication of Leibnitz's paper on friction in which he distinguished between rolling and sliding friction (Germany)
1707	Birth of Leonhard Euler (1707–83)		
1707	Union of Scotland and England effected		

Table 8.1 (*Continued*)

Date (A.D.)	Political and social events	General technical developments	Tribology
1710			**De Mondran's design of roller-disc bearings for carriages approved by the Academy of Sciences in Paris. Enabled one horse to do the work of two (France)**
1712		Newcomen's atmospheric beam engine built near Dudley Castle, Staffordshire	
1714	George I (1714–27)		
1715	Louis XV of France	Graham introduced his 'dead-beat' escapement	
1715	Jacobite Rebellion		
1716	Death of Leibnitz		**Henry Sully's prize-winning chronometer fitted with a roller-bearing (England)**
1723	Death of Christopher Wren		
1724			**Desaguliers' paper to the Royal Society on the Cohesion of lead (England)**
			Leupold's reference to the use of roller bearings in the hollow shafts of windmills (Germany)
1725		(onwards)–famous Dutch mill books published	
1727	George II (1727–60)		
1727	Death of Newton		
1729	John and Charles Wesley founded the Methodist Society		
1732	Colony of Georgia founded	Henri Pitot described his tube for measuring the velocity of fluids	
1733		Camus reported his studies of mechanisms which included an extension of gear contact analysis initiated by de la Hire	
1734		*A Course of Experimental Philosophy* published by J. T. Desaguliers	**Desaguliers introduced the concept of cohesion as a factor in friction. He showed how to allow for friction when calculating machine performance. Friction tests were described and measurements of friction**

Table 8.1 (*Continued*)

Date (A.D.)	Political and social events	General technical developments	Tribology
			displayed in tabular form ($\mu = 0{\cdot}15 - 0{\cdot}90$). Role of lubricant to fill up holes or surface imperfections, to facilitate motion by acting as rollers and to minimize wear (England)
			Jacob Rowe awarded patent by King George II for the use of 'friction-wheels' in carriages and fixed machinery. Refers to the use of cast-iron, brass and bell-metal in bearings. Mentions use of leather seals to prevent bearings clogging. Discusses economic advantages of his roller-bearings and calculates potential savings to the United Kingdom of £947,500 per annum (England)
		(1734) and (1736) Publication of *Groot Volkomen Moolenbock* by Van Natrus, Polly and Van Vuuren	Illustration of trunnion-mounted wooden rollers in Dutch roll-ring bearing in *Groot Volkomen Moolenboch* (Netherlands)
1735			Leupold's recommendation for the use of tallow or vegetable oil for lubricating rough surfaces (Germany)
1737		Publication of Bélidor's *Hydraulic Architecture*	Bélidor's analysis of friction based upon the representation of rough surfaces by spherical asperities. Predicted $F = P/\sqrt{8} = 0{\cdot}35P$ (France)
1738		Publication of Daniel Bernoulli's *Hydrodynamica*	
1743	France joins Spain to fight England in America and India		
1745	The 'Forty-five' Jacobite Rebellion		
1746	Culloden – rebellion crushed		
1750			Euler published two papers on friction in the Berlin Academy of Sciences

He introduced the symbol 'μ' for F/P and developed the expression |

Table 8.1 (*Continued*)

Date (A.D.)	Political and social events	General technical developments	Tribology
			$\mu = \tan\alpha$. He showed that asperities must have slopes of about 19° or 14° to satisfy Amontons' and Mr Bilfinger's (Leonardo da Vinci's) value of μ. He distinguished between static and kinetic friction (Germany)

A number of important accounts of industrial machinery similar to those discussed in Chapter 7 continued to appear in the early days of the seventeenth century and the posthumous publication of Vittorio Zonca's text *Teatro Nuovo di Machine et Edificii* in 1607 provided a good early example of this valuable type of record. The text can be viewed as an account of the form of machinery in the late Tudor period which was written at a time when another author, William Shakespeare, was in his prime.

Most of our seventeenth-century references are related to the development of plain bearings and an increasing recognition of the importance of brass and bronze as bearing materials. There were, however, two most important contributions by eminent scientists towards the end of the century. The Monarchy had been restored in England after the turbulent Civil War era and the Royal Society was well established. In 1685 Robert Hooke presented his discourse on carriage wheels which included an impressive section dealing with the nature of rolling friction, the advantages of steel-on-bell metal bearings, the importance of seals and lubrication and the virtues of 'truckles' or roller-disc bearings. Two years later Isaac Newton published his *Principia* which contained the essential hypothesis on viscous flow and which was to form the basis of classical studies of fluid-film lubrication almost exactly 200 years later. In the reign of James II English scientists had thus begun to lay the foundations of modern concepts in tribology before the dawn of the eighteenth century, although Newton's contribution was prompted not by a concern for lubrication but by an attempt to explain the motion of celestial objects.

In 1699 there started an exciting period of scientific studies of friction which merits special attention in the long history of tribology. The centre of activity of the work carried out at this time was Western Europe, with France and England playing important roles. Amontons' now classical studies of friction reported in 1699 were received with a certain amount of scepticism, but Philippe de la Hire soon confirmed their essential conclusions. Parent (1704) and Bélidor (1737) applied their mathematical skills to the concept that friction arose from the work done in moving interlocking asperities over each other in an impressive extension of the French activity.

Desaguliers (1725, 1734) made a valuable contribution when he drew attention to the role of cohesion in the friction process and in spite of his name the credit goes to England in this instance. This activity in studies of friction immediately before the Industrial Revolution was rounded off by contributions from Euler (1750*a*,*b*) when he was living in Berlin.

Perhaps the most significant contributions related directly to bearings, which were made in parallel with the studies of friction in the latter half of the seventeenth and first half of the eighteenth century, were those concerned with roller-disc bearings in machinery and carriages (Hooke, 1685; De Mondran, 1710 and Rowe, 1734). There was also ample evidence of an increasing interest in the influence of materials, lubricants and seals upon the wear of bearings.

Rowe's (1734) delightful early analysis of the economic significance of friction in wheel bearings leaves us in no doubt that the pre-Industrial Revolution period of one and a half centuries witnessed a great awakening of interest in many aspects of tribology.

8.7 Summary

The seventeenth and early eighteenth centuries represented a period of great advance in both practical and fundamental aspects of tribology. The major developments can be grouped under three headings: a greatly increased awareness of the importance of friction and wear in machinery; a more adventurous use of both plain and rolling bearings; and the first comprehensive scientific studies of the friction process.

The first point is illustrated by the number of contributions to the subject which contain some reference to the need to make full allowance for friction and wear in the design and operation of machinery. The papers by Hooke (1685), Amontons (1699) and Desaguliers (1734) bear witness to this point, while the quotation from Amontons' now classical paper which opens the chapter reflects the general position at the end of the seventeenth century. It is interesting to reflect upon the fact that tribologists are even now making only slightly modified observations on the extent to which their subject is neglected in machine design and operation.

The *ad-hoc* development of bearing materials for fixed machinery and wheeled vehicles is clearly seen in the writings of Zonca (1607), De Monconys (1666), Hooke (1685) and Plumier (1701). The willingness to experiment with the roller-disc bearing, which was known to Leonardo da Vinci and developed gradually in Renaissance times, marks the period under consideration as an era of innovation in bearing design. Practical forms of this bearing were mentioned by Hooke (1685), designed by De Mondran (1710), and exploited by Rowe (1734). The economic merit of improved bearings was noted with some emphasis by the two latter writers, for it appears that they anticipated a 50 per cent reduction in the motive power required for wheeled vehicles and possibly a 30 per cent reduction of losses in fixed machinery.

The meagre references to lubricants and lubrication in this period are swamped by the stirring attacks on the nature of friction, but they nevertheless show a steady interest in the subject. There is little to report apart from the brief mention of the use of oil in cart axles in China by Sung Ying-Hsing (1637), Hooke's (1685) plea for 'adequate' lubrication in carriage bearings, Amontons' (1699) use of pork fat in his friction experiments, De la Hire's (1699) reference to the practice of using lard or oil in machines, Desaguliers' (1734) speculation that the role of the lubricant was to fill up imperfections in the solid surfaces and to act as tiny rollers and Leupold's (1735) recommendation that tallow or vegetable oil should be used for lubricating rough surfaces. In retrospect the most valuable contribution to scientific studies of lubrication was provided by Newton's (1687) hypothesis on viscous flow; a contribution which arose from mighty considerations of the nature of the universe.

The last third of this period of 150 years is outstanding for the progress made through the new scientific method towards an understanding of the nature of friction. In this period the geometric concept of asperity interaction was established by Amontons (1699) and adopted by De la Hire (1699), Parent (1704), Leibnitz (1706), Desaguliers (1734), Bélidor (1737) and Euler (1750a). The role of surface texture in determining frictional resistance was readily accepted and the concept that the resistance was related to the force required to pull one rigid or elastic rough surface up the sloping sides of the protuberances of the mating component was entirely in accord with the current thinking in mechanics. In addition, two significant additional factors were mentioned by Philippe de la Hire and Desaguliers. The former thought that it may be necessary to detach some asperities to initiate sliding, while the later argued cogently that cohesion would contribute to the resistance to sliding. In these further suggestions we see the beginnings of the 'deformation' or 'mechanical' and 'adhesive' aspects of modern understanding of friction.

Two further aspects of these early scientific studies of friction are worthy of note. It will be seen that the geometric concept of friction between rough surfaces slowly lost ground to our modern understanding of the phenomenon, but the laws of friction enunciated by Amontons in 1699 still form an acceptable and convenient summary of a complicated process of surface interactions. It should also be remembered that Leonardo da Vinci was apparently aware of these laws and the case for some recognition of this fact has already been presented in Chapter 7.

The second feature of note is the didactic approach of Desaguliers (1734). He was plainly concerned with the application of the new-found knowledge on friction to engineering problems of the day. His book demonstrates an earnest desire to instruct as many persons as possible on both the fundamental nature of friction and the methods of taking it into account in machinery.

The emergence of the three important concepts of interlocking rough surfaces, cohesion and the shearing of asperities, in a short period of thirty-five years ensures a special place for the late seventeenth and early eighteenth centuries in the complete history of tribology. Workable rules had been established which were to provide an adequate, if far from complete, understanding of the important phenomenon of friction immediately prior to the Industrial Revolution.

Chapter 9

There is no subject in science, perhaps, on which there is a greater diversity of opinion, than in the laws which govern friction; and the previous experiments, though, perhaps, sufficient in many cases for practical purposes, yet by no means tend to bring the enquiry into any more settled state.
Nicholas Wood (1838)

The Industrial Revolution (c. A.D. 1750–1850)

9.1 Introduction

In many ways this span of time in which man so dramatically applied his ingenuity and energy to the utilization of natural resources for the manufacture and transportation of goods has proved, somewhat surprisingly, to be the most difficult to research from the point of view of the history of tribology. It is an enigmatic period since it is clear that the outstanding mechanical innovations of the time were posing new and urgent tribological questions, and yet, apart from a few well-documented and in some cases classical examples, the threads of both scientific inquiry and practical achievement in our subject are less evident and more confusing than in any previous or subsequent period. The underlying reasons for this state of affairs are both numerous and interesting, making it more important than ever to acquire an appreciation of the general historical, social and technological background of the times. No attempt will be made to present a comprehensive history of the origins or nature of the Industrial Revolution, but the commentary offered in Section 9.2 is intended to form a backcloth for the better appreciation of the tribological progress which was made in this hectic century. Patent specifications have formed a valuable source of historical records for this period, particularly in relation to rolling-element bearings and lubricants.

Studies of friction developed steadily throughout the Industrial Revolution with notable contributions coming from English, French and Russian scientists and engineers. The summit was reached by that most comprehensive and well-known work presented in response to an open competition in 1781 by the outstanding French engineer Charles-Augustin Coulomb. However, the significance of Coulomb's great contribution to the subject should not be permitted to obscure three further important trends which became evident as the history unfolded.

Coulomb was confirming the findings of previous workers on static friction and extending the studies to kinetic and rolling friction in such an authoritative manner that those concerned with the topic in the eighteenth century might well have felt that the subject had been completely unravelled. Yet at the very time when an account of the friction process was emerging in a form which enabled engineers to make reasonable predictions of energy losses in machines, a few lingering doubts about the fundamental nature of interacting surfaces were being bolstered by isolated studies in both tribology and the new science of therodynamics, primarily in the United Kingdom and France.

The second significant trend in eighteenth-century work was the increasing impact of practical men in posing, and in many cases finding solutions to, the problems of a society which was becoming ever more dependent upon the satisfactory functioning of machinery. The professional civil, as opposed to military, engineer was barely recognized until the end of the Industrial Revolution. The natural philosophers of the day were often men who combined great intellect with practical skills, but even the most highly skilled craftsmen in the body of artisans worked without any opportunity to develop an appreciation of the scientific background to their crafts. Nevertheless, the problems encountered by industry increasingly formed the basis for scientific studies elsewhere; the challenge offered by the French Académie des Sciences which led to Coulomb's studies being a good illustration. In due course the need to acquaint engineers with science and to combine the knowledge thus gained with the great practical skills of the craftsman was recognized in the industrial nations of the world, although the manner in which the goal was achieved differed greatly in different countries and still presents significantly different patterns of education and professional training. These questions are of great interest and importance, but in the present text we can do no more than comment that it was in the period of the Industrial Revolution that many problems concerned with the education and training of engineers were highlighted and the foundations of long-standing practices established.

Towards the end of the period now being discussed, the beneficial nature of friction was recognized. Previous studies had been centred upon the effort that had to be exerted to overcome friction, but the advent of the clutch or friction drive in power transmission systems focused attention on a somewhat different aspect of the problem. The exciting development of steam-driven locomotives introduced questions about the tractive effort which could be transmitted from wheel to rail. Vehicles had been running on rails for hundreds of years before the Industrial Revolution, but animals had provided the main source of tractive power. It was now quite a different problem when the wheel had to transmit both the load and the driving force to the rail.

The rolling-element bearing continued to attract attention throughout much of the Industrial Revolution, although many independent forms were in fact very similar. Many were built for horse-drawn carriages and pride of place will be given in section 9.3 to accounts of these fascinating inventions. Some were built for special purposes, like the masterly project of the building of a statue to Peter the Great in St Petersburg (Leningrad), and the support of post-mill structures like Sprowston near Norwich, while others found their way into general machinery. However, the modern ball- and roller-bearing industry was a child rather than a father of the Industrial Revolution, since the satisfactory functioning of such machine elements is exceptionally related to accurate machining, particularly grinding, which itself developed during this period.

Plain bearings, increasingly made of suitable combinations of metals, continued to satisfy the demands of industrialization without major changes, although locomotive and carriage axles posed a problem towards the end of the period which in due course promoted the classical studies of lubrication in the latter half of the nineteenth century. The most significant development was undoubtedly the introduction of the first widely employed bearing alloy by Isaac Babbitt in 1839. This contribution was more than the development of a suitable low-shear-strength alloy of tin, antimony and copper capable of running with either hardened or unhardened journals; it also reflected the rapid progress of the United States into the industrial and hence the tribological arena.

In previous chapters reference has been made to the important development of windmills and water-mills, but little has been said about their construction or their bearings. A section of the present chapter is devoted to these delightful and ingenious devices, since it was in this period that they reached their zenith as the major sources of power for drainage, water-pumps and milling of various kinds. Steam engines were introduced into flour-mills before the end of the eighteenth century and the liberation of such processes from the whims of natural streams of water or currents of wind proved to be an irresistible attraction. Many important improvements were made to both forms of mills long after the introduction of the steam engine and some are still functioning today, even though their contribution to the total demand for mechanical power is infinitesimal. Few people can fail to be intrigued by the aesthetic appeal and the mechanical ingenuity of the traditional windmill, but the author also hopes that the tribologist will share with him the pleasure and interest which results from a study of windmill bearings.

While the immediate achievements in tribology during the Industrial Revolution were undoubtedly attributable to the skills of practical men, important theoretical studies of fluid flow in Europe were providing the foundations for most spectacular progress in the science of lubrication in a subsequent period. The important elements of these studies will therefore be noted in this chapter.

These are the major topics to be discussed, but the reader might well feel that the impressive progress evident in many branches of engineering was not fully reflected in either the science or the art of tribology. In a curious way it appears that man's knowledge of bearings, lubricants and friction was, with a few exceptions, adequate and sufficiently adaptable to cope with the great upheaval in the form of machinery which took place throughout the Industrial Revolution. It is therefore important that we pose the question: what was the Industrial Revolution?

9.2 The Industrial Revolution

Terminology and background

The term 'Industrial Revolution' was first used by nineteenth-century writers and historians to describe the rapid technological, social and economic changes which were initiated in England, but which soon spread to Western Europe and North America. The expression is normally attributed to Arnold Toynbee (1815–83), who also proposed the time span 1760–1840, but the words can be misleading and the period cannot be specified with any precision. The word 'Revolution' is too emotive for the process under consideration. 'Industrial' and energetic it certainly was, but the transformation of England from an agricultural and mercantile nation in the mid-eighteenth century to the state of industrialization which fully justified the title '. . . the workshop of the world. . .' by the time of the Great Exhibition in 1851, though spectacular, was nevertheless 'evolution' rather than 'revolution'. The world did not wake up to the news that an Industrial Revolution had gripped the United Kingdom in 1750, 1760 or any other year, and it is interesting to note that the early writers who coined the term were the commentators on a shorter and more violent revolution which took place in France at the end of the eighteenth century.

Different authors have adopted various dates for the process of industrialization in England. The origins were usually dated close to 1760, but the termination is more obscure. In a sense the industrialization of England continues, while the appropriate dates for the same process in the United States, Germany, Australia, Russia and China are naturally quite different. The allocation of exact dates for an evolutionary process of this nature is less important than an appreciation of its nature and consequences. In this brief account emphasis will be placed upon the strict engineering aspects of the phenomenon, although the economic, social and political elements are exceedingly important and vital to a full appreciation of the period. General historical texts form a good starting-point, but the reader will also find that the books of Armitage (1961), Bernal (1954), Burstall (1963), Kranzberg and Pursell (1967), the two volumes of collected historical articles published by the Institution of Mechanical Engineers under the title *Engineering Heritage* (1963, 1966) and Volume IV of the massive work by Singer et al. (1958) provide much factual information and excellent appreciations of the period.

The textile industry

The progress of industrialization in England between 1750 and 1850 impinged upon most of her productive processes and services. The textile industry was one of the most obviously affected and it provided, for various technical and social reasons, the origins of the factory system at the expense of the eighteenth-century cottage industry. Both spinning and weaving were mechanized and it is often said that it was the imbalance between the rates of production in the two processes which prompted practical men to consider alternative methods of working. The spinning process in the cotton industry showed the earliest response, with notable innovations like Arkwright's water-frame in 1769, Hargreaves' spinning-jenny in the same year, Crompton's mule in 1779 and Thorp's ring spinning-frame in the United States in 1828. Early versions of many of these machines were driven by hand, animal power or by water-mills. The woollen industry had gradually moved to the wetter, higher regions of northern England, and particularly Yorkshire, to take advantage of the availability of streams for processing the cloth and producing power, while the cotton industry had centred round the Lancashire streams on the western side of the Pennines close to Liverpool, the port to which the raw cotton was delivered from America. The innovations in spinning originated in the cotton, and not the longer established woollen industry, probably because the cotton fibre was easier to spin by mechanical means. Arkwright's water-frame spinning-machine was so called because it was driven by a water-wheel and the rapid evolution of spinning equipment was established long before the steam engine became the dominant source of power.

In weaving, the development of a satisfactory power loom was a slower process. Important contributions were made by Cartwright between 1785 and 1792, Johnson in 1803 and Horrocks in 1813. The number of power looms increased by a factor of six[1] in about as many years and by 1850 they had superseded hand weaving for both cotton and worsted cloth in England and the USA. There are few references to tribological problems or advances in the rapidly developing textile industry during the late eighteenth and early nineteenth centuries. The bearings on the newly invented machines were on the whole simple in form and lightly taxed and, in association with the lubricants of the day, were entirely adequate. This state of affairs continued until recent times, when concern for contamination of the product by the lubricant and the lubricant by the environment, the introduction of synthetic fibres, the call for exceptionally high-speed spindles coupled with a reappraisal of the performance of traditional machinery, all combined to exercise the skill of the modern tribologist.

Agriculture

In agriculture the introduction of the portable steam engine through machines like the steam thresher was making its contribution to the very considerable increase in productivity which went hand-in-hand with urbanization and the growth of the factory system. A number of ingenious machines were introduced into the industry without disturbing too greatly the role of traditional farming equipment. A more impressive change can be detected in the materials of construction, with iron implements replacing practically all the working components of tools traditionally made of wood. Once again, however, the bearings were of very simple form.

1. See Julia de L. Nann's account of the development of textile machinery between 1760 and 1850 in Singer, C. et al. (1958), Vol. IV, p. 300.

Communications and transportation

The need for adequate communications and the transportation of goods between the new industrial centres, their sources of raw materials and their customers in the United Kingdom and abroad, was responsible for remarkable feats of construction – first of the roads, then of the canals and finally the railways.

In Chapter 8 some attention was paid to the degree of sophistication of coach and carriage construction, with particular reference to axle bearings. Coaching inns and stations were established in association with a fine network of roads which permitted coaches to change horses and maintain an average speed of 9 or 10 mph (14·5–16 km/h) by 1835. Such speeds were possible only because of advances in roadway construction and the design of better suspensions for vehicles. The excessive width of the rims of wagon wheels used for carrying heavy loads had persisted since Roman times (see Ch. 5) while these dual advances were awaited, with the widths on some wagons rising to 16 in (406 mm) by 1750. In due course the improved roads and vehicles permitted the width of flat rims to fall to 9 in (229 mm) and then to 6 in (152 mm) before the bevelled rim was introduced. The wheelwright was a highly esteemed craftsman who had, over the centuries, employed studs of wood, metal and stone (see Chs. 4 and 6), iron hoops and rims of adequate width to combat wear on the roads of the day. By 1850 both pneumatic and solid rubber tyres had been tried, with but little initial success, by R. W. Thomson and T. Hancock, respectively. In due course they were to be reinvented for both the bicycle and the motor car.

Roads

The evolution of networks of roads was not restricted to England alone, although in this and many other aspects of industrialization, she set the pace. Good roads had been built in Scotland by General Wade after the 1715 Rebellion. Telford (1757–1834), whose name is usually associated with the fine bridge which he built over the Menai Straits, contributed substantially to road-building techniques. The Scot John McAdam (1756–1836) achieved great success in the United Kingdom and North America with the introduction of his impervious road surface. There were two major factors in the secret of sound road construction: the building of adequate foundations and the provision of an impervious surface to keep dry the underlying soil and structure. The origins of the use of asphalt can be traced back to 1712, although its use was mainly restricted to Switzerland and France throughout the century under discussion. Tar was first used in Nottinghamshire in the 1830s and the term 'tar-macadam' is now standard terminology for the material used to seal metalled roads.

The technical improvements in roads were matched by a spectacular increase in mileage in support of military, commercial and industrial developments in the United Kingdom, continental Europe, Russia and North America. Although the first Turnpike Act was passed in 1663 few toll roads were constructed in England until the eighteenth century. The first turnpikes in America were not constructed until 1785, but by the early years of the nineteenth century the canal and then the railway booms slowed down the development of roads on both sides of the Atlantic.

The construction of good metalled roads and the related progress in road carriage design accounts for the numerous references to improved axle-tree bearings in the literature covering the early part of the period of industrialization. Some of these entries are indeed fascinating and important, particularly in relation to the future extensive use of rolling-element bearings.

Canals

The great expansion of inland waterways in response to the demands of industrialization called for little innovation in bearing design, although it promoted, primarily in France, some basic studies of fluid flow and resistance[2] which furthered the science of fluid mechanics and thus the basis for late nineteenth-century studies of lubrication.

The railways

The steam engine finally confirmed the railway as the major mode of transportation for goods. The great railway boom, which was centred on northern England in the first half of the nineteenth century, represented a remarkable corner in the canvas of engineering history. Apart from the mechanical innovations, the civil engineering feats in designing and building bridges, cuttings, embankments and tunnels to keep the gradients low enough to enable the tractive force between wheel and rail to propel the train, were outstanding. The earth-moving and the rate at which it was effected were phenomenal. Lecount (1839) put some meat on the bones of this statement in his account of the construction of the London to Birmingham line. It was, he said, the greatest public work executed in ancient or modern times. The Great Wall of China sank into the shade when compared with the railway, while the building of the Great Pyramid in Egypt involved a smaller total effort and a much lower rate of activity. The 100,000 to 300,000 men who built the Pyramid over a period of twenty years performed work equivalent to the lifting of $15 \cdot 7 \times 10^9$ ft^3 ($0 \cdot 4 \times 10^9$ m^3) of stone through a height of 1 ft ($0 \cdot 3$ m), but the effort of the 20,000 railway navvies working on the London to Birmingham line was equivalent to the raising of 25×10^9 ft^3 ($0 \cdot 71 \times 10^9$ m^3) material through the same height in less than five years. The material removed, if spread in a pile 1 ft ($0 \cdot 3$ m) wide and 1 ft ($0 \cdot 3$ m) high would encompass the earth three times.

The hard, dangerous, but skilful process of cutting the soil and rock, tunnelling and blasting, was undertaken by gangs of men who had earlier been employed in building sea walls, canals and roads. These 'navigators' or 'navvies' as they were soon called were men apart from the casual labourers employed in such operations. Their existence was exceptionally

2. See Tokaty, G. A. (1971).

hard, and a vivid insight into their way of life has been presented by Coleman (1965).

The railway age may be said to have started on the afternoon of Thursday 22 May 1822 when the first rail was laid on the Stockton and Darlington Railway. For several years the carriages were hauled by horses and some prefer to think that the Liverpool to Manchester line opened in 1830 has a substantial claim to this important title in transportation engineering. However, locomotives or no locomotives, both were public lines designed for the transportation of both goods and passengers.

Steam locomotives had been operating on purely industrial tracks for some years before any of the passenger-carrying lines opened. Yorkshire's connection with the railway dates back to 1758 when the first railway was sanctioned by Parliament for the purpose of conveying coal to Leeds from the Belle Isle, Middleton coal-mines owned by Charles Brandling of Middleton Hall. The wagons were originally drawn by horses but steam locomotives were introduced on 24 June 1812. The famous engineer Matthew Murray (1765–1826) was entrusted with the task of building these locomotives. It was Murray, who walked from Stockton-on-Tees to Leeds at the age of twenty-four and who is locally given the title 'Father of Leeds Engineering', who founded the steam engineering industry in the city in the 1790s. John Blenkinsop, agent for the Middleton Collieries, entrusted the design and construction of these early locomotives to Murray in 1810 and it is pleasing to note that public-spirited persons have ensured the continuation of this historic line to the present day.

Most of the early railway engineers were troubled by the idea that a smooth wheel on a smooth rail would not have sufficient traction to enable an engine to pull a heavy weight. Murray shared these doubts and his engines were therefore fitted with a fifth wheel which was cogged and worked in teeth on the outside of one of the rails. It was twenty two years later in 1834 that the more conventional form of railway truly came to Leeds with the opening of the Leeds to Selby line. Trevithick (1771–1833) built and employed a steam locomotive to draw a load of 10 tons (100 kN) over almost 10 miles (16 km) of cast-iron tramway connecting the Pendarren Ironworks with the Glamorganshire canal in 1804, while George Stephenson built his first locomotive for use at Killingworth Colliery near Newcastle upon Tyne in 1813.

The first railway boom took place in the early 1830s with a second one in the mid- and late 1840s. Vast lengths of lines were completed in the United Kingdom and overseas with the English navvy travelling the world in these and later years to participate in the frenzy of construction.

Two major tribological questions arose from this activity. Would the wheels be able to transmit the necessary tractive effort to the track to ensure starting, stopping and steady running?—and how could an adequate supply of lubricant be provided for the bearings of a machine which was itself moving? Static engines could be lubricated by careful and fairly continuous attention from the engineers, but these new vehicles provided an important spur for the development of self-contained lubrication systems.

Most of the major lines of the railway system in the United Kingdom had been completed by 1860 and it can be seen that the steam locomotive played a vital role in this development from the turn of the nineteenth century. This brings us to the outstanding engineering feature of the Industrial Revolution; the large-scale introduction of the steam engine in industry, on the railways and at sea.

Steam power in industry

In a later section of the present chapter wind- and water-driven mills will be discussed, but it is useful to consider in a little more detail the rapid evolution of the machine which ousted these traditional sources of mechanical power. The early developments have already been mentioned in Chapter 8, where it was noted that by the end of the seventeenth century, drainage had become one of the most serious problems in mines. Existing methods based upon simple pumps operated by men, animals or perhaps water-wheels, had become so inadequate that further development of the industry in terms of the depth of mining was in jeopardy.

Galileo's pupil Evangelista Torricelli, had claimed in 1643 that the atmosphere exerted pressure because of its weight. Some years later the value of this concept and the great force which could be created by allowing the differential between this pressure and a vacuum to act on a surface was demonstrated by Otto von Guericke of Magdeburg and Denis Papin, assistant to the Dutch scientist, Christian Huygens. In 1690 Papin condensed steam beneath the piston in a $2\frac{1}{2}$ in (63·5 mm) diameter cylinder and succeeded in raising a mass of 60 lb (27·2 kg) by means of the partial vacuum thus created. A Cornish military engineer Thomas Savery (*c.* 1650–1715) employed a large-scale steam-generated vacuum pump based on this principle in 1698 which became known as 'The miner's friend' following the publication of a pamphlet of this title in 1702. Savery's pump, which found application not only in mines but also for the supply of water to large buildings, had but a modest life, was extremely inefficient and fell into disuse within a decade or so. A similar but more elegant form of condensing engine was later built by Thomas Newcomen (1663–1729), an ironmonger from Dartmouth.[3] Newcomen's visits to the Cornish tin mines in his capacity as a dealer in iron tools brought the drainage problems to his attention and prompted him to devise his pumping engine. After many years of experiments Newcomen constructed an atmospheric steam engine which was erected at Dudley Castle in 1712 for the purpose of draining a coal-mine. The cylinder had a bore of 19 in (483 mm) and a stroke of 6 ft (1·8 m) and the distinguishing feature was a large trunnion-mounted rocking beam connected to the metal piston-rod at one end and the pump-rod at the other by means of chains. The majestic beam engine had emerged, and although Newcomen died in 1729 his engines were still being built on a large scale in 1750. A

3. See Rolt, L. T. C., 'Thomas Newcomen–father of the steam engine', *Engineering Heritage* Semler, E. G., ed. (1963) Vol. 1, Heinemann on behalf of the Institution of Mechanical Engineers, London.

number of these early engines can be seen in the Science Museum, London, while the reader with a general interest in the subject of industrial archaeology will find much to interest him in the beautifully illustrated book by Bracegirdle (1973).

In the early stages of the Industrial Revolution that outstanding engineer John Smeaton,[4] born in the village of Whitkirk near Leeds in 1724 and better known as the builder of the Eddystone Lighthouse, was to apply his skills to the study and improvement of water-wheels, wind-mills and the Newcomen beam engine. In the latter case major improve-ments were effected by improved cylinder-boring techniques and better sealing of the piston by means of rope packing. The original Newcomen engine had a thermal efficiency of about 0·5 per cent, but this had been raised by Smeaton to 1·4 per cent in 1774.

It was in the winter of 1763–64 that the Scottish instrument maker James Watt[5] (1736–1819) was asked to repair a model Newcomen engine in the University of Glasgow. He became aware of the excessive loss of heat through the cylinder walls due to the alternate steam entry and condensation, and after experimenting with other cylinder materials he conceived the idea of a separate condenser, to enable the cylinder to retain a more uniform temperature. The outcome of this novel arrange-ment was a considerable increase in efficiency (to 2·7 per cent) and power output. Watt also established the unit of mechanical power and he coined the term 'horsepower'. He was granted his famous patent in 1769 and after a brief financial link with John Roebuck he began a highly successful association with the entrepreneur and industrialist Matthew Boulton which continued until 1800. The firm of Boulton and Watt produced many fine engines of increasing power and efficiency in their Soho Works in Birmingham and by 1800 almost 500 had been completed.

A further important innovation was the development of the rotative-beam engine[6] in which the pump-rod link was replaced by a connecting-rod-crank or sun-and-planet gear arrangement. Watt further improved his engine by cutting off the steam inlet early in the stroke; the 'expansive' engine of 1792 having an efficiency of 4·5 per cent. Compounding was first successfully applied to beam engines and the Cornish mining engineer Arthur Woolf achieved a thermal efficiency of 7·5 per cent with such engines in 1816.

The early days of the eighteenth century also saw the inevitable but important development of the direct-acting steam engine in which the piston was connected through a cross-head and rod or directly by a rod to the crank mechanism. Henry Maudsley's table engine patented in 1807 confirmed this development. The beam engine, particularly in rotative form, continued to enjoy high popularity throughout the first half of the nineteenth century, but the direct-acting reciprocating engine which poses so many problems for the twentieth-century tribologist had, in

4. See Bowman, G., 'John Smeaton–consulting engineer', *Engineering Heritage* (1966), Vol. 2, pp. 8–12.
5. See Wailes, R., 'James Watt–instrument maker', op. cit. (1963), Vol. 1, pp. 57–61.
6. See Clark, R. H., op. cit. (1963), Vol. 1, pp. 62–9.

essence, arrived. The introduction of high-pressure steam technology by Richard Trevithick (1771–1833)[7] in the early years of the eighteenth century finally raised the thermal efficiency of the steam engine to the dizzy heights of 17 per cent by 1834. The basic development of steam engines for pumping, milling, general industrial purposes, locomotives and in due course steamships was now complete. Furthermore, the development of steam power by practical engineers was being paralleled in the development by others of the science of thermodynamics. In particular, the synthesis of Sadi Carnot's (1824) penetrating interpretation of the functioning of steam engines with Joule's (1884) careful measurement of the mechanical equivalent of heat and the views of Rankine (1881) led to a clear statement of the second law of thermodynamics by Clausius and Lord Kelvin in 1851. The story has been well told by Bradley (1966).

Machine tools and manufacture	The industrial progress between 1750 and 1850 was, not surprisingly, associated with a tremendous development of machine tools. The lathe and the drill, both central to manufacturing processes, were in 1750 made essentially of wood. Existing tools improved greatly, iron was used more extensively and most of the machine tools known today such as milling machines, shapers, planers, boring machines and grinders were developed in this energetic age. At the beginning of the period James Watt found that the diameter of a steam-engine cylinder varied by as much as $\frac{3}{8}$ in (9·5 mm), but by the end of the era the machine tools required for the precision bearing industry of the future were already available. Gilbert[8] has written '. . . In 1775 the machine-tools at the disposal of industry had scarcely advanced beyond those in the Middle Ages: by 1850 the majority of modern machine-tools had been invented.'

The development of machine tools, the introduction of the factory system and the ready availability of prime-movers provided fertile ground for the seeds of mass production. One of the earliest examples was the scheme proposed by Sir Mark Isambard Brunel (1769–1849), father of an even more famous railway engineer and builder of ships and bridges, Isambard Kingdom Brunel (1806–59) (see Rolt, 1957), for the production of pulley-blocks for the Admiralty at Portsmouth. When the plant became fully operational in 1808 the annual output of 130,000 blocks previously achieved by 110 skilled men was undertaken by 10 unskilled men with machinery involving a capital outlay of £54,000. Forty-three machines were driven by a 30 hp (22·4 kW) steam engine through overhead shafting.

The introduction of interchangeable parts became possible with the improved accuracy of manufacture, initially in muskets in Europe by 1785 and independently in the USA through Eli Whitney in 1798.

The education and training of engineers	While the process of industrialization in the United Kingdom and the USA between 1750 and 1850 was carried through by practical men and engineers

7. See Wailes, R., 'Richard Trevithick – engineer and adventurer', *Engineering Heritage* (1963), Vol. 1, pp. 70–4.
8. See Gilbert, K. R., 'Machine tools', in Singer, C. et al. (1958), Vol. IV, pp. 417–41.

who had no education in and little knowledge of science, the period represents a transition from a craft to a science foundation for technology. It is a curious and interesting fact, much discussed by historians, that while England was establishing a dominant position in practical engineering, the foundations for studies in engineering science were being laid across the English Channel in continental Europe.

The famous Ecole Polytechnique was officially established in 1795 after the French Revolution, but others had been opened at the dawn of the Industrial Revolution. The students of engineering science in France received a basic two-year course of instruction in mathematics and mechanics, physics and chemistry. This was followed by courses presented by practising engineers, including men of the stature of Coulomb, on design, structures, construction and machines. The German polytechnics soon followed, with courses containing more emphasis on machine design and the training of engineers for industry. Engineering, as an academic discipline, crept into British universities more slowly and much later. The first chair was created at the University of Glasgow in 1840, but this was soon followed by appointments at both King's and University Colleges, London and at Cambridge in 1875. At this stage there were no universities centred upon the cities of northern England and the midlands which nurtured the Industrial Revolution.

| **Scientific societies and professional institutions** | In England the Royal Society had united those interested in natural philosophy and its applications, but further steps were taken to extend an awareness of science and its potential to the adult population through the creation of the Royal Institution in 1799 and the British Association for the Advancement of Science in 1831. The latter held its first meeting in York, and while many of its sponsors were deeply interested in the application of science to the problems of society, like the prolific inventor now recognized as one of the fathers of aviation Sir George Caley, few would claim that it has played a dominant role in engineering. More important was the creation of the professional engineering institutions. A Society of Engineers had been formed in 1771 which in due course came to be known as 'The Smeatonian' in honour of the undoubted leader of the profession at the time of the Society's inception. The formation of the *Institution of Civil Engineers* was the result of a reaction against the Society. The first meeting of a group of prominent engineers was called by H. R. Palmer in 1818 with William Maudslay in the chair. Palmer's definition of the engineer is of some interest. . . '. . . An engineer is a mediator between the philosopher and the working mechanic, and, like an interpreter between two foreigners, must understand the language of both. . . .' |

The Institution provided a sense of unity among the early nineteenth-century civilian engineers and one of its objectives was '. . . the general advancement of mechanical science'.

A further revolt was to lead to the formation of the *Institution of Mechanical Engineers*; a story told in full by Parsons (1947). A group of railway engineers was watching some locomotive trials near the Lickey incline at Blackwell on the Bristol and Birmingham line one afternoon in 1846. Heavy rain obliged them to take shelter in a platelayer's hut where the conversation turned to the refusal of the Institution of Civil Engineers to admit to membership the doyen of their profession, George Stephenson, unless he submitted '. . . a probationary essay as proof of his capacity as an engineer.' Stephenson understandably refused, and after further informal gatherings the Institution of Mechanical Engineers was founded at a meeting held in the Queen's Hotel, Birmingham on 27 January 1847.

The present-day patterns in engineering education and professional training still reflect their origins during the period of the Industrial Revolution, although the national differences are much reduced. It is against this background of outstanding practical achievement in Britain and the United States, the creation of formal courses of education for engineers in France and Germany, the emergence of the professional engineering institutions and the late but firm acceptance of engineering science as an academic discipline in the United Kingdom that developments in tribology in the century prior to 1850 must be viewed.

9.3 Bearings in the Industrial Revolution

The steady development of both *plain* and *rolling-element* bearings noted in Chapter 8 was again in evidence during the Industrial Revolution. The word 'steady' is the preferred description, since most of the design concepts and materials had already emerged in earlier times. Nevertheless, some long-standing lines of development culminated in this century in notable achievements which can justifiably be identified as important landmarks in the history of bearing development. Achievements like Varlo's ball-bearing for road carriages, the Sprowston Mill ball thrust bearing and the introduction by Isaac Babbitt of a plain bearing material which, in various forms, still supports in an entirely adequate way much of our present-day machinery, may be cited in this category.

The major structural change which influenced bearings was the near-complete replacement of wood by iron in most existing and new forms of machinery. Metal-on-metal bearings thus became the normal rather than the exceptional arrangement, although the use of certain kinds of timber in bearing construction continued long after the Industrial Revolution.

Road carriages continued to provide the spur for improvements in both plain and rolling-element bearings, but towards the end of the period the bearings in locomotives and wagons were to pose increasingly severe problems. The major use of iron in the numerous new machines of the period provided few problems for the designer, since the bearings themselves were not severely loaded, but in due course the new materials of construction were to give rise to the development of special bearing materials like 'babbitt'.

It is convenient at this stage to consider separately the development of *plain* and *rolling-element* bearings, although the latter provide evidence of greater innovation.

Plain bearings

Whereas the *half-bearing* was adequate for many machines in which the predominant load acted vertically downwards, the *split-360° bearing*, introduced initially in Renaissance times and developed during the seventeenth and early eighteenth centuries, became increasingly popular during the Industrial Revolution. The great merit of the split-bearing was that it was adjustable for wear; a feature of immense value to the operators of the new class of factory equipment.

Both wood and simple iron bearings were used and it has already been noted that bronze was assuming a position of some importance in the list of potential bearing materials. The introduction of replaceable bushes before the end of the eighteenth century has been confirmed in an interesting article by West and Rathie (1972). The authors have clearly enjoyed their excursion into industrial archaeology, since their article contains a number of interesting illustrations of bearings from the period under review, together with notes on the locations of machinery containing examples of early bearing forms in the Luton Museum, the Birmingham Museum of Science and Technology, the Science Museum in London, the Manchester Museum of Science and Technology and in a rolling mill in Sheffield. Two of their illustrations are reproduced in Fig. 9.1.[9] The first shows a main drive-shaft bearing on a *rolling mill* at the Little Matlock Works, Loxley Valley, Sheffield which has the distinction of having operated with suet and water as a lubricant from the late eighteenth or early nineteenth century until the present day. The second illustration shows an early nineteenth-century *combination bearing* in which an iron shaft rotated in split-brass halves. This particular bearing was employed on the engine of the P.S. *Comet*, the first commercial steamboat, and is thus dated to 1812. The split-brass bushes were sometimes square as shown in Fig. 9.1(*b*), sometimes hexagonal and later plain cylindrical bushes. Occasionally, these early bushes displayed simple oil holes for manual feeding, but in due course the various forms of continuous lubrication systems discussed later in this chapter became more useful.

Interesting accounts of the use of *dissimilar metals* in bearings and the merits of various lubricants occur in the early texts on steam engines. Galloway (1831) and Treadgold (1838) wrote interesting contemporary accounts of this new source of power: the former indicating that the transition from rope-packing to brass piston rings was connected with the tendency for the former to lead to scoring and presumably greater power loss due to steam leakage. The trunnion-mounted beam of Newcomen's original atmospheric engine was supported in bearings mounted on the wall of the engine house as described by Stuart (1829), while the split-brass bearing construction on a majestic beam engine built for drainage purposes in the low-lying areas near Ely in 1831, which still survives, can be seen in Bracegirdle (1973, plate 29).

---

9. I am indebted to Dr G. H. West for his interest in this work and his ready assistance in providing the original photographs and additional information on these bearings.

Fig. 9.1
Early nineteenth-century journal
bearings.

(a) Main driving shaft on
a late eighteenth- or early
nineteenth-century rolling mill
lubricated by suet and water.
Little Matlock Works, Loxley
Valley, Sheffield.

(b) Split-brass bearing supporting
iron shaft from first commercial
steamboat, PS *Comet*, 1872.

Fig. 9.2
(a) Early nineteenth-century plain footstep bearing intended to support the weight of a vertical shaft. The body of the bearing is a heavy cylindrical casting having a flange on the bottom with four bolt-holes for fixing. The brass is fitted into a square recess in the middle of this casting and it is held in position by soft metal which was melted and run into the space between it and the casting.

(a)

(b)

(b) Early nineteenth-century sliding footstep bearing for the dual purpose of supporting a vertical shaft and disengaging two driving shafts geared together. The cup-shaped bearing brass is fitted into a casting which is arranged to slide in guides in a heavy foundation plate. A pivoted lever is also provided to enable the bearing to be moved sideways in the slides when required. The vertical shaft which rotated in the bearing had on it a spur wheel which was put into or out of gear with another wheel by the use of the sliding mechanism shown. The example was removed from a brewery in Watford.

The two fine illustrations of early nineteenth-century *footstep bearings* shown in Fig. 9.2 serve to remind us that the manufacturers of machines with vertical shafts or spindles had to accommodate thrust as well as radial loads. One of the bearings clearly shows a lubricant supply groove.

Perhaps the most important feature of thoughts about plain bearings in the last half of the eighteenth century was the growing appreciation in the minds of practical men of the advantages of *dissimilar metals* in sliding pairs. The wrought-iron gudgeon pins or cast-iron shafts mounted in bronze split-bearing bushes were dominant in medium to heavily loaded machines. In due course other plain bearing materials were sought and of all the materials available at reasonable cost in the early days of the nineteenth century only *tin* and *lead* emerged as realistic candidates. Neither material on its own was hard enough to support appreciable loads, although their very softness and low melting points meant that they were ideally suited to absorb abrasive dirt or wear particles and to operate for short periods with inadequate lubrication. Alloying presented a possibility of improving the mechanical properties of these potential bearing materials and in the USA in 1839 Isaac Babbitt proposed the addition of *antimony* and *copper* in his famous patent for *tin-based alloys*. A typical alloy of the type patented by Babbitt consisted of about 89 per cent tin, 9 per cent antimony and 2 per cent copper. At a later stage alloys of *lead* containing antimony, tin and possibly small proportions of other metals also found favour and both alloys became known as *babbitts* or *white metals*–a terminology which was retained in the Organisation For Economic Cooperation and Development, *Glossary of Terms and Definitions* (OECD, 1969).

White metal has remained a stalwart bearing material until the present day and it seems to be destined for use for many years to come. Wilcock and Booser (1957) estimated that three-quarters of the plain bearings then in use employed the material. The properties of white metal can be modified by using different proportions of other metals in the alloys; hardness and mechanical strength being increased with the addition of antimony and copper, but only with penalties in the form of a decrease in ductility and hence conformality and an increase in shaft wear rate. The mechanical weakness of traditional babbitts made it necessary to support the layer of bearing material on a more rigid backing of bronze or steel. Initially some babbitt linings were almost $\frac{1}{2}$ in (13 mm) thick, but the thickness was reduced dramatically over the years to about one-hundredth of this value to improve the fatigue characteristics of plain bearings. Useful accounts of the properties of bearing materials and the historical aspects of white metal bearings have been presented by Shaw and Macks (1949, Ch. 11: 466–70), Wilcock and Booser (1957, Ch. 13: 376–9) and Morris, in Braithwaite (1967, Ch. 7: 310–16).

Probably the first scientific study of journal bearings to be published was undertaken towards the end of the century under review by von Pauli (1849) of the Nuremberg carriage works. By this time the relatively poor performance of railway carriage and locomotive bearings compared with that of the longer standing and highly developed road carriage bearing was causing concern in a number of countries affected by the rapid growth of railways. As pointed out by Cameron (1966, Ch. 11: 263–75) in his excellent account of the history of journal bearings, the experiments were conducted by two fitters, Werder and Hävel, on two different 120° partial arc bearings having diameters of 2·36 and 2·64 in (59·9 and 67·1 mm). An important and representative feature of both bearings was the large length to diameter ratio of almost exactly 2.

The object of the experiment was to investigate the influence of bearing material upon friction with a journal speed of 153 rpm, equivalent to a forward speed of the train of 25 mph (40 km/h), and a bearing load of 3,870 lbf (17·2 kN), equivalent to a mean pressure of 515 lbf/in² (3,551 kN/m²). The bearings were lubricated with '. . . normal machine oil . . .' and the best coefficient of friction was obtained with an alloy consisting of 91 per cent tin, 6 per cent copper and 3 per cent zinc. The very low coefficient of friction recorded for this material on the larger bearing was 0·0033, thus indicating that the bearing was operating close to the region of transition from hydrodynamic to mixed lubrication. Evidence to support this view comes from von Pauli's observation that the increase in mean pressure of 36 per cent which accompanied a change to the smaller bearing produced a tenfold increase in friction.

While it is true that the impact of iron as the major material of machine construction and the building of both beam and direct-acting steam engines led to the progress in bearing metals recorded above, it has to be admitted that the Industrial Revolution had little impact upon the form of the simple sleeve or split-bush bearings developed in a previous age. *Carriage-wheel bearings* posed the major problems in the late eighteenth century and there is evidence of great ingenuity and elaboration in their design. Naylor (1965) has drawn attention to a particularly fine array of such bearings in use about 1800, which was shown in Felton's *Treatise on Carriages* (Felton, 1794), and which is reproduced here in Fig. 9.3. An indication that carriage bearings were not always entirely satisfactory has been attributed by Hersey (1966) to Turner (1832), the author of an early textbook on chemistry who wrote that, '. . . The axle-tree of carriages has been burned from this cause, and the sides of ships have taken fire by the rapid descent of the cable.'

This reference to frictional heating is a useful reminder that it was not only the practical achievement of the steam engine but also the development of concepts of the nature of friction which took place against the background of Lavoisier's (1743–94) imponderable element *caloric* rather than the modern science of *thermodynamics*. We shall return to this subject in Section 9.4.

Rolling-element bearings

(a) *General industrial applications with particular reference to carriages for road and rail.*   Important developments in the history of both ball- and roller-bearings undoubtedly took place during the Industrial

Fig. 9.3
Array of carriage bearings in use early in the nineteenth century – reproduced from Vol. 1 of Felton's *Treatise on Carriages*.

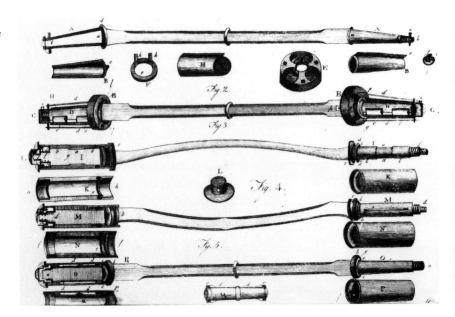

Fig. 9.4
John Ladd's chain-and-sprocket
wheel drive for waggons and
carriages (Ladd, 1757). Brit. Pat.
No. 714.

Revolution, although the early forms of the bearings themselves were but primitive versions of their twentieth-century offspring. Precision and mass production came with grinding, which was itself a product of the development of machine tools in the period under review. The decisive steps had in fact been taken before the end of the eighteenth century, with *free rolling-element* ball- and roller-bearings being made and used for both road carriages and industrial machinery.

Evidence that *friction-wheels*, patented and so enthusiastically described by Jacob Rowe in 1734 (see Ch. 8), were established features of road carriage construction at the opening of this energetic century is provided in British Patent No. 714 of 1757 awarded on 30 April to a Wiltshire surveyor and dealer in timber by the name of John Ladd (Ladd, 1757). The major part of Ladd's specification was concerned with a double reduction *chain and sprocket wheel* system connected to the axles of wagons and carriages and operated from '. . . a common winch, to be turned by a man, as is usually done in draw-wells and malt mills, who with common labour at such winch will have power to move great weights with much greater facility than has hitherto been practiced.'

The construction of the chain itself is not without tribological interest. The account of the chain, presented in association with the illustration shown here as Fig. 9.4 reads

> . . . *which links may be perhaps best joyned together by a smooth well-hardened spindle, rivetted or otherwise made fast at each end, on which all the said links may easily move, and which will operate better and more easily by having a roller of bell metall or other hard metall, running on each spindle or rivet in the middle, and between either side of each link at k, l, m, n, o, in the said Figure 1, by means of which rollers the friction will be greatly lessened. . . .*

Ladd's objective was to eliminate or greatly reduce the number of horses and cattle used to draw '. . . Weights in the common Roads and Highways; . . .' at the expense of '. . . a very moderate and almost inconsiderable Human Force'. He was at pains to point out that his man-powered chain-drive could be used on vehicles with or without friction-wheels on their axles. Indeed he included in his invention '. . . a new method of placing or applying friction wheels . . . to rollers, gudgeons of mill wheels, shafts, spindles, arbours, or other axes or axletrees that move horizontally, which will eventually take off or abate one eighth part at least of the friction . . . and save them much longer from wearing out.'

The fascination of friction-wheels was not restricted to those concerned with road carriages, for in 1766 Alexander Barclay built a *sugar mill* which utilized '. . . A New Method of Constructing Sugar Mills by the Application of Friction Wheels to Diminish the Resistance arising from Friction . . .' which had been '. . . found out and discovered . . .' by John Greenhill Yonge. The pair were awarded British Patent No. 862 on 16 March 1766 (see Yonge, J. G. and Barclay, A. (1766)), but the specification is brief and the details sparse.

One of the most interesting entries in the late eighteenth-century literature on carriage bearings is the delightful little book printed in

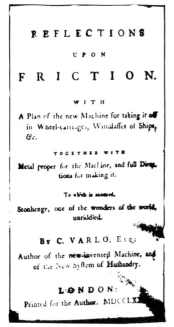

REFLECTIONS

UPON

FRICTION.

WITH

A Plan of the new Machine for taking it off in Wheel-carriages, Windlasses of Ships, &c.

TOGETHER WITH

Metal proper for the Machine, and full Directions for making it.

To which is annexed,

Stonehenge, one of the wonders of the world, unriddled.

By C. VARLO, Esq;

Author of the new-invented Machine, and of the New System of Husbandry.

LONDON:

Printed for the Author. MDCCLXX

The inside Machine for a Post Chaise or not side out for a Waggon

Fig. 9.5
C. Varlo's plan of his new machine for taking off friction in wheel-carriages (Varlo, 1772).

London for C. Varlo[10] (Varlo, 1772). The essence of Varlo's message is the great merit of rolling compared with sliding motion and the development of a bearing containing free, as opposed to trunnion-mounted, iron balls. This was undoubtedly one of the earliest practical ball-bearings and it is interesting to speculate on the factors which promoted the invention. Varlo was aware of the use of 'cannon balls rolling between bars of iron' in the transport of the 900-ton (9 MN) block of stone for the statue of Peter the Great some two years earlier (see subsection (*c*) below). In responding to those sceptics who considered his first advertisement erroneous, he probably overstated the case by proposing the concept of zero resistance to pure rolling between materials of the same hardness provided that the loads were not so immense '. . . as to cause an imperfection'. He recognized cohesion, but rejected the idea that it was involved in the friction process, and he also noted that the grease which reduced sliding friction by smoothing the surfaces could actually increase rolling friction by clogging the balls. Varlo claimed to have '. . . met with some mechanics by chance, that will assert there is a friction in a rolling globe, though it roll upon so hard a surface as not to penetrate; for, say they, were it not so, a ball would roll for ever without stopping'.

A somewhat lengthy discussion eventually led Varlo to attribute the retardation of freely rolling balls to a lack of coincidence of the geometric centre with the centre of gravity of the sphere. He recognized the finite elastic contact area between globes and crudely estimated the diameter of the contact by visual inspection to be about $\frac{1}{16}$ in (1·6 mm) when a load of more than 100 tons (1 MN) was applied to a $1\frac{1}{2}$ in (38 mm) diameter cast-iron ball. His two major claims for the bearing, namely that it could '. . . figuratively speaking . . .' go without either friction or the need for grease, led him to predict that it would '. . . doubtless in time . . . become general, as in every mechanical power it will save at least one third the strength, but in heavy weights much more . . .'. Not a bad prediction of the future of the ball-bearing industry over 200 years ago!

The details of the bearing design and construction are contained in Chapters III, IV and V of Varlo (1772) while Chapter II is devoted to an explanation of the potential for such bearings in preventing the rubbing of anchor cables on wooden ships and in improving the functioning of windlasses. Varlo dismissed any anxiety over excessive heating in rolling bearings by the simple, but enlightened, statement that '. . . where there is no friction, there can be no heat'.

Varlo's own plan of the machine (bearing) is shown in Fig. 9.5. The inner bush, marked B, was located on the non-rotating axle A by means of the octagonal mounting. Originally, while the inner bush was fixed continuously in one orientation relative to the load, on a square-section shaft, Varlo detected a wear groove about $\frac{1}{8}$ in (3 mm) deep after a mere 500 miles (805 km) of running in his post-chaise. The square and eventually

10.   I am grateful to Professor David Tabor of the Cavendish Laboratory, Cambridge, for drawing this work to my attention and kindly making a copy of the book available for study.

octagonal mounting of the inner bush on the stationary shaft facilitated periodic realignment and more even distribution of the wear.

The balls or globes (G) rolled freely between inner and outer bushes (B) and (F) in grooves formed by the side-plates (C) and (E). The outer bush (F) was made in two parts joined by a square bolt (I), rotation within the nave being prevented by the three projections (H). Bearings containing globes of three diameters of 1 in (25·4 mm), $1\frac{1}{4}$ in (31·6 mm) and $1\frac{1}{2}$ in (38·1 mm) were recommended for different duties in ships and carriages and several important practical suggestions are included in the book. For example, it was written that the hardest cast or case-hardened metal with the finest grain should be used for both the globes and bushes. It was further recommended that the side-plates should form a channel such that the lateral clearance with the globes was kept to a minimum, with the number of balls being the maximum that could be accommodated with freedom. It is interesting to note that Varlo attributed such friction as did exist in his bearing to the rubbing of adjacent balls, rather than to rolling friction at the contacts with the bushes.

Varlo's own description of his field trials will be used to close this account and to testify to the success of his invention. The journeys were undertaken from York and I will leave it to the reader to estimate the loading to be attributed to the passengers and their luggage. It is a colourful and impressive passage.

> *. . . In order to prove the machines and to find out wherein the difference of the temper of the metal lay, & induced me to take long journeys in my own post-chaise, which is a remarkably large heavy one.*
>
> *In it were two ladies, all their luggage, and myself, drawn only by two horses.*
>
> *My first journey was from York to Liverpool, taking round-about roads, and back again, through a great part of Derbyshire, to York.*
>
> *In this journey I only had the inside bushes of the hind wheels fixed, where they have been ever since, viz. from the second of July 1772, to the twenty-ninth of October following; and am now on a long journey at Edinburgh; so that they have run at least 700 miles, which is no bad trial.*
>
> *They are not yet wore much above the sixteenth of an inch deep; and the balls are very little smaller, and keep globular.*

Fifteen years after Varlo successfully introduced his ball-bearing into carriage wheels, a merchant from Redland, Gloucester, by the name of John Garnett (Garnett, 1787) introduced the essential features of modern roller-bearings. He was awarded a Patent No. 1580 of 1787 for a bearing containing rollers of various forms. Furthermore, the design shown here in Fig. 9.6 contained an early form of cage or retainer in the form of '. . . polygons of wood, metal or other substance, of any thickness . . .' such that '. . . the rollers, by means of their pivots or holes, will be kept asunder'.

The rollers, whether 'conic' or 'cylindric' or 'spheric', rotated within the space between the inner ring (AA) attached to the axle and the outer ring (BB) located in the wheel. Garnett emphasized the important role of the cage when he wrote '. . . the essential circumstance being the keeping the rollers separate, that is, from touching each other, without their being attached either to the axis or container, and without the encumbent load being borne by the separating principle'.

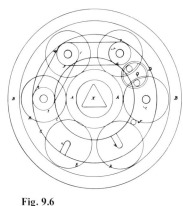

Fig. 9.6
John Garnett's roller-bearing for reducing friction in axles (Garnett, 1787). Brit. Pat. No. 1580. (Note the proposed forms of cages or retainers: (*a*) Slots (n) or holes (y) to locate pivots on the rollers; (*b*) Small separating rollers; (*c*) Bands of leather, silk or hemp connecting adjacent rollers.)

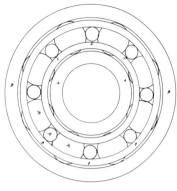

Fig. 9.7
One of Watkin George's proposed
bearings for 'destroying friction in
all kinds of axles and shafts'
(George, 1787). Brit. Pat. No. 1602.
(In a second arrangement more
external cylinders, cones or rollers
were involved.)

A number of suggestions for keeping the rollers apart were proposed and shown in the diagram. Slots (n) or holes (y) in the cage could locate pivots on the rollers, small separating rollers (J) could be used or chains, bands of leather, catgut, silk or hemp could connect adjacent rollers as shown in 1–7.

Garnett's roller-bearing represented a triumph for late eighteenth-century tribology. Allan (1945) has pointed out that a model of the bearing with a spherical surface on the outer ring for self-aligning purposes can be seen in the James Watt Collection of the South Kensington Science Museum, while Davison (1957*b*) has given further details of the apparatus from James Watt's workshop to which the bearing was fitted.

Later in the same year a further roller bearing patent was issued to a Somerset millwright, Mr Watkin George (George, 1787). The proposed designs for roller-bearings capable of '. . . Destroying friction in all kinds of Axles and Shafts . . .' were based upon the extensive use of intermediate contra-rotating rollers in a well-packed bearing as shown in Fig. 9.7.

An iron-founder from Carmarthen by the name of Philip Vaughan (Vaughan, 1794) was the first to patent a ball-bearing of great merit for carrying the loads on '. . . Certain Axle Trees, Axle Arms, and Boxes for Light and Heavy Wheel Carriages'. The drawing reproduced in Fig. 9.8 shows a complete axle with balls running in deep grooves. The balls were loaded at point A, and after packing, the section of the outer race labelled item '4th' was inserted to complete the bearing assembly. It will be noted that, like Varlo's earlier invention, the bearing was devoid of a cage or separator for the balls.

A year after Vaughan's ball bearing was patented James Edgell (Edgell, 1795) of Somerset returned to the concept of friction-wheels in a further patent. The major difference from earlier versions was the axle construction and the use of iron friction-wheels for the support of iron or steel shafts.

The French Army, which had shown increasing interest in the possibility of using rolling-element bearings throughout the 1780s and 1790s,[11] provides the last entry in the eighteenth-century section of our story, for a certain Captain Rebart of the Engineers attempted, apparently with little success, to install roller-bearings in French artillery vehicles in 1798.

An interesting feature of the rolling-element bearings introduced in the very early years of the nineteenth century was the widening range of their applications. Cardinet[12] was granted French Patent No. 263 on 8 June 1802 for rolling contact bearings for a roundabout. The footstep thrust bearing contained tapered rollers running on suitably shaped rings or, alternatively, balls running in grooved thrust washers. In both cases a cage was used to separate the rolling elements. In addition, cylindrical roller-bearings were used to carry the radial load.

Davison (1957*b*) has drawn attention to Horwitz's report of the use of roller-bearings in fast-running, lightly loaded machines like grinding-

---

11. See Allan, R. K. (1945), pp. 11–12.
12. See ibid., p. 12.

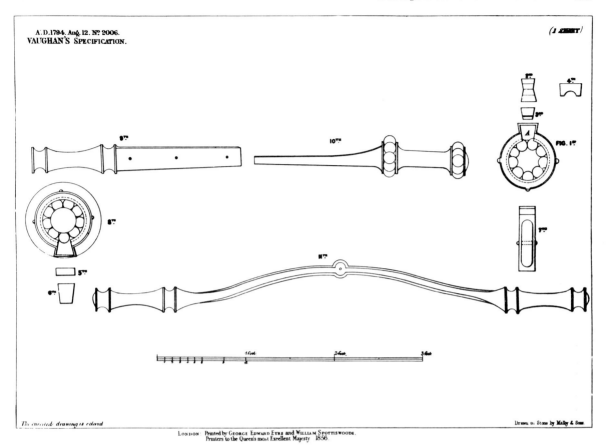

LONDON. Printed by GEORGE EDWARD EYRE and WILLIAM SPOTTISWOODE,
Printers to the Queen's most Excellent Majesty. 1856.

Fig. 9.8

Philip Vaughan's ball-bearing for 'certain axle-trees, axle-arms, and boxes for light and heavy wheel carriages' (Vaughan, 1794).

spindles by 1803. The Scottish civil engineer R. Stevenson, father of the novelist Robert Louis Stevenson, used roller-bearings in the lantern of the Bell Rock Lighthouse and ball-bearings in the cranes used for the lighthouse construction.[13]

The application of friction-wheels to railway carriages was patented by no less a person than a Knight of the Kingdom of Bavaria, Joseph de Baader (De Baader, 1815). The friction-wheels formed a small part of a somewhat curious proposal for either a single- or a narrow-gauge line railway in which flanged wheels ran on flat-topped rails with lateral stability of the vehicle being provided by horizontal or inclined wheels running on the vertical or inclined plates which supported the rail. Friction-wheels were well-established features of mechanical systems by this time, as evidenced by De Baader's view that '. . . friction wheels or rollers, which, being well known to mechanicians, need not be described. . .'.

In this quaint but fascinating patent De Baader suggests such things as long stretches of line of gentle slope so that the carriages will go entirely by themselves (there is no mention of the return journey, but in a later section the specification calls for the raising of carriages by means of an

---

13. See Allan, R. K. (1945), p. 12.

inclined plane after sloping lengths of line of 100 yd (91 m) or so), the possibility of man-propelled systems employing a crank-drive to the wheels, special carts or wagons with platforms to transport carriages from one stretch of line to the next, the use of long pointed poles like those used to propel punts on water, and the use of wind and sails or even nimble-footed dogs for the motive power on light carriages! It was indeed an all-embracing concept.

Diverse and numerous applications of rolling-element bearings were reported in the decade following 1818. A Saxon art master called Brendal designed a hollow winch shaft with a ball footstep bearing in 1818.[14] Furniture castors containing balls to take the thrust load were patented by James Harcourt of Birmingham in 1820 (British Patent No. 4,481) and Allan (1945) has also drawn attention to two patent applications in France by Dentillot in 1821 and Bridgman in 1822 for a most curious scheme involving road wheels and an endless chain to be operated on a principle similar to that of later tracked vehicles.

The concept of a most unusual form of roller-bearing and cage for carriage wheels was outlined in considerable detail in British Patent No. 4709. It was submitted in the joint names of a Middlesex mechanic, John Whitcher, a common carrier of London, Matthew Pickford and a Middlesex coachsmith, James Whitbourn (Whitcher et al., 1822). The idea arose from a genuine concern for the loading and resisting torques in friction-wheel pivots and some apparently erroneous ideas about the epicyclic motion of caged rollers in cylindrical roller or disc bearings. The complex arrangement embraced by this specification is shown in Fig. 9.9, but there is no evidence that it was ever employed by anyone other than Messrs Whitcher, Pickford and Whitbourn.

Joseph Resel, an Austrian inventor of Trieste, was granted a patent in 1829 for roller- and ball-bearings for carriages and machinery and it is interesting that he claimed for his bearings both a reduction in friction and the possibility of running without lubricant. Not all applications for patents for rolling-element bearings were successful. Allan (1945) cites the case of a machine designer named Staubes, who had his specification for a bearing in which hardened steel rollers located by a brass cage revolved between iron bars rejected by the Technical Trade Committee because the members felt that the rollers would become ground so badly as to lose their mobility.

A resident of Douglas in the Isle of Man, Mr Charles Greenway (Greenway, 1840), was awarded British Patent No. 8333 for a wide range of rolling-element bearings with cages to be employed in carriage wheels, or in a large number of industrial applications. He recommended that the components should be of cast iron to produce great hardness and durability. Details of the cages, called cradles, for both journal and thrust bearings are shown in Fig. 9.10(a). Applications mentioned in the specification include carriage-wheel bearings, gun carriages, swing bridges,

---

14. See Allan, R. K. (1945), p. 12.

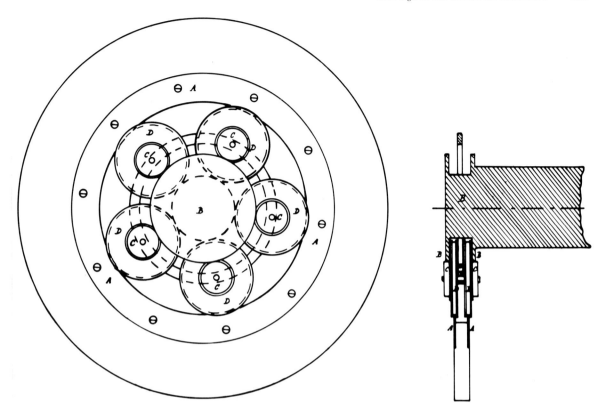

Fig. 9.9
Complex roller-bearing for wheels
of carriages–John Whitcher,
Matthew Pickford and James
Whitbourn (Whitcher et al., 1822).
Brit. Pat. No. 4709.

railway turntables, cranes, ships and slipways, but it is the gun-carriage problem which holds the greatest interest.

Mr Greenway recognized and stated the two different aspects of friction in gun carriages. Low friction was required when the cannon was being manoeuvred into its port-hole or embrasure, but high friction was required on firing, to prevent the gun from recoiling too far. He thus introduced his rolling bearings into the wheels shown in Fig. 9.10(*b*), but arranged for the carriage to run on all wheels only when raised at the front by means of the lever P. When in the firing position the front of the carriage was lowered to make contact with the deck or floor at KK. This simple mechanical arrangement utilized the high sliding friction between wood-on-wood in the firing position, and the merit of low rolling friction for returning the gun to its correct location after recoil.

I may appear to have dwelt for some time on these accounts of rolling-bearing inventions, but if the reader does not share my enthusiasm for the work of the late eighteenth- and early nineteenth-century engineers and tribologists I hope that he will at least recognize the point that all the essential requirements for effective rolling contact bearings had emerged by the end of the eighteenth century. The progress can be attributed to practical men and it is important to note that the bearings were made in embryo form well before adequate grinding and general manufacturing techniques became available. This thesis has been presented in greater detail by Dowson (1975).

Fig. 9.10
(a) Charles Greenway's bearing with cage (DDDDD).
(b) Charles Greenway's scheme for providing maximum friction between gun and deck (on KK) for firing, and minimum friction for running gun out into firing position by relieving contact on KK by means of lever PP.

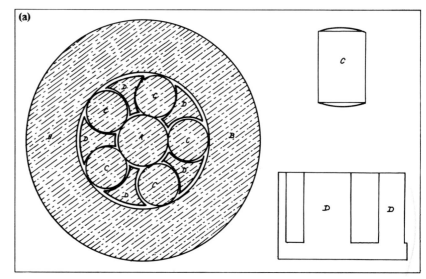

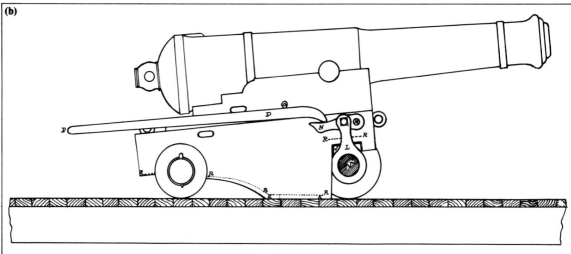

(*b*) *Weather-vanes: An early contribution to tribology from the United States* When the weather vane of Old Trinity Church, Lancaster, Pa., was being renovated in 1909 it was found that the vane had been supported on one of the oldest known roller-bearings in North America. The thrust bearing contained six copper rollers, having diameters of about $1\frac{1}{4}$ in (31·8 mm); running on bronze rings of approximately $5\frac{1}{2}$ in (139·7 mm) diameter. The rollers were pierced to enable them to rotate on brass spindles in a copper cage. The bearing, dated at 1794, has been variously attributed to the marine engineer Robert Fulton, the engraver of the Washington penny and the Great Seal of America, M. Gètz, or the rifle manufacturer Leman.[15]

---

15. See Allan, R. K. (1945), p. 12.

Fig. 9.11
Early roller-bearing (1770?). From Independence Hall, Philadelphia, USA.

A duplicate, and probably older, version of this interesting, and long-lasting bearing was later found to have been fulfilling a similar purpose on top of a building of some consequence in American history. When a young sales engineer of the Fafnir Bearing Company, Mr Charles Haefner, was called to examine a bearing failure in the weather-vane on the roof of Independence Hall, Philadelphia, he found the bearing shown in Fig. 9.11. It is thought that the bearing was constructed in the early 1770s and it established for rolling bearings a respectable reputation for durability by supporting a load of some 750 lb (3·3 kN) for a period of almost two centuries. The story has been reported in the Winter 1968–69 issue of a quarterly publication of the Fafnir Bearing Company known as *The Dragon* (Eaton, 1969), in an article appropriately entitled 'A trip down memory lane'.[16] In the article it is stated that all the components of this historic bearing, which may even pre-date the Declaration of Independence in 1776, were made of bronze. The structural forms of the Lancaster and Philadelphia bearing were, however, identical.

It appears that the eighteenth-century stimulus for rolling-element bearing development was provided mainly by road carriages in Europe and weather-vanes in the United States.

(c) The equestrian monument to Peter the Great – St Petersburg (c. 1769)
The founder of St Petersburg (Leningrad) was the first of three tsars of Russia to bear the name Peter between 1682 and 1762. Peter I (1672–1725),

---

16. I am grateful to Mr Ed Brown, a fellow enthusiast and active researcher into the history of tribology for drawing my attention to this historic bearing and to the Fafnir Bearing Company for kindly providing background information and the photograph reproduced as Fig. 9.11.

Fig. 9.12
Monument to Peter the Great
(1672–1725)–St Petersburg
(Leningrad) (*c.* 1769).

also known as Peter the Great, not only founded the city after driving the Swedes out of the area, but also established the Russian Navy. His grandson, Peter II, enjoyed but a brief reign from 1727 to 1730 before he died at the age of fifteen, while his maternal grandson, Peter III, was murdered at the age of thirty-four after being forced to abdicate six months after his succession in 1762. The murder might well have been carried out on the orders of his wife, the former Princess Sophia of Germany, who was subsequently acclaimed as Catherine II, Empress of Russia.

Catherine decided at an early stage in her thirty-four-year reign to create a statue to Peter the Great in the city which he had founded. The sculptor Etienne-Maurice Falconet, who was commissioned to execute the work, spent a full twelve years on the task before the unveiling in the summer of 1782. His desire to depict '. . . a living, pulsing, passionate nature . . .'[17] was to yield the fine equestrian statue depicted in Fig. 9.12. The statue shows the horse rearing and thrashing a snake with its foot, but restrained by its powerful rider, at the summit of a cliff. It overlooks the wide river Neva from a fine square, originally named Senate, but later

17. I am indebted to the USSR Novosti Press Agency and Mrs Sadie Alford, picture librarian of their London Office for descriptive material and photographs of the statue to Peter the Great in Leningrad.

renamed Decembrist's in honour of the 1825 uprising. The statue is important not only as a fine example of massive sculpture and a significant aspect of Russian history, but as a little-known tribute to the tribological skill of Count Marin Carburi.

Allan (1945) has referenced a full account to this enterprise by Ivan T. Aminoff, but in January 1972 in Cleveland, Ohio, Mr E. D. Brown, Jr drew my attention to a copy of a book in the New York Public Library by Count Carburi himself (Carburi, 1777). This fascinating text provided the basis[18] of our account of the transportation of a single granite block weighing some 3 million lbf (13·3 MN) from Finland to St Petersburg to form both the pedestal and the statue to Peter the Great.

Count Marin Carburi of Cessalonia, who was at one time a lieutenant-colonel in the service of Catherine II, later became a lieutenant of police and censor in charge of the noble body of the sons of the nobility of St Petersburg. He was known in Russia as the Chevalier de Lascary . . . but the '. . . Advertisement . . .' in his book published in Paris in 1777 openly declares this to be a false title adopted '. . . after a shameful act of violence in his youth . . .'. Catherine's pardon of this youthful misdemeanour and Carburi's desire to salve his conscience in the two-page advertisement not only add interest to our story, but enable us to know the true family name of our author. It also resulted in an interesting two-page eulogy to the Empress of all the Russias in an early section of the book.

The sculptor, Falconet, built a model to show how several large rocks could be joined together with iron or copper to provide a pedestal of suitable proportions for the statue to one so great, but Carburi saw disadvantages and enthusiastically proposed the use of a single large stone for the monument. Lieutenant Betski, Supervisor of Buildings and Art, at first told the Senate that such a scheme was impossible, but in due course his former aide, Count Carburi, in whom he had great confidence, persuaded him that it could be done.

On the advice of a peasant, Carburi pursued his search for a suitable stone by taking a journey to Finland to see a moss-covered block of granite lodged in marshy ground and surrounded by birch and pine trees. It was 42 ft (12·8 m) long, 27 ft (8·2 m) wide and 21 ft (6·4 m) high, but a mere 6 ft (1·8 m) of the rock was visible above the ground. The rock was located near a bay in the Gulf of Finland, some 6 versts[19] from the water and 20 versts from the town. It was at this stage that the question of transporting the rock, which even after preliminary cutting on the site weighed some 3 million lbf (13.3 MN), across the deep marsh and streams, the river Neva and/or the sea, ultimately yielded the elegant solution which justifies an honourable place for late eighteenth-century Russia in our history of tribology. The reader might care to note that the rock in question was more than twenty times heavier than the stone colossus moved over less troublesome terrain at El-Bersheh *c.* 1880 B.C.

---

18. I am pleased to acknowledge the translation of this work and helpful interpretations by Mrs Juliet White.
19. The verst was a unit of length equivalent to 3,500 English feet or about two-thirds of an English mile and just slightly greater than a kilometre.

During the winter of 1768/69 preparatory work in the form of clearing a swath through the forest from the site to the banks of the river Neva, digging the earth from around the rock, driving piles into the earth and eventually tipping the rock on to its side was undertaken by about 400 men. The forest must have vibrated with activity and in a note of gratitude to the workmen Carburi wrote '. . . nearly all Russian countrymen and soldiers are carpenters, which was a great help in our work, they are so skilled with a hatchet that there is no work of carpentry that they cannot do with this alone and a chisel . . .'.

These simple feats, carried through under very difficult conditions, involved the use of capstans and massive levers some 65 ft (19·8 m) long and 15–18 in (0·38–0·46 m) in diameter, cut from the forest. The rock was finally lowered on to a bed of hay and moss whose thickness was reduced from 6 ft (1·8 m) to a compact mass of as many inches. By the time this part of the operation had been completed in March 1769 the deep frost was yielding and the earth became too soft to support the heavy loads. The summer was thus devoted to the building of a mole of considerable strength out of the forest path into the river Neva. In due course this mole, which had to withstand the impact of enormous ice-floes, was to be used to pass the rock into a specially constructed vessel.

The work of moving the rock began in earnest as the winter of 1769/70 gripped the land. The system adopted was beautiful in its simplicity; a triumph for mechanics and tribology. The rock sat on a hefty wooden chassis containing longitudinal grooves which was allowed to rest upon similarly grooved beams on the ground. Within the grooves were placed 5 in (127 mm) diameter balls to form a 'free' linear ball-bearing as shown in Fig. 9.13(*a*). The chassis was 42 ft (12·8 m) long and 17 ft (5·2 m) wide; the longitudinal beams (marked cc in Fig. 9) being 18 in (0·46 m) wide and 16 in (0·41 m) deep, with each of the four cross-beams being 12 in (0·3 m) square. Six of the lower beams or rails were constructed, each 33 ft (10 m) long, 14 in (0·36 m) wide by 12 in (0·25 m) high and as each became free behind the rock, six men dragged them to the front to be used again, in much the same way as the Egyptians had used planks at El-Bersheh about 1800 B.C. and the Assyrians had used logs at Kouyunjik in 700 B.C. (see Ch. 4). The grooved beams were lined with 2 in (50·8 mm) thick copper channels as shown in section in Fig. 9.13(*b*). The disadvantages of rollers compared with balls were fully outlined by Carburi and it is interesting to note that the material eventually selected for the rolling spheres was the alloy brass to which some tin or pewter had been added. In earlier experiments Carburi found that ordinary iron balls broke or were flattened by the great weight, while cannon-ball-like cast-iron spheres splintered into several pieces. Only copper mixed with a little tin and calamine provided a satisfactory alloy, the success being attributed to the homogeneity of the parts of copper. The balls were placed at intervals of about 2 ft (0·6 m), making a total of thirty or thirty-two in use at any one time, but it is clear from the account that the load sharing was uneven. Indeed, at times some balls became entirely free and it was necessary to employ seven men on each side to keep the balls apart by means of iron rods. Even with perfect load sharing each ball would be subjected to a force

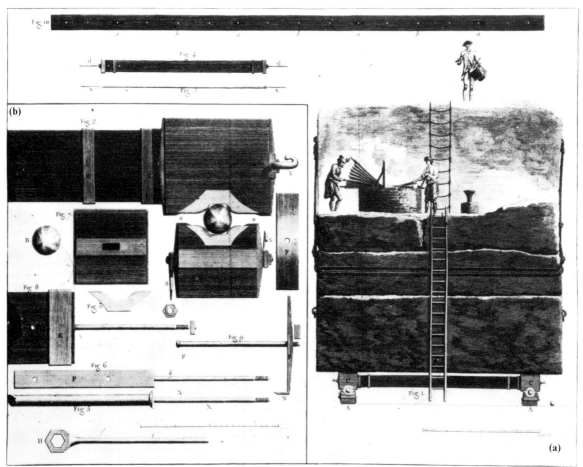

Fig. 9.13
Count Marin Carburi's system for transporting on linear rolling bearings the granite block for the equestrian monument to Peter the Great (*c.* 1770).
(a) Granite rock, forge and drummer on wooden chasis running on 5-in (127-mm) diameter spheres in linear bearings.
(b) Section of linear bearing showing grooves lined with copper and the spheres of copper and tin.

of about 10^5 lbf (0·44 MN). The maximum Hertzian compressive stress between a sphere and a flat surface of similar material is given by the expression $0·616(WE^2/d^2)^{1/3}$, and since the elastic modulus (E) for the copper alloy in question would be about 15×10^6 lb/in^2 (103 GN/m^2), each sphere would theoretically experience a contact stress of about $5·9 \times 10^5$ lbf/in^2 (4 GN/m^2). This exceeds the yield stress for the material by a factor of at least 10 and it is thus clear that permanent flattening of the balls would occur.

The complete system of transportation is illustrated in Fig. 9.14, where the capstan and towing cables are clearly seen on the left and a group of workmen can be seen bringing one of the free tracks from the rear to the front of the rock. A single workman responsible for keeping the spheres apart is shown on one of the seven pieces of matting towed along with the rock, while one of the sledges carrying a hut for stores can be seen bringing up the rear. Two further remarkable features of the process can be seen on top of the rock; one of two drummers mounted on top for synchronizing commands and, at the front, a complete forge for the day-to-day manufacture of miscellaneous devices and tools. The previous

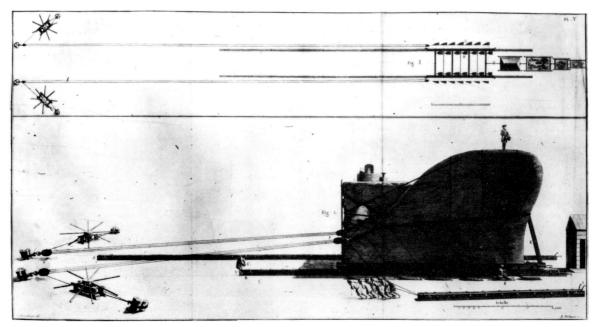

Fig. 9.14
Count Marin Carburi's arrangement of capstans and towing cables for the transporting of the large granite block on a linear roller-bearing (*c.* 1770).

arithmetic also suggests that the forge would be fully occupied in restoring the copper-alloy spheres to their spherical forms.

Two capstans each operated by thirty-two men were adequate for moving the rock over level ground, but four and sometimes six capstans were required for sloping ground. It was not possible to move the rock in a straight line from its original location to the banks of the Neva owing to the nature of the marshy ground, so Carburi built a 12 ft (3·6 m) diameter turntable in which fifteen cannon-balls ran in copper-lined grooves as shown in Fig. 9.15. In this truly remarkable early thrust bearing equal load sharing would ensure a load of 2×10^5 lbf (0·9 MN) on each ball. The turntable was employed on several occasions and during the winter of 1769/70 the rock was moved a distance of about 4 miles (6·4 km) over difficult terrain in a mere six weeks.

The rock was floated down the Neva between two guide-ships on a specially built barque constructed by the Admiralty. The loading and unloading were not without incident, but nothing further of tribological note appears to have occurred during this formidable exercise. Perhaps the most suitable commentary on a remarkable episode in the history of tribology was provided by the sceptics and the jokers, when they described the arrangement employed in the transportation of the massive pedestal for the monument to Peter the Great as '. . . the mountain on top of the egg'.

9.4   Further studies of friction

Sliding friction

It might well have been thought that the extensive studies carried out during the late seventeenth and early eighteenth centuries would have

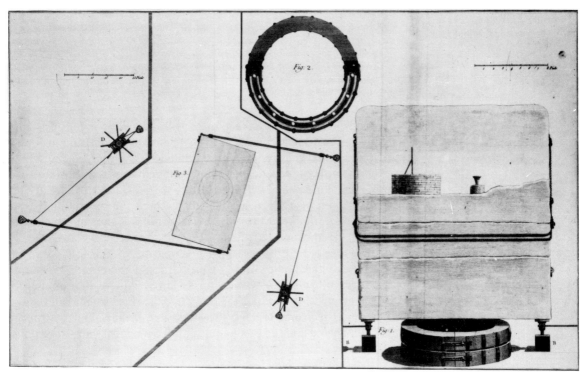

Fig. 9.15
Count Marin Carburi's 12-ft
(3·6-m) diameter turntable for
turning the granite block on
fifteen cannon-balls running in
copper-lined grooves (*c.* 1770).

exposed the true nature of such a commonplace phenomenon as friction, but it is, alas, as elusive and difficult to explain as it is easy to recognize. It is, in fact, a subject which exercises the minds of many present-day scientists and engineers.

The period opened with Johann Andreas von Segner's (1758) studies of friction. He is widely believed to have been the first person to draw a clear distinction between static and dynamic friction; a topic which formed a special feature of Coulomb's (1785) detailed experimental study.

The concept of friction established in Europe, and in France in particular, in the seventeenth and eighteenth centuries, dominated the thinking and teaching of most workers during the period of the Industrial Revolution. The writings of Richard Helsham[20] (Helsham, 1767) who was Professor of Physics and Natural Philosophy in the University of Dublin, exemplify this view. Lecture IX of the fourth edition of the book published after his death was entitled 'Of friction'–a title which is itself indicative of the significance attached to the subject by natural philosophers of the day.

The mechanical view of friction based upon opposing asperities riding over each other had great appeal and merit. It was beautiful in its simplicity, yet concordant with the laws of friction. Helsham explained it thus:

---

20. I am grateful to Professor C. Brockley of the University of British Columbia, for drawing my attention to the work and for permitting me to consult his personal copy of the text.

. . . When a body is moved upon a plane, the prominent parts of the body and plane must of necessity fall into each others' cavities, and thereby create a resistance to the motion of the body, inasmuch as the body cannot be moved unless the prominent parts thereof be continually raised above the prominent parts of the surface whereon it slides; and this cannot be done unless the whole body be at the same time lifted up, and as it were raised on an inclined plane equal in height to the forementioned protuberant parts; upon which account the moving power must sustain some part of the weight of the body, even in moving it along an horizontal plane.

Professor Helsham notes that there will always be some resistance to sliding since all surfaces are '. . . in some measure rough and unequal'. He further argues that the resistance offered by a given roughness to a given weight will be independent of the '. . . magnitude of the surfaces . . .' since '. . . the same weight will ever require the same force to raise it over the prominences of a given height, whatever be the magnitude of the surface whereon the weight rests . . .'. The concept is thus in accord with Amontons' second law.

He then argues that the resistance arising from friction will vary with the weights of the bodies, and be proportional thereto: '. . . for if a certain force be sufficient to raise a certain weight over prominences of a given height, it is manifest that a double or triple weight will require a double or triple force to raise it to the same height.' Once again the logic is convincing and the predictions in full agreement with Amontons' first law.

In a later section of the lecture devoted to wheel carriages,[21] Helsham refers to *friction-wheels* as late contrivances for the reduction of carriage bearing friction '. . . Whereby the axle, contrary to what is usual in most carriages, is made to revolve, and its arms, instead of pressing against the boxes, are made to bear on the circumferences of moveable wheels. . . .' This passage in a series of lectures printed in 1767 shows that the use of friction-wheels on carriages was quite well known, although the full extent of their usage was not established.

Two years before Count Carburi started to move the massive rock which was to form the pedestal for Peter the Great's equestrian statue in St Petersburg, Euler had returned to the city on the invitation of Catherine II. Euler had long since published his papers on friction in Berlin, and although he became totally blind soon after his return to St Petersburg he would no doubt have followed the exploits of Count Carburi with great interest. One of Euler's pupils, who in due course also turned his attention to the study of friction, was Semen Kirilovich Kotel'nikov[22] (1723–1806). He entered the faculty of higher mathematics of the Russian Academy of Sciences at the age of thirty-one and in 1761 Lomonosov promoted him as a member of the Chancellery of the Academy.

Kotel'nikov also lived in St Petersburg while Count Carburi was effortlessly moving his large rock to the city from Finland. Indeed, while the sculptor Falconet was busily working on the monument, Kotel'nikov (1774) published a book in St Petersburg dealing with the mechanics of

21.   See Helsham, R. (1767), Lecture IX, p. 142.
22.   See Kragelskii, I. V. and Shchedrov, V. S. (1956), pp. 33–8 for an account of Kotel'nikov's work on friction.

equilibrium and the movement of bodies. With Euler, Carburi and Kotel'nikov in the city, St Petersburg laid claim to a noteworthy role in the development of tribology, *c.* 1770.

Bowden and Tabor (1964) have described Kotel'nikov as the first serious Russian worker on friction. It is clear from his writings that his views on the role of asperity interactions were similar to those of earlier mathematicians like Euler and Belidor. The mechanistic view thus prevailed in London, Paris, Berlin and St Petersburg. Kotel'nikov did, however, identify the resistance arising from friction with a form of energy dissipation. His work added little to the established concepts of friction, but he clearly attempted to draw attention to the need to consider the phenomenon when studying machines. He noted that,[23] '... All those who have tried to experiment with friction show that it does not depend on the size of the body, nor its shape, nor on the area of the mutually rubbing surfaces; but only on the forces pressing on the body.'[24]

He then went on to point out that since Amontons declared that the force of friction was always 0·33 of the normal force, while Bilfinger preferred 0·25, it could be assumed that friction would always lie between these limits. He considered the discrepancy to be neither serious nor surprising and he even proposed a compromise of 0·29!

Kotel'nikov considered friction to be due to interlocking asperities and in a short but interesting section he clearly introduced the concept of a 'coefficient of friction', without apparently using this terminology. He wrote: '... If we denote the friction content F and the applied force P as unknowns, in the ratio $\mu:1$, then friction $F = \mu P$.'

Before leaving the last of our eighteenth-century tribologists in St Petersburg, it is worth noting that Kotel'nikov also studied the effect of friction upon general bearing performance. He drew attention to the displacement of the resultant force when friction was present, a concept with important implications in the study of both journal bearings and rolling friction. Kragel'skii and Shchedrov (1956) note in the section relating to Kotel'nikov that exceptionally large thrust bearings were constructed and applied, and since his writing was essentially contemporary with Carburi's exploits, it would be interesting to know if the two were acquainted.

The most comprehensive study of friction during this period was undoubtedly undertaken by Charles Augustin Coulomb (1736–1806). Like many of the great contributors to science, Coulomb was an all-rounder. He is perhaps best known for his contributions in the fields of strength of materials, electricity and magnetism, but to tribologists his name is practically synonymous with friction. Some think of him as a physicist, others as an engineer, but a brief summary shows that he achieved recognition as an outstanding military engineer long before embarking upon his great contributions to physics. At an even earlier stage, when his mother was anxious for him to study medicine, he chose to devote his time to mathematics. A more complete account of the life

---

23. See Kotel'nikov, S. (1774), Ch. xiv, pp. 161–5.
24. I am grateful to Mr V. H. Brix for his translation of this text.

of this great man is presented in the Appendix, Section A.5, but the interested reader is also recommended to turn to Professor Stewart Gillmor's (1971) most excellent book.

The Corps des ingénieurs du génie militaire, founded in 1675, had established at Mézières an Ecole du génie which developed a reputation as the best technical school in Europe by the time Coulomb reached his decision to train as an engineer. The creation of formal courses of instruction in engineering encouraged the preparation of comprehensive texts, of which Bernard Forest de Bélidor's *L'architecture hydraulique* (Bélidor, 1737) discussed in Chapter 8 was an example, and generally promoted the reputation of France in technical education and training. Coulomb entered the school at Mézières in February 1760 and graduated with the rank of *lieutenent en premier* in November 1761. After less than two years mapping coastal works between Brest and La Rochelle, Coulomb was posted to Martinique, where he undertook the largest constructional project of his career in the form of the grand Fort Bourbon on Mount Garnier. It was only after his return to France in 1772 that Coulomb was able to devote some time to writing and to change the emphasis of his work from engineering to physics. This intensely practical background, supported by mathematical skill, was the foundation of Coulomb's scientific studies. There was, no doubt, a backlog of basic questions resulting from his engineering experience which formed a reservoir for future work, but he responded with great success to the open contests promoted by the Paris Academy of Sciences.

The Academy announced in 1773 that it would offer a prize for the best contribution to the question of constructing magnetic compasses. There was considerable interest at the time in the use of magnetic compasses for navigation and for scientific studies of the earth's magnetism. No award was made in 1775, but Coulomb shared with Van Swinden the doubled prize in 1777. This success is of interest not only because it established Coulomb in French scientific circles, providing as it did a major impetus towards his election to the Academy, but also because it focused his attention on subjects like friction, torsion, electricity and magnetism, which were to prove such fruitful fields of study in later years. Prior to this time magnetic needles had been mounted on pivot bearings, but Coulomb discussed pivot friction and then described his new magnetic torsion balance in a memoir entitled, 'Investigations of the best method of making magnetic needles' (Coulomb, 1780). Friction was such an important aspect of magnetic compass behaviour that Coulomb's improved low-friction device was adopted by the Paris Observatory in 1780.

France and Britain were in competition for colonies, sea power and world trade throughout much of the eighteenth century. Halfway through the century Britain had a marked superiority in both the quality and number of her ships. The French Government understandably encouraged the expanding machine-building industry, the design of precision instruments and the construction of ships in order to counteract the superior experience of the British by improved design and the application of scientific knowledge to practical problems. The general background has been described by Gillmor (1971) and by Courtel and Tichvinsky (1963).

THÉORIE
DES
MACHINES SIMPLES,

*En ayant égard au frottement de leurs parties,
et a la roideur des Cordages.*

Pièce qui a remporté le Prix double de l'Académie des Sciences
pour l'année 1781.

La Raison a tant de formes, que nous ne sçavons à laquelle nous prendre,
l'Expérience n'en a pas moins.
Essai de Montaigne. Liv. III, ch. 13.

*Par M. Coulomb, Chevalier de l'Ordre de Saint Louis,
Capitaine en premier au Corps Royal du Génie, pour lors
Correspondant, & depuis Membre de l'Académie des
Sciences.*

Tome X. x

Coulomb's study of friction is a good example of high-quality scientific work prompted by practical problems of the day. The Navy was acutely aware of the influence of friction upon the operation of pulleys, capstans and the launching of ships on slipways. Tallow was widely used as a lubricant in the launching of ships, but Coulomb (1785) himself noted that excessive friction would often cause the ship to stick and even fall over when halfway down the slipway. In certain cases frictional heating caused the ship to catch fire, while decomposition of the tallow made it more like an adhesive than a lubricant. It was by no means easy to persuade workers to repair the damaged hull and slipway under these circumstances, and even more difficult to convince a crew that the hull had not been sprung. Gillmor (1971) has recounted the story of the ill-fated British ship, *Felicity*, which stuck halfway down the slipway, strained the hull and sank two months after launching with the loss of all hands.

In 1777 the Academy of Sciences announced that the prize of 1,000 *louis d'or* for 1779 would be offered for work on friction. In particular the Academy sought to encourage work which could be used in the calculation and design of machines and it required that '. . . the laws of friction and the effects resulting from the stiffness of ropes be determined from new experiments made on a large scale; it required as well that the experiments should be applicable to machines used in marine applications such as the pulley, the capstan and the inclined plane. . . .'

No winner was announced in 1779, but the Academy doubled the prize and reopened the competition for 1781. Coulomb had been posted to the arsenal at Rochefort on the west coast of France in 1779, and it was here, in the shipyards, that he carried out his experiments on friction. He had good facilities, the active support of the port commander, Latouche-Tréville, and the services of two men. He submitted his now classical friction memoir 'Théorie des machines simples' during 1780 and was judged the winner in the spring of 1781. This was his second consecutive success in the naval prize competitions and it ensured his election to the Academy in December 1781.

It is interesting to note that his entry defeated a paper by Lazare Carnot, father of Sadi, who had also submitted an entry in 1779.

Coulomb's (1785) friction memoir is a long and detailed essay which exemplifies his devotion to the scientific method and his attention to detail. In the introduction he outlines the current state of knowledge, with reference to Amontons' (1699) initial concepts and some later experimental indications of contrary behaviour in terms of time of repose, area of contact effects and the role of cohesion by de Camus (1724), Desaguliers (1734) and Musschenbroek (1762).[25] The latter attributed friction to surface asperities in these words, '. . . Le Frottement naît de l'aspérité des surfaces, qui sont remplit de petites monticules, et de petites cavités. . . les petites eminences de chacune de ces surfaces s'engage dans les cavités l'autre.'

Musschenbroek argued that sliding would be accompanied either

---

25. A valuable appraisal of these contributions to the science of friction and the physical studies of Coulomb has been presented by Sharp, G. R. (1934).

Fig. 9.16
C. A. Coulomb's apparatus for the
study of sliding friction
(Coulomb, 1785).

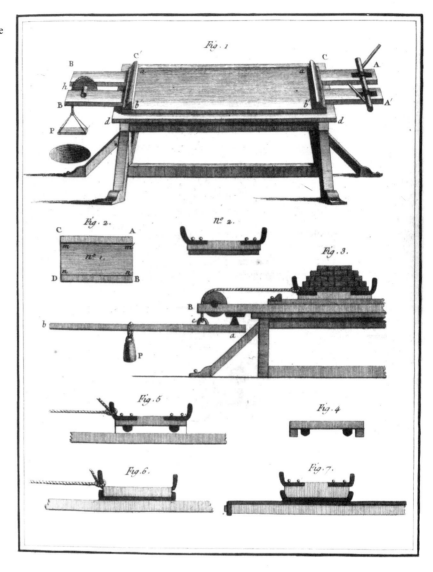

by a rising motion of one surface over the other, or by rupture of the
asperities. He was pessimistic about ever establishing a general law
of friction, yet many of his ideas are reflected in Coulomb's more detailed
study. Coulomb was critical of the neglect of deformation in earlier
studies of friction and he acknowledged his debt to a former professor,
M. l'Abbé Bossut, with this reference to his writings '. . . dans son
excellent *Traité de Mécanique*'.

The memoir is divided into two parts; the first dealing with the friction
of sliding plane surfaces and the second with the stiffness of ropes and
friction in rotating parts. His apparatus for the study of sliding friction
between plane surfaces is shown in Fig. 9.16. *Une table très solide* of
substantial proportions was carefully prepared to form the lower surface
of the sliding pair. *Traineau* or sledges of width 18 in (0·46 m) and various
lengths were connected either directly to a weight or indirectly via a

lever to a weight by means of a rope which passed over a 1 ft (0·3 m) diameter guaiac wood pulley mounted on a green oak shaft. Strips of various widths were nailed to the underside of the sledge as shown in Coulomb's fig. 2 and weights of various magnitudes were added to the sledges, as shown in his fig. 3 to investigate the effect of apparent contact area and load, respectively. The materials tested, in various combinations, were *oak*, *green oak*, *guaiac wood*, *fir* and *elm* together with *iron* and *yellow copper*. Both *dry* and *lubricated* conditions were studied, the lubricants being *water*, *olive oil*, *tallow*, *axle grease* and *soot*, Both *smooth* and *rough surfaces* were investigated and the *maximum speeds* and *pressures* of 11 mph (4·8 m/s) and 300 atm (30·4 MN/m^2) adequately covered the range of practical conditions. Periods of repose ranging from 0·5 s to almost 4 days provided data of considerable importance in the development of Coulomb's theory of friction.

Coulomb set out to investigate the influence of four main factors upon friction:
1. The nature of the materials in contact and their surface coatings.
2. The extent of the surface area.
3. The normal pressure (or load).
4. The length of time that the surfaces remained in contact
   (time of repose).

In later studies he considered the influence of ambient conditions like temperature, humidity and even vacuum. He was at pains to explain that the two major hypotheses which had been introduced by earlier workers to explain friction were related to asperity interactions (Amontons, 1699) and cohesion (Desaguliers, 1734). It is important to note that at the time these two concepts appeared to be mutually exclusive, since it was thought that cohesive forces would increase in proportion to the size of the surfaces in contact, while Amontons and others had demonstrated that the mechanistic view of resistance arising from asperities riding over each other did not require the sliding force to vary with the size of the surfaces. His determination to resolve this point on the basis of experimental findings was demonstrated when he wrote '. . . l'experience seule pourra nous décider sur la réalité de ces différentes causes.'

A wide range of friction experiments were reported and the results recorded in great detail. A discussion followed each major account of experiments and Coulomb then presented representative values of the coefficient of friction. He found, for example, that under dry conditions the friction between unlubricated wooden surfaces reached a constant value after periods of repose of one or two minutes and that the typical values were as follows:

Materials	Weight/Friction
Oak sliding on oak	2·34
Elm sliding on elm	2·18
Pine sliding on pine	1·78
Oak sliding on pine	1·50

Table 9.1

	T (time of repose, min)	$A + mT^\mu$ (static friction force, lbf)	mT^μ
I^{cre} observation	0	$A = 502$	0
II^c	2	790	288
III^c	4	866	364
IV^c	9	925	423
V^c	26	1,036	534
VI^c	60	1,186	684
VII^c	960	1,535	1,033

By presenting such data Coulomb exposed his essentially pragmatic approach and his desire to provide usable data. However, he went much further than this since he used his experimental results to construct empirical equations relating the force of friction to the variables listed earlier. The procedure is well illustrated by the report of his study of the influence of time of repose upon the static friction of two pieces of well-worn oak lubricated with tallow. His results were presented in tabular form as in Table 9.1.

It was apparent that the observed values of friction force could be represented with good accuracy by an equation of the form

$$F = A + mT^\mu \qquad \qquad \ldots [9.1]$$

–and Coulomb found that the constants were given by

$$A = 502, \quad m \approx 2700, \quad \mu \approx 0\cdot2.$$

A major disadvantage of this relationship is that it predicts very large friction forces as the time of repose extends indefinitely beyond the range of experimental conditions. Coulomb therefore found it expedient to suggest an alternative relationship of the form

$$F = \frac{A + mT^\mu}{C + T^\mu} \qquad \qquad \ldots [9.2]$$

–from which it can be seen that A/C represents the initial friction at zero time and that m indicates the static friction after exceedingly long periods of repose.

We will return later to Coulomb's interesting explanation of this significant effect of time of repose upon the friction of fibrous materials, but it is convenient to consider two further important observations at this stage.

Coulomb found that in most cases friction was *almost* proportional to load and independent of the size of the contacting surfaces. He concluded that *cohesion* had a very small influence upon friction, but he nevertheless retained a small constant (A) in the expression he used to relate friction to normal force. In the general case of a body of weight (P) resting upon an inclined plane of slope (n) to the horizontal and subjected to a tension (T) applied at an angle (m) to the plane as shown in Fig. 9.17, he expressed

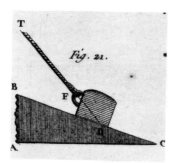

Fig. 9.17
C. A. Coulomb's study of the movement of a block on an inclined plane (Coulomb, 1785).

the force components normal and parallel to the plane as

$$P \cos n - T \sin m,$$

and

$$P \sin n - T \cos m.$$

The limiting friction force was written as

$$P \sin n - T \cos m = A + \frac{P \cos n - T \sin m}{\mu},$$

where A represents the effect of cohesion (*cohérence*) or surface films (*duvet*), and μ is the inverse of the coefficient of friction (a curiosity and impediment to modern readers!)

These equations enable the force (T) to be written as

$$T = \frac{A\mu + P(\cos n + \mu \sin n)}{\mu \cos m + \sin m}$$

or, in the case of frictional resistance to sliding of horizontal surfaces,

$$F = T = A + \frac{P}{\mu}. \qquad \qquad \dots [9.3]$$

Coulomb was the first to use this two-term expression for friction. Similar representations have become commonplace in recent times, with the first term (A) being attributed to *adhesive* or *cohesive* effects and the second (P/μ) to *deformation* or *ploughing* action.

The second important observation, discussed in Coulomb's Chapter II, was the relationship between kinetic and static friction. Once again the apparatus shown in Fig. 9.16 was utilized, the sliding speed being determined from the use of a pendulum designed to beat half-seconds and a graduated scale. In general, kinetic friction was found to be smaller than static friction, the difference being quite large for fibrous materials but almost imperceptible for metals.

The last 5 per cent of the ninety-eight-page first part of Coulomb's memoir is particularly interesting. It is his '... Essai sur la théorie du frottement'. He starts by summarizing the four principal features of his experimental findings which form the foundations for his theory. They are:

1. For wood sliding on wood under dry conditions the friction rises initially but soon reaches a maximum. Thereafter the force of friction is essentially proportional to load.
2. For wood sliding on wood the force of friction is essentially proportional to load at any speed, but kinetic friction is much lower than the static friction related to long periods of repose.
3. For metals sliding on metals without lubricant the force of friction is essentially proportional to load and there is no difference between static and kinetic friction.

4. For metals on wood under dry conditions the static friction rises very slowly with time of repose and might take four, five or even more days to reach its limit. With metal-on-metal the limit is reached almost immediately and with wood-on-wood it takes only one or two minutes. For wood-on-wood or metal-on-metal under dry conditions speed has very little effect on kinetic friction, but in the case of wood-on-metal the kinetic friction increases with speed.

Coulomb's explanation of these observations was simple and direct. He concluded first that friction could come only from the meshing of asperities, and that, since the resisting force was almost exactly proportional to load and independent of the size of the surfaces, cohesion must have but a small influence. He felt that cohesion, if present, would inevitably increase the resistance as the number of contacting points and hence the extent of the surfaces increased–a view which was to remain unchallenged until the exploration of real areas of contact between solids in the middle of the twentieth century. Coulomb concluded that although cohesive forces were not exactly zero, their contribution to friction could be neglected in practice.

The variation of friction with time of repose, which had been noted earlier by the Abbé Bossut, was readily explained by considering the surfaces of fibrous materials to have the form shown in Fig. 9.18, the so-called brush–bristle analogy of Musschenbroek. The surfaces of wood were considered to be covered by flexible, elastic fibres like the hairs on a brush. When such surfaces were brought together the bristles penetrated each other and since it took a finite time for meshing to become complete it was quite likely that the static friction would also increase with the time of repose.

Coulomb considered that the application of a tangential force would cause the elastic fibres to fold over as shown in his fig. 9 (Fig. 9.18). After a certain amount of deformation, which depended only upon the size of the asperities or fibres, sliding would take place as the opposing asperities slipped out of mesh. The slope of the effective inclined planes formed by the enmeshed asperities in this limiting situation would determine the frictional resistance, which would thus be dependent upon the nature of the surfaces but independent of their extent and directly proportional to load. Once sliding started the asperities on each surface

Fig. 9.18
C. A. Coulomb's representation of rough surfaces (Coulomb, 1785).

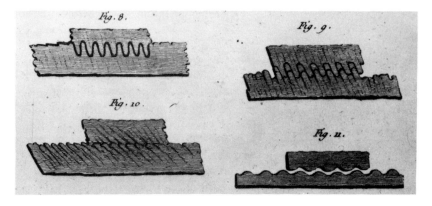

Fig. 9.19
C. A. Coulomb's apparatus for the study of rolling friction.

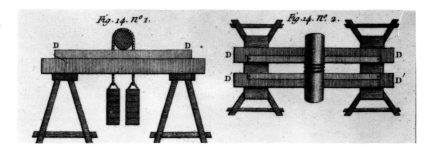

would fold over even further to fill up the voids previously occupied by the opposing surfaces, thus giving smaller slopes to the effective miniature inclined planes and a kinetic friction lower than the static friction for the surfaces. The situation in sliding is depicted in Coulomb's fig. 10 (Fig. 9.18).

The surfaces of metals were thought of as '. . . Angular, globular, hard, and inflexible parts, so that no degree of pressure nor of tension can change the shape of the parts which cover the surface of metals.'

This view explained why there was little discernible difference between static and kinetic friction between pairs of metals, since the folding over of asperities described earlier would not take place.

In this forthright, brief but ingenious and simple way, Coulomb explained the nature of friction. His results provided a workable theory for many generations of engineers, although twentieth-century studies of surfaces and contact have inevitably led to reservations about his conclusions. With hindsight, the major criticism is that much of his analysis adhered to the concept of frictional resistance arising from asperity interaction, without reference to the mechanism of energy dissipation. It should be remembered, however, that the science of thermodynamics had not yet been born and that a detailed knowledge of surface interactions had to await the sophistication of twentieth-century science.

A third law of friction (see Ch. 8) is often attributed to Coulomb, although the above text demonstrates that the investigator was probably more aware of its limitations than many of its subsequent users. It states:[26]
(3) Kinetic friction is independent of the sliding velocity.

Coulomb also investigated rolling friction using the apparatus shown in Fig. 9.19, which is broadly similar to that employed in his sliding experiments. The rollers were made of lignum vitae or elm, and the boards on which they rolled were of oak. Once again he found the force of friction to be directly related to the load, although its magnitude was much smaller than sliding friction. He also found that the resistance to rolling was inversely proportional to the radius of the roller. The remainder of his memoir was devoted to the effect of stiffness of ropes, friction in pulleys, capstans, etc. and a useful summary of his work is given in the first of a series of articles entitled 'Historic researches' which appeared in *The Engineer* (1944*a*).

There can be no doubt about the great contribution to tribology made by Charles Augustin Coulomb through his studies of friction. Here was a man with a training in engineering and a long, successful field career, applying in a most effective manner the scientific method to his investigations. He never lost sight of the need to provide results and understanding of practical importance, but he also received recognition as a physicist through his work in the Academy in later life in what might be called his second career. His work dominated the thinking of tribologists concerned with friction for over 150 years. Even now many of his views are respected and are defended in scientific debates.

On 25 November 1784 the Rev. Samuel Vince read a paper to the Royal Society on a subject which '. . . seems to be of a very considerable importance both to the practical mechanic and to the speculative philosopher.'

The title of the paper, was, 'On the motion of bodies affected by friction' (Vince, 1785). The work reported appears to have been carried out without any knowledge of Coulomb's work and it is particularly interesting because of the emphasis placed upon kinetic friction.

Samuel Vince (1749–1821) was a Fellow of the Royal Society, Archdeacon of Bedford and Professor of Astronomy and Physics at Cambridge. He was conscious of the contradictory findings of earlier workers and clearly felt that the laws of friction governing bodies in motion had not been truly established. At a later date, Nicholas Wood, a civil engineer concerned with railroads and colliery development, was to express similar doubts in the words which open this chapter. In setting out to redeem the situation by further experiments, Vince expressed confidence in his observations in no uncertain terms: '. . . The experiments, in which I was assisted by my ingenious friend the Rev. Mr. Jones, Fellow of Trinity College, were made with the utmost care and attention, and the several results agreed so very exactly with each other, that I do not scruple to pronounce them to be conclusive.'

In Vince's experiments the sliding body was connected to a falling weight by means of a string which passed over a pulley. The essence of his experimental procedure was to record the distance moved by the weight in intervals of time of one to four seconds recorded by a pendulum clock. He found that for a variety of materials the kinetic friction of hard bodies was independent of sliding speed, but that when the surfaces were covered by cloth the resisting force increased with velocity. The kinetic friction regained a constant value when the surfaces were coated with paper. His careful experiments also revealed that whereas the kinetic friction always increased with increasing load, the rise was less than would have been expected from direct proportionality. He also found that kinetic friction was not entirely independent of the apparent contact area. He emphasized the care which had to be exercised in retaining similar surface roughness when different faces of a body were brought into contact with the plane, particularly in materials like wood where knots and grain had to be contended with. In all cases he found that the smallest surface gave the least friction for a given weight.

In seeking to explain the discrepancies in earlier work Vince distinguished between *cohesion* between bodies at rest and the resistance of *friction* encountered by bodies in motion. He preferred his own '. . . most decisive experiments' and said of the findings of others: '. . . All the conclusions therefore deduced from the experiments, which have been instituted to determine the friction from the force necessary to put a body in motion, have manifestly been totally false; as such experiments only show the resistance which arises from the cohesion and friction conjointly.'

It seems strange that Vince attributed cohesion or adhesion to static situations alone, but it was not until 1829 that an anonymous author, writing about Vince's work in the Library of Useful Knowledge published in London,[27] asked the most pertinent question '. . . Does the motion of one body upon the other altogether destroy the cohesion?'

A generation after Coulomb and Vince had made their contributions to the science of friction, a scientist who was trying to expose the nature of heat made some interesting observations. The name of the scientist was John Leslie (1766–1832), sometime tutor to the Wedgwood family, Fellow of the Royal Society and Professor of Physics in the University of Edinburgh. A book published by Leslie (1804) described his careful study of radiation, including the procedure involving the 'Leslie Cube' which still forms an established feature of experiments on heat in physics courses. In a later part of the book he discussed the resistance to flowing fluids and the friction experienced by moving solids. The work is particularly interesting since it is written against the background of the *caloric* theory of heat developed by the French chemist Lavoisier (1743–94). The modern science of thermodynamics was promoted primarily by later studies of steam power,[28] although concern for the relationship between work done against friction and the heat generated in sliding was also a factor of some importance. According to Lavoisier the evolution of heat during combustion was related to the release of an imponderable element caloric. Leslie set out to determine '. . . What then is this calorific and frigorific fluid after which we are enquiring?'

After a lengthy discourse in which he associated caloric with air he concluded that '. . . heat is an elastic fluid, extremely subtle and active', and that '. . . heat is only light in a state of combination'.

In this brief account of Leslie's studies, our attention must be restricted to his work on the friction of solids, although his views on fluid resistance are equally interesting. He recognized a weakness in the mechanistic view of Amontons and Coulomb which attributed friction to the effort required to pull opposing surfaces over a series of inclined planes representing surface asperities, and it is not without significance that these reservations arose in the mind of a man concerned with studies of heat in the budding field of thermodynamics.

After noting that friction persisted after surfaces had been smoothed by '. . . the grinding and abrasion of their prominences', Leslie wrote:

27.   See Bowden, F. P. and Tabor, D. (1964), Pt II, pp. 509–10.
28.   See Bradley, D. (1966), Vol. 2, pp. 26–31.

. . . Friction is, therefore, commonly explained on the principle of the inclined plane, from the effort required to make the incumberent weight mount over a succession of eminences. But this explication, however currently repeated, is quite insufficient. The mass which is drawn along is not continually ascending; it must alternately rise and fall; for each superficial prominence will have a corresponding cavity; and since the boundary of contact is supposed to be horizontal, the total elevations will be equalled by their collateral depressions. Consequently, if the actuating force might suffer a perpetual diminution in lifting up the weight, it would, the next moment, receive an equal increase by letting it down again; and those opposite effects, destroying each other, could have no influence whatever on the general motion.

Leslie was equally sceptical about the role of adhesion proposed by Desaguliers and considered by Coulomb.

Adhesion seems still less capable of accounting for the origin of friction. A perpendicular force acting on a solid can evidently have no effect to impede its progress; and though this lateral force, owing to the unavoidable inequalities of contact, may be subject to a certain irregular obliquity, the balance of chances must on the whole have the same tendency to accelerate, as to retard, the motion.

Having dismissed the inclined plane and adhesive mechanisms of friction, Leslie expounded his own explanation of the process. He argued that deformation of the contacting asperities and the surrounding surfaces would continually change the surface topography during sliding. However, the flattening of opposing asperities was seen as a time-dependent process, such that '. . . In some cases, a few seconds are sufficient; in others, the full effect is not produced until after the lapse of several days.

Leslie saw in this time-dependent effect an explanation of the general observation that the force required to initiate motion after a period of repose was greater than kinetic friction. His summary of the nature of friction was colourfully related to the classical legend of Sisyphus, first King of Corinth who was punished in Hades for certain misdemeanours by the vain effort of continually rolling a block of marble uphill only to see it roll down again.

. . . Friction consists in the force expended to raise continually the surface of pressure by an oblique action. The upper surface travels over a perpetual system of inclined planes; but that system is ever-changing, with alternate inversion. In this act, the incumbent weight makes incessant yet unavailing efforts to ascend: for the moment it has gained the summits of the superficial prominences, these sink down beneath it, and the adjoining cavities start up into elevations, presenting a new series of obstacles which are again to be surmounted; and thus the labours of Sisyphus are realised in the phenomena of friction.

The role of lubricants was not overlooked by Leslie,

. . . The intervention of a coat of oil, soap or tallow, by readily accommodating itself to the variations of contact, must tend to equalize it, and therefore must lessen the angles, or soften the contour, of the successively emerging prominences, and thus diminish likewise the friction which thence results.
Such is apparently the real origin or friction.

I have felt it worthwhile to quote extensively from Leslie's writings, since the work represents an entirely new approach to the nature of friction. It is concerned with the *deformation loss* concept built into many modern theories of friction, although it cannot be claimed that Leslie worked out the details. Furthermore, as pointed out by Bowden and Tabor (1964, Pt. II: 510) he failed to appreciate the possibility of shear stresses and a resulting tangential force arising from adhesion. Nevertheless, John Leslie deserves recognition for introducing early in the nineteenth century, the concept, if not the fine detail, of a deformation aspect to the friction process.

The increasing tempo of industrialization in the early part of the nineteenth century soon led to a demand for further clarification of the nature and influence of friction. There were two major problems: the increasing diversity of opinions prevalent among philosophers and the lack of data for engineers concerned with the construction of bridges, arches and machines and the launching of ships on slipways. It was the latter problems which prompted George Rennie to undertake and report on an impressive series of experiments (Rennie, 1829). His work has rarely been adequately reviewed, apart from the account by Kragelskii and Shchedrov (1956), yet it is a most interesting indication of the growing recognition of the significance of tribology in an industrial society.

The resistance offered by ice to the motion of sledges and skates; the friction of leather seals in the pistons of pumps; the resistance experienced by wood used in pile-driving, carpentry and the launching of ships; the effect of friction upon the equilibrium of stone arches and buildings; the remarkable difference between the frictional characteristics of cloth and hard solids; and the friction of metals in their universal application to machinery, were all examples which formed the background to Rennie's investigations. It was, however, the agitation of the canal and railroad question in 1824 and 1825 and the tribological problems posed by wheeled vehicles and steam power which provided the main stimulus for his work. His careful survey and criticism of existing literature is particularly valuable to those wishing to trace the history of studies of friction.

Rennie used two sets of apparatus, the first, shown in Fig. 9.20(*a*), being for fibrous materials like cloth and wood, and the second, shown in Fig. 9.20(*b*) being for metals like brass and iron. Both sets of apparatus are essentially similar – though somewhat more elegant – forms of the equipment used by Coulomb. In the first arrangement a sliding block moved over a platform which could be elevated to any angle up to 30°.

The experiments on cloth revealed coefficients of friction varying in value from one-third to slightly in excess of unity. The coefficient was found to increase with surface area and time of repose, but to decrease with both load and velocity. The behaviour of five different cloths was recorded.

Eleven different combinations of wood were tested and Rennie found that the resistance was directly proportional to load but independent of the extent of the surfaces, time of repose or velocity. Hardness of the wood emerged as an important influence upon friction, yellow deal on yellow deal providing the greatest resistance and red teak on red teak the least.

The data derived from each set of experiments are clearly recorded in ten tables throughout the paper, but it is the 'Remarks' at the end of each section which are most informative. The naval background to Rennie's extensive studies of the friction of wood is clearly indicated by the following extract.

> *According to Mr Knowles of the Navy Office, F.R.S., the weight of the 'Prince Regent' of 120 guns on the slips previous to launching, was 2400 tons; which, divided by the area of the sliding surface of her bilge-ways (equal to 149, 184 square inches), gives a pressure of 36 lbs. per square inch.*
>
> *But the weight of the 'Salisbury' of 58 guns on the slips, according to the area of her bilge-ways, was 44 lbs. per square inch. Now, by the foregoing Table, the average force required to put in motion the three different kinds of oak under a pressure of 56 lbs. per inch, is about $\frac{1}{8}$th of the pressure, which proportion prevails even as high as 6 cwt. per inch area: and by Table IX we find that soft soap (the ingredient mostly used for diminishing the friction of bilge-ways under a pressure of 56 lbs. per inch) gives about $\frac{1}{26}$th of the pressure for the friction. Hence the angle at which a building slip should be laid can be easily determined. Coulomb even makes 49 lbs. per square inch, and $\frac{1}{27}$th for the pressure for hogslard.*

The three oaks mentioned in the extract were listed as English, Norwegian and American live. The direct value of these friction experiments in relation to the important exercise of launching ships is clearly seen. Furthermore, the beneficial effect of soft soap or hogslard in reducing the coefficient of friction to about one-third of the dry case is well demonstrated; the consistency between Coulomb's value of $\frac{1}{27}$ and Rennie's of $\frac{1}{26}$ being quite noteworthy.

Fig. 9.20
Apparatus used by George Rennie for his experiments on the friction and abrasion of the surfaces of solids.
(a) Apparatus used for studying the friction of solids – primarily cloth and wood.
(b) Apparatus for studying the effect of speed upon the friction of metals – primarily brass and iron.

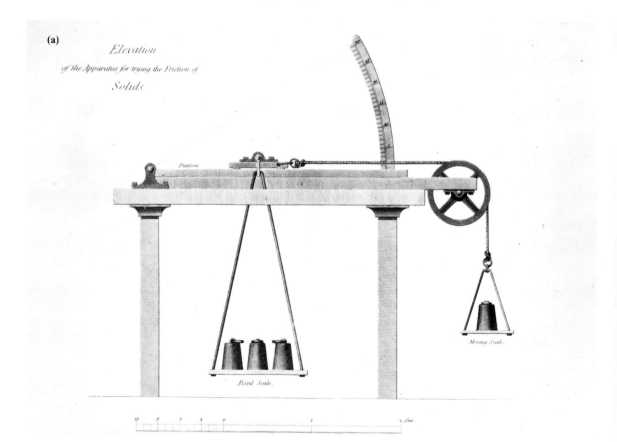

(a)

Elevation

of the Apparatus for trying the Friction of

Solids

Rennie's studies of metals encompassed the effects of load, surface area, sliding velocity and lubricants upon the friction of wrought and cast iron, steel, brass and tin in various combinations. In the absence of unguents he was able to confirm the findings of earlier workers with regard to both static and kinetic friction. He advised that the coefficient of friction for hard metals against hard metals was about one-sixth and that soft metals like tin have more friction than hard metals. The merit of steel was noted with the words: 'The remarkable property of steel in hardening, and its power to resist abrasion, render it preferable to every other substance yet discovered in reducing the friction of delicate instruments, as is exemplified in the different experiments on the pendulum, and the assay and other balances recently introduced at His Majesty's Mint and the Bank of England.'

The second apparatus was used to study the friction of axles at various speeds, with and without unguents. The list of lubricants used provides an interesting insight into lubrication practice in the Industrial Revolution. In addition to an *oil* of unspecified nature, *soft soap, tallow, hogslard, black-lead* and an *anti-attrition composition* of *black-lead and hogslard* are mentioned. The reference to the use of black-lead (graphite) represents one of the earliest indications of the use of solid lubricants. The tabulated results also show that many experiments on axles were carried out under lubrication regimes that would now be described as fluid-film or mixed.

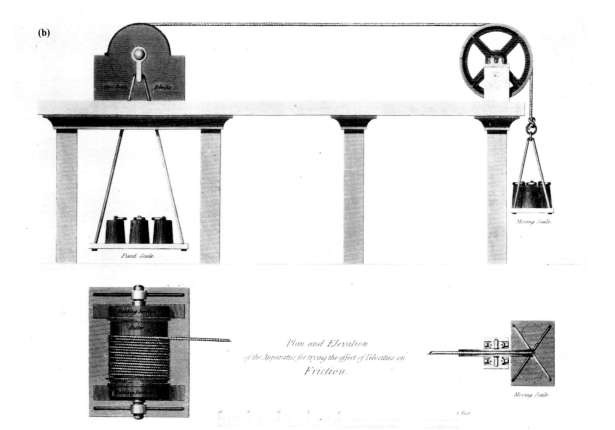

Plan and Elevation
of the Apparatus for trying the effect of Velocities on
Friction.

Perhaps the most significant finding was that the friction experienced by lubricated axles depended upon the nature of the lubricant rather than the nature of the bearing materials.

The wide-ranging nature of Rennie's experiments was confirmed in the last few pages of his paper where he discussed the friction of ice, hide-leather, stones and machinery. Experiments on ice at a temperature of $-2.2°C$ ($28°F$) revealed coefficients of friction which fell steadily with increasing load, the smallest recorded value being 0.0143. Leather was used extensively for sealing pistons in water-pumps and Rennie found that the coefficient of friction under dry conditions was influenced much more by surface area than by load. The coefficients ranged from 0.25 to 0.17. When the leather was soaked in water the friction increased greatly with both load and time of repose, thus enabling Rennie to write: 'This circumstance explains the enormous friction evinced in the pistons of pumps when first put in motion.'

Reference to the role of friction as a '. . . most powerful assistant in maintaining the equilibrium of arches . . .' reflects the importance of the phenomenon in the construction of arches, bridges, etc. employing either dry or mortared joints. The angles of friction determined by others for a variety of stones ranged from $33°$ to $40°$ (0.58–0.70 rad) under dry conditions. With beds of fresh and finely ground mortar interposed the angles were reduced to 25–$34°$ (0.44–0.59 rad). An interesting account of the use of copper sheets and tallow in the building of the New London Bridge reads:

> *The weight of the middle arch (of 151 feet 9 inches span) of the New London Bridge, together with the centres, is 4900 tons. This acting upon the surface of the striking wedges equal to 540 square feet, gives a pressure of 140 lbs per square inch. The angles of inclination of the wedges are equal to 8° 45', and their surfaces are covered with sheets of copper well coated with tallow. On removing the check pieces, the wedges commenced gliding back slowly and uniformly by the gravity of the arch and centres, and the motion was checked and continued until the arch was left in equilibrio.*

In a short section devoted to friction in machines Rennie draws attention to the beneficial effects of both lubricants and the process which would now be described as 'running in'.

> *It has been customary to deduct one fourth of the power expended for friction. This allowance may maintain in machines newly set in motion. When the bearings have been equalized and the rubbing surfaces extended by the abrasion of the irregularities, the friction will be diminished and the movements of the machine be more steady. But when the bearings are properly proportioned to the weight of the parts of the machine, and their surfaces kept from contact by unguents, a much less allowance may be made.*

Reference to the action of unguents in keeping the surfaces from touching each other suggests that Rennie was moving towards an appreciation of the process which was to be fully exposed some fifty or sixty years later in a decade of study of fluid-film lubrication.

To reinforce the view that a 25 per cent loss of power or force in machinery may be excessive in well-lubricated, run-in mechanisms a number of practical cases were cited. A corn-mill erected for His Majesty's

Table 9.2
Mr George Rennie's amounts of dry friction for different substances at a load of 36 lbf and within the limits of abrasion of the softest substance

Materials	Mean pressure (lbf/in²)	Coefficient of friction (μ)
Steel on ice (load 144 lbf)	171	0·014
Ice on ice	2·25	0·028
Hardwood on hardwood	28–728	0·129
Brass on wrought iron	6·09	0·136
Brass on steel	5·33	0·141
Brass on cast iron	6·10	0·139
Soft steel on soft steel	6·10	0·146
Cast iron on steel	6·10	0·151
Wrought iron on wrought iron	6·10	0·160
Cast iron on cast iron	5·33	0·163
Hard brass on cast iron	4·64	0·167
Cast iron on wrought iron	6·10	0·170
Brass on brass	6·10	0·175
Tin on cast iron	5·33	0·179
Tin on wrought iron	6·10	0·181
Soft steel on wrought iron	6·09	0·189
Leather on iron	0·66 → 28·44	0·25
Tin on tin	6·10	0·265
Granite on granite	–	0·303
Yellow deal on yellow deal	–	0·347
Sandstone on sandstone	–	0·364
Woollen cloth on woollen cloth	–	0·435

Victualling Department at Deptford required only one-tenth of the weight of the mass to overcome inertia and friction in the bearings. In another case two double purchased cranes indicated 0·11 and 0·13, respectively, for the coefficient of friction.

A summary of the results displayed in Mr Rennie's extensive tables of friction coefficients is reproduced in Table 9.2. It provided wide-ranging guidance on dry coefficients of friction for the engineers of the late Industrial Revolution period.

Rennie did not attempt to formulate a theory of friction, although his penetrating observations and general conclusions demonstrated an excellent early nineteenth-century appreciation of the nature of the phenomenon. He provided data covering a very wide range of materials which were of immediate value to the engineers of the day. Furthermore, he drew attention to the importance of avoiding abrasion of the surfaces, the merits of running in and the beneficial role of lubricants. The list of substances tested and the lubricants employed is of considerable historical interest since it reflects tribological practice in the early nineteenth century. The significance which Rennie attached to the subject is in evidence in his closing sentence. '. . . It only remains to conclude by expressing a hope, that the data now furnished will in some degree enlarge the bounds of our knowledge on this subject, interesting as one of philosophical inquiry, and intimately connected with every branch of the mechanical arts.'

Shortly after the publication of Rennie's results in England a young captain in the artillery of the French Army confirmed and extended Coulomb's findings in a long series of experiments carried out at the Engineering School of Metz. His name was Arthur Jules Morin

(1795–1880) and his work became so well known in Europe that the laws of friction were widely known as Morin's laws until the end of the nineteenth century. He was appointed Professor of Applied Mechanics at Metz at an early age[29] and subsequently became a Professor and later the Director of the Conservatoire des Arts et Métiers at Paris. He ultimately rose to the rank of General in the French Army, devoting his later years to problems of ventilation and the reconstruction of Paris and other French cities in connection with Napoleon III's grandiose schemes for the Second Empire. Morin's (1835) reports were concerned with both sliding and rolling friction, but it was the latter subject which led him into a great and bitter controvesy with a young French engineer named Dupuit.

Rolling friction

It was well known that rolling friction was considerably less than sliding friction, but interest in the force required to move freight wagons, passenger coaches and cannons necessitated a more detailed study of the subject.

Morin agreed with Coulomb's finding that the resistance to rolling was directly proportional to load and inversely proportion to the radius of the roller, according to the relationship

$$F = k\,\frac{P}{R}.$$

However, the young Arsène Dupuit reported a careful study of rolling friction (Dupuit, 1839) in which he found that the observations were more accurately represented by an expression of the form

$$F = k\,\frac{P}{R^{1/2}}.$$

The David and Goliath situation thus created between the young engineer Dupuit and the great General Morin was exacerbated by the latter's suggestion (Morin, 1840) that there was probably some interfacial slip in Dupuit's experiments. This precipitated a three-year running debate and dispute of increasing intensity before the French Academy as reported by Stone (1956).

Dupuit (1840) stood his ground and refuted the suggestion that the discrepancy could be attributed to interfacial slip. He also carried the attack to Morin by drawing attention to the fact that the latter's experiments were not carried out on road surfaces in a standard condition. His boldness further led him to point out that Morin's results contained many numerical errors and that when recalculated the data fitted quite well the inverse square root relationship with radius.

Twelve months later Morin (1841*a*) presented new data which, he claimed, supported conclusively the view that resistance was inversely proportional to the radius of the roller. On this occasion he also questioned

---

29. A brief biography can be found in 'Historic researches, No. 1–Friction: Coulomb and Morin's experiments', *The Engineer* (1944), July, 22.

the sensitivity of Dupuit's dynamometer. Naturally such a suggestion could not remain unanswered, but in the next response, Dupuit (1841) concentrated on the effect of the axial length of the cylinder upon rolling friction.

Dupuit found no effect of cylinder length upon rolling friction, whereas Morin had concluded that resistance was inversely proportional to the square root of the length. Dupuit pointed out with some clarity that if the length influenced friction as suggested by Morin, the result would be inconsistent with the linear relationship between friction and load. Clearly, if a cylinder was cut into four parts and each part carried one-quarter of the total load, the sum of the resistances of the four sections would be twice as great as that of the original cylinder if the relationship proposed by Morin was correct.

In his final response, Dupuit (1842) answered the criticism of his dynamometer with rhetoric. Imagine that a force of 100 kgf (981 N or 220 lbf) was required to move a wagon having wheels of diameter 2 m (6·6 ft). If smaller wheels of diameter 0·5 m (1·6 ft) were fitted, the new force required to cause movement would be 400 kgf (3952 N or 882 lbf) according to Morin and 200 kgf (1962 N or 441 lbf) according to Dupuit. Dupuit simply asked, what sort of sensitivity is required in a dynamometer to distinguish between these two? He also argued that if Morin's results were correct there would be no future for the new railway systems, since it would be possible to build road carriages with a rolling resistance very little greater than that of coaches running on rails.

Dupuit's (1839) theory of rolling friction was remarkably simple yet farsighted, particularly since it was presented long before the theories of contact stresses and plastic deformation were available. In essence Dupuit argued that the material behind a rolling cylinder would not fully recover after deformation and that this would cause the ground reaction to act in front of the rolling axis as shown in Fig. 9.21. The couple created by the rolling resistance (FR) would thus have to overcome the resisting torque ($P\delta$) before motion could take place.

Hence

$$F = \delta \left[\frac{P}{R} \right].$$

Fig. 9.21
Dupuit's concept of rolling friction.

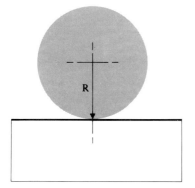
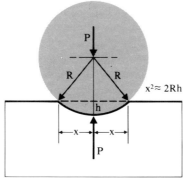
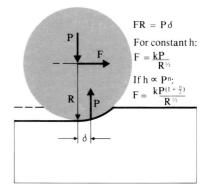

Now, for a rigid cylinder pressed into a softer solid with penetration h, the chord rule tells us that the width of the half-contact zone is given by

$$x^2 \approx 2Rh.$$

Dupuit considered the arc of contact on the front of the roller to be smaller than that on the rear part and it therefore appeared to be quite reasonable to assume that the location of the effective ground reaction would be related to the half-contact width under all conditions, such that

$$\delta \propto x \propto (Rh)^{1/2}.$$

Hence, for a constant depth of penetration,

$$F = k\left[\frac{P}{R^{1/2}}\right].$$

The main weakness in the argument is that the depth of penetration must, of course, be related to the load.

If $h \propto P^n$ the result becomes

$$F = k\frac{(P)^{1+n/2}}{R^{1/2}}.$$

Almost half a century later Hertz (1881) was to provide the contact stress analysis which enabled values of n to be established for various geometrical forms of elastic solids. For an elastic cylinder on a plane n is close to unity and it is interesting to note that Tabor (1955) found that the rolling resistance associated with elastic hysteresis at light loads was governed by an expression of the form

$$F = k\frac{P^{3/2}}{R^{1/2}}.$$

Further evidence that this modified Dupuit expression is remarkably consistent with modern theories of rolling friction is provided by the elegant analysis of Bentall and Johnson (1967). The problem is, of course, extremely complex and the exact relationship between rolling resistance (F), load (P) and roller radius (R) depends upon the conditions considered. Collins (1972) has shown by means of a rigid-plastic analysis that very high loads give rise to an expression of the form

$$F = k\frac{P^2}{R}.$$

A similar result is obtained if it is assumed that a rigid roller sinks into a soft material having a uniform and constant bearing pressure.

Johnson (1972) has provided a survey of the various regimes of rolling resistance varying from elastic through elastic-plastic to fully plastic conditions. The complex nature of the problem is now recognized, and if we find it hard to understand the bitterness of the debate between Morin and Dupuit, we can at least sympathize with their predicament.

However, it is clear that Dupuit was really suggesting a friction mechanism which would now be attributed to 'deformation' or 'hysteresis' losses. He wrote:

> *. . . Thus for rolling resistance to exist, one of the bodies must be compressible; that is to say, under a given load the molecules are deflected by a distance ε'. Further, the molecules must not be perfectly elastic; that is to say, the molecules do not return to their original position when the load is removed but remain displaced by a certain distance ε. These quantities being given the coefficient of rolling friction can be at once deduced. Thus friction of the second kind is not a special property of the surfaces and distinct from the compressibility and elasticity. It is indeed simply the consequence of these, so that if we know these two properties for the body we have at once its resistance to rolling.*

Dupuit was unable to persuade the French Academy of the validity of his views, although many of his concepts were remarkably close to present-day understanding of the subject and Morin (1841*b*) himself discussed deformation effects in his last paper in this series. He was, of course, in a difficult position, since General Morin had already achieved high distinction for his scientific work and his findings were, after all, in accord with those of the great Coulomb. In due course a Commission ruled in favour of Coulomb and Morin, but the unfortunate Dupuit can rightly claim a respected place in the history of tribology. The full story of this little-known episode in nineteenth-century studies of friction has been handsomely told by Tabor (1962).

9.5 The heyday of water-mills and windmills

Reference has already been made in earlier chapters to notable tribological developments associated with windmills and water-mills. It is nevertheless worth while to look closely at these magnificent machines in the present chapter since it was the period of the Industrial Revolution which witnessed their zenith and the origins of their ultimate demise. While some working water-mills and windmills can still be seen at the present time they are few in number and, alas, still in decline. Their contribution to the total power requirements of modern industrial countries is insignificant, although present concern for the conservation of fossil fuels in an avaricious world has reawakened interest in their future.[30]

The vertical water-wheel mounted on a horizontal shaft was a relatively simple arrangement compared with windmills in which the structure had to face the prevailing wind. It is for this reason that advanced forms of windmill structures tend to hold the greatest tribological interest, although the thrust bearings for vertical spindles of early water-wheels have a special fascination.

Water-wheels

These have been used for more than 2,000 years. They emerged in simple form long before it was recognized that power could be extracted from the wind as well as from flowing streams. They supplemented and in due course replaced the hand-operated querns and the stones worked by oxen

---

30. See *New Scientist*, **57**, No. 827, 24, Jan. 4th, 1973. Brackner, A. (1974) *New Scientist*, **61**, No. 891, 812–14; ASME (1974*a*), p. 55; and ASME (1974*b*), p. 29.

and horses for grinding corn. The significance of this mechanical contrivance in early society is nicely reflected in the epigram by Antipater of Thessalonica of 85 B.C. which forms the first known reference to a water-mill. 'Cease your work, ye who laboured at the mill. Sleep now, and let the birds sing to the blood red dawn. Ceres has commanded the water nymphs to perform your task; and these, obedient to her call, throw themselves on the wheel, force round the axle tree and so the heavy mill.'

The Greek mill with a horizontal wheel at the lower end of a vertical spindle which was also connected to the upper stone, represented the earliest form of water-mill. Similar structures appeared in Greece, Rome and China some 2,000 years ago. The vertical water-wheel rotating on a horizontal shaft emerged in Roman times. The Roman mill, which was described by Vitruvius *c.* 15 B.C. as a natural extension of the treadmill, was introduced into many European countries following the Roman conquest. It was the forerunner of the traditional water-mill in Western Europe throughout the period of the Industrial Revolution and the predecessor of the windmill.

Two examples of quite remarkable bearings on eighth–tenth-century horizontal water-wheels will serve to illustrate the importance attached to bearing performance in the early development of water power in the British Isles. In a fascinating and informative article Mr Robert MacAdam described an ancient wooden horizontal water-wheel found in the bog of Moycraig in the district of the Grange of Drumtullogh, Ireland (MacAdam, 1856). A sophisticated structure of some nineteen ladles or buckets of oak formed the horizontal wheel at the lower end of an upright axle of oak some 6 ft 6 in (2 m) tall. The Moycraig wheel, which together with an accurate working reconstruction, can be seen in the Ulster Museum, Belfast is shown in Fig. 9.22.[31] Perhaps the most remarkable feature of this ancient water-wheel, which has a date established by radio-carbon dating of A.D. 950 $\pm$ 110 years, is the stone pivot or gudgeon driven firmly into the oak axle. MacAdam described the arrangement as follows:

> *The whole mechanism was supported by a pivot or gudgeon secured by a wedge at the foot of the axle, where it still remains. This pivot, no doubt, revolved upon another stone hollowed to fit it. A stone of this kind was, in fact, found near the water-wheel at Killinchy, and is preserved along with it in the Belfast Museum, bearing evident marks of having been deeply perforated by some pivot constantly revolving in it.*

This outstanding evidence of pivot bearings in horizontal water-wheels is a further example of the use of stone as a bearing material in the Middle Ages (see Ch. 6).

A further and probably contemporary example of the pivot bearings in horizontal water-wheels involving an exceptional use of metal has emerged in recent years from Tamworth in Staffordshire. Rahtz and Sheridan (1972)

---

31. I am grateful to Dr W. A. McCutchen, Keeper of Technology and Local History of the Ulster Museum, Belfast, for kindly supplying this photograph and full details of the Moycraig wheel.

Fig. 9.22
Moycraig horizontal water-wheel,
Ireland (*c.* 950 A.D. ±110 years).

have described excavations of an eighth-century Saxon water-mill which
might well have formed part of the palace complex of Offa, King of Mercia
from A.D. 757 to 796. Among the wooden remains of the mill was a wooden
sole-plate some 75 ft (2·3 m) long containing a peg-hole at one end and a
remarkable inset metal bearing. The pivoted sole-plate could be raised or
lowered to provide control of the clearance between the millstones. The
bearing consisted of a piece of 'iron', approximately 3 in (80 mm) square
and 1·4 in (35 mm) thick set into the plank with no other means of fixing.[32]
There was a thick encrustation of rust surrounding the metal as shown
in Fig. 9.23, but most of the iron was uncorroded. Detailed examination
revealed a remarkable bearing form. The bearing socket had a diameter
of about 0·7 in (20 mm) at the top and was roughly conical in form as
shown in the section in Fig. 9.24. The wear pattern is also evident in this
figure. The iron had been forged and quenched such that the carbon
content and hardness were consistent with quite a fine-quality steel: a truly

32. I am grateful to Mr Philip Rahtz of the School of History and Dr E. M. Trent of the
Department of Industrial Metallurgy of the University of Birmingham for kindly
providing full details of this bearing and the photographs shown in Fig. 9.23.

Fig. 9.23
An iron pivot bearing from a
Saxon water-mill at Tamworth.

remarkable feature of early metal bearing construction. Metallurgical examination revealed a further and initially unsuspected feature of the bearing. On the underside of the metal plate an earlier bearing socket of similar form to the one shown in Fig. 9.24 had been plugged. It therefore appeared that the value of the metal bearing had been appreciated and the plate inverted to double its effective life.

In due course the relatively small horizontal-wheel mills gave way to vertical wheels of considerable size. Water-mills were initially used for grinding corn, but in due course they provided power for a number of industrial processes including fulling, forging, air-blowing through bellows, crushing, pumping, boring, sawing, metal-rolling and paper-making. Initially they were built of wood and the drawings of Agricola show that little iron was used in the construction of sixteenth-century mills. At the dawn of the Industrial Revolution the water-wheel was the most important prime-mover for driving rotating machinery. Small animal-driven mills were still in use, windmills were used for grinding corn and draining the fens and steam power was used to drive pumping engines in mines, but water-wheels provided most of the industrial power. The location of new industrial sites on the fast-flowing Pennine streams provides evidence of this fact which can still be seen. Expanding industry and the growth of the factory system rapidly used up the available water

Fig. 9.24
Sketch of wear pattern and cross-section of socket in an iron bearing from Tamworth water-mill.

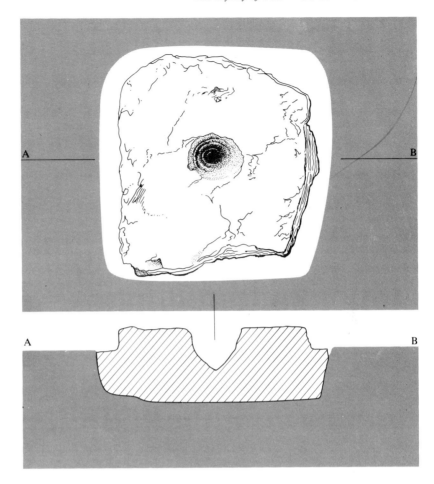

sources and drew attention to the need to maximize the efficiency of water-wheels.

John Smeaton, born at Austhorpe Lodge near Leeds on 8 June 1724 was destined to provide the answers to many of the questions of the day. He was to become one of the great engineers of the eighteenth century and although he is best known as the designer and builder of the Eddystone Lighthouse, he was highly respected as an experimenter and a designer of water-mills. He made two major contributions to the technology of water-mills. His experiments on models reported to the Royal Society in May 1759 demonstrated the marked advantage of overshot and breast wheels over undershot wheels. Smeaton (1759) found that the maximum overall efficiency of undershot wheels was about 22 per cent and of overshot wheels 63 per cent. He devoted much effort to the study and design of breast wheels.

Smeaton's second major contribution was the widespread introduction of cast iron to replace wooden wheels. His first cast-iron shaft was made in 1769 and fitted to the wheel of the Carron furnace-blowing engine. Several of the early cast-iron shafts broke, for the quality of the material and the understanding of stresses in rotating components were both inadequate.

The heavy cast-iron wheels presented rotating members of considerable inertia, thus providing a more stable speed for many industrial processes. Water-wheels provided a valuable source of power to support industrialization, although it was rare for individual wheels to produce more than 20 hp (15 kW). It is comforting to find that the remaining examples of these fine machines are increasingly coming into the care of preservation societies.

The weight of the vertical water-wheels which developed in the Industrial Revolution was usually large compared with the horizontal force arising from the impingent water. The resulting bearing force was thus essentially vertical and the journal bearings were simple 180° or 360° forms. The common material of construction was metal and animal fats were the main lubricants. That boundary lubrication prevailed in these water-wheels can invariably be seen from the location of the wear marks on the bearings.

Windmills

The windmill followed the water-mill as a source of mechanical power and immediately provided scope for the construction of alternative power units on sites remote from flowing streams. It is known that a windmill existed in Normandy almost 800 years ago[33] and the first known illustration of a windmill is probably the sketch incorporating a post-mill on the opening page of the English Psalter written in Canterbury about A.D. 1270. Initially such mills were used more widely for drainage and driving other forms of machinery.

In recent years interest in the generation of power from the wind has been reawakened, but if the modern wind-turbines now being studied by engineers in various countries ever make a significant contribution to the overall requirements of society, they will bear little resemblance to the graceful structures of the past. They will no doubt present a new range of bearing problems, but in this text attention will be focused upon the tribological features of the traditional windmill.

Windmills were constructed in different forms in different geographical locations and in different historical periods, but there are two basic types of structure: the post-mill and the tower-mill. *Post-mills* which were the first to be constructed, consisted of a wooden structure containing all the machinery, wind-shaft and the sails, mounted on a vertical post held firmly in the ground. It was necessary for the wooden structure and all its heavy contents to rotate on the fixed post to face the prevailing wind and this necessarily restricted the size of such mills. The second arrangement, known as a *tower-mill*, emerged in the fourteenth century. This ingenious structure allowed the subsidiary drive shafts, gears and grinding stones to be located in a stationary tower while the wind-shaft and sails alone turned in a dome or cap to face the wind. Both forms of mill are sketched in Ramelli's (1588) text as shown in Figs 9.25 and 9.26. These cutaway

33.   See reference to a deed dated *c.* A.D. 1180 which records the gift of land near a windmill in Wailes, R. 'A note on Windmills', in Singer, C. et al. (1956), Vol. II, p. 623.

Fig. 9.25
Post-mill for grinding corn (after
Ramelli, 1588).

drawings reveal not only the basic forms of windmills, but also the
adventurous engineering approach to the problems of grinding corn and
raising water by mechanical means in the sixteenth century. The development of wooden gears and the mechanical power transmission system
which was necessary for the successful operation of a tower-mill was
itself a notable landmark in engineering history. The two basic forms of
windmill are shown in simplified form in Fig. 9.27, together with the sites
of major bearings in each structure. Both arrangements called for satisfactory journal and thrust bearings to support the heavy wind-shaft
and this promoted the early development of interesting forms of wooden,

Fig. 9.26
Tower-mill for driving a
rag-and-chain pump (after
Ramelli, 1588).

Fig. 9.26
Tower-mill for driving a rag-and-chain pump (after Ramelli, 1588).

stone and metal bearings. In addition the post-mill relied upon a single
pivot or *pintle* bearing to support the rotating structure on the post,
together with a horizontal *steady* bearing wrapped round the post to
provide lateral stability to the structure. The caps of tower-mills had to
turn upon the fixed tower to face the prevailing wind and this called for
special forms of sliding or rolling-element thrust bearings to carry the
weight and in some cases to centre and retain the cap upon the fixed
structure. The basic forms of *wind-shaft*, *pintle* and *cap bearings* will be
outlined.

Wind-shaft bearing. Wind-shafts revolved in the tops of windmills and they were attached directly to the sails. They were normally inclined at a small angle of between 5 and 15° (0·087–0·26 rad) to the horizontal to ensure that the sails did not strike the structure of the mill and to facilitate bearing-support arrangements. An excellent account of wind-shafts and their bearings has been published by Wailes (1947).

The front or *neck bearing*, which supported most of the weight of the wind-shaft and the sails, was invariably a partial-arc journal having a wrap angle between 90° and 120°, made of wood, stone or metal. The early wooden wind-shafts were of great size and weight, being either square, round or quite frequently octagonal in section and up to 2 ft (0·6 m) in diameter. They were made of oak or pine and wear of the journals was resisted by longitudinal strips of wrought or cast iron in the wood. Nails and iron rings placed fore and aft of the bearing surface retained these inset strips of metal in the journal.

Various woods such as thorn, oak, elm, apple and lignum vitae have been used for wind-shaft neck bearings, but it also appears that stone bearings were used from an early date. Marble was a popular bearing material and Wailes (1954) has recorded the use of red granite at Thornton Mill in Lancashire. It is said that at one stage a miller had the right by law to lift any cobblestone which he considered suitable for a bearing. A neck bearing from Thornham Mill, near Hunstanton, Norfolk which now resides in the Bridewell Museum, Norwich, is reproduced in Fig. 9.28. The metal strips are bound to an oak wind-shaft and the roughly shaped blue marble bearing is retained in an oak block which would form part of the breast beam by lead and plaster. Visitors to the museum can see how the marble has been worn and polished to present a beautifully smooth bearing surface to the journal.

At Kingston Down, near Lewes in Sussex, a unique glass neck bearing was successfully employed in a six-sailed post-mill which was ultimately blown down in 1916.[34]

White metal bearings were also used, but hard brass or gun metal were confirmed as the most usual neck-bearing materials when the wooden wind-shaft gave way to cast iron following John Smeaton's work in 1754. It is interesting to note that the neck bearing of a wind-shaft is generally known as '*le marbre*' in France and the '*neck-brass*' in England. A selection of stone and hardwood neck bearings now preserved in The Falcon, Leiden and, photographed by Mr John Reynolds, is shown in Fig. 9.29.[35]

In East Anglia a trunnion mounting known as a *swing pot* was developed to make a self-aligning neck bearing. Simple forms of roller-bearings utilizing three to five rollers were also introduced, but they caused excessive wear on cast-iron windshafts.[36] Cast-steel wind-shafts or split-steel sleeves later mitigated this effect.

---

34. See Wailes, R. (1954), p. 101.
35. I am grateful to Mr Reynolds for providing the negative from which Fig. 9.29 was obtained and to the publishers, Hugh Evelyn Limited, for permission to reproduce the illustration from Reynolds (1970).
36. See ibid., p. 121.

The axial force arising partly from the inclination of the wind-shaft in its neck-bearing arrangement, but also from the force exerted by the wind upon the sails was taken by a thrust bearing, or more commonly by a combined thrust and journal bearing at the tail of the wind-shaft remote from the sails. The tails of wooden wind-shafts were invariably fitted with iron gudgeons, although separate journal and thrust bearings were sometimes used. A single housing was used to support both thrust and radial loads on the *tail bearings* of cast-iron wind-shafts. The thrust load was generally transmitted through a wooden or metal collar on the wind-shaft, but pivot bearings consisting of cast-iron balls in spherical seats have also been used as shown in Fig. 9.30.

One of the most remarkable tribological components of post-mills was the *pintle bearing*. It was the pivot on the top of the post which carried

Fig. 9.27
(a) Basic form of post-mill showing location of bearings.

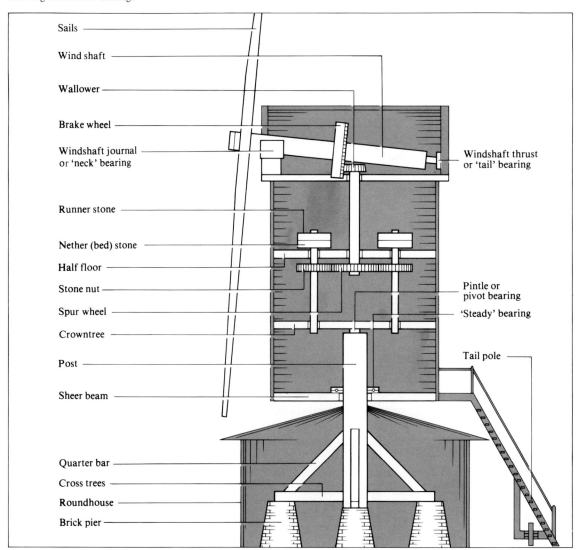

Sails

Wind shaft

Wallower

Brake wheel

Windshaft journal or 'neck' bearing

Windshaft thrust or 'tail' bearing

Runner stone

Nether (bed) stone

Half floor

Stone nut

Spur wheel

Crowntree

Post

Sheer beam

Pintle or pivot bearing

'Steady' bearing

Tail pole

Quarter bar

Cross trees

Roundhouse

Brick pier

the great weight of the mill superstructure, yet it was simple in form and generally made of wood. In some cases the top of the post was simply reduced in a single step to a cylindrical section some 7 in (177 mm) long and 7 in in diameter as shown in Fig. 9.31(*a*). Sometimes a two-step configuration was used, the first one having a diameter of 10–12 in (254–305 mm) and the inner one being about 7 in (178 mm) as before. In the case of the Fenstanton Post-mill in Huntingdonshire shown in Fig. 9.31(*b*) the height of each step was 5 in (127 mm).

It was customary to provide additional support to the top of the post by means of iron straps as shown in Fig. 9.31. The wear marks on the faces of the wooden collars are clearly seen, but it seems quite remarkable that such simple wooden bearings could be effective for such long periods of time. In most cases lubricant was supplied from the lower floor of the

(b) Basic form of tower-mill showing location of bearings.

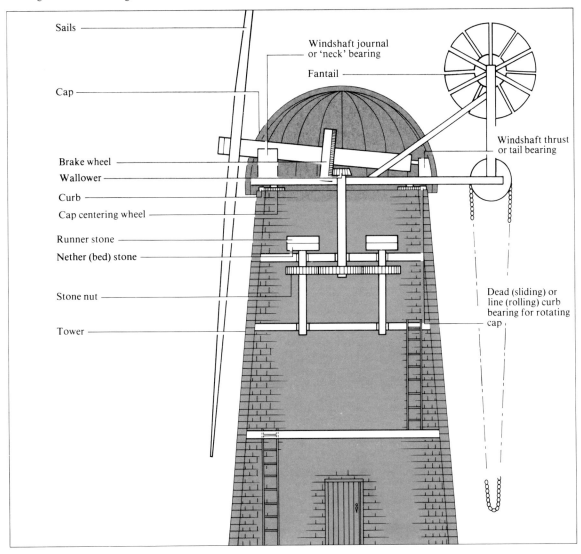

Fig. 9.28
Neck bearing from wind shaft of Thornham Mill, near Hunstanton, Norfolk.

post-mill through a hole cut in the bottom of the crown-tree or occasionally drilled in the crown-tree itself.[37]

The *steady bearings* used to provide stability to the delicately mounted post-mill superstructures were often simple wooden arrangements in the lower floor. There was, however, one outstanding example which deserves a place in the history of both windmills and rolling-element bearings.

A few miles to the east and slightly north of Norwich in Norfolk a post-mill was built about 1730 in the small village of Sprowston. It incorporated the highest level of achievement of the millwright's craft and established a reputation as a fine example of an English post-mill during its 200-year life. It attracted the attention of both artists and engineers. The founder of the Norwich School of Painting, John Crome, recorded it with some artistic licence in his famous picture of *A Windmill on Mousehold Heath* which now hangs in the Tate Gallery, London, and which is reproduced here as Fig. 9.32(*a*). The late Mr H. O. Clarke of Norwich made a fine and accurate 1 in 12 working model of Sprowston

37. I am grateful to Mr Rex Wailes for his comments on pintle bearings and for the illustrations shown in Fig. 9.31.

Fig. 9.29
Windmill neck bearings of stone and hardwood – preserved in The Falcon, Leiden.

Mill which now fascinates young and old alike as they visit the Science Museum in South Kensington, London. It is most fortunate that these artistic and engineering records still exist, together with an entertaining account of the history of the mill by the late Wing-Commander H. C. Harrison,[38] for the life of this famous mill was terminated in tragic circumstances in 1933. An illustration of the mill included in Wing-Commander Harrison's book (Harrison, 1949) is reproduced here as Fig. 9.32(*b*). Wing-Commander Harrison was born at the Mill House, Sprowston, in 1888 and his family had been connected with the mill since 1780. His father operated the mill from 1884 to 1920 and his younger brother, Mr H. H. Harrison, worked it during its last eight years of useful life until 1928. In due course, and after a sum of £100 had been subscribed to a preservation fund, it was arranged that the mill should be placed in the care of the Norfolk Archaeological Trust, the transfer being arranged for 25 March 1933. On the previous day, when the wind blew strong, some misguided person decided to burn off a copse of gorse brushwood on the windward side of the mill-hill. Showers of sparks and burning brush soon settled on the stationary sails such that the canvas-covered shutters and the entire wooden structure were soon ablaze. The millstones, the gears and the large cast-iron rings shown in Fig. 9.33(*a*) crashed into the round-house to complete the destruction of Sprowston Mill. Our main interest lies in the heavy cast-iron rings.

---

38. I am indebted to Mr F. R. Hall of Norwich for providing much interesting background information and a fine photograph of Sprowston Mill which he took in the mid-1920s.

Fig. 9.30
Windshaft thrust bearing in the
form of a cast-iron ball in a
spherical seat.

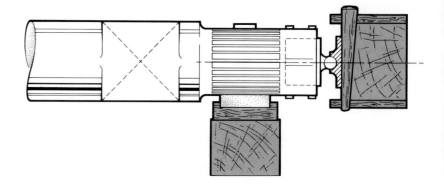

The cast-iron rings, which were presented to the Bridewell Museum, Norwich, where they can still be seen, proved to be the steady bearing from the sheer beams beneath the lower floor of the post-mill super structure. It was certainly built by 1780 and, being contemporary with Varlo's (1772) smaller carriage bearings, it ranks as one of the earliest known applications of metal ball-bearings in England. The cast-iron rings were large, having outer and inner diameters of 34 and 24 in (0·86 and 0·61 m), respectively. The rings were cast in halves, with bolts holding the sections together through lugs, and each ring had a thickness of $1\frac{5}{8}$ in (41 mm). Forty cast-iron balls of $2\frac{1}{4}$ in (57 mm) diameter were contained in grooves of $2\frac{3}{4}$ in (70 mm) diameter as illustrated in Fig. 9.33(*b*). The ratio of the groove to ball radius of 1·22 is somewhat larger than that encountered in the modern products of precision technology, but quite remarkable in concept and execution for the later eighteenth century. An outer hoop of wrought iron surrounding the lower ring (see Fig. 9.33(*a*)) was probably intended as a shield to exclude foreign matter from the grooves. It should be emphasized that this deep groove ball-bearing was employed to prevent wobble of the mill superstructure, and not to carry its weight, this being the function of the pintle bearing.

Clark (1938) has presented a detailed account of this historic ball-bearing and further descriptions can be found in the books by Harrison (1949) and Wailes (1954). This completes our examination of the special bearing features of post-mills and it now remains to look at *cap bearings* in tower-mills.

The essential form of tower-mills is the fixed tower made of stone, brick or even timber and a movable cap carrying the wind shaft and the sails. The track on which the cap turns on top of the tower is known as a *curb*. These curbs were initially made of wood and wooden skids attached to the cap provided simple sliding bearings. Similar skids arranged round the periphery of the cap centred the cap by running against the side of the curb. Such arrangements in which simple sliding took place between the cap and the curb on the tower were known as *dead curbs*. Wear and high friction must have posed serious problems, and in due course the tops and sides of curbs were faced with iron and iron or gun-metal skids were attached beneath the caps.

The advantages of rolling over sliding were quickly recognized for this bearing situation, and in due course rollers were inserted between

Fig. 9.31
Pintle bearings from post-mills.

(a) Single-step pintle from
a post-mill at Kirkby-Green,
Lincolnshire (nominal size of pintle
6 × 6 in (152·4 × 152·4 mm)).
(b) Double-step pintle from a
post-mill at Fenstanton,
Huntingdonshire (nominal size of
pintle and lower step 7 × 5 in
(177·8 × 127 mm) and 11 × 5 in
(299·4 × 127 mm), respectively).

(a)

(b)

(a)

(b)

Fig. 9.32
(a) John Crome's painting *A Windmill on Mousehold Heath*, showing the post-mill at Sprowston, near Norwich.
(b) Sprowston post-mill (*c.* 1730–1933).

the cap and the tower. Such an arrangement is described as a *live curb*. The rollers, normally made of iron, were generally located in the cap with an iron track or curb on top of the tower. Ramelli showed rollers located in a horizontal ring in the base of the cap of the rag-and-chain pump depicted in Fig. 9.26. In live curbs rollers or truckles located by vertical pins ran against the sides of the curb to centre the cap. In a variation of the live curb arrangement known as the *shot-curb*, metal rings face the top of the tower and the underside of the cap and an independent ring of rollers in a form of retaining cage is inserted between the two. Similar roller-rings were used in the base of the Dutch *paltrok mills* as explained in Chapter 8. In Holland the rollers in such mills were made of wood, whereas in England cast-iron rollers were usually employed in live curbs.

In this section it has been shown that millwrights had to face considerable bearing problems. The millwrights and craftsmen of former years achieved great success in constructing viable mills. An impressive feature of mill history is the complex nature of the structure and machinery and the effectiveness of simple bearings and mechanical drive systems, coupled with concern for pleasing and, many believe, beautiful features of the landscape. Stone and wood were used extensively in the bearings of both windmills and water-mills throughout most of their effective span of history, although metal bearings became standard in later years. There were numerous additional and interesting tribological problems associated with the gears and bearings on the drive shafts to the stones which cannot be covered in the present text. Animal fats were widely

Fig. 9.33
(a) Ball-bearing from Sprowston post-mill (*c.* 1780), (outer diameter 34 in (0·86 m); inner diameter 24 in (0·61 m)).
(b) Ball-bearing from Sprowston post-mill (*c.* 1780), (showing some of the forty cast-iron balls of $2\frac{1}{4}$ in (57 mm) diameter).

(a)

(b)

used as lubricants, particularly for neck bearings, but Mr Rex Wailes[39] has drawn my attention to the use of pitch in wind-shaft bearings in Finland. Harrison (1949) has described how the neck-brass supporting the cast-iron wind shaft in the Sprowston Post-mill was lubricated. A

39. Wailes, R., Private communication, Nov. 1974.

large piece of *tallow* was smacked on to the brass adjacent to the wind-shaft. As the bearing warmed up, the tallow would melt and lubricant would run in between the shaft and the brass.

The efficiency of a lubricant cannot be assessed without reference to the bearing materials. One firm of Lincolnshire millwrights[40] told me of an old mill in which the brass neck bearing lasted but a week or two. When the brass was replaced by wood (thorn) the bearing lasted for years. The general development of lubricants throughout this period is discussed in Section 9.6. The reader is advised to consult the papers and books already mentioned, together with the text by Freese (1971), for more detailed information.

The power losses in these early sources of mechanical power were considerable. It can be shown that the theoretical aerodynamic efficiency of windmills can achieve a maximum value of 59 per cent, but the overall efficiency rarely exceeded 10 per cent and was generally nearer to 5 per cent. Immediately after the Second World War Dutch engineers demonstrated that a significant improvement could be achieved by giving special attention to the aerodynamic performance of the sails and by using more efficient bearing systems. It is quite possible that further improvements could be made by applying current tribological expertise, but this will no doubt be considered as the modern wind-turbine replaces the aesthetically pleasing traditional windmill.

9.6   Lubricants and lubrication

It is convenient to review progress in the art of lubrication at this time for three reasons. In the first place the science of fluid mechanics was established and the equations governing continuum mechanics and viscous flow which were to form the basis of subsequent understanding of fluid-film lubrication were developed in the period under review. Secondly, industrialization and the demands of the textile industry and the railways in particular, were responsible for a great deal of empirical development of animal, vegetable and even mineral oil lubricants. The third and probably the most important reason for this review is that the animal and vegetable oils which had carried the burden of machine lubrication throughout the ages, were about to be largely replaced by mineral oil from shale or from flowing or pumping wells.[41]

Fluid mechanics – viscous flow and viscosity

Leonhard Euler (1707–83) is normally regarded as the founder of the science of fluid mechanics, although Daniel Bernoulli (1700–82) published a book (Bernoulli, D. (1738)) on the subject in 1738. Euler presented the equation of continuity and his well-known equations of motion for an ideal (non-viscous) fluid.

40. Thompson, J. E., Private communication, Nov. 1972.
41. See Dowson, D. (1974), pp. 1–8 for an account of the transition from animal and vegetable lubricants to mineral oils.

Continuity equation

$$\frac{\mathrm{D}\rho}{\mathrm{D}t} + \rho\left[\frac{\partial u}{\partial x} + \frac{\partial v}{\partial y} + \frac{\partial w}{\partial z}\right] = 0.$$

Euler's equations of motion

$$\frac{\mathrm{D}u}{\mathrm{D}t} = -\frac{1}{\rho}\left[\frac{\partial p}{\partial x}\right],$$

$$\frac{\mathrm{D}v}{\mathrm{D}t} = -\frac{1}{\rho}\left[\frac{\partial p}{\partial y}\right],$$

$$\frac{\mathrm{D}w}{\mathrm{D}t} = -\frac{1}{\rho}\left[\frac{\partial p}{\partial z}\right].$$

Western Europe dominated the next stage in the development of the subject through the work of great scientists like Lagrange (1736–1813), d'Alembert (1717–83), de Borda (1733–99) and Venturi (1746–1822).

Studies of the flow of water in a canal led the Frenchman Chezy to present a mathematical relationship governing steady, viscous flow, while a military engineer and fellow countryman by the name of du Buat (1734–1809) carried out systematic experimental studies of the role of friction (viscosity) upon the flow of fluids in pipes and canals. It was, however, a French physician Jean M. L. Poiseuille (1799–1869) who first established experimentally the law governing viscous flow in pipes through his thorough investigation of the flow of blood in narrow capillaries. Poiseuille (1846) introduced his study with the words;

> ... I began my investigations because progress in physiology demands a knowledge of the laws of motion of the blood, which is to say, a knowledge of the laws of motion of fluids in small-diameter pipes. Of course, Du Buat, Girard, Navier and others have already studied this problem: but it is desirable to have further study and experimental investigation, so as to have reliable comparison of theory with experimental data.

The name given to the unit of dynamic viscosity used in scientific work has acknowledged for a long time the contribution to studies of viscous flow by Poiseuille through the Poise (cgs system), proposed by R. M. Deeley and P. H. Parr in 1913, and the Poiseuille or Pascal second (SI system).

Hagen confirmed Poiseuille's findings in later independent work in Germany and the formula for viscous flow through a circular section tube is usually known as the Hagen–Poiseuille equation,

$$q = -\frac{\pi R^4}{8\eta}\left[\frac{\mathrm{d}p}{\mathrm{d}x}\right].$$

It was the great French mathematician Claude Louis M. H. Navier (1785–1836) who first introduced viscous terms into the general equations of motion. He used the symbol ε to represent viscosity and in the classical paper which he presented to the Académie des Sciences on 18 March 1822 he wrote: '... With this assumption, the constant ε represents in units of weight the resistance occurring from sliding to any two layers one against another for unit area.'

Navier (1823) was thus the first person to include the viscous terms in an expression of Newton's laws of motion for a flowing fluid. The same set of basic equations governing viscous flow were later obtained rigorously and independently by the English physicist and mathematician Sir George Gabriel Stokes (1819–1903) (Stokes, 1845) and thus they are universally known as the Navier–Stokes equations. They can be written as follows:

The Navier–Stokes equations of motion

$$\rho \frac{Du}{Dt} = \rho X - \frac{\partial p}{\partial x} + \eta \left[\frac{\partial^2 u}{\partial x^2} + \frac{\partial^2 u}{\partial y^2} + \frac{\partial^2 u}{\partial z^2} \right],$$

$$\rho \frac{Dv}{Dt} = \rho Y - \frac{\partial p}{\partial y} + \eta \left[\frac{\partial^2 v}{\partial x^2} + \frac{\partial^2 v}{\partial y^2} + \frac{\partial^2 v}{\partial z^2} \right],$$

$$\rho \frac{Dw}{Dt} = \rho Z - \frac{\partial p}{\partial z} + \eta \left[\frac{\partial^2 w}{\partial x^2} + \frac{\partial^2 w}{\partial y^2} + \frac{\partial^2 w}{\partial z^2} \right].$$

The term *coefficient of viscosity* appears to be of relatively recent origin. Navier did not use it and Stokes refers to 'μ' in the expression for viscous stress in a Newtonian fluid $\tau = \mu(du/dz)$ as the *index of friction*.

This background to the stirring developments in the field of viscous flow theory must necessarily be brief, but the interested reader will find a more complete account in Tokaty (1971). The stage was set for the application of these governing equations of viscous flow to the problems of lubrication in the 1880s, but in the meantime there was ample scope for empirical, but nevertheless impressive, developments in the art of lubricant blending and application by practical men of the day.

Lubricants

Traditional animal and vegetable sources satisfied the ever-increasing demand for lubricants throughout the Industrial Revolution, but the position changed dramatically in the mid-nineteenth century. Mineral oils, which had previously been available only in relatively small volumes by distillation from shale, emerged in large quantities from flowing and pumping wells to form the major source of lubricants. The 1850s witnessed a far-reaching transition in the origin of lubricants and the start of a petroleum industry which was to support and be vital to industrial expansion in the nineteenth and twentieth centuries.

It is both interesting and instructive to examine the list of lubricants most commonly employed during the Industrial Revolution. A selection is shown in Table 9.3.

Sperm oil was most highly prized for its ability to reduce friction and for its stability, but was also the most expensive lubricant of the day. It was obtained from the large cavity in the head of the sperm-whale and was particularly valuable for lightly loaded spindles and general machinery.

Table 9.3
Lubricants of the Industrial
Revolution era

Liquid lubricants			Solid lubricants
Animal	Vegetable	Mineral	
Sperm oil	Olive oil	Mineral oil	Graphite or
Whale oil	Rape-seed oil	(by distillation	plumbago
Fish oils	Palm oil	or other forms,	Soapstone or
Lard oil	Coconut oil	or refining from	talc
Neat's-foot oil	Castor oil	crude oil, shale,	Molybdenum
Tallow oil	Ground-nut oil	coal or wells)	disulphide

Whale oil, obtained by heating the blubber of any of the whales, was more disagreeably odorous, rarely used in lubrication, but extensively used in the making of soaps and illuminants.

The species was reprieved and the whaling industry almost ruined by the growth of the petroleum industry and the increasing use of paraffin as an illuminant. A century later, when the emphasis was largely, but by no means exclusively, on meat, the future of the whale was again to be seriously threatened. *Fish oils* found little application in machinery, although the oils of cod and porpoise found occasional use.

Lard oil was much cheaper and in consequence much more widely used than the excellent sperm oil. It was obtained from the fat of pigs and was held to be an excellent lubricant. *Neat's-foot oil*, obtained by boiling the feet of cattle, was similarly rated, particularly for low-temperature applications. It appears that *tallow oil*, which resulted from the application of pressure to the tallow of cattle, was also a good lubricant, but it was more widely used in fine soap.

The wide range of vegetable oils found round the world were normally extracted from the seeds or fruit of plants. By far the most common source of vegetable oil for lubricating purposes during the Industrial Revolution was *olive oil*, extracted from a member of the jasmine family, *Olea europea*. This plant provided fine table oil as well as illuminating and lubricating liquid. The best olive oil used for lubrication was said to be the top-ranging vegetable oil and fully equal to sperm oil. It has a much higher viscosity than the fine sperm oil, thus enabling it to be used for heavier duties.

In climates inhospitable to the olive, an alternative source of lubricant was provided by the seeds of various forms of brassica. The resulting *rape-seed-oil* found wide application and became an important contributor to the array of lubricants in industrial countries. *Palm oil* was imported in large quantities from the East and the West Indies by the industrial nations of Europe. It was widely used in general manufacture and to a limited extent as a lubricant, either alone or, more frequently, as a constituent of a special lubricant formulation. The nut of the *cocoa-palm* yielded a semi-fluid substance (melting point 69°F (21°C)) which was one of the more stable vegetable oils, but its use as a lubricant appears to have been limited.

Both *ground-nut oil* and *castor oil* tend to thicken with exposure to the air, but they are nevertheless good lubricants. The former, obtained from ground-nuts or pea-nuts, found similar application to olive oil, while the latter, from the *Ricinus* plant, rapidly developed a reputation as a viscous lubricant of great merit for lubricating severely loaded machinery.

Mineral oils, known throughout the ages in various countries, were first processed to provide illuminating fluid. Evans (1921) has noted that pitch, tar and oil had been obtained from a kind of stone in the late seventeenth century; the oil being sold as Betton's British Oil to cure rheumatism. Shale was the main source of the oil used for distillation in the eighteenth and early nineteenth centuries. Small distillation plants were built in Prague in 1810 and, on a larger scale, by Selligue in France in 1834. At the very end of our period James Young patented a process for producing paraffin oil from coal which was to place Scotland in the forefront of petroleum production in the latter half of the nineteenth century. The full story will be told in Chapter 10. The role of mineral oils in the industrial processes of the day was initially quite small, but from these insignificant beginnings was to grow a great industry capable of providing vast quantities of hydrocarbon lubricants at very low cost.

It is commonly thought that solid lubricants are a relatively recent phenomenon, but their use in the lubrication of heavy, slow-moving machinery was well established during the Industrial Revolution. An indication that solids were viable lubricants is provided by the references to the use of black-lead and an anti-attrition composition of black-lead and hogslard in the friction experiments of Rennie (1829). But what was this *black-lead* or *plumbago*?

In Roman times rods or discs of metallic lead, which had been known since prehistoric times, were used to draw guidelines for subsequent writing in ink. At a later date tin and possibly other metals were added to the lead to give a more intense mark, but the practice appears to have lapsed in the Middle Ages. In the sixteenth century a fine deposit of graphite, an allotrope of carbon, was discovered in Cumberland and the material was soon put to use as a writing material. The material provided a fine black line and its resemblance to the metallic lead used previously caused the term *black-lead* to be introduced. In 1779 the material was shown to be a form of carbon and although the name graphite was proposed ten years later, the terms 'black-lead' and 'lead' pencils are still in general use. A full account of the use of graphite in pencils has been given by Voice (1950), but it is the recognition of the virtues of the material as a lubricant which is now recorded.

Graphite of great purity was later found in Ticonderoga in New York State, Ceylon and Siberia. It was used for lubricating lightly loaded machinery and also for silk-looms. The various references in nineteenth-century patents to mixtures involving plumbago show that it was widely appreciated as one of the earliest forms of dry lubricant.

Campbell (1970) has suggested that *molybdenum disulphide* was first used as a lubricant in the great trek west in the United States in the 1880s.

Soapstone or *talc* in powder form also found application as a dry lubricant on rotating axles as described by Thurston (1885). Vogelpohl (1969) has noted in an interesting historical article that eighteenth-century books on mills contained references to solid lubricants. Talc was recommended as a lubricant in 1785, and in 1800 graphite was used in the Hungarian mines. In due course the men who built and maintained machinery, particularly in mills, developed the habit of carrying with them small quantities of graphite or even flowers of sulphur. Applications of small quantities of these secret ingredients to bearings which were running hot or experiencing other forms of distress sometimes had dramatic results, thus enhancing the reputation of the so-called mill doctors. This list, although by no means complete, throws light on the types of lubricant in common use throughout the Industrial Revolution. Greases, or semi-fluid lubricants, based upon either beef or mutton tallow, were still used extensively for lubricating machinery, but it is the patent specifications of the day which reflect the ingenious, though empirical and occasionally amusing developments of lubricants in the early nineteenth century.

On 6 August 1812 in the reign of King George III, British Patent No. 3573 was awarded to Henry Thomas Hardacre of the parish of Saint Mary-le-bone for his invention of 'A Composition to Prevent to a very Extensive Degree the Effects of Friction'. The specification read: '. . . One hundredweight of plumbago to four hundredweight of pork lard or beef suet, mutton suet, tallow, oil, goose grease, or any other kind of grease or greasy substance, but pork lard is the best, which must be well mixed together, so as to appear to the sight to be only one substance.'

Hardacre's (1812) specification demonstrates the use of heavy grease consisting of considerable quantities of the dry lubricant graphite mixed with lard or suet. Instructions were given on the methods of application and amounts of the composition to be used in bearings, steam-engine piston rods and the stone spindles of mills.

In the reign of King William IV, a gentleman from Liverpool bearing the name Henry Booth was awarded the British Patent No. 6814 for patent axle greases and lubricating fluids (Booth, 1835). In the former, solutions of common washing soda were mixed with '. . . three pounds of good clean tallow and six pounds of palm oil'. The mixture was heated to about 200°F (93°C), agitated then allowed to cool and to develop a consistency like that of butter. Such grease was said to be suitable for carriage axles and the axles of every description of railway carriages. The *lubrication fluid*, said to be applicable to the rubbing parts of machinery, was manufactured in a similar way, but with the addition of rape oil.

Nathaniel Partridge of Stroud had a similar composition approved in British Patent No. 6945 (Partridge, 1835). His anti-attrition paste was made primarily from olive oil and a solution of lime in water. For heavier duties, like the lubrication of '. . . cogs or teeth of wheels . . .', whale oil was preferred to olive oil and the paste thus formed could be thickened with additions of palm oil or tallow and small quantities of carbonaceous matter such as plumbago, black-lead or soot. The merits of carbon as a lubricant had been recognized long before the physical mechanism of lubrication had been resolved.

A strap manufacturer from Bermondsey by the name of Thomas Denne mentioned the use of an unctuous black substance known as vegetable black, or otherwise the ordinary lamp-black of commerce in his specification covered by British Patent No. 11674 (Denne, 1847) for a composition suitable for atmospheric pipes and for lubricating axles and the moving parts of machinery.

In the eleventh year of Queen Victoria's reign Joseph John Donlan of Staffordshire was awarded British Patent No. 12109 for 'Compounds for Lubricating Machinery' (Donlan, 1848) of quite an extraordinary nature. Both novelty and effectiveness were claimed for the compounds with the words:

> ... Nor do I believe any ordinary chymist, unless instructed so to do, could command sufficient knowledge to convert these three chemical kingdoms, namely, animal, vegetable, and mineral, into one pure and liquid amalgamation. And I do solemnly believe that this union of bodies is the first and only combination ever yet produced, competent or capable of preventing friction, heat, and vibration in locomotive machinery.

The compounds were intended to solve the tribological problems of the day associated with locomotive, marine and stationary engines. They included:

> ... The green leaves of the vegetable substance called laurel; ... the green leaves of the vegetable substance called ivy; ... the oxymuriate of mercury; ... subnitrate (or trisnet) of bismuth; ... sulphate of copper; ... sulphate of zinc; ... tartarized antimony; ... pulverized Roman alum; ... Greenland or southern whale oil; ... Gallipole or olive oil; ... sperm oil; ... palm oil; ... the vegetable substance called caoutchouc; ... brown soap common; ... minimum; ... white pepper; ... ceruss; ... common salt and powdered alum.

In addition it was noted that black-lead and celestial blue could be added in powder form to colour the compounds.

The mid-nineteenth century must truly have witnessed the zenith of the entrepeneur's skill in concocting lubricants before the large-scale introduction of mineral oils.

The use of mineral oils obtained by distillation was imminent. William Little obtained Patent No. 12571 for the preparation and use of such substances in England for lubricating machinery such as the axle-trees of locomotives (Little, 1849). The idea apparently came to him from '... a certain foreigner residing abroad'. Hompesch's (1841) Patent Specification No. 9060 was concerned with improvements to the methods then in use for the extraction of oils from bituminous materials such as schist or clay slate and asphalt. At the very end of the period under review James Young firmly established the process of extracting mineral oil from coal (Young, 1950), but the story belongs more conveniently with the history of the late nineteenth-century petroleum industry told in Chapter 10.

Lubrication systems

The need to supply lubricants to the surfaces of bearings had been recognized for hundreds of years before the Industrial Revolution. In most cases the lubricants of vegetable and animal origin had been applied intermittently by hand, but the new machine age called for more continuous feeding systems. There are, however, a few indications of previous concern for this subject.

A drill driven by a water-wheel had been found by de Caus (1615) to contain a bearing complete with a filling hole for the lubricant. Davison (1957*b*) has noted that a continuously lubricated bearing had been found on a lathe in 1792, but such indications were rare prior to the age of steam. On the Liverpool and Manchester Railway Stephenson lubricated the carriage axles by a wick-feed system. A tin box containing the oil was attached to the main frame of the carriage and oil was conveyed to the bearing surface by means of a cotton wick. Such systems, which are usually incapable of ensuring flooded lubrication, have been used in various forms until the present day. Lubricators soon became standard features of machinery. Henry Booth,[42] whose 1835 patent for axle greases and lubricating fluids was mentioned in the previous section, also conceived an ingenious supply arrangement. The grease was placed into a box on the carriage frame and allowed to pass to the bearing via a hole. When the bearing became dry and hot the grease melted and the bearing received a fresh charge of lubricant. The need to supply the bearings of moving carriages and locomotives with a constant source of lubricant undoubtedly promoted the development of lubricating systems during the early nineteenth century. Progress was such that for a while the bearings on moving vehicles were probably better protected than those on fixed machinery.

The next stage was to utilize the principle of viscous lifting, whereby a component loosely or rigidly fastened to the shaft was allowed to dip into a reservoir and thereby pump fluid to the top of the bearing by its rotation. In some cases the shaft itself served this function, but it was more usual to use rings, discs, chains or even ladles. Patents for both the ladle and loose ring forms of oil circulation were awarded to A. E. Jaccoud of Vienna in 1829 and a disc lubricator was patented by Baudelot of Harancourt in 1838. The chain lubricator was attributed to M. Decoster of Frankfurt in 1847 and exhibited at the Great Exhibition in London in 1848 by J. Hick.

Vogelpohl (1969) has given a valuable and complete account of these and other early nineteenth-century developments of lubricating systems. These elementary atmospheric pressure pumps were adequate for immediate requirements, but Mr Beauchamp Tower's observations in the 1880s were to place new emphasis on the importance of lubricant supply systems.

9.7 Wear testing

Scientific studies of wear developed little until the mid-twentieth century, but workers in the field were then quick to recognize the vagaries of experimental work. The lack of repeatability of tests on a given apparatus and

---

42. See Wood, N. (1838) p. 201.

the divergence of results from different forms of equipment presented both a challenge and a problem. The laws of wear are similar in number and simplicity to the laws of friction, although the complexities of the underlying surface interactions which govern both processes are widely recognized. One of the laws is the inverse dependence of wear upon hardness, and it is generally thought that the recognition of this relationship belongs to the present century. There is, however, in a paper by Mr Hatchett, F.R.S. dated 1803, rewarding reading and some little-known illumination on this point.

The problem arose from concern in the United Kingdom in the reign of King George III (1760–1820) for the coinage. On 10 February 1798 His Majesty established a powerful Committee of the Privy Council to '. . . take into consideration the state of the coins of this kingdom . . .'. The Committee was concerned with the '. . . considerable loss which the gold coin appeared to have sustained by wear within certain periods . . .'.

Two Fellows of the Royal Society, Mr Henry Cavendish and Mr Charles Hatchett, were invited to examine the problem and it is Mr Hatchett's voluminous paper, occupying no less than 151 pages of the *Philosophical Transactions*, which is the subject of our story.

Mr Hatchett clearly recognized the mechanical nature of wear, for in the opening pages of his paper he wrote: '. . . The wear of coin is an effect produced by mechanical causes, subject to be modified by certain physical properties, such as ductility and hardness, . . .'.

Much of his investigation was devoted to a study of the general characteristics of alloys of gold. Emphasis was placed upon the specific gravity of the alloys used in coins of the realm and he was able to demonstrate the impropriety of estimating the value of small numbers of coins on this basis.

It is the last third of Mr Hatchett's lengthy paper which is devoted to wear, and his experiments had two main objectives. In the first case the effect upon diminution, or wear, of the gold coin resulting from rubbing upon similar gold, copper or silver pieces was explored and in the second series of tests the ability of various alloys of gold to resist abrasion by sand, metal filings and gritty powders was investigated.

The very fine hand-driven reciprocating wear-testing machine shown in Fig. 9.34(*a,b*) was used to study the diminution of various coins sliding together. This must have been one of the earliest wear-testing machines. It could accommodate up to twenty-eight specimens or coins, had a maximum stroke of $\pm\frac{3}{8}$ in (9·5 mm) and a static loading system. Lead weights were used to press the coins together, the normal load on each coin being 21,429 grains (about 13·6 N). This relatively large load was employed in order to achieve measurable changes in weight in reasonable times. The argument for accelerated wear tests, coupled with a recognition of the disadvantage of such procedures, is evident in one of Mr Hatchett's footnotes.

> . . . This weight may appear to be very considerable; but it was not employed until repeated trials had proved the extreme difficulty, and almost impossibility, of producing any perceptible effect within a moderate period of time; and, even with this weight, the experiments were found to be exceedingly tedious. The only evil

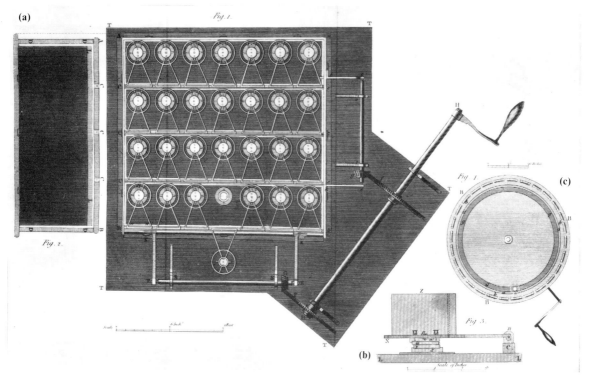

Fig. 9.34
Hatchett's wear-testing machines (Hatchett, 1803).
(a) Reciprocating machine for rubbing together twenty-eight coins.
(b) Enlarged view of one pair of coins (u, y) and dead-weight (z) in machine.
(c) Abrasive wear-testing machine.

which resulted from such a pressure was, that the comparative wear of the fine gold appeared much more considerable than would have been the case, if a small weight could have been employed. . . .

One of the most remarkable aspects of the reciprocating wear-testing machine shown in Fig. 9.34(a) was the design feature which caused the direction of rubbing to vary continuously to prevent the formation of '. . . little furrows or gullies . . .' in the test coins which promoted rapid wear. This was achieved by moving the frames containing the upper and lower layers of coins in perpendicular directions and at different rates. The gear wheels (F) and (f) attached to the drive shaft (H) had 90 and 75 teeth respectively, and the mating wheels (G) and (g) 20 each. The shafts (E–E) and (e–e) thus rotated in their fixed bearings at speeds in the ratio 6 : 5. The ends of each shaft were turned to provide eccentrics for driving the supporting tables through links (EK) and (ek). This simple yet ingenious mechanism ensured that the coins did not rub continuously over the same wear path, a facility which many investigators build into current wear-testing machines.

A second machine evaluated the resistance of coins to both rubbing and impact. It was delightfully simple, being an oak box of cubic form having sides 8 in (0·20 m) long, which was filled with 200 coins and tumbled about the fixed axis (E–E) of the apparatus shown in Fig. 9.34(a).

The machine used for the abrasive wear tests is shown in Fig. 9.34(c). It consisted of a turntable in which coins were loaded against the 29 in (0·74 m) diameter rim with various abrasive powders being applied to a shallow groove forming the wear track.

Experiments were carried out on smooth, unstamped specimens as well as upon stamped coins and wear was measured by recording the initial and final weights of the pieces. Many of the problems of wear testing familiar to those who have engaged in the exercise are reported in the paper. Mr Hatchett states that although the tests were carried out '... with every possible precaution ... perfect accuracy could not be attained, nor indeed expected; for, various minute and unavoidable circumstances contributed to produce very sensible effects; even a few particles, collected and retained between the pieces during the operation, frequently prevented the loss by friction from being correctly ascertained ...'.

He also found that the fine gold coins has their embossed parts obliterated not so much by abrasion as by having their protuberant parts pressed and rubbed into the mass. The parallel roles of plastic flow and wear of soft materials were thus noted at an early stage.

It is interesting to note that the term 'friction' is used throughout the text in a manner which would now be thought to be ambiguous. It was clearly seen to be intimately linked with, and indeed the source of, mechanical wear.

The tests were carried out for varying numbers of revolutions of the handle up to 229,000. Since it took about 40 hours to turn the drive shaft to the tumbling box some 72,000 revolutions, the long-term nature of effective wear testing was evident. It is also noted that the 1 in (25.4 mm) thick sides of the oak box were nearly half-worn through in this period, an indication that in many instances wear testing can tax the machine almost as much as the specimens.

It is also interesting to note that in some experiments the weights of the test pieces actually increased, an observation which must have been made on inumerable occasions and with varying degrees of irritation since the dawn of the nineteenth century.

The dominant findings reported by Mr Hatchett were as follows:

1. Fine gold (23 carats $3\frac{3}{4}$ gr) rubbing against like material suffered more wear than alloyed gold (with copper and/or silver). This implies that when like materials rub together the amount of wear is inversely proportional to the degree of ductility.
2. When metals of different hardness slide together the more ductile metals always wear more than those which are harder.
3. Metal transfer takes place from the softer to the harder material. The coating being '... commonly spread thinly over the surface; but, in some few instances, (especially when a very hard metal rubbed against one which was very soft), the particles of the latter, instead of being spread over the whole surface, became accumulated, so as to form little protuberances or knobs'.
4. The wear of copper sliding against copper was much greater than for the other metals tested.
5. Earthy powders and metallic filings used in the abrasive wear tests tend to wear the different materials (alloys of gold) in proportion to the respective ductility.

Mr Hatchett concluded that most of the wear of gold pieces was attributable to abrasive action by extraneous and gritty particles encountered by the coins in circulation rather than the wear resulting from the rubbing of adjacent coins. He also decided that the total wear due to both mechanisms could not account for the great and rapid diminution observed in the gold coins of the land.

The deterioration of coinage has been the subject of various studies since and even before Mr Hatchett considered the problem. When Spurr (1965) considered the matter again he was able to report his findings in a mere three pages.

9.8 Tribology and the railways early in the nineteenth century

The stimulus afforded to tribology by the development of the railways during the Industrial Revolution inevitably came to the surface in the early sections of this chapter. It would, however, be less than just to this history if a special, albeit brief, section was not devoted to this topic.

Wheel–rail adhesion

There are innumerable texts devoted to the history of railways, but some of the books written prior to 1850 provide the most instructive and fascinating references to tribological matters. Few excel those descriptions of practical matters written by Wood (1838) and Bourne (1846). In his history of the Great Western Railway, John Bourne drew attention to *gravity* and *friction* as the two forces which always resisted the natural efforts of mankind to move themselves and their commodities to and fro upon the earth. He further distinguished between *sliding* or *dragging friction* and *rolling friction* and noted some of the factors which affected each. *Rolling friction* was attributed *to some cohesion between the surfaces* and *the presence of dust, and other small bodies, which the cylinder crushes or displaces.* He argued that on a good *road* the *rails* should be *level*, *smooth* and *hard*. Permissible inclinations on the new railroads were much lower than those on existing turnpikes, the former being at least an order of magnitude smaller than the latter and preferably about 1 in 300.

The ability to transmit adequate tractive effort from the moving engine to the track loomed large as a problem for this new mode of transportation. Bourne's view of the situation was well told in the following extract from his book.

> . . . *One of the limits of the propelling power of the locomotive engine will of course be the adhesion or friction between the driving wheels and the rail, since, should the engine be required to draw too great a load, or to ascend too steep a hill, or in anyway to exert a power too great for this adhesion, even if it be capable of exerting such a power, the wheels would slide or slip round upon the rail without advancing the engine. The adhesion of the wheels depends partly upon the weight which is thrown upon them, and partly upon the state of the rails. It is possible to increase the power of adhesion of the engine by coupling or connecting its other wheels with the driving wheels, so that all turn round together, and the engine thus receives the benefit of the adhesion of the whole. Engines whose power of draught is intended to be considerable, to draw luggage, or ascend steep inclinations, are so constructed.*

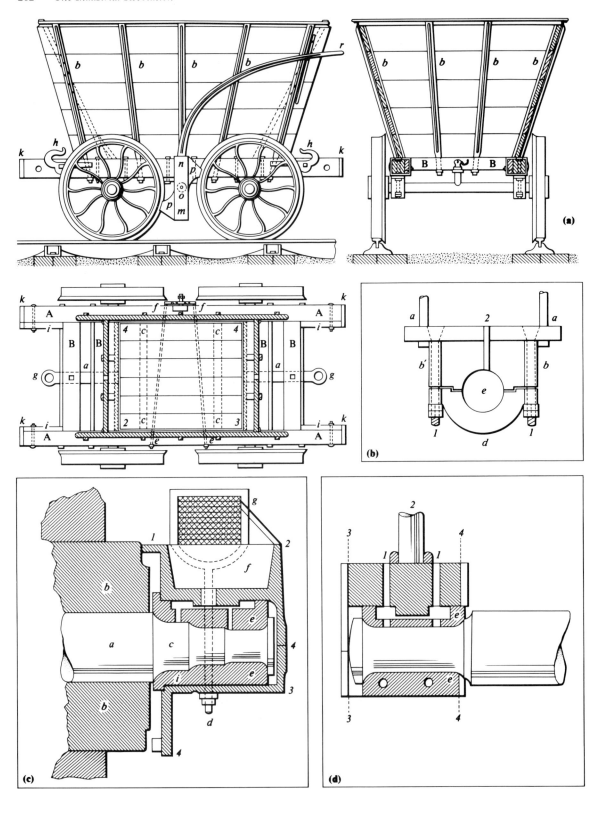

Fig. 9.35
Wood's illustrations of typical
forms of railway waggons and
bearings in the 1830s.
(a) Representative waggon used
for carrying coal.
(b) Typical 'inside' half-bearing of
brass showing the oil hole '2' and
dust-cover 'd'.
(c) Typical 'outside' brass bearing
showing lubricant reservoir 'f' and
two feed holes.
(d) 'Outside' bearing with two
lubricant feed-holes used on the
carriages of the Newcastle and
Carlisle line.

The limitations imposed by variable and often inadequate wheel–rail
adhesion still represent a problem to the railways. Diesel locomotives
are designed to make use of an adhesion coefficient of about 0·24,[43]
but values slightly less than 0·15 and as high as 0·70 have been recorded
in laboratory and track studies. This wide range of adhesion coefficients
well reflects Bourne's view that much depends upon the state of the rails.
It has been shown that the coefficient depends greatly upon small
quantities of debris or contaminants, including grease and oil from the
engine, autumn leaves, clay, cement, sawdust, coal and rust. Moisture
plays a significant role and hence the traction coefficients fluctuate with
the humidity and variations from dry to rainy days.

Bearings for railway wagons

By the time of the great railway boom early in the nineteenth century,
the bearings for road vehicles like the stagecoach and heavy wagon had
reached, through a long period of evolution, a satisfactory stage of
perfection and reliability. It was, therefore, quite natural that railway
engineers should initially think in terms of a simple extension of such
bearings to the new mode of transportation.

On 23 September 1826 George Stephenson wrote to one of the most
celebrated mining engineers of the day that,[44] '. . . We are fitting up some
Bearings for Waggon axles on the principle of the Mail Coach; which
plan from your experience you must be aware is much wanted in Coal
Waggons.'

The simple wooden and iron bearings had given way to the use of
better combinations of bearing materials including bronze, brass and, in
due course, the tin-based alloys introduced by Isaac Babbitt.

The form of bearings on wagons used for carrying coal in the 1830s
and the method of their lubrication was well described and illustrated
by Wood (1838). A typical wagon, together with various bearing arrangements, is shown in Fig. 9.35. For wheels carried outside the framework
of the wagon or carriage, simple half-bearings made of solid brass to the
form shown in Fig. 9.35(*b*) were used. Lubricant was supplied through
the hole marked '2', and 'd' represented a dust cap.

When heavier loads had to be transported the bearings were mounted
outside the wheels as shown in Figs 9.35(*c*) and (*d*), the latter being an
example from carriages employed on the Newcastle and Carlisle line.
In both cases lubricant reached the brasses through two holes as shown.

In all these examples of early wagon bearings the lubricant supply
holes were located on the load line in the loaded half of the bearing,
an arrangement which the scientific studies of lubrication in the 1880s were
to expose as being most unfortunate. Correct location of the oil holes is a
lesson which has been hard to learn and designers were to develop a
mystique of favoured arrangements for holes and grooves which has
lingered on to the present day. Dissatisfaction with the relatively poor

43.   See Broster, M. et al. (1974), **29**, 309–21.
44.   See Skeat, W. O. (1973), p. 102.

performance of railway wagon and locomotive bearings compared with that of the longer standing and highly developed road carriage bearings mounted in the next decade and prompted von Pauli's (1849) studies in the Nuremberg carriage works. It was also, in due course, to provide the stimulus for scientific studies which resulted not only in better bearings, but the exposure of the physical basis of fluid-film lubrication in the 1880s.

Grease was widely used as a lubricant on the railways and Dowson (1974) has observed that the steam locomotive promoted interest in automatic lubricant feeding systems, since it was neither convenient nor possible to employ a man to walk round the bearings applying grease or oil to a machine moving through the countryside at speeds approaching 25 mph (40 km/h).

Yes, the railway boom in the early half of the nineteenth century gave birth to problems in lubrication, bearing design and surface traction which were to be investigated and in most cases solved to the great advantage of tribology as a whole and not just for the benefit of a new mode of transportation.

9.9 Chronology

The Industrial Revolution is recognized as a period of rapid and impressive development of the machinery of production. At the start of the period engineers relied almost entirely upon castings, forgings, the lathe, the drill and their skill with hand tools, but by 1850 most of the machinery now seen in factories had developed to a stage of great utility. Once the merit of the steam engine had been recognized it became necessary to develop machine tools capable of producing in adequate quantities components of high precision. The role of steam power and the subsequent development of the railways in promoting manufacturing skills is evident in the chronology of the period shown in Table 9.4. The textile industry was largely responsible for the collapse of cottage industries and the creation of the factory system. It also created a fertile ground for seeds of numerous ingenious inventions by practical men.

Table 9.4
Chronology for the Industrial Revolution (1750–1850)

Date (A.D.)	Political and social events	General technical developments	Tribology
1750		Lathe and drill major machine tools (made of wood). Wagon wheels achieve widths up to 16 in	
1751	New style calender adopted in England		
	Clive seized in Arcot in India		
1752		Franklin's lightning conductor	
1753	French drove English traders out of Ohio Valley		
1754	Royal Society of Arts		

Table 9.4 (*Continued*)

Date (A.D.)	Political and social events	General technical developments	Tribology
1756	Start of Seven Years War between England and France	John McAdam (1756–1836) (road building)	
1757	Clive's conquest of Bengal	Thomas Telford (1757–1834) (roads and bridges)	**John Ladd's patent for friction-wheels and construction of chains**
1758	English capture Louisburg	First commercial railway sanctioned by Parliament–Bell Isle, Leeds to Middleton Colliery	
1759	Wolfe's ascent of the Heights of Abraham and capture of Quebec from the French	John Smeaton's Royal Society paper on water-wheels Wedgwood Potteries founded	
1760	Accession of George III (1760–1820)	Industrial Revolution (1760–1840) Arnold Toynbee Smeaton's Eddystone Lighthouse completed Carron Ironworks founded	
1762	Catherine II, Empress of Russia (1762–96)		**Musschenbroek's work on friction**
1763	Peace of Paris ends Seven Years War		
1764		Hargreaves spinning-jenny invented	
1766			**Alexander Barclay and John G. Yonge's patent on use of friction-wheels in sugar-mills**
1767		Robert Helsham's book on natural philosophy published Rails cast at Coalbrookdale	
1768	Royal Academy founded, London. Captain Cook started his first voyage to the Pacific		
1769		James Watt granted patent for his steam engine with separate condenser Arkwright's water-frame and Hargreaves' spinning-jenny patented. John Smeaton introduced cast-iron shafts for mills	**Transport of granite block for equestrian monument to Peter the Great on a large linear ball-bearing by Count Carburi (copper-alloy balls)**
1770	Captain Cook landed in Botany Bay	Ramsden's screw-cutting lathe	**Roller-bearing fitted to weather-vane in Independence Hall, Philadelphia (early 1770s)**
1771	Society of Engineers, later known as the Smeatonian Club, founded		

Table 9.4 (*Continued*)

Date (A.D.)	Political and social events	General technical developments	Tribology
1772			**C. Varlo's ball-bearing for road-carriages (cast-iron globes 1–1½ in dia.)**
1773	Boston Tea Party		
1774	Louis XVI, King of France (1774–92)	Wilkinson's boring-mill Newcomen engine thermal efficiency raised to 1·4%	**Semen Kirilovich Kotel'nikov's work on friction**
1775	War of American Independence (1775–83)	Coulomb shared Paris Academy of Sciences prize for magnetic compass with Von Swinden	
1776	American Declaration of Independence. Adam Smith's *Wealth of Nations* and Gibbon's, *Decline and Fall of the Roman Empire* published		
1778	The Maritime War (England and France)		
1779	Spain declared war on England	Crompton's mule – England	
1780			**Large ball-bearing fitted to post-mill at Sprowston, Norwich by 1780**
1781			**Charles Augustin Coulomb awarded Academy of Sciences Prize for his memoir on friction (published 1785)**
1783	Peace of Versailles ended War of American Independence	Steamboat *Pyroscaphe* on river Sâone	
		First balloon ascents – France	
1784		Meikle's threshing machine	**The Rev. Samuel Vince's work on kinetic friction**
1785		Steam used in cotton industry	
		Edmund Cartwright invented the power loom	
		First American turnpikes constructed	
1786	Death of Frederick II of Prussia		
1787			**John Garnett patented roller-bearing**
			Watkin George patented roller-bearing
1788	First fleet of convicts arrived in Australia under Captain Arthur Philip		

Table 9.4 (*Continued*)

Date (A.D.)	Political and social events	General technical developments	Tribology
1789	French Revolution began – Declaration of the Rights of Man		
1791	Ordnance Survey established	Metre defined by French Academy of Sciences	
1792	France declared war on Austria and Prussia	James Watt's expansive engine achieved 4·5% thermal efficiency	
1793	Louis XVI executed	Whitney's cotton gin	
	France declared war on England		
1794	Lavoisier executed		**First patent on ball-bearing – Philip Vaughan, England. Roller-bearing fitted to weather-vane of Old Trinity Church, Lancaster, Pa. – Robert Fulton?**
	French conquered Belgium		
1795	Ecole Polytechnique established		**James Edgell's patent for friction-wheels made of iron**
1796	Napoleon conquered Italy		
1797			**Rev. Edward Cartwright proposed use of metal piston rings, rather than hemp**
1798	Nelson defeated Napoleon's fleet at the mouth of the Nile	Invention of lithography (Senefelder). Eli Whitney's musket with interchangeable parts (USA)	**Roller-bearings introduced into French artillery vehicles by Captain Rebart**
1799	Royal Institution founded	Sir George Caley's design of a glider	
1800		Trevithick's high-pressure steam engine	**Felton's *Treatise on Carriages* showing forms of bearings**
1801	Union of Great Britain and Ireland	Trevithick's steam road carriage	
	Nelson destroyed Danish fleet at Copenhagen		
1802	Treaty of Amiens ended war between England and France	Mechanized manufacture of pulley blocks for Royal Navy	**Cardinet's patent for roller-bearing in roundabout – France**
1803	England resumed war with France	Johnson's power loom Mr Wise produces first steel pen nib (London)	**Horwitz's report on use of roller-bearings for grinding spindles**
			Mr Hatchett's study of the wear of gold coins and the development of wear-testing procedures
1804		Caledonian Canal begun (Telford) Trevithick's railway locomotive	**John Leslie's writings on friction (deformation loss) and the role of lubricants**

Table 9.4 (*Continued*)

Date (A.D.)	Political and social events	General technical developments	Tribology
1805	Nelson defeated French fleet at Trafalgar	Iron Railway–Surrey, England	**R. Stevenson used ball-bearings in the cranes and roller-bearings in the lantern of the Bell Rock Lighthouse**
1806	Napoleon blockades British Isles Death of Pitt		
1807	Slave trade abolished France invaded Portugal	Henry Maudslay's table engine Steamship *Clermont* on Hudson River	
1808	England launched Peninsular War in Spain (1808–14)	John Dalton's atomic theory	
1810		Distillation of shale (Prague)	
1811	Luddite Riots (1811–16)	Bell Rock Lighthouse	
1812	Failure of Napoleon's Russian campaign. America declared war on England (1812–14)	P.S. *Comet*–first commercial steamboat–on the Clyde. Steam locomotives used on Middleton Colliery Railway, Leeds	**Henry Thomas Hardacre's patent for a heavy grease**
1813		G. Stephenson built his first locomotive for use at Killingworth Colliery	
1814	Treaty of Ghent concluded war between United States and England		
1815	Napoleon defeated by Wellington at Waterloo		**Joseph de Baader patented use of friction-wheels for railway carriages**
1816		Davy lamp	
1818	Institution of Civil Engineers formed		**Ball footstep bearing used by Brendal**
1819	Gold standard established Peterloo massacre		
1820	Accession of George IV (1820–30)		**Simple ball-bearing used in furniture castors by James Harcourt**
1822		First rail laid on Stockton and Darlington Railway Equations of motion for a viscous fluid–Claude Louis M. H. Navier	**Roller-bearing with cage patented by J. Whitcher, M. Pickford and J. Whitbourn–for carriage wheels**
1825		Opening of Stockton and Darlington Railway	

Table 9.4 (*Continued*)

Date (A.D.)	Political and social events	General technical developments	Tribology
1826		Telford's Menai Bridge	
1827	University College, London, founded	Fourneyron's water-turbine	
		First friction matches	
1828		Thorp's ring–spinning-frame–USA	
1829		G. Stephenson's *Rocket*	George Rennie's extensive studies of friction
1829			Joseph Resel's patent for roller- and ball-bearings for carriages and machinery
			A. E. Jaccoud's patents for ladle and loose ring forms of oil circulation systems
1830		Liverpool–Manchester Railway opened. First use of tar in road-building–Nottingham–'tar-macadam'	G. Stephenson used cotton-wick lubricators for carriage axles
1831	British Association for the Advancement of Science founded	Faraday demonstrated electromagnetic induction	Galloway's reference to use of brass piston-ring rather than hemp-packing in steam engines
1832	Durham University founded	Horse-trams in New York	Edward Turner's reference to frictional heating (axle-trees and anchor chains)
1833	Slavery abolished in British Empire First successful Factory Act in UK	Iron smelting with anthracite in USA	
1834		Steam-engine efficiency 17% Distillation of shale (France)	
1835		Average speed of coaches between coaching inns 9–10 mph	Arthur James Morin's studies of rolling and sliding friction
		Brussels–Malines Railway	Henry Booth's patent for axle greases and lubricating fluids
		Colt revolver patented	Nathaniel Partridge's patent for an anti-attrition paste (olive oil, lime and water)
1836	Patent office established	Galvanized iron	Baudelot of Harancourt patented disc lubricator
1837	Reign of Queen Victoria (1837–67) New Zealand settled		

Table 9.4 (*Continued*)

Date (A.D.)	Political and social events	General technical developments	Tribology
1838	Dickens' *Oliver Twist* published Royal Agricultural Society founded	Atlantic crossings started by SS *Great Western* Screw propeller introduced N. Wood's treatise on railways London–Birmingham Railway opened Telegraph on railway	**Treadgold's reference to the use of dissimilar metals and lubricants in steam engines**
1839		Nasmyth's steam hammer	**Arsène Dupuit's work on rolling friction and dispute with Morin (1841–42)** **Isaac Babbitt's patent for tin-based alloys involving antimony and copper**
1840	Penny post introduced First UK Chair of Engineering established at Glasgow University	Electroplating	**Charles Greenway's patent for rolling-element bearings with cages for carriage-wheel and industrial applications** **Joseph John Donlan's patent for compounds for lubricating machinery**
1841		Vulcanization (Goodyear) of rubber to withstand temperature changes	**Hompesch's patent for improvements to methods for the extraction of oils from bituminous materials such as schist clay, slate or asphalt**
1844		Jean M. L. Poiseuille's work on viscous flow in narrow capillaries	
1845	Royal College of Chemistry founded	Equations of motion for a viscous fluid – Sir George Gabriel Stokes	
1846	Ether established as an anaesthetic	J. C. Bourne's account of the Great Western Railway Standard railway gauge introduced T. Hancock's solid rubber tyres fitted to road vehicles	
1847	Institution of Mechanical Engineers founded Karl Marx's *Communist Manifesto* published – London		**Thomas Denne's patent for lubricant for axles and machinery** **M. Decoster of Frankfurt introduced chain lubricator – exhibited at Great Exhibition in London by J. Hick, 1848**

Table 9.4 (*Continued*)

Date (A.D.)	Political and social events	General technical developments	Tribology
1848	California gold rush. Publication of Macaulay's *History of England* and Mills' *Political Economy*	McCormick reaper produced in Chicago factory	
1849			**von Pauli's study of journal bearing friction and materials, Nuremberg carriage works William Little's patent for use of mineral oils obtained by distillation for lubricating machinery**
1850	Public Libraries Act	Pneumatic (R. W. Thomson) and solid (T. Hancock) rubber tyres tested for road vehicles (By 1850 the lathe and the drill had been supplemented by milling machines, shapers, planers, boring and grinding machines.)	

Exploration, colonization, wars between leading European nations and the emergence of an independent United States of America dominated the political scene. In this expansive period a number of important organizations which were to promote the development of science and technology were created. The formation of the Society of Engineers (Smeatonian Club) in England in 1771, the Institution of Civil Engineers in 1818 and the breakaway Institution of Mechanical Engineers in 1847 reflected the need for a new form of engineer, standing between the natural philosopher and the mechanic, to further the progress of industrial societies. The Royal Institution (1799) and the British Association (1831) encouraged popular interest in science and its application to practical problems, while in the field of education, science and technology were introduced or expanded through the creation of such institutions as the Ecole Polytechnique (1795) in France, new universities throughout Europe and professors of engineering in the United Kingdom. These developments took place against an increased knowledge of the earth itself. Captain Cook reached Botany Bay in 1770 and a penal settlement had been established in the new continent of Australia by 1788. New Zealand was settled in 1837. It was not only a time of expansion but in some fields of turbulence; a period in which science and technology impinged increasingly upon society.

In tribology three major threads of development can be traced: the development of sound forms of plain and free rolling-element bearings, extensive studies of friction and the increasing attention to lubricants and lubrication. In addition, the remarkable development of wear-testing procedures, albeit for the restricted but important concern for the coinage, heralded late nineteenth-century and particularly twentieth-century work in this field.

The rolling bearings were required primarily for vehicles and a miscellaneous range of lightly loaded machinery, while plain-bearing development was linked to windmills and water-mills, the steam engine and the railways. The introduction of dissimilar metals and the tin-based alloys or babbitts in the late 1830s represented important progress in this field. The impact of patents is seen in the chronology. Patents for bearings and lubricants in this period provide fascinating reading and an important record of events.

Studies of friction were numerous and impressive, with an increasing interest being shown in kinetic and rolling friction. Charles Augustin Coulomb's work is perhaps the best-known essay on the subject (Coulomb, 1781), but the slightly earlier work by Kotel'nikov (1774), the Rev. Samuel Vince's studies of kinetic friction (Vince, 1784), John Leslie's speculations on the fundamental nature of sliding friction (Leslie, 1804), George Rennie's extensive and undervalued studies of the subject (Rennie, 1829) and the illuminating dispute between Arthur James Morin and Arsène Dupuit on the nature of rolling friction (Morin, 1835; Dupuit, 1839) all provide significant signposts to the development of understanding of this basic aspect of tribology.

Lubricants used throughout practically the whole of the period of the Industrial Revolution were of animal and vegetable origin, with the patent applications reflecting great ingenuity in the mixing of various naturally occurring solid and liquid substances. Towards the end of the period, mineral oil was extracted from shale and coal and the scene was set for a complete and rapid upset of the traditional sources of lubricants. This will represent one of the major stories of Chapter 10.

9.10   Summary

Many have experienced difficulties in tracing the progress of a particular branch of technology through the period known as the Industrial Revolution. The threads of progress in the long period of recorded history prior to 1750 were reasonably clear, but the trends became more numerous and complex in the century under review. In tribology I have chosen to highlight the development of both plain and rolling-element bearings, studies of friction and the growing concern for lubricants and lubrication, even though some topics have been given little space and in some cases omitted completely.

The calls upon tribology were essentially practical. This was evident not only in the need to provide adequate bearings for the new sources of mechanical power which emerged during the period, but in the factors which promoted more fundamental studies of friction and, to a lesser extent, lubrication and wear. Was it not such practical marine problems as the stiffness of ropes on pulleys and capstans and the difficulties occasionally encountered in trying to launch ships on slipways which prompted the French Academy of Sciences to offer the 1779 prize ultimately awarded to Coulomb for work on friction?

- the effect of friction upon the motion of sledges, the behaviour of stones in arched bridges, the sealing of pistons in pumps and the tribological problems posed by wheeled vehicles and the steam engine which provided the stimulus for Mr George Rennie's impressive experiments on friction?
- and concern for the coinage in the United Kingdom which led to Mr Hatchett's important studies of wear and wear-testing procedures?

It was undoubtedly the energetic development of new mechanical devices and the great advance of the methods of production which dictated the pattern of tribological progress. The ceaseless activity of practical engineers in the United Kingdon and America was supplemented by the development of important patterns of technical education, first in France and then throughout Europe. These advances were followed by the growth of societies and institutions for engineers in the United Kingdom which have formed the mainstay of professional activities to the present day. In presenting this background I have endeavoured to convey an appreciation of the expansive nature of the times and to describe some of the political, social and technological factors which nurtured the Industrial Revolution.

If a single field of application has to be selected to form the background to bearing development between 1750 and 1850 it must be that of mechanical power. This fact alone justifies the inclusion of a section on water-wheels and windmills, since both had reached their zenith and were giving way to steam by 1850. Wood was rapidly being replaced by iron as the major material of construction, even though it continued to provide 90 per cent of the energy production in the United Kingdom up to the middle of the nineteenth century. However, the metal bearings were in most cases similar to their wooden forerunners and it was the development of accommodating bearing materials and an array of patent solid, liquid and semi-liquid lubricants or greases which deferred the need for basic changes in bearing configurations. Even the new railway locomotives and wagons failed to create, at least initially, the severity of operating conditions which might have been expected to lead to a basic revision of plain bearings. As late as 1826 George Stephenson was reporting that he was fitting bearings on the principle of the mail-coach. In general there was little need for innovation in bearing design to meet the demands of the Industrial Revolution. It is for this reason that many of the most interesting aspects of tribology in the period are found in peripheral activities like Count Carburi's exercise in moving a large granite block on a self-made, large-scale linear ball-bearing in Finland; the design and construction of unusual bearings for windmills (Sprowston); weather-vanes (Philadelphia); personal road-carriages (Varlo), and the ever fascinating patents for curious mixtures of lubricants (Donlan).

In the continuing debate on friction there was certainly much work reported and many important concepts tentatively proposed, but few definitive advances in the understanding of the phenomenon. Existing skills in designing and making bearings were, on the whole, equal to the changing industrial situation, and the skilful use of naturally occurring vegetable- or animal-based lubricants overcame most residual deficiencies in bearing performance. Perhaps the most important feature of tribological history in this period of rapid progress towards an industrial world was the legacy of problems of growing severity. The physical and chemical nature of lubrication had still to be unravelled, a satisfactory theoretical model for friction had yet to be constructed from the mass of individual experimental investigations, while the complex nature of surface interactions associated with wear were not to be studied in any depth for a further century. The general state of confusion after a century of considerable activity and inquiry into matters tribological is well reflected by the words of Nicholas Wood about the subject of friction which opened the chapter.

Chapter 10

Mineral oil and scientific studies of lubrication
A.D. 1850–1925

10.1 Introduction

It is an interesting and curious fact that in the long history of tribology prior to 1850 very little attention was paid to the nature of lubrication or even lubricants. Some experimenters had measured the magnitude of friction between lubricated solids, but most natural philosophers had addressed themselves to the problem of understanding the physical nature of friction rather than lubrication. The discovery of mineral oil, the development of the steam engine and the railways, together with the increasing concern of professional engineers for the reliability, safety and efficiency of lubricated machinery, all contributed to the pressures which were responsible for the spectacular redress of this situation during the period under review. By 1925 Nature was to yield to science the broad range of her secrets surrounding the physical and chemical nature of lubrication, leaving many aspects of dry friction and the whole basis of wear for consideration in the mid-twentieth century. It is for this reason that lubricants and lubrication dominate our story in the present chapter.

Mineral oil was found in large quantities in several parts of the world late in the nineteenth century. Once its value as a lubricant had been confirmed and its competitiveness established on the basis of quantity and price, it rapidly established itself as the premier lubricant of the age. The entry of mineral oil into the machine age and the organizations for its exploitation will be discussed in Section 10.2.

The beautiful mechanism of fluid-film lubrication was exposed by careful experimental work and impressive mathematical analysis in the 1880s. This was indeed a vintage decade in the history of tribology. However, the full impact of Reynolds' theory of hydrodynamic lubrication upon bearing design was not felt for about twenty or twenty-five years and is thus an early twentieth-century phenomenon. The elegance of the almost simultaneous solutions to the aggravating problem of poor thrust-bearing performance in hydraulic and marine applications by Michell and Kingsbury early in the twentieth century, which represents one of the finest examples of sound engineering analysis and design, is the subject of Section 10.7. It is an indication of the substance of the theory of hydro-dynamic lubrication from the 1880s, and the engineering principles of thrust-bearing design developed between 1900 and 1910, that neither have needed much modification to the present day. Many of the major and well known manufacturers of both plain and rolling-element bearings were established early in the twentieth century.

Patent specifications continue to form a valuable source of information for the developments recorded in this chapter. This is particularly evident in the late nineteenth century, when, for example, the requirements of the velocipede played no small part in promoting the development of the ball-bearing described in Section 10.7. The outstanding contributions of men of science and professional engineers are recorded in the proceedings or transactions of learned societies which are readily available. The difficulty experienced in locating the sources of information recorded in earlier chapters was much relieved for the present period, but the rapid growth in the quantity of relevant published material meant that interpretation and the selection of significant milestones was more onerous.

Towards the end of the span of seventy-five years covered by this chapter, the biologist Sir William Bate Hardy turned his attention to the question of friction between lubricated surfaces. The outcome was the masterly exposure of that form of lubrication known as *boundary*. The two major modes of lubrication were thus revealed by 1925 and this has helped to define the period under review. It is a most exciting and important period in the history of tribology, finding its foundations in the evaluation of lubricants, machines and science itself in previous centuries and providing the building blocks for modern studies and applications of the science of lubrication.

10.2 Mineral oil

Early knowledge

Although it was known in various parts of the world throughout the ages, mineral oil was first produced commercially as a lubricant in the eighteenth and early nineteenth centuries. Seepage of natural gas and oil fuelled the sacred fires of the Zoroastrians at Baku in the Caucasus from the sixth century B.C., the Romans used petroleum as an illuminant and there are references to its use in ancient writings from Greece, India and China. In the late seventeenth century a patent was awarded in England for the production of pitch, tar and oil from stone; the oil being sold as Betton's British Oil to cure rheumatism. Illuminating oil was produced by distillation in Prague as early as 1810 and on a larger scale by Selligue in France in 1834. A good account of the early knowledge of mineral oils has been presented by Forbes (1958). Three countries featured in the foundations of the petroleum industry in the 1850s – Scotland, the United States of America and Canada.

Scotland (James Young)

In the middle years of the nineteenth century the son of a Glasgow cabinet-maker developed a process for producing paraffin oil from coal which was to place Scotland in the forefront of petroleum production for many years to come. This marked the true beginnings of commercial exploitation of petroleum,[1] although small quantities had previously been collected by digging wells and skimming the surfaces of streams at sites of natural seepage from the earth. It seems particularly appropriate to retell this part of the history of mineral oil exploitation at a time when Scotland is once again establishing a position of some importance in connection with petroleum production.

James Young was born in 1811 and initially apprenticed to his father. He attended chemistry classes at Anderson's College with Lyon Playfair and David Livingstone as fellow pupils. In 1832 he abandoned carpentry to become a lecture assistant to Professor Graham and he accompanied Graham to London when the latter was appointed to the Chair of Chemistry at University College in 1837. Two years later, on the recommendation of Graham, he was appointed chemist at the Muspratt Alkali Works near Liverpool. In 1844 he became manager of Charles Tennant's Ardwick Chemical Works in Manchester and it was while he was in this position that his old school friend, then Sir Lyon Playfair, M.P., wrote to him about his discovery of the first British source of mineral oil in Ridding's Colliery at Alfreton in Derbyshire. Tennant showed no interest in the discovery, but Young was soon exploiting the find with financial support from a lawyer named Binney. With a young chemist named Meldrum in charge, production of mineral oil rose to 300 gal (1,364 l) a day, with much of the output finding a convenient outlet in the Lancashire cotton-mills.

The Derbyshire spring had a short life, but a new plant was erected at Bathgate in the centre of the Torbane Hill Fields in 1851. Messrs Binney and Meldrum joined the venture with Young and in 1866, when Young sold out, the annual production of mineral oil in Scotland had reached over 6 million gal (27 million l) a year. Twenty years later it was over 50 million gal (227 million l) a year. In the reign of Queen Victoria, Young (1850) obtained a now historic patent for his process which was enrolled on 17 April (1851) under the title, '. . . Treating Bituminous Coals to Obtain Paraffine and Oil containing Paraffine therefrom'.

Young says little about the application of his product, apart from noting that the more volatile component could be used as an illuminant and the remainder as a lubricant, either by itself or mixed with an animal

1. The word 'petroleum' comes from the Latin and Greek word *petra* for rock, and the Latin word *oleum* for oil.

or vegetable oil. The treatment of the bituminous coal was described in some detail and the raw materials deemed to be best fitted for the process were parrot, cannel and gas coals. The coal was broken into small pieces about the size of a hen's egg and treated in a common gas retort containing a water-cooled pipe attached to a refrigerator.

By 1858 Young's company was the world's largest manufacturer of coal oil, with successful markets not only in the United Kingdom but also in the United States and Europe. He produced mineral oil and grease for lubrication, illuminating oil, naphtha, and paraffin wax for candles. Young became a rich man and he was a liberal supporter of his lifelong friend Dr Livingstone. He was also a Fellow of the Royal Society and founder of the Chair of Technical Chemistry at Anderson's College. Nineteenth-century industrialists were quick to appreciate the advantages of mineral oils, but it was perhaps the illuminating fluid which left a bigger impact on society. David Livingstone wrote: '"Paraffin" Young raised himself to be a merchant prince by his science and art, and has shed pure light in many lowly cottages and in some rich palaces.'

In due course the supply of mineral oil from flowing and pumping wells was to supersede the supply from shale and coal. Goldblatt (1973) has reminded us that Young delivered an uncharacteristic and totally erroneous judgement after visiting the early wells in the United States in 1861, 'I dinna think it will amount to much; it's ephemeral, it won't last.'

Canada
(James Miller Williams)

It is generally accepted that Drake's Well in the USA marked the start of the oil industry in 1859, but several other countries have claimed this distinction. Forbes (1958) has pointed out that Hunäus struck both oil and gas at Wietze, Hanover, two years earlier when drilling a series of five wells for the specific purpose of producing crude oil. The fact that Germany is usually left out of the reckoning in terms of the origins of the oil industry is no doubt due to the fact that the quantities of oil extracted were limited. The industry never attracted the frenzied activity typical of North America and European production was small in the nineteenth century. The Canadians issued a commemorative stamp and celebrated the centenary of oil in June 1958 on the basis of the exploits of Mr James Miller Williams, an Ontario business-man. In 1856 Williams bought the properties of Charles N. Tripp, who had been producing asphalt from some gum beds near Oil Springs, Ontario. Williams' first well was dug to a depth of 27 ft (8·2 m) on the banks of the Thames River near Bothwell. In his second well at Enniskellin he found petroleum at a depth of 65 ft (19·8 m). There appears to be some uncertainty about the date of his drilling operations between 1857 and 1859, but if the date of June 1858 is correct it certainly preceded Drake's success. Williams went on to build Canada's first oil refinery on the banks of Black Creek, before selling out his interests in oil as a result of increasing American competition in 1880. A useful perspective on the Canadian contribution to the early exploitation of petroleum is contained in an article devoted to 'The Drake centenary' in the *Institute of Petroleum Review* (1959b).

**The United States
of America**
(Edwin L. Drake)

In 1846 on Prince Edward Island, Gesner, who qualified as a physician in London in 1827, introduced an illuminating oil obtained from the distillation of coal which he called kerosene.[2] In this respect the activities of Gesner in the United States and Young in Scotland were similar in both nature and timing, but the exploitation of mineral oil in the United States soon extinguished the flame of small-scale production of kerosene from coal.

The North American Indians had known about the presence of mineral oil long before the white settlers sought to extract the fluid from the earth. The Indians found that the wells they dug frequently produced mixtures of brine and oil, the former being valuable and the latter a contaminant. They called the viscous fluid *seneca oil* and used it medicinally.

Evans (1921) has told how Samuel Kier, a pharmacist from Pittsburg, improved the kerosene lamp and sold a special illuminating fluid produced by the distillation of mineral oil. His father had operated salt wells at Tarentum for several years and initially Samuel was content to market the mineral oil contaminant obtained from the diggings as a patent medicine under the label, *Kier's Petroleum or Rock Oil*. He later sold the 'carbon oil' obtained from a still in the main street of Pittsburg as a somewhat smoky illuminant. He adorned his bottles of kerosene with a label showing a derrick, and although this was totally unrelated to contemporary methods of petroleum production, it did lead to a colourful version of the origins of oil wells. It is said that the pictures of derricks on Kier's bottles of kerosene gave an enterprising lawyer from New York called Bissell the idea that it might be possible to drill for oil in much the same way that artesian wells were bored for water. More sober accounts have been proposed in the Drake centenary article in *The Institute of Petroleum Review* (1959*b*) and the book by Tugendhat and Hamilton (1968).

When the village of Titusville in north-western Pennsylvania was settled in 1809 the new residents soon heard about the local oil springs. One settler named Nathaniel Carey sold the oil as a medicine, while a local physician, Dr Francis Brewer, arranged the lease of land from a lumber firm owned partly by this father for the purpose of obtaining oil for illuminants and lubrication. Mr J. D. Angier, a resident of Titusville, was employed to work the springs and he obtained 3 or 4 gal (11–15 l) a day from the lease.

In 1853 Dr Brewer took a sample of the Titusville oil to his friend Dr Dixi Crosby of a Dartmouth College Medical School in New Hampshire. The sample aroused much interest and it was during the winter of 1854 that the law graduate from New York Mr George H. Bissell returned to Dartmouth, saw the sample and recognized in the excitement the commercial possibilities of large-scale production of the fluid. The college professors who had been experimenting with the sample of oil told Bissell that if it were suitably refined it could provide a better lamp

---

2.   From the Greek word *keros* meaning wax.

light than kerosene derived from coal. Bissell and his business partner, Jonathan G. Eveleth, enthusiastically agreed to form a company to develop the springs and to market petroleum. Mr James M. Townsend, President of the City Savings Bank, and other business-men in New Haven showed interest in the venture, but decided that second opinions should be sought on the scientific analysis and economic potential of the oil. In due course a Boston chemist and the Professor of Chemistry at Yale, named Luther Atwood and Benjamin Silliman, Jr, respectively, confirmed the views of the Dartmouth scientists. Silliman concluded his report of 16 April 1855 with the words: '. . . In conclusion, gentlemen, it appears to me that there is much encouragement in the belief that your Company have in their possession a raw material from which, by a simple and not expensive process, they may manufacture very valuable products.'

The land at Titusville was transferred from Dr Brewer to Bissell and Eveleth to launch the Pennsylvania Rock Oil Company of New York in November 1854. There was no rush by financiers to support the new enterprise and it was not until Professor Silliman presented his favourable report in April 1855 that the new company started to attract attention. Bissell decided that it was desirable for tax reasons to reorganize the company in Connecticut and so the Pennsylvania Rock Oil Co. of Connecticut was formed in September 1855.

It was at this stage that Edwin Drake came on the scene. He was appointed by a director of the company, Mr James M. Townsend, to investigate the site and to initiate production. Drake was born at Greenville, NY. He had little formal education and worked successively as a clerk, farm labourer, express agent and conductor on the New York and Newhaven Railroad. He was given the courtesy title 'Colonel' by Townsend to impress the inhabitants of Titusville and he was warmly received when he first visited the town in December 1857. Drake reported favourably on the prospects for the venture and the Seneca Oil Company was formed to succeed the Pennsylvania Rock Oil Co. Colonel Edwin Drake was elected Chairman and General Agent of the new company.

The initial attempts to boost production by digging trenches and then pits were thwarted by flooding. Drake then considered the possibility of drilling for oil. He consulted the salt-well drillers at Tarentum, installed drilling equipment at Titusville and hired a blacksmith and experienced salt-well driller by the name of William A. Smith. The townsfolk of Titusville were much amused by this novel approach to oil production and the derrick built in the early summer of 1859 became known as *Drake's yoke* or *Drake's folly*. A fine picture of Colonel Drake and his oil well is shown in Fig. 10.1.

Drilling commenced in June 1859 and Drake drove an iron pipe down to bedrock at a depth of 32 ft (9·8 m) to keep water out of the hole. This was the first application of the well-casing technique. Subsequent drilling of the rock proceeded at a rate of about 3 ft (0·91 m) per day, but the numerous delays and lack of favourable results since the formation of the company caused the stockholders to become a little nervous. They finally refused to advance more money, instructed Drake to settle outstanding bills and advised him to return to Newhaven.

Fig. 10.1
Drake's Well (1859)–Titusville, Pennsylvania, USA. (The photograph shows the bearded 'Colonel' Edwin L. Drake talking with Peter Wilson, a Titusville druggist. On the extreme right in the background is 'Uncle' Billy Smith, Drake's head driller.)

Colonel Drake had not received the stockholder's instructions by the time the drilling crew ceased work with the drill at a depth of about 69 ft (21 m) on Saturday, 27 August 1859. As 'Uncle' Billy Smith and his crew were closing down for the day, the drill slipped into a crevice at a depth of $69\frac{1}{2}$ ft (21·18 m), Smith visited the well again late on the afternoon of Sunday, 28 August and found that oil had flowed to the surface. Excitement grew as the news spread through neighbouring districts and crowds of people flocked to the well-head. Such was the beginning of a vast industry.

Drake failed to recognize the full implications of his find and the oil-strike left him comparatively unmoved. He never bought any of the oil-rich land for himself even though the frenzy associated with this first oil-rush grew daily. He rebuilt the first well after a fire and then proceeded to build two more before becoming a buyer of oil on commission to a New York firm. For a while he was Titusville's Justice of the Peace. He subsequently lost most of his money through speculation on Wall Street, but the people of Titusville subscribed about $5,000 to provide him with a home and an income. He also received an income of $5,100 from the Pennsylvania Legislature. He died, some twenty-one years after his historic oil-strike, in November 1880 at Bethlehem, Pennsylvania. Records of the stirring times of the exploitation of mineral oil at Titusville are preserved in the Drake Well Museum just outside the town.

The New York lawyer, Mr Bissell, quickly leased or purchased surrounding farmland after the 1859 strike and in a matter of a few months he made a fortune. Drake's Well reached but a modest output of 400–1,200 US gal (1514–4542 l) per day, but further wells were constructed rapidly. A tanner from England called William Barnsdall built the second well and cleared $16,000 in five months, while a well called 'Mapleshade' eventually made $1·5 million for its owner Mr Charles Hyde, a local storekeeper.

The growth in demand for the new product was remarkable and by 1861 kerosene made from oil had entirely displaced the variety produced from coal in the USA. It was at this stage that James Young visited the American oil wells and uttered his erroneous forecast for the industry. Exports of kerosene commenced in December 1861 when the brig *Elizabeth Watts* sailed for London from Philadelphia with the fluid stowed in barrels. By the end of 1865 substantial quantities of kerosene were being shipped across the Atlantic to Britain, France and Germany. The remarkable impact of mineral oil upon America can be judged from a paragraph in a book written by Bone (1865) which was published in Philadelphia a mere six years after Colonel Drake struck oil at Titusville.

'From Maine to California it lights our dwellings, lubricates our machinery, and is indispensable in numerous departments of arts, manufactures and domestic life. To be deprived of it now would be setting us back a whole cycle of civilization. To doubt the increased sphere of its usefulness would be to lack faith in the progress of the world.'

10.3   Origins of the major oil companies

In the early days of mineral oil exploitation in the United States production, refining and distribution were quite separate activities in the hands of different companies. As the various groups tried to preserve and expand their interests a number of bitter commercial and industrial disputes arose which were to have a long-standing influence upon America and the industrial world. Fortunes were made by a few and lost by many, large companies and trusts were created and concern for the effect of ruthless business procedures led to the introduction of the Sherman Act of 1889. The dominant personalities were the philanthropist John D. Rockefeller in the United States, the inventive Nobel brothers from Sweden and the banker Baron Alphonse de Rothschild in Paris. By the end of the nineteenth century the American scene was dominated by the giant Standard Company which had been nurtured from the outset by Rockefeller.

John D. Rockefeller was born on 8 July 1839 on a farm near Moravia in western New York State. He was the eldest of six children, brought up in an environment of dedication to religion and private enterprise. His father had no fixed occupation and was apparently indicted for rape, though never charged, before the family moved from Moravia to Cleveland. His mother was deeply religious. John helped to bring up the family during his father's frequent absences from home and his Baptist background, family life and the influence of his mother were no doubt dominant factors in his forceful business yet philanthropic life.

In 1855 John Rockefeller became a book-keeper in a firm of commission agents and produce shippers. Three years later, after a dispute over his salary, he went into partnership with an immigrant from England called Maurice Clark. The firm of commission agents known as Clark and Rockefeller prospered. Business boomed during the American Civil War as the firm supplied food, clothing and, significantly, kerosene to the Northern armies.

With oil production centred in Pennsylvania, Cleveland was well placed as a refining and distribution centre. It not only had rail links with the petroleum-producing areas and the east coast, but ready access to the Great Lakes waterways. Trading in kerosene aroused Rockefeller's interest in the oil business, and by 1863 Clark and Rockefeller had joined another English immigrant by the name of Samuel Andrews to form a refining company in Cleveland. Andrews, who had previously worked in a refinery in Cleveland, brought both skill and enthusiasm to the new venture. By 1865 it was the biggest refinery in Cleveland, but the need to borrow considerable sums of money to finance the rapidly expanding, capital-intensive business led to a difference of opinion between Clark and Rockefeller. Clark was cautious, whereas Rockefeller was ready to raise all the money that was necessary. The partnership broke up by agreement on 2 February 1865, with Rockefeller withdrawing from the commission agency and Clark from the refining side of the business. The agreement required Rockefeller to pay Clark $72,500, and this he managed to do at the age of twenty-six.

Andrews and Rockefeller were joined by a grain merchant named Mr Henry M. Flagler, who suggested that the best way to finance their expanding refinery was to form a joint-stock company. It was this proposal that led to the creation of the Standard Oil Company of Cleveland on 10 January 1870. The President of the new company was John D. Rockefeller. Rockefeller was convinced that profitability depended upon a reduction in the number of small competitive refineries and the creation of a large consolidated company. This was achieved by Standard taking over many small refineries with impressive speed. By 1871 Standard was probably the largest refining company, not only in Cleveland but throughout the whole of the United States.

The purpose of consolidation was not only to ensure that the refining side of the oil business was profitable, but also to place the refining activity in a strong position in negotiations with the oil producers and distributors. The main form of distribution of refined petroleum products was by rail, and before Standard achieved its pinnacle of power in the refining world the railway companies snatched the initiative from Rockefeller and his associates. The railroads were revolutionizing transport and communications in America, but they also had much of industry in their grip.

The three dominant oil-transporting railroad companies suggested to thirteen refinery companies that the two industries should be formed into two associations. The Association of Oil Refiners would then agree to transport their products on the three railroads, and the railroads would offer advantageous rates to the refineries concerned. In this way rival refiners and transporters of petroleum products were to be squeezed out of the market. The refining companies agreed to the proposal in principle and in January 1872 they decided that an association known as the South Improvement Company should be formed.

News of the scheme leaked out before it became effective, and the oil producers of Pennsylvania reacted vigorously. They formed a Petroleum Producers Union to protect their position after a mass meeting in the Titusville Opera House. It was agreed that oil should not be supplied to the members of the South Improvement Company and demands for an investigation into this apparent restriction on free trading were dispatched to the Pennsylvanian Legislature and to Congress. Proposals were tabled for the construction of pipelines to break the monopoly of the railroads, and public indignation arose when it was admitted at the Congressional Investigation in Washington that the real purpose of the agreement was to increase prices of petroleum products.

The oil embargo was effective and the South Improvement Company and the railroads were soon negotiating with the Oil Producers Union. An accommodation of the various interests was achieved by March 1872, and although some restrictions on sales by oil producers continued for a few weeks the 'Oil War' of 1872 was over by the end of April. If consolidation of the oil refiners through the South Improvement Company had been stillborn in 1872, it was to be given new life in spectacular fashion through the persistence and opportunism of John D. Rockefeller within the next few years. It is necessary to appreciate the general economic conditions in the United States in the 1870s in order to understand how Rockefeller was able to achieve such dominance in the aftermath of the Oil War of 1872. Oil production had exceeded demand since the end of the Civil War in 1865 and the viability of many companies became precarious during the general economic depression of 1873. Over-production and violent fluctuations in prices produced a turbulent business environment. It also gave rise to the phrase, 'The bottom fell out of the market'. Rockefeller was able to buy up many refineries on the verge of bankruptcy and by 1879 Standard controlled about 90 per cent of American refining capacity.

Although Bone (1865) had said of mineral oil that '. . . it lights our dwellings, lubricates our machinery, and is indispensable in numerous departments of arts, manufactures and domestic life, . . .' it seems clear that the refining companies were initially interested only in kerosene. Other products were discharged as waste and in due course were to injure the environment of Cleveland. It was the harsh economic climate of the 1870s and the fight for business which promoted additional applications of the by-products of kerosene production. In the 1870s Andrews and others encouraged industry to use oil as fuel and to replace the animal- and vegetable-based lubricants by mineral oils. The advantages of fuel oil over coal for firing ships' boilers had already been noted in an article in *The Times* of 6 April 1865, and in 1878 the forerunner of the present-day paraffin oil stove for domestic use proved to be a great success. Mineral oils had been used as lubricants in small quantities in many countries since the middle of the nineteenth century. Most oil refineries in Europe and the United States listed lubricating oils among their products, but the large-scale movement away from animal and vegetable lubricants started in the USA in the 1870s and developed rapidly in Russia and Europe in the 1880s.

Once Standard had achieved an effective monopoly in refining activities, the issue of transportation became the next commercial battle-ground. Initially mineral oil and its products were transported in wooden barrels, iron drums or tin cans. In due course Standard started to manu-facture their own cans and barrels, but the real battle was between the railways and pipelines. The Empire Transportation Company bought out the two largest pipeline companies in order to control the distribution of oil to Standard's refineries. Standard responded by buying its own pipelines, Empire then bought its own refineries and a bitter struggle was fought between 1873 and 1877. The result was once again a victory for John D. Rockefeller and the Standard Oil Company. In 1877 Standard purchased both the Empire Transportation Company's oil interests and the largest remaining pipeline company. It thus held a monopoly in both the transportation and refining of mineral oil. A despairing attempt to break Standard's monopoly over oil transportation was mounted by the Tidewater Pipe Company at the end of the 1870s. A pipeline some 120 miles (193 km) long was built over the Allegheny Mountains, and this forced Standard to construct its own lines and eventually to absorb the rival company. The construction of these pipelines across the mountains to the eastern seaboard firmly established direct pumping as a viable technical and commercial system. Initially, attempts had been made to construct wooden pipelines, but cast iron and steel provided the real basis for construction.

The law prevented the Standard Oil Company based in Ohio from owning companies in other states, yet the organization was assuming nationwide proportions. In 1882 Rockefeller overcame the problem by creating a trust in which the activities of separate Standard Oil companies in various states would be controlled by a single board of Trustees in New York. In due course the Ohio Supreme Court ruled that this new arrange-ment violated the original company charter and the trust was dissolved in 1892. By this time some of the states had recognized the advantages of accommodating petroleum enterprises within their boundaries and new company laws were introduced. The new legislation enabled a holding company known as Standard Oil (New Jersey) to be created.

Public and political hostility to the Standard monopoly and very large profits was reflected in various articles and books of which Tarbell (1904) is the best known. In 1890 Senator John Sherman's Anti-Trust

Act was signed by President Harrison, but it was not until Theodore Roosevelt's time that the full impact of the legislation was felt. After a prolonged legal battle Chief Justice White pronounced in the Supreme Court in May 1911 that Standard Oil must dispose of its subsidiary companies within six months. Sampson (1975) has written a fascinating account of the impact of the dissolution of the original Standard Oil Company.

Exxon

Standard Oil of New Jersey, now known as Exxon in America and Esso in Europe, was to become the biggest oil company in the world. In 1972[3] its assets still exceeded those of other oil companies and other industrial giants including General Motors, IBM and ITT. Multinational companies, whose annual budgets exceeded those of many nations in which they operated, were creating a new feature of international trade.

Socony (Mobil)

The *Standard Oil Company of New York* became Socony. In 1931 Socony merged with another company known as Vacuum which specialized in lubricants to become Socony-Vacuum. The company name later changed to Socony-Mobil and is now known simply as Mobil.

Socal (Chevron)

In California Standard had been involved in battles with smaller independent companies similar to those earlier encounters in the east. In due course *Standard Oil of California* became known as Socal and Chevron.

Gulf

There were two major rivals to the offshoots of the original Standard Oil Company. At Spindletop in Texas close to the Gulf of Mexico a rotary oil rig was built by an engineer named Anthony Lucas. Backed by the banking firm of T. Mellon and Sons of Pittsburgh, Lucas struck oil at a depth of 1,000 ft (304 m) on 10 January 1901. The now legendary gusher released a jet of oil which rose to twice the height of the derrick and the link between Texas and oil was established. The company formed to exploit the new oilfield was named Gulf.

Texaco

A former employee of Standard Oil known as Joseph Cullinan moved to Texas to establish the Texas Fuel Company after the Spindletop gusher of 1901. Additional finance and a partnership with Arnold Schlaet led to the formation of a larger organization known as the Texas Company. Initially the company traded by buying oil from Spindletop and selling it to sugar planters along the Mississippi and to customers in the East,

3. See Sampson, A. (1975), p. 189.

but it gradually expanded its share of oil production. In 1952 the *Texas Company* became known simply as Texaco, one of the largest oil producers in the world.

The proportions of various types of hydrocarbons found in the crude oil produced in different parts of the United States varied considerably. In the East the Appalachian fields produced a crude oil which was relatively rich in paraffins (about 40 per cent), the Gulf of Mexico fields in Texas provided crude oil in which naphthenes and aromatics were dominant, while the proportion of naphthenes in some Californian oils was as high as 75 per cent.

It is interesting to note how soon international trade in mineral oil was established. The brig *Elizabeth Watts* carried a cargo of petroleum across the Atlantic in 1861, two years after Drake found oil at Titusville. The first tanker, named the *Zoroaster*, was built for the Caspian Sea trade in 1877, and Armstrong-Whitworth built the first European tanker, the *Glückaup*, on the Tyne in 1885.[4] There were over fifty tankers sailing under American, British and Russian flags by 1890, and about three-quarters of the business of the American Standard Oil Company was overseas by 1885. This inevitably brought Standard Oil into competition with the expanding Russian and European oil interests, and it is to these regions that our study now moves.

Russia

Small quantities of mineral oil were being produced in Russia as early as 1856 and in Roumania by 1857, but it was not until 1873 when the Tsarist regime allowed private prospecting and exploitation that the Russian industry became firmly established. Three brothers from Sweden known as Ludwig, Robert and Alfred Nobel established a dominant position in the Baku region, with interests in both the production and transportation of oil. The Nobels transported oil up the river Volga and then by train to the Baltic. The competitive but more convenient outlet by tanker through the Black Sea had to await the construction of a railway from the Caucasus to the port of Batum. The banker Baron Alphonse de Rothschild of Paris provided the capital, and the growth of the Russian oil industry after completion of the railway in 1883 was both assured and remarkable. It will be seen in Section 10.4 that the year 1883 was significant in Russia not only because of the rapid growth of the oil industry, but also because of the scientific studies of lubrication by a professor of steam engineering, Nicolai Petrov.

The Russian wells yielded vast amounts of oil from a relatively small area at Baku. By 1900 output exceeded the total production of the United States, although this position was to be reversed within a decade. Europe became the meeting point and the market place for commercial competition, with price wars between the Standard Oil Company, the Nobels, the Rothschilds and smaller Russian and American companies. By the 1890s oil was being shipped through the Suez Canal and competition between Russian and American crude oil spread to the Far East.

---

4. See Forbes, R. J. (1958), Vol. V, pp. 102–23.

A valuable account of the history of the Russian petroleum industry was published by Redwood (1885).

Royal Dutch/Shell

Prominent European trading nations like Holland and Britain had to look to remote parts of the earth for the oil which could retain their trading and political positions. They were in a good position to exploit their interests in worldwide transportation and this was to be a feature of both the large European oil companies. In 1890 a company was formed in The Hague, Holland under the title Koninklijke Nederlandsche Maatschappÿ tot Exploitatie van Petroleumbronnen in Nederlandsch–Indië, N.V. (Royal Dutch Company for the Working of Petroleum Wells in the Netherlands Indies). It was initially managed by J. B. August Kessler, but in 1900 Henri Deterding assumed control of the company.

In London the Shell Company was to grow out of the Far Eastern trading interests of a Jewish merchant named Samuel. The family firm was established in 1830. The trader was fascinated by seashells brought from the East by sailors and his business in the East End of London benefited when such items became fashionable in Victorian society. By 1892 Marcus Samuel was well established as a trader in petroleum. In the previous year the Suez Canal Authority had approved plans prepared by Mr Fortescue Flannery, a British marine engineer, for the construction of a tanker for use by the Samuels in the transportation of Russian oil to the Far East. Previously, oil passing through the canal had been carried in tins within wooden containers, but Marcus Samuel had visited the Russian oilfields and recognized the great potential of tankers for the transportation by sea of products like kerosene. A nine-year contract from the Rothschilds, the construction of storage depots in the East and the building of eleven tankers by 1893 firmly established the Samuels' position in the Far East. In 1897 the Shell Transport and Trading Company was formed to take over the oil interests of M. Samuel and Company. The name 'Shell' reflected Marcus Samuel's father's earlier trading interests and all the tankers were identified by the names of different seashells.

The merger between the 'Royal Dutch' and 'Shell' companies was initiated in 1902 and completed in 1907 by a transfer of assets to two new companies based in London and The Hague. Royal Dutch took 60 per cent and Shell Transport 40 per cent of the shares. The Royal Dutch/Shell group of companies ranks second in the list of oil producers whose history has been told in refreshing style by Sampson (1975).

British Petroleum

Reports of oil seepages in the Middle East became more frequent in the 1870s, although historical records suggest that bitumen had been known in Mesopotamia for about 5,000 years. German geologists and railway engineers made increasing reference to indications of significant oil deposits in the area. The Sultan became concerned about German mineral rights adjacent to the railways and turned to a twenty-one-year-old graduate in civil engineering from King's College, London named

Calouste Sarkis Gulbenkian for his advice. Gulbenkian produced an encouraging report, but there was no immediate response. Gulbenkian moved to Cairo for a while and then to London in 1897. In 1900 he declined an invitation to take an oil concession in Persia, the opportunity passing to the founder of British Petroleum, William Knox D'Arcy. D'Arcy made his fortune in gold mining in Australia before returning to England in 1900. He was granted the original tax-free oil concession to 480,000 mile^2 (1,243,200 km^2) of territory in Persia on 28 May, 1901. The concession cost £20,000 in cash, 20,000 £1 shares and a promise to pay the Persians 16 per cent of the net profits.

A self-taught geologist, sometime Indian Rubber Works Department official and oil driller in Sumatra named George Bernard Reynolds started the drilling in 1902 and a small amount of oil was located in January 1904. As more dry wells were sunk, the capital ran out at a time when the British Admiralty was becoming convinced about the future for fuel oils in ships and concerned for the lack of British-owned sources of production. Britain's oldest oil company, Burmah, was encouraged to finance further drilling operations and on 26 May 1908 a gusher 50 ft (15 m) high at a place called Masjid-i-Salaman relieved British anxiety.

On 14 April 1909 the Anglo-Persian Oil Company was formed, the name being changed to Anglo-Iranian in 1935 and to British Petroleum in 1954. Delays and difficulties in pipeline and refinery construction produced a further financial crisis and the refinery at Abadan did not function successfully until 1913.[5] The First Lord of the Admiralty since 1911 had been Winston Churchill and he became convinced of the need to change the Navy's fuel from coal to oil. This in turn led him to ensure security of supplies and he persuaded his Cabinet colleagues to agree to a 50 per cent Government participation in the company. The necessary Bill was passed by the House of Commons and received Royal approval six days before the outbreak of the First World War.

The major oil companies were thus established before the dreadful holocaust in Europe between 1914 and 1918. Exxon (Esso), Socony (Mobil), Socal (Chevron), Gulf, Texaco, Shell and BP established dominant positions and attracted the name 'The Seven Sisters' (Sampson, 1975). In due course further national and private companies developed in various parts of the world. The growth of state control of oil production extended from the Soviet Union to the Middle East and the North Sea, vast reserves of oil were located and exploited in areas remote from the original fields, but the Seven Sisters have continued to dominate world trade. It is beyond the scope of this book to trace the development of the oil industry in recent times, but the origins of the mineral oil industry in the period covered by this chapter deserve special recognition in the history of tribology. Within a short space of time animal and vegetable fats and oils which had been used as basic lubricants for thousands of years gave way to mineral oil. The transition had a profound effect upon economic and technical developments in the twentieth century, but it

5. See 'The Golden Jubilee of B.P.', *Inst. Petrol. Rev.*, **13**, No. 149, May 1959, 135–7.

also gave a spur to the scientific studies of lubrication which are discussed in Section 10.4. Before proceeding to a detailed consideration of the science, it is useful to recognize the changing role of petroleum products at the end of the nineteenth and in the early years of the twentieth centuries as illuminants, fuels and lubricants.

Illuminants

From 1859 until the dawn of the twentieth century kerosene was by far the most important product from the refineries. Sperm and rape-seed oil were used as illuminants before the drilling of Drake's Well. There is no doubt that the widespread use of the kerosene lamp along with coal gas improved the level of illumination in homes and workplaces enormously in the latter half of the nineteenth century. The supremacy of coal gas and kerosene as sources of light continued until the extensive generation and distribution of electricity was established early in the twentieth century. Michael Faraday's invention of the dynamo at the Royal Institution in London in 1831 and Thomas Edison's commercial exploitation of the device in 1878 laid the foundations for the competition from electricity. Edison went on to develop the incandescent electric lamp in 1879 and by 1882 he had established an electricity generating station.

Fuels

The economic depression of the 1870s in the United States did much to promote interest in the use of some of the by-products of kerosene manufacture. In due course fuel oil was to compete strongly with coal in firing the boilers of factories, steam-driven ships and locomotives, but it was to be the road vehicles which opened up the greatest market for a fuel based upon petroleum. Steam-driven road carriages had been running in France since 1769 and in England since 1784. The first electric road vehicle ran in 1839 and the internal combustion engine powered cars in 1860. The early internal combustion engines used a mixture of gas and air, and Lenoir's spark-ignition system of 1860 facilitated rapid development of such engines. A commercial plant for the manufacture of coal gas had been established in Birmingham, England, by William Murdock in 1795 and a company was formed to supply gas to London in 1813.

A chemist named Eugene Carless started to distil mineral oils obtained from Scottish shale in Hackney, London, in 1859. He produced a spirit of specific gravity 0·68 for enriching gas which he called *carburine* in 1864. When this highly volatile spirit was used in the internal combustion engine its original role in enriching *gas* was recalled through the adoption of the word *gasoline*, a term still used in America. In Germany, Benz produced his first car in 1885 and Daimler in 1886.[6] Daimler-powered cars were produced in France in 1891 and Diesel took out his patent for a compression-ignition engine in 1892. When a young engineer in England named Frederick Richard Simms decided to import Daimler gas-oil

6.   See Woodbridge, E. (1959), pp. 1–3.

engines to power launches on waterways he turned to the firm of Carless, Capel and Leonard for the supply of a deodorized petroleum spirit of 0·68 specific gravity which became known as *launch spirit*. In 1893 Simms suggested that the spirit should be renamed *petrol*. The full story of the origin of the term *petrol* in the United Kingdom has been told by Liveing (1959).

The rapidly expanding petrol-driven road-vehicle market was controlled by legislation with its origins in the competitive field of railways. The Red Flag Regulations of 1831, whereby locomotives had to be preceded by a red flag carried at walking pace, were deemed to apply to motor vehicles in 1865. The provision was rescinded by the Highways and Locomotives (Amendment) Act of 1878, and by 1896 the Locomotives on Highways Act introduced a speed limit of 14 mph (22·5 km/h). Most local government boards moved cautiously and restricted the upper speed to 12 mph (19·3 km/h), but on 1 January 1904 the limit was raised to 20 mph (32·2 km/h). It will be a surprise to many to know that this limit persisted until 1930, and it has been noted[7] that some believe the laws to have had a limiting effect upon both the speed of road vehicles and the expansion of the British motor industry. Nevertheless, the rapid expansion of the motor trade did much to establish, 'petrol' or 'gasoline' as a petroleum product of equal importance to kerosene by the start of the First World War.

| Lubricants | The transition from the widespread use of vegetable and animal oils to mineral oils was short in duration and dramatic in effect. Understanding of the optimum usage of the range of vegetable and animal products had developed over 2,000 or 3,000 years, whereas the change to mineral oils was effected in but a few decades. A detailed consideration of nineteenth-century lubricants has been presented by Dowson (1974). |

Lubricants

The transition from the widespread use of vegetable and animal oils to mineral oils was short in duration and dramatic in effect. Understanding of the optimum usage of the range of vegetable and animal products had developed over 2,000 or 3,000 years, whereas the change to mineral oils was effected in but a few decades. A detailed consideration of nineteenth-century lubricants has been presented by Dowson (1974).

The change was promoted by several factors. Growing industrial demand and the requirements of the railways was forcing up the cost of vegetable and animal oils and fats; many of the existing lubricants were unstable; systematic evaluations of the relative performance of oils from different sources were showing the readily available mineral oils to be entirely satisfactory for most applications, and scientific studies of lubrication by Hirn in France, Petrov in Russia and Beauchamp Tower in England were providing a firm base for the utilization of the new range of lubricants. All these factors contributed to the ready acceptance of mineral oils in industry and the creation of a strong market for another product of the rapidly expanding petroleum industry.

Detailed accounts of the range of lubricants and lubrication procedures employed in the nineteenth century can be found in Thurston (1885) and a historical review by Vogelpohl (1969).

The increased openings for petroleum products with rapid expansion of the markets for fuel oils and mineral oils at a time when kerosene was

---

7. See Woodbridge, E. (1959), pp. 1–3.

facing competition from both gas and electricity provided stability and opportunity for the new oil industry. In due course many new companies were formed, but the early years were dominated by the seven big companies whose histories have been outlined in this section.

10.4   Scientific studies of fluid-film lubrication

At the opening of the nineteenth century understanding of the mechanism of lubrication was primitive compared with knowledge of the friction process, but by the time it closed the science of fluid-film lubrication represented the pinnacle of tribological progress. The manner in which this transition took place was as sudden as it was important to technological progress, and a brief background to the stirring events of the 1880s helps to place them in perspective.

The background

Observations by Leupold (1735), Leslie (1804) and Rennie (1829) neatly summarize the development of ideas on lubrication prior to the middle of the nineteenth century. The concept that friction was intimately related to the roughness of opposing surfaces led naturally to the idea that any substance used to reduce friction must act effectively through its influence upon surface topography. Leupold (see Ch. 8) had the idea that the lubricant reduced the surface roughness such that '. . . the bearing surface is rolled over as if it had small balls on it'. Leslie concluded that lubricants reduced friction by filling up the depressions in rough surfaces (see Ch. 9). In 1829 Rennie gave a tantalizing hint that he was moving towards the view that the effectiveness of lubricants was related to their ability to separate sliding surfaces. He commented that when bearings were properly proportioned '. . . and their surfaces kept from contact by unguents, a much less allowance [for friction] may be made'.

Arthur Jules Morin published a number of papers on friction in the 1830s and his work on journal bearings dominated thinking for many years to come. He represented his findings as the ratio of the force of friction to the applied load and provided coefficients of friction for a wide range of bearing materials. He found that Coulomb's laws of friction were valid, and the simple concept of the laws of dry friction supported by Morin's coefficients provided the basis for much calculation and most of the teaching until the end of the nineteenth century. Morin also noted that vibration reduced static friction.

Morin certainly noted the influence of lubricants like oil, tallow and lard upon friction, but the effect was recorded simply as a modification to the friction coefficient. For example, he found that wrought- or cast-iron axles mounted in cast-iron or bell-metal bearings and lubricated by oil, tallow or lard exhibited a coefficient of friction of 0·075 with 'ordinary lubrication' and 0·054 when the lubrication was 'well maintained'. He appears to have given little consideration to the mechanism of lubrication or the possibility that well-lubricated surfaces might exhibit totally different relationships between friction, load, speed and bearing area to those apparent from dry contact studies.

Sporadic references to the nature of lubrication typified the situation which existed between 1850 and 1880, yet by 1890 all the basic features of fluid-film lubrication had been revealed. The exposure, achieved primarily by Petrov in Russia and Tower and Reynolds in England, represents one of the most interesting examples of simultaneous and yet totally independent achievements in the history of engineering science. The nature of these now classical investigations will be discussed shortly, but we must first address ourselves to the question of the causes and the timing of the studies.

What prompted this major work on fluid-film lubrication in the 1880s?

There appear to have been two main influences. The first was the discovery and widespread exploitation of mineral oil discussed in the previous sections. This new source of lubricants not only supplemented but in due course almost entirely replaced the existing supply of animal and vegetable products. The transition has been described by Dowson (1974). The second, and perhaps major influence, was the energetic growth of the machine age in general and the railways in particular. The poor performance of bearings on locomotives, trucks and carriages on the railways provided the necessary stimulus for studies of lubrication.

It is also important to recognize the quite sizeable empirical base which supported the great leap forward in the 1880s. The empirical testing of lubricants and bearing materials was by no means uncommon in the United States, Europe and Russia in the mid-nineteenth century. Von Pauli (1849) is credited with some of the earliest studies of bearing materials, as outlined in Chapter 9, but the assessment of lubricants by Hirn (1854) was a development of great significance.

France

Gustav Adolph Hirn's[8] *studies of journal bearing friction* (Hirn, 1854) Hirn was one of the founders of the science of applied thermodynamics. His interest in the relationship between work and heat led him to consider the nature of friction. He confirmed the laws of dry friction established by Amontons and Coulomb and then studied the influence of several lubricants upon the friction of a bronze bearing of fixed composition. His experiments were completed at the end of 1847, but it was 1854 before the work was published. Hirn submitted his findings first to the Académie des Sciences and then the Royal Society in London, but neither body felt moved to publish the paper.

Hirn loaded a half-bearing against a polished cylindrical drum of 9 in (230 mm) diameter and 8·6 in (220 mm) length. The hollow, water-cooled, cast-iron drum dipped into a lubricant and rotated at either 51 or 92 rpm (0·85–1·53 Hz). The half-bearing was made of bronze, consisting of one part of tin and eight parts of copper. The bearing was loaded by dead weights and friction was measured by adding balance weights to the lever shown in Fig. 10.2. The friction balance was said to be an extremely delicate and accurate brake. The temperature rise was also recorded

8. See Appendix, Sect. A.7 for biographical details.

Fig. 10.2
Hirn's friction balance.

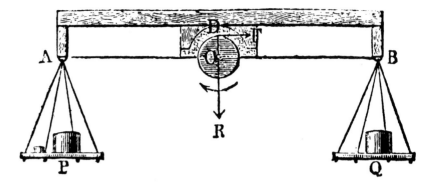

and the general arrangement of the testing machine was to form the basis of equipment used by several workers in subsequent decades.

Hirn found that in the absence of a lubricant the coefficient of friction followed Coulomb's law and was independent of load, speed and apparent contact area. For lubricated bearings the coefficient of friction was approximately proportional to the square root of the load, speed and apparent contact area, although the power of speed in the relationship was more nearly equal to 1 than $\frac{1}{2}$ when the lubrication was adequate and the loads light. This variation in the power of speed in relation to the coefficient of friction was explored and resolved most satisfactorily by Hirn. He was able to pass water through the hollow cast-iron drum and hence to exercise control over its temperature. He found that the coefficient of friction was directly proportional to speed when the temperature, and hence viscosity of the lubricant, was held constant, but that the power fell to 0·76 or even 0·5 if the machine was allowed to drift to an equilibrium temperature. He also noted that the coefficient of friction was directly related to the viscosity of the lubricant. He tested fats and twelve natural oils as well as water and air and wrote that '. . . For water and air to act as lubricants it is necessary for the drum to turn sufficiently rapidly to drag them into the bearing. When the speed reduces to a certain value the two non-viscous fluids are expelled by the pressure and the surfaces come into direct contact, and the friction at once becomes enormous.'

He found that air could reduce the friction to one-hundredth of its normal (liquid lubricated) value.

Gustav Adolph Hirn well deserves an early and respected place in the history of lubrication research and is considered by many to be the father of the subject. His study of friction in journal bearings revealed all the essential features of fluid-film lubrication although it lacked theoretical justification. He was certainly aware of the advantages of copious lubrication and he established the linear relationship between journal-bearing friction and speed at light loads. He discovered the effect of running-in upon bearing friction and pointed out that lubricated bearings must be run continuously for a certain time before a steady value of friction, lower than the initial value, is attained. He not only tested the current range of animal and vegetable oils like sperm, olive and rape but he also pioneered the use of mineral oil. Colonel Drake had not yet drilled his oil-well in Titusville when Hirn experimented with mineral oil obtained from oil-sands and it was widely believed that mineral oil was unsuited as a machine lubricant. Hirn demonstrated that distilled mineral oils could provide most satisfactory lubricants in the form of heavy machine oils and light spindle oils alike. He was also the first to demonstrate that air could be used as a lubricant and the origin of gas-bearing studies can be placed firmly in the middle of the nineteenth century.

His results on the friction of lubricated surfaces were so contrary to the well-established laws of dry friction proposed by Coulomb and confirmed by Morin that they received little support and recognition. This could well have accounted for his publication problems. It is only with hindsight that the excellence of his experimental work and the validity of most of his findings can be recognized. In due course Petrov was to refer extensively to the experiments of Hirn in the development of his theory of viscous friction in bearings. Robert Henry Thurston, who was to become the first President of the American Society of Mechanical Engineers, had no doubt about Hirn's contribution to lubrication research. His famous book, *A Treatise on Friction and Lost Work in Machinery and Millwork* (1885), was dedicated to 'The Memory of The Engineer, Physicist and Mathematician G. A. Hirn, one of the earliest workers in this field'.

England	*A pragmatic approach to lubricants and lubrication in the 1850s* While Hirn was conducting his revealing studies of the friction of lubricated half-bearings in France, the members of the newly formed (1847) Institution of Mechanical Engineers in England were demonstrating their practical interest in the subject. It was commendable that the Council of the Institution periodically presented a list of subjects on which they invited communications from members and their friends. In January 1850, and on various other occasions, the list included, '. . . Friction of various bodies–facts relating to friction under ordinary circumstances–friction of iron, brass, copper, tin, wood &c–proportion of weight to rubbing surface–best form of journals, &c.–lubrication, best materials and means of application, and results of practical trials–best plan for oil tests.'

The members responded, particularly in the 1850s, through a number of short papers of considerable tribological interest. In January 1851 Mr Joseph Barrans described his new axle-box for railway engines and carriages and indicated the developing interest in replacing fats and greases by oils with the following words (Barrans, 1851): '. . . There is another material advantage attending this part of the invention, inasmuch as oil, which is generally admitted to be a better lubricator, and more certain in its action than grease of any kind, has been mostly kept out of use, by reason of the great waste attending it in ordinary axle boxes. . . .'

The following year Mr Paul R. Hodge compared the performance of an oil-filled axle box of American design with the traditional tallow lubricated boxes on the London and North Western Railway (Hodge, 1852). When the new axle boxes were fitted to tender No. 182 and charged with 10 quarts (11·4 l) of oil by Mr R. McConnell and his assistants, they worked beautifully. Furthermore, the new axle-boxes weighed but 50 per cent of the old boxes and the costs per day were 9*d.* and 1½*d.* for the tallow and oil lubricated sets of six axle-boxes, respectively. Mr Hodge made the interesting observation that grease was never used as a lubricant on the railways of the United States. He also discussed the evils of hot axles resulting from high friction in the bearings. '. . . Scarcely a train passes over our roads (in the summer more particularly) but some one or more of the axle journals heat. In one instance that the writer experienced, the whole train had to be passed into a siding for more than two hours, before it could again proceed on its journey.'

The year 1853 saw the presentation of two papers which neatly convey the outlook of the day on the lubrication of axle-boxes. Mr John Lea of London extolled the virtues of a special lubricant based upon '. . . carefully refined southern whale oil as a basis, to which are added India-rubber and levigated white and red leads, to constitute a kind of metallic soap' (Lea, 1853). It was in fact the composition patented by Mr Donlan in 1848 and described in the previous chapter.

In a delightful paper describing a new form of axle-box designed to transmit an adequate amount of lubricant to the journal, Mr W. Bridges Adams drew attention to the disadvantages of soap or grease. It appears that the early design of railway axle bearings was based upon the sizes of bearings used to support fixed shafting in factories. A fluid oil was used in the factory bearings, but a viscid soap was introduced into the railway bearings to accommodate the concussion from the rails. However, as Adams (1853) noted, '. . . A strong objection to soap lubrication is, that it requires a considerable amount of friction in the winter time to make it fluid; and it is sometimes difficult to start a train into motion when the grease has been frozen . . .'.

He also reiterated the concept of lubrication propounded by Leupold in 1735.

> . . . *Oil, or grease, or soap, interposed between two metallic bodies moving one upon the other, is composed of a series of small globules, which keep the bodies separated, and serve as rollers. The surfaces of metallic bodies, however apparently smooth, are composed of salient and re-entrant angles of larger or smaller size, according as the metal is hard and polished, or soft and rough. Therefore the more imperfect the structure of an axle and bearing, the more viscid must be the lubricating material to keep them from contact.*

The role of lubricant viscosity in effective fluid-film lubrication was thus expressed from different points of view by Adams in England in 1853 and Hirn in France in 1854.

The disadvantages of grease-like lubricants for trucks and locomotives continued to be the dominant issue on railway lubrication problems in the 1850s. There were two main problems. Grease or tallow was normally packed in a metal box above the bearing which itself sat upon the axle. It began to flow through vertical holes into the clearance space between the journal and bearing only when it was warmed. In cool weather the sole source of heat was friction in the bearings and hence the lubricant was inactive on starting up the train or on very cold days. If the grease became too hot due to excessive heating of the bearings, or the warmth of a summer's day, the lubricant became too fluid and escaped from the bearing. The latter point was emphasized by Mr William G. Craig when he addressed a meeting of the Institution of Mechanical Engineers in Birmingham on 22 April (Craig, 1855) '. . . In numerous instances, from the metal becoming overheated, the grease is rendered fluid, and runs out and is wasted'.

The advantages of liquid lubricants for the railways thus became apparent long before mineral oil became available in substantial quantities. The immediate impact of this development in the 1850s was to stimulate ideas on the way in which liquid lubricant might be contained in axle-box bearings. The longer-term effect was to promote studies of the physical nature of lubrication itself in the 1880s.

References to axle-box lubrication diminished for a while after the 1850s, but there is one remarkable landmark in the history of lubrication from the year 1860 related to the use of water. On Thursday 9 August Mr Sampson Lloyd of Wednesbury described an invention of M. Aerts of Belgium known as *The Water Axlebox* (Lloyd, 1860). The use of water rather than oleaginous matter as a lubricant at this stage is most remarkable, although Hirn (1854) had already demonstrated the possibility of using water as a lubricant. The stimulus for Aerts' invention was the inherent disadvantages of fatty substances and particularly the need to provide heat to greases before they became effective.

A metal disc attached to the axle was dipped into a bath of water as shown in Fig. 10.3. A brass scraper removed water from the top of the rotating disc and directed it into a spout, thus providing an effective, self-contained lubrication system. Leather seals retained the water in the

Fig. 10.3
M. Aerts' water axle-box for the railways.
(A) Cast-iron disc dipping into water. (B) Water reservoir. (C) Brass scraper. (D) Spout. (E) Cast-iron collar. (F) Leather collar. (G) Annular cast-iron disc. (H) Back plate. (I) India-rubber band or steel spring. (K) India-rubber ring. (L) Plug. (M) Lid.

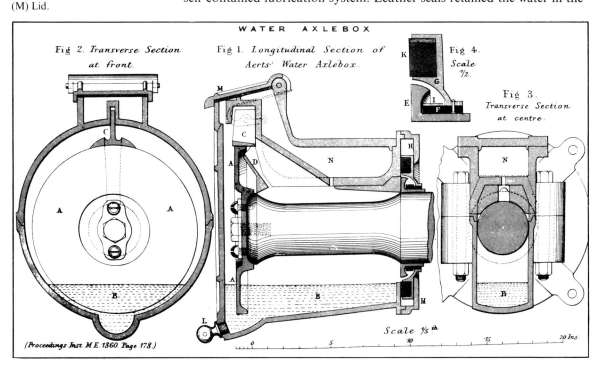

WATER AXLEBOX

Fig 2. *Transverse Section at front.*

Fig 1. *Longitudinal Section of Aerts' Water Axlebox.*

Fig 4. *Scale 1/2.*

Fig 3. *Transverse Section at centre.*

(Proceedings Inst. M.E. 1860. Page 178.)

Scale 1/5 th

axle-box. A similar arrangement was proposed for the bearings supporting line-shafting. Oxidation of the metal was prevented by greasing both the brass bearing and the axle before fitting the water-boxes. The effect was said by Mr Lloyd to be that '. . . the grease is found to solidify after a short time of running, and forms a sort of varnish of a dark brown colour, thus proving the absence of actual contact between the two metal surfaces; for were it otherwise, the surface of the axle would become polished'.

Trials on the Great Northern and Eastern Counties Railways demonstrated that less wear took place in the bearings of both first- and second-class carriages when conventional grease-boxes were replaced by M. Aerts' water axle-boxes. A large shaft which previously caused great trouble by overheating at Mr Leech's mill at Stalybridge was caused to run without trouble for at least a year by fitting water-lubricated bearings.

M. Aerts, who was present at the meeting, explained that the use of water as a lubricant was not new, but that '. . . the principle aimed at in the water axlebox was to have greased surfaces for the bearing, with a constant stream of water running over them, so as to interpose a film of water constantly between the rubbing surfaces, preventing contact of the two metallic surfaces and thereby removing the cause of heating'.

During the discussion of this most interesting and early application of water-lubricated bearings, objections were raised on the grounds that the lubricant would freeze when the train had to stand for long periods in the sidings. M. Aerts pointed out that in such cases the water could readily be drained through the hole provided. He thought this might be necessary in cold countries like Russia or Canada, but not England where there had been no problems with the freezing of boxes on the Eastern Counties Railway during the previous two winters. He also pointed out that existing boxes had to be recharged with grease after running only 40 or 50 miles (64–80 km), whereas the water axle-boxes would run for 1,000 miles (1,609 km) before needing a fresh supply of water. The saving in cost of attendants at various stations along the line would be considerable, and since wear in the new boxes was much reduced, the cost of repairs, then estimated at four and a half to ten times the cost of grease, would also diminish. The third economic argument in favour of the new axle-boxes was, of course, that water was cheaper than grease.

From these background notes it is clear that axle-boxes were providing severe problems for those responsible for the efficient running of the railways in the 1850s and 1860s. Numerous practical solutions were proposed to the lubrication problems of the day. These included design changes, the optimum selection of bearing materials and ingenious proposals for lubricants. Not only did Hirn (1854) demonstrate that air and water could be used as lubricants, but practical men like Lloyd (1860) showed that the Aerts water axle-box could be effective in mills, on line-shafting and on railway wagons. It is widely believed that the concept of fluid-film lubrication, in which the bearing surfaces could be separated by a film of lubricant, arose suddenly in the 1880s, but there is strong evidence that the concept emerged gradually and naturally. In support of this thesis I would refer to the works of Leupold (1735), Leslie (1804),

Rennie (1829), Adams (1853), Hirn (1854) and finally M. Aerts and Lloyd (1860). Some of these men were scientists and others were practical engineers concerned with the provision of adequate lubrication for machinery. In their own way each contributed to the growing belief that the role of a lubricant was to separate the bearing surfaces.

The timing of the account of M. Aerts' water axle-box is interesting since it occurred almost exactly one year after Colonel Drake struck oil at Titusville. The increasing availability of mineral oil in America and its impact upon the problems of the railways affords a useful opportunity to switch attention to the work of that great contributor to lubrication science, Robert Henry Thurston.

The United States

Robert H. Thurston and the testing of lubricants (Thurston, 1879)   Robert Thurston[9] was a man of great energy who became the first President of the American Society of Mechanical Engineers in 1880. He graduated at Brown University in 1859 and spent two years working in his father's steam-engine manufacturing business before joining the Navy. His engineering philosophy was simple and convincing. If machines were correctly designed and constructed they would not experience permanent or plastic distortion in performing their functions and would not therefore cause energy to be dissipated. Friction thus emerged as the source of *lost work* and *inefficiency* in machinery. '. . . The study of the laws of friction, the construction of its theory, and the experimental investigation of the conditions which determine the loss of efficiency in machinery by friction, are thus obviously of supreme importance to the engineer who designs, the mechanic who constructs, and the operator or manufacturer who makes use of machinery.'

He argued that it was to the engineer '. . . a vitally important branch of applied science, and it is coextensive with the applications of mechanical science'.

The message is as clear and valid today as it was almost 100 years ago; yet it appears that it has to be restated and emphasized for successive generations.

Thurston backed up his general remarks with painful facts. '. . . The loss of power in mills ranges, with different machines, from 5 to 90 percent, averaging for cotton and flax mills about 60 percent with good management, and in woollen mills about 40 percent. . . .'

The loss of 50 per cent of the available power in overcoming friction between lubricated surfaces in mills, together with figures ranging from 3 to 16 per cent in steam engines and about 15 per cent in iron-working tools provided an adequate motivation for his subsequent evaluation of lubricants.

Thurston delivered a series of lectures on *friction* and *lubrication* before the Master Car Builders Association and elsewhere, which were later published in the *Railroad Gazette* (1879). A more extensive account

9.   See Appendix, Sect. A.9 for biographical details.

of the subject formed the basis of his famous book dedicated to Hirn and entitled *A Treatise on Friction and Lost Work in Machinery and Mill-work* (Thurston, 1885). The book ran to seven editions and was reprinted as late as 1907. Accounts are given of the current understanding of sliding and rolling friction under dry conditions and the friction between lubricated surfaces. The concept of dry friction was linked to the inter-locking of asperities with adhesion receiving scant attention. The laws of friction are presented and applied to numerous machine elements. This no doubt accounted, at least in part, for the great popularity of his text, for he demonstrated how the principles of *pure mechanics* could be modified to account for the actual behaviour of mechanical devices. Subjects covered included journals, pivots, belts, ropes, pulleys, friction-wheels, road vehicles, railway trains, earthen retaining walls, gears and brakes. The book contained a most valuable collection of coefficients of friction and this, together with an excellent account of lubricants and machines for their evaluation, represents the greatest merit of the text.

After describing a number of other machines Thurston gives an account of his own lubricant tester shown in Fig. 10.4. The test bearings marked G and G′ on the overhung journal F were loaded by means of a helical spring. The test shaft was mounted in slave bearings B and B′ and driven at variable speeds by a pulley at C. Temperature was recorded by a thermometer housed at Q. Friction caused the pendulum H to rotate to some equilibrium position and readings of angular displacement and compression of the spring enabled the coefficient of friction to be calculated directly. A similar machine was later constructed for testing railroad bearings at realistic loads and speeds.

Thurston's results were fully reported in tables of coefficients of friction under realistic operating conditions for a wide range of lubricants. Such findings must have been of immense value to those concerned with the lubrication of machinery at the end of the nineteenth century.

Fig. 10.4
Robert H. Thurston's lubricant testing machines (Thurston, 1885.)
(a) Initial pendulum machine.
(b) Railroad bearing machine.

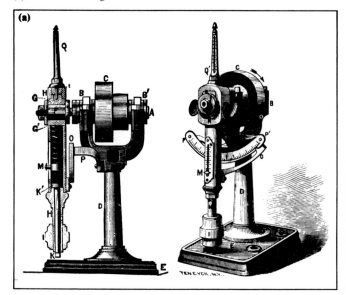

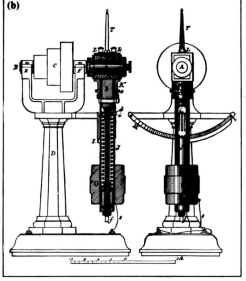

Robert Henry Thurston made significant contributions in many fields of engineering activity, but his work on lubricants, lubrication and friction alone has brought him great renown. His major contributions were threefold: he unceasingly drew attention to the importance of studying and applying tribological concepts, he presented coefficients of friction for materials encountered in a wide range of machine elements and he firmly established empirical testing of lubricants. He appears to have been the first person to report that the coefficient of friction passed through a minimum as the load increased, a transition which is now recognized as the change from full fluid-film to mixed or even boundary lubrication. He wrote:

> *... The general conclusion that the coefficient of friction decreases with increasing pressure must evidently be qualified by the undoubted proposition that, with any given condition of the rubbing surfaces, and with all other conditions unchanged, there must always be ultimately reached a point at which, with increasing pressures, the limit of bearing power is attained or approached, and the friction must exhibit a change of law, the coefficient increasing, beyond that limit, as the intensity of pressure is augmented.*

Robert Thurston's lectures and his book (Thurston, 1885) had a great impact upon late nineteenth-century attitudes to the subject we now call tribology. His discourse on the nature of friction and lubrication, his comprehensive accounts of the properties of lubricants, his emphasis on the need for testing lubricants on appropriate instruments and machines and his powerful economic arguments for consideration of the effect of friction upon machine performance established his book as a most important text. His crusade against the evils of friction was recognized not only in the United States but also in Europe and Russia.

Russia

Mineral oil, the railways and the scientific work of Nikolai Pavlovich Petrov (Petrov, 1883)   Whereas Thurston was concerned mainly with the effects of friction and lubrication upon the performance of machinery in mills and factories, and engineers in Western Europe were struggling to improve bearings on the railways, the two major influences which combined to foster the pioneering work of Petrov in Russia were the need to find an application of Caucasian petroleum products and the inadequate performance of axle-boxes on the railways. In a previous section of this chapter the origins and growth of the Russian oilfields were discussed. The nature of the Russian crude oil was somewhat different from that of the American product. About 70 per cent of the lighter fractions produced from the American crudes could be sold for lighting and heating, whereas the Russian yield was only 30 per cent. As the rate of production of Russian crude oil increased, the accumulation of a 70 per cent residue caused the industry to give serious consideration to possible uses of the waste product.

The residue of *residuum* was the tarry substance remaining after distillation of crude petroleum to form naphtha and illuminating oils. At Baku it was called *astatki* by the Russians and *mazut* by the Tartars. *Astatki* was in great demand for fuel and the inducement to produce lubricating oils was initially small. In due course the *astatki* was made into lubricants by distillation with superheated steam and marketed in competition with the long-established animal and vegetable oils. Production was started by V. I. Ragosine in 1876 and Russian mineral oils were introduced into England in 1878.[10] The conservative attitude of the engineers was eventually overcome by the considerable economic advantage of the new mineral oils, but Petrov felt that it was necessary to clarify the situation by means of a comprehensive and well-conducted series of experiments.

The second factor was the problems experienced with bearings on the railways, the very issues which featured in the meetings of the Institution of Mechanical Engineers in England from the 1850s onwards. Petrov was well acquainted with these problems through his professional interest in applied mechanics and the railways. He held the Chair of Steam Engineering and Railway Vehicles in the Technological Institute of St Petersburg (Leningrad) from 1871 and he became a member of the Engineers of the main Society of Russian Railways in 1873. A full account of his life and work is given in a biographical note in section A8 of the Appendix, but at this stage it is adequate to appreciate that it was the combination of his background with the Russian railways and the interest in the possibility of using mineral oil from Russian crude to lubricate the bearings of wagons and machinery which directed Nikolai Pavlovich Petrov's attention to the question of friction and lubrication in journal bearings.

Petrov was sensitive to the economic implications of introducing the new mineral oils. They were certainly much cheaper than the vegetable oils then in use, the direct savings being about 70 per cent. However, it was known that different lubricants could affect the friction in bearings operating under identical conditions by factors up to one order of magnitude, and the use of a cheap oil might result in much greater friction and fuel consumption. Petrov pointed out that Russia spent millions of roubles on fuel for machinery, and imperfect lubrication could therefore cost the country gigantic sums of money if it involved a 5 or 10 per cent increase in fuel consumption.

The direct effect of the lubricant upon the friction of journals led Petrov to undertake numerous experiments in which the measured friction could be related to viscosity. To appreciate the importance of this work it is necessary to realize that at this stage mineral oils were generally selected on the basis of measurements of density, flash points and freezing points, together with observations on transparency and the absence of injurious matter. Petrov was to show that it was *viscosity* and not *density* which governed the lubricating characteristics of an oil at a given temperature.

Petrov wrote at length and on several occasions on the influence of lubricant upon mediate friction in journal bearings, but he came to the heart of the matter very early in his first paper (Petrov, 1883). He wrote:[11]

---

10. See Archbutt, L. and Deeley, R. M. (1900), p. 95.
11. See Parr, P. H. (1912), p. 244.

'. . . I have noticed when approaching this problem that those experimenters who have studied mediate friction up to the present have never paid sufficient attention to the fact that, in all cases where the friction is effectively mediate, the liquid film completely separates the two solid surfaces.'

The term 'mediate' was used to indicate that the friction between the solids was generated through an intermediary–the lubricant. Petrov realized that the answer was to be found in simple hydrodynamics rather than the contemporary concepts of interlocking asperities, and he proceeded to analyse the problem within the framework of three assumptions.

1. That the liquid film was bounded by two cylindrical surfaces.
2. That each particle of the film revolved at a constant speed round the common axis of the enveloping cylinder.
3. That the hydrodynamical pressure, in the greater part of the liquid film, varied so little that it could be considered to be constant.

The question of slip between the lubricant and the surfaces of the solids complicated the situation studied by Petrov. Navier (1823) had discussed the question and proposed that if such *external friction* existed it could be represented by

$$\tau = \lambda v, \qquad \qquad \dots [10.1]$$

where τ is the shear stress at the interface between the fluid and the solid, λ the coefficient of external friction and v the relative sliding speed between the fluid and the solid at the interface.

The relationship for *internal friction* in a viscous fluid proposed by Newton (1687) could be written as,

$$\tau = \eta \frac{du}{dn}, \qquad \qquad \dots [10.2]$$

where τ is the internal shear stress in the fluid, η the coefficient of internal friction (or viscosity) and du/dn the normal velocity gradient.

Petrov repeated the hydrodynamic theory of Margules (1881) for the analysis of flow of a viscous fluid between concentric cylinders, applied the relationship [10.2] to the inner (λ_1) and outer (λ_2) solid surfaces bounding the oil film in order to satisfy the equilibrium equation for moments on the lubricant, and concluded that the torque could be written as

$$M = 4\pi\eta l U \left[\frac{r_1^2 r_2^3}{r_1 r_2 (r_2^2 - r_1^2) + 2\eta(r_2^3/\lambda_1 + r_1^3/\lambda_2)} \right], \qquad \dots [10.3]$$

where l is the axial length of the bearing.

With c representing the mean and small thickness of the lubricating film ($r_2 - r_1$), the expression for mediate friction force (F) reduces to

$$F = \frac{2\pi r_1 l \eta U}{[c + (\eta/\lambda_1) + (\eta/\lambda_2)]} = \frac{\eta U A}{[c + (\eta/\lambda_1) + (\eta/\lambda_2)]}, \qquad \dots [10.4]$$

where A is the bearing surface area $2\pi r_1 l$.

The coefficient of mediate friction (μ) can thus be written as

$$\mu = \frac{F}{P} = \frac{\eta U A}{P[c + (\eta/\lambda_1) + (\eta/\lambda_2)]}, \qquad \dots [10.5]$$

where P is the load on the bearing.

Petrov did not have to come to any firm conclusion about slip at the interface between the lubricant and the solids for the simple reason that he was concerned with the relative performance of different lubricants. He was, of course, intending to compare the friction of mineral oils with that of the well-known vegetable and animal oils such as olive, rape and sperm. He did, however, present reduced forms of equation [10.4] for situations in which the internal and external friction coefficients adopted quite dissimilar magnitudes. For example, he argued that if the ratio of internal to external friction could be neglected (no slip), as might be the case with water and glass, the relationship [10.4] reduced to

$$F = \frac{\eta U A}{c} \qquad \dots [10.6(a)]$$

or

$$\mu = \eta U \frac{A}{c} \frac{1}{P} = \frac{2\pi r_1 l \eta U}{cP} = \frac{4\pi^2 r_1^2 l}{c} \left(\frac{\eta N}{P}\right). \qquad \dots [10.6(b)]$$

The relationships [10.6(a)] and [10.6(b)] are variously known as *Petrov's law*, and they refer to the configuration shown in Fig. 10.5.

The main difference between Petrov's studies of journal-bearing friction and previous work was that he constantly related his findings to the viscosity of the lubricant. He constructed an apparatus similar to that used by Poiseuille (see Fig. 10.6) and measured the viscosity of castor, olive, rape and sperm oil, water and eleven mineral oils from Caucasian petroleum over the temperature range 17·5–65°C. It is interesting to note that he wrote down the equation for flow of a viscous fluid in a pipe which included a term for slip at the wall. He then rejected the term on the grounds that it was negligible without apparently feeling confident enough to make the same assumption for lubricated bearings.

Fig. 10.5
Petrov's law

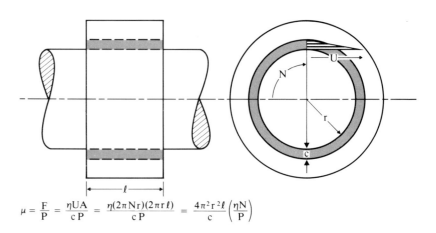

$$\mu = \frac{F}{P} = \frac{\eta U A}{c\,P} = \frac{\eta(2\pi N r)(2\pi r\,\ell)}{c\,P} = \frac{4\pi^2 r^2 \ell}{c}\left(\frac{\eta N}{P}\right)$$

Fig. 10.6
Nikolai Petrov's apparatus for
determining the viscosity of
lubricants.

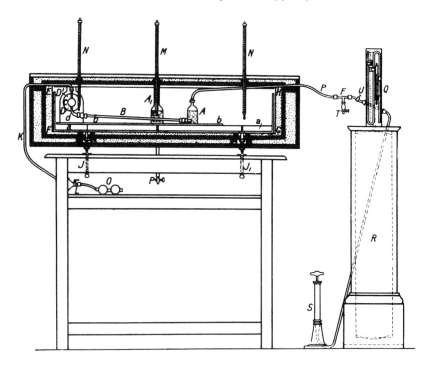

Petrov analysed the experimental results of Hirn and then carried out a comprehensive study of journal friction on both a pendulum machine made by Klein, Schanzline and Boecker and a specially constructed wagon-axle machine for testing Laouenstein axle-boxes of the type used on the St Petersburg–Varsovie Railway. These machines are shown in Figs 10·7 and 10·8, respectively. The experiments were carried out in the workshops of the St Petersburg–Varsovie Railway and one of the wagon axles (G) on the second machine was driven by a belt from a steam engine. Friction caused the wheels (L, L$_1$) on the test axle (K) to rotate and friction in the axle-boxes tended to disturb the equilibrium of beam MM. Balancing weights on the loading cradles (N$_1$) enabled the friction in the bearings to be measured. The rigours of nineteenth-century experimental work on bearings was indicated by the range of ambient temperatures recorded by Petrov–in one case the air temperature was $-10\cdot6°$C ($12\cdot9°$F). In these experiments the surface speed of the journal varied from 0·9 to 7 ft/s (280–2,200 mm/s) and the pressure per unit area varied from 9 to 90 atm (0·91–9·1 MN/m^2). The experimental work was both extensive and convincing. The number of experiments completed was 627 and the most impressive conclusion was that the term $[c + (\eta/\lambda_1) + (\eta/\lambda_2)]$ in the denominator of equation [10.4] was essentially independent of both η and U. This confirmed the somewhat novel view that mediate friction was dependent upon the viscosity and not the density of the lubricant.

It was Petrov who established the hydrodynamic nature of journal-bearing friction and the relationships [10.6(a)] and [10.6(b)] known as Petrov's law are among the best-known relationships in lubrication theory.

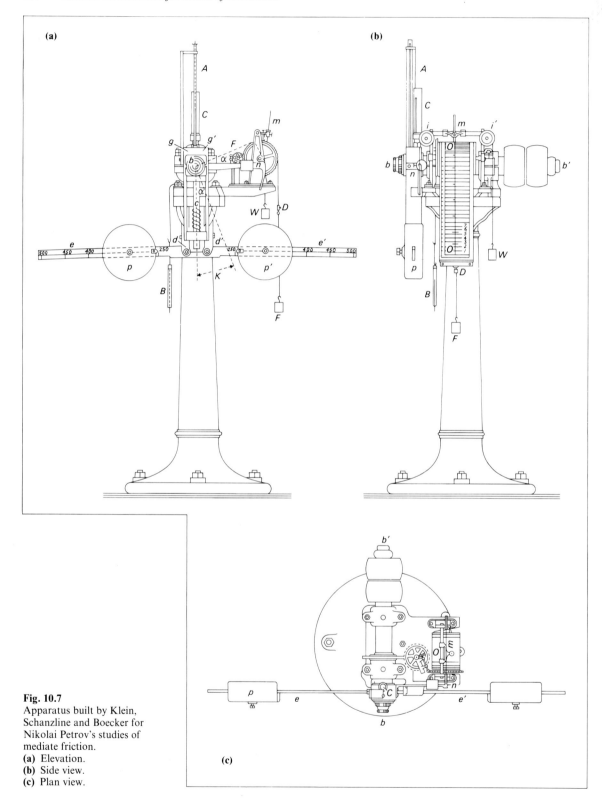

Fig. 10.7
Apparatus built by Klein,
Schanzline and Boecker for
Nikolai Petrov's studies of
mediate friction.
(a) Elevation.
(b) Side view.
(c) Plan view.

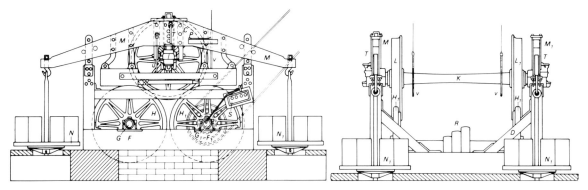

Fig. 10.8
Nikolai Petrov's machine for testing the friction of wagon axles.

It is therefore surprising that he never measured pressure in the bearing and that he missed the opportunity to expose the action of pressure generation in fluid-film bearings. This was left to Beauchamp Tower in England. Petrov did recognize that the effective average clearance $(c + (\eta/\lambda_1) + (\eta/\lambda_2)]$ might be influenced by load and he proposed the relationship,

$$\left(c + \frac{\eta}{\lambda_1} + \frac{\eta}{\lambda_2}\right) \propto \frac{1}{P^{1/2}} = \frac{k}{P^{1/2}}$$

Such a relationship would enable the coefficient of friction (μ) to be expressed as

$$\mu = \frac{F}{P} = \frac{\eta U}{k} \frac{A}{\sqrt{P}}. \qquad \ldots [10.7]$$

The equation was empirical and entirely lacking in theoretical justification.

Petrov's first papers on lubrication were published in the *St Petersburg Engineering Journal* (Petrov, 1883). An authorized translation in German was prepared by Wurzel (1887) and a useful summary appeared later in Ostwald's (1927) *Klassiker*. Petrov completed over eighty scientific papers and many are contained in the USSR Academy of Sciences collection of selected works edited by Leibenson (1948). Many of his papers published after 1890 were written in French and a valuable English translation of a summary of his 1883 papers was published by Parr (1912) in *The Engineer*. Many of the papers are difficult to trace, but Cameron (1966) has given an excellent guide to the literature. Some of Petrov's original notebooks can be seen in the Polytechnic in Leningrad.

In 1900 Petrov repeated Reynolds' analysis of journal bearing. It was a surprising paper and one which added little to the theory of lubrication. In the binomial expansion of some of the integral terms up to (e^{28}) were retained and the penultimate section of the paper carried the title 'L'erreur dans les calculs d'Osborne-Reynolds'. There are, however, further errors in Petrov's work and the contribution is of little significance and rarely mentioned. In 1920 Martin[12] was to comment, somewhat

---

12.   I am grateful to Dr P. B. Neal for access to this letter.

harshly, to Professor Goodman about this paper as follows, '. . . I am simply astounded at the contents . . . He (Petrov) seems to have made no contribution of any importance whatever to the study of lubrication unless it be a few experimental results.'

Even if this evaluation of Petrov's (1900) theoretical paper is accepted, the judgement of his few (627!) experimental results cannot be supported. Nicolai Pavlovich Petrov was undoubtedly the first person to confirm by careful experimental study and great physical insight the hydrodynamic nature of mediate friction in journal bearings. The idea had been nurtured by Leupold (1735), Leslie (1804), Rennie (1829), Adams (1853) and particularly Hirn (1854), but none had developed the concept or its experimental verification to the stage at which it could be widely recognized. This distinction undoubtedly belongs to Petrov (1883), and his work published in 1883 received the Lomonosov Prize of the Imperial Russian Academy of Sciences.

The United Kingdom

The Institution of Mechanical Engineers' Committee on Friction at High Velocities and the experimental work of Mr Beauchamp Tower (Tower, 1883)   The accidental discovery of substantial pressures in the oil films in journal bearings by Mr Beauchamp Tower is one of the best-known landmarks in the history of tribology. It provided the experimental basis and conceptual stimulus for the famous analytical study of Osborne Reynolds and is seen by many as the foundation of understanding of fluid-film lubrication.

What is less well known is the background to this discovery. Why was the work undertaken? What was the role of the Institution of Mechanical Engineers? What was the composition of the Committee guiding the work? Who was Beauchamp Tower and why was he appointed to undertake this investigation? It is the answers to these questions which provide the necessary background for a full appreciation of the great leap forward in the understanding of lubrication in the 1880s.

The work was prompted by a growing state of confusion in the second half of the nineteenth century about the laws governing friction between moving bodies. The laws of static friction were well established and Coulomb's view that the friction of motion was independent of speed had been supported by the highly respected General Morin. Yet Hirn (1854) had concluded that the mediate friction of axles in their bearings was directly proportional to speed when the temperature remained constant. A contrary view was expressed by Bochet (1861) after experiments in which he attached skids of different materials to the wheels of railway wagons, thereby converting them into sleds. A spring dynamometer indicated that the kinetic friction decreased with sliding speed for all materials.

The continuous nature of friction during transition from static to dynamic conditions was demonstrated by Jenkin and Ewing (1877) while Kimball (1877) not only confirmed this finding but showed that kinetic friction could be greater or less than the static friction. He further reported that at very low speeds the friction increased as the velocity rose, but that the trend was reversed at higher speeds. The important studies of friction between brake-blocks and wheels and between wheels and rails undertaken by Westinghouse and Galton (1878*a,b*) are reviewed in Section 10.9. The results clearly indicated that the coefficient of friction between the brake-blocks and the wheels decreased as the sliding speed increased.

In the United States Thurston (1879) was obtaining results from his journal-bearing experiments which appeared to be in opposition to those of Kimball. Thurston found that as the speed increased the coefficient of friction fell from the static value, reached a minimum and then rose steadily. He also noted that the speed at which the minimum occurred was influenced by the load on the bearing. It is to Thurston[13] that the discovery of this important feature of the coefficient of friction–speed characteristic of lubricated bearings must be attributed.

This accumulation of apparently contradictory evidence on the characteristics of kinetic friction in the third quarter of the nineteenth century was, of course, related to the wide range of different experimental conditions. Nevertheless, the confusion was so great and the question of such importance to technological progress, that the subject came to the attention of the Research Committee of the Institution of Mechanical Engineers. One of the praiseworthy activities of the young Institution was the encouragement of studies of matters of immediate concern to engineers. The lists of 'Subjects for papers' which appeared in each volume of *Proceedings* indicated the desire of the Council to '. . . invite communications from the Members and their friends on the preceding subjects, and on any Engineering subjects that will be useful and interesting to the Institution . . .'.

The list invariably included 'Friction of various bodies', but the matter was advanced by the scheme for *experimental research* adopted by the Council at its meeting on 20 December 1878.[14] The scheme called for the appointment of five members of Council to form a *Research Committee*, the first duty of the Committee being to '. . . prepare a list of subjects, on which further research is desirable . . .'.

The Research Committee was also charged with the task of appointing '. . . a sub-committee for each of the selected subjects and invite gentlemen (not necessarily Members of the Institution) to give assistance to such sub-committees'.

The Research Committee moved quickly and it presented its list of recommended subjects to the meeting of the Council on 20 February 1879. It was resolved that the following subjects be first investigated:

(1) The conditions of the hardening, tempering, and annealing of steel.
(2) The best form of riveted joints to resist strain, in iron or steel, or in combination.
(3) Also, should time and money be found to be sufficient; Friction between solid bodies at high velocities.'

---

13. See Hersey, M. D. (1933), pp. 411–29.
14. I am grateful to Professor Cameron and the Librarian, the Lyon Playfair Library, Imperial College, for access to extracts from the Council of the Institution of Mechanical Engineers Minutes provided by the Librarian of the Institution in 1961.

Fortunately, both the time and the money became available and in due course a talented investigator by the name of Beauchamp Tower[15] was commissioned to investigate journal-bearing friction.

The members of the *Committee on Friction at High Velocities* were: E. A. Cowper, Capt. Douglas Galton, C.B., Hon. D.C.L., F.R.S., Dr John Hopkinson, F.R.S., Prof. H. C. Fleeming Jenkin, F.R.S., John Ramsbottom, Lord Rayleigh, F.R.S., Prof. A. B. W. Kennedy (Reporter).

The terms of reference of the Committee were to study '. . . friction at high velocities, specially with reference to friction of bearings and pivots, friction of brakes, &c.'

It is interesting that the membership included men who had already contributed so much to tribology, like Galton of railway braking fame and Ramsbottom whose piston-ring design had become established practice, and Fleeming Jenkin who had skilfully studied friction, together with eminent scientists and engineers like Lord Rayleigh, John Hopkinson, E. A. Cowper and Professor Kennedy. The last two gentlemen both served as Presidents of the Institution.

Each sub-committee of the Institution Research Committee was instructed to first collect and collate all the records of experiments and other information already existing on the subject in hand. The first report to the Council of the Committee on Friction at High Velocities was made in November 1879 by the Reporter, Alexander B. W. Kennedy, Professor of Engineering at University College, London. The report, which was subsequently published in the *Proceedings* of the Institution (1883), gave a fine, succinct review of previous work and concluded with a translation of General Morin's response to an inquiry sent out to a number of persons. Kennedy noted that '. . . General Morin's letter is specially interesting, as coming from such a veteran worker in the subject of friction as its writer'.

On 30 December 1881 The Committee informed the Council of the Institution that it had received indications of interest in constructing the testing machine from a Mr Paget and others, including Mr H. Fell. It received a grant of £100 for the purpose of experiments from the Council on 15 August 1882 and a further sum of £200 on 22 December of the same year.

According to the *Proceedings* of the Institution (1882) a Mr W. W. Hulse made an inquiry on 20 April 1882 about the progress of the investigation as follows:

> *. . . Mr W. W. Hulse said if he was not out of order he should like to ask what progress was being made by the Research Committee on the question of Friction at different surface velocities. The subject – including as it did the friction of revolving shafts, the safe and the unsafe load per square inch to put on the bearings at various velocities, with various materials, and with various lubricants; and also the question of the friction of worm gear as compared with other gearing – was an interesting and important one to the Institution, and was anxiously awaited.*

---

15. See Appendix, Sect. A.12 for biographical details.

The President, Mr Percy G. B. Westmacott of Newcastle upon Tyne said, '. . . a gentleman had that day been appointed, who was in every way admirably fitted to carry out the experiments proposed by the Committee and who would soon be in a position to take the subject in hand'.

The man appointed on 20 April was Beauchamp Tower, who had been in practice on his own account since 1878 developing some of his inventions. His background was well suited to the task in hand. He had worked as an assistant to Mr William Froude, whose investigations of the resistance to motion of ships had established his reputation in hydrodynamics, and also for a few months in the winter of 1875–76 as assistant to Lord Rayleigh in carrying out experiments on sound. Lord Rayleigh first met Beauchamp Tower in the summer of 1875 when he was visiting his aunt, Mrs Christopher Tower, at Weald Hall, Brentwood. They remained firm friends and Lord Rayleigh would no doubt be able to speak with confidence and personal knowledge of Tower's skill as an experimenter when the Research Committee on Friction at High Velocities was considering the appointment of a gentleman to take charge of its experiments.

Tower entered into the work with enthusiasm and he was to be engaged in making experiments for the Institution of Mechanical Engineers from 1882 to 1891. The journal-bearing experiments were carried out on the specially constructed testing machine shown in Fig. 10.9. The tests were carried out at the Edgware Road Station, Chapel Street Works of the Metropolitan Railway in London. The steel journal was 6 in (152·4 mm) long and of 4 in (101·6 mm) diameter, with its axis horizontal. A partial journal bearing made of gunmetal rested on top of the shaft, thus simulating the conditions found in railway axle-boxes. The angle of wrap was about 157° (2·74 rad) in most of the experiments, but in some cases it was as small as 68° (1·18 rad). The lubricants included vegetable oils (olive and rape), animal oils (lard and sperm), mineral oils and a mineral grease. Unlike Petrov, Tower failed to measure and record the viscosities of his lubricants. The speeds ranged from 100 to 450 rpm and the loads per unit projected area from 100 to 625 lbf/in^2 (0·69–4·31 MN/m^2). The bearing assembly (ABC) supported a loading frame (DD) on a knife edge, and friction between the bearing and journal resulted in a displacement of the end of a lever (LL). The shaft was driven in both directions and the net movement at P was used to determine the coefficient of friction.

Tower soon found that '. . . all the common methods of lubrication are so irregular in their action that the friction of a bearing often varies considerably'. He found, for example, that when rape oil was used as a lubricant at a speed of 150 rpm, the coefficients of friction with a pad under the journal, a siphon lubricator and an oil-bath were 0·009, 0·0098 and 0·00139, respectively. In order to achieve consistent results, the majority of Tower's investigations were carried out with the journal immersed, or at least dipping into, a bath of oil. This recognition of the importance of adequate lubrication was one of the major results of Tower's investigations.

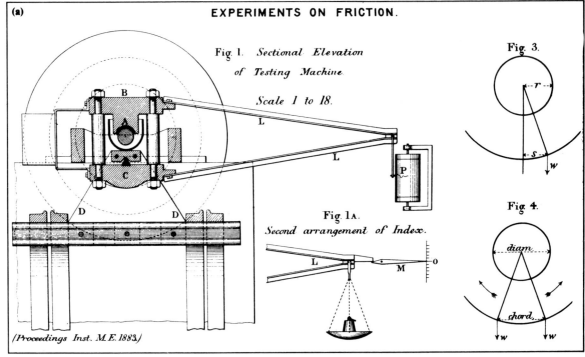

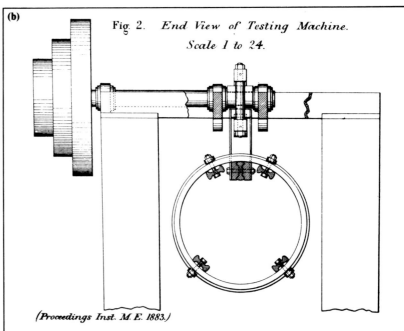

Fig. 10.9
Beauchamp Tower's testing machine for experiments on friction (Tower, 1883).
(a) Elevation.
(b) End view.

The above details were all contained in Tower's (1883) 'First report on friction experiments (friction of lubricated bearings)' adopted by the Committee on Friction and presented to the Council on 28 September 1883. It is a report which should be read by all serious students of tribology and those interested in experimental investigations. The account of the accidental discovery of oil-film pressures can best be told in Tower's own words.

> *A very interesting discovery was made when the oil-bath experiments were on the point of completion. The experiments being carried on were those on mineral oil; and the bearing having seized with 625 lbs. per sq. in., the brass was taken out and examined, and the experiment repeated. While the brass was out, the opportunity was taken to drill a $\frac{1}{2}$-in. hole for an ordinary lubricator through the cast-iron cap and the brass. On the machine being put together again and started with the oil in the bath, oil was observed to rise in the hole which had been drilled for the lubricator. The oil flowing over the top of the cap made a mess, and an attempt was made to plug up the hole, first with a cork and then with a wooden plug. When the machine was started the plug was slowly forced out by the oil in a way which showed that it was acted on by a considerable pressure. A pressure-gauge was screwed into the hole, and on the machine being started the pressure, as indicated by the gauge, gradually rose to above 200 lbs. per sq. in. The gauge was only graduated up to 200 lbs., and the pointer went beyond the highest graduation. The mean load on the horizontal section of the journal was only 100 lbs. per sq. in. This experiment showed conclusively that the brass was actually floating on a film of oil, subject to a pressure due to the load. The pressure in the middle of the brass was thus more than double the mean pressure. No doubt if there had been a number of pressure-gauges connected to various parts of the brass, they would have shown that the pressure was highest in the middle, and diminished to nothing towards the edges of the brass.*

This open yet graphic account conveys the sense of excitement which must have surrounded the remarkable discovery. It truly marks the beginnings of detailed experimental and analytical studies of hydro-dynamic lubrication, although Petrov had correctly interpreted the nature of mediate friction.

Tower found that the normal practice of feeding oil through a single hole to an axial groove on the loaded side of the bearing was quite in-adequate. Indeed he said, '... This arrangement of hole and groove, instead of being a means of lubricating the journal, was a most effectual one for collecting and removing all oil from it.'

The oil-feed arrangement is shown in Fig. 10.10(*a*) together with a successful modification in Fig. 10.10(*b*). The usual groove arrangement and twin-hole supply used in locomotive axle-boxes and shown in Fig. 10.10(*c*) was also found to be quite ineffective. Tower concluded that with bath lubrication and correct bearing proportions, it was possible to reduce the coefficient to 1/1000. He also detected the minimum in the coefficient of friction–speed relationship discovered by Thurston.

> *... Observations on the behaviour of the apparatus gave reason to believe that with perfect lubrication the speed of minimum friction was from 100 to 150 feet per minute; and that this speed of minimum friction tended to be higher with an increase of load, and also with less perfect lubrication. By the speed of minimum friction is meant that speed in approaching which, from rest, the friction diminishes, and above which the friction increases.*

This initial study of journal-bearing friction and lubrication by Beauchamp Tower dispelled the confusion which had accumulated

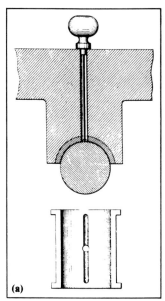

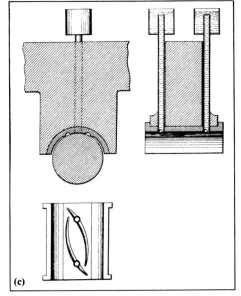

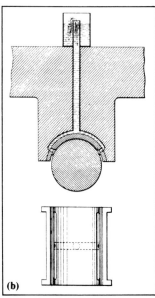

Fig. 10.10
Beauchamp Tower's experiments
on friction-lubricant supply
arrangements (Tower, 1883).
(a) Arrangements for ordinary
lubrication – needle lubricator.
(b) Modifications to feed lubricant
to two axial grooves.
(c) Usual arrangement of grooves
in locomotive axle-boxes.

over many decades. Many of the observations are now regarded as
self-evident, but in order to appreciate their impact in the 1880s, it should
be remembered that practically all the earlier studies were related to
friction and that little attention had been given to the mechanism of
lubrication. Tower's major findings were:

1. That repeatable friction measurements could be achieved only when
   the bearing was provided with adequate quantities of lubricant.
   This was best achieved by allowing the rotating shaft to dip into an
   oil-bath.
2. Lubricant should be supplied on the unloaded side of the bearing
   and not through an oil hole or groove on the load line. (The latter
   arrangement had previously been adopted on most steam engines for
   locomotive and stationary applications.)
3. The tangential force of friction acting on the journal when the lubrication was adequate, increased with velocity but was nearly independent
   of load (i.e. it followed the laws of liquid friction much more closely
   than those of solid friction).
4. There was a minimum on the friction–speed relationship such that
   the friction initially diminished and then increased again as the speed
   of rotation increased from rest.
5. In a well-lubricated bearing the brass actually floated on a film of oil
   such that pressures within certain parts of the film considerably
   exceeded the mean pressure due to applied load.

Tower's 'Second report on friction experiments' (1885) was devoted
to the question of pressure distribution in a partial journal bearing
embracing slightly less than 180° of arc. The diameter and length were
once again 4 in (101·6 mm) and 6 in (152·4 mm), respectively, the lubricant
was a heavy mineral oil, with the journal bearing half immersed in an
oil-bath and the speed of rotation was 150 rpm.

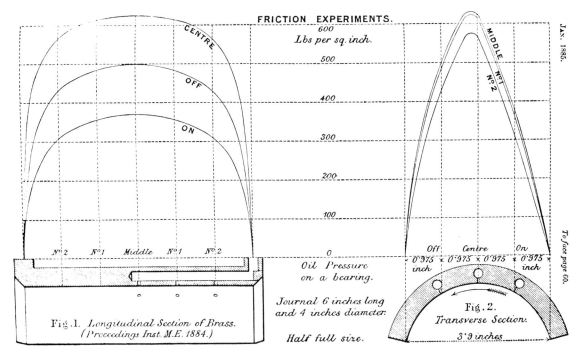

Fig. 10.11
Beauchamp Tower's experiments on the pressure distribution in a partial arc journal bearing (Tower, 1885).

Pressure was measured by connecting a Bourdon gauge to one of three $\frac{1}{4}$ in (6·3 mm) axial holes drilled in the brass in locations which divided the arc of contact into four equal regions. A $\frac{1}{16}$ in (1·6 mm) diameter hole was drilled from the bearing surface into the bigger longitudinal holes, and once the pressure had been ascertained the hole was stopped and another hole drilled until readings had been recorded at nine points. There was no switching and automatic recording of pressure in bearing experiments in the 1880s!

This remarkable study yielded the illuminating axial and circumferential pressure distribution shown in Fig. 10.11. The sides of the bearing on which the rotating shaft entered and left were designated the 'on' and 'off' planes, respectively–a designation which has been retained to the present day. Tower's comment on the recorded pressures was that, '. . . Their most clearly marked feature is seen to be that the place of greatest pressure is on the "off" side of the centre, Fig. 2, the pressure at the holes in the "on" side being in every case considerably less than that at the corresponding holes on the "off" side.'

A remarkable testimony to the skill of Mr Tower and the accuracy of the experiments was provided by the fact that the integrated pressures yielded a load-carrying capacity of 7,988 lbf (35·5 kN) compared with the applied load of 8,008 lbf (35·6 kN). Tower attributed the difference of 20 lbf (89 N) to '. . . errors of observations'!

It appears that the cost of these experiments was carried not by the Institution, but by the current Vice-President and Chairman of the Friction Committee, Mr Joseph Tomlinson.[16]

16. See *Tribology News*, Dec. 1975, pp. 4–5.

Fig. 10.12
Institution of Mechanical
Engineers Research Committee on
Friction ('Experiments on the
friction of a collar bearing')
(Tower, 1888).
(a) Side and plan views of the
apparatus.
(b) Transverse section showing
'serpentine oil grooves'.

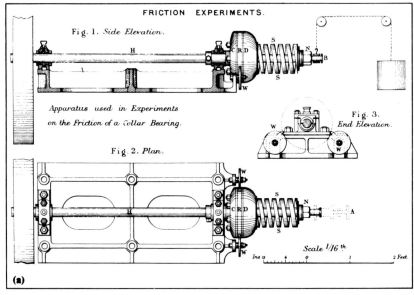

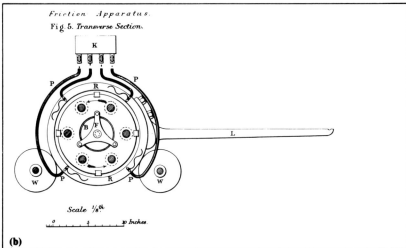

Tomlinson, who was to become the President of the Institution in
1890–91, was the Resident Engineer and Locomotive Superintendent of
the Metropolitan Railway in whose Edgware Road Works the experiments
were carried out. He not only covered most of the cost of the experiments,
he was devoted to the subject and also gave a considerable amount of
his time to the project. It is for this reason that at the end of the meeting
Mr Arthur Paget proposed a vote of thanks, not to Beauchamp Tower,
but to Joseph Tomlinson.

A further remarkable feature of the experiments conducted by
Beauchamp Tower was that the journal-bearing apparatus was driven
by a spherical engine. This early rotary engine, described by Heenan
(1885), was invented and designed by none other than Mr Beauchamp
Tower.

The 'Third report of the Research Committee on Friction' (Tower, 1888) was devoted to 'Experiments on the friction of a collar bearing'. The bearing consisted of an annular steel ring (R) coated with gun-metal and loaded between two cast-iron discs (CD) by means of a spiral steel spring (S) as shown in Fig. 10.12(*a*). The accepted method of lubricating such parallel surface bearings was to admit oil to sealed radial grooves which communicated with *serpentine grooves* of the form shown in Fig. 10.12(*b*). The general conclusions were that such a bearing was '. . . very inferior to a cylindrical journal in its power of carrying weight'. Mean pressures of only 75–90 lbf/in² (517–620·5 kN/m²) could be supported compared with values almost an order of magnitude greater on well-lubricated journals. The coefficients of friction, ranging from 0·0264 to 0·0646, were at least ten times greater than those recorded in well-lubricated journal bearings.

If the substance of this third report is of lesser interest, the same cannot be said of its timing. It was presented two years after the magnificent formulation of the theory of fluid-film lubrication by Osbourne Reynolds, yet Tower declined to incorporate the physical wedge concept in his presentation. Professor Archibald Barr of the Engineering Department in the Yorkshire College, Leeds, which later became the University, suggested that it would be interesting if experiments could be conducted on a bearing incorporating a tapered clearance which would be consistent with the '. . . beautiful theory . . .' published by Professor Osborne Reynolds (Barr, 1888). Barr's suggestions, which show clearly the tapered wedge in both journal- and thrust-bearing arrangements, are seen in Fig. 10.13. It is interesting and somewhat surprising that Tower made no response to this suggestion, although he did comment on other aspects of Professor Barr's contribution to the discussion.

The fourth and final report of the Research Committee on Friction was presented in March (Tower, 1891) when Joseph Tomlinson, current President of the Institution, was in the chair. The report was concerned with the friction of pivot bearings and the experiments were conducted

Fig. 10.13
Professor Archibald Barr's proposal for a tapered clearance footstep bearing (Barr, 1888).

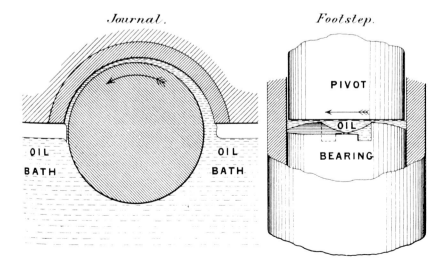

by Beauchamp Tower, Joseph Tomlinson and Mr Mair-Rumley. The experiments covered by the third and fourth reports were carried out in the works of Mr Mair-Rumley's company, Simpson and Co., of Pimlico.[17]

This complete series of experiments conducted by Beauchamp Tower and promoted by the Institution of Mechanical Engineers revolutionized the state of knowledge on lubrication. It was not to be long before the genius of Osborne Reynolds crowned the golden decade of the 1880s with his brilliant mathematical analysis of the physical situation revealed by Petrov and Tower.

United Kingdom

Osborne Reynolds and the theory of fluid-film lubrication (Reynolds, 1886)   The theory of fluid-film lubrication was established in splendid style after the publication of Beauchamp Tower's first report to the Institution of Mechanical Engineers. The development represents one of the best examples of the way in which careful experimental work can provide the stimulus for mathematical analysis which in turn confirms the physical nature of the phenomenon and provides a springboard for further engineering development.

An interesting editorial in *The Engineer* of 22 April 1884 carried the title 'What is friction?' It deplored the way in which busy engineers accepted the results of General Morin and worked with them, rather than investigating further the nature of friction. The article reviewed the background studies from Coulomb onwards, before turning to the work of Beauchamp Tower. It chided the Institution of Mechanical Engineers for its leisurely approach to such an important issue, since it was some six years since the Committee on Friction at High Velocities had been established, yet recognized the significance of the findings with the words: '. . . Such was the state of the case when the Institution of Mechanical Engineers took up the question. Their progress in determining it has certainly been of the slowest; but they have lately issued a report which consolidates and advances our knowledge of the question in a remarkable degree.'

The writer clearly felt that Beauchamp Tower had opened the door for a fresh approach to the question of friction, particularly in relation to lubricated surfaces. In a remarkable plea for a scientific approach to the subject *The Engineer* stated: 'Any physicists, therefore, who would put forward a good working hypothesis on the question of friction, or rather on the two questions of solid and liquid friction, would probably deserve well of the engineering profession and the world at large.'

The day before this editorial was published the Lucasian Professor at Cambridge, G. G. Stokes,[18] had written to Beauchamp Tower as follows: 'I thought I would try what would be the results to which we should be conducted by the application of the known equations of motion of a viscous fluid, which in the present case are very much simplified, since the inertia of the lubricant does not sensibly come into account.'

---

17. See *The Engineer* (1944b), **178**, July, 40–2.
18. See Larmor, J. (1907), pp. 246–8.

With great insight into the problem of lubricated journal bearings, Stokes stated that the position of minimum film thickness was not on the load line, but advanced in the direction of motion. He also commented on the question of *fitted bearings*, in which the journal and bearing had equal radii and wrote: '. . . I am led to regard the side portions of the brass as very important. I have not however as yet worked this out, though I see the way to do it . . . I thought it well to communicate to you these preliminary indications of theory, as I cannot at present work the thing fully out on account of lectures.'

It appears that Stokes had responded quickly to the intriguing experimental findings of Beauchamp Tower, although there is no evidence that he finalized his theory. Not everyone recognized the potential of Tower's work, as the following initial response to the editorial in *The Engineer* by a reader from Birmingham with initials J. C. H.[19] indicated.

> *I anticipated that Mr Tower's researches would have had some practical value; but as far as I can see they might just as well never have been undertaken.*
>
> *To certain minds I have no doubt that it may prove valuable to know that friction between dry surfaces is not the same thing as friction between oiled surfaces; but to the great body of engineers the statement is simply useless. It is really a matter of no importance whatever that frictional coefficients vary with the speed, because the fact cannot be usefully applied. Again, we gain absolutely nothing of any value from the discovery, if such it be, that the use of an oil bath diminishes friction enormously. We cannot use oil baths, and the fact is therefore of no importance.*
>
> *It seems to be the fate of the Institution of Mechanical Engineers to always carry out investigations of no practical value to anyone. If Mr Tower had told us something about the relative values of different materials for bearings he would have done good service. . . .*

In retrospect this represented one of the most erroneous judgements on the merits of tribological research ever reported.

The harsh views on the Institution of Mechanical Engineers and its sponsorship of the work on journal bearings which appeared in *The Engineer* in April 1884 were not shared by Lord Rayleigh, Professor of Experimental Physics at the University of Cambridge. This was the year in which the British Association for the Advancement of Science decided, for the first time, to hold its summer meeting in the colonies. The venue was Montreal, Canada, and it fell to Lord Rayleigh as the incoming President to present his address on the evening of 27 August. Before giving his interesting review of advances in physics he said, '. . . The History of science teems with examples of discoveries which attracted little notice at the time, but afterwards have taken root downwards and borne much fruit upwards.'

Viewed in the light of this remark, his reference to the work of Beauchamp Tower immediately after a commendation of Professor O. Reynolds' study of laminar and turbulent flow in pipes was both prophetic and historically significant.

> *. . . As also closely connected with the mechanics of viscous fluids. I must not forget to mention an important series of experiments upon the friction of oiled surfaces, recently executed by Mr Tower for the Institution of Mechanical*

---

19. See *The Engineer* (1884), **57**, 29 Feb., 164.

Engineers. The results go far towards upsetting some ideas hitherto widely admitted. When the lubrication is adequate, the friction is found to be nearly independent of the load, and much smaller than is usually supposed, giving a coefficient as low as 1/1000. When the layer of oil is well formed, the pressure between the solid surfaces is really borne by the fluid, and the work lost is spent in shearing, that is, in causing one stratum of the oil to glide over another. In order to maintain its position, the fluid must possess a certain degree of viscosity, proportionate to the pressure; and even when this condition is satisfied, it would appear to be necessary that the layer should be thicker on the ingoing than on the outgoing side. We may, I believe, expect from Professor Stokes a further elucidation of the processes involved. In the meantime, it is obvious that the results already obtained are of the utmost value, and fully justify the action of the Institution in devoting a part of its resources to experimental work. We may hope indeed that the example thus wisely set may be followed by other public bodies associated with various departments of industry.

The reference to Stokes is interesting, but we can only speculate on the extent to which Lord Rayleigh and Professor Stokes had discussed the work of Beauchamp Tower. In the event it was to be Professor Osborne Reynolds who first successfully applied hydrodynamic theory to the journal-bearing problem.

On Thursday 28 August the President of the Mathematical and Physical Science Section, Professor Sir William Thomson, delivered an address which was recorded in full. It was also noted that Professor Osborne Reynolds, F.R.S. read two papers, one of which was entitled, 'On the action of lubricants' (Reynolds, 1884).

Five days later, on Tuesday 2 September, Reynolds (1884) read a further paper, 'On the friction of journals'.

These brief titles appear to represent our total knowledge of the work reported by Reynolds in Montreal, but on 3 September the Montreal *Daily Witness*[20] carried an account of the meeting which stated that, 'Professor Osborne Reynolds gave a paper on "The Friction of Journals" which was entirely theoretical and only of interest to the initiated'.

This report suggests that Osborne Reynolds was probably reporting to the British Association in Montreal in 1884 the essence of his famous theory of fluid-film lubrication. It is interesting to note that Robert H. Thurston also attended the meeting and presented a paper 'On the theory of the steam engine'.

Sixteen months after the Montreal meeting of the British Association for the Advancement of Science, Reynolds (1886) submitted his paper on the theory of lubrication to the Royal Society. It was read to the Society on 11 February 1886 and was to become the most venerated publication in the history of tribology. In the introductory paragraphs Reynolds remarked that the equation relating pressure and velocity, which appeared to explain the existence of the film of oil at high pressure in Mr Tower's experiments, was first mentioned in a paper read before Section A of the British Association at Montreal in 1884. He also referred to the paragraph in Lord Rayleigh's presidential address from which it appeared that Professor Stokes and Lord Rayleigh and simultaneously arrived at a similar result.

---

20.   See Cameron, A. (1966), pp. 270–1.

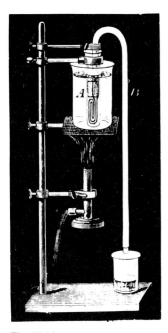

Fig. 10.14
Apparatus used by Professor
Osborne Reynolds to determine
the viscosity of olive oil
(Reynolds, 1886).

It thus appears that the *Reynolds equation* of fluid-film lubrication
was first revealed to the world, but not recorded, in Montreal in 1884.

Reynolds himself remarked that he had at that time '. . . no idea of
attempting its integration . . .', but he recognized how this might be
achieved in approximate form '. . . on subsequent consideration'. The
classical paper of 1886 not only contained the basic differential equation
of fluid-film lubrication and its approximate solution for restricted situations, but also a direct comparison between the theoretical predictions
and the experimental results obtained by Beauchamp Tower. A brief
review of Reynolds' rather lengthy paper appeared in *The Engineer*
(1944).

Like Petrov (1883), Reynolds recognized the need to measure viscosity
and to relate it to temperature. The fact that he included in the title
of his paper the words '. . . including an Experimental Determination of
the Viscosity of Olive Oil', emphasizes this point and also indicates the
dominant role of vegetable oils in the 1880s. It is no accident, but a
matter of some historical interest, that the founder of the modern science
of lubrication should consider it equally important to determine the
viscosity of the vegetable oil which had sustained the bearings of
machinery in general and locomotives in particular throughout the
nineteenth century, as it was to expose the beautiful mechanism of fluid-
film lubrication.

According to Reynolds, his theory drew attention to two circumstances of particular significance: the role of radial clearance and the fact
that the point of nearest approach of the journal to the brass was always
on the *off* side of the load line. He recognized the possibility that negative
(sub-ambient) pressures might lead to rupture of the oil film, and also
that the theory could accommodate dynamic circumstances like those
encountered in crank-pin bearings.

Two questions were still actively under discussion in the field of hydrodynamics when Reynolds applied slow viscous flow theory to the problem
of lubrication. One was the possible dependence of the property known
as viscosity upon motion, or speed, and the other was the possibility of
slip at the boundaries with solids. Reynolds concluded that both issues
could be ignored for the problem in hand.

The viscosity of olive oil was determined as a function of temperature
by means of the simple apparatus shown in Fig. 10.14. The exponential
relationship which was used to represent the results was:

$$\eta_1 = \eta_2 \exp[-\alpha(T_1 - T_2)] \qquad \qquad \text{. . . [10.8]}$$

Reynolds quoted empirical viscosity–temperature relationships for
olive oil based upon his own experiments and for water and air based
upon the work of Poiseuille and Maxwell, respectively. It is interesting to
note that the table of experimental results for olive oil and water obtained
by Reynolds is dated 11 April 1884, almost five months before the presentation of this Montreal paper.

The basic mechanism of fluid-film lubrication was explained with
great clarity in Section III of the paper. Reynolds discussed the development of pressure in lubricating films by considering the necessity to

introduce Poiseuille (due to pressure gradient) flow in addition to Couette (1890:433–510) (due to surface motion) flow to preserve continuity of the rate of mass or volume flow. It was in this section that the important *physical wedge* concept was explained in relation to plane surfaces, a revolving cylinder near a plane and a partial arc journal bearing as illustrated in the graphic pictures reproduced in Figs 10.15–10.17. The exposure of the fundamental requirement that the film thickness should decrease in the direction of surface motion if load-carrying pressures were to be generated within the film, was to provide the foundation of sound twentieth-century bearing design procedures. The student of lubrication will find much to interest him in this classical paper. It contains within its numerous pages many gems, particularly in relation to physical interpretations of phenomena. For example, to ascertain the directions in which the viscous stresses act on the bounding solids, Reynolds advised the reader to imagine the broken lines joining the surfaces in Fig. 10.16 to be distorted to the shapes of the solid lines and to be in tension.

The basic equations of hydrodynamics were swiftly reduced to match the conditions of slow viscous flow between conforming solids. Integration then yielded the well-known *Reynolds equation*, which can be written as

$$\frac{d}{dx}\left[h^3 \frac{dp}{dx}\right] + \frac{d}{dy}\left[h^3 \frac{dp}{dy}\right] = 6\eta\left\{(U_0 + U_1)\frac{dh}{dx} + 2W_1\right\}, \qquad \dots [10.9]$$

where p is the pressure in the lubricant, h the film thickness, x, y are coordinates within the 'plane' of the film, U_0, U_1 are velocities of lower (plane) and upper (curved) bearing surfaces in the x-direction, W_1 the velocity of separation of the upper surface relative to the lower (plane) surface $[= dh/dt]$ and η the lubricant viscosity.

With this form of the equation it is only W_1 and not the surface velocity U_0 or U_1 which causes local values of film thickness to change with time.

Reynolds presented two analytical solutions to equation [10.9] before tackling approximate solutions to the journal-bearing problem. In the first case he showed that the squeeze-film problem for elliptical plates of semi-major and minor axes a and b, respectively, which had previously been considered by Stefan (1874) had a solution:

$$p - p_a = \frac{12\eta}{h^3}\frac{a^2b^2}{(a^2 + b^2)}\left[\frac{x^2}{a^2} + \frac{y^2}{b^2} - 1\right]\frac{dh}{dt}. \qquad \dots [10.10]$$

Further integration revealed a squeeze-film time for the film thickness to reduce from h_1 to h_2 under constant load W given by

$$t = \frac{3\eta\pi a^3 b^3}{(a^2 + b^2)W}\left[\frac{1}{h_2^2} - \frac{1}{h_1^2}\right]. \qquad \dots [10.11]$$

The great merit of this example was that it was '. . . the most complete as well as the simplest case in which to consider the important effect of normal motion in the action of lubricants'.

The second example was the plane inclined slider bearing of infinite width. In this case Poiseuille flow in the y-direction was neglected, the

Fig. 10.15
Professor Osborne Reynolds'
general view on the action of
lubrication of plane surfaces
(Reynolds, 1886).
(a) Parallel surfaces in relative
tangential motion.
(b) Parallel surfaces approaching
with no tangential motion.
(c) Parallel surfaces approaching
with tangential motion.
(d) Surfaces inclined with
tangential movements only (i).
(e) Surfaces inclined with
tangential movements only (ii).

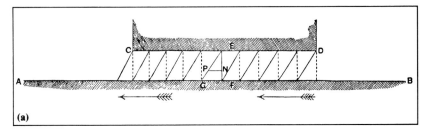

(a)

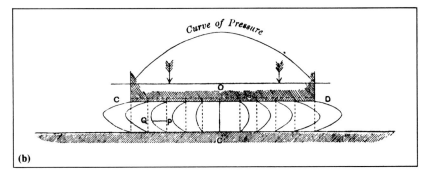

(b)

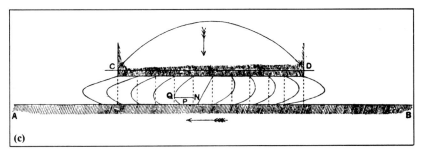

(c)

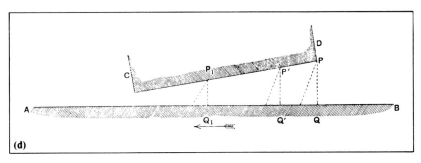

(d)

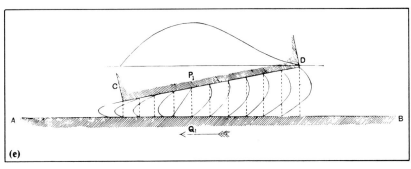

(e)

Fig. 10.16
Professor Osborne Reynolds'
general view on the lubrication
of a rotating cylinder near a plane
(Reynolds, 1886).
(a) Revolving motion with plane
surface symmetrically placed.
(b) Revolving motion with plane
surface asymmetrically placed.
(c) The effect of a limiting supply
of lubricating material.

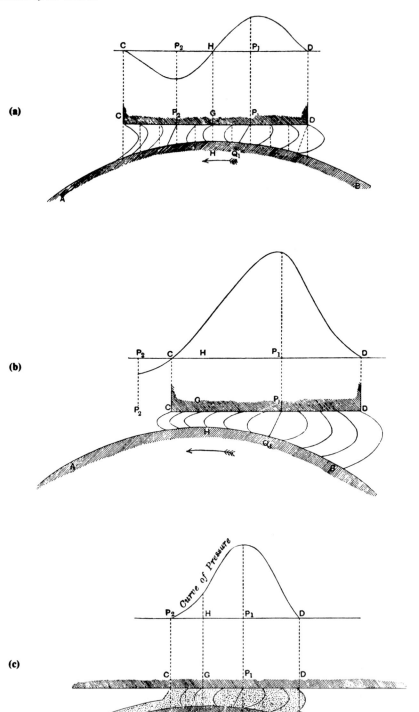

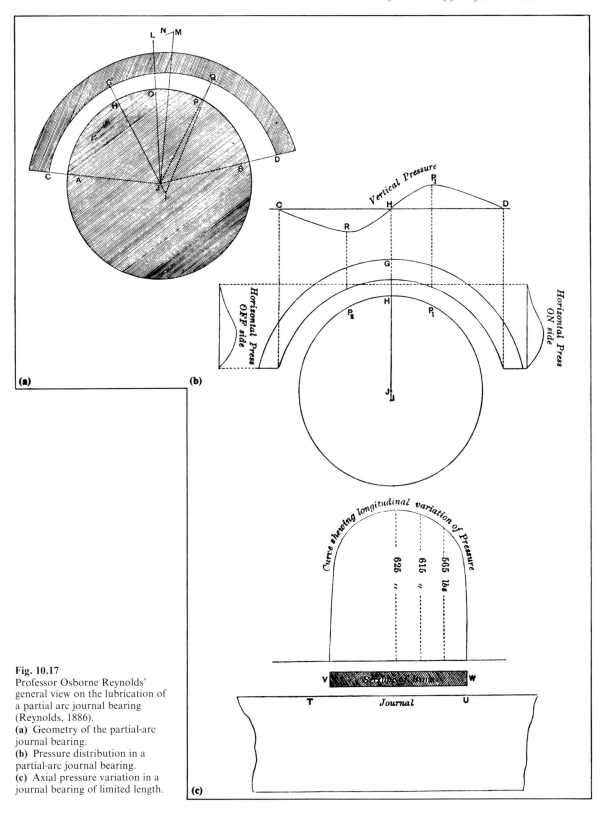

Fig. 10.17
Professor Osborne Reynolds'
general view on the lubrication of
a partial arc journal bearing
(Reynolds, 1886).
(a) Geometry of the partial-arc
journal bearing.
(b) Pressure distribution in a
partial-arc journal bearing.
(c) Axial pressure variation in a
journal bearing of limited length.

situation now frequently described as one of *no side-leakage* or two-dimensional flow.

Reynolds derived analytical solutions for pressure and load-carrying capacity per unit width for a bearing of length *l* and then proceeded to show that the maximum load-carrying capacity per unit width (W/b) was attained when the inlet film thickness was about 2·2 times the minimum film thickness (h_0). The approximate expression for W was shown to be

$$\frac{W}{b} = 0.16\eta U \left[\frac{l}{h_0}\right]^2.$$
...[10.12]

For partial-arc journal bearings Reynolds chose to seek approximate solutions for pressure and surface tractions through the use of trigonometric series. Unfortunately, the numerical coefficients associated with the series expressed in terms of increasing powers of eccentricity ratio failed to converge, although the series themselves converged with adequate accuracy if the eccentricity ratio was less than 0·6 and ten or twelve terms were considered.

Reynolds found that for a bearing of total arc similar to that used by Mr Tower, the best agreement between theory and experiment occurred when the eccentricity ratio was 0·5. He further found that the pressure reached a minimum about 2° (0·035 rad) before the 'off' side of the bearing for these conditions. He recognized that solutions for higher values of the eccentricity ratio would introduce negative pressures, that is pressures lower than ambient, and hence concluded that an eccentricity ratio which rendered the pressure negative at the 'off' extremity of the brass was the largest value for which lubrication could be considered certain. At higher eccentricity ratios and greater loads the continuity of the film would be maintained by the pressure of the atmosphere.

These statements are significant since they indicate Reynolds' concern for the effect of oil-film rupture or cavitation. The phenomenon had been observed by Newton (1709) using water as a lubricant, and it continues to be a subject of physical interest, mathematical complexity and in some cases of practical importance. Reynolds' description implies an outlet or cavitation boundary condition of the form,

$$p = \frac{\mathrm{d}p}{\mathrm{d}x} = 0.$$
...[10.13]

Reynolds did not state this condition explicitly, but it is nevertheless referred to by several workers as the *Reynolds cavitation boundary condition*. The same condition was used by Gümbel (1921) and given independent theoretical justification by Swift (1931) and Stieber (1933). For this reason the statement is also known as the *Swift–Stieber cavitation boundary condition*.

Reynolds' ability to combine mathematical analysis with sound physical reasoning was further emphasized in this truly wide-ranging paper by an interesting, but necessarily speculative, discourse on the influence of thermal and elastic action upon journal-bearing performance.

Tower measured neither the radial clearance nor the eccentricity in his experiments on a 4 in (101·6 mm) diameter journal bearing, but by matching theoretical and experimental solutions for both pressure and friction, Reynolds was able to deduce that the radial clearance in the two bearings must have been about 0·0008 in (0·02 mm). This deduction was to provide a useful quantitative guide for journal-bearing designers for many years to come.

In his conclusions Reynolds noted that his theory had, '. . . been tested by experiments throughout the entire range of circumstances to which the particular integrations undertaken are applicable'.

He discussed wear in relation to his theoretical findings and noted that continuous running in one direction would lead to bedding in over part of the arc. This action was consistent with Tower's observation that the journal, after running for some time in one direction, would not run at first in the other. Reynolds noted that wear would take place on starting as a result of initial metallic contact, and that the resulting abrasion would introduce metallic particles into the oil and blacken it.

The last two paragraphs of this long paper neatly summarize the wide-ranging nature of the study and the impact of the theory upon late nineteenth-century understanding of fluid-film lubrication.

The verification of the equations for viscous fluids under such extreme circumstances affords a severe test of the truth and completeness of the assumptions on which these equations were founded. While the result of the whole research is to point to a conclusion (important in Natural Philosophy) that not only in cases of intentional lubrication, but wherever hard surfaces under pressure slide over each other without abrasion, they are separated by a film of some foreign matter whether perceivable or not. And the question as to whether this action can be continuous or not turns on whether the action tends to preserve the matter between the surfaces at the points of pressure as in the apparently unique case of the revolving journal, or tends to sweep it to one side as is the result of all backwards and forward rubbing with continuous pressure.

The fact that a little grease will enable almost any surfaces to slide for a time has tended doubtless to obscure the action of the revolving journal to maintain the oil between the surfaces at the point of pressure. And yet, although only now understood, it is this action that has alone rendered our machines, and even our carriages possible. The only other self-acting system of lubrication is that of reciprocating joints with alternate pressure on the separation (drawing the oil back or a fresh supply) of the surfaces. This plays an important part in certain machines, as in the steam engine, and is as fundamental to animal mechanics as the lubricating action of the journal is to mechanical contrivances.

The final sentence shows that the founder of fluid-film lubrication theory was also the first to comment on the possibility of hydrodynamic action in synovial joints.

Reynolds' theory took the understanding of lubrication to the heights of the tribological landscape. It has retained its dominant position and all successful fluid-film bearing design is based upon it. There have been many refinements, but no major modifications for almost a century. This in itself reflects the importance of Osborne Reynolds' contribution to our subject.

United Kingdom

John Goodman's estimate of film thickness (Goodman, 1886)   In November 1885 a young man who was working as an assistant to William Stroudley in the Brighton Works of the London, Brighton and South Coast Railways read a student's paper to the Institution of Civil Engineers. The man was John Goodman[21] and the paper was entitled 'Recent researches in friction'.

Goodman later became Professor of Civil and Mechanical Engineering in the Yorkshire College in Leeds and by the time the College was transformed into the University of Leeds by the award of a charter on 25 April 1904, he had established a firm reputation for his studies of bearings, friction and lubrication.

One interesting aspect of Goodman's (1886) student paper was that it confirmed the existence of a lubricating film between journal and brass, and recorded for the first time a quantitative measure of its magnitude. Goodman noted that the brass with its load was raised bodily as the shaft began to rotate. He then mounted the micrometer screw shown in Fig. 10.18 on the brass and used a weak voltaic cell and galvanometer to indicate contact between the screw and the shaft. The micrometer was capable of measuring to 0·0001 in (0·0025 mm) and was insulated from the brass by paraffined paper. The difference between readings when the shaft was rotating and at rest gave an indication of the lift of the brass. Goodman wrote, '. . . The amount was found to vary with the efficiency of the lubrication. With a pad-lubricator it was raised to 0·0013 inch on first starting, and as the films became better developed the friction decreased and the brass was ultimately raised 0·0029 inch.'

This was the first recorded measurement of film thickness in a running journal bearing. It came just before Reynolds' (1886) paper was published and provided additional experimental backing for the concept of fluid-film lubrication which had emerged from Petrov's (1883) friction measurements and Tower's (1883) pressure measurements.

The United States

Albert Kingsbury's experiments on air-lubricated journal bearings (Kingsbury, 1897)   While the experimental results obtained by Tower and Goodman and Reynolds' analysis of fluid-film lubrication were being digested in Europe, Albert Kingsbury performed a most remarkable experiment in the United States.

Kingsbury[22] studied under Thurston at Sibley College, Cornell University between 1887 and 1889, before accepting a teaching post at the College of Agriculture and the Mechanic Arts at Hanover, New Hampshire for one year. He left this position after one year, but returned as Professor of Mechanical Engineering in 1891. While developing laboratory apparatus at Hanover, he stumbled across a most remarkable feature of the behaviour of a piston in a cylinder.

---

21. See Appendix, Sect. A.15 for biographical details.
22. See Appendix, Sect. A.16 for biographical details.

Fig. 10.18
Professor John Goodman's micrometer screw used to measure film thickness in a journal bearing (Goodman, 1886).

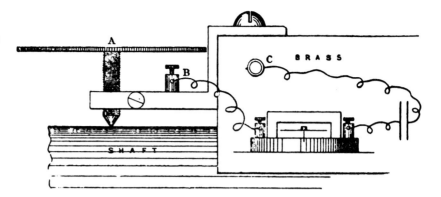

About 1892 Kingsbury built a torsion–compression machine to study the friction of screw threads. The compression part of the machine consisted of a piston of diameter 6 in (152·4 mm) and length $6\frac{1}{4}$ in (158·8 mm) within a heavy cylinder. With the cylinder vertical, Kingsbury found that the piston could be spun rapidly, apparently without contact with the cylinder wall. His machine-shop instructor and excellent mechanic, John Brown, then found that the same result was obtained when the 22 lbf (97·8 N) piston was rotated in a horizontal position. Kingsbury concluded that the air was acting as a lubricant and he proceeded to build the special air-lubricated journal bearing shown in Fig. 10.19. The radial clearance was about 0·0016 in (0·002 mm) and the diameter of the 50 lbf (222·4 N) cast-iron rotor 6 in (152·4 mm). A slot in the end of the shaft allowed a screwdriver driven by a hand or breast drill to produce a high speed of rotation.

Kingsbury measured the pressure at various places in the air film in both circumferential and axial directions and studied the effect of speed. His demonstration of the apparatus before the meeting of the Association of Agricultural Colleges in Washington in November 1896 and before the engineers at the Bureau of Steam Engineering at the Navy Department

Fig. 10.19
Albert Kingsbury's air-lubricated journal bearing (Kingsbury, 1892).

attracted great interest. Kingsbury tested the bearing with both air and hydrogen and it was at the latter meeting that Kingsbury was shown Osborne Reynolds' paper for the first time by John H. MacAlpine. He studied the paper, extended his experimental work on the air-lubricated journal bearing and subsequently published his outstanding paper on 'Experiments with an air-lubricated bearing' (Kingsbury, 1897) in the *Journal of the American Society of Naval Engineers*.

The form of the measured pressures was later to be compared convincingly with solutions of the Reynolds equation for a compressible fluid by Harrison (1913). These experimental and theoretical results will be presented and discussed later in this section.

Germany

Richard Stribeck's experiments on journal friction (Stribeck, 1902) *and Ludwig Gümbel's rationalization of the results* (Gümbel, 1914a)   Robert Thurston was probably the first person to draw attention to the existence of a minimum point on the curve of coefficient of friction plotted against load, but it was a famous German engineer by the name of Richard Stribeck who first investigated the minimum point in some detail.

Stribeck was Professor für Maschineningenieurwesen at the Technische Hochscule in Dresden in the mid-1890s. In 1898 he was appointed one of the directors of the newly founded Centralstelle für wissenschaftlich-technische Untersuchungen, Berlin-Neubabelsberg,[23] and in the next four years he completed and published the findings of some outstanding experiments on both ball-bearings and plain journal bearings. His work on rolling-element bearings will be considered in Section 10.7, but his celebrated study of the frictional characteristics of plain journal bearings has special interest in this account of scientific studies of lubrication.

Stribeck (1902) reported the results of a carefully conducted and wide-ranging series of experiments on journal bearings, in which not only the load but also the speed was varied. The results clearly showed the minimum point reported earlier by Thurston and which is now recognized as the demarcation between full fluid-film lubrication and some asperity interaction. It is important to realize that confusion over the nature of journal-bearing friction had steadily increased in the latter half of the nineteenth century. Stribeck's systematic and definitive experiments ended the confusion, although the debate about the true nature of lubrication on each side of the minimum in the friction curve was to continue for many decades.

The characteristic form of the curve of the journal-bearing coefficient of friction is frequently referred to as 'the Stribeck curve' in recognition of this convincing experimental work in Germany early in the twentieth century.

In 1914 Dr Ludwig Gümbel (1914a), who had devoted his earlier years to mechanical and hydrodynamic problems associated with ship

---

23. I am grateful to Prof. Dr-Ing. H. Czichos of the Bundesanstalt für Materialprüfung for providing background information on Stribeck's career.

performance, analysed the experimental results published twelve years earlier by Stribeck. He showed that the separate curves to the right of the minimum in the friction traces could be condensed into a single graph if the coefficient of friction was plotted against

$$\left[\frac{\eta\omega}{\bar{p}}\right], \qquad\qquad \dots [10.14]$$

where η is the lubricant viscosity, ω the angular velocity of the shaft and $\bar{p}$ the load per unit area.

This important dimensionless grouping, which emerges clearly from hydrodynamic theory, was later to be known as the Gümbel number in relation to friction data. A different dimensionless grouping which is sometimes used to represent bearing-load capacity is also referred to as the Gümbel number in German literature.

Gümbel (1921) later considered the question of boundary conditions to be applied to the solution of Reynolds' equation for journal bearings. He recognized the problem of cavitation and the inability of lubricants to withstand sub-ambient pressures and proposed, without theoretical support, three alternative cavitation boundary conditions.

1. $p = p_{max}$ at $\phi = \pi$ and $p = 0, \pi < \phi < 2\pi$;
2. $p = 0$ at $\phi = \pi$ and $p = 0, \pi < \phi < 2\pi$;
3. $p = dp/dx = 0$ at the rupture point and $p = 0$ beyond this position in the divergent region.

The first condition is untenable on the grounds of continuity of pressure, the second is known as the half-Sommerfeld solution since it merely neglects the Sommerfeld predictions in the region $\pi < \phi < 2\pi$ and is equally untenable on the basis of flow continuity, and the third is a mathematical statement of the physical situation outlined by Reynolds. This was the first explicit statement of a widely used cavitation boundary condition, but it is rarely attributed to Gümbel since he merely presented it without theoretical support or physical justification.

Germany

Arnold Johannes Wilhelm Sommerfeld (Sommerfeld, 1904)   From 1900 to 1906 the nuclear physicist Arnold Sommerfeld[24] held the Chair of Technical Mechanics in the Technische Hochschule in Aachen. It was during this period, when he was closely involved with technical problems, that he completed his solution to the problem of journal-bearing lubrication. The great merit of this work was that it presented an analytical solution to the problem tackled by Reynolds and represented the most important theoretical study of lubrication during the twenty years from 1884 to 1904. During this period the designers of journal bearings had gained little from the work of Osborne Reynolds. The mathematics was complex, the numerical solutions limited to small eccentricity ratios and

---

24. See Appendix, Sect. A.18 for biographical details.

no attempt had been made to produce a range of solutions suitable for design. In short, it was elegant analysis rather than synthesis.

Whereas Reynolds used a tedious and restricted series solution for a journal bearing of infinite width, Sommerfeld found that the integrated form of the Reynolds equation could be solved in closed form.

If side-leakage is neglected the Reynolds equation [10.9] can be integrated once with respect to x or ϕ without any knowledge of the relationship between the film thickness (h) and the coordinates x or ϕ. The resulting integrated form of the Reynolds equation becomes

$$\frac{dp}{d\phi} = 6\eta Ur\left[\frac{h - h_0}{h^3}\right],\qquad\qquad \dots [10.15]$$

where h_0 is the film thickness at the location of maximum or minimum pressure, where $dp/d\phi = 0$. The problem of determining the pressure distribution in the bearing thus resolves itself to a question of evaluating $\int d\phi/h^2$ and $\int d\phi/h^3$.

Sommerfeld (1904) expressed the relationship between h and ϕ as $h = c + e \cos \phi$, where c is the radial clearance and e the eccentricity. He then introduced the ratio $c/e = \alpha$ and defined three integrals,

$$J_1 = \int_0^{2\pi} \frac{d\phi}{(\alpha + \cos \phi)},$$

$$J_2 = \int_0^{2\pi} \frac{d\phi}{(\alpha + \cos \phi)^2},\qquad\qquad \dots [10.16]$$

$$J_3 = \int_0^{2\pi} \frac{d\phi}{(\alpha + \cos \phi)^3}.$$

The first integral can readily be solved by means of the standard substitution $\delta = \tan(\phi/2)$ to yield

$$J_1 = \frac{2}{(\alpha^2 - 1)^{1/2}} \tan^{-1}\left[\left(\frac{\alpha - 1}{\alpha + 1}\right)^{1/2} \tan\left(\frac{\phi}{2}\right)\right]_0^{2\pi} = \frac{2\pi}{(\alpha^2 - 1)^{1/2}}$$

$$\dots [10.17]$$

and this was the form in which Sommerfeld expressed the integral.

With the solution for J_1 available, Sommerfeld noted that reduction formulae could be used to determine J_2 and J_3, since,

$$J_3 = -\frac{1}{2}\frac{dJ_2}{d\alpha} \quad \text{and} \quad J_2 = -\frac{dJ_1}{d\alpha}.$$

Sommerfeld utilized these relationships to show that if the lubricating film did not rupture due to cavitation, the attitude angle (ψ) between the load line and the line of centres would be 90° ($\pi/2 \times$ rad) and the load-carrying capacity per unit axial length (P) given by

$$P = \frac{12\pi\eta r^2 U}{c^2}\left[\frac{\alpha^2}{(\alpha^2 - 1)^{1/2}(2\alpha^2 + 1)}\right]. \qquad\qquad \dots [10.18]$$

Furthermore, the friction torque on the shaft per unit axial length (M) was given by

$$M = \frac{4\pi\eta r^2 U}{c}\left[\frac{\alpha(\alpha^2 + 2)}{(\alpha^2 - 1)^{1/2}(2\alpha^2 + 1)}\right]. \qquad \ldots [10.19]$$

Hence the coefficient of friction (μ) = M/Pr was given by

$$\mu = \frac{c}{r}\left[\frac{\alpha^2 + 2}{3\alpha}\right]. \qquad \ldots [10.20]$$

The coefficient of friction exhibited a minimum when $\alpha = (2)^{1/2}$ and Sommerfeld displayed the variation of μ, with both U and P. The minimum value is

$$\mu = \frac{2\sqrt{2}}{3}\left(\frac{c}{r}\right)$$

and it is important to note that in the concentric case ($\alpha = \infty$), the viscous force on the shaft reduces to the Petrov solution

$$F = \frac{\eta U A}{c}$$

(see eqn [10.6(a)]).

The derivation of a set of simple relationships for load-carrying capacity (P), friction moment (M) and coefficient of friction (μ) from the differential equation presented by Reynolds was a great achievement. The equations provided for the first time a basis for the rational design of journal bearings, although the utility of the work was not immediately recognized. Furthermore, the neglect of cavitation in the divergent clearance space and the prediction of sub-ambient pressures in this region was to be a matter of concern. The pressure distribution predicted by the Sommerfeld analysis was skew-symmetric about the line of centres, with the gauge pressures in the divergent clearance space being numerically equal but of opposite sign to those in the convergent film. Sommerfeld himself recognized the problem, but the determination of a physically acceptable and mathematically convenient cavitation boundary condition for lubricating layers has exercised the minds of research workers until the present day.[25]

It was noted that the coefficient of friction predicted by the Sommerfeld analysis achieved a minimum when plotted against eccentricity, and the fact that Thurston (1885) and Stribeck (1902) had recorded an apparently similar effect experimentally engendered confidence in the theory. It was later recognized that the minimum normally recorded by experimenters marked a transition from fluid-film to mixed or even boundary lubrication.

Perhaps the most amazing finding which emerges from a careful study of this classical paper, is that the mathematical transformation attributed to Sommerfeld was apparently never used by him. He simply used the standard substitution $\delta = \tan(\phi/2)$ and reduction formulae to solve the integrals which yielded expressions for pressure, load-carrying capacity

---

25. See Dowson, D. et al. (1975).

and viscous traction in his solution of the full, 360°, journal bearing problem.

In his analysis of the half, 180°, journal bearing, Sommerfeld introduced the arctan function (γ) in the form,

$$\tan \gamma = \left[\frac{\alpha - 1}{\alpha + 1} \right]^{1/2} \tan\left(\frac{\psi}{2} + \frac{\pi}{4} \right)$$

This can be rearranged to appear in a form similar to the relationship later described as the 'Sommerfeld Transformation', but it is by no means certain that Sommerfeld himself adopted this approach. It would certainly not be necessary to do so, and in his own paper there is no explicit statement of the relationship associated with his name in modern texts on lubrication.

Boswall (1928) certainly rationalized the mathematical procedure in his excellent book, and he showed that the integrals could readily be solved by means of the now familiar 'Sommerfeld Transformation'.

$$(1 + \varepsilon \cos \phi) = \frac{1 - \varepsilon^2}{(1 - \varepsilon \cos \gamma)} \qquad \qquad \dots [10.21]$$

where (γ) is known as the Sommerfeld variable and (ε) is the eccentricity ratio (e/c).

If Sommerfeld had used the transformation written here as equation (10.21), the solution (J_1) would not have emerged in the form shown in equation (10.17).

Boswall (1928) wrote, '. . . Sommerfeld showed that the mathematical difficulty could be got over, and that the series could be replaced by definite integrals which could be calculated without any serious expenditure of labour. . . . The mathematical solution, although closely following the method given by Sommerfeld, will therefore be given in a perfectly general form. . . .'

It appears that this is a further case in which a contribution to science was subsequently recognized by naming a useful relationship, equation or substitution after the original investigator – rather like the story of the Bernoulli equation in fluid mechanics.[26]

I have discussed this interesting aspect of the history of the Sommerfeld transformation with a number of people, including Professors Blok, Cameron and Booker[27] and my colleague Dr Taylor, but so far I have been unable to find any evidence to show that Sommerfeld ever used the mathematical transformation which we all attribute to him.

Sommerfeld (1921) wrote one more paper on lubrication, but this was more in the nature of a review and a restatement of his earlier work than a new contribution to the subject. Again there was no reference to the use of the 'Sommerfeld substitution' as we now recognize it.

The next stage in the development of journal-bearing theory was to be the introduction of more realistic cavitation boundary conditions and,

---

26.   See Tokaty, G. A. (1971), pp. 73, 80.
27.   See Booker, J. F. (1965*a*).

in due course, an allowance for side-leakage. However, there is no doubt that Arnold Sommerfeld played a major role in developing journal-bearing theory. Many of the essential characteristics of bearing behaviour emerge from his work and the penetrating yet beautifully direct nature of the analysis rarely fails to excite the interest of students of the subject.

Australia	*Anthony George Malden Michell* (Michell, 1905*a,b*) While Arnold Sommerfeld was exercising his analytical skills on the journal-bearing problem, a man sometimes described as Australia's most famous engineer was tackling the thrust-bearing question. Anthony George Malden Michell[28] was a first-generation Australian even though he was born in Islington, London, England. His parents came from Devon, had emigrated to Australia and were visiting relatives in England when George was born. George's elder brother John Henry Michell studied mathematics at Cambridge and it was during this period that George attended lectures on mechanics in the new School of Engineering. He also attended lectures on physics by J. J. Thomson, Osborne Reynolds' most distinguished student. George Michell graduated as an engineer from the University of Melbourne in 1895 and obtained his practical experience with a firm of hydraulic engineers. It was the severity of thrust-bearing problems in the centrifugal pumps used in the Murray River Irrigation Scheme and the water-turbines employed in the early Australian hydro-electric schemes which promoted Michell's analytical studies and his invention of the tilting-pad thrust bearing. The bearing invention is described in Section 10.7, but it is the background analytical work which is of interest here.

Reynolds (1886) produced an analytical solution to the governing equation for pressure in a plane inclined surface bearing of infinite width (no side-leakage) and an approximate numerical solution for journal bearings with the same restriction on side-leakage. Sommerfeld (1904) neglected side-leakage in his analysis of journal bearings. Michell's (1905*b*) analysis of the plane inclined slider bearing was remarkable in that it included for the first time the effects of side-leakage on the slider-bearing problem. The analysis was completed during the period 1902–4 and the paper was published in English in the *Zeitschrift für Mathematik und Physik* (Michell, 1905*b*).

Michell assumed the lubricant to be isoviscous and he sought a solution for pressure (p) which could be written as

$$p = p_1 + p_3 + p_5 + \cdots + p_m + \cdots,$$

where the general term $p_m = \omega_m \cdot (\sin my)/mx$; ω_m being a function of x only and m a positive odd number.

The solution included Bessel functions and the coefficients were selected to make the pressure vanish along the boundaries of rectangular

---

28. See Appendix, Sect. A.19 for biographical sketch.

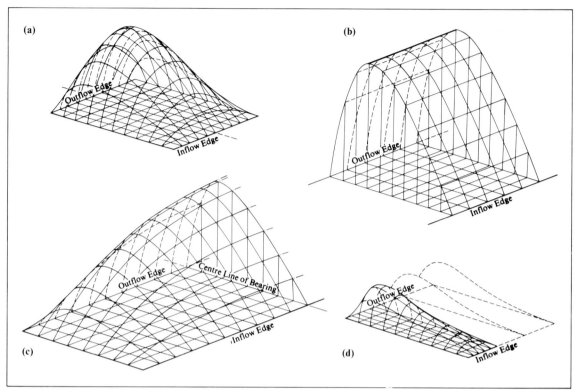

Fig. 10.20
A. G. M. Michell's predictions of pressure distributions on plane-inclined pads (Michell, 1905). (Inlet film thickness = 2 × outlet film thickness. All pressures shown to the same scale except (*d*) which is exaggerated by a factor of five.)

pads. Numerical solutions were obtained for a square pad ($b/l = 1$), a wide pad ($b/l = 4$) and a narrow pad ($b/l = \frac{1}{3}$). The three-dimensional displays reproduced in Fig. 10.20 showed for the first time the distribution of pressure in realistic bearing arrangements. It was a masterly piece of work, combining physical insight with impressive analytical skill. In 1943 Michell was awarded the James Watt International Medal and the Vice-President of the Institution of Mechanical Engineers, Professor Andrew Robertson, remarked that even then the work was regarded by some as '. . . the only really important extension of Reynolds' theory yet effected.' It is interesting to note that Michell had some personal acquaintance with Sommerfeld and Kapitza in Europe and Kingsbury in the United States.

In the minds of most people the name of Anthony George Malden Michell is normally and justifiably associated with the invention of the tilting-pad thrust bearing. The triumph of his analytical studies alone would have merited recognition in a book on the history of tribology. He was a remarkable engineer from a remarkable family. In due course both he and his elder brother John Henry became Fellows of the Royal Society.

United Kingdom

W. J. Harrison's analysis of gas bearings (Harrison, 1913), *and dynamic loading* (Harrison, 1919)   A lecturer in mathematics in the University of Liverpool rounded off a period of exciting analytical work in the field of lubrication. The work begun by Osborne Reynolds in Manchester, extended and applied by Sommerfeld in Germany and Michell in Australia, returned to Liverpool to be compared with experiment and extended to gaseous lubricants by Harrison (1913).

Hirn (1854) had mentioned the advantages of using air as a lubricant under suitable circumstances, but it was Kingsbury who first presented firm experimental evidence that load-carrying pressures could be generated in an air film between a bush and a rotating journal. Harrison (1913) first solved Reynolds equation for a 360° (2π rad) journal bearing lubricated by an incompressible fluid. He was unaware of Sommerfeld's (1904) treatment of the problem at the time that he completed his analysis, but his more direct approach has merit and is worthy of study. An additional bonus from this study was the revelation that there must necessarily be a divergence between the viscous torques on the journal and bearing at all non-zero values of eccentricity. He rightly drew attention to the practical significance of this result for experimenters, who normally measured the torque on the bearing yet really sought the torque on the journal. The theory predicted that journal friction would always be greater than bearing friction at finite values of eccentricity. This important result was to play an important role in explaining differences between journal torques based upon power loss measurements and direct readings of bearing torques.

Harrison went on to discuss the essential features of the flow of a compressible fluid between parallel plates. He then considered continuity of mass flow rather than volume flow, integrated the modified Reynolds equation for zero side-leakage numerically by means of Runge's method and compared the findings with the theory for an incompressible lubricant and the results of Kingsbury's beautiful air-lubricated journal-bearing experiments. The comparisons are evident in Fig. 10.21. In each case curve 1 represents the solution for an incompressible lubricant, curve 2 Harrison's solution for a compressible lubricant obeying the relationship for an isothermal, compressible fluid, $p/\rho = k$, and curve 3 the distribution of pressure observed by Kingsbury. The general agreement between Harrison's theory (2) and Kingsbury's experimental results (3) was excellent, particularly at the lowest and highest speeds considered. His analytical skill was matched by his perseverance with numerical work, for he recorded the fact that a single complete solution of the journal-bearing problem involved the writing down of 12,000 digits.

Harrison's work was quite outstanding, coming as it did long before the emergence of gas-bearing technology. He even went on to consider plane inclined surface bearings and at a later stage was to consider dynamically loaded bearings. In his 1913 paper he wrote, '. . . I have obtained some results and have work in hand treating of cases in which the influence of variable speed and variable load on the lubrication of a cylindrical bearing is taken into account'.

Harrison's (1919) analysis of dynamically loaded journal bearings was to be the first serious contribution to a complex problem which exercises the minds of bearing designers and analysts alike to the present day. By this time, Harrison was at Cambridge, where he was a Fellow of Clare College. His analytical skill enabled him to predict, within various

Fig. 10.21
Harrison's comparison between
incompressible (1) and
compressible (2) lubricant
solutions and Kingsbury's (3)
experimental values for pressure
in an air-lubricated journal
bearing (Harrison, 1913).

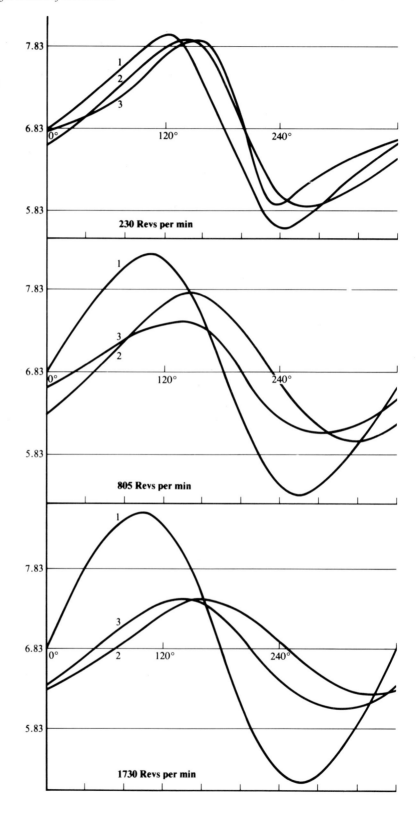

assumptions, the orbital motion of a journal carrying a steady load yet suffering some initial displacement from the position of static equilibrium. He also considered the question of response of the journal to small fluctuations of the load. The paper had little practical significance, yet it was a pioneering venture in the field of dynamically loaded bearing analysis. Once again Harrison's analytical work was published in advance of the awareness of bearing designers of the significance of the problems under discussion. His second paper, like the first, was double-barrelled, in that it also contained a theoretical study of pivot bearings. The analysis was prompted by the results of the fourth report of the Institution of Mechanical Engineers Committee on Friction at High Velocities (Institution of Mechanical Engineers, 1891), and the approach was based upon the likelihood of a chamfer on the bearing surface adjacent to the groove providing a physical wedge in an otherwise parallel surface-bearing situation.

United States

Mayo D. Hersey's use of dimensional analysis in the study of journal-bearing friction (ZN/P) (Hersey, 1914).   Mayo D. Hersey, affectionately known in the United States as 'the Father of Lubrication,' carried out a large number of friction measurements on full journal bearings in the Department of Mechanical Engineering at the Massachusetts Institute of Technology in 1909. He plotted the coefficient of friction against load, speed, temperature, viscosity and rate of oil supply, but found it difficult to condense the data into a form suitable for publication. The answer came from the application of dimensional analysis.

Prior to the First World War a number of workers had utilized dimensionless groups and the principle of dynamical similarity in an *ad hoc* way.[29] Dr Edgar Buckingham of the United States National Bureau of Standards clarified the subject and introduced his famous *Pi* theorems (Buckingham, 1914a,b). The concepts appealed to engineers and Hersey (1914) was the first to apply dimensional theory to the problem of journal-bearing friction. He showed that hydrodynamic friction should be a unique function of,

$$\left(\frac{\eta n}{p}\right) \qquad \qquad \qquad \ldots [10.22]$$

which differs from Gümbel's number (see eqn [10.14]) only by replacing the angular velocity (ω) by the rotational speed (n).

Many years later Wilson and Barnard (1922) introduced a lower-case z to stand for *Zähigkeit* (viscosity) and replotted all the available friction data from Stribeck onwards against

$$\left(\frac{zn}{p}\right). \qquad \qquad \qquad \ldots [10.23]$$

---

29.   See Hersey, M. D. (1966), p. 131.

It was S. A. McKee, another worker at the National Bureau of Standards, who finally promoted the lower-case to an upper-case representation of this familiar dimensionless group; thus emerged the familiar form

$$\left(\frac{ZN}{P}\right).$$
. . . [10.24]

Gümbel's posthumous book (Gümbel and Everling, 1925) made no mention of Hersey's work and it must therefore be assumed that it was not known in Germany. Professor Blok[30] has suggested that this delay in communication of ideas between 1914 and 1922 or so could have resulted from the isolation of Germany during and immediately after the First World War.

Some workers now refer to the well-known curve for journal-bearing coefficient of friction as the *Stribeck–Hersey curve*, to honour the total contributions by the experimenter and the man who rationalized the data by means of dimensional analysis.

England

Martin's analysis of gear lubrication (Martin, 1916) *and Stanton's approach to the experimental problem* (Stanton, 1923)   Once Reynolds' equation had been solved so convincingly for journal (Sommerfeld) and thrust (Michell) bearings, attention turned to the more complicated problem of gear lubrication. In an unsigned article in *Engineering*, Martin (1916) presented a direct and impressive analytical approach to the problem. He recognized that the conjunction between spur gears could be represented with adequate accuracy by two cylinders and proceeded to solve the Reynolds equation for this geometry on the understanding that the cylinders were rigid and the lubricant isoviscous. Martin deduced a simple relationship between the minimum film thickness (h), effective radius of curvature (R), load per unit width (P), viscosity (η) and mean surface speed (u) of the form,

$$\frac{h}{R} = 4\cdot9\left(\frac{\eta u}{P}\right).$$
. . . [10.25]

Calculations based upon this relationship revealed film thicknesses which were small compared with the known surface roughness of gears, and hence it had to be concluded that fluid-film action could not account for the satisfactory lubrication of gears. The quandary was not finally resolved for a further half-century, when the theory of elasto-hydrodynamic lubrication became firmly established.

Dr T. E. Stanton worked at the National Physical Laboratory and was a member of the 'Lubricants and Lubrication Inquiry Committee' set up in England by the Department of Scientific and Industrial Research. The Committee reported its findings in 1920 and among its 'Recommendations for future research on lubricants and lubrication' related to physical and

---

30. Blok, H., Private communications, June 1976.

engineering subjects was the simple statement, '. . . (b) . . . (viii) To carry out a research on the friction of surfaces in rolling contact, and in combined rolling and sliding contact (such as gear teeth) in the partially lubricated, completely lubricated, and in the dry states.'

Dr Stanton and Mr Hyde had already carried out a number of experiments at the National Physical Laboratory on behalf of the Committee on a Lanchester worm-testing gear, but Stanton (1923*b*), went on to study further the form of lubrication encountered in worm gears. The Committee had felt that the very high stresses encountered in gears would prevent fluid-film action and that the friction would be governed by interactions between molecules of the lubricant and the solid. It was certainly known that the mechanical efficiency of gears was little affected by lubricant viscosity, but Stanton was reluctant to be carried away completely by the wave of support for boundary lubrication. He wrote:

> *. . . That this evidence is conclusive cannot, in the opinion of the writer, be regarded as certain, since it is conceivable that in these tests the thickness of the film and the area of it which is under pressure should vary in such a way as to produce changes in the resistance of the order of those due to variations of viscosity of the lubricant, and further, that the flow of the material of the worm wheel, which undoubtedly takes place under the extremely high pressures, may account for the observed effect of the nature of the material.*

Stanton's perception of the essential ingredients of elasto-hydrodynamic lubrication has remained in obscurity, but it is well to recognize his penetrating thoughts on the subject before reviewing his experimental work of 1923.

Stanton could not envisage direct observation of the actions between worm gear teeth and he turned instead to a simulation of the high pressures in a heavily loaded journal bearing having a large radial clearance. A single pressure tapping and a Bourdon gauge was used to measure pressure and the circumferential distribution was measured by rotating the 360° bush. The large radial clearances of up to 6 per cent on a 1 in (25·4 mm) diameter shaft produced small effective load-bearing arcs and maximum pressures as high as 3·5 tonf/in² (54 MN/m²). A friction balance was used to measure torque on the overhung bearing and temperature was measured in the pressure-tapping hole by means of a thermojunction. Lubricants tested included sperm, rape, castor and a mineral oil.

A typical result from Stanton's work is shown in Fig. 10.22. The pressure distribution shows that the maximum pressure attained was about 3·5 ton/in² (54 MN/m²) and that the effective load-carrying arc of contact was only about 15° (0·26 rad). This was a truly remarkable demonstration of fluid-film lubrication under very severe conditions and it probably represents the first record of elastohydrodynamic action. It not only demonstrated the great potential of hydrodynamic action for the protection of bearing surfaces under very severe operating conditions, it added to the understanding of journal-bearing performance under cavitated conditions. Reynolds (1886) thought that cavitation would limit the safe operating condition in a bearing to about 0·5, but Stanton was able to demonstrate safe performance at eccentricity ratios of about 0·99. There were no significant sub-ambient pressures reported in

Fig. 10.22
T. E. Stanton's measurements of
pressure in a journal bearing at
high eccentricity ratios (Stanton,
1923*b*). (Diameter of journal 1 in
(25·4 mm), diameter of bearing
1·06 in (26·9 mm); load 690 lbf.
(3·07 kN); lubricant–castor oil;
speed 1000 rpm (16·67 Hz);
temperature of film 49·4°C.)

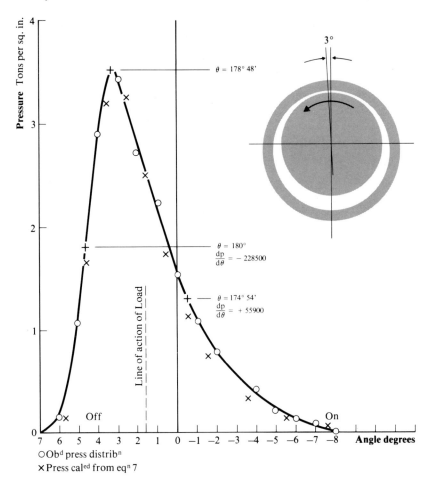

○Ob^d press distrib^n
×Press cal^ed from eq^n 7

Stanton's paper and he deduced both the eccentricity ratio and the
lubricant viscosity from the two points of inflection on each of the
pressure curves. He then recalculated the theoretical pressure distribution
and obtained impressive correlation over the full arc of the film, as shown
in Fig. 10.22.

Stanton's experimental work did much to promote confidence in
hydrodynamic action under conditions which had previously been
thought to be so severe as to require protection for the bearing surfaces
from boundary action. His physical insight and outstanding experimental
work provided a much neglected and oft forgotten springboard for subsequent studies of gear lubrication and elastohydrodynamic lubrication.

United Kingdom

*Lord Rayleigh's early analysis of externally pressurized (hydrostatic)
thrust bearings* (Lord Rayleigh, 1917*a*) *and his determination of the
optimum profile of thrust bearings* (Lord Rayleigh, 1918*a*)   As the virtues
of fluid-film lubrication generated by motion of the bearing surfaces
became recognized, attention was naturally diverted to the limitations
imposed by excessive loading or low surface speeds. Lord Rayleigh (1917*a*)

was the first to present an analysis of a hydrostatic thrust bearing. He derived expressions for the logarithmic radial decay of pressure from a central supply in the space between parallel circular discs, the resisting torque and the limiting load-carrying capacity of the bearing. His brief yet direct and complete analysis was published in *Engineering* in December 1917, less than nineteen months before his death. He pointed out the practical importance of the situations which could be accommodated by such bearings and emphasized that with '. . . the utmost accuracy in fitting together the two opposed surfaces, . . . as well as the removal of all suspended solid matter from the lubricant . . . there should be no wear of the solid surfaces, which should never come into contact.'

In a characteristic footnote he mentioned that he had defaced the coinage of the realm in the cause of experimental work. 'P.S.–I may perhaps mention that I have made a small model, in which the opposed surfaces are those of two pennies ground to a fit, and the "lubricant" is water supplied from a tap.'

The following month Lord Rayleigh (1918*a*) published his 'Notes on the theory of lubrication' in the *Philosophical Magazine*. An accurate and fair account of earlier contributions by others to fluid-film lubrication theory preceded a most intriguing study of the optimization of thrust-bearing profiles. He neglected side-leakage and repeated the solution for the plane inclined slider bearing ($h = mx$), before extending the range of solutions to a general curved surface ($h = mx^n$), and an exponential profile ($h = e^{\beta x}$). He then posed the problem of determining the bearing profile which would yield a maximum load-carrying capacity per unit width (P) for a given bearing length (l) and a given minimum film thickness (h_0). The answer, revealed by a neat application of the calculus of variations, turned out to be the stepped parallel surface bearing. Lord Rayleigh's results for various profiles are summarized in Table 10.1.

Lord Rayleigh's suggested arrangement for the optimum form of thrust bearing is shown in Fig. 10.23. He again chose to demonstrate the efficacy

Table 10.1

Values of (h_i/h_0) and dimensionless load-carrying capacity $(P/\eta U)(h_0/l)^2$ For various thrust-bearing profiles

Profile		$k = \left(\dfrac{h_i}{h_0}\right)$	$\dfrac{P}{\eta U}\left(\dfrac{h_0}{l}\right)^2$
Plane		2.2	0.160
Parabolic		2.3	0.163
Exponential		2.3	0.165
Stepped-parallel		1.87	0.206

of his proposed bearing by defacing the coinage and once more relegated the delightful account of the device to a postscript.

> *... P.S.–Dec. 13 (1917).–In a small model the opposed pieces were two pennies ground with carborundum to a fit. One of them–the stationary one–was afterwards grooved by the file and etched with dilute nitric acid according to (Figure 10.23), sealing wax, applied to the hot metal, being used as a resist. They were mounted in a small cell of tin plate, the upper one carrying an inertia bar. With oil as a lubricant the contrast between the two directions of motion was very marked.*
>
> *Opportunity has not yet been found for trying polished glass plates, such as are used in optical observations in 'interference'. In this case the etching would be by hydrofluoric acid, and air should suffice as a lubricant.*

Summary

In the forty-year period from about 1880 the physical nature of fluid-film lubrication was revealed by inspired experimental work, the theory was established by classical analysis, the governing equations solved for both journal and thrust bearings and the basic ideas accepted with such confidence that they were employed in the rationalization of bearing friction data, extended in a preliminary way to the problem of gear lubrication, applied to gas and externally pressurized bearings, adapted to time-dependent situations and utilized in optimization procedures to predict the most effective thrust-bearing profiles.

There are three important aspects of this most creative period; the outstanding calibre of the analytical and experimental investigations, the possible influence of communication of the written and spoken word in speeding up scientific inquiry and the grouping of four main thrusts in periods concerned with practical developments, basic concepts, the solution of the equations for the major forms of journal and thrust bearings and, finally, the more adventurous extension of the concepts to fields outside those originally envisaged by the early investigators.

This grouping together of major developments at intervals of about ten years shows clearly enough in Fig. 10.24. Throughout this Section it has been necessary to examine all the known scientific and technical developments in the field of fluid-film lubrication, but to refer only to those of outstanding merit. A more complete bibliography can be found in Hersey's (1933, 1934) notes on the history of lubrication and his later comprehensive text (Hersey, 1966).

We can see in this era the increasing effect of international mobility and communications upon scientific research. Although Petrov (1883) and Tower (1883) appeared to know nothing of each other's work, since it was practically simultaneous, both knew of the work of Thurston (1878, 1879) and others. Reynolds' (1886) analysis resulted directly from Tower's (1883) experiments, but there is nothing to suggest that Reynolds was acquainted with Petrov's work. Sommerfeld (1904) certainly knew of Reynolds' analysis, but Harrison (1913) had completed his own solution to the problem of a journal bearing lubricated by an incompressible fluid before he heard of Sommerfeld's (1904) earlier analysis. Gümbel (see Gümbel and Everling, 1925) seemed to be unaware of Hersey's (1914) use of dimensionless groups to represent journal-bearing friction data, but the interruption to the exchange of ideas caused by the First World War might well have accounted for this.

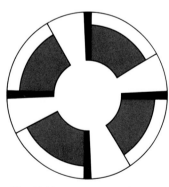

Fig. 10.23
Lord Rayleigh's suggestion for an optimum stepped parallel surface thrust bearing (Rayleigh, 1918*a*). (White parts represent an original plane surface, shaded parts are slight depressions of uniform depth. The four wide black radial lines represent grooves for the easy passage of lubricant.)

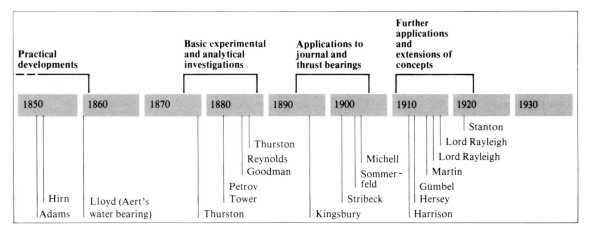

Fig. 10.24
Major eras in scientific studies of
fluid-film lubrication (1850–1930).

Many of the contributors discussed in this section travelled to each others' countries, but few of them appear to have been introduced to each other. Both Tower and Petrov went to the USA, Thurston travelled in Europe, Reynolds visited Canada and Michell visited both Europe and the USA. The era was indeed characterized by a great mobility of men of science and their ideas.

10.5   Viscometry

Several instruments which could compare the *fluidity* of lubricants were established in Europe and North America in the second half of the nineteenth century. It might be thought that the introduction of such instruments was a consequence of the scientific studies of the 1880s which confirmed the pre-eminent role of viscosity in the physical process of fluid-film lubrication. An examination of the literature shows that this is not entirely accurate, although the 1880s not only accelerated but also provided a sound basis and sense of purpose to the process.

The steam age had already exposed the limitations of unstable and increasingly expensive animal and vegetable lubricants and generated interest in the testing of lubricants when Charles Dolfuss addressed the Société Industrielle de Mulhouse on 29 June, 1831. He described and demonstrated to the Société an instrument consisting of a vessel filled with test fluid with a small hole in its base. By noting the time taken for a given volume of lubricant to leave the vessel Dolfuss derived an *index for its liquidity*. The inventor called the device a *viscomètre* and it was widely acclaimed as a most useful instrument.[31]

In the early 1840s a professor in the Paris medical schools chose to study the passage of blood through fine capillaries by investigating the flow of pure water through the bores of thermometer tubes. Determination of viscosity by means of the measurement of flow in tubes was thus established by Jean M. L. Poiseuille. Initially, horizontal tubes were used

31.   See Forbes, R. J. (1958), pp. 102–23.

as described by Barr (1931), with improvements being introduced in subsequent years to avoid mixing of the test fluid with the surrounding fluid and to improve the knowledge of capillary bore, length and end conditions. Perhaps the best-known variation on this arrangement in lubrication circles is the viscometer introduced by Slotte (1881).

If the horizontal capillary viscometer is made longer to improve accuracy it becomes somewhat fragile, difficult to mount and awkward to contain in a constant-temperature chamber. Stone (1915) therefore introduced a vertical tube viscometer which can be seen as the forerunner of the modern U-tube viscometer.

In parallel with the ever increasing refinement of the long capillary viscometer, various forms of robust short-tube viscometers were introduced in industry in the 1880s. Three, the Redwood, Saybolt and Engler, developed in England, the USA and Germany, became particularly well known and can still be seen in lubricant-testing laboratories. While they all differ in detail these commercial viscometers employ the same principle. The test fluid is placed in a central, open cup to a definite depth, while being surrounded by a water- or oil-bath to provide some degree of temperature control. The centre of the base of the cup contains a small orifice or short length of capillary controlled by a simple valve. The time required for a definite volume of the test fluid to flow through the jet into the atmosphere and hence into a collecting vessel gives a measure of the viscosity in arbitrary units known as *Redwood seconds*, *Saybolt seconds* or *Engler degrees*. It can be seen at once that these commercial instruments were essentially developments of the Dolfuss *viscosimètre*, and all were known in English-speaking countries as viscometers.

The *Engler viscometer* was the first to receive official recognition,[32] having been adopted for a comparison of lubricants by a German railway committee as early as 1884. Engler described the viscometer in a paper (Engler, 1885), although it was subsequently subjected to a number of minor modifications designed to standardize dimensions and thus to improve comparability between individual viscometers. The *Engler degree* (E) was merely the ratio of the times taken for 12·2 in^3 (200 cm^3) of the test fluid and water to flow from the cup. The kinematic viscosity, (v) defined as the ratio of the absolute viscosity (η) to the density (ρ) of the lubricant, could then be calculated from an empirical relationship between v and E.

Mr Boverton Redwood designed his viscometer in 1885 and in (1886) he gave a full account of the instrument. It was, apparently, a modification of an instrument developed by a chemist at Price's Patent Candle Company at Battersea by the name of C. Rumble. At the same lecture Redwood also displayed an instrument developed for the same purpose by an Inspector of the Standard Oil Company of New York, Mr G. M. Saybolt. Redwood used agate as the material for his jet and expressed the viscosity of lubricants as a percentage of that of rape oil at 15·6°C

---

32. See Barr, G. (1931), p. 99.

(60°F). He appears to have done this deliberately to enable those engineers using the new mineral oils to appreciate the viscosity of their new lubricants against a background of something within their experience. The *Redwood second* was merely the time of efflux of a given quantity of lubricant and the instrument achieved wide acceptance in England. Like the Engler viscometer it was to be subjected to a long period of standardization and in due course the National Physical Laboratory was to issue certificates to instruments which met agreed specifications. An empirical equation enabled Redwood seconds to be converted to kinematic viscosity.

In due course this early type of Redwood viscometer was designated No. 1, because a later instrument for more viscous liquids was designated Redwood No. II viscometer. By this time Boverton Redwood had been knighted and he was investigating fuel oil for the Admiralty when the No. II instrument was introduced. It was widely known as the 'Admiralty' type for many years.

In the *Saybolt viscometer* the volume of efflux was measured within the cup itself. The instrument was clearly available in the mid-1880s, for it was exhibited by Redwood in England, but a full description did not emerge until Herschel (1918) published a standard set of dimensions agreed by Saybolt. A later agreement between Saybolt and the Bureau of Standards was adopted by the American Society of Testing Materials (ASTM) and the Standard Saybolt universal viscosimeter became as familiar and well established in the United States as the Redwood viscometer was in England and the Engler viscometer in Germany.

Since all these instruments enable viscosity to be calculated from empirical equations related to the characterstics of the form of viscometer there are, of course, relationships between the basic units of measurement such as Redwood and Saybolt seconds and Engler degrees. It is beyond the scope of this book to present such relationships, but they can be found in Barr (1931), Cameron (1966) and publications of various professional societies and standards laboratories throughout the world.

Other forms of viscometer involving falling spheres and rotational and oscillatory motion between annular discs or concentric cylinders have developed since the nineteenth century. Many of the ideas have had to be rethought as knowledge of the complex rheological behaviour of lubricants has increased. The relationship between viscosity and temperature, pressure and rate of shear is often a matter of great importance to the present-day tribologist, but contemporary testing procedures still reflect the urgent requirements of the early decades of the mineral oil age.

It was mentioned in Chapter 9 that Deeley and Parr (1913) proposed that the cgs unit of viscosity should be called the *poise*. The American H.A.S. Howarth later proposed the *reyn* for British units (Howarth, 1934) and Max Jakob recommended the *stoke* for kinematic viscosity (Jakob, 1928). In the now widely accepted SI system of units it has been proposed that the units Ns/m^2 or Pa s should be known as the Poiseuille, abbreviated to Pl to distinguish it from the pose, but this does not yet enjoy international acceptance.

In this section attention has been focused upon the early history of the accurate measurement of that most important single property of most lubricants–viscosity. It should be remembered, however, that it was not until the 1880s that scientists and engineers developed a sound appreciation of the importance of viscosity. Prior to that time 'fluidity' was discussed and several other properties were measured and related to lubricating ability. In particular, density or specific gravity was often the basis of comparison. The books by Thurston (1885) and Archbutt and Deeley (1900) give a fascinating insight into the background of lubricant property measurement. Finally, it should not be forgotten that 'viscosity' and the *coefficient of viscosity* were not firmly defined until Maxwell (1860) discussed the matter in his Bakerian Lecture the year after Colonel Edwin L. Drake struck oil at Titusville.

10.6 Boundary lubrication

The scientific studies of fluid-film lubrication in the 1880s appeared to reach such definite and complete conclusions that engineers and physicists of the era must have felt that the basic problem of the nature of lubrication had been solved. There was every justification for this view, since not only were mathematical predictions of bearing friction and pressure distribution in accord with experimental findings, but engineers like Michell and Kingsbury applied the new-found knowledge to the design of highly effective bearings in a most convincing manner.

In due course it was recognized that *fluid-film* was but one of at least two distinct modes of lubrication. It was the biologist and Secretary of the Royal Society, Sir William Bate Hardy, who clarified the position and introduced the term *boundary lubrication* in a report to the Lubrication Committee of the Department of Scientific and Industrial Research which was published in the *Proceedings of the Royal Society* (Hardy and Doubleday, 1922a,b). An account of Hardy's life is given in the Appendix, Section A.17, together with an appraisal of the circumstances which directed his scientific curiosity towards the subject of lubrication after the First World War. It was in fact a natural extension of his earlier work on colloids and surface tension, and it is in these areas that the foundations of boundary lubrication were constructed in the nineteenth and early part of the twentieth centuries. Studies of surface tension, gas release at liquid–solid interfaces, capillarity, floatation and the spreading of surface films of contaminants on liquids carried out without reference to lubrication were to provide the essential background.

In his discussion of the inhibiting effect of oil upon the motion of camphor on water, a lecturer on physical science at King's College School, London, by the name of Charles Tomlinson noted that as early as 1787 Volta had written (Tomlinson, 1863): '. . . If the water be defiled with any foreign substance, or its surface only slightly fouled with oily matter, if only the dust of the room be upon it, the looked for motions of camphor and of benzoin will not take place, or will be so feeble as to be scarcely sensible.'

Young (1805) considered the laws governing the spreading of liquids on solids and presented his well-known equation for the equilibrium of a drop of liquid on a solid surface in terms of the surface tensions of the respective interfaces and the wetting angle between the liquid and the solid. The influence of surface cleanliness upon gas release between liquids and solids was discussed in terms of *adhesive forces* by Tomlinson (1867, 1875) who was by this time a Fellow of the Royal Society.

A notable step forward in the study of thin films on liquid surfaces was reported by Pockels (1891). Miss Agnes Pockels, who was not a professional physicist, wrote to Lord Rayleigh in German from Brunswick on 10 January 1891. Lord Rayleigh forwarded a translation of her letter to the journal *Nature*, on 2 March and the article, 'Surface tension', appeared on 12 March 1891 with the following introduction: '. . . I shall be obliged if you can find space for the accompanying translation of an interesting letter which I have received from a Germany lady, who with very homely appliances has arrived at valuable results respecting the behaviour of contaminated water surfaces. . . .'

In a delightfully simple yet elegant experiment she found that very small quantities of oil on the surface of water did little to disturb the surface tension of the water. As the amount of oil was increased, the surface tension fell quite rapidly from that of water to oil once the quantity of oil per unit area reached a well-defined value. Lord Rayleigh (1899) repeated the experiments and drew attention to the importance of the finding. The effect was attributed to the spreading of oil on the surface of water, with the fall in surface tension coinciding with the complete formation of a monomolecular layer of oil of thickness slightly in excess of 39×10^{-9} in (10^{-9} m). The narrow brass Pockels–Rayleigh trough filled with water, with a glass strip forming a surface barrier between the clean water surface and that contaminated with oil, together with the associated surface tension balance, has formed the basis for numerous studies of thin films on liquids until the present day.

Lord Rayleigh (1918*b*) referred again to these experiments in his paper dealing with the lubricating properties of thin oily films which provided the opening quotation of this chapter. It is a delightful paper in which Lord Rayleigh measured the friction between two solids and found that the presence of a thin, imperceptible film of grease from the hands reduced the friction to levels below those detected when the surfaces were made moist, or even immersed in water or paraffin oil. When the surfaces of glass were thoroughly cleaned with an alcohol flame the friction attained very high values.

Clearly the increased friction associated with moisture or ample amounts of water or paraffin oil raised a number of questions in relation to established views on lubrication. The observations prompted Lord Rayleigh to write:

> . . . *In view of the above estimate and of the probability that the point at which surface-tension begins to fall corresponds to a thickness of a single layer of molecules, we see that the phenomena here in question probably lie outside the usual theory of lubrication, where the layer of lubricant is assumed to be at least many molecules thick. We are rather in the region of incipient seizing, as is perhaps not surprising when we consider the smallness of the surfaces actually in contact.*

The role of thin surface films upon lubrication was thus emerging as a question of some importance towards the end of the First World War. Across the Atlantic Langmuir (1917, 1920) of the General Electric Company in Schenectady and Harkins et al. (1917) were studying the theory of thin surface films. They developed the concept of spreading of oil and water and discussed the physical and chemical forces of interaction between the molecules in thin films and the surfaces on which they were deposited. Langmuir's (1917) general paper on the fundamental properties of solids and liquids is a particularly valuable contribution to early twentieth-century literature on this subject. In his fascinating paper to the Faraday Society, Langmuir (1920) explained that the spreading of oil on water was due to attraction between the water and some active group in the oil molecule. This further explained why the surface tensions of various long-chain hydrocarbons were essentially the same, since the active ends of the molecules were drawn inwards leaving the methyl groups to form the true surface with the environment.

Sir William Bate Hardy and boundary lubrication	Hardy graduated as a zoologist at Cambridge in 1888 and became a lecturer in physiology in the University in 1913. He studied various aspects of histology prior to 1898 and in due course developed an interest in the role of ionization upon the coagulative power of electrolytes. He developed his own appreciation and knowledge of physical chemistry and molecular physics and presented papers to the Royal Society on colloid solutions and the tension of composite fluid surfaces just two years before the First World War. He distinguished himself on Government committees concerned with food during the war and was instrumental in the formation of important food research centres.

In his first publication on friction and lubrication, Hardy (see Hardy and Hardy, 1919) drew special attention to the role of adsorbed layers on surfaces. The property of colloids in forming concentrations on the surfaces was noted by Willard Gibbs and known as 'adsorption'. Hardy found that very thin films, perhaps only one millionth of a millimetre thick, were sufficient to cause two glass surfaces to slide over each other with minimum friction. He concluded that under such circumstances the lubrication depended wholly upon the chemical constitution of the fluid and that a true lubricant was a fluid which was adsorbed by the solid surface.

In a most impressive report to the Department of Scientific and Industrial Research (DSIR) on lubricants and lubrication (DSIR, 1920) there appeared a penetrating 'Review of existing knowledge of lubrication' by a Special Committee chaired by Mr S. B. Donkin. More will be written about the report and the work of the Committee in Section 10.11, but it is relevant to the present discussion to note the clear distinction drawn between three stages of lubrication. The statement is summarized in Table 10.2 and it is evident that the concept of modes of lubrication other than fluid-film was already emerging.

The kind of lubrication in which very thin films of molecular proportions are effective in reducing friction and wear between opposing

Table 10.2
DSIR Committee, 'Review of
existing knowledge of
lubrication' (DSIR, 1920)

Stages of lubrication	Laws	Coefficient of friction
1. Unlubricated surfaces	Dry friction	0·10–0·40
2. Partially lubricated surfaces	Greasy friction	0·01–0·10
3. Completely lubricated surfaces	Viscous friction	0·001–0·01

solids was first described as *boundary lubrication* by Hardy and Doubleday (1922) in a report to the Lubrication Committee of the DSIR which was also published in the *Proceedings of the Royal Society*. Hardy (1920) had earlier discussed some of the essential features of lubrication when 'boundary conditions' are fully operative in his address on 'Problems of lubrication' to the Royal Institution of Great Britain.

It was Hardy and Doubleday's (1922a,b) joint papers which firmly established the basic concepts of boundary lubrication. He was aware of the work of Lord Rayleigh and that of both Langmuir and Harkins reported in the *Journal of the American Chemical Society* in 1917, which emphasized the orientated and monomolecular nature of surface films, but his own experiments led him to the view that orientation effects extended well beyond the range of a single molecular layer. The debate on mono- versus multi-molecular layers in boundary lubrication continues to the present day.[33]

The Hardy and Doubleday (1922a,b) papers are now seen as a classical contribution and many reviews of boundary lubrication start from this point.[34] A quotation from the early part of the opening paragraph will provide the background and perspective on this important development.

> . . . *In what is often called complete lubrication, the kind of lubrication investigated by Towers and Osborne Reynolds, the solid surfaces are completely floated apart by the lubricant. There is, however, another kind of lubrication in which the solid faces are near enough together to influence directly the physical properties of the lubricant. This is the condition found with 'dry' or 'greasy' surfaces. What Osborne Reynolds calls 'boundary conditions' then operate, and the friction depends not only on the lubricant, but also on the chemical nature of the solid boundaries. Boundary lubrication differs so greatly from complete lubrication as to suggest that there is a discontinuity between the two states.* . . .

Hardy's reference to the use of the term 'boundary conditions' by Osborne Reynolds is slightly curious, since it suggests that he was associating his thoughts on the role of molecular films attached to solid surfaces with the need to specify values of velocity or even pressure along the *boundaries* of hydrodynamic films. Reynolds (1886) wrote, '. . . The fluid is subject to boundary conditions as regards pressure and velocity . . .' and he proceeded to adopt the condition of no-slip, namely, '. . . At the lubricated surfaces the fluid has the velocity of these surfaces. . . .'

The condition of no-slip can now be given an explanation in terms of physical adsorption, or molecular interactions between the fluid and the bounding solids of the van der Waals type, and in this sense the process

---

33. See Allen, C. M. and Dragulis, E. (1969) pp. 363–84.
34. See Bowden, F. P. and Tabor, D. (1950); Godfrey, D. (1965) pp. 283–306; Rowe, G. W. (1966) pp. 100–11; and Campbell (1969a) pp. 87–117.

is just as important to fluid-film as it is to boundary lubrication. However, it seems more likely that Hardy was really referring to that section of Reynolds' paper in which the importance of an adequate amount of lubricant ascertained in Beauchamp Tower's experiments was discussed. Reynolds wrote '... in the case of the oil bath the film of oil might be sufficiently thick for the unknown boundary action to disappear, in which case the results would be deducible from the equations of hydrodynamics'. Perhaps Hardy had these 'unknown boundary actions' in mind when he used the term 'boundary conditions'.

An impressive feature of Hardy's studies was the simplicity of his apparatus. In some cases a watch-glass loaded with lead sliding on a sheet of glass was adequate. The experiments simply involved the measurement of friction in the presence of various quantities of a range of lubricants for three materials: glass, steel and bismuth. The lubricants were paraffins, their related acids and alcohols. Great care was taken with the preparations for the experiments, attention being focused upon the importance of cleanliness.

Hardy and Doubleday found that the coefficient of friction gradually fell from the value corresponding to clean, dry surfaces as the concentration of lubricant increased. The effect, shown in Fig. 10.25, was similar to that observed by Lord Rayleigh (1899) in relation to the surface tension of films on water. The limiting values reached at the highest concentrations shown in Fig. 10.25 were identical to those measured when the faces of the solid were in equilibrium with saturated vapours of the lubricant. Furthermore, it was noted that these limiting values of friction were identical for saturated vapour, flooded and what was termed *primary-film*

Fig. 10.25
The influence of concentration of ethyl alcohol vapour upon the coefficient of friction of glass and steel (after Hardy and Doubleday, 1922b). (Ordinate (μ). Abscissa: gram-molecules of C_2H_5OH per litre of air.)

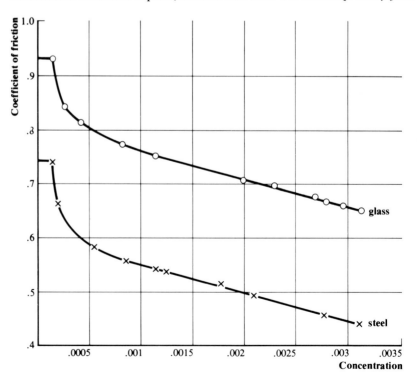

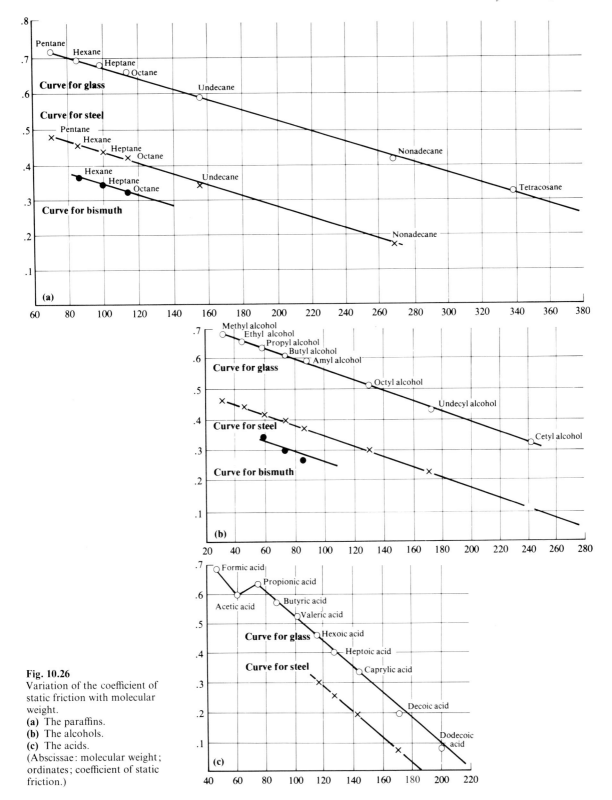

Fig. 10.26
Variation of the coefficient of static friction with molecular weight.
(a) The paraffins.
(b) The alcohols.
(c) The acids.
(Abscissae: molecular weight; ordinates; coefficient of static friction.)

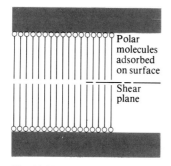

Polar
molecules
adsorbed
on surface

Shear
plane

Fig. 10.27
Diagrammatic representation of
Sir William Bate Hardy's concept
of boundary lubricating films.

conditions. The latter represented the situation achieved, often after waiting for an hour or so, by the spreading of an invisible *primary* film of lubricant over the surface of the solid. It was the measurement of the coefficients of friction corresponding to these fully developed primary films which formed the basis of Hardy and Doubleday's work.

Typical findings, showing the coefficients of friction of these primary films for glass, steel and bismuth plotted against the molecular weight of the paraffins, alcohols and acids are shown in Fig. 10.26. The outstanding feature of these results is that they can be well represented by linear relationships of the form,

$$\mu = b - aM, \qquad \qquad \dots [10.26]$$

where μ is the coefficient of (boundary) friction, M the molecular weight of the lubricant, a a constant independent of the nature of the solid but dependent upon the chemical composition of the lubricant and b a constant dependent upon the nature of both the solid and the lubricant.

These results and subsequent understanding of the phenomenon explained for the first time how the friction between sliding surfaces could be influenced by the properties of both the lubricant and the solid. This was, of course, quite different from fluid-film lubrication, which was dictated by the properties of the lubricant alone. The laws of friction for boundary lubrication are essentially the same as those for dry solids, but with a reduced value of the coefficient of friction.

Hardy summarized the experiences of many investigators of friction when he wrote '. . . The relations disclosed by the experiments are of surprising simplicity, whilst their interpretation is difficult.' He did, however, offer a qualitative picture of the way in which molecular weight might influence lubrication by suggesting that the molecules of the lubricant were orientated by the fields of attraction of the solids so that their long axes were at right angles to the solid surfaces. A given tangential force applied in the surfaces of the solids would then be expected to have greater influence on the plane of slip as the molecular weight and hence the length of the carbon chain increased. He recognized that such a concept, depicted in Fig. 10.27, was not entirely satisfactory owing to the simplified representation of the molecular chains, and further advanced the view that the atomic structure might experience rearrangement or crystallization. Before the end of the period under review Hardy and Doubleday (1922*a,b*, 1923) were to further elucidate the phenomenon of boundary lubrication.

The major concepts of both *fluid-film* and *boundary* lubrication were thus established by scientific inquiry during the period of seventy-five years covered by this chapter. Bearing designers were afforded a full account of the advantages of fluid-film lubrication early in the period, but a cautionary note and illuminating description of another and indeed limiting form of lubrication known as boundary was issued by Sir William Bate Hardy close at the end of the period. Between these two limiting modes of lubrication is a region in which both boundary and fluid-film lubrication mechanisms contribute to the tribological performance of many machine elements – a third regime now known as *mixed lubrication*.

10.7   Plain bearings

It is reasonable to inquire about the impact of the studies outlined in Section 10.4 upon the design and construction of plain journal and thrust bearings in the period under review. The work on boundary lubrication described in Section 10.6 came too late to have a direct impact on the formulation of lubricants in the period, but it inspired the mid-twentieth century activity on additives. It is quite evident that many recent advances in plain-bearing design are based upon firm foundations dating back to the 1880s, but there are also notable examples of the direct application of the new-found knowledge prior to 1925.

In the period between 1850 and the mid-1880s the development of plain bearings and lubrication systems was entirely empirical. A fair indication of the unsatisfactory nature of bearings and lubrication at the beginning of the period is revealed by the opening words of the paper presented to the Institution of Mechanical Engineers by Mr Paul R. Hodge of London (Hodge, 1852). 'No part of the machinery of a railway requires constant lubrication more than the axle journals of locomotives, tenders, and carriages; as the heating of one journal in the whole train is sufficient to produce the most serious results, not only by delaying the traffic, but also by endangering the lives of the passengers in the train.'

Mr Hodge's solution to the problem, together with the recommendations of Lea (1853) and Adams (1853) have already been mentioned in Section 10.4. Most of the proposals involved the use of alternative lubricants or lubricant supply systems, but the inventors were working in the dark, quite unaware of the basic concepts of fluid-film lubrication to be revealed by Osborne Reynolds in 1886. But solutions to bearing problems were not always sought in the use of alternative lubricants or lubrication systems. It was in the 1850s that wooden bearings, and that remarkable, naturally occurring bearing material, lignum vitae, achieved some prominence.

Lignum-vitae

This is the common name given to the *Guaiacum*, a genus of trees important for their exceptionally hard, resinous wood, belonging to the family Zygophyllaceae.[35] The lignum-vitae tree is a native of the West Indies and the northern coast of South America. It is found at altitudes up to about 2,600 ft (800 m) and attains a height of 20–30 ft (6–9 m). The colour of the heartwood varies from a dark greenish brown to almost black, while the sapwood is thin, pale and yellow. The wood has a fine, even texture and an oily feel due to resin which accounts for about 25 per cent of its weight. The timber is heavier than water, with a specific gravity normally in the range 1·1–1·3. It is hard, difficult to split and one of the most durable woods in the world.

The wood had been used in the Caribbean for various purposes, but it was introduced into Europe for medicinal purposes by the Spaniards in 1508. It acquired a strong reputation as a cure for many vile ills such

---

35. See Record, S. J. and Mell, C. D. (1924); Record, S. J. and Hess, R. W. (1943).

as syphilis, and in 1517, when it had reached Germany, it was thought to '. . . have cured French pockes, and also helpeth the gout in the feete, the stoone, the palsey, lepree, dropsy, fallynge euyll (epilepsy) and other diseases'. In Germany the word is still known as *pockholz*.[36] The reputation of *Guaiacum* and the name lignum-vitae (wood of life) were well established in Europe early in the sixteenth century, but it was first mentioned in the *London Pharmacopoeia* in 1677. The medicinal role of lignum-vitae diminished rapidly after the publication of a book questioning its efficacy by John Pearson, a Fellow of the Royal Society, Senior Surgeon of the Lock Hospital and Asylum and Surgeon of the Public Dispensary (Pearson, 1807).

In 1717 the clockmaker John Harrison used lignum-vitae roller pinions which '. . . move so freely as never to need any Oyl.'[37] The wood also found application in furniture, mallets, spools and pulleys, castor wheels, woods for the game of bowls, chisel blocks and small bearing bushes. Common features of the requirements for most of these applications were durability and a resistance to splitting and wear. In light industry, the self-lubricating property of the wood attracted attention, and it was used in situations where conventional lubricants might affect the product, or where conventional lubrication systems would be inconvenient. Such applications have included the use of rollers and thrust collars in the food industry, bearings in pumps, tanks and agitators and rollers for conveyors in the munitions industry.

In heavy industry the major applications were in shipbuilding and rolling-mills. In the latter case it was found that correctly cut lignum-vitae lubricated by water sprays had marked advantages over both plain and rolling-element roll-neck bearings lubricated by oil or grease. An article published in *The Metallurgist* 1932 claimed a 20 per cent reduction in power loss and a 15 fold increase in the life of roll-neck bearings when lignum-vitae replaced white metal. Hobson (1955) has given details of the application and it appears that mean loads of 2,000 lbf/in^2 (13·8 MN/m^2) were supported at surface speeds up to 400 ft/min (2 m/s) by bearings made from strips of lignum-vitae in which the rubbing face was formed by the end grain. In their book, *Timbers of the New World*, Record and Hess (1943) noted that lignum-vitae was being used in increasing amounts to replace brass and babbitt metal bearings in roller-mills and pumps. It was claimed that the initial cost of the wood was less than that of the metal, the life several times longer and lubrication unnecessary.

By far the most important tribological application of lignum-vitae was its underwater use on steamships. Mr John Penn of Greenwich, London did much to promote this application through his papers to the Institution of Mechanical Engineers (Penn, 1856, 1858). The first paper, read in Birmingham on Wednesday 30 January 1856 by Mr F. P. Smith, owing to Mr Penn's unavoidable absence due to illness, described the

---

36. See the *Sunday Times*, 28.2.1971.
37. See Lloyd, H. A. (1953) and Ch. 8.

results of a preliminary study of the utility of wood bearings for both merchant and naval vessels. Severe problems had been encountered with corrosion of the wrought-iron screw shaft and subsequent injury to the brass bearings in both wooden and iron ships. The corrosion was particularly severe in the former, which were clad with copper, as a result of galvanic action in the sea. A solution to the corrosion problem was sought by casing the iron shaft in a brass tube, but the subsequent wear of brass journals on brass bearings was naturally tremendous. In some cases the wear amounted to a penetration of about 1 in (25 mm) in a few months.

Penn (1856) described the basic construction of wooden bearings designed to overcome these problems. The brass bush had longitudinal dovetailed grooves machined within it to receive strips of wood as shown in Fig. 10.28(a). The strips were normally about $2\frac{1}{2}$ in (63 mm) wide, with $\frac{3}{4}$ in (19 mm) spaces between them, the initial bearing surface standing proud of the brass by about $\frac{1}{4}$ in (6 mm). By the time John Penn presented his first paper to the Institution of Mechanical Engineers a large number of merchant ships and 200 of the latest ships built for the British Navy had been fitted with wood bearings.

The transport *Himalaya* of 3,500 tons (35 MN), 700 hp (522 kW) engines and a screw weighing 11 tons (110 kN) had experienced severe wear of its initial metal bearings. When wood bearings were fitted, the vessel went on to complete 20,000 miles (32,187 km) in ten months '. . . without exhibiting the slightest signs of wear'. The propeller-shaft diameter was 18 in (0.46 m) and the length of the initial brass bearing and the replacement strips of wood was 4 ft (1.2 m). Similar success was reported for the 200 hp (149.1 kW) steam sloop *Malacca*, when the brass bearings were replaced by lignum-vitae.

Fig. 10.28
John Penn's investigation of wood bearings (Penn, 1856, 1858).
(a) Transverse section of shaft and bearing showing dovetailed inserts of lignum-vitae in the brass.
(b) Testing machine for submerged wood bearings.

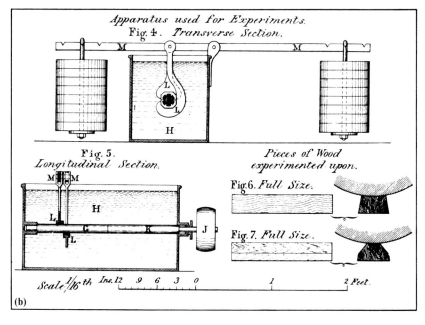

(a)

(b)

Penn found that the unit loading could be increased by a factor of about ten above the nominal 60 lbf/in^2 (0.41 MN/m^2) adopted for brass bearings, working pressures of about 2,000 lbf/in^2 (13.8 MN/m^2) being acceptable for lignum-vitae. Penn, who was a well-known marine engine builder, constructed in his Greenwich works the bearing testing machine shown in Fig. 10.28(*b*). The shaft or axle (g) was driven by a pulley (J) at 700 rpm and various test bearings (L) were loaded by dead weights on the lever (M). The bearings, of diameter 1½ in (38.1 mm) and length 2 in (50.8 mm), were submerged in a bath of salt water and the temperature of the bath was recorded. Penn emphasized the role of water as a coolant, '... The two rubbing surfaces appear in reality to run without any lubricating material between them, and the water acts merely as a conductor to carry off the heat as rapidly as it is produced.'

At the end of his paper Penn (1856) introduced a subject which he was to amplify later (Penn, 1858). It was not only the stern-tube bearings, but also the thrust bearings which transmitted the propelling effect of the screw to the hull, which suffered excessive wear. Such bearings consisted of a number of collars on the shaft which ran in corresponding grooves in the brasses as shown in Fig. 10.29(*a*). The transport ship the *Himalaya* mentioned earlier suffered longitudinal wear of about ¾ in (18 mm) on the brasses before an emergency repair became necessary at sea. The engineer produced a set of sawn half-rings of lignum-vitae and slipped them into the recesses created by wear of the brasses. No perceptible wear was evident after the homeward journey and the vessel put to sea again with no disturbance to the oil-lubricated lignum-vitae rings. In his paper Penn (1858) was able to report the successful replacement of multiple-collar thrust bearings by a single brass thrust ring working against end-grained hardwood bearing pads as shown in Fig. 10.29(*b*). This was indeed a remarkable demonstration of the value of wooden bearings in ships. It was a development bettered only by the invention of the tilting-pad thrust bearing almost half a century later.

Penn tested various timbers including softwoods like poplar, box, snakewood, cam-wood, elm and lignum-vitae. He found that end-grain orientation generally gave the greatest resistance to wear, but doubted whether bearings made of wood could ever find application in rolling-mills because of the enormous pressure occurring momentarily as the strip or bar went through the rolls. Lignum-vitae was also used in the pintle bearings of ships' rudders as described by Economou (1973).[38]

John Penn was a remarkable man and an outstanding engineer. He was the son of a Somerset millwright who had moved to London to open an engineering works in Greenwich in 1800. His father's background could well have influenced his trials of wooden bearings, although the work was conducted well after 1843 when his father died. The firm started building marine engines in 1825 and Penn lived from 1805 to 1878. He was a Vice-President of the Institution of Mechanical Engineers for seven years and

---

38. I have been fortunate to have access to a full survey of the timber lignum-vitae prepared by Philip Economou as part of a project for the degree of M.Sc. in tribology in the University of Leeds.

Fig. 10.29
End thrust-bearings for screw propeller shafts.
(a) Original multiple-collar thrust bearing showing excessive wear on the brass.
(b) John Penn's single plate bearing showing lignum-vitae thrust pads on both external (flooded) and internal (water-bath-lubricated) arrangements (Penn, 1858).

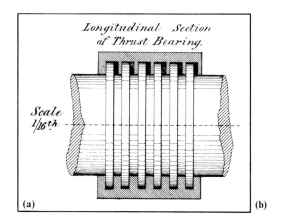

(a) (b)

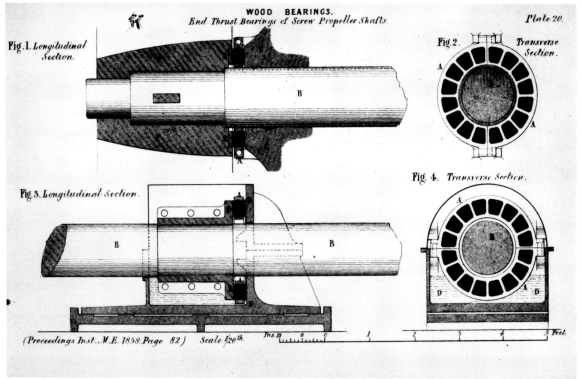

President for the 1858–59 and 1867–68 sessions. As President of the Institution he was also Chairman of the meeting at which his second paper was presented in 1858. A useful biography of Penn can be found in Parsons (1947).

This naturally occurring material had quite a long run in engineering in general and stern-tube bearings in particular. In 1871 some 1,500 tons were shipped from Santo Domingo to Europe alone, with London and Hamburg being the most important ports of entry. The timber was actively used until the start of the Second World War, and it has been estimated that between 1939 and 1945 no fewer than 1,000 Liberty and Victory ships were fitted with lignum-vitae bearings. It was the 1960s when the size of new ships and the loadings on stern-tube bearings increased enormously, sealing techniques and bearing design procedures improved and a new range of polymeric materials became available that the use of lignum-vitae declined. It is also interesting to note the long gap in studies of the tribological properties of the material since the days of John Penn. In recent times McLaren and Tabor (1961) investigated the frictional properties of the material, while Dowson and Economou (1977) studied its wear characteristics.

Vulcanite, glass and papier mâché

Metallic bearings lined with low-friction materials increasingly dominated the plain bearing field towards the end of the nineteenth century. However, lignum-vitae and other woods were not the only non-metallic bearing materials proposed for industrial duty.

On 12 November 1855 provisional patent protection was afforded to Mr J. H. Johnson for 'Vulcanite bearings and linings'. In his application he stated that, '. . . A suitable substance, such as oyster shells, after being calcined and powdered, is added to a mixture of india-rubber and sulphur and the mixture is then vulvanized. The bearings may be made of metal covered with the rubber composition.'

There also seemed to be a great fascination with glass as a bearing material for lightly loaded bearings. Typical applications which won provisional patent protection included a proposal from Mr J. S. Hendy in 1857 for glass tubes to be used as bearings for the moving parts of chimney and ventilating cowls, and another from Mr W. Tice in 1862 for tubes of glass, porcelain or like materials to be used as bearings for all gas regulators and like instruments.

For heavier duty Mr J. Green proposed, and in 1856 was awarded patent cover for, a glass bearing in which molten glass bearings were cast into metal frames.

Attempts to form suitable self-lubricating bearings for the lace-making industry included a proposal by Mr W. Ashton in 11 July 1862 under the heading 'Papier-mâché bearings' that, '. . . Tubes, bushes . . . and other frictional surfaces for the spindles and other parts of braiding and spinning machinery are made of a mixture of paper pulp, plumbago or lubricating material, and gum. Steatite or French chalk, or other earthy or metallic substances, may replace the plumbago.'

In general, however, the only serious contender to the metallic plain bearing in this period was lignum-vitae, and we shall see that by the 1920s white-metal-lined plain journal and thrust bearings were being produced in vast quantities by specialist firms.

The wear of brasses

It will be evident from the quantity and diversity of the discourse on the background to the subject of this chapter that excessive friction in machinery provided the major stimulus for scientific studies of lubrication. The wear of journal bearings was a nuisance, but excessive friction and the development of *hot-boxes* was viewed with even greater concern. The main objective was to deliver adequate power to the driving wheels of locomotives, the propellers of ships or the spindles of factory equipment, and this necessarily focused attention on the friction losses encountered in machines. The practitioners of the day were more concerned with the debilitating effect of friction than with the insidious and destructive action of wear. It was only in the middle of the twentieth century, when the importance of reliability and life of machinery and the economic significance of excessive wear were fully recognized, that the emphasis moved from studies of friction and lubrication to the question of wear.

There was, however, one observation on wear by Goodman (1887) which is of interest not only because it represented one of the few direct commentaries on wear, but because it interlocked with the developing concepts of lubrication.

Goodman noted that when the brass, which by now was a well-established synonym for bearing, was bored to a diameter considerably greater than that of the shaft such that line contact prevailed under static conditions, rotation caused the brass to creep round the periphery of the shaft and to develop contact on the 'on' side of the bearing. This would cause wear to take place in the arc O–B shown in Fig. 10.17(a). With a well-fitted brass the opposite took place and wear was observed on the 'off' side between O and A. The explanation is, of course, to be found in the effect of the relative diameters upon the ability of the bearing to develop a satisfactory fluid film: in the former case dry or mixed friction yields surface tractions which cause the shaft to sit in equilibrium towards the 'on' side of the bearing, while in the latter case the formation of Osborne Reynolds' physical wedge and the resulting equilibrium of forces requires the line of minimum film thickness to be located towards the 'off' side of the bearing.

These somewhat strange effects were described by Goodman as *peculiar*, but he recognized the significance of bearing geometry and the role of different surface tractions in determining the location of the line of minimum film thickness and hence the region of wear. He also investigated the role of oil grooves and condemned the current practice of cutting axial grooves along the crown of such bearings. This wear pattern was hard enough to explain when the theory of fluid-film lubrication was new and barely appreciated, but I remember well the effect of contrary conditioning in an age when we tend to think that hydrodynamics always prevails. I viewed the wear of bearings on a large but then disused water-wheel in a mill near my home in Adel and almost persuaded myself that water flowed uphill by automatically interpreting the results in terms of fluid-film lubrication. All became clear when I realized that the lubrication had been far from perfect and that wear had inevitably taken place on the 'on' side of the bearing.

Tilting-pad bearings

Plain bearings have such simple forms that the science and technology upon which they are based is rarely appreciated. Yet their very simplicity can represent the pinnacle of ingenious design, and no better justification for this statement can be found than the *pivoted-shoe* or *tilting-pad bearing*. The story of tilting-pad bearing development excites students of both machine design and tribology.

Such bearings, known as *Michell* and *Kingsbury* in Europe and America, respectively, are remarkable for four reasons:

1. They were the direct results of engineering applications of Osborne Reynolds' theory of fluid-film lubrication.
2. They were conceived and developed almost simultaneously and quite independently by Michell in Australia and Kingsbury in America.
3. Problems with hydraulic machinery promoted the inventions in both countries, but marine applications soon followed.
4. They represented a plain-bearing development which has probably never been excelled.

Reference is made to the theoretical analysis of thrust bearings by Michell in Section 10.4 and details of the inventors' lives are recorded in the Appendix Sections A.16 and A.19, but it is the historical background to the bearings themselves which concerns us here.

Towards the end of the 1880s Kingsbury worked as a student under Dr Robert H. Thurston[39] at the Sibley College of Cornell University. He was given a project on half-journal bearings for the Pennsylvania Railroad Company and hence developed an interest in lubrication. Kingsbury's skill yielded such unusually low coefficients of friction for well-fitted brasses that neither he nor Thurston could explain the observations. The question remained in Kingsbury's mind while he was engaged in other pursuits, but in 1891 he returned to the academic world as Professor of Mechanical Engineering in the New Hampshire College of Agriculture and the Mechanic Arts, Hanover, and was again confronted by the intriguing behaviour of lubricating films. While making a torsion–compression machine for the laboratories of the new college premises at Durham, he accidentally found that the accurately made piston would float on a film of air when spun rapidly in the cylinder. This discovery caused him to build an air-lubricated journal-bearing demonstration model which attracted much attention at the November 1896 Washington meeting of the Association of Agricultural Colleges. It was only at a subsequent demonstration of the apparatus to the engineers of the Bureau of Steam Engineering at the Navy Department that Reynolds' (1886) paper was shown to Kingsbury. He subsequently obtained on loan from the Boston Library the appropriate volume of the *Philosophical Transactions* and reported more fully on his air-bearing experiments in Kingsbury (1897). It was during his detailed examination of Reynolds' paper that the idea of the tilting-pad bearing came to Kingsbury (1950) and the story is of such importance and simplicity that it is best told in his own words.

In reading the section of Reynolds' paper dealing with flat surfaces, it occurred to me that here was a possible solution of the troublesome problem of thrust bearings. Reynolds showed that if an extensive flat surface rubbed over a flat surface slightly inclined thereto, oil being present, there would be a pressure between the surfaces distributed about as sketched in [see Figure 10.30(a)]

---

39. See Appendix, Sect. A.9 for biographical details.

Fig. 10.30
Sketches of tilting-pad bearings by Kingsbury and Michell.

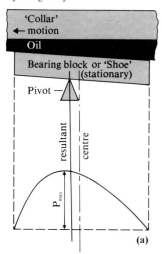

(a) Kingsbury's sketch of pressure distribution in an offset shoe.

(b)

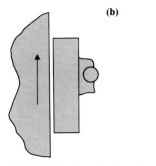

(b) Michell's thrust bearing from British Patent No. 875 of 1905.

The maximum pressure would occur somewhat beyond the center of the bearing block in the direction of motion and the resultant would be between that maximum and the center line of the block. It occurred to me that if the block were supported from below on a pivot, at about the theoretical center of pressure, the oil pressures would automatically take the theoretical form, with a resulting small bearing friction and absence of wear of the metal parts, and that in this way a thrust bearing could be made, with several such blocks set around in a circle and with proper arrangements for lubrication.

In 1898 Kingsbury and his respected mechanic John Brown built such a bearing with the blocks pivoted on the centre line on small spherical bosses. The tests were entirely successful, as can be judged from the fact that unit loads of 4,000 lbf/in² (27·6 MN/m²) were supported at 285 rpm, compared with typical values of 40–50 lbf/in² (0·28–0·34 MN/m²) in the conventional bearings for ships propeller shafts and 400–500 lbf/in² (2·76–3·45 MN/m²) for the first long vertical shafts at Niagara Falls.

From June 1903 Kingsbury worked for the Westinghouse Electric and Manufacturing Company in East Pittsburgh. When the company made a large electric motor with a vertical shaft in 1904 they introduced, at Kingsbury's suggestion, one of his bearings. The bearing ran hot during a shop test carried out in Kingsbury's absence and was replaced by a ball-bearing. In 1909 the Westinghouse Company again built a large vertical motor provided with a ball thrust bearing for the Minidoka irrigation project in Idaho. Kingsbury thought that he detected signs of incipient failure and suggested that one of his bearings be used in its place. Perhaps it was the 1904 experience that caused H. P. Davies, vice-president and manager of engineering, to insist that if this was done Kingsbury should pay for all the parts! The designer's confidence swept aside such minor frustrations, the bearing was made from a few notes and freehand sketches within a week and it operated perfectly under test. The company still lacked confidence in such a revolutionary device and the machine was finally equipped with a roller-bearing. A pleasing postscript to the tale is that the US Army engineers operating the Idaho plant applied to Kingsbury fourteen years later for one of his bearings to replace the roller-bearing. He gladly obliged.

In 1907 Kingsbury applied for a patent. The initial application was rejected by the Patent Office on the grounds that Michell had patented the idea in England in 1904, but affidavits by Baker and Putney, former students at the New Hampshire College, and Graffam and Traill who graduated under Kingsbury at Worcester Polytechnic Institute in 1900 established beyond doubt that Kingsbury had built and operated tilting-pad bearings before the Michell patent award in England. The now famous patent No. 947242 was finally awarded three years later on 25 January (Kingsbury, 1910). This placed Kingsbury's tilting-pad bearings firmly on the map and for a number of years the Westinghouse Machine Company made the bearings for customers to Kingsbury's designs. A most instructive paper by Howarth (1919) outlined some of the successful and spectacular applications which followed the 1910 patent. In 1911 these included Kingsbury bearings to take the bevel-pinion thrust on the horizontal drive shaft of a billet-mill and the large vertical spindle loads on a plate-glass grinding and polishing machine. Kingsbury thrust bearings were also fitted to vertical and horizontal hydroelectric machinery, horizontal steam turbines and centrifugal pumps. One of the most memorable and early successes was the bearings built in 1912 for the Pennsylvania Water and Power Company's hydroelectric plant at McCall's Ferry, now called Holtwood, on the Susquehanna river. The bearing replaced a roller-bearing and carried a load of 405,000 lbf (1·8 MN) on a 48 in (1·2 m) diameter thrust collar at 94 rpm (1·57 Hz). At a celebration dinner attended twenty-five years later in July 1937 by Professor and Mrs Kingsbury it was found that the maximum wear on the blocks occurred near the trailing edges and was between 0·005 and 0·006 in (0·13–0·15 mm). The plant engineers found by extrapolation that the probable life of the bearing was 1,300–1,700 years.

Further successful applications included the bearings for hydroelectric generators at Koekuk on the Mississippi and Cedar Rapids on the St Lawrence. In 1917 the bearings were first fitted to ships of the US Navy and rapidly became standard items of equipment for major ships down to and including the four-stack destroyers and harbour tugs. A notable demonstration of the advantages of tilting-pad bearings over current thrust collars on propeller shafts was reported by Howarth (1919). The German ship the *Prince Eitel Friedrich* had been interned in the United States and the bearing experiments were carried out at Hoboken by the Navy Department. The twin screws were driven by reciprocating engines and the thrust bearings had twenty-one collars, fourteen for the thrust when moving ahead and the other seven for the astern thrust. When one of the collars was removed and replaced by a Kingsbury bearing the single collar took the full thrust in either direction! In the postwar years Kingsbury bearings found applications in passenger and cargo ships, centrifugal pumps and steam turbines.

The Westinghouse Machine Company soon found that it could not keep up with demand and Kingsbury started to manufacture on his own account about 1921. In 1924 the Kingsbury Machine Works was founded in Philadelphia, Pennsylvania, USA.

Kingsbury's invention of the tilting-pad bearing was typical of the majority of engineering inventions. All the ingredients are evident in the story – a chance observation, awareness of a current problem, consideration of available theory, the birth of an idea, early success tempered by later frustrations, persistence, confidence and great engineering skill leading to a sound device. The approach was largely empirical, whereas Michell followed a more scientific and theoretical route to a very similar conclusion.

The motivation for Michell's studies of bearings came from problems encountered with large thrust loads in hydraulic machinery. He was a consultant to the Melbourne company which made and installed the pumping machinery for the Murray Valley Irrigation Scheme in Australia and to the Tasmanian company responsible for one of the first Australian hydroelectric schemes.

Michell's fundamental investigations of methods of supporting the large thrust loads in such machinery have been reported in Section 10.4.

His work was apparently effected in the short period between 1902 and 1904 immediately prior to the publication of his now classical paper in Germany (Michell, 1905a). Almost forty years later it was still being described[40] as the only really important extension of Reynolds' theory, yet effected. Michell's theory demonstrated that the resultant hydrodynamic force in a plane inclined slider bearing was always located between the pad centre and the trailing edge, and that maximum load-carrying capacity was achieved at a particular ratio of inlet to outlet film thickness. Furthermore, the centre of pressure was also determined by this ratio, and Michell quickly recognized that a pad pivoted about a suitable position would adjust its tilt as the load and minimum film thickness changed to provide optimum hydrodynamic load support at all times. The tilting-pad bearing conceived by Michell was firmly based upon sound scientific principles and was a masterly engineering development stemming from Osborne Reynolds' theory of lubrication.

Michell was quick to recognize the engineering simplicity yet great potential of such a bearing, and he was awarded British and Australian patents in England and Australia on 16 January 1905. The sketch of his bearing which would establish its own inclination by means of such a pivot which appeared in British Patent No. 875 of 1905 is shown in Fig. 10.30(*b*). The initial and final specifications were issued on 16 January and 20 June, respectively, and the complete specification was accepted and published on 20 July 1905.

The first successful tilting-pad bearing built to Michell's instructions was constructed by George Weymouth (Proprietary) Ltd of Melbourne for a centrifugal pump at Cohuna on the Murray river, Victoria, in 1907. Michell entered a partnership with Mr H. T. Newbigin for the manufacture of tilting-pad bearings and by 1913 their advantages were recognized by shipbuilders. The first merchant ship thrust blocks[41] were fitted to two ferries plying between Buenos Aires and Montevideo in 1913. The engines generated 5,000 hp (3·73 MW) and the mean pressure on the pads or blocks was 300 lbf/in^2 (2·07 MN/m^2). The first English ship to be fitted with the bearing was the cross-Channel steamer the *Paris*, but many naval vessels were similarly equipped during the First World War.

Marine engineers' interest in pivoted pad bearings expanded rapidly during the First World War in Europe and the United States. This was not attributable solely to the increased volume of shipping but to the need to accommodate greater propulsive effort for larger or faster ships. Ironically, it is said that it was the use by Krupps in Germany, who had declined to take licences from Michell, which stimulated interest elsewhere. Kingsbury (1950) also recalled that it was the use of Michell bearings by the British Navy that caused the US Navy to adopt his bearing. The steam turbine played a significant role in this development. It was introduced by Parsons in 1884, and by the turn of the century had been

---

40. See Appendix, Sect. A.19.
41. I am grateful to Mr Advani of Vickers Limited Engineering Group (Michell Bearings) for this information.

adopted for certain naval applications. Many tilting-pad bearings were introduced into naval vessels during the First World War, the first being HMS *Caledonian* and HMS *Mackay* in 1914, and when Michell successfully applied for an extension of his patent in 1919 it was stated that naval applications alone had saved the country coal valued at £500,000 in 1918. A comprehensive and valuable account of Michell as a man and an inventor can be found in a biographical memoir prepared by Cherry (1962) for the Royal Society and in the special publication of the Institution of Mechanical Engineers (1943) issued on the occasion of the award of the James Watt International Medal.

In the post-war years the use of tilting-pad bearings in large ships became almost universal and a company, Michell Bearings Ltd, was formed in Newcastle on the banks of the river Tyne by the inventor in 1921 to meet the demand. Both Michell and Kingsbury started their own manufacturing companies early in the 1920s, and since there is always much interest in the simultaneous and entirely independent developments of tilting-pad bearings by these exceptional engineers it is convenient and instructive to compare the detailed chronology of events shown in Table 10.3.

Table 10.3
Chronology of tilting-pad bearing invention and development by Albert Kingsbury and Anthony George Maldon Michell

	Kingsbury		Michell
1898	Kingsbury and Brown built and tested a tilting-pad bearing in which the blocks were pivoted on spherical boses. (New Hampshire College, Durham)		
		1902	Michell completed his basic analysis of
		1904	plane inclined slider bearings of finite width
1904	Westinghouse Electric Company built a tilting-pad bearing to Kingsbury's design for a large electric motor with vertical shaft. The bearing ran hot in shop tests and the machine was delivered to the customer with a ball-bearing		
		1905	Publication of 'The lubrication of plane surfaces' in *Z. Math. Phys.*
		1905	Michell bearing patented in England (No. 875) and Australia
1907	Kingsbury unsuccessful with US patent application	1907	Tilting-pad bearing made to Michell's design by George Weymouth (Pty) Ltd, Melbourne and fitted to centrifugal pump at Cohuna on Murray river, Victoria
1909	Westinghouse Electric Company again built a tilting-pad bearing to Kingsbury's design. Shop tests were entirely satisfactory but the company decided to release the machine for the Minidoka irrigation project with roller-bearings.		
1910	Kingsbury bearing awarded US Patent No. 947242		

Table 10.3 (*Continued*)

	Kingsbury		Michell
1912	Roller-bearing at the Pennsylvania Water and Power Company's hydroelectric plant at McCall's Ferry (Holtwood) on the Susquehanna river successfully replaced by tilting-pad bearing built for Kingsbury by the Westinghouse Machine Company		
		1913	First merchant ships fitted with Michell thrust blocks–ferries plying between Buenos Aires and Montevideo
		1914	First use of tilting-pad bearings by the Royal Navy
1917	US Navy used Kingsbury bearings on propeller shafts		
		1919	Michell's British patent extended
		1921	Michell Bearings Ltd, formed in Newcastle upon Tyne, England
1924	Kingsbury Machine Works founded on Tackawanna Street, Frankford, Philadelphia, USA		

It should not be thought from this comparative chronology that there was any dispute between the two men as to the propriety of their views of the respective inventions, although Grieve (1944) tries to establish the precedence of Michell. In fact there appeared to be mutual respect between the inventors and in a review paper presented to the Metropolitan Section of ASME in New York on 17 May 1929 Michell (1929) generously recorded the following statement.

> *... The author avails himself of the opportunity to explain that the pivoted thrust bearings known in America and Europe respectively as the 'Kingsbury' and 'Michell' bearings are in principle the same, and to state that the development of these bearings was effected in its early stages by Professor Kingsbury and himself independently in their respective countries without knowledge of each other's work. Professor Kingsbury's work was commenced a few years earlier, though his first publication on the subject was later than the author's. The author also takes the occasion to note the improper omission of Professor Kingsbury's name in some recent European publications, in which are described not only the principle of the bearing but actual constructions effected by the latter and his associates.*

A number of bearing manufacturing companies and individuals have made models to illustrate the principle of the tilting-pad bearing. One such model made in 1926 by the Kingsbury Company is shown in Fig. 10.31. When the flat annular disc sits at rest on the three pivoted pads it closes an electrical circuit and causes the flashlight bulb to glow. When the disc is spun by hand the pads tilt and the circuit is broken by a thin but coherent lubricating film of air which persists for a very long time and simply and vividly demonstrates the validity of the tilting-pad bearing principle.

Fig. 10.31
Air-lubricated tilting-pad
demonstration model produced by
Albert Kingsbury's company (*c.*
1926).

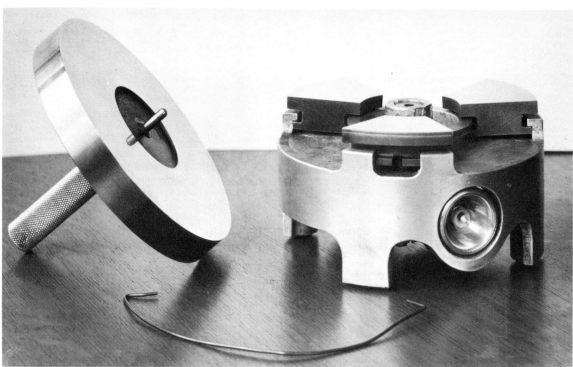

In general terms the improvement in thrust-bearing performance brought about by the replacement of plain thrust collars by tilting-pad bearings was an increase in mean load-carrying capacity from approximately 50 to 500 lbf/in^2 (0·34–3·45 MN/m^2) and a reduction in the coefficient of friction from 0·04 to 0·005 or even 0·002.

Origins of specialist plain bearing manufacturers

It had become established nineteenth-century practice to make plain bearings by keying the soft, tin-based, anti-friction bearing metals into harder casings made of bronze or iron. The term 'babbitt' became synonymous with anti-friction bearing materials of various compositions, even though the original American invention consisted of 89 per cent tin, 9 per cent antimony and 2 per cent copper.

Machine manufacturers invariably made their own bearings, and when excessive wear took place the process of relining bearings was not infrequently carried out by the machine user or the original manufacturer. There was, therefore, a large and growing demand for anti-friction bearing material towards the end of the nineteenth century and this led to the formation of companies who concentrated on this role.

The Glacier Metal Company Limited[42] was founded in 1899 by two Americans for the purpose of manufacturing anti-friction metal. One of the founders soon returned to the USA, but Cuyler Findlay remained in England and gradually developed both the range and quantity of the products. When the First World War broke out the young company was initially called upon to manufacture die-cast parts for hand grenades, but as the war effort developed and industry was required to supply fighting vehicles in ever increasing quantities, the Glacier Company started to manufacture plain bearings. This growth of a specialist plain-bearing manufacturing company was essentially spawned by the rapid development of the internal combustion engine about the time of the First World War, but the process gathered momentum as the motor industry developed in the post-war years.

The significant move to bearing manufacture was complementary to, rather than at the expense of, the manufacture of anti-friction bearing metals. *Findlay's motor metal* became widely known and is still sold on a worldwide basis.

Growth of the Glacier Company was inevitably linked to a rapid development of the general motor industry, but an approach from Morris Motors for the supply of bearings for its new products firmly established a link with the automobile industry which has influenced company development to the present day. Expansion was rapid in the 1920s and an important development was the introduction of steel-backed bearings and power presses for their manufacture in 1929. This was an important move towards automated continuous processing and mass production of the simple yet essential machine element, the modern journal bearing.

42. I am grateful to Mr F. A. Martin and other members of the Glacier Metal Company for kindly providing background information.

While companies like Glacier were developing their potential for producing white-metal anti-friction alloys and enlarging the scope of their manufacturing facilities to make complete plain bearings, two companies were being formed solely for the manufacture of that extraordinary thrust bearing known as the tilting-pad. The background to the invention of these most effective bearings has been presented in Section 10.6, but it is convenient at this stage to comment on the development of appropriate manufacturing facilities in England and the United States of America.

It was in 1921 that Anthony George Malden Michell formed the company Michell Bearings Ltd,[43] in Newcastle upon Tyne. The fact that the company was established in the heart of the shipbuilding industry on the north-east coast was no accident, since the marine industry had done for the tilting-pad bearing what the automobile industry had done for the white-metalled journal bearing. In 1962 the company moved to its existing premises in Scotswood Road, on the banks of the river Tyne and in 1971 it became part of Vickers Limited Engineering Group. Although marine applications, including such famous passenger liners as the *Queen Mary* and *Queen Elizabeth* built for the Cunard Company, still account for a large proportion of the output there has been a steady development of the role of tilting-pad thrust and journal bearings in general industry for turbines, electric motors, pumps, fans, mixers and large crushing equipment. White-metal reclamation and bearing repair also plays an important part in the overall function of the firm.

In the United States Albert Kingsbury had licensed a number of firms to manufacture the tilting-pad bearing which he had invented. He was essentially in business on his own account soon after the successful patent award of 1910, and by 1912 he was signing contracts for the supply of large thrust-bearing units. He was not, however, manufacturing his own bearings at this time, since he had agreements for this purpose with the Westinghouse Electric and Westinghouse Machine Companies. It was only after 1921 when the Westinghouse Machine Company found that it could no longer keep up with his work that he decided to expand the limited workshop facilities which he owned on Cherry Street in Philadelphia.

Kingsbury bought an old factory on Tackawanna Street, Frankford, in the north-eastern part of Philadelphia, installed new and second-hand machinery and started to manufacture tilting-pad bearings on his own account. The Kingsbury Machine Works[44] was incorporated and founded in 1924 and it expanded several times on the original Tackawanna Street site before moving in 1966 to a new site in Drummond Road on the Philadelphia Industrial Park established near the North Philadelphia Airport in 1958. The range of applications of tilting-pad bearings has grown on each side of the Atlantic and now embraces propeller shafts and propulsion turbines in marine situations, steam turbines and hydro-

---

43.   I am grateful to Mr Advani of Vickers Limited Engineering Group (Mitchell Bearings) for background information.
44.   I am grateful to Mr R. S. Gregory of Kingsbury, Inc., for background information.

electric units, compressors and gas expanders, process plant equipment, electric motors and various items of equipment in nuclear power plants on land and at sea.

10.8   The velocipede, rolling-element bearings, surface contact and rolling friction

The modern forms of rolling-element bearings finally emerged during the period 1850 to 1925, although many refinements in design, manufacture and analysis were to occupy the central period of the twentieth century. The history of the period represents more of a gradual evolution than an explosive revelation of concepts and the creation of completely new designs which chart the path of plain-bearing progress and lubrication in the same period, yet the overall results were just as important for industrial progress.

Allan's (1945) painstaking listing of patent applications is a valuable record of rolling-element developments throughout the ages. In the last half of the nineteenth and first quarter of the twentieth century the Patent Office received applications for novel ball- and roller-bearings designed to ease the movement of capstans, turntables, axles and railway rolling-stock axle-boxes, mills and millstones, lifting tackle, footstep bearings, marine propeller shafts and big-ends and gudgeon pins in reciprocating engines. Nor were the associated lubrication systems overlooked.

Not all the applications were accepted and few of the proposals achieved widespread recognition. There were, however, several notable devices which emerged during the period, like the use of hollow balls and rollers in thrust bearings for turntables proposed by W. H. Ward in 1858, which attracted the award of British Patent 2585 of that year. In recent years the clock has turned full circle and much sophisticated analysis and ingenious manufacture has been devoted to the prospects for hollow balls to reduce both centifugal loading and slip in very high-speed gas-turbine bearings.

The first use of the word 'cage' in a patent application to denote the structural element used to separate the loaded rolling elements appeared in British Patent No 2855 awarded to Bailly, Durand, Mesnard and Porier in 1864.

A most valuable record of the state of the art of ball-bearing design and construction at the close of the nineteenth century was reported by Benjamin (1898). He pointed to the rapid extension of the use of steel balls for reducing the friction in bearings and noted that the monthly production of one company alone ranged from 15 million to 20 million balls with diameters ranging from $\frac{1}{16}$ in (1·6 mm) to 4 in (101·6 mm). The article stressed the '. . . almost entire absence of accurate knowledge in regard to the strength and durability of steel balls. . .'. Perhaps the most telling comment appeared at the very end of a closing summary on the principles governing the design and construction of this species of bearing '. . . It is generally better in machine design not to be too radical. Let the other fellow do that.'

The velocipede

While the range and nature of the bearings developed for industrial applications in the latter half of the nineteenth century is of interest, the real impetus for ball-bearing development was provided by that familiar vehicle the bicycle.[45]

The history of two-wheeled, self-propelled vehicles stretches back well beyond the nineteenth century, but it was the *Draisine* invented by the German, Freiherr Drais, in 1818 which started the development of the modern bicycle. Two wheels were attached to a frame and as the rider sat forward on a pad he pushed or kicked on the ground to provide forward movement. The Scotsman K. Macmillan invented the two-wheeled, self-balancing machine with treadles linked to the rear wheel in 1839, and he wisely added a brake. A tricycle with pedals and cranks attached to the front wheel was made and exhibited at the 1862 Exhibition by Messrs Mehew and Chelsea, but it was the Frenchman, Pierre Michaux, who put cranks and pedals directly on the front wheel of a bicycle in the 1860s. He promptly formed a company to manufacture his velocipedes[46] and the bicycle industry thus began in France. It appears that the proposals for cranks on the front wheels came from Monsieur Lallement of the Michaux concern who promptly patented the idea when he arrived in America in 1866. Bicycle manufacture spread to Coventry in England in 1869, but general progress was delayed as long as the rider sat astride a large front wheel of the penny-farthing arrangement designed to provide a maximum distance of travel for one revolution of the crank. The problem was resolved in 1876 when the geared-up chain drive to the rear wheel made it possible for the rider to sit in a better, more stable position between the wheels on a *safety-bicycle*. Better balance, steering and braking were crowned by comfort provided by the invention of the pneumatic tyre in 1888 by J. P. Dunlop,[47] a Belfast veterinary surgeon.

The great technological development known as the velocipede did much to enhance leisure pursuits in the 1880s and 1890s, but it also focused attention on frictional resistance in the bearings of such machines. The early machines employed plain bearings and the first patent for ball-bearings was No. 1485 issued to Mr A. L. Thirion on 16 May 1862. His application argued that, '. . . The direct weight of the body on the axles of railway and ordinary carriages, carts and velocipedes is diminished by the interposition of spheres or cones with or without a bridge.'

The date of the application is more significant than the details of the proposed device, since the former came very early in the history of velocipedes and the latter turned out to be unsuitable. A great deal of experimental work on ball-bearings for bicycle wheel hubs, crank spindles and steering sockets took place in England and France in the fifty years following 1860. A count of the British patent specifications for ball-bearings for velocipedes for the period is interesting as shown by the

---

45.   I am grateful to Mr F. R. Whitt, former President of the Southern Veteran Cycle Club, for interesting discussions on this topic.
46.   Velocipede: *velox-ocis* (swift) + *pes pedis* (foot).
47.   Derry, T. K. and Williams, T. I. (1960), p. 392.

Fig. 10.32
British patent applications for velocipede ball-bearings.

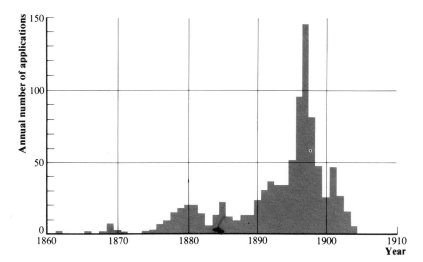

distribution in Fig. 10.32. It is clear that the frenzy of innovation reached its climax at the very end of the nineteenth century.

The efforts were well worth while. Frictional resistance was reduced, maximum speeds increased and bicycles fitted with ball-bearings started to win road races. Whitt (1970) has recalled these events and noted that in 1868 an Englishman by the name of George Moore won the Rouen–Paris race on a Michaux velocipede fitted with ball-bearings. In these machines the bearing friction was probably a fifth or even a tenth of the comparable resistance in the earlier plain-bearing machines.

Landmarks in the history of ball-bearing development for the velocipede included A. C. Cowper's new machine of 1868, F. J. Suriray's velocipede of 1869 which formed the basis of the machine raced by James Moore when he completed 130 km (80·7 miles) in 10 hours 25 minutes, Dan Rudge's British Patent No. 526 of 1878 for a complete cycle wheel hub with cup-and-cone bearings, W. Brown's patents for cup-and-cone bearings sealed to keep out the dust and yet permit oiling in 1879, his Aeolus bearing of 1886 and the Humber Company's double-row ball-bearing for cycles in 1889.

Cheaper forms of Victorian bicycles used plain bearings until the 1890s, but the ball-bearing machine soon dominated the market. By this time the bicycle had achieved great popularity and the ball-bearing undoubtedly reduced the propelling effort required to initiate and maintain motion and lessened the fatigue of the rider. The demand for steel balls of adequate quality and reliability placed new demands upon manufacturing industry. If one ball was larger than others in a bearing it carried a disproportionate load and was liable to fail. The situation was much relieved between 1887 and 1900 when better grinding machines were invented for the manufacture of steel balls.

Materials and manufacture of steel balls

The impetus given to the widespread introduction of ball-bearings by the velocipede was tempered by the difficulty of making metal balls of adequate strength and geometrical perfection. A sizable industry grew up in

England for the manufacture of steel balls for bicycles, with export markets developing on the Continent. When Friedrich Fisher[48] opened his factory for the manufacture of cycles in Germany in 1883, he originally imported steel balls from England, and when he tried to make his own iron balls he experienced great difficulties.

It was undoubtedly the lack of suitable materials that delayed the development of ball-bearings at the end of the nineteenth century. In former times the balls were made of cast iron, as in the case of the Sprowston windmill bearing described in Chapter 9, but the brittleness and lack of homogeneity caused manufacturers to look for alternative metals. It was soon recognized that it was not just the crushing strength of the balls, but also their ability to resist plastic deformation and surface fatigue failure that governed the material requirements. With contact stresses not infrequently in the range 100–300 tons/in^2 (1·54–4·63 GN/m^2), the requirements placed upon ball-bearing materials soon became more exacting than in almost any other engineering endeavour. It is no exaggeration to say that the requirements for special steels and precision manufacturing processes in the ball-bearing industry greatly extended materials science early in the twentieth century.

It is normally necessary to harden rolling-element bearing components to Rockwell 60–65 C to windstand the severe surface contact conditions outlined earlier. The advent of *case-hardening* at the turn of the century considerably extended the scope of ball-bearing steels. In this process the steel components containing about 0·15 per cent carbon are heated in an environment of carbonaceous material for a sufficiently long period to enable additional carbon to be absorbed and to react to form iron carbide in a thin skin or case. Extreme care and skill has to be exercised to achieve homogeneity, and variations in hardness over the surface of the component have a deleterious effect upon bearing life. In due course most manufacturers turned to a steel containing about 1 per cent carbon and 1·5 per cent chromium which exhibited a fine-grained structure, great strength and the ability to be hardened to much greater depths. This *through-hardened* steel, with a hardness in the range Rockwell 62–65 C, is used extensively, but not quite exclusively, in the manufacture of ball-bearings, but case-hardening is widely used for roller-bearings.

The introduction of stainless steel just before the First World War attracted the attention of bearing manufacturers interested in producing rolling-element bearings for corrosive environments. Such steels normally contained very high proportions of chromium, in the range 12–15 per cent, but their fatigue lives were so low that the material had little immediate impact upon rolling-element bearing technology. In many cases the material showed little improvement over bearings made of hard bronze.

The life of ball-bearings was greatly influenced by the surface quality and geometrical accuracy of the balls. The long-established methods of producing balls were casting and turning in a lathe from rod or bar.

---

48.   See Allan, R. K. (1945), p. 17.

Much ingenuity was in evidence in the development of lathe work in which balls of remarkable accuracy could be turned and parted off from a bar with barely a vestige of a machining pip. In due course the sheer demand called for more rapid production techniques and grinding became the established practice. Rough balls in the smaller size ranges were first produced from steel wire or bar in automatic cold-heading presses, while larger balls were made by hot pressing. In either case a thin flash of metal remained on the equator of the ball where the two hemispherical dies met, this being removed by rough grinding or tumbling. Further grinding between cast-iron discs and flat grinding wheels arranged in such a way that the axes of rotation of the balls were continuously varying, supplemented by long periods in rotary tumblers containing such items as abrasives, lubricants, sawdust and strips of soft leather, and various forms of grading and final inspection yielded the precision products which mark the excellence of this particular twentieth-century form of a tribological component whose history can be traced over thousands of years.

By 1890 balls could be made to within ± 0.001 in (25.4 μm) of a nominal size and this limit was halved by 1892. Finer tolerances could be achieved only at prohibitive cost, but the scene was set for the emergence of specialist rolling-element bearing companies to meet the needs not only for the cycling fraternity but machine manufacturers in general and the automobile builders in particular.

The birth of precision ball- and roller-bearing companies

Prior to the end of the nineteenth century the manufacturers of equipment tended to make complete machines, including their own rolling-element bearings. Some firms concentrated on the manufacture of balls, but the majority of these organizations were small in both size and life. On the whole the modern rolling-element bearing industry developed from firms already involved in the manufacture of bearings in substantial quantities for their own products, but the following outlines, which are illustrative rather than comprehensive, show how some of the well-established companies originated.

The taper roller-bearing   Attempts to utilize some form of rolling bearing to reduce axle friction and hence to minimize the *draught* or force required to move four-wheeled vehicles over roads have a long history. Speculation surrounding the Celtic cart hub bearings has been mentioned in Chapter 5, Jacob Rowe's account of his attempts to convince the English of the vast savings to be made by introducing roller-disc bearings on the wheels of carriages and wagons was published in 1734 and Varlo's (1772) prediction for the future of devices similar to the impressive cast-iron ball-bearings fitted to his post-chaise was that they would '. . . doubtless in time . . . become general, as in every mechanical power it will save at least one third the strength, but in heavy weights much more'.

The story was continued and brought to fruition when a German immigrant by the name of Henry Timken, whose portrait is reproduced in Fig. 10.33, became an apprentice in a carriage shop in the United States

Fig. 10.33
Henry Timken–founder of the
Timken Roller-Bearing Axle
Company, St Louis, Missouri,
1898.

in 1830. He later established his own successful carriage-building business and eventually devoted much of his time to the question of axle bearings. In 1898 Timken was awarded an American patent for his taper roller-bearing and in the same year he formed the Timken Roller Bearing Axle Company in St Louis, Missouri. The integral shaft and taper roller-bearing assemblies incorporated numerous advantages for wagons. They greatly reduced both the static and kinetic friction of the vehicle, typically by 25 to 50 per cent, accepted troublesome thrust loads as wagons turned or moved along ruts in bumpy roads and enabled bearing wear to be accommodated by the simple device of tightening the end locating nuts.

If it was the horse-drawn vehicle that nurtured the initial development of the taper roller-bearing, it was the automobile that ensured the rapid growth of the industry in the twentieth century. The company moved to Canton, Ohio, in 1902 to be near the developing automobile industry and in 1909 the Bearing and Axle Divisions separated. Axle manufacture moved to Detroit, but the Timken Roller Bearing Company remained in Canton. A licensing agreement resulted in the manufacture of Timken bearings in Great Britain in 1909 and the company opened its first steel plant in Columbus, Ohio, in 1919.

It is interesting that the need to carry combined radial and thrust loads in road axle bearings promoted the development of a relatively sophisticated form of rolling-element bearing at such an early stage. The cup-and-cone-arrangement of the lightly loaded bicycle ball-bearing could accommodate the gradual yet inevitable wear of the balls over long periods of time, yet when heavy loads had to be carried and roller-bearings were deemed to be necessary in four-wheeled vehicles, it was difficult to see how slackness produced by the progressive wear of a cylindrical roller could be resisted. This problem, coupled with a recognition of the relatively high axial loading on axle bearings, caused Henry Timken to invent his taper roller-bearing.

Present-day taper roller-bearings have four elements: a cone or inner ring which fits on to the shaft, the taper rollers, a cage or retainer to maintain some separation and even spacing of the rollers, and an outer tapered cup to contain the complete bearing in a housing. It is interesting to follow the early steps towards the present-day arrangement. In the first bearing of 1898 no cage was used and the space around the cone was filled with taper rollers each grooved roughly equidistant from the ends as shown in Fig. 10.34(*a*).[49] The grooves fitted over projections on the inner cone to prevent the rollers from skewing. Within a year a cage was introduced to ensure alignment of the roller and small *pintles* or *stub-shafts* at the ends of the tapered rolling elements were located in the retainer rings. The rollers still retained their circumferential grooves, but it is even more interesting to compare the *pintle*-ended roller formation shown in Fig. 10.34(*b*) with the stub-shafted cylindrical bronze and wooden rollers used by the Romans on Laki Nemi 1,850 years earlier and reproduced in Figs 5.12 and 5.13.

The early taper roller-bearing axles found application in fine carriages, light road wagons, buggies, surreys, delivery wagons and trucks as indicated by company catalogues issued before the First World War.[50] These documents confirmed the advantages of such bearings, particularly the 25 per cent to 50 per cent reduction in draught or, as one unsolicited testimonial expressed it, '. . . I used to drive three horses with the same loads that I now easily take with two small horses'.

By the end of the First World War the motor-driven road vehicle was rapidly replacing horse-drawn carts and a Timken publication relating to adjustable taper roller-bearings for motor cars and other services

---

49. I am grateful to Mr E. F. Instone of the Physical Laboratory of British Timken, Mr Walt E. Littmann of the Research Department and Mr D. E. Eagon, Jr, Public Relations Manager of the Timken Company, Canton, Ohio, for background information and permission to publish these illustrations.

50. I am grateful to Mr Instone of British Timken for the opportunity to inspect copies of the *Company Catalogue* (1910) and *Making Horse-Haulage More Profitable* (1914).

Fig. 10.34
Late nineteenth-century forms of tapered roller-bearings.

(*c.* 1919) throws an interesting light on some of the early problems faced by motor manufacturers.

> ... *Motor Car Manufacturers are continually experiencing trouble, and Motor Repairers throughout the country are carrying out a steady trade in renewing worn and broken ball bearings on English, as well as foreign cars, to the detriment of manufacturers reputations.*
> ... *When turning a corner, 25 feet radius at 15 m.p.h., the end thrust on the front wheel bearing which gets the maximum thrust is just about the same amount as its normal radial load.*

Fig. 10.35
Arthur Charles Barrett – founder of the Preston Davis Ball Bearing Company and joint founder of the Hoffman Manufacturing Co. Ltd (1898).

Ball-bearing manufacture   A number of small companies were established to manufacture steel balls for bicycles being built in England and in Europe at the end of the nineteenth century. A remarkable gentleman whose name is rarely mentioned in accounts of the development of the ball-bearing industry enters the story at this stage. He was a partner in a family business of iron and brass founders known as R. Barrett & Sons, of Beech Street in the Barbican. Charles Arthur Barrett, whose photograph is shown in Fig. 10.35, was an entrepreneur who recognized the growing demand for steel balls and bearings early in the 1890s. He formed the Preston Davis Ball Bearing Company as an offshoot of a company managed by his cousin Mr G. F. Barrett, which traded under the name of the Westminster Engineering Company, and commenced the manufacture of cycle bearings and carriage hubs.

The new company soon experienced difficulty in making, or obtaining, steel balls of adequate accuracy and quality for bearings. Mr Barrett thus journeyed to America to explore the claims of the Hoffmann Machine Company on 143rd Street, New York. In 1897 Mr Barrett met Mr E. G. Hoffmann, a naturalized American who held, jointly with the company, a number of patents for the manufacture of steel balls of high accuracy. Mr Barrett was impressed, and since the processes involved appeared to be capable of resolving the fortunes of the English company, he bought the English and European Hoffmann patent rights through R. Barrett & Sons. Mr Hoffmann came to England and a new company known as the Hoffmann Manufacturing Co. Ltd, was formed and registered in January 1898 with a capital of £100,000. Mr C. A. Barrett, Mr E. G. Hoffmann and Mr C. F. Barrett formed the board of directors, the two former being joint managing directors.

A small new factory was constructed in Durrant's Meadow next to the Great Eastern Railway Goods Yard in Chelmsford, the ground floor of the main building being completed and equipped with machinery during 1899. The initial staff totalled twenty, but within ten years the company was producing over 1 million steel balls each day. Initial development of the company was plagued with technical problems, since there was no satisfactory procedure for grinding the balls once they had been turned from a bar.

Hoffmann's original invention was based upon a lathe having two chucks revolving together which facilitated the turning of steel balls from a bar with an insignificant parting blemish and an overall diametral accuracy in the range 0·001–0·002 in (25·4–50·8 μm). A drawing of this remarkable lathe, bearing Mr Hoffmann's signature, is shown in Fig. 10.36(*a*). A photograph of this elegant machine tool which was some 20 in (0·51 m) long is shown in Fig. 10.36(*b*).[51] A German foreman in the Chelmsford factory by the name of Otto Schmidt happened to read an article in the technical press about Christian C. Hill's method of grinding spheres. Mr Hill was an American and his English patents were about to lapse, so this time Mr G. F. Barrett travelled across the Atlantic, bought the patent rights and returned to Chelmsford to work out their technical content. In due course this development enabled Hoffmann's to produce balls of high accuracy and sphericity and the move provided the company with a sound technical base for future expansion. Mr C. A. Barrett suffered a breakdown, retired from the board for about fifteen months, had a second breakdown in 1913 and died in 1916. In the meantime the company was guided by Mr G. F. Barrett and Mr P. C. Low, Managing Director of the London company of R. Barrett & Sons. In 1903 they bought out Mr Hoffmann, but retained his name in the company title.

The company rapidly developed a fine reputation for the manufacture of ball-bearings of high precision early in the twentieth century. British Patent No. 15131 was awarded to E. G. Hoffmann in 1902 for a single-row ball-bearing having both an inner groove and an outer spherical

51. I am grateful to Dr S. Y. Poon, Manager, Engineering Research, RHP Bearing Research Centre, Chelmsford, for unearthing these illustrations and the photograph of Mr Charles Arthur Barrett.

Fig. 10.36
(a) Reproduction of engineering drawing of Hoffman ball turning machine and parting-off cutter (1899) carrying E. G. Hoffmann's signature.

(b) E. G. Hoffman's original machine (length approximately 20 in (0·5 m) for turning and parting-off balls from 12 ft (3·6 m) lengths of rod.

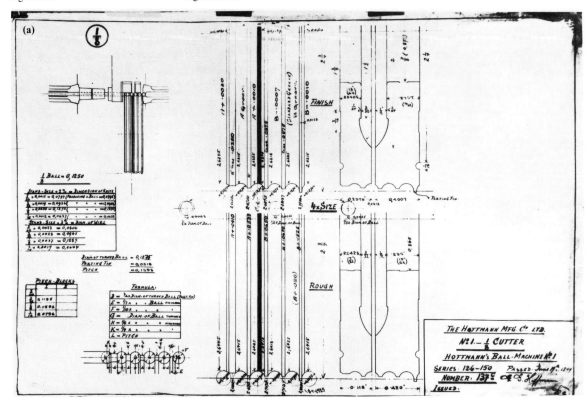

track. A cylindrical roller-bearing with grooved inner and outer rings was introduced to meet the ever increasing duties required of bearings, and it was patented by the British company in 1910. The lengths of the rollers were made equal to their dimensions, this being the origin of the so-called square rollers.

At the beginning of the twentieth century a company which manufactured woodworking machinery moved from its initial site in Chelsea to a new, enlarged location in Newark in Nottinghamshire. The company had always made its own bearings and after the move to Newark the demand for its rolling bearings from other concerns increased. This commercial incentive and difficulties with overseas supplies during the First World War led to the creation of a separate company known as Ransome and Marles Bearing Co. Ltd. Wartime demand and the ever increasing requirements of the automobile industry established the company as a major United Kingdom bearing manufacturer.

In Sweden a plant engineer Sven Wingquist with the Göteborg textile company of Gamlestadens Fabriker became concerned with unnecessary stoppages and production losses resulting from bearing failures. The failures were frequent and the delivery times of replacement bearings were often long. It appears that most of the bearing failures in the belt-driven machines arose from misalignment and deflection of the line shafting, the misalignment being attributable in part to settlement of the factory premises after construction.

The problem led Wingquist to consider bearing arrangements which would accommodate misalignment. He first considered a deep-groove ball-bearing, but late in 1906 and early in 1907 he developed the concept of a self-aligning, double-row, ball-bearing. Wingquist[52] was anxious to form an independent company and on 16 February 1907, a board meeting was held in the Gamlestadens Fabriker offices and the company A. B. *Svenska Kullagerfabriken* (SKF) was established with a share capital of SKr110,000 issued as 110 shares. Sven Wingquist held ten shares and became Managing Director of the new company. A new factory having a floor area of 12,000 ft^2 (1,115 m^2) was built in Göteborg and in 1908 seventy employees were producing 45,000 bearings per annum. By 1913 these figures had risen to 3,200 employees and 1,325,000 bearings.

Sven Wingquist was awarded Swedish Patent No. 25406 in 1907 for his double-row ball-bearing with a spherical outer track. His invention of the self-aligning ball-bearing capable of supporting essentially radial, but also some axial, load in misaligned shafts was a great step forward in ball-bearing development. It was followed by the award of British Patent No. 1617 in 1917 for a double-row, self-aligning bearing with barrel-shaped rollers. The first factory outside Sweden was built in England in 1911, but overseas expansion was rapid and worldwide production was ensured by the construction of factories in Germany in 1914, the United States in 1916 and France in 1917. The first research laboratory, itself a novelty in the engineering industry, opened in Göteborg in 1912.

---

52. See *Ball-Bearing Journal* (1977), No. 190, Feb.

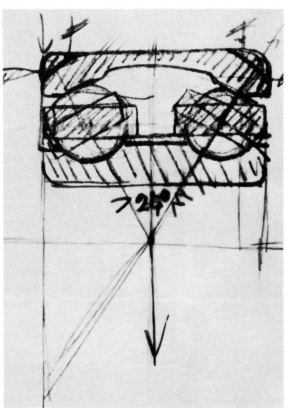

Fig. 10.37
(a) Sven Wingquist–founder of
A. B. Svenska Kullagerfabriken
(SKF).
(b) Sven Wingquist's first sketch
of his revolutionary self-aligning
ball-bearing–drawn on Easter
Day, 31 March 1907.

The founder of SKF, Sven Wingquist was born in Hallsberg, Sweden on 10 December 1876. He graduated in engineering, but specialized in textile engineering and gained practical experience in Sweden and the USA. He joined Gamlestadens Fabriker in 1899. Problems associated with the rapid expansion of the company and wartime difficulties took their toll and Wingquist was ill for a long period after the First World War. He resigned as Managing Director in 1919, but remained on the Board until his death in 1953. He was Chairman of the Board during the last fifteen years of his life. Worldwide recognition as the father of the Swedish ball-bearing industry was accorded to Wingquist and he was awarded an honorary doctorate of the Stevens Institute of Technology, Hoboken, USA, in 1951. He was also an honorary member of the Swedish Society of Inventors and the Swedish Academy of Engineering Sciences.

A portrait of Sven Wingquist is reproduced in Fig. 10.37, together with a print of the first sketch of his self-aligning ball-bearing drawn in the evening of Easter Day, 31 March 1907.[53]

By the end of the first decade of the twentieth century, or at least by the time of the First World War, most industrial nations were well advanced towards the creation of specialist ball- and roller-bearing

53. I am grateful to SKF for permission to reproduce the portrait of Sven Wingquist from Steeckzen, B. (1957) and the first sketch of his self-aligning bearing from the article in the *Ball-Bearing Journal*, (1977), No. 190, Feb.

companies. In Germany the Deutsche Waffen und Munitionsfabriken played an important role, while the combination of an expanding automobile industry and precision manufacture provided fertile ground for the growth of rolling-element bearing companies in the United States. One such firm was the Fafnir Bearing Company, created in a corner of a New Britain factory in Connecticut with a staff of seven under the General Manager, Elisha H. Cooper, in 1911.[54]

A recognition of the advantages of rolling compared with sliding motion dates back to antiquity and many of the principles of operation of modern bearings were inherent in devices developed during classical Greek and Roman times, but four major ingredients appear to have been responsible for the firm foundation and impressive growth of the rolling-element bearing companies about 1900.

The first was the rapid development of the bicycle or velocipede in the latter half of the nineteenth century, coupled with impressive demonstrations of the merits of ball-bearings in such machines. The second was the concomitant advance in precision manufacture of balls and rollers, first through ingenious turning and then in highly skilled grinding and polishing techniques. Accurate manufacture facilitated the production of rolling elements of adequate geometrical perfection, but their ability to resist the exceptionally high contact stresses called for the introduction of both case-hardened and through-hardened steels. The fourth factor which really provided the spur for mass production of precision components was the development of road vehicles and the automobile in particular. The use of such vehicles in the First World War provided an additional spur, but by 1920 the merits of rolling-element bearings in motor cars and a wide range of manufacturing equipment was well established.

It now remains to consider the scientific studies of contact stresses, load-carrying capacities and the nature of rolling friction which accompanied these spectacular practical innovations.

The scientific studies of Hertz, Stribeck and Goodman

These three men were undoubtedly the outstanding investigators of contact mechanics throughout the early stages of development of the rolling-element bearing industry. It is often assumed that a recognition of the high contact stresses encountered in ball-bearings must have promoted Heinrich Hertz's theoretical studies of the subject (Hertz, 1881), but this was not the case.

Hertz[55] was working with Helmholtz in Berlin on electrical phenomena about 1880. He attended a meeting of the Physical Society of Berlin at which Newton's rings were discussed and began to wonder about the influence of changes in geometrical form of the glasses which were pressed together in demonstrations of such optical phenomena. He proceeded to solve the problem of contact stresses between elastic solids

54. I am grateful to Mr T. J. Rosinski, Advertising Manager of the Fafnir Bearing Company for background information.
55. See Appendix, Sect. A.13 for biographical details.

during the Christmas vacation of 1880 and his now classical paper was read to the Physical Society of Berlin on 31 January 1881. He was twenty-three years old at the time and although it was contact between glass lenses which initiated the problem, he concluded his paper by calculating that the contact time between two steel spheres the size of the earth moving with initial speeds of 0·4 in/s (10 mm/s) would be almost 27 hours!

The expressions developed by Hertz (1881) for the deformations and contact stresses within the elliptical regions of intimate contact between arbitrarily shaped elastic solids when loaded together, have formed the foundations of modern studies of non-conforming tribological components like rolling-element bearings and gears.

Professor Richard Stribeck[56] undertook most of his basic studies of bearings between 1898 and 1902 while working in Berlin. He was head of the Metallurgical Division of the Centralstelle für wissenschaftlich-technische Untersuchungen when he was commissioned by Deutsche Waffen-und Munitionsfabriken to carry out experiments to ascertain the safe loads on balls and complete bearings at various speeds. This was, and in some respects still is, the most basic issue facing the manufacturers of rolling-element bearings.

Stribeck first ascertained the mechanical properties of bearing steels, hardened and tempered in various ways. He then tested the predictions of Hertzian contact analysis and measured by means of a mirror apparatus and a compression testing machine the deformation of three equal steel balls, two steel balls with a flat plate sandwiched between them, a similar arrangement in which the flat plate was replaced by a grooved one to accommodate the balls, and finally a single sphere contained between the cup-shaped ends of two short, solid cylinders. One of the early findings was that *plastic deformation* and *permanent set* could be produced by very small loads. Stribeck found that the relationship between the load on a ball (P) required to produce a given permanent deformation, or to just avoid plastic flow, was related to the diameter of the ball (d) by the simple expression,

$$P = kd^2. \qquad \qquad \dots [10.27]$$

The constant (k) depended on the geometrical configuration and the type and condition of the steel.

Stribeck also considered the manner in which the total load (W) would be shared between the balls in a complete bearing and demonstrated that if the number of balls was Z, and the load on the most heavily loaded ball was P_0, the ratio ZP_0/W exhibited little variation and could be taken as 4·37. He then derived the celebrated and extremely useful *Stribeck equation* for static load-carrying capacity by writing the more conservative figure of 5 for the theoretical value of 4·37 and noting that,

$$W = \frac{ZP_0}{5} = \frac{kZd^2}{5}. \qquad \qquad \dots [10.28]$$

---

56. See Appendix, Sect. A.14 for biographical details.

The great merit of Stribeck's work was that it not only confirmed the essential features of Hertzian contact theory for elastic solids, but it also provided a sound basis for manufacturers to predict the general form of bearings for particular applications. The constant (k) had to be determined empirically and as laboratory testing and operating experience developed it soon became possible to select rolling-element bearings of adequate capacity and life for specified applications.

Professor John Goodman[57] made significant contributions to the subject of friction and plain bearings (Goodman, 1887), before being appointed to the Chair of Engineering in the Yorkshire College of Science, forerunner of the University of Leeds, in 1890 at the age of twenty-eight. He continued his interest in plain bearings for many years to come, but in 1887 he was already writing about the friction of '. . . an ordinary 'Rudge' bicycle bearing . . .' and comparing it with plain bearing performance. He thus started his work on rolling bearings before Stribeck, but he also published his wide-ranging and most notable paper on the subject as late as 1912. Goodman's (1912) paper is a fascinating commentary on the evolution of both the science and technology of rolling-element bearings. He described the testing machines, the wide range of ball- and roller-bearings tested, and recorded in commendable fashion the essential features of his measurements of friction, wear and life. A selection of his writings will illustrate the character of his studies. He was familiar with Hertzian theory and the confirmation provided by Professor Stribeck and Dr Heerwagen. He emphasized the importance of procedures adopted by '. . . reputable makers of ball bearings . . .' for the production of balls of uniform hardness and geometrical conformity. He wrote,

> . . . *The argument that large balls will wear more than the rest, so that after a short time every ball will take a fair share of the load, is altogether fallacious, since the overloaded balls do not simply wear down. It is well known to those who have used such bearings that the balls very soon develop specks and flakes, which damage the ball-races and quickly bring about the failure of the whole bearing.*

Goodman sought a procedure for the early detection of bearing failure and tried careful diametral measurements and periodic accurate weighing before finding that microscopic examination of the surfaces gave the most satisfactory results. In this connection he supplemented the above commentary on wear with reference to such terms as '. . . pitting . . . flaking . . . scratching . . . and peeling'.

Goodman's major contribution was a clear demonstration that the load-carrying capacity of a rolling-element bearing diminished with speed. In speculating on this effect he commented that, '. . . The speed-effect on the balls has possibly some relation to the well-known effect of very rapid reversals of stress.'

This reference to the phenomenon of 'fatigue' was to be prophetic as far as Goodman's later researches were concerned, as those readers

---

57. See Appendix, Sect. A.15 for biographical details.

familiar with his general engineering work and stress analysis in particular will be aware.

Goodman found that quite considerable reductions in bearing load-carrying capacity below the Stribeck static level could be introduced by speed, or more correctly, by the number of stress cycles. In general terms the results of his wide-ranging experiments showed that the load (P) on a single ball of diameter (d) was related to the shaft rotational speed (N) and the ball-race path diameter (D) by an expression of the form,

$$\left[\frac{ND}{d} + A\right]P = B, \qquad \qquad \ldots [10.29]$$

where A and B are constants depending upon the bearing geometry.

Now the load-carrying capacity of a ball at any speed depends upon the square of the ball diameter (d), as illustrated by Hertzian analysis for static conditions according to equation [10.27]. Hence,

$$P = \left[\frac{Kd^3}{ND + Ad}\right]. \qquad \qquad \ldots [10.30]$$

This general form of the *Goodman equation* proved to be exceedingly valuable in predicting the load-carrying capacity of bearings at various speeds and it has found application to the present day with little modification.

But John Goodman had a few other words of wisdom to deliver on the subject of early twentieth-century rolling-element bearings. He attributed the curious observation that the smallest balls in a bearing became scratched while the larger ones showed little distress to *slipping* or *skidding*. He concluded that many failures could be attributed to dirt and emphasized the need to keep dust and grit out of the bearings. On the question of lubrication he was ambivalent, but finally concluded that a little oil or grease was probably desirable. He was impressed by the very low friction under both static and kinetic conditions in ball-bearings, and noted that the friction in roller-bearings was also considerably less than in plain bearings. He was concerned, but slightly mystified, by the large and not infrequently deleterious effects of end thrust in roller-bearings.

The paper itself was forty-five pages long, excluding illustrations, and it attracted no less than thirty-eight pages of discussion from manufacturers, investigators and bearing users. It contains a wealth of information on the development of understanding of the basic mechanics of early twentieth-century rolling-element bearings and anti-friction materials, and is a fine tribute to the sound engineering approach to bearing problems adopted by John Goodman. His summary left the reader in no doubt about his assessment of, or realistic attitude towards, rolling bearings.

> . . . *The author believes that there are very few instances in which roller or ball bearings might not be used to advantage, and he considers that the extra first cost is very soon repaid by the reduction of friction and the saving of oil. Such bearings require far less attention; they are more 'foolproof'; and the wear is extremely small and therefore does not upset the alignment of the shaft; but to ensure success*

ball bearings must not be greatly overloaded, and they must be carefully fixed in the first instance.

Few would disagree with Allan's (1945) assessment that, '. . . the contributions of Heinrich Hertz, Professor Stribeck, and Professor Goodman were doubtless the most outstanding in the early days of the industry. . .'.

Rolling friction

The euphoria which accompanied the rapid introduction of the rolling-element bearings into lightly loaded machines like the velocipede at the end of the nineteenth century might well have obscured the fact that the resistance to rolling motion, although much reduced from that encountered in sliding, was still finite. The advantages inherent in replacing sliding by rolling motion had been recognized since antiquity and the subject had attracted the attention of Leonardo da Vinci (Ch. 7), Hooke (Ch. 8), Coulomb, Morin and Dupuit (Ch. 9). However, it was in the present period that the first substantive scientific studies of the mechanics of rolling friction were reported. It was Reynolds (1875), the father of fluid-film lubrication theory, who also initiated the studies of rolling friction for cylinders, and Heathcote (1921) who extended the concept to a ball rolling in a groove.

But why did Reynolds the hydrodynamicist become involved with this issue? The answer is that he came to it from a consideration of belt drives long before he examined the question of fluid-film lubrication. He argued that the transmission of power between two parallel shafts by means of a belt wrapped round two pulley-wheels required the formation of tight and slack sides of the connecting belts. The belt moving on to the driving pulley would experience tension which gradually relaxed until it left the wheel on the slack side. Since belts are elastic, the change in length associated with this relaxation in stress represented a relative movement or *creep* between the pulley and the belt, and the combined effort caused the driven pulley to move less quickly than might otherwise by imagined. Reynolds (1874) reported this work in *The Engineer*, but it clearly promoted the wider considerations of rolling friction discussed in his Royal Society paper published in the following year (Reynolds, 1875). In the latter paper he argued that in rolling motion there would be an '. . . analogous slipping when a hard roller rolls on a soft surface, or when an india-rubber wheel rolls on a hard surface'.

This relative movement between surfaces in the region of elastic deformation between rolling objects is now widely known as *micro-slip* and Reynolds attributed rolling friction to the integral of local sliding friction throughout the contact zone. In a characteristically simple experiment he demonstrated that when an iron cylinder rolled over india-rubber the cylinder moved forward less than its circumference in one revolution. In fact he reported that, '. . . an iron roller rolled through something like three-quarters of an inch less in a yard when rolling on india-rubber than when rolling on wood or iron'.

The problem is that the differential slip within the deformed track considered by Reynolds is normally minute and, more seriously, quite

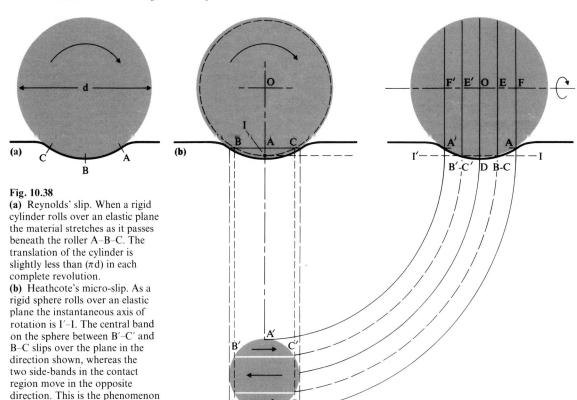

Fig. 10.38
(a) Reynolds' slip. When a rigid cylinder rolls over an elastic plane the material stretches as it passes beneath the roller A–B–C. The translation of the cylinder is slightly less than (πd) in each complete revolution.
(b) Heathcote's micro-slip. As a rigid sphere rolls over an elastic plane the instantaneous axis of rotation is I′–I. The central band on the sphere between B′–C′ and B–C slips over the plane in the direction shown, whereas the two side-bands in the contact region move in the opposite direction. This is the phenomenon of micro-slip within the contact zone.

incapable of providing a contribution to rolling resistance unless some dissipative action accompanies the effect. Reynolds himself recognized the difficulty and tried to reconcile his overall view of the essential mechanics of the problem with the requirement for some form of *hysteresis* through references to successive heating and cooling and a possible viscous effect within the solids.

This was one of Osborne Reynolds' less authoritative contributions to tribology, yet his views on the nature of rolling friction were widely accepted for the next half century. His work has been widely discussed in more recent times (Johnson, 1962; Bowden, and Tabor, 1964; Bisson, and Anderson, 1964; and Barwell, 1970*b*), and even if his tentative conclusions appear less appropriate in the 1970s than in the 1870s, they established a line of reasoning on the subject of *micro-slip* which was characteristically novel.

Heathcote (1921) extended the work of Reynolds to the action of a ball rolling in a groove under Hertzian contact conditions. His work was directly linked to efforts to appreciate the nature of friction in rolling-element bearings. The difference between Heathcote's approach and that of Reynolds is illustrated in Fig. 10.38. Reynolds' considered the differential stretching of the elastic plane as it passed beneath a loaded roller from A through B to C as shown in Fig. 10.38(*a*). Heathcote considered a ball rolling on an elastic plane and realized that there would be two lines of zero sliding motion bounding regions in which the direction of sliding of the elastic 'plane' relative to the ball adopted different directions, as shown in Fig. 10.38(*b*). The reason is simple: each point in the mid-plane of the contact on the surface of the *rigid* ball will have a velocity in the direction of sliding given by its angular velocity about the axis O multiplied by its radius from that axis (*R*). The largest velocity occurs at the bottom and the smallest at the edge of the groove, with points of zero relative motion lying somewhere in between. The resultant rolling friction was deemed to arise from the net tractive action between regions (1) and (2).

By 1925 the concepts of micro-slip established by Reynolds and Heathcote were widely accepted, but in recent years refinements to the analysis and further elegant experimental work has demonstrated that hysteresis effects are perhaps of even greater significance in the overall action known as *pure rolling*.

10.9   Dry friction and brakes

While studies of friction between dry solids continued during the period under review they were generally overshadowed by the work on fluid-film and boundary lubrication. If an exception can be found it is in the steady accumulation of data on kinetic friction. Once again the motivation for such studies arose primarily from the railways, where the problems of traction and braking were assuming some significance.

There was during this period a good deal of discussion about the relationship between static and kinetic friction. Some considered the two coefficients to be quite different, with a sharp transition from one value to another as sliding commenced or ceased. The experimental evidence presented by several workers gradually confirmed the view that the friction of motion and the friction of rest could be represented by a continuous function. The emergence of this concept can be traced through the writings of Bochet (1861), Jenkin and Ewing (1877), Kimball (1877), Galton (1878*a,b*), and Wellington (1884). The work of Jenkin and Ewing clearly established that although the coefficient of friction depended upon speed, there was no abrupt change in value as the sliding speed was reduced until the surfaces became stationary. Morin, whose results from an elaborate series of experiments executed between 1830 and 1834 had been reported to the Paris Academy of Sciences, had found that the static friction was usually greater than, but sometimes sensibly equal to, the kinetic friction. He also noted that in many cases a slight shock was sufficient to destroy the distinction between the two.

Professor Jenkin of the Engineering Department at Edinburgh University and Ewing thought that instead of an abrupt change there might be continuity between the two kinds of friction. They therefore measured the coefficient of friction in a delightfully simple experiment. They mounted a 2 ft (0·6 m) diameter by $\frac{3}{4}$ in (19 mm) thick disk of cast iron on a horizontal axle fitted with small (0·1 in (2·5 mm)) diameter stub shafts which sat in rectangular grooves cut in appropriate bearing

Table 10.4
Coefficient of friction recorded
by Jenkin and Ewing at very
low speeds

Materials	μ (dry contact at 0.002 ft/s)
Steel on steel	0.351
Steel on brass	0.195
Steel on agate	0.200
Steel on greenheart	0.215
Steel on beech	0.366

materials. The disc was set in motion and the coefficient of friction was deduced by monitoring its retardation. Air resistance was neglected and the influence of the recorder used to measure surface speed was also thought to be negligible. The recorder was in fact based upon a pendulum device used by Sir William Thomson in his studies of electrical impulses in long submarine cables. The dry contact results obtained at very slow speeds are shown in Table 10.4.

There was little variation in μ over the speed range 0.0002 ft/s (0.00006 m/s) to 0.0089 ft/s (0.0027 m/s), and the high values of μ recorded strongly supported the view that transition from kinetic to static friction was continuous.

The Italian scholar Conti (1875) working in Alexandria measured the kinetic friction of different materials by allowing specimens to slide down a very large inclined plane. The motion was carefully monitored by allowing the slider to close electrical contacts placed within the plane at known locations.

In a series of experiments reported in the *American Journal of Science* Kimball (1877) demonstrated that kinetic friction could be either greater or less than the static friction, depending upon the nature of the surfaces, the loads, sliding speeds and lubrication. In all cases, however, he found a smooth transition from the static to the kinetic situation. In the case of a leather belt on a cast-iron pulley the kinetic friction at very low sliding speeds exceeded the static friction, increased initially with increasing speed until a maximum was reached and thereafter decreased steadily as the speed reached higher values. He concluded that the conflicting views of earlier workers like Coulomb, Morin, Bochin and Hirn could be attributed to the different experimental conditions. Such views would appear to be self-evident to the present-day tribologist, but it emphasizes the state of confusion which scientists and engineers faced in the field of friction and lubrication at the end of the last century. Galton's (1878*a,b*) studies of the friction of railway brakes in England, Wellington's (1884) experiments on the friction of journals at low velocities and Goodman's (1886) investigations are particularly good examples of the way in which sound experimental evidence on the characteristics of dry friction or friction between lubricated surfaces at very low sliding speeds was established in the latter half of the nineteenth century.

Captain Douglas Galton, F.R.S. studied the friction between brake-blocks and wheels and between wheels and rails by means of sophisticated instrumentation on a specially constructed van on the London, Brighton and South Coast Railway. The van was equipped with Westinghouse pneumatic brakes and that remarkable engineer from London Mr George

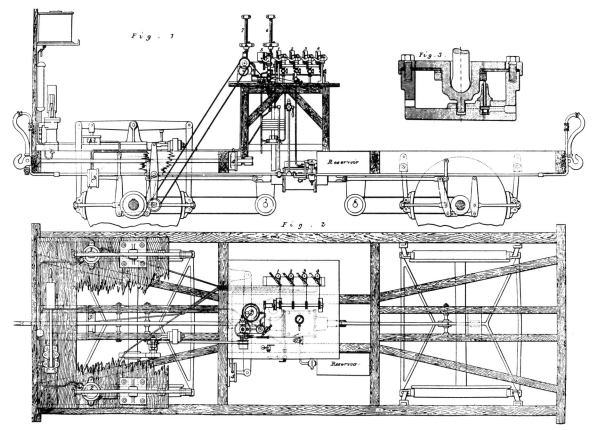

Fig. 10.39
Experimental van used by Captain Douglas Galton for the determination of the action of railway brakes (Galton, 1878*a*). (Designed by George Westinghouse.)

Westinghouse designed hydraulic dynamometers which recorded the normal force between the brakes and the wheels, the friction moment or force between the brakes and the wheels, the traction on the draw-bar and the speed of both the wheels and the van. A drawing of this ingenious van is shown in Fig. 10.39 and the records of a number of tests appear in Fig. 10.40. One of the advantages of the hydraulic dynamometer was that

Fig. 10.40
Typical record of braking force (*A*), braking pressure (*B*), wheel speed (*C*, *C'*), and retarding force on the draw-bar (*D*) obtained by Captain Douglas Galton for a van moving with an initial speed of 21 mph (33·7 km/h) (Galton, 1878*a*).

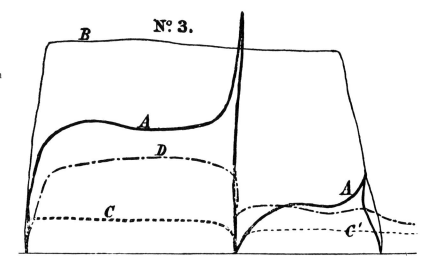

Table 10.5
Captain Douglas Galton's
coefficients of friction for
brake-blocks on wheels and
wheels on rails

Speed		Coefficient of friction			
mph	ft/s	0–3s	5–7s	12–16s	24–25s
(a) Friction of cast-iron brake-blocks on steel tyres of wheels					
<5	7	0·360	–	–	–
10	14	0·320	0·209	–	–
20	29	0·205	0·175	0·128	0·070
30	43	0·184	0·111	0·098	–
40	58	0·134	0·100	0·080	–
45	65	0·125	–	–	–
50	73	0·100	0·070	0·056	–
60	88	0·062	0·054	0·048	0·043
(b) Friction of wrought-iron brake-blocks on steel tyres of wheels					
18	–	0·170	–	–	–
31	–	0·129	0·11	0·099	–
48	–	0·110	–	–	–
(c) Friction of steel tyres of wheels skidding on steel rails					
10	–	0·110	–	–	–
15	–	0·087	–	–	–
25	–	0·080	0·074	–	–
38	–	0·057	0·044	0·044	–
45	–	0·051	–	–	–
50	–	0·04	–	–	–

remote recording could be adopted. In this case the instruments were placed upon a table in the centre of the van and the recording drums were rotated at constant speed by the ancient device of a water-clock. Galton's preliminary findings were first presented to a meeting of the Institution of Mechanical Enginers in Paris and subsequently published in *Engineering* on 14 June 1878 (Galton, 1878*a*). A subsequent paper presented to Section G of the British Association meeting in Dublin was reported in *Engineering* on 23 August 1878 (Galton, 1878*b*). In the latter paper average coefficients of friction from a large number of tests were recorded as shown in Table 10.5.

The three major conclusions to be drawn from the results displayed in Table 10.5 are;

1. The coefficient of kinetic friction between the brake-blocks and the wheels decreased as the sliding speed increased.
2. The coefficient of kinetic friction between the brake-blocks and the wheels decreased with time at a fixed speed.
3. The coefficient of kinetic friction between the brake-blocks and the wheels was greater than that between the rails and the wheels when the latter were skidding.

Captain Galton drew attention to the coefficients of friction recorded by Morin (0·44), Rennie (0·30) and Jenkin (0·351) for very low sliding speeds or static conditions, thus emphasizing the continuity between static and kinetic friction conditions. The second conclusion was attributed to temperature changes during sliding and the third one to differences in surface conditions. With the brake-block rubbing against the wheel, contaminants were continuously cleaned from both surfaces, thus leading to high friction between relatively clean solids. It is for this reason that dissimilar metals (e.g. cast iron or steel) were used. When the wheel skidded the relatively clean tyre continuously met contaminated surface of the rail and the coefficient of friction was much reduced.

The rolling resistance of the van was variable but generally between 0·24 and 0·28, and hence frictional forces up to about one-quarter of the weight of the vehicle could be applied at the brakes before the wheels would skid. It can be seen from Table 10.5(c) that once skidding occurred the resistance to motion was much reduced. This revelation that a train could be brought to rest most effectively by applying braking forces which permitted rolling rather than skidding had a major effect upon subsequent designs of brakes.

A further consequence of this finding followed from the increase in coefficient of friction of the brakes as the sliding speed fell (see Table 10.5(a)). If a fixed normal force was applied to the brake of a train moving at high speed, the coefficient of friction and hence the force of friction responsible for retarding the train would increase as the forward velocity decreased. If this braking force exceeded the force of adhesion between the wheels rolling on the rail, the wheels would then skid with a much reduced overall braking effect. The effect is shown with great clarity in the record of Test 3 in Fig. 10.40. Trace B shows the near constant braking pressure and trace C′ the small but steady retardation of the van from an initial speed of about 21 mph (34 km/h). The effective tangential braking force (trace A) started to rise as the speed fell in accordance with the evidence of Table 10.5(a). When the speed fell to 18 mph (29 km/h) the wheels locked (trace C) and the subsequent retarding force recorded on the draw-bar (trace D) was much reduced.

The essential principles of effective braking on the railways were succinctly stated by Captain Galton as follows:

> . . . *In order to obtain the maximum retarding power on the train, the wheels ought never to skid; but the pressure of the brake blocks on the wheels ought just to stop short of the skidding point. In order that this may be the case, the pressure between the blocks and the wheels ought to be very great when the brakes are first applied, and gradually to diminish until the train comes to rest.*

Wellington (1884) added to the growing literature on journal friction with his paper presented to the American Society of Civil Engineers. Like Bochin before him, Wellington had previously deduced the magnitude of friction from the motion of trucks on inclined tracks, but these experiments were designed to study the starting and low-velocity friction of journals as used on the American railways. Lubrication was by means of pads of waste material saturated by West Virginia mineral oil and placed on each side of the journal. The bearings operated under conditions which would now be recognized as boundary or perhaps mixed lubrication and many of the findings were overshadowed by the studies of fully lubricated journals reported by Petrov (1883) and Tower (1883). The paper is nevertheless interesting, particularly in view of its emphasis on low-velocity conditions.

A full discussion of the paper attracted contributions from Beauchamp Tower and Robert Thurston. It was Thurston, who had presented his inaugural address as first President of the American Society of Mechanical Engineers four years earlier (Thurston, 1880), who commented on the continuity of the friction process during the transition from kinetic to static conditions. He said, '. . . I think that there can now be no doubt that the friction of motion, at low speed, gradually increases as the velocity of the rubbing is decreased, until it becomes a maximum at the moment of coming to rest, and that the friction of motion and the friction of rest are a "continuous function".'

A young engineer by the name of John Goodman[58] had his interest in friction and lubrication awakened when he became an assistant to William Stroudley on the London, Brighton and South Coast Railway. It was Stroudley who, in his capacity as locomotive superintendent, made provision for Douglas Galton's work on the friction of brakes, and since Goodman was to become a professor in the University of Leeds with an international reputation for his studies of bearings, there seems little doubt that Stroudley initiated some of the major practical investigations of tribological problems on the railways.

Goodman (1886) presented his account of 'Recent researches in friction' to the students of the Institution of Mechanical Engineers on 20 November 1885. His survey of the laws of friction for dry surfaces and his explanation of the phenomena involved provide a useful glimpse into the late nineteenth-century understanding of the subject. The behaviour was still related to the concept of interlocking asperities, with the projections on one surface mounting up and riding over those on the other surface. The influence of Coulomb's studies described in Chapter 9 was much in evidence. The force required to pull one body up the numerous inclined planes on the other constituted friction, and heat was generated as the body fell down the far side of the projections.

While recognizing – and this in itself is interesting – that the theory of dry friction was of a much more complex nature than that of lubricated surfaces, Goodman suggested that many, if not all, of the difficulties could be overcome if '. . . it be assumed as a basis that all the surfaces of solid bodies resemble in structure the surface or pile of velvet.'

This proposal was illustrated by Goodman as shown in Fig. 10.41(*a*). His explanation that the coefficient of friction was equal to the tangent of the angle which the loaded fibres made with the surfaces was depicted in relation to the inclined plane as shown in Fig. 10.41(*b*). The argument was extended to take account of the observation that friction was always greater on the reversal of the direction of sliding by reference to Fig. 10.41(*c*). It was argued that the pile, already flattened in the original direction, had to describe an arc of a circle before it could be flattened in the reverse direction, thus increasing the vertical lift of the upper body from $(\alpha - \beta)$ to $(m - n)$. Finally, the lower friction observed between dissimilar metals compared with similar metals, was attributed to dif-

58.   See Appendix, Sect. A.15 for biographical sketch.

Fig. 10.41
Professor John Goodman's
explanation of dry friction in
terms of an analogy between
surface structure and the pile of
velvet (Goodman, 1886).
(a) In sliding (A) over (B), (A)
rises repeatedly through a height
$(\alpha - \beta)$.
(b) Tan ϕ is equal to the
coefficient of friction – note that
the fibres in the pile are horizontal.
(c) When the direction of sliding
is reversed, the pile, already
flattened in the original direction,
must describe the arc of a circle
before it can be flattened in the
reverse way.
(d) Geometrical explanation of
the observation that friction is
always higher with similar, than
dissimilar metals.

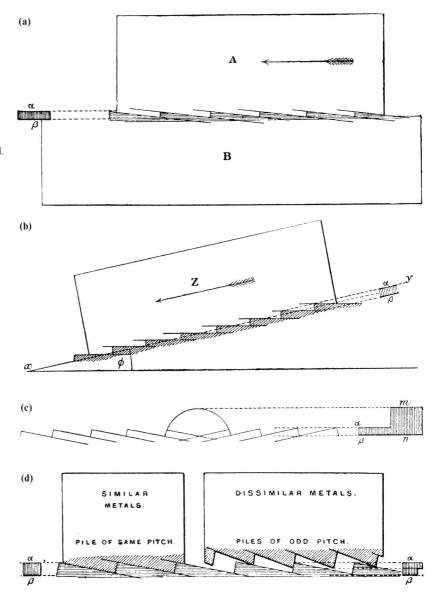

ferences in the surface 'piles' and a decrease in the effective lift $(\alpha - \beta)$
as shown in Fig. 10.41(*d*).

A number of valuable contemporary accounts of friction appeared
in books published in the period under review. One of the most interesting
was written by Archbutt and Deeley (1900), respectively Chemist and
Inspector of Motors and Boilers to the Midland Railway Company in
Derby. Later texts by Stanton (1923*a*) and Gümbel and Everling (1925)
also make rewarding reading. The latter writer, who had contributed
extensively to lubrication theory in earlier years, expressed the view that
dry friction arose from the force necessary to displace material surrounding penetrating asperities – the so-called deformation element
in modern friction theory.

10.10 Piston rings

Early piston seals in reciprocating machinery

Piston rings have once again been the centre of attention in recent years as a result of a crop of problems associated with the scuffing of diesel engine cylinder liners. However, it is worth noting that the modern form of piston ring is essentially a product of steam rather than internal combustion engine development. The development of a satisfactory seal for reciprocating machinery has a long history. Water-pumps probably provided the earliest demands for such components, although the severity of the operating conditions was slight. Leather discs or rings were normally adequate in hydraulic equipment.

The famous demonstration of atmospheric pressure by Ottonis de Guericke, Burgomaster of Magdeburg, in 1650 was achieved by means of a piston within a cylinder. The piston was made of hard oak and it was provided with a circumferential groove filled with *flax* or *hemp*. When air was withdrawn from the cylinder the seal was adequate to hold a pressure difference between the atmosphere and the partial vacuum capable of overcoming '. . . the efforts of twenty or more men'.

Robert Boyle built a similar device in 1667 which used leather seals on the piston. The leather was kept 'turgid and plump' by means of a layer of water above it. Attempts to develop reciprocating machines driven by the explosion of gunpowder within cylinders, or by the subsequent condensation of the products of combustion, in the latter half of the seventeenth century were largely unsuccessful and Economou (1976) has speculated on the role of inadequate sealing of the piston in such devices.

Leather discs or flaps were favoured by Papin and Newcomen for sealing the early steam engines. The cylinder bores were far from accurate and the ability of leather or rope to conform to the clearance was an essential feature of these early developments. Thermal efficiencies were still less than 1 per cent early in the eighteenth century and leakage of steam past the piston was a major nuisance. The discs or flaps of leather were screwed to the piston and caused to turn up in the cylinder, like the seals on bicycle pumps, to contain a small amount of water. *Bridle* or other *soft rope* was sometimes preferred to leather and the cylinder wall was usually smoothed by rubbing sand against it. John Smeaton introduced improved techniques for the manufacture of engine cylinders and for sealing pistons by means of rope packing. Both developments contributed to the improved thermal efficiency of steam engines of about $1\frac{1}{2}$ per cent achieved by 1774.

Steam-engine efficiency improved dramatically to about 2·7 per cent when James Watt introduced the separate condenser in 1769. The firm of Boulton and Watt dominated the development and manufacture of the steam engine in the latter half of the eighteenth century. Almost 500 engines of increasing power and efficiency were produced in the Soho Works in Birmingham. The cylinders were bored by a new technique by John Wilkinson.[59] They were clamped securely on a stationary work

---

59. See Town, H. C. (1973), pp. 73–8.

support as the boring bar moved forward. This produced a much improved cylinder bore, such that Watt was able to promise his customers that '. . . a 72 in cylinder would not be further from absolute truth than the thickness of a thin sixpence in the worse part'.

The association between Boulton and Watt and Wilkinson broke up when the latter used his manufacturing skills to make and sell Boulton and Watt engines without permission of the inventors.

Watt devoted considerable effort to the development of piston seals. He initially used rings of varnished cloth since, '. . . cloth has a great reputation to adopt itself to a bad liner'.

As the circularity of cylinders improved, the need to use such compliant material diminished. However, the increased steam pressures associated with engines of greater power and the higher cylinder temperatures which resulted from the introduction of the separate condenser created new problems for the piston seals. Watt's original single ring was soon augmented by a second and a third so that, '. . . if steam goes past one of the rings of cloth, the other will take it up'.

In his search for alternative materials Watt considered *wood, hat, paper-chewed, and a mixture of paper-pulp and flour paste*. He eventually found *oakum* to be the most suitable material. It was formed from loose fibres obtained by untwisting and picking old rope. Oils and animal fats were used to lubricate and to improve the seal. Prior to the introduction of the separate condenser it was customary to retain a layer of water on top of the piston. This assisted sealing and preserved the flexibility of the seal. The higher cylinder temperature associated with the condensing engine caused the water seal to evaporate (see Robinson and Musson, 1969), and this at once made it essential to provide adequate lubrication for the ring. Tallow replaced oils and continued to be used until the advent of heavy mineral oils in the 1860s as told by Fowle (1974).

As steam pressures increased, longer and heavier pistons became necessary. Early in the nineteenth century an 18 in (0·46 m) diameter piston required an axial length of packing of about $2\frac{1}{2}$ in (63·5 mm) and the piston had to be fabricated to enable the packing material to be compressed. The manufacturers issued detailed instructions on the methods of inserting oakum, applying tallow, beating the packing into shape and tightening the springs against lead segments which secured the rings of oakum. The use of bolts, nuts, cotters and pins provided plenty of scope for loosening and the release of damaging metallic components within the cylinder. The mechanical complexity was later reduced by the introduction of a single *junk ring* bolted to the piston to secure the hemp gasket or junk packing in place.

Metallic piston rings

A contemporary of James Watt, the Rev. Edward Cartwright, proposed in 1797 that alcohol vapour should be used instead of steam in reciprocating engines. He recognized that such a change would raise difficulties for existing piston seals and lubricants, and this led him to the idea of using a metallic ring. He built such a piston with unlubricated brass rings pressed against the cylinder by steel springs.

The metallic ring did not oust the hemp-packed seal until the latter half of the nineteenth century. It was necessary to press the rings, often of segmental form, against the cylinder wall by means of spiral or leaf springs in order to minimize leakage. The forces involved and the state of manufacture contrived to cause numerous failures of the early metallic rings, but the situation was resolved in 1854 by the genius of John Ramsbottom, of the London and North Western Railway.

John Ramsbottom was born in Todmorden, Lancashire, in 1814 and he died on 20 May 1897 at the age of eighty-two.[60] He was a founder member of the Institution of Mechanical Engineers, a member of Council for seven years, Vice-President for thirteen years and President in 1870–71. He was also a member of the Institution of Mechanical Engineers Committee on Friction at High Velocities which guided the work of Beauchamp Tower. He devoted his life to the railways and is remembered particularly for inventions which included his piston rings, the duplex safety-valve, the sight-feed displacement lubricator and the water-trough between rails which enabled running locomotives to pick up water.

Ramsbottom (1854) described his new piston and piston ring shown in Fig. 10.42 at a meeting of the Institution of Mechanical Engineers held in Birmingham on 25 January 1854. The piston consisted of a single casting (A) held on to the conical end of the piston rod by means of a nut. Three grooves (BBB) $\frac{1}{4}$ in (6·4 mm) wide, $\frac{1}{4}$ in (6·3 mm) apart and $\frac{5}{16}$ in (7·9 mm) deep were turned in the circumference of the piston to receive the rings. Rings made of brass, steel or iron were drawn to a suitable section to fit the grooves as shown in the small section in Fig. 10.42, and bent into a circular form in rollers such that their diameter was about 10 per cent greater than the cylinder bore. The rings were forced against the cylinder wall by their own elasticity and Mr Ramsbottom found that an initial 10 per cent oversize was quite sufficient to keep the rings steam-tight.

The four advantages claimed for Ramsbottom's piston and rings were *lightness*, *simplicity*, *security* and *reduced friction*. The normal piston weight for a cylinder of 18 in (0·46 m) bore was reduced from 260 to 121 lbf (1,156 to 538 N), a reduction of some importance in relation to balancing and the magnitude of inertia forces in reciprocating machinery. The old fabricated pistons were replaced by a single component plus three rings. The only machining required was turning of the rim and grooves and boring of the central tapered hole. The piston rings were manufactured at a cost described as '. . . little more than nominal'.

The rings could not become loose, unless they broke, and the absence of loose parts like nuts, bolts, cotters and pins which might become deranged was a great advantage. Ramsbottom attributed the reduction in friction partly to the smaller piston weight resting on the bottom of a horizontal cylinder and partly because of the great reduction in '. . . the amount of elastic surface pressed against the cylinders'.

Mr Ramsbottom exhibited rings from service and noted that the first had been fitted some sixteen months earlier. A set of rings would run for

---

60. See Parsons, R. H. (1947), pp. 262–4.

Fig. 10.42
John Ramsbottom's improved piston fitted with metal rings made of brass, steel or iron in grooves $\frac{1}{4}$ in (6·3 mm) wide, $\frac{5}{16}$ in (7·9 mm) deep and $\frac{1}{4}$ in (6·3 mm) apart (Ramsbottom, 1854).

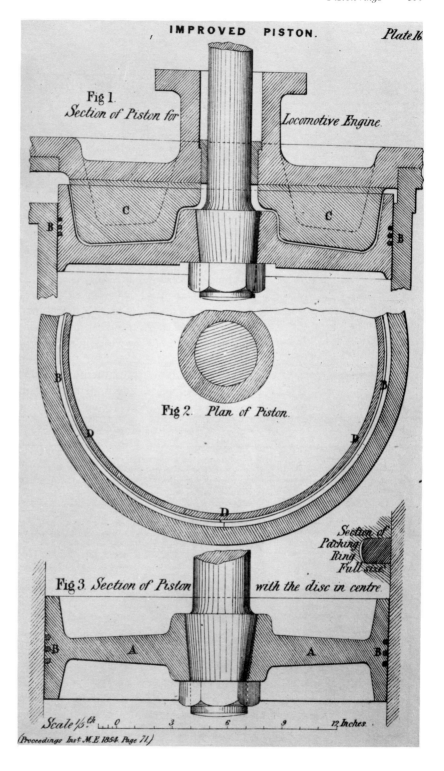

3,000–4,000 miles (4828–6437 km) and cost, when new, 2s. 6d. (12·5p). The average coal consumption of the first fifteen engines fitted with the new rings was compared with the consumption of the same engines in the previous four years. The average savings of 5·7 lb (2·3 kg) per mile over an aggregate distance of 269,800 miles (419,736 km) prompted Ramsbottom to claim '. . . a result which has been carefully arrived at, and which goes to show that this piston, either from greater average tightness, or reduced friction, or both combined, is greatly superior to those which it has superseded'.

The steam tightness of the Ramsbottom piston ring arose directly from the elastic compression of the ring against the cylinder. To provide adequate strength the rings were necessarily rigid and quite unable to follow any irregular contours of the cylinder. This gave rise to dynamic loading, wear and occasional breakage. A solution to this problem was proposed by Mr David Joy of Worcester (Joy, 1855). He introduced a spiral ring of cast iron or brass into the grooves to give a longer effective spring and less pressure on the cylinder wall as shown in Fig. 10.43. The arrangement allowed the ring to compensate more adequately for wear, but although its merit was confirmed from installation in several engines, it was more complex than the simple Ramsbottom ring. The *coil ring* never achieved the widespread popularity of the simple *circular rings*.

Ramsbottom (1855) tackled the question of uneven wear on his piston rings by determining in a practical and most effective manner the optimum geometrical form of the undeformed component. He first made a split

Fig. 10.43
David Joy's spiral cast-iron piston ring (Joy, 1855).

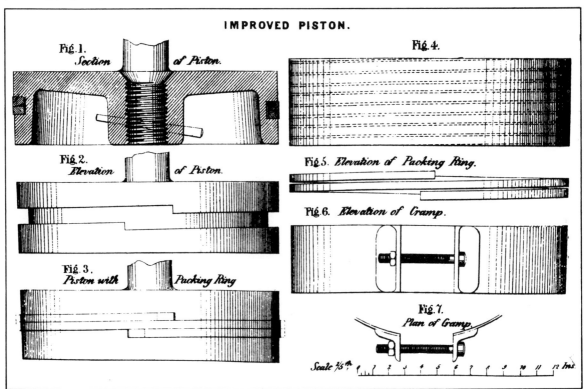

Fig. 10.44
John Ramsbottom's apparatus for determining the optimum geometrical form of unloaded, split, metal piston rings (Ramsbottom, 1855).

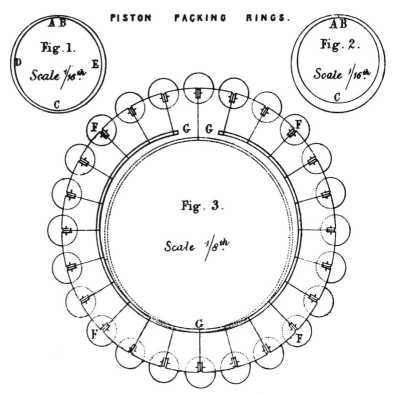

ring truly circular of a diameter equal to the bore of the cylinder. He then attached twenty-four strings carrying equal radial loads to regularly spaced locations round the periphery of the ring and noted the distorted shape GG shown in Fig. 10.44. He argued that a ring having an initial shape GG would deform into a circle when subjected to uniform pressure round its periphery and that this would equalize the wear. This proved to be the case and rings thus bent were found to last much longer than rings having an initial circular shape.

It was also found by accident during these experiments that the widely held view that sealing pressures should at least equal the steam pressure was quite erroneous. In Ramsbottom's experiments a sealing pressure of $3\frac{1}{2}$ lbf/in^2 (24,132 N/m^2) was adequate to seal steam pressures up to 100 lbf/in^2 (689,476 N/m^2).

The sealing methods utilized by Ramsbottom and Joy relied upon the inherent elasticity of the rings. Miller (1862) introduced an ingenious modification which overcame the disadvantages of rigid rings. In a double-acting piston he drilled small holes between the back of the ring groove and the piston crown, thus allowing cylinder pressures to act on the back face of the ring. The sealing pressure thus reflected the pressure in the cylinder and it became possible to use very light, flexible rings having all the advantages of the spiral coil, yet low wear and efficient sealing. With a few modifications this principle has remained in operation to the present day. The holes have been dispensed with and the pressure of the working fluid is allowed to act on the back of the ring by making the axial length of the ring somewhat less than the groove width.

The lubrication of piston rings followed an essentially empirical approach for well over half a century after metal rings became accepted practice. Stanton (1925) carried out experiments on two coupled pistons in horizontal cylinders and concluded from observations of the motion of a driving pendulum that the friction was consistent with boundary lubrication. Castor oil was used as the lubricant and the recorded coefficients of friction ranged from 0·023 to 0·028. However, the speed was only 15 cycles/min and the possibility of fluid film lubrication at more realistic speeds remained. In more recent times, detailed analytical studies and careful experimental investigations have suggested that fluid-film lubrication can normally be achieved throughout most of the stroke, but that boundary lubrication frequently occurs near the limit of the piston excursion at top dead centre. The piston ring-cylinder liner still presents a formidable challenge to the tribologist, but this short history demonstrates that most of the basic design concepts emerged in the last century.

10.11   The influence of professional institutions and government committees

There is one feature of progress in tribology in the period 1850 to 1925 which distinguishes it from earlier history. It was the era in which the newly formed institutions of professional mechanical engineers and government laboratories recognized the significance of specific aspects of the subject and actively encouraged research in the field. Illustrations of such trends from earlier times, like the promotion of work on accurate chronometers by the Board of Longitude in London, early in the eighteenth century, on friction by the French Académie des Sciences in the latter half of the same century and on wear of coins of the realm by Fellows of the Royal Society at the request of a powerful Committee of the Privy Council in the reign of King George III, have already been mentioned, but these were less frequent and more specific than the general trends being discussed here.

The Institution of Mechanical Engineers Research Committee and its Sub-committee on Friction at High Velocities – 1878

It was common practice for the Institution of Mechanical Engineers, formed in 1847, to invite members to submit lists of subjects that would be of interest to its members. Such lists ensured the relevance of meetings and debates to current industrial problems. Furthermore, at its meeting on 20 December 1878, the Council of the Institution took the enlightened view that a *Research Committee* should be established to determine '. . . which further research is desirable'. It was this active promotion of research on matters of immediate industrial concern which led directly to the classical experiments of Beauchamp Tower on plain bearings and the subsequent development of the theory of fluid-film lubrication by Osborne Reynolds described in Section 10.4.

Membership of the Committee on Friction at High Velocities, many of whom will be recognized immediately by readers of the present chapter as notable contributors to the subject of tribology, included: E. A.

Cowper, Captain Douglas Galton, Dr John Hopkinson, Professor H. C. Fleeming Jenkin, John Ramsbottom, Lord Rayleigh, Professor A. B. W. Kennedy.

The American Society of Mechanical Engineers Research Committee on Lubrication – 1915

An appreciation of the importance of lubrication to the rapidly developing industry of the United States was soon developed by the professional society established to concern itself with mechanical engineering in that country in 1880. In 1915 the ASME established its *Research Committee on Lubrication* (RCL) with aims outlined by Flowers and Hersey (1932) as follows: '. . . to investigate the fundamental problems of lubrication phenomena, to formulate the results of investigations previously made and to keep in touch with contemporary research in this field'.

The aims were consistent with the zealous approach to lubrication of the first President of the ASME, Robert H. Thurston, but they once again illustrated an awareness of the impact which research could have upon an essential aspect of industrial development.

The list of eleven members of the first Research Committee on Lubrication again makes interesting reading to the student of tribology: Dr A. E. Flowers, Professor E. Buckingham, Mr L. P. Alford, Professor L. J. Bradford, Dr H. C. Dickinson, Mr M. D. Hersey, Mr H. A. S. Howarth, Professor G. B. Karelitz, Professor A. Kingsbury, Dr B. L. Newkirk, Professor A. E. Norton.

Barwell (1956) has noted that although the First World War interrupted Hersey's work for this Committee, it enabled him to visit the National Physical Laboratory in England where he found that a high priority was being afforded to research on lubrication. The Research Committee on Lubrication of the American Society of Mechanical Engineers still actively identifies subject areas requiring further elucidation and, furthermore, seeks sponsorship for work on such topics. Its *Appraisal of World Literature on Boundary Lubrication* (Ling et al., 1969) is but one example of the fruits of its recent labours.

The Department of Scientific and Industrial Research Advisory Council, Lubricants and Lubrication Inquiry Committee – 1917

In February 1917 a Special Committee of the Department of Scientific and Industrial Research (DSIR) was established under the Chairmanship of Sir Maurice Fitzmaurice to report on a grant application from the Bradford Association for Engineering Research in support of '. . . proposed researches to determine the relation between the viscosity of lubricants and the load on a bearing, and the action of lubricants at high temperatures as applied to commercial methods of oil testing'.

The Committee responded to its task with incredible speed, reported back to the Advisory Council of the DSIR on 1 March 1917, '. . . deprecated an attempt to obtain partial results in any section of the field for research in lubrication, . . .' and rejected the application.

It was, however, impressed by the need for a thorough investigation of the problems of lubrication and suggested that a memorandum should be prepared '. . . on the field for reasearch on Lubricants and Lubrication, which would contain an analysis of the problems involved, together with a suggested scheme for research'.

Financial justification for the recommendation was also introduced in the exceptionally short time available and the wording adopted to persuade those in authority of the economic significance of the subject is well worth repeating.

> ... *The Committee pointed out that before the war the annual expenditure on lubricants in this country was six million pounds, and it was estimated that an annual saving of one to two millions could be effected if a systematic investigation of the subject were undertaken, and the results made freely available to the public. Furthermore, the loss caused by the improper lubrication, whether arising from undue friction or from unduly rapid wear of bearings or from the wastage of oil used in excess, would represent a very large addition to the figure given above.*

The ingredients of the argument are interesting. The ghosts of all who had studied the elements of tribology from Leonardo da Vinci onwards would recognize the merit of the technical arguments, but emphasis was also placed upon the economic significance, excessive use of lubricants, the losses associated with excessive friction and wear and, equally significant, the need to awaken public consciousness. The word *tribology* might well have been introduced in 1917, but it took another bite of the apple half a century later to confirm the situation.

Fortunately the Advisory Council accepted the recommendations of the Special Committee and the *Lubricants and Lubrication Inquiry Committee* was set up with the following terms of reference: '... To prepare a memorandum on the field for Research containing an analysis of the problems involved, together with a suggested scheme of research which would be likely to lead to valuable results.'

Membership of the Committee was as follows: Mr S. B. Donkin, Mr L. Archbutt, Mr A. S. Barnes, Mr A. E. Berriman, Professor C. V. Boys, Mr R. M. Deeley, Professor F. G. Donnan, Mr H. M. Martin, Mr W. W. F. Pullen, Captain D. R. Pye, Dr T. E. Stanton, Mr T. C. Thomsen.

Consultative members included The Rt Hon. Lord Rayleigh and The Hon. Sir Charles Parsons, with Dr C. H. Lander and Mr A. Richardson as Technical Officer and Secretary, respectively.

The membership is once again of interest to students of the subject since it not only contains well-known names of scientists and engineers from the field of tribology but representatives of industry, the armed services, universities and, most significantly, Barnes and Pullen who were HM inspectors of schools. The need to introduce awareness of the subject and education at an early age had clearly been recognized. A Sub-Committee on Lubrication was also formed.

The prodigious work of the Committee was fully reported in a report (DSIR, 1920) submitted to the Advisory Council. Some twenty-two meetings of the main Committee were held, five interim reports submitted, valuable bibliographies prepared, five meetings of the Sub-Committee on Lubrication were held, research work was instituted at the National Physical Laboratory, bulletins on lubricants and cutting fluids were published, existing knowledge on lubrication was summarized and reviewed and recommendations were made for future research on lubricants and lubrication.

The report still forms a good starting point for those wishing to appreciate the significance of tribology and there are two clear impressions which remain. In the first place the Committee sifted all the previous, and in many cases conflicting, reports on friction and, secondly, it classified the various modes of lubrication and their laws with a clarity which had not previously existed, in the form shown in Table 10.2.

Coming as it did towards the end of a most energetic period in the history of tribology the DSIR (1920) Report is a document of wisdom and merit which is only now being recognized. It represents a sound lesson for scientists, engineers, tribologists and those responsible for administration of research at national and even international levels.

10.12 Chronology

It is evident that political, social, scientific and technological events were becoming altogether more international in nature in the period under review. This was not entirely attributable to improvements in communications and transportation, although both were remarkable, but also to the growth of strong nations on most of the continents of the world. Perhaps it was the growing strength of the United States–evident in technological terms from the moment of the International Exhibition at Crystal Palace, London, in 1851, and in political terms in her influence on world events–which was most dominant. While colonial clashes and rebellions occurred in Africa and the East, the European nations engaged in bitter disputes and provided the tinder-box for the first calamitous conflict of worldwide proportions from 1914 to 1918.

Monarchies gave way to republics in numerous states and ferocious civil wars afflicted many nations. The bloody fighting experienced by the young North American States from 1861 to 1865 was balanced in 1917 by the upheaval of the Russian Revolution. Yet science and technology not only survived these political changes, but expanded at an increasing rate and even contributed, as in the past, to military endeavours. It became possible to manufacture high-quality steel and other metals of importance to industry in vast quantities. Parallel developments of machine tools provided the equipment necessary for the production of a new range of man-made devices ranging from the bicycle to the steam turbine. The internal combustion engine entered the scene and impinged upon the domination of steam power. This was particularly evident in the development of propulsion systems for the motor car. Railways continued to expand, impressive tunnels were cut through mountains, rolling-stock improved greatly in quality and electric traction was introduced. Yet the inventions of the age which had even more spectacular impacts upon individual mobility and life-style were the *velocipede*, the *motor car* and the *aeroplane*. All influenced tribology as well as society, and were to introduce social problems whose full impact are only now apparent.

In physical science the great strides forward in atomic physics dominated the period. Thomson's discovery of the electron and Rutherford's work on the structure of the nucleus of the atom revealed and confirmed an atomic form of matter of great beauty and deceptive simplicity.

In tribology pride of place must be given, as indicated by the title of this chapter, to the discovery and commercial exploitation of *mineral oil*, together with the revelation of the nature of both *fluid-film* and *boundary lubrication*. This was as much the age of scientific studies of lubrication as the previous periods had been the eras of consideration of friction. Furthermore, it was not just the science of lubrication which emerged, but the most impressive nature of its application to bearing design. The independent development of *tilting-pad* bearings by Michell in Australia and Kingsbury in the United States is one of the most exciting stories of engineering development based upon sound tribological principles.

The studies of elastic contact by Hertz, the investigations of rolling friction by Reynolds and Heathcote and the studies of bearing load-carrying capacity by Professors Stribeck and Goodman provided the scientific background which supported the emergence of a strong *rolling-element bearing* industry. New manufacturing and mass production techniques, together with the preparation of suitable steels, were the direct result of the demands of the ball- and roller-bearing industry.

All these trends are evident in the chronology shown in Table 10.6. The reader wishing to consult more complete chronologies of scientific, technological and general world events will find rewarding reading in Fisher (1935), Derry and Williams (1960), de Vries (1971), Russell and Goodman (1922), Beazley (1975) and Gunn (1975). The various developments discussed separately throughout the chapter can now be seen in time-sequence and in relation to general political, scientific and technological progress. Such presentations facilitate an appreciation of the interaction of events which might otherwise be lost through studies of particular aspects of the subject. The record of non-tribological developments in the world are necessarily sparse and random in nature, but I hope that they help to provide a useful framework for the reader. The total number of entries, with events listed in almost every year, is but a reflection of the increasing pace of tribological progress.

Table 10.6
Chronology for (1850–1925). Mineral oil and scientific studies of lubrication.

Date (A.D.)	Political and social events	General technical and scientific developments	Tribology
1792	John Russell–later 1st Earl Russell–born. Prime Minister 1846–52		
1819	Alexandrina Victoria born–later Queen Victoria 1857–1901 (UK)		
1850	Millard Fillmore (1850–1853) President of the USA Death of Sir Robert Peel T'ai P'ing rebellion (1850–1864), China Universal suffrage established in France	Robert Stephenson's wrought-iron Britannia Bridge built across Menai Straits Henry Bessemer developed interest in glass-making Rudolph Clausius–second law of thermodynamics	**British patent No. 13292 application by James Young for 'Treating bituminous coals to obtain paraffine . . .' (enrolled 17 April, 1851)**

Table 10.6 (*Continued*)

Date (A.D.)	Political and social events	General technical and scientific developments	Tribology
1851	Great Exhibition–Crystal Palace–London Gold found in Australia Hsien Fêng (1851–61) Ruler of Ch'ing (Manchu) Dynasty, China	Isaac M. Singer's domestic sewing machine, Boston, Massachusetts McCormick reaper exhibited in England Kelly's steel-making converter, USA Submarine telegraphic cable established between England and France Donisthorpe–Lister wool combing machine advances worsted trade in Bradford	**Scottish mineral oil plant opened at Bathgate in the Torbane Hill Fields** **J. Barran's new axle-box for railway engines and carriages**
1852	Louis Bonaparte–Napoleon III, French Emperor (1852–70) (Second Empire) Death of Duke of Wellington Lord Derby Prime Minister UK Lord Aberdeen Prime Minister (1852–55) UK	American Society of Civil Engineers instituted Henri Giffard's steam-driven airship in France Great Northern main line between London (King's Cross) and York completed Sir William Siemens discussion of the physical properties of steam and invention of water-meter	**P. R. Hodge demonstrated success of oil-filled axle-boxes**
1853	David Livingstone began his exploration of the Zambezi M. F. Maurey's *Physical Geography of the Sea* published Franklin Pierce (1853–57), President of the USA Melbourne University founded	Samuel Colt established his armoury (1,400 machine tools) USA Railway built through the Alps between Vienna and Trieste Edge-wheel surface grinder Hypodermic syringe	**J. Lea's patent lubricating composition** **W. Bridges Adams discussion of axle-box lubrication** **Dr Brewer took Titusville oil sample to Dartmouth College for analysis (USA)**
1854	Crimean War between Britain and Russia (1854–56) Siege of Sebastopol (1854–55) Florence Nightingale's reformation of nursing US Republican Party formed University College, Dublin, founded	Water-driven impulse turbine invented in California SS *Brandon*, first ship with compound expansion steam engines	**Pennsylvania Rock Oil Company of New York founded (November) (USA)** **G. A. Hirn's studies of journal-bearing friction published–tests on animal, vegetable and mineral oils, water and air as lubricants** **John Ramsbottom's proposals for piston rings**

Table 10.6 (*Continued*)

Date (A.D.)	Political and social events	General technical and scientific developments	Tribology
1855	Alexander II (1855–81), Russia Professorship of Technology established at Edinburgh University Lord Palmerston Prime Minister (1855–58) UK	Bunsen burner Turret lathe – R. S. Lawrence, Vermont, (USA) Tungsten steel developed, F. Köller, Austria Metal naval 'mines' used by Russians in Crimean War Rayon patented – G. Audemars.	**Pennsylvania Rock Oil Co. of Connecticut formed (September) (USA)** **J. H. Johnson's patent for vulcanite bearings** **David Joy's spiral piston ring**
1856	Alexander II of Russia Treaty of Paris ended Crimean War	Sir Henry Bessemer's converter for the manufacture of steel Frederick Siemens British patent for heat regeneration in engines Synthetic dye 'mauve' created by William Henry Perkin, UK Borden's condensed milk (USA)	**James Miller Williams bought asphalt producing properties at Oil Springs, Ontario from C. N. Tripp** **Small quantities of mineral oil produced in Russia** **Glass bearing patent applications from J. Green (1856), J. S. Hendy (1857); W. Tice (1862)** **John Penn's use of lignum-vitae as a journal-bearing material – particularly marine applications**
1857	Indian Mutiny (sepoy regiments of Bengal Army at Meerut) James Buchanan (1857–61), President of the USA	Safety elevators (hydraulic lifts) – Otis (USA)	**James Miller Williams first (dry) well dug to depth of 27 ft at Bothwell (Canada)** **Small quantities of oil produced in Roumania**
1858	Secretary of State for India appointed to supersede the East India Company Lord Derby Prime Minister for 15 months (UK)	G. F. Göransson perfected Bessemer steel-making process (Sweden) South Foreland lighthouse fitted with arc lamps supplied with electricity from steam-engine-driven generator	**James Young's company world's largest manufacturer of coal oil (Scotland)** **James Miller Williams second well found petroleum at a depth of 65 ft at Enniskellin (Canada)** **John Penn's successful use of lignum-vitae as an improved marine thrust-bearing material** **W. H. Ward's British Patent No. 2585 for hollow balls and rollers in thrust bearings**
1859	Franco-Austrian War in Italy (1859–60) Lord Palmerston Prime Minister (1859–65) UK	Steam-roller invented (France) Iron steamship the *Great Eastern* completed (combined paddle and screw propulsion) Wooden frigate encased by $4\frac{3}{4}$ in armour plate (France) Darwin's *Origin of Species* published	**(Colonel) Edwin Drake struck oil at a depth of $69\frac{1}{2}$ ft at Titusville. 28 August (USA)** **Eugene Carless produced carburine (gasoline) for internal combustion engines in Hackney, London** **Bochet's studies of friction**

Table 10.6 (*Continued*)

Date (A.D.)	Political and social events	General technical and scientific developments	Tribology
1860	Anglo-French Commercial Treaty	Lenoir constructs first practical spark-ignition internal combustion engine Wheatstone patented printing telegraph Maxwell's discussion of viscosity and coefficient of viscosity Traction engines used on farms	**M. Aerts' 'water axlebox' described by Sampson Lloyd**
1861	Victor Emmanuel first King of United Italy Abraham Lincoln (1861–65) President of the USA American Civil War began Legislative Council established in India Emancipation of serfs in Russia	First 'all-iron' warship HMS *Warrior* (380 ft) constructed	**Brig *Elizabeth Watts* carried cargo of petroleum across the Atlantic** **Kerosene exported from USA to Europe**
1862	Companies Act established principle of limited liability International Exhibition (London) Bismarck in power in Prussia T'ung Chih (1862–75) ruler of Ch'ing (Manchu) Dynasty, China	Universal milling machine constructed by Joseph Brown of the company Brown & Sharpe (USA) Steel rails introduced R. J. Gatling's hand-cranked multi-barrelled gun patented	**W. Ashton's patent application for papier-mâché bearings** **A. L. Thirion's application for Patent No. 1485 for use of ball-bearings in a velocipede** **Miller's proposals for pressure-sealed piston rings**
1863	Battle of Vicksburg, General Grant, USA Slavery declared illegal by Lincoln in USA, 1 January	Steam-operated London Underground Railway–cut and cover basis of construction First New Zealand Railroad–Christchurch to Ferrymead	**Clark, Rockefeller and Andrews formed a refinery company in Cleveland, Ohio**
1864	German war against Denmark	Torpedo (self-propelled) invented by the Scotsman, Robert Whitehead, in Austria James Clerk Maxwell's theory of electromagnetic radiation	**Bailly, Durrand, Mesnard and Porier's British Patent No. 2855 first to use the term *cage***
1865	End of American Civil War Assassination of President Abraham Lincoln by John Wilkes Booth Andrew Johnson (1865–69), President of the USA	Electrolytic refining of copper Carbolic acid (phenol) used as antiseptic–Joseph Lister (UK) First 'Pullman' Railway sleeping car (USA)	**First oil pipeline (6 miles) in Pennsylvania (USA)**

Table 10.6 (*Continued*)

Date (A.D.)	Political and social events	General technical and scientific developments	Tribology
	Red Flag Regulation deemed to apply to motor vehicles. Light car speeds restricted to 4 mph in the country and 2 mph in town with a man carrying a red flag in front of each mechanical vehicle		
	Foundation of Massachusetts Institute of Technology (MIT)		
	Lord John Russell Prime Minister (UK)		
1866	Austro-Prussian War	England linked to USA by submarine telegraphic cable	**Scottish mineral oil production 6 million gall per annum**
	Lord Derby Prime Minister (UK)	Aeronautical Society of Great Britain founded	
		Siemens–Martin open-hearth steel	
1867	Dominion of Canada established	Pierre Michaux's velocipede	**C. Tomlinson's discussion of surface forces**
	Reform Bill passed		
	Paris Exhibition	Alfred Nobel discovered dynamite	
	Das Kapital published by Karl Marx		
1868	B. Disraeli Prime Minister (UK)		
	W. E. Gladstone (Liberal) Prime Minister (1868–74) UK	Westinghouse compressed air brake	**J. H. A. Bone's book, *Petroleum and Petroleum Wells*, published (Philadelphia)**
		Disc harrow introduced	**Englishman George Moore won Rouen–Paris road race on a Michaux velocipede fitted with ball-bearings**
			A. C. Cowper's ball-bearings for velocipedes
1869	Ulysses Simpson Grant (1869–77) President of the USA	USA spanned by rail; golden spike ceremony at Promontory Point, Utah, May	**F. J. Suriray's velocipede with ball-bearings**
		Suez Canal opened	
		Bicycle manufacture started in Coventry, England	
		Periodic table–Dmitri Mendeleyev–Russia	
1870	Napoleon III declared war on Prussia	Twine binder for corn	**John D. Rockefeller became President of newly created Standard Oil Company of Cleveland**
	Lenin (1870–1924)		

Table 10.6 (*Continued*)

Date (A.D.)	Political and social events	General technical and scientific developments	Tribology
1871	German Empire established France – the Third Republic Great Chicago fire Trade Union Act (UK) Paris surrendered	1,500 tons of lignum-vitae imported into Europe alone from Santo Domingo Mnt Cenis Tunnel opened (linking Lyons to Turin) G. Westinghouse automatic railroad air brake	
1872	Yellowstone – first National Park in USA	World oceanographic survey commenced by HMS *Challenger* (1872–76) Nicholas Otto's four-stroke internal combustion gas engine	**South Improvement Company formed by dominant refining and railroad companies**
1873	General economic depression	Remington Company markets early typewriter Brayton's oil engine (USA)	**Nobel brothers developed production and transportation of oil in Baku region of Russia**
1874	B. Disraeli's Conservative Administration (1874–80) UK	Telegraph – T. A. Edison, USA	
1875	Disraeli purchased shares in the Suez Canal Company Kuang Hsu (1875–1908), Ruler of Ch'ing (Manchu) Dynasty, China	Gelatine for blasting, A. Nobel Main drainage system completed in London Combination harvesters introduced in USA	**Osborne Reynolds' study of rolling friction**
1876	Philadelphia's Centennial Exhibition Queen Victoria declared Empress of India	Alexander Graham Bell patented telephone with electromagnetic microphone Nicholas Otto's silent gas engine	**Mineral oil produced by V. I. Ragosine in Russia**
1877	Patent protection established in Germany Russia declared war on Turkey Bulgaria autonomous Rutherford Birchard Hayes (1877–81), President of the USA	First tanker (*Zoroaster*) built Joseph Monier's patent for reinforced concrete beams Phonograph invented by T. A. Edison, USA	**'Standard' purchased Empire Transportation Company**
1878	Red Flag law rescinded by the Highways and Locomotives (Amendment) Act, UK Treaty of Berlin Second Afghan war with Russia	Domestic paraffin oil stove invented Joseph Swan's successful carbon-filament lamps Thomas Edison made the dynamo a commercial success	**Russian mineral oil imported into England** **Westinghouse and Galton's studies of friction between brake-blocks and wheels and rails and wheels**

Table 10.6 (*Continued*)

Date (A.D.)	Political and social events	General technical and scientific developments	Tribology
	Salvation Army founded		**D. Rudge's British Patent No. 526 for complete cycle wheel hub with cup-and-cone bearings**
			Institution of Mechanical Engineers Research Committee Established (20 December)
1879	Board of Trade authorizes use of steel in bridges Trotsky (1879–1940) Pacific war (Chile v. Bolivia and Peru).	Electric railway – Berlin T. Edison's carbon-filament incandescent lamp S. G. Thomas and P. Gilchrist's method of producing basic steel	**Institution of Mechanical Engineers Committee on Friction at High Velocities formed** **W. Brown's sealed cup-and-cone bearings**
1880	*Progress and Poverty* by Henry George published. W. E. Gladstone's second Liberal Ministry (1880–85) (UK)	American Society of Mechanical Engineers founded. Robert H. Thurston elected first President of the Society Carnegie's first big steel furnace built in Pittsburgh Hans Renold's roller chain	
1881	Alexander III (1881–94), Russia James Abram Garfield, President of the USA Chester Alan Arthur (1881–85) President of the USA		**Heinrich Hertz's analysis of contact between elastic bodies**
1882	Triple Alliance of Germany, Austria-Hungary and Italy International agreement on 3-mile limit for territorial waters (Hague Convention)	Thomas Edison established a central electricity generating station which provided electric power throughout an area St Götthard Tunnel (28·5 miles) completed	**Beauchamp Tower appointed by the Institution of Mechanical Engineers to investigate friction (20th April)** **Board to Trustees created in New York for separate Standard Oil companies**
1883	International Patents Convention	F. Fisher opened factory for manufacture of bicycles in Germany – steel balls imported from UK Brooklyn Bridge completed	**Beauchamp Tower's first report on friction of lubricated bearings and deductions regarding fluid-film lubrication** **N. P. Petrov's paper on mediate friction in journal bearings published** **Slotte viscometer. Railway from Caucasus to Batum on Black Sea completed – financed by Baron Alphonse de Rothschild**

Table 10.6 (*Continued*)

Date (A.D.)	Political and social events	General technical and scientific developments	Tribology
1884		Charles A. Parsons built first successful steam turbine – Gateshead, UK	**British Association for the Advancement of Science meeting in Montreal, Canada, venue for first presentation of Osborne Reynolds equation of fluid-film lubrication**
		Discovery of ionic dissociation – Svante Arrhenius	
		Photographic roll-film patented by G. Eastman, USA	**A. M. Wellington's study of bearing friction from the motion of trucks on inclined tracks**
			Engler viscometer
1885	Gordon killed at Khartoum	Gottlieb Daimler carburettor, petrol engine and motor cycle, Germany	**R. H. Thurston's *Treatise on Friction and Lost Work in Machinery and Millwork* published**
	Grover Cleveland (1885–89), President of the USA	Karl Benz of Mannheim built petrol-driven automobile, Germany	**Saybolt viscometer (mid-1880s described by Herschel, 1918)**
	Lord Salisbury Prime Minister (UK)	Rover *safety bicycle*, England	
		First skyscraper (10 storeys), Chicago, USA	**Redwood viscometer**
1886	Gladstone Prime Minister (UK)	C. M. Hall (USA) and P. L. T. Héroult (France) independently developed an electrolytic process for manufacturing aluminium	**Osborne Reynolds paper on fluid-film lubrication published by the Royal Society, London**
	First Irish Home Rule Bill		
	Lord Salisbury's (1886–92) Conservative Administration (UK)	First hydroelectric station built at Niagara Falls	**John Goodman's first publications on tribology, *Recent Researches in Friction***
			First direct measurement of film thickness in journal bearings by John Goodman
1887	Clipper *Cutty Sark's* record voyage from Sydney to London in 75 days.	Heinrich Hertz – discovery of radio waves	**Wurzel's translation of Petrov's paper**
			John Goodman's studies of friction in both plain and ball-bearings and his observations on the wear of brasses
		AC motor – N. Tesla (USA)	
1888	Discovery of gold and formation of De Beers' diamond monopoly in South Africa	J. P. Dunlop invented pneumatic tyre for bicycles	
		Cordite – Alfred Nobel	
		First 'Kodak' box camera by G. Eastman (USA)	
1889	London dock strike – expansion of trade unionism	C. G. P. de Laval invented impulse steam turbine – Sweden	**Humber Company's double-row ball-bearings for cycles**

Table 10.6 (*Continued*)

Date (A.D.)	Political and social events	General technical and scientific developments	Tribology
	Cecil Rhodes charter for the British South Africa Company	Eiffel Tower – Paris	
	Benjamin Harrison (1889–93) President of the USA		
1890	John Sherman's Anti-Trust Act signed by President Harrison	About 50 tankers sailing under American, British and Russian flags	**Royal Dutch Company for the working of petroleum wells in the Netherlands Indies formed**
	Slaughter of Sioux at Wounded Knee, South Dakota	First electric tube railway with tunnels bored through the clay	**Oil production in Russia exceeds that in the USA**
		Disc ploughs come into use on farms	**Steel balls for bearings manufactured to within ± 0.001 in of nominal size**
		Daimler motor company founded	
1891		Trans-Siberian Railway begun	**Fourth and final report of the Institution of Mechanical Engineers Research Committee on Friction**
			Miss Agnes Pockles study of thin films on liquid surfaces
1892	Independent Labour Party founded	Rudolf Diesel's engine patented in Britain	**Standard Oil Company Trust dissolved**
	Gladstone (1892–95) Prime Minister (UK)		
1893	Grover Cleveland (1893–97), President of the USA	Henry Ford built his first car	**Petroleum or launch spirit named *petrol* by Simms**
1894	Urban, rural district and parish councils established	Gramophone disc	
	Nicholas II (1894–1917), Russia		
1895	Lord Salisbury (1895–1902) Prime Minister (UK)	Rubber tyres for motor cars by André and Edouard Michelin, France	
		X-rays discovered by Wilhelm von Röntgen – Germany	
1896	Locomotives on Highways Act restricted speeds to 14 mph	Stationary petrol engines used on farms	
		Antoine Becquerel discovered radioactivity – France	
1897	William Cleveland (1897–1901), President of the USA	First successful manufacture of the diesel engine in the Augsburg Works	**Albert Kingsbury's experiment on air-lubricated journal bearings**

Table 10.6 (*Continued*)

Date (A.D.)	Political and social events	General technical and scientific developments	Tribology
		First steam-turbine-driven ship, the *Turbina* demonstrated at Cowes Diamond Jubilee Naval Review	**Shell Transport and Trading Company formed – Marcus Samuel**
		J. J. Thomson discovered the electron, Cavendish Laboratory, England	
1898	Philippines annexed by USA during Spanish-American War		**Albert Kingsbury and John Brown built and successfully operated a tilting-pad thrust bearing**
			Henry Timken awarded US patent for taper roller-bearing
			Timken Roller Bearing Axle Company formed in St Louis, Missouri
			Hoffmann Manufacturing Co. Ltd, formed
			Benjamin's account of ball-bearing design and construction published
1899	Second Boer War (South Africa (1899–1902)	Lord Rayleigh's study of the surface tension of oil films on water	**The Glacier Metal Company Ltd, founded, England**
1900	Labour Party founded, UK	Quantum theory – Max Planck	**Publication of Archbutt and Deeley's (1900) book on lubrication and lubricants**
	Boxer Rising in China	Count von Zeppelin's first rigid airships	**Henry Deterding assumed control of Royal Dutch Company**
1901	Death of Queen Victoria, UK	Rayon – Arthur D. Little, USA	**Richard Stribeck's studies of elastic distortion and static load-carrying capacity of rolling-element bearings – Germany**
	Edward VII (1901–10), UK	Transatlantic wireless telegraphy – Marconi	
	Theodore Roosevelt (1901–9), President of the USA		**Spindletop gusher – 10 January, Texas, USA**
	Nobel Prize Foundation and award of first prizes		
1902	Treaty of Vereeniging ends Boer War	Petrol (gasoline) engine-driven farm tractors introduced – USA	**Richard Stribeck's experiments on friction in plain and rolling-element bearings**
	Mr Balfour Prime Minister (1902–5) UK		**British Patent No. 15131 awarded to E. G. Hoffmann for single-row ball-bearing with inner groove and outer spherical track**

Table 10.6 (*Continued*)

Date (A.D.)	Political and social events	General technical and scientific developments	Tribology
			Timken Company moved to Canton, Ohio
			Merger between Royal Dutch and Shell initiated
1903	Russian occupation of Manchuria	Atomic nucleus discovered by Ernest Rutherford, UK	
	Alaskan frontier defined by USA and Canada	Orville Wright's flight in 12 hp 'Flyer', at Kitty Hawk, North Carolina, (12 s flight), 17 December	
	Universities of Liverpool and Manchester founded		
1904	*Entente Cordiale* between France and England	Thermionic valve invented by Ambrose Fleming–England	**Arnold Johannes Wilhelm Sommerfeld's theoretical solution of the Reynolds equation for journal bearings**
	Speed limit for road vehicles raised to 20 mph (UK)	Rolls-Royce Company founded	**Westinghouse Electric Company built a tilting-pad bearing for large vertical electric motor to A. Kingsbury's design–not installed**
	University of Leeds founded	Work begun on Panama Canal	
1905	Swedish–Norwegian Union dissolved	Theory of relativity–Albert Einstein	**Anthony George Malden Michell's solution of the Reynolds equation for plane inclined slider bearings of finite width**
	Mrs Pankhurst began militant action in support of women's suffrage		**A. G. M. Michell awarded British Patent No. 875 for tilting-pad bearings**
1906	Anglo-Russian *entente*		
	Self-government in South Africa		
	Sir Henry Campbell-Bannerman's new Liberal Administration 1906–8 (UK)		
1907	Agreement between Britain and Russia on Persia	Cunard ships *Mauretania* and *Lusitania* launched	**George Weymouth (Pty) Ltd built and fitted Michell tilting-pad bearing to centrifugal pump on Murray river, Victoria, Australia**
	Vast immigration to the USA (1·2 million)		**Albert Kingsbury applied for US patent for tilting-pad bearing**
			Sven G. Wingquist awarded Swedish Patent No. 25406 for a double-row ball bearing with spherical outer track
			Royal Dutch/Shell Company founded

Table 10.6 (*Continued*)

Date (A.D.)	Political and social events	General technical and scientific developments	Tribology
1908	First Old Age Pensions Act, UK Mr Askquith's Liberal Administration (1908–11) UK Hsuan Tung (1908–12), Ruler of Ch'ing (Manchu), Dynasty, China	Bakelite invented by L. H. Baekeland, USA	**Masjid-i-Sulaiman gusher in Persia**
1909	William Howard Taft (1909–13) President of the USA R. E. Peary (USA) reached North Pole	First aviation meeting – Rheims, France Louis Blériot flew English Channel from Calais to Dover Model T Ford mass produced (15 million produced between 1909 and 1927) Bakelite – first major industrial polymer (plastic) produced	**Anglo-Persian Oil Company formed (Anglo-Iranian 1935– British Petroleum 1954)**
1910	George V (1910–36) UK Union of South Africa Halley's comet observed	Milking machine invented	**Albert Kingsbury awarded US Patent No. 947242 for tilting-pad bearings** **Cylindrical roller-bearing with ground inner and outer rings patented (Hoffmann) in UK**
1911	Parliament Act and Coalition Government (UK) Winston Churchill appointed First Lord of the Admiralty Sun Yat-Sen (1911–12), President of the Republic of China R. Amundsen reached South Pole Philanthropic Carnegie Corporation founded	Cloud chamber – C. T. R. Wilson	**Aktiebolaget Svenska Kullagerfabriken (Swedish Ball Bearing Company) formed** **Fafnir Bearing Company formed in Connecticut, USA** **SKF awarded British Patent No. 1617 for double-row self-aligning bearing with barrel-shaped rollers** **Standard Oil Company dissolved after pronouncement in Supreme Court by Chief Justice White**
1912	China became a republic Yüan Shih-k'ai (1912–13), President of the Republic of China First Balkan War (Serbia, Bulgaria, Greece, Montenegro v. Turkey) R. F. Scott reached South Pole	The *Selandia* became the first large ship (tonnage 4,950) to be fitted with diesel engines. Speed 12 knots	**John Goodman's wide-ranging study of rolling-element bearings and development of dynamic load capacity equations** **P. H. Parr's publication of English translations of N. P. Petrov's papers in *The Engineer***

Table 10.6 (*Continued*)

Date (A.D.)	Political and social events	General technical and scientific developments	Tribology
1913	Woodrow Wilson (1913–21), President of the USA Yüan Shih k'ai (1913–16), President of the Republic of China Irish Home Rule Act	Niels Bohr model of the atom Henry Moseley's discovery of atomic numbers	**W. J. Harrison's analysis of bearings lubricated by compressible (gas) lubricants** **Deeley and Parr proposed the term *poise* for the cgs unit of viscosity** **Abadan Refinery became operational**
1914	Start of the First World War, 4 August Russian invasion of Prussia Battle of the Marne stops German advance on Paris Irish Home Rule Bill passed but suspended during the war		**Ludwig Gümbel's rationalization of bearing friction data** **Mayo D. Hersey's use of dimensionless analysis in the study of journal-bearing friction [ZN/P]** **Ransome and Marles Bearing Co. formed** **Oklahoma oil-field**
1915	Italy joins the Allies, May British Dardanelles campaign fails *Lusitania* sunk off Irish coast	Albert Einstein's general theory of relativity Mills bomb – hand grenade – UK Ford produced one-millionth car	**ASME Research Committee on Lubrication (RCL) formed** **Stone – vertical tube viscometer**
1916	Naval battle of Jutland Lloyd George War Cabinet formed (UK) Li Yüan-hung (1916–17), President of the Republic of China		**H. M. Martin's analysis of gear lubrication**
1917	Germany opens submarine warfare The USA joins the Allies in the First World War *Introduction to Psychoanalysis* published – Sigmund Freud Russian–German Armistice – December Russian Revolution (October revolution led by Lenin) (1870–1924) British captured Jerusalem from Turks Gen. Feng Kuo-Chang (1917–18), President of the Republic of China IQ test instituted (USA)	Trans-Siberian Railway completed	**Lord Rayleigh's analysis of externally pressurized thrust bearings** **Harkins et al.'s study of surface films and molecular orientation** **I. Langmuir's study of thin surface films**

Table 10.6 (*Continued*)

Date (A.D.)	Political and social events	General technical and scientific developments	Tribology
1918	End of the First World War – Armistice concluded, 11 November		**Lord Rayleigh's determination of the optimum profile of thrust bearings**
	General election – Lloyd George's coalition defeated Labour and Liberals (December)		
	Hsü Shih-Ch'ang (1918–22), President of the Republic of China		**Lord Rayleigh's study of the lubricating properties of thin oily films**
1919	Paris Peace Conference and Treaty of Paris	Lord Rutherford discovered proton, Cambridge, England	**W. J. Harrison's study of dynamically loaded bearings**
	Soviet Rupublic established	First transatlantic flight by Sir John Alcock and Sir Arthur Whitten Brown	
1920	Peace declared (January)		**DSIR Advisory Council report of the Lubricants and Lubrication Inquiry Committee**
	Palestine established		
	First meeting of League of Nations		**I. Langmuir's further studies of thin surface films**
	Prohibition – Eighteenth Amendment (USA)		**California oil-field**
1921	Warren Gamaliel Harding (1921–23), President of the USA		**Ludwig Gümbel's discussion of cavitation boundary conditions in fluid-film bearings**
			H. L. Heathcote's recognition of micro-slip in ball-bearings
			Michell Bearings Ltd, formed, Newcastle, England
			Introduction of copper–lead bearing alloys
1922	Li Yuan-Hung (1922–23), President of the Republic of China	Technicolour films	**Sir William Bate Hardy and Miss Ida Doubleday introduced the concept and term 'boundary lubrication'**
	USSR established Rise of Mussolini – Italy		
	A. Bonar Law Prime Minister (UK)		**T. E. Stanton's experimental study of the lubrication of gears and low conformity journal bearings**
1923	S. Baldwin Prime Minister (UK)	Air conditioning developed by Willis Carrier, USA	
	Turkish Republic proclaimed		
	Mikhail I. Kalinin, Chairman of Presidium of the Supreme Soviet of the USSR		

Table 10.6 (*Continued*)

Date (A.D.)	Political and social events	General technical and scientific developments	Tribology
	Calvin Coolidge (1923–29) President of the USA Tsao Kun (1923–24), President of the Republic of China		
1924	Civil war in China (1924–28) Death of Lenin – replaced by Stalin Stalin v. Trotsky First Labour Government in Britain Ramsay MacDonald (Jan.) and S. Baldwin Prime Ministers (UK)	Wave mechanics – L. de Broglie and E. Schrödinger Sir Edward Appleton used 'radar' to detect electrically charged layers in upper atmosphere (ionosphere) H. Eckener flew his airship Z-R-3 across the Atlantic	**Kingsbury Machine Works founded in Philadelphia**
1925	Paul Von Hindenburg, President of Germany	Wolfgang Pauli's exclusion principle Tungsten carbide cutting tools	**Ludwig Gümbel's view that the major source of dry friction arose from deformation of material by penetrating asperities** **T. E. Stanton's deduction of boundary lubrication between piston rings and cylinder walls**

10.13 Summary

The essential thesis of this chapter is that the outstanding features of the period 1850–1925 in terms of tribology were the discovery of mineral oil and the penetrating scientific studies of the process known as lubrication. The emergence of vast quantities of mineral oil in the USA and Russia had an immediate effect upon illuminants, but once it was recognized that part of the naturally occurring fluid formed an excellent lubricant and fuel, the oil industry expanded rapidly. The story of the discovery of mineral oil has been linked with brief accounts of the emergence of the early major oil companies to provide a more complete understanding of some of the general technical and tribological developments.

In previous times the source and nature of friction in machine elements had been the main concern of scientific investigations in the field of tribology, but in the present period scientists and engineers turned their attention to lubrication. For the first thirty years or so following 1850 progress towards a better understanding of the nature of lubrication was essentially empirical, but thereafter experimental and analytical approaches yielded spectacular results. The experimental background developed in the USA (Thurston), Russia (Petrov) and in England (Tower) in the late 1870s and early 1880s. The simultaneous and totally independent studies by Petrov and Tower in two different countries which were reported in 1883 provides a good subject for debate on the subject of historical coincidence. In this case it appears to have been the twin spurs of problems with bearings on railway rolling-stock and the availability of a new and potentially valuable lubricant in the form of mineral oil which accounted for the complementary work in the two countries.

The theory of fluid-film lubrication, first presented by Osborne Reynolds at a meeting of the British Association in Canada in 1884 and published in full by the Royal Society in London in 1886, is probably the most significant single feature in the whole history of tribology. It has therefore been afforded some prominence, but accounts of the further studies of lubrication which rippled through the scientific world between 1886 and 1925 have also been included. The revelation of the beautiful mechanism of fluid-film lubrication, which is still the finest example of slow viscous flow theory for the student of fluid mechanics, was matched in 1922 by Sir William Bate Hardy's classical work on boundary lubrication. It is for this reason that the story of the present chapter spans the period 1850–1925, for within that time span the two major modes of lubrication were exposed and clearly recognized. It suddenly became clear that the successful operation of sliding machine elements depended upon the existence and persistence of lubricating films having thicknesses in the range 10^{-9}–10^{-5} m.

The scientific achievements, attributable incidentally to an engineer (Reynolds) and a biologist (Hardy), were matched by equally impressive engineering work. The elegance, simplicity and outstanding effectiveness of the tilting-pad bearings devised independently by Michell (Australia) and Kingsbury (USA) form illustrations of engineering at its best. The growth of specialist plain-bearing manufacturing companies to exploit the demand for the more conventional and the revolutionary tilting-pad bearings has been outlined to complete the picture. It is doubtful if many of the general technical developments of the period, like the high-powered, fast, steam-turbine-driven ships, or even the automobile could have taken place at all without the corresponding progress in lubrication and bearing technology.

The rolling-element bearing also developed with impressive speed. The velocipede provided a major impetus for the development of the ball-bearing, while horse-drawn, four-wheeled road carts and the early automobile had a similar effect on cylindrical and taper roller-bearing progress. The background scientific work on elastic contact (Hertz), bearing load-carrying capacity (Stribeck and Goodman) and rolling friction (Reynolds and Heathcote) is reported. Likewise, to facilitate an appreciation of overall progress, an account is given of the emergence of some of the early specialist ball- and roller-bearing manufacturing companies.

Basic studies of friction were dominated by a consideration of the relationship between static and kinetic friction. It was only in this period that it was finally recognized that there was normally a steady, continuous transition from one state to the other. The major factors promoting such studies were the interest in traction between locomotive wheels and the rails and the need to provide effective braking for trains.

There were equally important and impressive developments in the sealing of pistons in steam and internal combustion engines. The inventions of John Ramsbottom were outstanding and there can be no finer tribute to his work than to record that present-day piston rings look remarkably like the original Ramsbottom rings of 1854.

Finally, it was felt to be desirable to draw attention to the increasing role of committees, the professional institutions and government laboratories upon the forward march of scientific research in general and tribology in particular. Such groups drew attention to subjects requiring further elucidation and actively promoted research work in these fields. It was a pattern which was to be extended throughout the twentieth century in many lands.

Chapter 11

. . . *A ring is worn thin next to the finger with continual rubbing. Dripping water hollows a stone, a curved plough-share, iron though it is, dwindles imperceptibly in the furrow. We see the cobble stones of the highway worn by the feet of many wayfarers. The bronze statues by the city gates show their right hands worn thin by the touch of all travellers who have greeted them in passing. We see that all these are being diminished since they are worn away. But to perceive what particles drop off at any particular time is a power grudged to us by our ungenerous sense of sight.*
Lucretius (95–55 B.C.)

Towards tribology
1925–the present

11.1 Introduction

Since starting to write this book I have developed a strong affinity for those historians who fight shy of assessing recent, by which I mean the past half-century, events. In the first place it is extremely difficult to assess the long-term relevance of current and recent work, and secondly the very volume of it makes such a task doubly difficult.

There seems little hope of achieving any valid historical perspective of the half-century at this stage, yet I am conscious of the fact that those engaged in the subject of tribology frequently seek to connect the emergence of contemporary events with the past. I had intended to curtail my history with the year 1925 reached in Chapter 10, but in order to meet this understandable desire of some readers and to avoid a complete gap between 1925 and the present day, I propose to highlight some of the work undertaken in the recent past.

This chapter will be quite different from the earlier ones. The approach will be selective rather than comprehensive, an outline of problem areas rather than a detailed account of their solution, and in some cases inevitably reporting rather than appraisal.

In chronological terms the developments fall into two periods of time of almost equal duration, with the end of the Second World War marking the demarcation. In terms of subject-matter there seems little doubt that if the previous seventy-five years stood for *lubrication*, and the time before that for *friction*, the present period must be labelled *wear*. It is for this reason that I have adopted an opening quotation on this most intriguing, important, yet evasive subject.

Apart from *wear*, a number of other topics in which progress has been made will be mentioned, but no attempt will be made to record a detailed chronology. The serious student should have little difficulty in locating the relevant literature, completing a review and forming some judgement of particular topics. This is another sense in which the lack of a full history can be justified: published books and papers for the period are readily available, whereas those from earlier times are more difficult to locate. Perhaps the most useful recent texts for supplementary reading on the history of tribology are those by Kragelskii and Shchedrov (1956) and Hersey (1966). Furthermore, many individual books on tribological topics like those by Allan (1945), Bowden and Tabor (1964), Rabinowicz (1965), Kragelskii (1965), Cameron (1966), Dowson and Higginson (1966), Moore (1972) and Halling (1975) include sections devoted to historical developments in several fields. A number of papers on historical aspects of tribology have been published in recent years including those by Hersey (1933, 1934), Davison (1957a,b), Tabor (1962, 1969), Courtel and Tichvinsky (1963), Muijderman (1966), Furey (1969), Dowson (1973, 1974a,b, 1975, 1976) and Wilcock (1974a).

It would, of course, be inappropriate to complete the story without giving some immediate background to the Department of Education and Science (1966) Report and the emergence of the word *tribology*. In this section of the chapter the discussion will be broadly divided into the *economics* and the *philosophy* of the situation, with some reference to *education*. I hope that the reader will find that the previous chapters have provided firm indications of the foundations upon which the developments of the 1960s were built.

11.2 Surface topography

An understanding of dry contact between engineering surfaces, friction, wear and some forms of lubrication depends upon a valid representation of surface topography. It was Bowden (1950) who said that, '. . . Putting two solids together is rather like turning Switzerland upside down and standing it on Austria – the area of intimate contact will be small.'

The sentiment is valid, even if the ratio of vertical and horizontal scales is somewhat magnified. It is now known that most tribological surfaces are covered by asperities having slopes almost entirely within the range 0–25° (0–0·44 rad), with the vast majority in the narrow band of 5–10° (0·09–0·17 rad). Perhaps a better analogy would have been the Yorkshire Wolds turned over on to the South Downs of England.

Various physical methods have been developed to yield information on surface topography, including *oblique sectioning, optical interferometry, optical* and *electron microscopy* and *profilometry*. Good introductions to these procedures can be found in Bowden and Tabor (1950), Halliday (1955), Barwell (1956), Williamson (1967), Ling (1969, 1973), Moore (1975a) and Halling (1975).

Of all these physical methods used to study surfaces, profilometry, in which a fine diamond stylus is drawn across the surface and its vertical excursions magnified and recorded has found the widest application. The procedure was first introduced by Abbott and Firestone (1933) of the University of Michigan and the *talysurf* record has proved to be a valuable if not entirely adequate print of the nature of tribological surfaces. The procedure known as *micro-cartography*, in which large numbers of closely spaced parallel talysurf traces are used to generate contour maps of surfaces, was developed in the 1960s by Greenwood and Williamson.

In recent years there has been evidence of greater vigour in the search for better surface quantification. Advances in friction, wear and studies of the lubrication of rough surfaces call for much greater information than the *centre line average* or *arithmetical mean deviation* and the *height of irregularities*. In particular it is necessary to know the distribution of asperity heights, slopes, spacing and the radii of peaks. The importance of these studies is evident when it is realized that most manufactured surfaces are very rough on an atomic scale, with asperity heights being 10^2-10^4 times as large as the atoms which form the features.

11.3 Friction

Sliding

In the long history of studies of dry friction four major approaches can be recognized. They are:

1. The geometrical or mechanistic view.
2. The molecular or adhesive concept.
3. The deformation or ploughing approach.
4. Combined adhesive and ploughing actions.

The first view, which dominated most of the analysis and interpretation of experiments throughout the seventeenth, eighteenth and nineteenth centuries, has been discussed in earlier chapters. In essence it attributes friction to the force required to pull a solid up a series of inclined planes formed by the asperities on the opposing surface.

Adhesion was first proposed as an element in the friction process by Desaguliers (1734), but it was Prandtl (1928) who returned to the issue and Tomlinson (1929) who considered the question of sliding friction under dry conditions in terms of energy dissipation within the molecular fields of force between opposing surfaces. The *molecular* aspect of friction tentatively introduced by Desagulier 200 years earlier was thus revived, although certain aspects of the resurrection remained empirical and necessarily conceptual rather than quantitative. It nevertheless provided the springboard for a remarkable development of the molecular theory of friction by Deryagin (1934), Holm (1938), Bowden and Tabor (1939) and Ernst and Merchant (1940).

Tomlinson (1929) assumed that both the normal load and the tangential force of friction were linearly related to the number of interacting atoms, but in the absence of more detailed knowledge of surface deformation and intermolecular forces he was unable to quantify the phenomenon.

Deryagin made a bold attempt to quantify some of the concepts outlined by Tomlinson in a statistical approach involving detailed representations of intermolecular forces and crystal structure. He derived a two-term or binomial expression for friction of the form,

$$\mu = \frac{F}{a(p + p_0)}, \qquad \qquad \dots [11.1]$$

where μ is the coefficient of friction, F the total force of friction, a the real area of contact and p and p_0 the mean pressures or stresses introduced by external forces and molecular adhesion, respectively. Details of the work have been questioned and debated but there is little doubt that it was a valuable step forward from the tentative proposals put forward by Tomlinson. A useful and more detailed account of Deryagin's work can be found in Akhmatov (1966).

Holm carried out a comprehensive study of the manner in which electricity was conducted from one solid to another in such devices as terminals, relays and circuit-breakers and by 1938 he was convinced that clean metal surfaces would deform plastically at asperity contacts and *cold-weld*. He concluded that one element of the force of friction must be attributable to the sum of the shearing strengths of the asperity contacts.

Bowden and Tabor's approach was also based upon the recognition that surfaces in contact touched only at points of asperity interaction and that the very high stresses induced in such regions of small area would lead readily to local plastic deformation. The penetration of an asperity into the opposing surface could be likened to a miniature hardness test and the mean normal stress (p) over the real areas of asperity contact (a) could be represented to all intents and purposes by the hardness (H) of the softer material. Likewise, if s represented the shear stress of the asperity junctions, the normal load (P), friction force (F) and coefficient of friction (μ) could be expressed with appealing simplicity by the relationships,

$$P = aH,$$
$$F = as, \qquad \qquad \dots [11.2]$$
$$\mu = \frac{F}{P} = \frac{s}{H}.$$

This expression for the coefficient of friction in terms of established mechanical properties of materials represented a great step forward in the theory of friction, although it is clearly incomplete. It neglects the detailed, more complex nature of asperity interactions and deformation and, of course, accounts only for the *adhesive* element of friction. The limitations are apparent when it is recognized that, for metals, $s \approx 0.5\sigma_y$ and $H \approx 3\sigma_y$, where σ_y is the yield stress in tension. All clean metals should thus exhibit a universal coefficient of friction of $\frac{1}{6}$, which is qualitatively and satisfyingly consistent with the statements of Leonardo da Vinci and Amontons, but unfortunately unrepresentative of more sensitive experimental findings.

In their studies of the metal-cutting process Ernst and Merchant (1940) attempted to take account of surface roughness and the fact that the real areas of contact between asperities might be inclined at an angle (θ)

to the direction of overall sliding. The resulting expression for μ was similar to equation [11.2], but involved additional terms in $\tan \theta$ on the right-hand side of the equation.

In the third approach to the nature of friction it was argued that a force would be required to move hard asperities through or even over another surface and that this micro-cutting motion represented the friction process. The idea had received its first serious recognition from Gümbel and Everling (1925) and it is still the subject of detailed studies in the field of plasticity theory. The idea is simple enough. If the sum of the projected areas of the indenting asperities perpendicular to the direction of sliding is a_0, and the mean stress resisting plastic deformation of the softer material which is being cut is equal to the hardness (H), the total force of friction (F) will equal $a_0 H$. Likewise, if the applied load (W) is carried on a number of asperity contacts of real area (a), W will equal aH. The coefficient of friction (μ) thus becomes

$$\mu = \frac{F}{W} = \frac{a_0}{a}.$$

For conical asperities having sides sloping at a mean angle θ to the direction of sliding,

$$\mu = \frac{a_0}{a} = \frac{2}{\pi} \tan \theta, \qquad \qquad \dots [11.3]$$

while for hemispherical asperities of radii (R) and small penetration,

$$\mu = \frac{a_0}{a} = \frac{4\theta}{3\pi}. \qquad \qquad \dots [11.4]$$

Since θ is small, typically 5–10° as seen in Section 11.2, it can be seen from equations [11.3] and [11.4] that alternative specifications of asperity geometry have little effect upon the magnitude of the friction force attributable to plastic deformation.

If it is assumed that both molecular (adhesive) and deformation (ploughing) actions are effective, then the resulting expression for μ is simply a combination of equation [11.2] and either [11.3] or [11.4]. Thus,

$$\mu = \frac{s}{H} + \frac{2}{\pi} \tan \theta. \qquad \qquad \dots [11.5]$$

In a series of experiments carried out in Melbourne, Australia during the Second World War, Bowden et al. (1943) used sliding contacts of differing geometries to ascertain the relative importance of *adhesion* and *ploughing*. The sliders included a *sphere*, a circular section *spade* and a *cylinder* with its axis parallel to the direction of sliding.

The findings demonstrated that adhesion played the major role in determining the friction between metals, and although this conclusion has to be treated with some caution and related to specific material combinations, it is still regarded as a valid observation.

Good overall reviews of present understanding of the alternative views of friction can be found in Bowden and Tabor (1950, 1964), Holm (1958),

Kragelskii (1965), Rabinowicz (1965), Merchant (1968), Moore (1975*a*) and Halling (1975), together with a valuable background text on the hardness of metals by Tabor (1951).

Before leaving this brief review of recent work on sliding friction reference should be made to a most interesting extension of studies of adhesion. It has already been noted that simple theory suggests that the coefficient of friction due to this process should be constant and equal to about ⅕. In practice the coefficients are nearer to unity for clean metals and dependent upon the combination of materials. Courtney-Pratt and Eisner (1957) drew attention to the intriguing mechanism of junction growth. The simple relationship (eqn [11.2]) was based upon the assumption that the area of real contact was determined by the normal force alone, but when the combined normal and shear stresses are considered and a yield criterion introduced, it becomes clear that the real area of contact can increase many times before sliding occurs. The net result of this is that the force and hence the coefficient of friction increases considerably above the predictions of equation [11.2]. The concept of junction growth is one of the most exciting developments in the field of friction studies in recent years. However, it remains clear that the continued application of plasticity theory to the friction problem will further enhance our understanding of the process.

Rolling

Thoughts about the mechanism of rolling friction in the period 1850 to 1925 had been dominated by the writings of Reynolds and Heathcote on micro-slip as outlined in Chapter 10. The subject was advanced considerably between the mid-1950s and 1970s.

In two valuable contributions to the subject Eldredge and Tabor (1954) and Tabor (1955) reported on the influence of both *plastic* and *elastic* deformation in the rolling process. In the first case a hard steel ball was rolled between parallel surfaces of softer metals including lead, tin, copper, mild steel, nickel, Hiduminium, phosphor-bronze and Duralumin. The authors concluded that on the first traverse the major source of resistance was plastic displacement of the softer metal ahead of the ball, and that it could be calculated from the area of the grooved track and a flow pressure of the metal which was approximately equal to the hardness of the material. Subsequent traverses caused the groove to enlarge and the rolling resistance to reduce until, in some cases, constant conditions were achieved. It was found that the presence of a lubricant had little effect upon the rolling resistance.

Tabor (1954) later considered rolling in the elastic range in some detail. He studied the results of rolling a steel cylinder over rubber, a sphere on flat and initially grooved rubber and steel and finally of rolling two spheres together in a pendulum machine. The merit of the latter experiment, which was first proposed by Tomlinson (1929), is that slip of the Reynolds type and differential micro-slip of the Heathcote type is eliminated.

The outcome of all these experiments was the conclusion that rolling resistance due to Reynolds micro-slip was negligible, that Heathcote differential slip also contributed very little unless the conformity between ball and groove was great and that since the presence of a lubricant had a negligible effect, the interfacial adhesion mechanism proposed by Tomlinson could not be recognized as an important factor. It was concluded that the major source of resistance to rolling under purely elastic conditions was a hysteresis loss.

At slightly higher loads plastic deformation occurs beneath the surface in the contact zone, but Merwin and Johnson (1963) showed that if the plastic zone was small and fully contained by material within the elastic range, the rolling resistance could be expressed as a function of the ratio of the maximum Hertzian contact pressure to the yield stress in shear. The analysis also revealed that residual stresses could attenuate the extent of plastic deformation on subsequent traverses of the rolling element and could even eliminate it in due course – a condition described as *shakedown*. Insight into rolling resistance in this elastic–plastic range was greatly advanced by the analysis. A valuable collection of papers representative of knowledge on rolling contact in 1960 was edited by Bidwell (1962).

For fully developed plastic flow with the region of plastic deformation extending to the surface of the solid both in front and behind the rolling component, Collins (1972) provided an approximate but valuable solution to the problem of rolling friction. He also drew attention to the interesting result that the coefficient of rolling friction in this plastic range was different if motion was caused by a tangential force applied to the axis of rotation or by a torque applied to the rolling element, the former being approximately twice as large as the latter.

By this time a fairly clear picture had emerged of the major factors responsible for rolling friction in elastic, elastic–plastic and fully plastic contact conditions. A neat non-dimensional representation of the complete findings was provided by Johnson (1972).

11.4   Plain bearings

Materials

The reciprocating automobile and aircraft engine did much to promote the development of plain-bearing materials in the past half-century. It is worth recalling that as the century opened, Archbutt and Deeley (1900) discussed metallic bearing materials under the headings of *cast iron*, *bronze* (consisting of copper, tin and zinc with the two latter elements in minor proportions), *white* metal or *babbitt* (being an alloy of tin, copper, antimony, lead and zinc with tin being the major constituent) and to some small extent *aluminium*. White metal was seen to be the most important material and the solid iron and bronze bearings had been generally supplanted by bearings made of the same hard metals but lined with anti-friction metal. The cost of replacing white-metal bearings was much less than that of changing the complete 'solid' bearings and the soft metal could readily be *scraped* and *bedded in*. Bochvar (1918, 1921) completed a wide-ranging study of babbitts and proposed the use of lead-based plain-bearing materials.

Non-metallic materials in frequent use in the 1900s included hard woods such as *crabtree*, *hornbeam* and *beech* fitted into an iron frame to form *gears* and *box*, *beech*, *holly*, *elm*, *maple*, *oak*, *hickory*, *canewood*, *snakewood* and *lignum-vitae* for shaft bearings. An interesting reference in Archbutt and Deeley (1900) extolled the virtue of *rhinoceros hide* gear wheels because the material '. . . wears well with small loads and runs silently'. Other non-metallic bearing materials included *stone*, *agate* and *jewels*.

Nearly sixty years later Wilcock and Booser (1957), in their excellent book, noted that babbitts still accounted for more than 75 per cent of sliding-bearing materials. By this time, however, the constant up-rating of bearing load-carrying capacities in reciprocating engines had induced more detailed studies of the desirable characteristics of bearing materials. These included *compatibility* with the rubbing counterface material, *embeddability* for dirt particles and wear debris, *conformability* to enable the bearings to accommodate misalignment, geometrical errors and deflections in the structure, *strength*, *corrosion resistance* and acceptable *cost* and *availability*.

Most of these desirable properties are self-evident, but some are conflicting. Embeddibility and conformability features generally call for a low modulus of elasticity and hardness and these properties are not usually consistent with high strength. In many applications in which the load is steady it is the simple compressive strength that is important, but when the load changes in magnitude and direction, as in reciprocating machinery, fatigue life of the material is a dominant factor.

The range of bearing metals has increased enormously in the past half-century, with the metallurgist making vital contributions. An early development, dating back to the 1920s, was to retain the desirable bearing characteristics of tin and lead and yet provide adequate strength by incorporating one of these materials in a harder metal matrix. Copper–lead alloys were typical of this two-phase system, although the increased hardness made it necessary to increase accuracy and alignment of the bearings and to use harder steel shafts. The relatively poor resistance of lead to corrosion presented a problem which was alleviated by the addition of tin and antimony and, in the case of copper–lead bearings, by forming a thin surface film of lead–tin or lead–indium alloy by electroplating. Such *overlays* were introduced early in the Second World War, about 1940, and remain an important feature of many modern bearings. Aluminium–tin alloys, having excellent corrosion resistance, were introduced in the 1950s and were described by Forrester (1968) as the most important development in soft-metal bearings in the previous decade.

At the beginning of the century babbitt linings were rarely less than $\frac{1}{8}$ in (3 mm) in thickness and not infrequently at least $\frac{1}{4}$ in (6·4 mm) thick. The need to provide adequate compressive and fatigue strength gradually brought the thickness of the bearing layer down to a few thousandths of an inch, at the expense of other desirable features such as embeddability and conformability. The thin-walled bearing in which a thin layer of soft bearing metal was mounted on a stiff steel shell was developed and was, of course, ideal for mass production. The optimum thickness of the bearing layer depended upon the engine and bearing ratings, but was generally between 0·001 and 0·005 in (0·02 and 0·12 mm). The development of thin-walled bearings is described in Section 11.5. Good accounts of the development of bearing metals during the twentieth century have been presented by Neave (1937), Wilcock and Booser (1957), Love (1957), Radzimovsky (1959), Holligan (1966), Morris (1967) and Wilson (1972).

Non-metallic bearings such as *rubber* and *graphite* have found increasing application, and early accounts of their use were presented by Brazier and Holland-Bowyer (1937) and Fogg and Hunwicks (1937). It is, however, the *polymeric* or *plastic* bearings which have made the greatest impact in this field. They fall into two categories: *thermosetting* and *thermoplastic* materials. The fabrics or non-orientated fibres were generally set in phenolic or very occasionally cresylic resins in the thermosetting materials introduced about 1930. The bearings were generally lubricated by water and were readily applied to rolling-mill bearings in the United States. Eyssen (1937) and Richardson (1937) were among those who talked about this important class of bearings before the Second World War and the subject has grown enormously in the postwar years.

In the thermoplastic field *nylon* was soon recognized as a valuable bearing material, followed by that remarkable low-friction polymer *polytetrafluoroethylene* (PTFE) early in the 1950s. Attention was drawn to the promising features of polyamides by Al'shits (1959). The *polyacetals* were introduced in the 1960s followed by *polypropylene* and *polyethylene*. The high-density, ultra-high-molecular weight version of the latter material has become an established component of most total replacements for synovial joints in the 1970s.

The great merit of these materials is that they can operate effectively without lubricant, although their mechanical properties generally limit their application to lightly loaded conditions, and often low speeds, such as those encountered in domestic equipment, toys, instruments, clocks and electrical switches. Sometimes solid lubricants like molybdenum disulphide (MoS_2) and graphite are added to improve the tribological characteristics of the materials, and silicone-fluid-impregnated polymers which provide a self-lubricating facility are now attracting attention. An important class of bearings which combines the good tribological characteristics of certain polymers with the strength of metals is the sintered bronze or iron bearing impregnated with PTFE.

Good general accounts of the history, tribological properties and application of these newer bearing materials can be found in Bowden and Tabor (1950, 1964), Love (1957), Wilcock and Booser (1957), Kragelskii (1965), Pratt (1967), Salomon (1968), Lancaster (1972), Clauss (1972), Tabor (1972), and Lee (1974).

Manufacture

The most general change in plain bearings relevant to manufacture in the past fifty years has been the substantial move away from solid bearing blocks, with or without thick white-metal linings, towards *bushes* or *thin-walled* bearing shells in reciprocating machinery. As early as 1919[1] a small company known as Dann Products in Cleveland, Ohio, was manufacturing bushes from brass strip by indenting the brass and filling the depressions with a mixture of graphite and pitch. The company changed hands, became known as Cleveland Graphite Bronze and shortly afterwards formed an association with a press engineer by the name of Klocke who was interested in bearing manufacture. The idea of continuously coating metal strip and mass producing bearings emerged and was patented (e.g. British Patent No. 409289 issued on 7 January 1932). Mr G. A. Vandervell was granted a licence to manufacture bearings in England by this means on 28 November, 1932. In the next five years various motor-car manufacturers introduced the engine design changes necessary to allow the introduction of interchangeable, precision, thin-walled, mass-produced bearings. The Glacier Metal Company also began to manufacture thin-walled bearings about 1938, and although challenged by Vandervell for infringement of the patent, the legal action was held in abeyance during the Second World War and only resolved in the House of Lords, after the patent had expired, in 1950.

Application of thin-walled bearings to motor cars was interrupted by the Second World War, but during the hostilities they were adopted in aircraft engines and military vehicles, including tanks. The bearing coatings were extended to *cast lead bronze* and *copper lead* on steel strip and in due course General Motors patented the use of *indium* plating on such materials to alleviate the problem of lead corrosion. The link between Vandervell and the Cleveland Graphite Bronze Company, which had by then become the Clevite Corporation, was broken in 1964, with Vandervell Products Ltd, continuing as an independent company in Maidenhead, England.

The Glacier Metal Company expanded rapidly in the interwar years. It became a public company in 1935 with the original owner as its first managing director. In 1938 Wilfred Brown took over, and during the Second World War production was expanded by the acquisition of additional premises in London and in Ayr, Scotland. Immediately after the war the company decided to retain production facilities in Scotland and a new factory was opened in Kilmarnock. In 1970 the company acquired the Glacier Metal Company of Richmond, Virginia, USA, which had been created by Mr John Battle, one of the original founders of the British company. In 1964, the company merged with, and became the Bearings Division of, Associated Engineering Limited.

Attention has been focused upon the development of two specialist plain-bearing manufacturers to illustrate the development of manufacturing procedures since 1925. Some large engineering companies still make their own bearings, but the emergence of mass production and companies specializing in bearing manufacture has greatly assisted general machine designers. Dowson (1937) has summarized the state of the art in the field of design, lubrication and selection of materials for both tilting-

---

1.   I am grateful to Mr D. F. Green, Research Manager of Vandervell Products Ltd, for kindly providing background information.

pad and plain-collar thrust bearings just before the start of the Second World War. The opportunity to select from catalogues readily available reliable plain bearings has relieved the machine designer of many problems, although the situation has only been reached as a result of great scientific and technological progress in lubrication theory, materials science, manufacturing techniques and the development of bearing-design procedures. The latter point is of sufficient merit to warrant special consideration in Section 11.8.

11.5   Rolling contact bearings

Ball- and roller-bearings must surely represent the pinnacle of twentieth-century accuracy of manufacture combined with mass production. We have seen how the companies formed early in the present century, or even at the end of the 1800s, relied upon the inventiveness of men skilled in engineering manufacture, the scientific studies of Heinrich Hertz and the experimental investigations of Professors Richard Stribeck and John Goodman. Stribeck considered load distribution and related static load-carrying capacity to bearing geometry, while Goodman confirmed the findings and related load-carrying capacity to speed, or life, in empirical equations which are not unlike those in current use. The industry was thus well established and supported by sound, if somewhat empirical, knowledge by 1925.

The major problem facing the industry was the question of bearing *life* or *endurance*. The elastic contacts in rolling-element bearings are subjected to high and repeated stresses, the levels of which are affected by such factors as load, material properties, lubrication and particularly geometrical perfection. It is not surprising that the life of such bearings under given conditions – like that of many manufactured articles and man himself – shows a spread or distribution characteristic of the species. The major mode of failure in rolling bearings is fatigue, and the probability of survival for a particular bearing varies from 100 per cent for zero revolutions to 0 per cent for an infinite number of revolutions. Bearing manufacturers have to exercise some caution in recommending suitable bearings for given loads and endurance, and in recognition of representative life-dispersion curves tend to refer to one or two meaningful features of the statistical nature of the problem. The L_{10} figure represents the fatigue life which 90 per cent of a given population of bearings will achieve, while the L_{50} value is the median life which 50 per cent of the bearings will endure. The L_{10} life is approximately one-fifth of the L_{50} life.

Dr Arvid Palmgren joined the technical staff of Aktiebolaget Svenska Kullagerfabriken (SKF) in Göteborg in 1917, and in due course he was charged with the task of finding a practical solution to the problem of rolling-bearing endurance. He built upon the earlier work of Stribeck carried out for the Deutsche Waffen-und Munitionsfabriken (DWF) and of Goodman in the Engineering Department at the University of Leeds and developed over many years a series of formulae of great utility. He quickly recognized the difficulty of relating static load-carrying capacity to fatigue life and initiated the publication of life formulae of increasing perfection from 1923. He was particularly gratified when, on 30 June 1930, Professor Stribeck wrote to him[2] saying, '... In the endurance tests, which were undertaken at my instigation during the last few months and which are still in progress, martensitic steels of great hardness were tested and as yet I have not been able to find any clear relationship between the static strength properties and the resistance to fatigue under alternating stress.'

Palmgren felt that Stribeck had finally accepted his earlier assertions. It became widely accepted that the life of a rolling-element bearing (L) was related to the applied load (W) with fair accuracy by the relationship,

$$W^3L = \text{a constant.} \hspace{4cm} \ldots [11.6]$$

The powerful influence of load upon life evident from this relationship emphasizes at once one of the major limitations of such bearings and the need for careful manufacture, handling and assembly.

Since load-carrying capacities based upon applied and centrifugal action are related to bearing size, or bore (D), another concept for estimating limiting speeds of rotation (N) which takes account of both load, and to some extent efficacy of lubrication, known as the *DN* factor, emerged. It is customary to record D in mm and N in rpm and products of several million are representative of certain bearing types and lubricants.

Palmgren's (1945) extensive work was well summarized in his book. Other valuable accounts of rolling-element bearing history, analysis, design and application have been presented by Allan (1945), Jones (1946), Shaw and Macks (1949), Wilcock and Booser (1957), Hall (1958), Bisson and Anderson (1964), Harris (1966), Hersey (1966), and Tallian (1969). The American Society of Mechanical Engineers published a useful booklet of *Life Adjustment Factors* (1971). Tallian (1969) distinguished between three eras of modern rolling-element bearing development: the *empirical* stage extending to the late 1920s, the *classical period* lasting until the early 1950s and the *modern period*. The classical period covered the development of sound engineering concepts of rolling bearings based upon Hertzian elasticity theory and Weibull fatigue statistics, matched by progress in manufacturing technology and the emergence of published standards from the Anti-Friction Bearing Manufacturers Association (AFBMA), the German DIN association and others.

The influence of operating conditions upon bearing friction is, of course, different for the two major classes of bearings, rolling and fluid-film. The friction of ball- and roller-bearings is impressively low and essentially steady whether the bearing is in motion or at rest. Nevertheless, many forms of fluid-film bearings, like externally pressurized and even self-acting gas- or liquid-lubricated systems, can exhibit even lower friction coefficients under certain conditions, and the use of the term *anti-friction*

---

2. I am grateful to Allan Palmgren, son of Arvid who also works for SKF in Göteborg, for kindly drawing this to my attention in 1976 and for providing copies of the correspondence.

for rolling-element bearings alone is unnecessarily restrictive. There has fortunately been an indication in recent years that the use of the term has diminished as a wider appreciation of the full spectrum of bearing characteristics has emerged. Furthermore, it can always be argued that the practice merely followed the earlier use of the term 'anti-friction metals' for babbitts by the plain-bearing manufacturers!

The period since 1950 is characterized by a prodigious amount of analytical work, even greater concern for geometrical accuracy and adequate surface finish, the availability of better steel and the emergence of improved lubricants, supported by an enhanced appreciation of the role of elastohydrodynamic lubrication. Excellent summaries of analytical progress can be found in the books by Jones (1946), Bisson and Anderson (1964) and Harris (1966). Topics which have received special attention include bearing kinematics, stresses, deformations, load-carrying capacity, friction and lubrication. The latter topic is reviewed in more detail in Section 11.13.

Vacuum-melted steels with fewer inclusions and material defects have greatly improved the fatigue life of rolling-element bearings, and special materials like high-temperature tool steels, martensitic stainless steels, stellites, molybdenum bearing steels and even ceramics have extended the use of rolling-element bearings to very high and cryogenic temperatures and other hostile environments.

11.6 Externally pressurized (hydrostatic) bearings

An alternative form of effective fluid-film bearing to the *self-acting* or *hydrodynamic* arrangement normally associated with slider bearings is that known as the *externally pressurized* or *hydrostatic* system. The minimum film thickness in a self-acting fluid-film bearing decreases as the applied load increases and as the sliding speed or fluid-entraining capability decreases. There are thus upper bounds on the loads and lower bounds on the speeds which self-acting bearings can tolerate if lubricating films of satisfactory thickness are to be maintained. In such circumstances safe fluid-film conditions can be preserved if the lubricant is pressurized before it enters the bearing, so that it effectively prizes open and maintains a space between the opposing solids.

The Frenchman Girard demonstrated the principle and took out a patent on hydrostatic bearings as early as 1865. The *paliers glissants* or externally pressurized water-lubricated bearings remained little more than curiosities for three-quarters of a century, although there were a few interesting applications along the way. An impressive demonstration model known as *Le Chemin de fer de glace* (ice railroad) attracted the attention of visitors to the 1878 Paris Industrial Exposition. A heavy block with four lion-like feet rested solidly on a steel plate, but when oil was pumped down each leg the block floated on films of lubricant and could be moved with astonishing ease. It is interesting to note that the principle has been revived in recent years for moving heavy plates and machines in shipyards and factories alike. Thurston (1885) referred to externally pressurized water-lubricated bearings, Lord Rayleigh (1917*a*) provided the first analysis of a simple form of thrust bearing based upon the hydrostatic principle and Gümbel and Everling (1925) contributed further to the subject.

The first detailed report on the successful application of hydrostatic bearings to machinery was presented by Welter and Brasch (1937), who were respectively a professor at the Technische Hochschule, Warsaw, and Chief Engineer at Frankfurt-am-Main. Lubricant was supplied to the crown of a 4·7 in (120 mm) diameter passenger-coach journal bearing on a large bearing testing machine in large quantities and at pressures in the range 100–200 atm (10·1–20·3 MN/m^2). It is an interesting paper in which the friction characteristics and temperature rise with and without external pressurization were plotted against time and compared with the performance of journal bearings. The main feature was, of course, the tremendous reduction in starting friction when external pressurization was used, but the longer-term results also gave the advantage to externally pressurized, rather than to plain or roller-bearings. Nevertheless, the bearing was never adopted for large-scale use on the railways.

In October 1947 the great 200 in (5,080 mm) diameter Halle optical telescope was completed on Mount Palomar, some 66 miles (106 km) north of San Diego in California. It was to provide the most spectacular demonstration of the merits of hydrostatic bearings. The structure had to rotate very slowly and smoothly to track celestial objects and hydrodynamic bearings were completely out of the running for diurnal motion. The moving mass weighed about 1 million lb (4·4 MN) and although ball- and roller-bearings were first considered, it became apparent that the resulting friction and its variation would be unacceptable. This became an ideal application for hydrostatic support bearings as described by Ormondroyd (1936), Karelitz (1938) and Fuller (1947, 1956). The coefficient of friction is of order 10^{-6} and the 1 million lb weight moves smoothly with a small fraction of the output from the $\frac{1}{12}$ hp (62 W) drive motor provided.

In the 1950s the use of hydrostatic bearings extended to machine tools, and in this field alone applications include the support of grinding machine, lathe and milling machine spindles, the tables of large vertical turning-mills and slideways which move without stick-slip or chatter. Other machines for which externally pressurized, liquid-lubricated bearings have been designed include dynamometers, turbogenerators, large rotating air-preheaters in power stations, tumbling-mills for ore crushing and liquid-sodium coolant pumps for nuclear reactors. The principle of hydrostatic bearings and their ability to provide high stiffness or accuracy of location depends upon careful attention to the design of the associated hydraulic circuits. External flow restriction by means of orifices or capillary tubes is normally important, and Royle et al. (1962) described an automatic flow-control valve of great potential for high-stiffness hydrostatic bearings.

Mathematical analysis and the development of design procedures has accompanied the increased interest in hydrostatic bearings in the past forty years, providing confidence in both the design and manufacture of such bearings at the present time. Good general introductions to the subject have been written by Shaw and Macks (1949), Fuller (1956), Wilcock and Booser (1957), Pinkus and Sternlicht (1961), while detailed design guides have emerged from Raimondi and Boyd (1957), Ripple (1963), Opitz (1968), Rowe and O'Donoghue (1971) and Stansfield (1970). The Institution of Mechanical Engineers (1972) published the proceedings of a conference devoted entirely to the subject of externally pressurized bearings and there have been many research papers on particular topics.

The externally pressurized bearing has undoubtedly established itself in a small but vital corner of the spectrum of bearing applications. The designs are usually required on an individual basis, although there are signs that manufacturers will offer standard arrangements of both hydrostatic bearings and associated spindles. The bearings will operate with liquids or gases, but consideration of the latter fluid is reserved for Section 11.7.

11.7 Gas bearings

It took about a century for the gas-lubricated bearing demonstrated by Hirn in 1854 to enter the lists as a serious form of bearing with several outstanding examples of its use. The investigations of Hirn (1854), Stefan (1874), Kingsbury (1897) and Harrison (1913) have been noted in Chapter 10, but it is interesting to note that both Klein and Sommerfeld (1903) and Charron (1910) decided independently early in the twentieth century that gaseous lubrication interfered with studies of dry friction between solids.

The need to operate high-speed spindles or *ring flyers* in the textile industry almost established the self-acting air bearing in the machine designers armoury very early in the twentieth century. In 1903 Dr S. Z. de Ferranti experimented with gas bearings for cotton-spinning machinery and Munday (1974) has traced the subsequent developments through patents dating from 1904. Many of the features of modern air bearings are evident in Ferranti's (1904) patent for a high-speed, vertical, rotating spindle, including radial clearances of only 0·0016 in (40 μm) and a rubber mounting sleeve to damp out the effects of rotor imbalance. F. J. de Ferranti took out an American patent for high-speed air lubricated bearings for textile machinery (Ferranti, 1904), and in the same year Sawtelle (1909) described a lathe attachment in the form of an air-lubricated traverse spindle grinder.

An interesting use of interferometry in the study of tilting-pad bearings by Stone (1921) confirmed the existence of effective self-acting air bearings between plane quartz supporting pads and a rotating glass disc. It appears that it was non other than A. G. M. Michell of tilting-pad bearing fame who suggested the scheme to Stone, and the determination of both film thickness and pad inclination from interference fringes not only demonstrated the validity of gas-film bearings but also confirmed in a most direct manner many of the features of hydrodynamic bearings. Kingsbury provided further proof of both the tilting-pad and gas-bearing effectiveness in the demonstration models built about 1922 and shown in Fig. 10.31. Dr Ferranti's widow (Ferranti, 1934) took out further patents which dealt mainly with materials to be used in gas-bearing construction and certain design details.

Full exploitation of the remarkable characteristics of self-acting gas bearings did not occur until the second half of the twentieth century. By this time surface speeds in bearings had advanced to such an extent that advantage could be taken of the low viscosity of gases to support relatively light loads with minute friction. The reason follows from the fact that load-carrying capacity depends upon the product of speed (U) and viscosity (η). However, since the viscous traction is also proportional to viscosity, the friction forces and torques exhibit spectacular reductions while the coefficients of friction are little different from those of liquid-lubricated hydrodynamic bearings. Inertial gyroscopes, which run at high speeds and low loads, were fitted with self-acting gas bearings in the late 1950s and the reduced friction led to a new generation of precision navigational aids.

Another postwar requirement which promoted the use of gas bearings was the need to operate lubricated equipment in very hot, cryogenic or radiation environments. In gas-cooled nuclear reactors like Calder Hall, conventional liquid lubricants would deteriorate due to irradiation and hence the concept arose of using carbon dioxide as both a coolant and a lubricant. The application was described by Ford et al. (1957). In early self-acting gas bearings, hard materials, accurately formed with a mirror finish, were the order of the day. Chromium, alone or as a plating on copper, and other hard polished metals were used, together with carbon. More recently ceramics and various polymers have become popular, although chromium and carbon are still widely used in externally pressurized gas bearings. The foil or tape bearing arranged in various configurations has been found to be flexible yet stable in a limited range of applications. The spiral-groove bearing discussed by Whipple (1949, 1951), utilized by Ford et al. (1957) and developed greatly by Muijderman (1966) has become firmly established. It was, however, briefly mentioned by Gümbel and Everling (1925) as noted in Chapter 10.

The principle of external pressurization has been exploited in the field of gas bearings to an even greater extent than self-acting behaviour. They overcome low-speed operating problems and permit satisfactory stiffness while providing low friction and hence small heat generation and minimal thermal distortion. Details of experiments, analysis and design have been given by Pinkus and Sternlicht (1961), Gross (1962), Grassam and Powell (1964), Ausman (1965), Hersey (1966) and Powell (1970). Applications include grinding machine spindles, feed-screws, slideways, sensitive and accurate metrology equipment, dynamometers, electric motors, flow-meters, air turbines and compressors. Their use in lifting heavy equipment in the manner of the oil-lubricated bearings at the Paris Exposition of 1878 has also attracted attention in recent years.

Hybrid bearings which combine the desirable low-speed characteristics of externally pressurized bearings with the advantages of low friction and desirable high-speed performance of self-acting bearings have been used in textile applications, ultra-high-speed centrifuges, gyroscopes, high-speed grinding spindles and turbines. Perhaps the device in the latter category best known to the general reader is the turbine-driven dentists drill which runs at speeds in the range 250,000–500,000 rpm (4,167–8,333 Hz). Air bearings have been used to support rotors running at speeds well in excess of 1 million rpm (16,667 Hz). The development of gas bearings has promoted a great surge of analytical, experimental and design effort in the past twenty years and there seems little doubt that they will be intimately linked with the history of mid-twentieth-century tribology. Their susceptibility to instability has alone attracted much attention (Marsh, 1965) while their operation in the turbulent region has been studied by Tipei (1957) and Constantinescu (1963, 1968b). They have found a firm, steadily expanding, but still small corner of the field of bearing applications.

11.8 Bearing selection and design (steady loading)

Selection

An important feature of twentieth-century bearing development, which emerges clearly from a study of the history of the subject, was the way in which new materials and new concepts were translated into practical realities. In the early days of the century the designer found his choice restricted to oil-lubricated, self-acting, metallic plain bearings or various forms of rolling-element bearings. Furthermore, both types were the outcome of largely empirical development.

In Section 11.4 attention was drawn to the development of non-metallic bearings and, in due course, roughly in the middle of the century, that led to the manufacture of polymeric bearings which could perform effectively without additional lubricant. These *dry* or *rubbing* bearings provided a third class of bearings and new scope for the machine designer.

A fourth class of bearings termed *partially lubricated*, which were also *self-contained*, included porous solids impregnated with lubricants. Morgan (1958, 1966, 1969) has described the characteristics of such bearings and outlined design criteria. Nor were the more traditional, intermittently lubricated, drip, felt and wick arrangements entirely eliminated. The oil ring and disc lubricators, whose early history was so well described by Vogelpohl (1969), continue to service various items of equipment such as pithead winding gear, and pumps, either as the sole or standby lubrication systems.

The possibility of separating load-carrying surfaces in relative motion in small instruments and large, high-speed trains by means of *magnetic levitation* has also received consideration.

In any event the advance of reliable and versatile *rolling-element* and *plain metallic* bearings, coupled with the introduction in impressive manner in limited but important fields of *externally pressurized* and *gas bearings*, the emergence of *dry-rubbing bearings* and the introduction of novel forms of *self-contained*, *partially lubricated* bearings in the middle of the twentieth century, increased both the range of bearing types available to the machine designer and his difficulty in selecting and then designing the most suitable device for a given application. Even today it is not uncommon in diagnostic studies of tribological failures to find that the problem rests in the use of an inappropriate bearing type rather than in errors of detailed bearing design.

Fortunately, men of experience and foresight recognized the spreading jungle and developed guidance paths which have introduced a measure of logic into *bearing selection* procedures.

Wilcock and Booser (1957) showed due recognition and respect for the situation by devoting Part I of their well-known book on bearings to the subject of the selection of bearing class. The authors concluded with a valuable list of examples in which consideration was given to the selection of suitable bearings for electric motors, motor-vehicle engines and chassis, steam turbines and aircraft gas turbines.

I well remember the impact which a brief, practical yet thoroughly useful approach to this problem by De Hart (1959) had upon my own deliberations as an engineer and a teacher of the mechanics of machines, lubrication theory and bearing selection and design. De Hart reviewed the basic forms and operating characteristics of ball-, roller-, self-acting and externally pressurized liquid- and gas-lubricated bearings and proceeded to draw up a general guidance chart of relative ratings to facilitate bearing selection. Six forms of ratings were applied to twenty-six bearing types in relation to eighteen different selection factors.

Several years' experience in lubrication research, bearing design and direct contact with operating situations enabled Neale (1964) to provide the first steps towards what are now widely recognized as logical and reliable bearing selection and design procedures. He started with a survey of the loads and speeds of operation of bearings in a wide range of machinery and discovered that the loads spanned about seven and the speeds six orders of magnitude. He then considered on logarithmic scales the general nature of the load and speed for *dry rubbing, rolling, externally pressurized, pressure-fed-self-acting, self-contained* and *porous bearings* of various sizes. Superposition of all these relationships for bearings of given sizes at once provided a bearing selection chart of immeasurable value.

The approach also contained caveats which took account of special environmental conditions and operating requirements. At a later stage Neale chaired an Institution of Mechanical Engineers Committee responsible for the production of bearing selection documents for both journal and thrust bearings which were published by the Engineering Sciences Data Unit (ESDU, 1965a, 1967). A full account of the basis of selection chart construction was presented by Neale (1968).

Design

A recognition of the importance of correct bearing selection and the provision of adequate guidance to machine designers could well be recognized as one of the major developments in the field of tribology in the third quarter of the twentieth century. Once the most appropriate form of bearing for a given application had been chosen, it then became necessary to select or design the bearing in detail. There is now a vast reservoir of basic design knowledge for all the major forms of bearings, based upon analytical studies, experimental research, the development of better materials and manufacturing processes, machine-operating experience and, of course, the long and steady evolution of the design process itself. In many cases the designer finds it possible by simple means to choose a standard bearing which will meet his requirements from manufacturers' catalogues. This situation arose firstly in the rolling-element bearing industry, but has now extended to plain bushes and thrust washers, tilting-pad bearings, dry-rubbing bearings, porous bearings and even to a limited extent to externally pressurized liquid- and gas-lubricated bearings.

When a machine designer wishes to design bearings without recourse to the facilities of the specialist, he can turn to readily available, versatile and reliable design manuals. These have also incorporated the fruits of experimental and analytical studies in many laboratories and a wealth of experience. Typical examples of such manuals are those produced by Hays (1960); the Institution of Mechanical Engineers and the Engineering Sciences Data Unit for steadily loaded pressure-fed hydrodynamic journal bearings (ESDU, 1965b), fixed inclined (ESDU, 1973) and tilting-pad thrust bearings (ESDU, 1975) and dry rubbing bearings (ESDU, 1976). Martin (1970) and Martin and Garner (1976) have shown great ingenuity in adopting slide-charts for this purpose.

These manuals enable the machine designer to follow a logical well-documented procedure and to produce sound proposals for the essential features of a satisfactory bearing. They have provided a new dimension to the designers' ability to deal satisfactorily with bearing design problems without the need for detailed and expert knowledge of lubrication theory or materials science and without the need for access to sophisticated computers.

Computers have played a major role in the development of satisfactory design procedures for both fluid-film and rolling-element bearings. There have been three discernible stages in the process. Initially, computers were used to yield approximate yet highly accurate solutions to the governing differential equations, information then being incorporated into hand or graphical design procedures and the design manuals mentioned earlier. Representative cases which can be cited include the solution of the finite-length journal bearing problem by Cameron and Wood (1949), the plane-inclined or tilting-pad bearing configuration by Jakobsson and Floberg (1958b) and the externally pressurized bearing problem by Raimondi and Boyd (1958).

Computer programs were then developed to yield design solutions to a particular problem by storing the information produced by previous solutions of the governing equations, or by pursuing an iterative procedure in which the equations were solved successively until a satisfactory match was achieved between the external, applied loads and the support capabilities of the bearing. Woolacott and Singh have illustrated how this can be achieved for journal bearings and their work formed the basis of the Engineering Sciences Data Unit, Item 69002 (ESDU, 1969).

The third and final stage involved the development of programs capable of yielding optimum bearing designs. The underlying philosophy of this final step is interesting, since it is a real attempt to encourage the computer to take on the full role of the bearing designer. The concept is simple, the execution of the task difficult and challenging. If the designer can produce a satisfactory bearing by taking a series of decisions in well-defined sequence and with adequate guidance, why not relieve him of the task and ask the fast but somewhat unimaginative computer to perform the task? The procedure has some disadvantages, most of which arise from the lack of versatility of computers compared with the human mind, but it has the merit of speed and efficiency. If a single solution can be obtained with great speed on the computer, it opens up the possibility of producing an array of solutions and optimizing on the basis of some previously defined criteria. The designer in a drawing office rarely has the luxury of adequate time for bearing optimization, but we appear to be on the verge of such developments in the field of digital computing.

The philosophy and the practice of this exciting possibility have been explored in recent years by Seireg and Ezzat (1969), Singh (1969), Moes and Bosma (1971), Dowson and Blount (1976), Dowson and Ashton (1976) and Dowson et al. (1977).

There have been immense strides towards the goal of sound bearing selection and design during the past fifty years. With a modicum of effort a designer can now prepare recommendations for bearings for a machine in a manner which was totally unthinkable in 1925.

11.9 Lubricants

The development of additives for mineral oils, the increasing use of process fluids as lubricants, the emergence of synthetic lubricants and the extended application of solid lubricants are the four most important factors which distinguish the past fifty years from the previous and long history of lubricant development.

Additives

The addition of small quantities of substances to enhance certain features of the performance of lubricants is by no means solely a twentieth-century development. Nineteenth-century patent literature abounds with enthusiastic suggestions for lubricating oils and greases containing additives. Nevertheless, it was only when mineral oils became the most common form of lubricants and the internal combustion engine became a mass-produced article that the additive industry really started to assume its present measure of importance. In fluid-film lubrication the most important physical property of a lubricant is its viscosity–a property which decreases rapidly with increasing temperature for liquids and which increases slightly for gases. Many of the late nineteenth-century studies of lubrication included investigations of the effect of temperature upon viscosity, as in the case of Petrov (1883) and Reynolds (1886). Lubricants of vegetable and animal origin are less susceptible to the effect than are mineral oils, but since the ability of a bearing to sustain load and avoid interference between the sliding members is proportional to viscosity, it is clearly important to be able to specify the viscosity–temperature characteristics of a lubricant.

A *viscosity index* (VI) intended to provide a useful yet simple measure of the extent to which the viscosity of a lubricant varied with temperature was proposed by Dean and Davis (1929, 1940). The lubricant which showed the least variation in viscosity with temperature at that time came from the Pennsylvanian fields and was arbitrarily given a VI of 100. Likewise, the oil which suffered the greatest decrease in viscosity with temperature came from the Gulf Coast and the sample which had the same viscosity as the Pennsylvanian oil at 210°F was assigned a VI of 0. The viscosity index of any other lubricant with the same viscosity at 210°F as the datum oils was defined by the relationship

$$\text{VI} = 100 \left[\frac{L - U}{L - H} \right], \qquad \qquad \dots [11.7]$$

where L is the viscosity of the standard Gulf Coast oil (VI $= 0$) at 100°F, H the viscosity of the standard Pennsylvania oil (VI $= 100$) at 100°F, and U the viscosity of the unknown oil, with the same 210°F viscosity as the standard oils, at 100°F.

In general, paraffinic and aromatic components of mineral oils have high and low VIs, respectively, with naphthenic components exhibiting intermediate values.

The great merit of the VI system is its simplicity, but for some purposes it is too simple and even inappropriate. The extension of exploration, the introduction of solvent-refining which removes aromatic components, in 1934, the use of additives and the introduction of synthetic lubricants yielded fluids with VI values well outside the original scale. This in itself did not matter, but meaningful extrapolation often became difficult. Most lubricants in large-scale use have a VI close to 100 and typical motor-car crankcase oils may have VIs of 150 or 160.

Another system of classification peculiar to the motor-car industry and proposed by the Society of Automotive Engineers (SAE), must possess some merit since it continues to be used commercially despite its growing inability to deal with the range of present-day lubricants and its total inadequacy for scientific work. In this system a single number designates the fact that the lubricant has a viscosity within a specified, but rather wide, band at either 0°F ($-11.8°C$) or 210°F ($98.9°C$). Two ranges of numbers are defined, one for engine crankcase lubricants and the other for gear-boxes. As an indication of cold-starting characteristics, crankcase oils intended for winter operation have a letter 'W' immediately after the SAE number and the temperature to which it is related is always 0°F ($-17.8°C$). The original SAE classification was related to viscosities in Saybolt universal seconds, (SUS), but the approximate ranges in SI units are also shown in Table 11.1.

The 5W, 10W and 20W oils were introduced to assist the starting of engines under very cold conditions about 1950, but of course the oil consumption at normal engine-operating conditions was excessive. It

Table 11.1
SAE classification for crankcase oils (Saybolt universal seconds (SUS) and SI ($\mu m^2/s$).

SAE grade	Viscosity range							
	0°F (255·4 K)				210°F (372·0 K)			
	SUS		$\mu m^2/s$		SUS		$\mu m^2/s$	
	Min	Max	Min	Max	Min	Max	Min	Max
5W	–	4,000	–	871	40	–	3·9	
10W	6,000	12,000	1,307	2,614	40	–	3·9	–
20W	12,000	48,000	2,614	10,460	40	–	3·9	–
20	–	–	–	–	45	58	5·8	9·6
30	–	–	–	–	58	70	9·6	12·9
40	–	–	–	–	70	85	12·9	16·8
50	–	–	–	–	85	110	16·8	22·7

was found that the addition of polymers like *polymethacrylates* and *poly-isobutylenes* led to an increase in viscosity which was more pronounced at high temperatures. These VI improvers yielded lubricants with a high enough viscosity, and hence SAE grade, for adequate lubrication at normal engine-running speeds, and yet a low enough viscosity corresponding to an entirely different SAE grade to ease cold starting. These new *multigrade* oils were introduced about 1952 and were designated by two sets of numbers, like 10W–30, corresponding to the low- and high-temperature SAE grades, respectively.

For scientific work neither the VI value nor the SAE grade of an oil is very useful. Full viscosity–temperature characteristics or empirical relationships between viscosity and temperature are required.

There are, of course, many other additives, like *rust inhibitors*, in oils apart from the VI improvers. It was also recognized at an early stage that mineral oils oxidized steadily by a somewhat complex chemical process and that the rate of oxidation increased with temperature. The products of oxidation are acidic and possibly corrosive and can lead to varnish formation, deposits of sludge and an increase in viscosity. *Oxidation inhibitors* or *anti-oxidant* additives appear to work by breaking down organic peroxides formed at an intermediate stage of the process or by preventing the action of metallic catalysts. They normally consist of compounds of sulphur, or phosphorus, or both, or oil-soluble amines and phenols. The amount of additive used to control oxidation normally ranges from about 0·25 per cent to 5 per cent and they came into commercial use between 1934 and 1938.

Additives used to control the formation of sludge and the deposition of products of combustion upon internal engine surfaces are usually oil-soluble, surface-active materials capable of holding material in suspension in the lubricant. These *detergent additives* are normally used in concentrations ranging from 0·5 per cent to 30 per cent. *Polymeric dispersant additives* may also be used in amounts ranging from 1 per cent to 5 per cent. The first detergent-dispersive additives were used to combat the problem of piston deposits in diesel engines in 1938.

At low temperatures dissolved wax separates from the oil to form crystals, although the effect is alleviated by solvent extraction. Eventually the crystals might form a gel which prevents the lubricant from flowing. The temperature at which this occurs is called the *pour point*, and for temperatures lower than about $-12°C$ (161 K) it is necessary to combat the action if a free-flowing lubricant is required. Polymeric *pour-point depressant* additives like polyalkylnaphthalenes, polyalkylphenol esters and polyalkylmethylmethacrylates in small amounts ranging from 0·01 per cent to 0·3 per cent are usually very effective.

The additives considered so far were nearly all concerned with improvements to basic lubricant characteristics such as viscosity, viscosity index, rate of oxidation and pour point. Other additives are used to control friction and wear directly whenever sliding contact is likely. These *anti-wear* additives generally fall into two categories (1) boundary lubricants which act by chemical adsorption, and (2) extreme pressure additives. The former were investigated by Hardy in the 1920s and consist essentially of the fatty acids. The use of the latter arose entirely from the adoption of the *hypoid gear* in rear-axle drives in motor cars in 1927. The development permitted the lowering of the chassis and hence the centre of gravity of the vehicle, but it was soon found that neither straight mineral oils nor those containing fatty acid additives could prevent severe wear of the hypoid gears. Being intermediate between worm and spiral level arrangements, hypoid gears experience both rolling and sliding interactions, with the latter being similar to the severe sideslip encountered in the less highly stressed worm gears. The widespread use of the hypoid gear not only led to the introduction of special lubricant additives, it also promoted the introduction of a new range of lubricant-testing machines to overcome the prohibitive cost of full-scale rear-axle tests. The proliferation of lubricant-testing machines and the debate on the value of data obtained from them was to continue for many years.

It has been argued that the lubricant chemist with his array of additives can disguise and sometimes overcome the effects of bad engineering, but it seems that motor-vehicle development in the 1930s could not have followed the now familiar path without his intervention. When the General Discussion on Lubrication and Wear was held by the Institution of Mechanical Engineers in 1937 the continuing cooperation between motor-car companies and the petroleum industry was evident in many papers like those by Evans (1937), Miller (1937) and Southcombe et al. (1937). An excellent account of the impact of motor-car development upon lubricants and bearings has been prepared by Greene (1967). One of the major features of the meeting was the concern for the development of adequate lubricants for machinery. Hypoid gears were responsible for the new forms of additives, but it soon became clear that other forms of gearing, and also cams and their followers, required the protection of these substances as the severity of loading increased. Miller (1937) noted that in most differentials in cars the mean pressures or unit loads on gears rose from 1,180 lbf/in^2 (8·14 MN/m^2) to 1,450 lbf/in^2 (10·0 MN/m^2) between 1924 and 1930. Dudley (1969) quoted figures for a wider time span by suggesting that hardwood gear-teeth in water-powered mills in 1832 sustained 75 lbf/in^2 (0·52 MN/m^2), cast-iron teeth in the same application 300 lbf/in² (2·07 MN/m²), machined cast-iron teeth in steam engines about 1900 moved up to 1,100 lbf/in² (7·58 MN/m²), and improvements in materials and manufacturing accuracy then increased the figures for electric motors, steam turbines, aircraft gas turbines and rocket motors to 5,000 lbf/in² (34·5 MN/m²), 6,000 lbf/in² (41·4 MN/m²), 9,000 lbf/in² (62·1 MN/m²), 15,000 (103·4 MN/m²) and finally 30,000 lbf/in² (206·8 MN/m²) in 1968.

Because the new range of chemicals were associated with conditions of high load or *extreme-pressure* they became known as EP additives. They are based upon chlorine, sulphur and phosphorus and include such materials as dialkyl dithiophosphates as outlined by Elliott and Edwards (1957) and sulphur in association with lead soaps as described by Campbell (1957). The general concept is that the EP additives react with metals if the local temperatures achieve values in the approximate range of 150–200°C. In this sense the term 'extreme pressure' is misleading, since the reaction process is controlled by temperature, although the latter might well reflect the severity of loading and general operating conditions.

Bowden and Tabor (1950, 1964) have pointed out that the highly successful EP additives become effective at just about the temperatures at which the metal soaps formed by the range of fatty acid additives desorb from the solid surfaces. A measure of continuity of protection over a wide range of temperatures is thus provided by adding both fatty acids and EP additives to lubricants.

Most of the films formed by chemical reaction, a process which has been called controlled corrosion, provide protection to sliding surfaces up to several hundreds of degrees C, and sometimes up to about 1,000°C (Godfrey, 1965). Another approach to the problem of surface protection at high temperatures and severe operating conditions is to use the range of solid lubricants like graphite or molybdenum disulphide.

This outline of additive development is by no means complete, but certain trends of historical significance can be identified in this most energetic, competitive and in some respects confusing area. The impact of motor-car development, the increasing collaboration between the motor-vehicle and the petroleum industries, the dominant role of the lubricant chemist or technologist, the alleviation of engineering problems by chemical means were all important. Perhaps the less evident, yet in the long run the most important features of the exercise, were the emergence of combined approaches to wear problems by engineers, chemists and physicists, with the recognition that the subject called for a multi-disciplinary effort, and the fact that the lubricant properties should be an integral part of the overall considerations in machine design.

Rewarding general and detailed discussions of lubricant additives and their roles can be found in Bowden and Tabor (1950), Bisson and Anderson (1964), Akhmatov (1963), Kragelskii (1965), Cameron (1966), Molyneux (1967), Godfrey (1968), Campbell (1969a), and the *Proceedings* of the General Discussion on Lubrication and Lubricants (Institution of Mechanical Engineers 1937), the Conference on Lubrication and Wear (Institution of Mechanical Engineers 1957) and the Conference on Lubrication and Wear: Fundamentals and Applications to Design (Institution of Mechanical Engineers, 1967), the American Society of Lubrication Engineers *Standard Handbook of Lubrication Engineering* (O'Connor et al., 1968), and Tabor (1972).

Process fluids	In the textile, food and chemical industries contamination of the product by lubricating oils can be at best a nuisance and at worst a disaster. The possibility of using the ambient or process fluid as the lubricant, thus obviating the need for seals and eliminating the contamination problem, is essentially a twentieth-century development.

The best-known example, *gas-film lubrication*, has already been discussed. In the applications of gas bearings considered previously the engineering advantages of low friction and high-speed operation were emphasized, but the great merit of using such bearings in textile spindles to produce clean rather than oil-stained yarn is clear. Likewise, the use of carbon dioxide lubricated bearings in gas circulators in nuclear reactors and air in dentists' drills form further illustrations.

A second important example of the use of process fluids as lubricants is the *water-lubricated bearings* in feed pumps for boilers. Water has also been used in marine (Newman, 1958) and sub-marine applications. In 1879 Thurston referred to the use of externally pressurized, water-lubricated bearings (Thurston, 1879), and as early as 1854 Hirn demonstrated that the fluid could be used effectively in self-acting journal bearings (Hirn, 1854). In some forms of nuclear reactor it is more appropriate to use water than air as the lubricant and this possibility was explored by Wepfer and Cattabiani (1955). Hays (1962) has provided an illustration of the use of water-lubricated bearings in oxygen compressors and this raises a further advantage of the use of water—the reduction of fire and explosion risks.

When water is used as a lubricant the bearing materials are usually non-metallic. Lignum-vitae, resin-bonded fabrics, rubber and carbon have all been employed. The subject of water-based lubricants and bearing materials was considered at the Institution of Mechanical Engineers' General Discussion on Lubrication and Lubricants (Institution of Mechanical Engineers, 1937), but the most wide-ranging studies have been reported by Hother-Lushington and Sellors (1964) and Hother-Lushington (1976).

In self-acting bearings all that is necessary is that the product of speed and viscosity should be high enough to ensure the formation of a film of adequate thickness, as noted by Gümbel and Everling (1925). However, the propensity of water to encourage corrosion of machine components, coupled with its relative inability to form protective boundary lubricating films, means that filtration and the correct choice of bearing materials is most important.

For high-temperature applications it may be possible to use *steam* rather than water, and the use of *liquid sodium* as a heat-exchanger fluid has led to studies of its potential as a lubricant by Campbell (1958). The use of liquid metals and suitable bearing materials in space vehicles has been considered by Anderson (1964) and by Anderson and Glenn (1968).

These examples illustrate the increasing need to consider at the design stage lubricant properties alongside those of the solids used in constructing machines. If the ambient or process fluid can be used effectively, then it is not just the design of the bearing, but the whole lubrication system, pumps, seals and the machine itself which may be influenced in an advantageous manner.

Synthetic lubricants

One of the major disadvantages of the most readily available and versatile mineral oil lubricants is that they oxidize rapidly at temperatures above about 90–120°C and show great reluctance to flow at temperatures below about −20°C. These limitations were not serious until the middle of the twentieth century, but advances in aviation in the Second World War – in particular the introduction of the gas turbine and the development of rocket motors and space vehicles which exposed moving parts of machinery to greater extremes of temperatures – the concern for fire hazards in hydraulic equipment and ovens, and the demands of the process industries from chemical plants to steel mills, all promoted the recent interest in synthetic lubricants.

The major categories of current synthetic lubricants are the *diesters*, *phosphate esters*, *silicate esters*, *polyglycols*, *polybutenes*, *silicones*, *polyphenyl ether* and *fluorocarbons*. The properties and applications of these newer types of lubricants are discussed fully by Wilcock and Booser (1957), Gunderson and Hart (1962), Moreton (1964), Goddard (1967), Blanchard (1968), Fowle (1968) and Gunther (1972). As a broad summary of the extent to which synthetic liquid lubricants have extended the useful range of fluid-film lubrication, it should be noted that some have pour points as low as −70°C, flash points between 180°C and 285°C and typical upper limits on normal operating temperatures of about 300°C, with 500°C being possible under some circumstances. There is, of course, a price to pay, since the cost of synthetic lubricants varies between about 3 and 300 times that of mineral oil. Nevertheless, they are firmly established for special applications and Wakelin (1974) has noted that one major oil company has already introduced a synthetic lubricant for reciprocating motor-vehicle engines. A wide-ranging review of the requirements of liquid lubricants has been published by Lansdown (1973).

Solid lubricants

If lubrication is required at even higher temperatures or lower pressures than those covered by mineral oils, process fluids or synthetic lubricants, solids with layer-lattice structures and some low shear strength solids without a layer-lattice structure might be used. By far the best known in the first category are *graphite* and *molybdenum disulphide*, while *lead oxide* and the oxides, sulphides and fluorides of other elements come under the latter heading.

Graphite and molybdenum disulphide have been known for centuries, and their tribological applications in the nineteenth century are recorded in Chapters 9 and 10. They can be applied as loose powders, dispersed in liquids or as a powdered metal mix. Many of the solid lubricants can also be used as self-lubricating bearing materials in solid form, impregnated into metals or as bonded coatings.

There is much to be done in the field of fundamental studies and applications of solid lubricants, and the very rate of their introduction in recent years makes it difficult to offer a judgement on their likely long-term roles. Their mode of operation is a good deal more complicated than was first realized, being particularly sensitive to humidity and general environmental conditions and chemical reactivity with other components of the system. They nevertheless extend the range of lubrication to harsh environments and can, in general, provide satisfactory lubrication up to about 500°C. Good introductions to the subject appear in Bisson et al. (1958), Deacon and Goodman (1958), Bisson and Anderson (1964), Braithwaite (1964), Lancaster (1966), Peace (1967), Campbell (1969*b*), the ASLE *Proceedings of International Conference on Solid Lubrication* (Boyd, 1971) and Clauss (1972).

Summary

The extended use of mineral oils by means of additives, the small but impressive use of process fluids as effective lubricants and the special use of synthetic and solid lubricants for hostile environments has enabled machinery in this technological world to function under conditions which would have been thought to be beyond all reason at the start of the century. The pace has been hectic and sometimes bewildering, with a mixture of empiricism and science, even in such a scientific age.

The engineer turned to the lubricant chemist and received support, sometimes when sound engineering alone might have yielded better solutions, then the lubricant technologist increasingly dictated the form of engineering progress. The initial impetus came from the reciprocating engine in road vehicles, then from marine and aircraft reciprocating engines and turbines, and in more recent years from man's desire to operate machinery in hostile environments of very high and cryogenic temperatures and within radiation fields in nuclear reactors. At the beginning of the chapter it was argued that the past fifty years was the tribological era of wear, and this account of the development of lubricants to combat the ever increasing challenge of wear control might ultimately be seen as one of the most significant practical developments of the century.

Perhaps the less evident, yet in the long run the most important features of lubricant development, were the involvement of physicists, chemists, materials scientists and engineers in separate and combined studies which played no small part in the progress towards a recognition of the multi-disciplinary nature of the problems in hand, and the growing recognition that a knowledge of lubricant properties should form an integral part of the overall approach to machine design. Both were to encourage the emergence of the philosophy and the reality of tribology in the 1960s.

11.10   The development of fluid-film lubrication theory for steadily loaded bearings

There has been an immense increase in the literature dealing with lubrication since 1925, and particularly since the Second World War. Much of it has dealt with the development of lubricant additives for both the mild and extreme-pressure conditions of boundary lubrication disclosed in Section 11.9 and will not be mentioned further. Other work yielded ingenious solutions to the governing Reynolds equation of fluid-film lubrication which provided the basis of the modern design procedures outlined in Section 11.8 and a brief history will be included here.

Some of the studies of lubrication were peripheral, but by no means unimportant, in the sense that they extended the original work to take account of such factors as thermal effects, lubricant compressibility, rate of shear, oil flow, cavitation and surface tractions in a much more realistic manner in bearing analysis. These refinements for both incompressible and compressible fluids have all contributed to the quite remarkable degree of sophistication achieved in current fluid-film lubrication theory.

This is by no means a complete list of the major developments in fluid-film lubrication in the past half-century, but as explained previously, the account must necessarily be selective rather than comprehensive. I can only hope that, when viewed alongside other developments outlined in this chapter, they will give a reasonable indication of mid-twentieth-century progress and some of the major problems yet to be resolved.

Solutions of the Reynolds equation for bearings of finite width

Reynolds developed his famous equation of fluid-film lubrication in 1886 to take account of pressure variations within the gently converging film of lubricant. His solutions for both thrust and journal bearings, like Sommerfeld's (1904) solution for journal bearings, neglected pressure variation and hence flow in the axial direction. In 1905, Michell's mathematical skill enabled him to take account of pressure variations in the two principal directions in thrust pads in approximate form, but it is not generally recognized that side-leakage solutions for bearings of various forms are of relatively recent origin.

The full Reynolds equation is by no means the easiest differential equation to solve and very few analytical solutions are available for realistic bearing forms. The approaches fall into four categories:

1. *Analytical solutions* to the complete or restricted equation.
2. Solutions in the form of *mathematical series*.
3. The use of *analogue* devices.
4. *Approximate solutions* by *numerical procedures*.

A number of solutions to slow viscous flow problems akin to lubrication were included in a splendid little book written by Purday (1949). However, to illustrate the type of rare analytical solutions in category (1), reference will be made only to the lubrication of spheres. Kapitza (1955) demonstrated that if a non-rotating sphere of radius R carrying a load (P) was floated above a plane surface sliding with velocity (U) at a minimum separation (h_0) in a sea of lubricant of viscosity (η), the solution of the Reynolds equation which considered the convergent, physical-wedge action alone yielded the relationship

$$\frac{P}{\eta UR} = \frac{6\sqrt{(2)}\pi}{5} \sqrt{\left(\frac{R}{h_0}\right)}. \qquad \dots [11.8]$$

Likewise, if the force squeezed the sphere through the fluid towards a stationary plane, it can be shown that the instantaneous velocity of approach (W) is defined by the expression

$$\frac{P}{\eta WR} = 6\pi\left(\frac{R}{h_0}\right). \qquad \dots [11.9]$$

The latter result was derived and used by Professor G. I. Taylor to determine the time taken for a sphere to sink through an oil film and thus to assist in Sir William Bate Hardy's interpretation of friction measurements on spherical sliders in his classical experiments on boundary lubrication described in the previous chapter.

Analytical solutions to squeeze-film problems for rigid solids exhibiting axial symmetry are generally straightforward and Osborne Reynolds' own analysis of squeeze-film action between parallel elliptical plates has been mentioned in Chapter 10.

These analytical solutions to the full Reynolds equation are indeed rare, but if the governing equation can legitimately be reduced new possibilities arise. In 1886 the problem in which side-leakage was neglected was tackled for both thrust and journal bearings by Reynolds himself (Reynolds, 1886) and when Sommerfeld (1904) made a similar assumption it appeared reasonable since the axial lengths of journal bearings of the day often exceeded their diameters by factors of two or three. In such cases axial variations in pressure were generally much smaller than those in the circumferential direction.

As engine ratings advanced and bearing materials were developed to accommodate higher mean pressures, the axial lengths of reciprocating engine bearings were gradually reduced until they were about half the diameter of the journal. This led to a reconsideration of the relative importance of the various terms in the full Reynolds equation in the 1950s. It was Fred W. Ocvirk[3] who recognized that such configurations permitted the neglect of pressure or Poiseuille flow in the circumferential direction compared with that in the axial or side-leakage sense. This facilitated the generation of an analytical approach to the restricted Reynolds equation which became known as the *Ocvirk* or *short journal-bearing theory*. The findings were published by Ocvirk (1952) and DuBois and Ocvirk (1953), and for load-carrying capacity of a cavitating film (half-Sommerfeld solution) take the form,

$$\frac{P}{\eta UL^2}\left(\frac{c}{L}\right)^2 = \frac{\varepsilon}{4(1-\varepsilon^2)^2}[\pi^2 + (16-\pi^2)\varepsilon^2]^{1/2}. \qquad \dots [11.10]$$

---

3. See Appendix, Sect. A.24 for biographical details.

Short bearing theory offended some purists and there is no doubt that its predictions have to be used with due caution and a knowledge of the limitations involved, but it provided estimates of bearing performance which were in close agreement with observed characteristics of contemporary bearings. Furthermore, it immediately involved all aspects of bearing geometry and provided a firmer base for the physical appreciation of bearing operating characteristics. The theory formed the foundation for many subsequent bearing design procedures.

It appears that Michell (1929) was conceptually in advance of Ocvirk since he had already applied the general approach to thrust-bearing analysis more than twenty years earlier. However, the extension to journal bearings in the 1950s was undertaken independently and considerably developed by DuBois and Ocvirk (1953).

An interesting analytical solution which was developed in terms of the comparatively unknown Heun functions was developed by Tao (1959). It is a highly mathematical paper, and although its immediate utility to the designer is limited it is a bold and interesting approach which can account for both the finite length of the bearing and variable viscosity of the lubricant.

The Reynolds and Sommerfeld analyses in which side-leakage is neglected and the Michell–Ocvirk short bearing theory in which axial Poiseuille flow was assumed to predominate over the comparable circumferential effect can be viewed as limiting forms of solutions to the Reynolds equation. They are colloquially known as *long*, or *infinite*, and *short* bearing solutions, respectively.

Perhaps the best-known and certainly the earliest example of a *series solution* to a side-leakage problem in lubrication theory was that performed by Michell in 1905. Froessel (1938) followed a similar approach in his combined theoretical and experimental attack on the tilting-pad journal-bearing problem. Products of circumferential and axial functions were employed by Boswall and Brierly (1932) to yield approximate solutions to the finite-width journal-bearing problem, and Fränkel (1944) assumed that the axial variation of pressure was parabolic with pressures being generated only in the convergent film of 180° extent in bearings of similar geometry.

The third category of work, in which analogues were established for the solution of the governing Reynolds equation, can be attributed to those pioneers of the tilting-pad bearing, Michell (1905b)[4] and Kingsbury (1931)[5]. Michell used an *elastic–solid* analogy in which the similarity between the equations governing viscous flow and the deformation of an elastic solid were used to ascertain, from a relatively simple experimental arrangement, the load-carrying capacity, the force of friction and the location of the resultant load on a plane thrust bearing of finite width. The apparatus was first used to provide an approximate check on the predictions of Michell's (1905) mathematical analysis, and a full account

---

4. See Appendix, Sect. A.19 for biographical details.
5. See Appendix, Sect. A.16 for biographical details.

with a physical interpretation based upon Clerk Maxwell's paper, 'The dynamic theory of gases', appeared in Michell's (1950) book.

Kingsbury's *electrical analogy* was based upon the close similarity between the Reynolds equation and the differential equation governing flow of electricity in a conducting material of resistivity (r) and variable thickness (H). A bath of electrolyte of identical plan shape, but normally scaled up for convenience, was provided with a variable depth (H) such that (H/r) was everywhere equivalent to (h^3/η) in the Reynolds equation (see eqn. [10.9]). The analogy between the flow of lubricant and electricity and lubricant pressure and electrical potential readily enabled solutions to be obtained for bearing forms of inconvenient geometry as far as mathematical approaches were concerned. It should be noted that not only variable film thickness but also variable viscosity can be accommodated and the approach has great appeal. It went further than Michell's elastic–solid analogy in that it provided details of the pressure distribution as well as the total load.

The electrolytic tank has been used to great effect in more recent studies. Needs (1934, 1935) obtained solutions which he presented in graphical and tabular form for various length to diameter and eccentricity ratios for 120°, centrally loaded, partial-arc journal bearings. Morgan et al. (1940) used an electrolytic model to ascertain the effect of side-leakage upon the performance of a full 360° (6·28 rad) journal bearing operating with an eccentricity ratio of 0·1. They also demonstrated the versatility of analogue procedures by studying the effect of lubricant supply holes and grooves. Kettleborough (1955, 1956) also used an electrolytic tank to study stepped-thrust bearings, including an interesting variation known as the *hydrodynamic pocket-thrust bearing* first proposed by Wilcock (1955). The fourth class of solutions, which ultimately led to the generation of data upon which modern design procedures are based, emerged in the 1940s. One of Sir Richard Southwell's young research students by the name of Christopherson (1941) demonstrated in a most impressive way how the *relaxation method* of solving sets of linear equations arising from the finite difference representation of differential equations could be used to solve problems in fluid-film lubrication. Furthermore, he was able by this means to incorporate the widely accepted *Swift–Stieber* or *Reynolds cavitation boundary condition* (see eqn [10.13]) with relative ease, even for bearings of finite width.

By the time Cameron and Wood (1949) obtained approximate solutions for the finite-width journal bearing by means of the same relaxation procedures, numerical methods had effectively become the main source of reliable data of great utility in bearing design. The numerical data upon which this important paper was based was computed by the Mathematics Division of the National Physical Laboratory. Cameron (1966) has generously recognized the improved solution based upon a similar approach by Sassenfeld and Walther (1954). This German research work was sponsored by the British Department of Scientific and Industrial Research after the Second World War and appeared first in DSIR Reports 11 and 12 before becoming more readily accessible in the *VDI Forschungsheft*. The Sassenfeld and Walther solutions achieved a new standard of accuracy, covered a wide range of eccentricity ratios and yet introduced innovations in that they considered both parallel and tilted journals. Solutions were also sought in terms of a variable ($\phi = ph^{3/2}$) which had been proposed earlier by Vogelpohl. The great merit of this function is that it damps down the rapid rise and decay in the independent variable at large eccentricity ratios, since p is normally small when h is large and vice versa.

The advent of high-speed digital computers greatly advanced the use of finite-difference methods in the solution of bearing problems, as noted briefly in Section 11.8. A clear trend evident in the massive list of papers in which the use of computers in the solution of the lubrication equations is concerned is that each step has achieved greater accuracy and realism. It became possible to consider more carefully lubricant supply arrangements such as the location of grooves and holes, the effect of film rupture due to cavitation and variation of lubricant properties. Some of the additional solutions which deserve special mention and which were, for one reason or another, of great utility, included those by Pinkus (1958) and Raimondi and Boyd (1958). A most valuable series of reports which emerged from Chalmers Tekniska Högskolas and Lund Technical University between 1957 and 1971 included Floberg (1957), Jakobsson and Floberg (1957), Jakobsson and Floberg (1958a,b, 1959), Floberg (1961), Håkansson (1964, 1967), and Floberg (1969). These reports presented with great clarity details of both thrust- and journal-bearing solutions and included graphical and tabular displays of basic bearing operating characteristics. Most solutions arbitrarily located axial lubricant supply grooves at the point of maximum film thickness or at $\pm 90°$ to the line of centres, but an extension to the situation in which the oil groove is located on the load line in the thicker film region has recently been supplied by Dowson and Ashton (1976). Yet many of the results of these refinements turn out to be second- or even third-order effects. The early numerical solutions of the 1940s and 1950s, coupled with the short bearing theory and sensible approximations to effective operating temperatures and oil-flow rates, essentially broke the back of the steadily loaded journal-bearing problem.

The finite-difference representation of the Reynolds equation and the use of relaxation methods has also been extended to other bearing configurations. Some machine elements, like roller-bearings and involute spur gears, can be represented by cylinders near planes, and although side-leakage is much less significant in such cases than in the conformal configurations of thrust and journal bearings, the effect has been examined theoretically for rigid solids by Dowson and Whomes (1967).

An alternative approach to the solution of finite bearing problems of considerable appeal is the use of the calculus of variations. This was pioneered and developed to good effect by Fränkel (1944) in Zurich, Hays (1959) in Detroit and Tanner (1961) in Manchester.

In recent years the popularity of finite-element techniques has impinged upon fluid-film lubrication problems as outlined by Reddi (1969), Fujino (1969), Argyris and Scharph (1969), Reddi and Chu (1970), Wada et al. (1971) and Wada and Hayashi (1971). Allan (1972) demonstrated the value of the finite-element approach in his analysis of pocket bearings, and an authoritative contribution stressing engineering aspects has been presented by Booker and Heubner (1972). At the present time it is probably fair to say that the considerable advantages of finite-element techniques over finite-difference approaches evident in solid mechanics have not been quite so evident in fluid mechanics in general and fluid-film lubrication theory in particular. However, it is very early days for this new approach and it can already be seen that it possesses certain virtues in dealing with complex film shapes. Comprehensive introductions to the finite-element method have been written by Zienkiewicz (1970) and Heubner (1975).

Initially, many research workers and designers preferred to use *side-leakage* factors to multiply the predictions of infinitely wide bearing solution characteristics in order to achieve sensible and useful design data. The early side-leakage factors emerged from graphical methods or from one of the major mathematical approaches outlined here, and were sometimes modified by operating experience. There is now little need for such an approach, although a somewhat more sophisticated solution was presented by Warner (1963), since there is a profusion of solutions of the Reynolds equation available for various bearing forms. It has been a happy hunting ground for the mathematically or computer-oriented engineer, and it appears that further refinements would move beyond the realm of realism for most engineering situations. It is of little use refining theoretical design data if the analysis assumes geometrical exactness of the bearing elements and the use of lubricants of the utmost cleanliness and rheological characteristics which can be specified with limited certainty.

Perhaps it is as well to close this sub-section by recalling the essentially recent advent of reliable solutions of the Reynolds equation for bearings of finite length. An outstanding book by Boswall (1928) was undoubtedly the major work of its kind between the publication of Gümbel and Everling's book in 1925 and the advent of the Second World War. Excellent summaries of the state of lubrication research and journal-bearing work in the 1930s were published by Hersey (1933, 1934). Another text of equal clarity by Norton (1942), but published posthumously, noted that, '. . . Boswall's *The Theory of Film Lubrication* is, perhaps, the best single statement of the hydrodynamic theory that exists today . . .'. Shaw and Macks (1949) were able to say very little on the subject of side-leakage and even when Michell (1950) added his classical text, side-leakage effects for all but tilting-pad bearings received little attention. All these four texts had a profound impact on the author, but of greater significance is the recognition that most side-leakage solutions of analytical or approximate form have emerged only in the third quarter of the twentieth century. Boswall's book demonstrated to good effect how graphical procedures could be used in the solution of the Reynolds equation.

It took a very long time to learn how to solve the basic differential equation presented to us by Osborne Reynolds in 1886 and the youthful nature of the solutions is not always appreciated. A valuable survey of journal-bearing literature which illustrates the path of progress in this field was prepared by Fuller (1958) and published by the American Society of Lubrication Engineers.

Thermohydrodynamic lubrication

In Section 11.9 attention was drawn to the pronounced effect of temperature upon viscous fluids, and since viscosity is the single most important property of lubricants, it is of some importance to take it into account in lubrication analysis. The subject has received much attention since 1925, and particularly since the 1940s. The state of usable solutions to the Reynolds equation for the purposes of bearing design at the time of the Institution of Mechanical Engineers' General Discussion on Lubrication and Lubricants in 1937 was well portrayed by the late Professor Swift when he wrote (Swift, 1937*b*): '. . . At the present time all theories of journal lubrication suffer because they neglect side leakage. Generally they also neglect changes in viscosity of the lubricant, but it has been found that the use of a representative value for this viscosity does not materially affect the results.'

The manner in which the paucity of side-leakage data was corrected in the ensuing years has just been described and it would appear from Swift's observations that thermal effects might not be as important as suspected, at least in the case of journal bearings. It has, nevertheless, been necessary to devote some time to the latter question, partly because Swift's observation raised the question, '. . . What is a representative value for the viscosity?' and partly because there arose about 1940 an interesting speculation about the role of thermal effects in the functioning of nominally parallel surface bearings. There was also the direct recognition by Williams (1937) of the influence of temperature upon bearing materials and wear, and a most interesting all-embracing attempt by Brillié (1937) to consider the effect of temperature, oiliness and bearing surface irregularities upon viscosity.

By far the easiest approach to the question of viscosity variation within a fluid-film bearing is to adopt the representative or mean value for viscosity, or temperature as proposed by Swift. Lasche and Kieser (1927) found that the rise in temperature in the direction of sliding in journal bearings could be represented with fair accuracy by a straight line. This and other indications have provided many suggestions for calculations of the effective viscosity in a bearing. However, the matter is really quite complicated, since it depends upon the rate of heat loss or supply to the oil film by conduction from the boundary solids, the rate of heat generation by viscous dissipation within the film and the overall transfer of heat by the lubricant by the mechanism of convection. An interesting alternative representation based upon the observation that film temperatures tend to be highest, and hence lubricant viscosities lowest, when the film is thin has been proposed by Tipei (1957). He adopts the relationship $\eta/\eta_1 = (h/h_1)^a$, where a is a constant between 0 and 1. For gas film and isoviscous liquid lubrication a is equated to zero and for most fluid-film bearing problems a is taken to be unity. Lubricant flow and side-leakage calculations are required and the development of an attack on the general problem can be found in the books by Norton (1942), Shaw and Macks

(1949), Fuller (1956), Barwell (1956), Wilcock and Booser (1957), Radzimovsky (1959), Tipei (1957), Cameron (1966) and Hersey (1966).

The variety of bearing configurations, lubricant supply systems and environmental conditions makes generalization of procedures difficult, but the only valid approach is to construct a complete heat balance for the bearing system. The merits of this approach were stressed by Professor Vogelpohl on numerous occasions and a comprehensive statement of his philosophy on bearing design was published in 1965. Whether or not such complications as outlined by Vogelpohl (1965) and the Engineering Sciences Data Unit, Item 66023 (ESDU, 1965*b*) are justified depends upon the degree of sophistication demanded of the design. Some relatively crude guides, such as the use of the mean temperature determined from a known inlet temperature and a lubricant temperature rise based upon adiabatic calculations in which it is assumed that all the heat produced by viscous dissipation remains within the oil, which yield a conservative design, can be defended on the important consideration of simplicity alone. In general, however, it is now widely accepted that rational bearing design, if based upon the use of effective temperatures or viscosities, must be derived from an overall heat balance approach.

A number of experimental studies of bearing and journal temperature distributions in both circumferential and axial directions have provided a qualitative understanding of heat flow in bearings which has been most helpful. Typical results were incorporated in an early publication by Clayton and Wilkie (1948) and at a later stage by Dowson et al. (1967). The latter paper provided experimental confirmation of a prediction made by Christopherson (1957) that the circumferential temperature fluctuation on the journal of a steadily loaded bearing would be small. The experimental measurements indicated changes smaller than 2°C.

In some interesting experiments carried out during the Second World War, but reported later, Fogg (1946) found that a grooved thrust plate intended to operate as a parallel surface bearing carried about 80 per cent of the load attributable to tilting pads under similar operating conditions. The friction was about twice as big as that expected from tilting pads, but Fogg suggested that the load-carrying capacity of the parallel surfaces arose from the expansion of the lubricant as it was heated during its passage through the bearing, a mechanism which became known as the *thermal wedge.*

The proposal was analysed by Cameron and Wood (1946) who also took account of the variation of viscosity along the bearing and side-leakage. In the United States Shaw (1947) and Charnes et al. (1953) all tackled the problem of density variation with varying degrees of additional sophistication for infinitely wide bearings. All the investigations demonstrated the reality of the density-wedge mechanism, for reasons which can readily be understood physically, but equally all agreed that the effect was far too small to account for the observed healthy performance of Fogg's bearings.

The Fogg effect was to exercise the minds of research workers for another twenty years and a factor normally neglected in the solution of the Reynolds equation which it was thought might account for the wartime experimental observations became known as the *viscosity wedge*. Cameron (1951) first applied it to contra-rotating discs and later (Cameron, 1958) to the parallel-surface bearing. The essence of the concept lies in a consideration of temperature and hence viscosity variations, both in the direction of sliding and across the thin oil film, it being assumed that the temperature of the moving bearing solid rises from inlet to outlet while that of the stationary member remains essentially constant.

Zienkiewicz (1957) first pointed out that if attempts at this class of problems were to be realistic they would have to incorporate not only solutions of the *Reynolds equation* which took account of the variation of lubricant properties along and across the film, but also the *energy equation* for the lubricant. In this and a subsequent paper by Hunter and Zienkiewicz (1960) numerical procedures for this more complex class of problems were developed. In these approaches to the problem it was necessary to impose known temperatures on the bounding solids, but it was shown by both Zienkiewicz (1957) and Cameron (1958) that the viscosity changes associated with the viscosity wedge were potentially more significant than the density changes involved in the thermal-wedge concept.

In the 1960s Dowson and Hudson (1964a, 1964b) introduced the term 'thermohydrodynamic' for this class of problem and developed the ideas proposed by Zienkiewicz and Cameron. In particular, they added the heat-conduction equations for both the moving and stationary solids of the bearing components in order to achieve more realistic predictions of the surface temperatures bounding the lubricant film. Both parallel-surface and plane-inclined surface bearings of infinite width were considered, and the findings for the former unexpectedly eliminated the viscosity wedge as a serious contender in the explanation of the performance of nominally parallel-surface bearings. It had previously been assumed that the major temperature rise would take place on the moving rather than on the stationary bearing surface, whereas the reverse effect normally occurs. As the moving surface dashes through the region of heat generation it has little time to acquire a high temperature, although it normally receives the most heat, as predicted by Christopherson (1957), whereas the stationary member steadily accumulates heat until it exhibits both a higher temperature and normally a greater temperature variation along its length, while receiving a smaller fraction of the total heat of viscous dissipation. The net result was that it was found that the surface temperatures of the solids originally assumed were reversed, and that the viscosity wedge was more likely to lead to negative than to positive load-carrying capacities!

A further important aspect of thermal effects in thrust bearings, known as *hot-oil carry-over*, which is now incorporated in realistic design procedures, was considered by Neal (1970). As lubricant which has been heated by viscous shearing emerges from a pad and passes over the groove or space before entering the following pad it encounters fresh and generally much cooler air. The mixing is not as straightforward as might at first be expected, since the hot oil within an enlarging boundary layer tends to contribute disproportionally to the oil supply at inlet to the next pad. The thermofluid mechanics of flow within an oil-groove is still the subject of research investigations.

From the historical point of view it is again relevant to draw attention to the recent advent of such sophistications as the simultaneous solution in numerical form of the combined Reynolds, energy and heat-conduction equations. It was, of course, linked with the increasing availability of digital computers, for the general concepts had been recognized, discussed and converted into firm predictions by ingenious simplifications in earlier years, but the late 1950s and early 1960s marked an important era of expansion in the application of digital computers to the solution of fluid-film problems.

By the mid-1960s it was realized that neither the thermal- nor viscosity-wedge concepts could account fully for the observed load-carrying capacity of apparently parallel surface bearings. Indeed the latter seemed to make matters worse. It was then that *thermal distortion* emerged as the most likely explanation of the Fogg effect. Swift (1946) suggested that this might be the case in a little-known discussion of Fogg's paper, but the idea was revived after all the abortive approaches mentioned above by Cameron (1960), Neal, Wallis and Duncan (1961), Dowson and Hudson (1964*b*) and Cameron (1964). It is now widely accepted that the creation of a convergent or convergent–divergent film by the thermal distortion of stationary, nominally parallel surface thrust pads, is the most likely explanation of the limited, but by no means negligible, load-carrying capacity of such bearings. The effect is greatly assisted by the presence of radial grooves carrying relatively cool oil in circular thrust rings. Indeed, it seems most likely that even if truly parallel surface bearings were produced with good accuracy, they would not remain parallel when running. The so-called parallel-surface bearing appears to be more a figment of imagination than a reality. Incidentally, a number of workers have concluded that small radii or bevels at entry to such films play a minor role. The influence of elastic as well as thermal distortion was also considered, but on the whole this is of small importance in conventional plain bearings, although it is a central feature of the highly successful mode of lubrication known as elastohydrodynamics, which is mainly associated with non-conformal machine elements as discussed in Section 11.13.

Oil flow

Beauchamp Tower impressed upon us the absolute necessity for adequate lubricant supply if fluid-film bearings were to operate successfully. At the end of the nineteenth century plain bearings were relatively lightly loaded and once the concept of fluid-film lubrication had been exposed the question of supplying adequate quantities of lubricant appeared both obvious and straightforward.

The development of quite sophisticated solutions to Reynolds equation for various bearing forms throughout the twentieth century enabled lubricant flow rates to be calculated, yet we find that once again the importance of this aspect of modern bearing performance has dawned upon us very slowly. As late as 1966 Cameron was to write in his book, *The*

Principles of Lubrication (Cameron, 1966), '. . . A quantity whose importance is only slowly being recognized is the oil flow.'

The matter is of real importance for two main reasons. In the first place it is necessary to be able to ascertain the flow rate through a bearing in order to determine the temperature rise and hence the viscosity changes which have to be taken into account in detailed bearing design. Secondly, the information is vital if the bearing is viewed as an integral part of the hydraulic circuit in a lubrication system involving sumps of adequate proportions, filters, valves, coolers and pumps. Space does not permit a full description of lubrication systems in the present text, but it is a practical matter of great importance. Many bearings, particularly those operating at high speeds, or remotely from the main lubricant supply pump, have experienced unnecessary distress owing to lubricant starvation. The subject has received special attention in a book by Nica in 1967 which was later translated into English and revised by Nica and Scott (Nica, 1969).

It is customary to distinguish between the flow which escapes readily from the supply holes or grooves in a bearing to the environment–the so called zero-speed or direct pumping flow–and that which is entrained within the bearing and sheared before emerging as side-leakage or being mixed with fresh oil to form the basis of a renewed supply. This division was proposed long ago when Barnard (1925) noted that the heat generated within a bearing had to be removed by side-leakage and that additional oil flow was necessary owing to direct leakage from the pressure-fed system. Wilcock (1957) has provided an excellent outline of the development of oil-flow studies in bearings, together with valuable design charts for the interpretation of basic data by bearing designers. Most modern texts like those by Wilcock and Booser (1957), Pinkus and Sternlicht (1961), Cameron (1966) and Barwell (1978) devote adequate attention to this important characteristic of film bearings, and design procedures like the Engineering Sciences Data Unit, Item 66023 (ESDU, 1965*b*) provide guidance of great utility.

Cavitation

One of the perennial aspects of the theory of fluid-film lubrication that has intrigued many people since Osborne Reynolds first introduced the subject in 1886, is the response of the lubricant to sub-ambient pressures. Reynolds demonstrated that a physical wedge, convergent in the direction of surface motion, was necessary for the creation of load-carrying hydrodynamic pressures, but equally the existence of divergent wedges in loaded journal bearings and several other convergent–divergent oil-film geometries, tend to yield predictions of pressures below ambient, probably below the saturation pressure of the lubricant, maybe below the saturated vapour or boiling pressure at ambient temperature and possibly of tension within the liquid.

It has been recalled that Osborne Reynolds discussed the possibility of film rupture and curtailed his calculations for partial-arc journal bearings at eccentricities which caused the pressure to return to ambient with zero gradient. For this reason this became known as the 'Reynolds cavitation boundary condition', although Sommerfeld later neglected the possibility of cavitation. By the time Gümbel considered the subject in 1921 he considered the full, non-disrupted film or *Sommerfeld solution*, an arbitrary cutting-off of sub-ambient pressures in the divergent film known as the *half-Sommerfeld condition* and the boundary conditions represented by equation [10.13] known as the *Reynolds boundary condition*.

The next significant steps took place independently and along different routes in England and Germany early in the 1930s and yet reached the same conclusion. Swift (1931) invoked stability considerations and showed that solutions based essentially upon the condition represented by equation [10.13] yielded a position of minimum potential energy of the journal relative to the load line. Stieber (1933) considered flow stability and also arrived at the conclusion that the condition [10.13] was valid. Since both Swift and Stieber were specifically concerned with the problem of film rupture, some writers prefer to describe the conditions represented by equation [10.13] as the *Swift–Stieber cavitation boundary conditions*.

The account so far suggests that if there was any doubt in the 1920s, the matter of cavitation boundary conditions in steadily loaded bearings had been well and truly resolved in the 1930s. Yet in due course experimental determinations of sub-ambient pressures in fluid-film bearings began to accumulate and could not be dismissed. An extensive experimental approach carried out at the American Bureau of Standards by McKee and McKee (1931) indicated a band of sub-ambient pressures just beyond the line of minimum film thickness, and a similar result was reported by Dowson (1957) for a totally different geometry. The latter work was based upon a research programme supervised by Professor Sir Derman Guy Christopherson at the University of Leeds and was the subject of the writer's first research in the field of lubrication.

The form of these sub-atmospheric pressure loops, normally amounting to a fraction of an atmosphere, prompted Birkhoff and Hays (1963) to propose that the cause was related to flow separation and that the appropriate cavitation boundary condition should be labelled the *Prandtl–Hopkins incipient counterflow condition*. Coyne and Elrod (1970) gave further detailed consideration to the form of the gas–liquid interface under the flow-separation condition, while Floberg (1968) concluded that the sub-ambient pressure was determined by the tensile strength of the lubricant, which was in turn related to surface tension and the size of minute gas bubbles within the fluid.

The subject of oil-film rupture by air entrainment from the environment or by the release of dissolved gas within the lubricant has received much attention. The gas, normally air, which collects in the divergent clearance space generally does so in the form of discrete finger-like bubbles, but comb and sheet formations have also been reported.

Cole and Hughes (1956, 1957) provided an excellent description of the form of 'finger' cavitation observed in journal bearings by means of a series of carefully conducted experiments. Further visual studies were reported for a range of bearing geometries by Dowson and Taylor (1975). The question of which boundary condition is the most consistent with experimental observations is the subject of continuing research, one of the most recent studies being reported by Taylor (1975).

Another form of cavitation in which the liquid actually boils at ambient temperature and reduced pressure is also encountered in some bearings, particularly under conditions of dynamic loading. The effect can also be caused by inertia action upon the flow from lubricant entry grooves and holes. Subsequent collapse of the vapour-filled cavities can cause severe damage to plain bearing materials, and there is a spate of activity in this field at the present time. The hydrodynamics of collapsing cavities was first analysed by Lord Rayleigh (1917*b*). The nature of this troublesome phenomenon has been vividly illustrated in recent times by Wilson (1975) and Conway-Jones (1975).

From time to time reports arise of measurements of liquids in tension within bearings, but it is probably fair to say that the question is still open to a certain amount of debate. There is no dispute about the ability of liquids to withstand tension under static conditions, as Berthelot demonstrated in 1850, and a survey of the literature has been presented by Temperley (1975). Osborne Reynolds (1882) himself subjected columns of mercury to tensions in excess of 1 atm (0·1 MN/m^2).

The problem of film rupture or cavitation in lubricating films in its various forms remains a challenge for physicists and engineers alike, although there is little effect upon the predictions of overall bearing performance if any of the currently discussed conditions are adhered to. It nevertheless remains an irritation to those concerned with studies of fluid-film lubrication that the factors governing oil-film rupture and the development of two-phase flow in bearings are imperfectly understood. A General Motors symposium (1964) dealt with the topic and the First Leeds–Lyon Symposium on Tribology was devoted entirely to the subject of cavitation and related phenomena in lubrication (1975).

Taylor vortices, inertia effects and turbulence

One of the corner-stones of Osborne Reynolds' theory of lubrication was the assumption of ordered or laminar flow in the thin film of fluid between the bearing solids. As twentieth-century technology advanced, the size and speed of bearing surfaces increased considerably, particularly in steam turbine–alternator electricity generating sets and in gas turbines. Furthermore, the use of unconventional or process fluids like air, water and liquid metals introduced a range of lubricants of much lower kinematic viscosity than conventional mineral oils. This combination of high speeds and reduced viscosity contrived to jeopardize the assumption of stable, laminar flow adopted by Osborne Reynolds.

Fortunately, fundamental research in fluid mechanics had provided an adequate base for the resolution of many of these practical problems. For example, in a now classical study, Taylor (1923) investigated both theoretically and experimentally the stability of a viscous fluid contained between concentric rotating cylinders. His masterly attack on the problem demonstrated that if the inner cylinder rotated while the outer remained fixed, the flow could exhibit an instability in the form of a series of regularly spaced, toroidal vortices. For a fluid of kinematic viscosity (v) small values of the radial clearance (c) relative to the inner cylinder or journal radius (R), and a rotational speed (Ω), a critical dimensionless group, now known as the *Taylor number* (T) must exceed about 1,700 according to the following relationship for the onset of this instability:

$$T = \frac{Rc^3\Omega^2}{v^2} \geq 1,700.$$

... [11.11]

The vortices can readily be illustrated by using a lubricant consisting of a suspension of aluminium paint pigment or by the injection of dyed fluid as described by Castle and Mobbs (1968). Their appearance is most appealing and although much additional work has been carried out since about 1960, Taylor's analysis has stood the test of time. For journal-bearing situations in which the journal runs in an eccentric location the critical number tends to exceed that given by equation [11.11], as deduced from the theoretical studies of Di Prima (1959, 1963) and the experimental studies of various investigators summarized by Castle and Mobbs (1968). The use of the critical Taylor number provided by equation [11.11] thus offers a conservative guide to the likelihood of encountering vortex flow.

Wider aspects of fluid-inertia effects in lubrication theory have been reconsidered from time to time in the past fifty years. Brillié (1928) included an inertia term in his analysis of thrust bearings, but a most interesting experimental study with the electrolytic tank by Kingsbury (1931) considered both the effect of centrifugal inertia forces on the flow of lubricant within the bearing and the magnitude of the ram-pressure effect at entry to the pad. The former effect was deemed to be quite small, but the latter was judged to be large; the estimate being thought to be excessive in the light of experience.

If it is assumed that the inertia effect is secondary to viscous action an iterative approach can be employed and this was the procedure favoured by Kahlert (1948). Brand (1955) considered the matter again in relation to thrust bearings, while Osterle and Saibel (1955) re-examined the role of inertia in hydrodynamic lubrication. The specific role of inertia effects in journal bearings was considered by Osterle et al. (1958). Both Osterle and Hughes (1958) and Dowson (1961) have considered theoretically and experimentally the significance of centripetal inertia effects in externally pressurized bearings and Dowson and Taylor (1967) studied the role of inertia flow in the somewhat unusual configuration of an externally pressurized spherical bearing.

In gas-lubricated externally pressurized thrust bearings the effect of inertia as fluid enters a small central hole can be significant and embarrassingly important in its effect on pressure distribution and load-carrying capacity. A pressure depression of considerable magnitude develops near the supply hole, as convincingly demonstrated by Gross (1962) and Mori (1961). Tipei et al. (1961) outlined the effects of inertia flow in bearings.

It can be seen that the role of inertia effects in lubrication was reconsidered spasmodically from 1928 to the late 1940s and early 1950s, with an intensive activity evident in the 1960s which in certain respects is still present today. Perhaps the most comprehensive account of inertia effects in lubrication was presented by Milne (1965) who had earlier contributed significant papers on the subject. Milne discussed the fundamental issues; entrance effects, both in terms of stagnation pressures and the development of merging boundary layers; transit effects involving the effect of acceleration and deceleration of the lubricant; centrifugal action; the consequences of bearing acceleration and oscillation; and finally the general role of inertia in limiting laminar flow. His work was particularly valuable in its effective demonstration of the use of various approaches, including averaged inertia, momentum integral, iteration, series expansions and marching techniques. The use of stream functions by Purday (1949) and Milne (1965) in lubrication problems exposes a little-explored area, and Milne's (1965) concern for lubricant flow at entry to a bearing was of direct interest to bearing design considerations.

Milne's overall conclusions were that, '. . . In general, it may be stated that inertia effects have been found to be of greater importance in externally pressurized than in self-acting bearings, and to be of greater importance in promoting instability of flow than in modifying directly the magnitude of pressure gradients.'

In some cases modern machinery operates at such high surface speeds or with fluids of such low kinematic viscosity that the lubricating films become fully or partially turbulent. Very often a journal bearing will run through a Taylor vortex regime before encountering fully developed turbulence. The first record of turbulence in journal bearings was reported by Wilcock (1949) and similar deductions were made by Wilcock (1955) and Abramovitz (1956) for hydrodynamic pocket and tilting-pad bearings, respectively. A most valuable experimental study of journal bearings operating in the turbulent regime was reported by Smith and Fuller (1956).

Mention was made of the gradual accumulation of research and operating evidence of turbulence in bearings in the Institution of Mechanical Engineers' General Discussion on Lubrication and Lubricants in 1957 and in the past twenty years the subject had advanced considerably. It proved to be a happy hunting ground for analysts with Constantinescu (1959) using Prandtl's mixing length approach, and Ng (1964), Ng and Pan (1965) and Elrod and Ng (1967) the eddy viscosity concept, respectively, to represent the turbulent theories. The former had the merit of simplicity, but the eddy viscosity approach was deemed by Taylor (1973) to be preferable. In both cases similar forms of the Reynolds equation for turbulent conditions were deduced of the general form,

$$\frac{\partial}{\partial x}\left[\frac{h^3}{k_x\eta}\frac{\partial p}{\partial x}\right] + \frac{\partial}{\partial y}\left[\frac{h^3}{k_y\eta}\frac{\partial p}{\partial y}\right] = \frac{U}{2}\frac{\partial h}{\partial x}, \qquad \ldots [11.12]$$

where both k_x and k_y could be related to the Reynolds number (R_e) by general expressions of the form,

$$k_{x,y} = 12 + f(R_e). \qquad \ldots [11.13]$$

For laminar flow the functions $f(R_e)$ become zero and equation [11.13] reduces to the original Reynolds equation [10.9].

More recently Hirs (1972) has adopted a bulk-flow approach which appears to enjoy both novelty and utility in bearing analysis. The various theories have been compared by Taylor and Dowson (1974).

When bearings operate in the turbulent regime several important operating characteristics are affected. The volume rate of flow for a given film geometry and pressure drop is reduced and the power loss or heat generation are increased as noted by Christopherson (1957). The effective viscosity tends to rise and this increases both the load-carrying potential and the friction or power loss. Both effects are quite marked, although the general consensus of views is that friction is affected more than load-carrying capacity, leading to an increase in the coefficient of friction.

In any event, bearings operating in the turbulent regime produce more heat and consume more power than bearings operating under laminar flow conditions. Cogent advice offered by Wilcock and Booser (1957) was, '. . . A good rule of design is to avoid the turbulent regime', although the accumulation of knowledge and experience modified this viewpoint. In 1974 Wilcock was able to write: '. . . Turbulence, as a phenomenon occurring in a bearing, has no intrinsic harmful effect on bearing behaviour. Thus the designer need not fear or avoid turbulent operation' (Wilcock, 1974*b*).

Nevertheless, a measure of the increasing concern for and interest in the subject of turbulence in bearings was provided by the fact that the Lubrication Division and Research Committee on Lubrication of the American Society of Mechanical Engineers (ASME, 1974) held a special and most valuable conference on the subject of turbulence and related phenomena at Northwestern University, Evanston, Illinois in June 1973. The Second Leeds–Lyon Symposium on Tribology (Mechanical Engineering Publications 1977) held in Lyon in 1975 was devoted to the subject of superlaminar flow in bearings and the *Proceedings* of these recent international events provide good surveys of present problems and the state of the art in this particular aspect of lubrication theory and bearing performance.

Body force: gravity, magnetohydrodynamics and magnetic levitation

It is evident from the full Navier–Stokes equations governing the flow of fluid that in addition to surface stresses associated with viscosity and pressure, forces distributed throughout the bulk of the material known as *body forces* might contribute to the motion under some circumstances.

The best-known body force in fluid mechanics is the *gravity effect*, and it can readily be shown that this is negligible in lubricating films. There are, however, some peripheral flow problems associated with self-contained bearings involving oil-rings and discs dipping into baths of lubricants, where gravity and viscous actions dictate the *viscous-lifting* mechanism by which lubricant is raised to the bearing and distributed along the shaft.

The subject is of considerable importance since the satisfactory design of self-contained bearings requires a knowledge of the ability of practical arrangements to deliver lubricant to the bearing. Some preliminary analytical studies of the fluid mechanics of viscous lifting have been reported by Blok and van Rossum (1948), van Rossum (1958) and Groenveld (1970), but the subject is still largely empirical. Valuable guidance for the designer has been provided by Martin (1973) in the *Tribology Handbook*.

A casual glance at the lubrication literature of recent years show that considerable analytical effort has been devoted to the subject of *magneto-hydrodynamic lubrication*. However, unlike the subject of viscous lifting, the field has not been adopted in practice and very few experimental studies have been reported.

The subject arose from a desire to improve the effectiveness of fluid-film, liquid-metal lubricated bearings. The advantages of using a process or heat transfer medium as the lubricant in nuclear reactors has already been noted, but one inescapable consequence is the very low viscosity of the fluid. If the lubricant is an electrically conducting fluid and it is caused to flow through a transverse magnetic field, it is possible to develop electromagnetic pressure gradients greatly in excess of the conventional hydrodynamic pressure gradients at the same flow rate. Good introductions to the subject were provided by Snyder (1962), Osterle and Young (1962) and Elco and Hughes (1962). Theoretical studies of journal, squeeze-film and slider bearings were reported by Kuzma (1963, 1964, 1965) of the General Motors Corporation Research Laboratories, and Shukla (1963), Shukla and Prasad (1966) and Shukla (1970) have considered both the general principles and theoretical formulations of both the magnetohydrodynamic, hydrostatic and slider-bearing problems.

Rare but refreshing evidence of experimental work in this field has been reported for squeeze-films by Kuzma et al. (1964), and for hydrostatic bearings by Krieger et al. (1966). The agreement between theory and experiment is good, particularly when the important effect of inertia is considered, but at the present time the practical potential of magneto-hydrodynamic bearings appears to be severely limited. The effect is real, but for the generation of realistic load-carrying capacities the vast bulk and expense of the electromagnetic facility seems to exclude contemporary applications.

Magnetic levitation in bearing applications has been reviewed by Taylor (1972). He points out that stability problems can arise, but that the problem is much reduced if superconducting materials or time-varying magnetic fields are used. Applications, real or contemplated, vary from the exotic frictionless cryogenic gyroscope, centrifuge bearings, high-accuracy balances and high-speed tracked vehicles to the mundane electricity supply meter.

Lubrication of real (rough) surfaces

The importance of adequate surface finish for the successful performance of fluid-film bearings has long been recognized and is largely self-evident. The dramatic improvement in performance of the journal bearings which received the careful attention and excellent workmanship of Albert Kingsbury made the point with some force.

Nevertheless, some workers felt that a theory of lubrication which took no account of gross geometrical deficiencies or the existence of surface roughness, could never be complete, and some went as far as to say that many of the features of bearing performance attributed to boundary lubrication might emerge from a more exact analysis of the hydrodynamic lubrication of rough surfaces. The present discussion will be limited to the development of *rough surface* lubrication studies rather than the alternative – but often just as significant – question of gross geometrical errors.

Michell (1950) presented one of the earliest and characteristically thorough analyses of the problem, at least as far as plain thrust pads are concerned, in an appendix to his book in which he explored mathematically the *effects of rugosities*. The somewhat quaint terminology was in reality reflected by a consideration of the hydrodynamic performance of pads covered by small corrugations in the form of a cosine function. For the purpose of analysis the corrugations were assumed to run in the direction of sliding and to be representative of turned and milled surfaces.

Various ratios of the amplitude of the cosine wave to the film thickness were considered and the general conclusion that was reached was that the load-carrying capacity decreased and the friction coefficient increased continuously as the ratio increased.

Burton (1963) also studied the influence of sinusoidal roughness on bearing load support and Tipei and Pascal (1966) discussed both micro- and macro-geometrical influences upon pressure generation. Dowson and Whomes (1971) extended the range of axial profiles to include rectangular, triangular and cosine ridges in an analysis of hydrodynamic effects upon rigid rollers on inclined planes. A simple experiment provided partial support for the theoretical predictions.

Tzeng and Saibel (1967) indulged in the processes of random analysis in their study of rough slider-bearing lubrication, but Christensen (1970) enlarged the scope of these investigations by linking them to the emerging appreciation of surface topography discussed in Section 11.2. He viewed the film thickness as a *stochastic* feature characterized by a number of statistical parameters. The approach was later applied by Christensen and Tonder (1971, 1972) to the analysis of both thrust and journal bearings of finite width. The concept provided a new impetus to film-lubrication studies of rough surfaces and the manner in which current studies are being linked even more closely to the expanding knowledge of bearing-surface topography suggests that investigations of hydrodynamic lubrication of real surfaces will represent a significant feature of late twentieth-century work on tribology.

11.11 Reciprocating machinery

Rapid development of internal combustion engines and reciprocating compressors has done much to promote studies of tribological components operating under dynamic conditions. Indeed, it is no exaggeration to say that much of the evident and impressive increase in engine ratings has been possible only with concomitant progress in lubricant technology, tribological materials, and a concerted theoretical and experimental attack on problems of dynamically loaded bearings and piston rings and liner lubrication and wear. In the present section attention will be restricted to progress in the fields of dynamically loaded bearings and piston rings.

Steadily loaded fluid-film bearings were the preoccupation of earlier eras, but in the past fifty years the analyst has opened up the possibility of solving dynamically loaded bearing problems to such an extent that design procedures for reciprocating machinery are now almost as well established as those for steadily loaded situations. The manner in which this difficult area of lubrication has developed in recent times certainly merits recognition.

Boswall's (1928) book was entirely, and Norton's (1942) text almost completely, concerned with steadily loaded thrust and journal bearings. Indeed, the latter was so involved with the intricacies of solving the steady-state problem, that he was moved to make the following partly erroneous but interesting comment on the state of thrust-bearing analysis. '. . . The majority of slider bearings operate with intermittent or reciprocating motion. The hydrodynamic analysis requires a steady state, and it is therefore impossible to give a numerical analysis for many of these bearings.'

A new class of problems arises when non-steady-state conditions are considered in lubricating films. Osborne Reynolds recognized that both *entraining action* with a physical wedge-shaped lubricating film, or *squeeze-film* action with the opposing solids approaching each other, could lead to the formation of load-carrying hydrodynamic pressures. He included both terms in equation (17) of his paper (Reynolds, 1886). The development of steady-state solutions which neglect changes in film geometry with time have received considerable attention and the next but merely intermediate step towards a full analysis of dynamically loaded bearings neglects entraining action and considers squeeze-film effects alone.

Squeeze-films

Osborne Reynolds provided a solution for the approach of parallel plates of elliptical form as an illustration of analytical solutions to restricted forms of his equation for fluid-film lubrication. Stefan had even solved the same problem twelve years before Reynolds, as noted in Chapter 10.

Good reviews of the not inconsiderable range of solutions to squeeze-film problems have been presented by Fuller (1956, Sects 19, 22), Archibald (1956), Hays (1961, 1962), and Moore (1965). One of the most comprehensive accounts of squeeze-film lubrication is contained in Chapter 7 of the book by Pinkus and Sternlicht (1961).

The lubricating film in bearings responds in two important ways when subjected to pulsating or dynamic loads. If the load tends to move the bearing surfaces towards each other the viscous fluid shows great reluctance to be squeezed out of the sides of the bearings. The tenacity of a squeeze-film is remarkable, and the survival of many modern bearings depends upon the phenomenon. It can readily be shown that for hydrodynamic thrust bearings of representative form, having a length (l) and minimum film thickness (h_0), the load-carrying capacities (P) due to separate squeeze and entraining actions involving velocities of approach (W) and sliding (U), are in the proportion,

$$\frac{P_{(squeeze)}}{P_{(entraining)}} = \frac{W}{U}\left(\frac{l}{h_0}\right). \qquad \qquad \ldots [11.11]$$

Since (l/h_0) is typically 10^3 and often 10^4 it is evident that a relatively small velocity of approach will provide a very large load-carrying potential.

The second important action is associated with separating surfaces. Very small velocities of separation can yield extremely low pressures and give rise to cavitation in the form of gas release or boiling at reduced pressure as outlined in Section 11.10. One of the most common possibilities of severe cavitation damage in bearings is, of course, associated with the likelihood of changes in the operating conditions which might tend to pull the bearing surfaces apart in the presence of the lubricant.

Dynamically loaded bearings

The hydrodynamics of dynamically loaded bearings was first considered by Harrison, who had earlier presented a basic analysis of gas bearings (Harrison, 1919) as outlined in Chapter 10. Major contributions to journal-bearing studies for both steady and dynamic loading emerged in the 1930s from the University of Sheffield, where Professor Swift[6] considered stability and cavitation (Swift, 1932), journal-bearing design (Swift, 1935) and the fundamental aspects of the response of bearings to dynamic loading (Swift, 1937a). The latter work was extensive and penetrating and although it was based upon many of the assumptions adopted by earlier investigators of fluid-film lubrication, it remains as a most significant milestone in the development of understanding of a complex problem.

Swift established the basic differential equations for dynamically loaded bearings and then proceeded to develop solutions within the broad framework of Sommerfeld's analysis. This meant that side-leakage was neglected and that there was no restriction placed upon the formation of sub-ambient pressures. The latter assumption was not taken lightly, and Swift went to the extent of arguing that since the change from maximum to minimum pressures took place so very quickly in reciprocating machinery, it was most unlikely that the physics of the situation would permit the repeated breakdown and re-establishment of the film.

Orloff (1947) pointed out that bearings subjected to rotating centrifugal loads could be analysed in terms of known solutions to steadily loaded bearing problems, and in the same year Stone and Underwood (1947) introduced a similar procedure which, although now seen to be of limited value in design, was a step of some importance in historical terms. The importance of the work was that it provided the designer with an

---

6. See Appendix, Sect. A.21 for biographical details.

approach of great simplicity and at the same time nudged the subject away from the strictly research field towards the practical considerations. The procedure became known as the 'equivalent speed concept' and was based upon the observation that the development of hydrodynamic pressures in journal bearings depends upon the term

$$K = [2\omega - (\Omega_b + \Omega_s)],$$

where Ω_b and Ω_s are the bearing and shaft rotational speeds and ω is the rotational speed of precession of the shaft centre about the journal centre. For a steady, rotating load the latter is merely the rotational speed of the load vector. If squeeze-film action is absent, it is clear that the locus of the shaft centre and the minimum film thickness depends directly upon K and that realistic situations can thus be related to known steady-state solutions. Clearly a difficulty arises when $K = 0$, and since this condition is often encountered, the equivalent speed concept predicts zero film thickness. In reality the all powerful squeeze-film effect frequently provided protection of the bearing surfaces during such moments of adverse motion and supporters of the approach thus argue that its predictions are conservative.

Burwell's (1947, 1949, 1951a) most valuable extension of the analysis of dynamically loaded bearings, including the introduction of side-leakage factors which had by then emerged in the steadily loaded bearing literature, with his accounts of the way in which the analysis could be applied to diesel engine connecting-rod bearings and radial aircraft engine master-rod bearings, was crowned in 1951 by his wide-ranging review of the situation at entry to the second half of the twentieth century (Burwell, 1951b).

It is significant that Shaw and Macks (1949) chose to devote so much of their excellent book to procedures for determining polar-load diagrams for a variety of engines, before tackling the question of dynamically loaded bearings in Chapter 6. Load calculations of this nature were by no means novel, as admitted by the authors, but the emphasis and care devoted to the presentation of this work was most timely in relation to studies of dynamically loaded bearings. Their consideration of the relative importance of gas forces and both rotating and reciprocating inertia forces was also to influence later workers.

In the 1960s there emerged new approaches of great practical and conceptual significance. Booker (1965b) introduced a graphical procedure for charting the locus of a shaft subjected to an arbitrary load which he described as a *mobility* (being a ratio of dimensionless velocity to force) *method of solution*. The approach considers the journal movement to arise from a combination of velocities of the journal in a stationary bearing along squeeze-paths, together with a component related to the angular velocities of the journal, bearing and load. The graphical procedure was later described in more detail by Booker (1971) and at about the same time Blok (1973) and his colleagues were developing a somewhat similar *impulse-whirl* or *iso-impulse* approach. In these latter procedures *impulse* replaces *time* as the independent phasing variable in the equations governing the motion of the journal centre. Both methods can take advantage of existing solutions for infinitely long, short or finite journal bearings, and in due course these initial and essentially graphical procedures were computerized by, for example, Herrebrugh and Moes.[7] They opened up both the ease and accuracy of dynamically loaded bearing calculations and at the same time provided the designer with a *physical* feel for events.

In Section 11.10 it was noted that the question of cavitation boundary conditions in bearings subjected to steady loads was still a subject of debate. The more complex problems of moving cavitation boundaries in dynamically loaded bearings was examined to good effect by Olsson (1965). He found that in time-dependent situations, cavitation boundary conditions based upon continuity of flow depended upon the relative magnitudes of the cavitation boundary and journal surface velocities. An important conclusion was that if the cavitation boundary moved at a speed less than half the journal surface speed, the more conventional, steady-state cavitation boundary conditions could be used.

Comprehensive theoretical, experimental and design studies of dynamically loaded bearings were reported by Lloyd et al. (1967*a,b*), Middleton et al. (1967), and Hahn and Radermacher (1967) at an Institution of Mechanical Engineers meeting held in Nottingham in 1966 and devoted to the subject of journal bearings. This range of papers indicated much progress and sophistication, particularly in the application of digital computers to dynamically loaded bearing problems.

The relative importance of inertia and gas forces was examined by Martin and Booker (1967) with the interesting conclusion being reached that the latter could often be neglected in studies of minimum film thickness in contemporary engines. A most comprehensive survey and comparison of alternative approaches to the determination of shaft orbits in reciprocating machinery was presented by Campbell et al. (1968).

There is a sense in which the degree of sophistication now reached in analytical procedures is impinging upon physical reality and some of the basic precepts upon which the theories are based. When the complications of approaches to dynamically loaded bearings are considered it is perhaps as well to remember that neither shafts nor bearings are round or smooth, that thermal and elastic distortions abound in reciprocating machinery, that lubricant supply arrangements are often inadequately understood and that the question of cavitation and lubricant rheology requires further study. The prospect of developing further refinements to solutions is indeed daunting and in some cases might even be misguided. In many cases it is not the numerical value of minimum film thickness determined by some theoretical approach for a particular bearing that is most important, but the ability to compare such predictions for other bearings in the engine and to explore the effect of engine design changes. The qualitative picture which emerges from such an approach can be most informative.

---

7. Private communication.

Piston rings

The Ramsbottom ring developed for steam engines in the nineteenth century was not only adequate for the demands of the day, but it provided the essential form of seal for the modern internal combustion engine. There have, of course, been changes in materials, an improvement in the accuracy of manufacture of cylinders, pistons and rings and a recent flurry of activity related to basic studies of ring and liner friction, wear and lubrication.

Ricardo (1922) expressed the view that friction between piston rings and cylinder liners accounted for a significant proportion of total engine friction. Stanton (1925) concluded that the friction followed dry or boundary lubrication laws, whereas Hawkes and Hardy (1936) found evidence of hydrodynamic action throughout most of the stroke. The general interest in ring and liner friction promoted much of the subsequent work on lubrication. Castleman (1937) contributed an early hydrodynamic theory of piston lubrication of both technical and historical interest.

Apart from the interest in friction, and hence lubrication, the other major considerations were the effectiveness of the ring as a seal for the combustion space, and wear. It is generally recognized that the top ring in a pack is all-important in the sealing process. Nevertheless, Illmer (1937) considered that efficient packs shared the overall pressure drop between the individual rings. Piston-ring dynamics is a complex business, but Williams and Young (1939) came to the conclusion that the forces and accelerations to which a ring was subjected near to top-dead-centre (tdc), could contrive to lift it off its bottom land, thus allowing gas to *blow by* and hence to let the ring collapse into its groove. The phenomenon of blow-by and its detrimental effects had been known to Taub (1939), and Dykes (1947) confirmed the phenomenon after the Second World War.

The form of wear most generally encountered in the ring and liner situation is known as *scuffing*, and Teetor (1938) drew attention to the importance of metallurgical considerations in this process. Williams (1936) concluded that cylinder wear was strongly influenced by the availability of an adequate supply of lubricant, and there was a paper presented to the Institution of Mechanical Engineers General Discussion on Lubrication and Lubricants in which Bouman (1937) stressed the relationship between wear and the amount of lubricant supplied to the cylinder wall.

Prior to the late 1940s and the 1950s, fundamental knowledge on piston-ring functioning had developed little since 1925. Immediately after the Second World War, Courtney-Pratt and Tudor (1946) confirmed experimentally prewar suggestions that hydrodynamic lubrication between ring and liner prevailed throughout most of the stroke, but that oil-film breakdown often occurred near the dead-centre positions. An important combined theoretical and experimental study by Eilson and Saunders (1957) was more optimistic about the potential of a continuous hydrodynamic film, but Faro Barros and Dyson (1960) worked on the same apparatus and counselled caution. Their suggestion that oil-film breakdown might occur during the first 30 per cent or so of the piston stroke has been largely confirmed by later work.

By about 1960 a concerted attack on the question of ring dynamics, lubrication and wear was under way. It was generally agreed that the major radial loading on rings arose from the gas pressures behind them, and since these varied, together with the velocities, throughout the stroke, the piston ring and liner was really a form of dynamically loaded bearing. In fact the subsequent studies of ring lubrication are similar in form, but compressed in time scale, to those of dynamically loaded bearings in general. Clearly, if squeeze-film effects are neglected, the mid-stroke conditions will predict, on the basis of simple hydrodynamic entraining action, a fluid film whose thickness is in reasonable accord with experiment, while at the ends of the stroke unrealistic and non-existent films are anticipated by an analysis comparable to that of Stone and Underwood (1947) for dynamically loaded bearings.

A high degree of sophistication has now been achieved in theoretical ring-lubrication studies by Furuhama (1959, 1960, 1961), Lloyd (1969), Baker et al. (1973, 1976), Hamilton and Moore (1974*b*), Baker et al. (1976) and others. Equally important experimental studies of friction, oil flow, piston-ring profiles, inter-ring pressures and even the pressure and film-thickness distributions have been reported. The friction studies of Faro Barros and Dyson (1960) were particularly useful, while Greene (1969) illustrated the fascinating but far from uniform behaviour of ring-lubricating films in a piston-cylinder simulation apparatus; Wakuri et al. (1970) studied oil loss; McCallion et al. (1972), Baker et al. (1973) and Hamilton and Moore (1976) considered the developed geometrical profile presented to cylinder liners by rings. Baker also provided valuable experimental data on inter-ring pressures, while Hamilton and Moore (1974) tackled the extremely difficult task of measuring the oil-film thickness by means of a capacitance technique. More recently Brown and Hamilton (1976) have reported the use of a piezoelectric pressure transducer to record oil-film pressures. In their impressive approach to piston-ring lubrication and wear, Ting and Mayer (1973) incorporated a procedure based on gas dynamics for calculating inter-ring pressures in cases where this important information might otherwise not be available.

It seems clear that theory and experiment in studies of piston-ring lubrication now form an active and promising part of current tribological investigations. There is still much to be done, but the improved understanding of piston-ring behaviour is already leading to proposals for innovation in design, as in the case of the *humped groove* described by Butler and Henshall (1976). A wider appreciation of the importance of ring dynamics, the influence of ring profile on lubricant transport (Baker et al. 1973) and the need to consider the behaviour of the complete ring pack is gradually emerging. There was also a somewhat dramatic development of interest in ring and liner scuffing in the 1970s.

Piston-ring scuffing in reciprocating internal combustion engines and gas compressors was identified by the Tribology Research and Liaison Committee as a subject of growing technical and economic significance late in the 1960s. Mr M. J. Neale was invited by the Ministry of Technology to determine the incidence of scuffing in terms of engine features and operating conditions. His report (Neale, 1970) clearly indicated that scuffing of piston rings and liners had assumed some importance and demonstrated the complex metallurgical, thermal and lubrication factors involved. He also concluded that there was no clear understanding of how piston rings worked and proposed various research projects to remedy the situation.

A meeting devoted entirely to the troublesome problem of piston-ring scuffing was held in May 1975 by the Institution of Mechanical Engineers. The high attendance of some 260 persons and the material included in the bound volume (Institution of Mechanical Engineers et al. 1976a) suggests that piston rings are likely to dominate the later 1970s and 1980s as much as studies of dynamically loaded bearings featured in research programmes in the 1950s and 1960s. Most modern texts on lubrication and tribology include sections on piston rings and a particularly valuable survey of fundamental work can be found in Hersey (1966).

The recent history of work on piston rings is interesting for many reasons. It calls for a multi-disciplinary approach; it is an indication of the widening scope of tribology to additional machine elements beyond the well-trodden ground of plain journal and thrust bearings; and it illustrates concisely the process of engineering development through problem identification, project specification, theoretical and experimental research and finally corrective and perhaps innovative design, coupled with materials and manufacturing progress.

11.12   Bearings and rotor dynamics

The 1920s witnessed a considerable development of the steam turbine and it was during this period that the special influence of fluid-film bearings upon rotor dynamics was clearly recognized. At the present time, when steam turbine–alternator electricity generating sets of 500 MW capacity having total lengths of about 164 ft (50 m) are quite common and future generations of machines of capacity in excess of 1,000 MW are being planned, the subject has once again attracted much attention. Other forms of machinery in which the influence of fluid-film bearings upon rotor dynamics is of considerable importance include fans, compressors, centrifuge spindles, gas turbines, electric motors, pumps, gyroscopes and textile spindles.

The subject is one which is rich in impressive research in rotor dynamics, much of it of long standing, with ample and sophisticated studies of bearing characteristics. The difficulty is to put the two facets together, since rotor dynamicists often adopt a simplistic and inadequate approach to bearings, and bearing analysts and designers too often allow the wider issues of system dynamics to escape them.

Rotor dynamics

Rankine (1869) first drew attention to the vibrations which could be encountered in rotating machinery at certain *critical speeds*. For a rotor of mass M concentrated at a radius (ε) from the undeformed axis of rotation on a shaft of radial bending stiffness (k) and negligible weight, rotating at an angular velocity (ω), the critical speed (ω_n) and amplitude of vibration (r) of the undamped system are given by the expressions:

$$\omega_n = \sqrt{\frac{k}{M}}, \qquad \qquad \dots [11.12(a)]$$

$$\frac{r}{\varepsilon} = \frac{1}{[(\omega_n/\omega)^2 - 1]}. \qquad \qquad \dots [11.12(b)]$$

Long before this time, Euler (1744) and Bernoulli (1751), had considered the lateral vibrations of bars supported in various ways, but it appears that Rankine extended the concept to out-of-balance rotating machinery. The frequency of the vibration is the same as the rotational speed of the shaft and is known as *synchronous whirl*. It can be seen from equation [11.12(b)] that a serious condition of large-amplitude vibration occurs at shaft running speeds close to ω_n.

The basic conditions for stability were established by Routh (1877) and still form the basis for many approaches to this important aspect of rotor behaviour.

Dunkerley (1894) carried out a wide range of experiments suggested by Osborne Reynolds and developed an approximate method for finding the *natural frequencies* or *critical speeds* of more complicated rotor systems. In the same year Lord Rayleigh (1894) introduced his approximate *energy method* for continuous beams. Jeffcott (1919) did much to establish the present-day concepts of shaft whirling. In the 1920s, steam-turbine development accelerated work in the field, with Holzer (1921) considering torsional vibrations and Stodola (1925) developing an iterative technique in which the mode shape of the vibration was assumed and successively improved, and which later formed the basis for *matrix iteration* procedures. In the 1930s Smith (1933) analysed non-symmetrical rotors, considered both bearing and rotor flexibility and found four critical speeds where there had been one for the symmetrical case.

The prolific literature on rotor dynamics which arose following the advent of the digital computer has, since the 1940s, been largely concerned with improved modelling of the rotor system and alternative procedures for predicting rotor response and stability. Support flexibility and damping, distributed-mass systems, gyroscopic and inertia effects, external forcing, environmental and internal damping were all considered. Various mathematical procedures ranging from developments of the initial *lumped-mass* concept to *transfer matrices* associated with distributed parameter models and *finite-element* techniques built upon consistent mass representations have been evaluated. Dynamic response has been approached through developments of the early numerical methods, modal analysis, direct integration, finite-element and perturbation techniques.

Balancing of rotating machinery has become a most important aspect of the successful operation of present-day equipment, with considerable interest being shown in the possibility of improving the balance of running machinery.

Bishop (1959) has surveyed the literature dealing with the response of unbalanced rotors and more recent reviews of critical speeds, response and balancing of flexible rotor systems have been proposed by Eshleman (1972), Shapiro and Rumbarger (1972) and Rieger (1973) and published by the American Society of Mechanical Engineers. General vibration texts like those by Den Hartog (1934), Bishop and Johnson (1960) and McCallion (1973) form a good basis for foundation studies of rotor behaviour.

Bearing stiffness and Damping

While the understanding of critical speeds and the behaviour of flexible-rotor systems was advancing steadily, in the 1920s there emerged evidence of a different form of vibration intimately linked to the operating characteristics of the supporting bearings. The link with the bearings was neither clear nor well understood in the early days, but some classical investigations by engineers concerned with steam turbines and blast-furnace compressors soon transformed the state of knowledge.

Professor Stodola contributed greatly to the steam-turbine literature in general and accounts of studies of vibrations in particular. His textbook (Stodola, 1927) was to influence subsequent generations of turbine designers, but before its publication he had suspected, from his wide experience of turbine testing, the existence of a critical speed which appeared to be unrelated to shaft flexibility and unbalance. He associated the vibration with the bearing stiffness, and introduced his *oil spring* concept. In this paper he suggested that fluid-film bearings acted like non-linear springs, with a definite stiffness associated with each eccentricity.

In the United States Newkirk (1924) had already drawn attention to the influence of friction–associated with slip between a wheel or disc and the shaft upon which it was mounted–upon vibrations. He described the phenomenon as *cramped shaft whirl*, the effect being similar to that of internal friction or elastic hysteresis of the shaft material. Kimball (1924) demonstrated and explained this effect of internal friction.

Newkirk's studies were related to disturbances of the rotors of some blast-furnace compressors running at speeds above twice the first critical speed. It was initially thought that the phenomenon was related to the *hysteresis* or *cramped* action mentioned in the previous paragraph, but the study was to reveal the phenomenon of *oil-film whirl*. A model of the rotors was prepared and whirl or vibration orbits were recorded photographically by means of a spherical mirror system described earlier by Newkirk (1924). Newkirk and Taylor (1925) reported a slight increase in amplitude of a synchronous vibration when running at the shaft critical speed, but a remarkable and totally unsuspected increase in the severity of the vibration when the rotational speed of the shaft was about twice the first critical speed. The vibration was not synchronous, but almost equal to the first critical speed, or half the shaft running speed. At higher running speeds the amplitude of the vibration increased slightly, but the frequency of vibration remained close to the first critical speed. Newkirk and Taylor used the descriptive terms *resonant whip* or *resonant whirl* to describe this oil-film induced, self-excited vibration of a flexible rotor. In

due course the terms *half-speed whirl* or *half-frequency whirl* were preferred, since it was found that the particular form of vibration considered
could also occur at running speeds below twice the first critical. The terms
'resonant whip' or 'resonant whirl' are generally reserved for the particular
condition in which the vibration frequency coincides with the first critical
speed.

It is interesting to recall that Harrison had already recognized the
possibility of such vibrations analytically, as mentioned in Section 10.4. In
essence the argument is that the mean velocity of flow of the lubricant in a
lightly loaded journal bearing in which only the shaft rotates with surface
speed (U) is approximately $U/2$. If the shaft centre precesses round the
bearing centre in the direction of shaft rotation at a rate equal to half the
angular speed of the shaft, the lubricant entraining action is lost and
equilibrium can thus be achieved by allowing the shaft to sit in a fixed
location or by permitting half-speed whirl. In practical situations, disturbances will inevitably cause the latter half-speed whirl possibility to be the
preferred motion.

One of Stodola's pupils was C. Hummel, who undertook an analytical
and experimental investigation of the suggestion that oil-film stiffness
could be important in some rotor vibration problems (Hummel, 1926).
He found two critical speeds and evidence of instability in the half-bearings
employed when the effective eccentricities were less than 0·7.

Once the phenomenon of half-speed whirl had been recognized,
empirical remedial procedures for the avoidance or alleviation of the
damaging vibration ensued. It was recommended that operating speeds
should not approach twice the first critical speed, that whirl could be
diminished by increasing the load, bearing supply pressure and radial
clearance, or by decreasing the bearing length and lubricant viscosity.
All these effects caused an increase in operating eccentricity ratio, but the
recommendations were qualitative rather than quantitative and in some
cases solutions to particular problems were found to be contrary to these
guidelines.

Newkirk (1931) confirmed experimentally the existence of half-speed
whirl with very stiff shafts running at speeds well below twice the first
critical speed, and in due course it was recognized that half-speed whirl
could occur over a wide speed range.

Robertson (1933) extended the theoretical analysis of stability by
Harrison for bearings of infinite extent, by considering journal and rotor
mass, but the most comprehensive theoretical study prior to the Second
World War was presented by Swift (1937*a*). His general analysis of non-
steady conditions in journal bearings was mentioned in Section 11.11, but
it undoubtedly provided the foundation for later analytical studies, even
though it applied only to bearings of infinite width and excluded the
possibility of cavitation.

Kapitza (1939) studied the transition of rotors through critical speeds
while he was working on the development of a high-speed expansion
turbine, and suggested that practical machines survived the ordeal only if
the oil film in the bearings provided adequate damping. It is interesting
to note that, prior to 1939, emphasis had been placed upon the elasticity

of the film, but Kapitza's proposals were to receive active consideration in the postwar years.

Studies of bearing-influenced rotor dynamics

There has been a great increase in interest in rotor dynamics and the effects of the bearings since the end of the Second World War. This appears to be attributable to the increase in size and speed of rotor installations, the availability of digital computers, the increased use of air bearings and the firm establishment of gas-turbine power units.

Hagg (1946, 1948) included oil-film damping in his studies of the behaviour of a rigid shaft supported by various forms of bearings. In an associated experimental investigation Hagg confirmed the stabilizing characteristics of tilting-pad journal bearings. Cameron (1955) and Cameron and Solomon (1957) confirmed the possibility of whirl of a rigid shaft in half-bearings operating at eccentricity ratios less than 0·7, and Hori (1956) presented a comprehensive study of the behaviour of a flexible rotor in full journal bearings. Korovchinskii (1953, 1956, 1960) gave detailed consideration to journal stability and the same subject was reviewed by Tipei (1956). Poritsky (1953) and Hagg and Warner (1953) both considered the question of stability of flexible rotors in bearings and derived simple stability criteria involving both rotor and oil-film stiffness.

In the 1950s it became clear that further information was required about oil-film stiffness and damping for a variety of bearing forms to provide the essential ingredients of an analysis of rotor dynamics and to assist in the design and development of whirl-resistant bearings. Sternlicht (1959) provided numerical solutions for the eight bearing stiffness and damping coefficients for cylindrical journal bearings required to represent the restoring forces in small perturbation analysis. Holmes (1960) applied short-bearing theory in his study of the stability and response of a rigid rotor, which marked the beginning of a series of papers on bearing-influenced rotor dynamics.

The rapid development of gas bearings in the 1960s was responsible for detailed studies of rigid rotors in bearings having elasticity and damping characteristics and Lund (1965) extended the analysis to the case of a flexible rotor mounted on flexible-damped supports. Lund considered the characteristics of various bearing forms in the 1960s and made outstanding contributions to the prolific literature on rotor dynamics. Hahn (1975) also considered the stability of flexible rotors mounted upon short sleeve bearings. Morton (1966, 1968, 1970, 1972) has studied, both theoretically and experimentally, the dynamic characteristics of bearings and their influence upon rotor dynamics. An interesting comparison between bearing stiffness and damping coefficients obtained experimentally on a 20 in (0·5 m) diameter bearing running in the laminar regime and current theoretical predictions in the 1970 paper showed considerable discrepancies. The theory tended to underestimate stability and to overestimate stiffness. Furthermore, in 1974 he extended the analysis to the case of rotors supported upon several bearings. Such arrangements are common in the electric power generating industry and Ettles et al. (1974) drew attention to the special problem of bearing misalignment in such applications. Glienicke (1967) had previously determined bearing dynamic characteristics experimentally by applying two different sinusoidal forces displaced from each other by $90°$ ($\pi/2$ rad). The journal diameter was 4·7 in (120 mm) and four different bearing types were studied. Black (1973) and Black and Lock (1976) have paid particular attention to the problems of lateral vibrations and the stability of pump rotors.

An interesting extension of these studies of particular importance to turbine and compressor applications is the investigation of squeeze-film dampers. The rotors are mounted in rolling-element bearings which in turn sit in hydrodynamic squeeze-film bearings. The arrangement introduces necessary damping while retaining the otherwise desirable features of rolling-element bearings. Successful applications have been reported in aero and automobile gas-turbine power units and the general system has been studied by Lund (1965), Gunter (1966, 1970), Mohan and Hahn (1974), Hahn and Simandiri (1974) and Rabinowitz and Hahn (1975).

Various *anti-whirl* bearings have been introduced into rotating machinery. The first, in which special grooving ending in a dam in the unloaded half of a bearing applied a stabilizing force to the journal, was proposed by Newkirk and Grobel (1934) and described in detail by Newkirk (1937). It appears that Rumpf (1938) was the first to report favourably on the use of *lemon-shaped* bearings. *Split bearings* in which the two halves are displaced at right angles to the load, *tilting-pad*, *Rayleigh-step*, *floating-ring* and *multi-lobe* bearings all have their advocates. It appears, however, that generalizations on performance are dangerous and that the only satisfactory procedure is to incorporate reliable bearing stiffness and damping characteristics in an analysis of the complete rotor system.

Summary

This is a most important aspect of tribology which is attracting much attention at the present time. It is impossible to discuss all facets of the subject in their historical context, but the above discussion highlights some of the major developments during the past half-century. Good general accounts of the subject were provided in books by Shaw and Macks (1949), Tipei (1957), Wilcock and Booser (1957), Pinkus and Sternlicht (1961), Freeman (1962), Gross (1962), Constantinescu (1963), Sternlicht (1965), Tondl (1965), Cameron (1966), Hersey (1966), Trumpler (1966) and Smith (1969). The reviews by Newkirk (1957), Stenbricht and Rieger (1968), Shapiro and Rumbarger (1972), Rieger (1973), Lund (1975) and Smalley and Malanoski (1976) also provide comprehensive and clear statements on the developing state of knowledge.

Steam-turbine development drew attention to the potentially important role of fluid-film bearings in rotor dynamics in the 1920s. The initial understanding of the phenomenon of half-speed whirl remained largely qualitative until after the Second World War. An appreciation of the effect of bearing elasticity upon rotor vibration and critical speeds has been supplemented in recent years by more detailed considerations of foundations. The response of a gas-turbine rotor on an airframe might well be different to that on a firm test-bed base, and large rotating machinery on oil-drilling rigs at sea cannot be expected to exhibit characteristics identical to those of similar components on land.

It is convenient and meaningful to distinguish between the response of a *rigid rotor on flexible bearings* and a *flexible rotor on rigid bearings*, although there is an important intermediate region in which *flexibility and damping of both the rotor and its supports* has to be considered.

The three major areas of interest in the field of bearing-influenced rotor dynamics are the effect of bearings and their supports upon:

1. Stability of the system.
2. Critical speeds.
3. Rotor response (or amplitude of vibration at various speeds).

It is now usually possible to predict for conventional bearings and simple rotor systems the thresholds of unstable operation. This allows necessary design changes to be effected and it facilitates decisions on remedial action in unstable arrangements. It is now recognized that the whirl frequency is not restricted to half rotor speed, but that it may be a smaller fraction, depending upon system characteristics and operating conditions. For this reason it seems preferable to refer to 'fractional frequency (or speed) whirl' and to retain the term 'resonant half-speed whirl (or whip)' for the situation in which the rotor speed is close to twice the first critical speed.

The effect of bearing and support stiffness upon rotor critical speeds can be quite significant and it is most desirable that the bearing characteristics should be incorporated into critical speed calculations. Many rotors operate above the first but below the second critical speed, but in some current designs of machinery rotational speeds carry the rotor beyond as many as six criticals.

There is an increasing interest in specifying the rotor response as a function of speed for a wide range of equipment. This is important, not only from the overall design point of view but also on grounds of safety and reliability. A high degree of sophistication has been reached in rotor dynamics, but adequate system modelling can still be a major problem. In some cases the users are imposing specifications on the response of rotating machinery, not only in the normal operating condition but in damage or emergency situations. For example, if a blade flies off a gas turbine or an axial-flow compressor, it is comforting to know that the machine can tolerate the sudden introduction of large centrifugal forces without experiencing catastrophic failure.

There is also an increasing recognition of the need to explore the stiffness and damping coefficients of components other than bearings which constitute large rotor systems. The range includes seals, glands and couplings. Aerodynamic effects and large rotor systems with multibearing supports are also attracting attention.

The subject of bearing-influenced rotor dynamics represents a good example of the transition from a recognition of the existence of the phenomenon, through an active period of diagnostic work, to monitoring and prognosis in the field of tribology, as outlined in Section 11.15.

11.13   Elastohydrodynamic lubrication

Background

Perhaps, in the fullness of time, the recognition and detailed study of a special form of fluid-film lubrication known as *elastohydrodynamic*, will be seen to be one of the major developments in tribology in the twentieth century. It not only revealed the existence of a previously unsuspected mode of lubrication in highly stressed and generally non-conformal machine elements like gears and rolling-element bearings, but it brought order to the understanding of the complete spectrum of lubrication, ranging from boundary to fluid-film, which was, to say the least, exhibiting a few irritating inconsistencies.

Once the Reynolds equation had been solved and the results applied so successfully to conformal journal and thrust bearings, it was entirely reasonable to extend the analysis to counterformal machine elements like gears and rolling-element bearings. It was, in fact, the gear problem, represented by Martin (1916) as the lubrication of rigid circular cylinders, which first attracted attention. The findings were negative, as outlined in Chapter 10, in the sense that the predicted film thicknesses were minute and considerably less than the surface roughness of the best machined gears.

The question was reopened in the middle years of the twentieth century, when it became clear from experimental evidence and operating experience that some gear sets enjoyed fluid-film lubrication, even though current theory denied the possibility. Some of the most persuasive evidence in support of hydrodynamic action arose from the persistence of machining marks, long after they might have been expected to disappear if surface contact or boundary lubrication had prevailed.

It was suggested that the high local stresses resulting from load transmission through non-conformal contacts like gears and rolling bearings could result in two effects beneficial to fluid-film formation. In the first place it was recognized that local elastic deformation would extend the effective length of the region in which hydrodynamic pressures could be generated, thus encouraging fluid-film lubrication. It was also appreciated that load-carrying capacity in hydrodynamic bearings was proportional to lubricant viscosity and that the latter might rise appreciably in regions of high pressure.

Analytical

These actions, which were additional to the normal considerations in bearing analysis, were investigated theoretically and separately in the 1930s and 1940s. The effect of elastic distortion was the first to be considered and this was followed by a study of the role of viscosity–pressure relationships in promoting fluid-film formation. Both effects were found to enhance the magnitude of the predicted oil film thickness in gears, but the increase was much too small to allow a reconciliation between theory and practice.

A notable breakthrough occurred when Grubin (Grubin and Vinogradova, 1949) managed to incorporate both the effects of elastic distortion and viscosity–pressure characteristics of the lubricant in an analysis of the inlet region to lubricated non-conformal machine elements. The most valuable result of this approach, which combined mathematical skill with sound physical insight, was the development of a new relationship for film thickness. The equation at once yielded values which were one or two orders of magnitude greater than the predictions of the Martin theory, comparable to the surface roughness of gears and at last consistent with experimental indications of satisfactory hydrodynamic films of lubricant. Shortly afterwards, Petrusevich (1951) provided three numerical solutions to the governing elasticity and hydrodynamic equations which confirmed the essential features of Grubin's analysis and yielded additional information on the details of film shape and pressure distribution throughout the conjunction. Two of the most intriguing features of Petrusevich's solutions were the local restriction in film thickness near the outlet end of the conjunction, immediately after a relatively long region of essentially constant film thickness and the appearance of an associated, very local, second maximum in the pressure distribution. The latter, quite extraordinary feature, soon became known as the *Petrusevich pressure peak* or simply the *pressure spike*.

Various procedures for the solution of the complex elastohydrodynamic lubrication problem were reported in the 1960s. Dowson and Higginson (1959) described an iterative procedure which not only yielded a wide range of solutions during the next decade but also enabled them to derive an empirical minimum film thickness formula for nominal line contacts. Archard and Cowking (1966) applied elastohydrodynamic theory to nominal point contacts of the form encountered between two spheres and thus extended the classical study of Kapitza (1955) as Grubin had extended that of Martin. The more general case of point contacts which give rise to elliptical contact zones under dry contact was analyzed by Cheng (1970) and more recently by Hamrock and Dowson (1976a, 1976b, 1977a, 1977b).

The essential features of film shape and film thickness for both nominal line and point contacts can now be determined theoretically. The results have proved to be of great utility in the analysis and design of highly stressed, lubricated machine elements like gears, cams and tappets, roller- and ball-bearings and stepless traction drives.

Experimental

Confirmation of the film thickness in elastohydrodynamic line contacts was offered by Crook (1958, 1961a, 1961b, 1963) in an outstanding collection of papers. He employed a capacitance technique and a knowledge of the dielectric constant of the lubricant to estimate the film thickness. His results were in close accord with theoretical predictions produced before and after his experiments and capacitance measurement remains a valuable technique for the experimental determination of film thickness in elastohydrodynamic conjunctions. The method was refined and used convincingly by Dyson et al. (1966) in a study which covered a wide range of lubricants. Valuable results were also obtained by an X-ray transmission technique pioneered by Sibley et al. (1960). A number of investigations have used transparent components and interferometry to study elastohydrodynamic film shapes. This revealing procedure has produced information of great clarity and beauty and has been exploited to good effect by Professor Cameron and his colleagues at Imperial College. The procedure has been outlined in some detail by Foord et al. (1970).

Pressure

Once agreement had been obtained, after nearly half a century, between theory and experiment in relation to film thickness in highly stressed, lubricated machine elements, attention turned to other aspects of elasto-hydrodynamic lubrication. *Pressure distribution* was revealed by the use of fine strips of manganin, an alloy of copper, nickel and manganese, whose resistance is influenced by pressure, deposited on the insulated surfaces of discs. The technique was developed by Kannel et al. (1964), extended by Cheng and Orcutt (1966) and utilized to reveal the presence of attenuated pressure spikes in an impressive study by Hamilton and Moore (1971).

Temperature

Temperatures within, or adjacent to, elastohydrodynamic contacts have been recorded by embedded and trailing thermocouples, films of platinum (Cheng and Orcutt, 1966) and nickel (Hamilton and Moore, 1971) on insulated discs and by the direct measurement of infrared radiation. The latter technique was developed and utilized in a most valuable study by Professor Winer and his colleagues in the Georgia Institute of Technology (see Turchina et al., 1974; Ausherman et al., 1976 and Nagaraj et al., 1977).

Effective lubricant temperature is particularly important in sliding elastohydrodynamic contacts because of its influence upon film thickness, traction and, in the limit, failure of the lubrication mechanism. About forty years ago Blok (1937*a*, 1937*b*, 1937*c*) established the foundations of his *flash-temperature theory* of film breakdown, in which it was postulated that the lubricant ceased to provide effective protection of gear-teeth and similar components when the conjunction temperature achieved a critical value which was normally thought to be in the range 200–250°C. The conjunction temperature concerned is the sum of the bulk temperature of the solids and the surface temperature rise associated with frictional heating in the regions of asperity interactions. The concept is appealing in its simplicity as a guide to failure prediction. It has been challenged, and supported, on numerous occasions, but it still represents one of the important twentieth-century concepts in tribology.

Rheology and traction

The lubricant in an elastohydrodynamic conjunction experiences rapid and very large pressure variations, a rapid transit time, possibly large temperature changes and, particularly in sliding contacts, high shear rates. The great severity of these conditions has called into question the normal assumption of Newtonian behaviour. While the assumption appears to be acceptable as far as film-thickness prediction is concerned, it may be unrealistic when details of the pressure and temperature distribution are considered and is certainly inadequate when traction has to be determined. For this reason lubricant rheology is a subject of great current interest. The dominant trend is to recognize the viscoelastic nature of most lubricants and to utilize a model of a Maxwell fluid in rheological analysis. Crook (1963) and Dyson (1970) suggested that viscoelasticity could account for the failure of Newtonian fluid analysis to predict observed traction behaviour, and Johnson and Roberts (1974) confirmed this view in an elegant experimental study involving skewed discs. Barlow et al. (1967) subjected samples of high-molecular-weight fluids to small-amplitude, high-frequency shear waves and deduced their viscoelastic characteristics from measurements of their complex impedence. This work complemented direct measurements on disc machines, and viscoelastic features of lubricant behaviour in elastohydrodynamic conjunctions is now widely accepted.

Gear and rolling-element bearing lubrication

Once the fundamental features of elastohydrodynamic lubrication had been established in the 1950s, 1960s and 1970s, it became possible to utilize the new-found knowledge in the analysis and design of lubricated, deformable machine elements. The first book devoted to elastohydrodynamic lubrication was published by Dowson and Higginson (1966), but the number of papers on the subject has grown enormously in the past decade. There has been spectacular progress in the understanding of gear and rolling-element bearing lubrication, but the subject has impinged upon such topics as seals, tyres on wet roads, conventional plain bearings, foil bearings, synovial joints and metal forming.

It has been possible to improve the life and reliability of both gears and rolling-element bearings by providing an effective elastohydrodynamic film thickness a few times greater than the composite surface finish of the mechanism. Dawson (1962) first drew attention to the important role of the ratio of these quantities in determining the fatigue life of rolling discs, while Wellauer (1967) has extended the concept to gear failure and Tallian (1976a, 1976b) has confirmed the vital role of elastohydrodynamic lubrication in the life and performance of rolling bearings.

A mode of lubrication

It is evident that the mode of fluid-film lubrication known as elastohydrodynamic has firmly established itself in the full spectrum of recognizable lubrication regimes. It is now clear that fluid-film lubrication applies not only to bearings exhibiting a high degree of geometrical conformity, but also to many highly stressed, non-conforming machine elements. As our knowledge of this interesting and remarkable form of lubrication is extended and consolidated, we can see with greater clarity than ever before where fluid-film lubrication ends and boundary lubrication begins. Indeed, many machine parts previously thought to be reliant upon boundary lubrication are now known to function under elastohydrodynamic conditions.

The protective elastohydrodynamic films are inevitably thin and often only slightly greater than the sum of the roughnesses of the surfaces they lubricate. It has thus become necessary to introduce the concepts of surface topography outlined in Section 11.2 into the analysis of elasto-hydrodynamic lubrication of real surfaces. Once again we see how the traditional boundaries between component subjects which comprise tribology have disappeared in the wider and more realistic approaches to problems of interacting surfaces which have emerged during the twentieth century.

11.14 Bio-tribology

This term was introduced quite recently by Dowson and Wright (1973) to cover '. . . all aspects of tribology related to biological systems'. Many of the studies in this field are of long standing, but the subject has received exceptional attention in the last two decades.

Perhaps the best-known aspects of the subject are the studies of those remarkable self-acting bearings, synovial joints, and the design and development of total joint replacements. There are many other examples of tribological studies of biological significance, including the abrasive wear characteristics of human dental tissues, fluid-transport problems ranging from the fluid mechanics of the ureter to the lubrication by plasma of red blood cells in narrow capillaries, the wear of bone implants like plates and screws, the wear characteristics of replacement heart valves and, on a slightly different front, the influence of micro-organisms on lubricants employed in industry.

Synovial joints

The load-bearing synovial joint forms a remarkable self-contained, dynamically loaded bearing. The analogy with the plain bearing in machinery is simply that the bone corresponds to the hard metal backing, the articular cartilage to the softer lining of low-friction bearing material and the synovial fluid to the lubricant.

This biological system attracted the attention of Osborne Reynolds in 1886, when he concluded his classical paper (Reynolds, 1886) on the theory of fluid-film lubrication with the words:

> . . . *The only other self-acting system of lubrication is that of reciprocating joints with alternate pressure on and separation (drawing the oil back or a fresh supply) of the surfaces. This plays an important part in certain machines, as in the steam engine, and is as fundamental to animal mechanics as the lubricating action of the journal is to mechanical contrivances.*

Reynolds does not appear to have written anything further on the subject, but forty-six years later the anatomist MacConaill (1932) studied the structure of joints and from a comparison of his findings with the wedge-shaped films in successful engineering thrust bearings, he concluded that the mode of lubrication must be hydrodynamic. In the same decade, and exactly half a century after Reynolds' observation, Jones (1936) detected an exponential decay in the amplitude of a pendulum in which the fulcrum was formed by a human interphalangeal joint. He interpreted this evidence of viscous damping as confirmation of hydrodynamic lubrication in the joint.

The hydrodynamic concept of synovial joint lubrication remained essentially unchallenged for almost three-quarters of a century, but then Charnley (1959, 1960) used cadaveric angle joints in further pendulum experiments, observed a near-linear decay and suggested that the conditions under which joints had to operate were most likely to preclude hydrodynamic and to favour boundary lubrication.

Charnley suggested that the evidence of viscous dissipation in Jones' experiments could be attributed to the use of intact joints, whereas his own studies related to the frictional characteristics of the articular cartilage surfaces alone, since the ligaments and tendons had been severed. Barnett and Cobbold (1962) confirmed this suggestion in another series of experiments in which the amplitude of a pendulum with a canine ankle joint was monitored before and after the ligaments were severed. However, their study left the subject in some disarray, since they also reported an essentially linear decay of amplitude when the synovial joint was replaced by a hydrostatic bearing in which, presumably, full fluid-film lubrication existed.

Further pendulum experiments in which the amplitude of motion was recorded as a function of time by Little et al. (1969) appeared to confirm that synovial joints worked in the boundary lubrication regime and that the detailed behaviour was controlled by lipid-rich surface layers of cartilage. Unsworth et al. (1975) demonstrated that experiments based upon observations of a decrease in amplitude of the pendulum were probably insensitive to any viscous action in the fulcrum, and an alternative form of pendulum machine was constructed in which the friction torque could be measured directly. Experimental results indicated that synovial joints experienced *boundary*, *fluid-film* (in conventional hydrodynamic, elastohydrodynamic or squeeze-film forms) and *mixed* lubrication, depending upon the magnitude and nature of the loading and the instant in the cycle of motion. Indeed, most experiments now seem to indicate, and probably have been suggesting to us all along, that synovial joints experience all the modes of lubrication familiar to the engineer and that there may be some variations on the technological concepts which have evolved to make the synovial joint such a remarkable bearing.

While the pendulum machines were producing the background data, other experiments and theories of considerable interest emerged. McCutchen (1959) proposed an intriguing self-pressurizing hydrostatic concept known as 'weeping lubrication'; Dintenfass (1963), Tanner (1966) and Dowson (1967) concluded that *elastohydrodynamic action* was likely at some stage in the cycle of events; Maroudas (1967) drew attention to the formation of gels on the surface of loaded cartilage which could well dominate frictional behaviour; Walker et al. (1968) proposed a modified form of squeeze-film action known as *boosted lubrication* in which the molecular structure of synovial fluid and the elasticity, porosity and surface topography of articular cartilage all contrived to prolong the life of effective lubricating films and Fein (1967), Dowson et al. (1970) and

Gaman et al. (1974) all drew attention to the important role of squeeze-film action in the lubrication of synovial joints.

Whatever the dominant form of lubrication might be in synovial joints, it is remarkably effective. It protects the self-contained bearing surfaces for about three-score years and ten and provides a dynamically loaded bearing with coefficients of friction which are frequently about 0·02 and normally in the range 0·002–0·10.

Total joint replacements	For patients who suffer the discomfort, stiffness and pain of some rheumatic disorders it is now possible for surgeons to contemplate the replacement of the complete joint by a man-made bearing.

Two major material combinations have emerged for total joint replacements: metal-on-metal and metal-on-plastic. In the former case it is generally considered to be necessary to use identical metals in order to avoid electrochemical corrosion in the very hostile environment within the body, but this is, of course, tribologically undesirable, since like metals tend to cold-weld at asperity junctions and to experience high friction and wear. In practice it is the high friction which is more troublesome than the rate of wear, since the former can readily cause loosening of the joint, while the latter can be controlled by careful selection of the type and hardness of the materials.

A combination of metal and plastic materials is most popular in current forms of total joint replacement. Corrosion-resistant metals like stainless steel (En58J) or chromium, cobalt and molybdenum alloys are normally used in association with the ultra-high-molecular weight, high-density form of polyethylene (UHMWPE). In the early stages of development of metal-on-plastic prostheses the low-friction polymer polytetrafluoroethylene (PTFE) was used, but the wear rate was excessive and few replacement hip joints survived for more than three years. The change to UHMWPE has extended the wear-life of prostheses enormously and current laboratory tests and extrapolation of clinical data suggest that successful implantation of present forms of prostheses should lead to wear-lives of at least twenty years. Another aspect of material selection which needs careful attention is the possibility that wear debris might lead to short- or long-term tissue reactions.

The hip-joint has received most attention, particularly in the 1950s and 1960s, but the more complex problem of finding a satisfactory knee prosthesis is the subject of much research at the present time. There are also replacement joints for the ankle, shoulder, elbow and fingers. Good accounts of studies of joint lubrication and the development of total joint replacements can be found in the proceedings of conferences organized by the Institution of Mechanical Engineers and the British Orthopaedic Association on lubrication and wear in living and artificial human joints (Institution of Mechanical Engineers, 1967), total knee replacement (Institution of Mechanical Engineers, 1976) and total replacement joints for the upper limb (Institution of Mechanical Engineers, 1978). Several books now deal with these topics including those edited by

Wright (1969), Kenedi (1971), Freeman (1973) and Dowson and Wright (1978).

The abrasive wear of teeth

An interesting study of the abrasive wear of teeth which occurs when they are cleaned by brushing has been reported by Wright (1969). It is necessary to consider the characteristics of the tooth surface, the machine used to remove deposits (toothbrush), the abrasive (dentifrice) and the environment, in such studies. One of the problems is that a dentifrice which is suitable for the outer layer of enamel might be too severe for the softer, underlying dentine which may be exposed in middle age.

Wright (1969) demonstrated that the abrasive wear resistance of dental tissue was proportional to hardness, as in the case of metals, but that the wear rates of the dental material were much greater than those of metals of comparable hardness. This result is typical of mineral materials and normally attributable to the formation of additional wear particles by brittle fracture. In all these experiments, which were directed towards a better understanding of the wear processes involved, the development of reliable test procedures and ultimately the formulation of recommendations for toothbrush and dentifrice materials, the abrasive was silicon carbide in various particle size ranges and the lubricant was water.

Fluid transport in the body

In some cases the flow of fluids like blood and urine can be treated as viscous-flow problems, and several authors have utilized the fundamental equations of lubrication in their studies.

In an interesting investigation of the fluid mechanics of the ureter, Lykoudis and Roos (1970) showed that the governing equations reduced to those of the theory of lubrication. The study represented a significant step forward in the understanding of peristaltic pumping. The flow and viscosity of sputum, a non-Newtonian fluid containing protein, water and salt, have also been studied in relation to chronic bronchitis and muco-viscidosis. Aspects of blood flow connected with the formation of thrombi have been actively studied in recent years, and although not directly concerned with bio-tribology, some of the earliest work on the related subject of bio-rheology emerged in this field.

One of the most interesting aspects of blood circulation concerns the movement of red blood cells in small capillaries. In the unstressed state the human red blood cell consists of a biconcave disc of diameter about 315 μin (8 μm). Since the smallest capillaries have diameters in the range 197–394 μin (5–10 μm) it is clear that the red cells, and possibly the less elastic walls of the capillaries, experience elastic distortion during the process of blood transport. The red cells occupy about 45 per cent of the blood volume and they pass along the capillaries in single file, being separated by blood plasma.

Lighthill (1968) and Fitz-Gerald (1969) have analyzed the problem in some detail and it appears that the passage of each red cell down a slightly bulging capillary is assisted by elastohydrodynamic lubricating films of thickness about 7·9 μin (0·2 μm). Lighthill (1968) also drew attention to that characteristic feature of elastohydrodynamic lubrication, the necking of the film towards the trailing edge of the cell.

The influence of micro-organisms on lubricants

This effect is almost entirely restricted to lubricants with an aqueous phase, and bio-degradation of cutting fluids in industry is recognized as a serious health, economic and technological problem. Infection of the lubricant can occur from aerobic or anaerobic organisms, but since most oils contain considerable amounts of dissolved oxygen the initial infection tends to be aerobic. Aerobic infections are usually associated with bad smells, particularly from hydrogen sulphide. Above neutral pH the dominant organisms are bacteria, but at lower pH, yeasts and fungi play a dominant role.

The complex processes of bio-degradation of lubricants and their consequences have been described by Hill (1969) and Littler and Purkiss (1970). Apart from the problems of hygiene and atmospheric pollution, the process can lead to the formation of corrosive fluids and acid-staining of bearings and workpieces.

Summary

The science of tribology has made a substantial contribution to the better understanding of, and in some cases the solution of, biological problems. The cooperative field is extensive and increasing. The list of topics discussed here is illustrative rather than comprehensive and it is interesting to note how many of the examples are concerned with the common ground between tribology and rheology.

11.15 The medical analogy–machine health, diagnosis, monitoring and prognosis

Health

Recognition that a substantial proportion of machine failures have a tribological basis and that reliability and safety of complex, large and expensive equipment calls for more attention during design and operation than has previously been the case, has prompted a considerable and growing interest in knowing the state of health of running machinery.

There is, of course, an analogy here with the medical field, where many of the doctors and a substantial proportion of the available facilities, are involved in diagnosis and treatment of illness rather than preventive medicine. The criticism that most nations have a 'sickness' rather than a 'health' service, is unfortunately equally true of many traditional engineering services. While it is true that much can be learned from parts salvaged from a breakdown or failure, and errors corrected in modifications to the reconstructed machine, the concept of monitoring the health of a system so that preventive action can be taken before an accident, or collapse, has much to recommend it. It is interesting to note that the dental profession tends to be more insistent upon this latter approach, and if regular inspection reduces the chances of severe damage or decay and wear of teeth, perhaps the same is true of tribological components.

The specification of routine service procedures at intervals which are often quite arbitrary is widespread and not infrequently the only means of attempting to retain a machine in good health. Yet many faults develop soon after, and possibly because of, such interruptions to the working life of a mechanical device.

Diagnosis

There has been much interest and marked success in the field of failure investigations in recent decades. Examinations of the worn or broken parts of bearings, seals, gears and brakes can often indicate the cause of the initial fault. An encyclopaedic approach to the nature and frequency of faults can lead to an improved understanding of defects in initial designs, or evidence of inadequate knowledge of the true operating conditions. A bearing failure frequently draws attention to a lack of knowledge of the actual loads, temperatures or even amounts and nature of the lubricant rather than to a detailed design fault, and it is for this reason that designers should be adequately equipped to deal with tribological as well as strength and other aspects of machine design.

The diagnosis of tribological faults is undoubtedly a most important aspect of technological progress and many experts are involved in this kind of work. It is quite amazing what careful physical and chemical analysis of wear debris, broken components or minute traces of lubricants can reveal. Mere visual inspection and a background of fundamental knowledge combined with experience can often and most speedily reveal the causes of failure.

The role of diagnostic work has assumed greater significance as a result of the move towards a smaller number of large machines in many industries. The trend has been evident in the process industries like chemical, steel-making and liquefaction plants; the electrical power generating industry; and transportation, particularly in the marine and aircraft fields. Modern equipment is often very large, frequently complex and inevitably expensive. The consequences of unnecessary breakdowns and shut-down time can be serious and even disastrous in both the economic and technological sense. The ever increasing emphasis on reliability of equipment has also promoted interest in the diagnosis of faults, but there is evidence of some impatience with this approach alone. The possibility of monitoring, on an intermittent or continuous basis, the running condition of a machine, such that potential failures can be detected early and remedied at a convenient stage in the life-cycle of the machine, is increasingly attractive. In short, we are witnessing the introduction of machinery health monitoring and the adoption of a prognostic approach to supplement, but not to replace entirely, the successful diagnostic methods of the past.

Monitoring

There are several features of machine behaviour which can be monitored to provide the basis for tribological prognosis. These include: the quantity and quality of the lubricant; thermal and elastic distortions of the structure or dimensional changes resulting from wear; the generation of excessive heat or temperature within the fluid or the solid elements of a device; vibration and noise; friction; and, of course, the magnitude and nature of wear.

A series of articles was devoted entirely to this topic early in the 1970s and progress has been rapid in recent years. The articles dealt with the concept of monitoring and tribological prognosis (Dowson, 1970*a*); details of particular methods for intermittent checks on the quantity and quality of the lubricant and evidence of incipient wear (March, 1970); vibration monitoring and the development of electromechanical systems for continuous recording of dimensions and their changes (Hother-Lushington, 1971). The system known as 'ferrography', in which samples of the lubricant circulating through bearings and other tribological components of a machine are encouraged to deposit wear particles on a transparent plate in a high-gradient magnetic field to facilitate subsequent analysis, is receiving much attention at the present time (see Seifert and Westcott, 1972 and Scott, 1975). Information on the concentration, size and morphology of particles is particularly useful in ascertaining the wear mechanisms operating in a machine.

It is obvious that it is necessary to select the most appropriate system or groups of systems for monitoring the running performance of a machine, but it is not always easy to make the best selection. More work is needed in this field, but machinery health monitoring has become a well-established practice.

Prognosis

The prediction of future development of machine faults is a complex process in which a number of difficulties have to be recognized. It is first of all necessary to know what is the characteristic of a healthy machine and this is often a function of the individual as well as the species. A medical doctor would find the measurement of body temperature of little value in his analysis of the condition of a patient if he did not know the average value for a healthy subject was 37°C. The establishment of 'norms' is an essential first step towards prognosis, yet to quantify such judgements in terms of 'hot' or 'noisy' bearings or quantities like rotor vibration amplitudes is rarely straightforward.

The second problem is to ascertain acceptable variations from the 'norm' of some operating characteristic, a task which is far from easy. If temperature or even temperature rise of lubricants with highly non-linear temperature characteristics is monitored on the propeller shaft thrust bearing assembly of an oil-tanker, acceptable limits might well change on a voyage from the Persion Gulf to the Arctic.

Many of the most useful procedures adopted so far involve periodic checks on lubricant characteristics, or the detection of both the nature and rate of production of wear debris. Particularly striking examples have been reported by March (1970). It seems that in some applications, prognosis is already well established.

Summary

It has at last been recognized that it is unnecessary and undesirable to wait for tribological failures to occur and then to try to take preventative action on future occasions. The cost of failures and the consequent loss of production in large-scale process plants confirm this view.

Nevertheless, the move towards prognosis must be taken carefully. It can be both a difficult and an expensive procedure and in many cases might be ruled out on the grounds of cost-effectiveness. There are, however, influences wider than the economic and technological which might dictate events, such as safety, reliability and environmental factors. In some cases the control is governed by legislation.

The introduction of machinery health monitoring and the gradual improvement of prognostic approaches must inevitably have an impact upon the designer. At the present time the assessment of performance of tribological devices is often qualitative, and the designer who devotes care and skill to the preparation of overall and detailed recommendations hears only of the failures!

It appears that we are entering a new era in the field of studies of machinery stoppages associated with tribological failures. In the past the art and science of diagnosis has been most valuable, but it is now recognized that it is sometimes necessary to know more about machine performance before a failure is reported. The transition from 'diagnosis' to 'monitoring machine health' and 'prognosis' in tribology is under way.

11.16   High-friction materials

The literature on tribology abounds with papers on lubrication, bearings and ingenious systems for reducing friction and hence improving the efficiency of machinery. But it is equally important to be able to offer high-friction materials to devices like tyres on roads, wheels on rails, shoes on floors, belt drives, clutches and, of course, brakes.

The development of rail, road and air transportation systems provided the stimulus for much of the work on high-friction materials, but the machine tools used in a wide range of industrial applications also influenced events.

Wood, leather and fabrics

Wood, leather and felt were widely used as friction materials prior to the present century. While their coefficients of friction were often quite acceptable, they failed to withstand high temperatures. The wood became charred, the surface of leather carbonized and braking was often accompanied by smoke and sometimes by fire.

Woven cotton fabrics, impregnated with resin, asphalt or bitumen provided satisfactory friction, but once again they were limited by poor thermal characteristics. The introduction of metal wires to improve the mechanical strength and to conduct heat away from the conjunction proved to be a useful development, but it was the introduction of the asbestos-based materials early in the twentieth century which dominated progress in friction materials. In the field of belting, fabrics woven from cotton or man-made fibres, leather and rubber materials with and without reinforcement, have all been developed for industrial drives.

Asbestos

The mineral asbestos is mined in opencast or underground workings in North America, Africa, the Soviet Union and Cyprus. It is normally white, off-white or blue in colour and occurs in the form of very fine hollow fibres about 4×10^{-8} inches in diameter. It is an inert material, being resistant to chemical reactions and, of course, high temperature. Its fibrous nature allows it to be spun into yarn and woven, either alone or with metal wires. For clutches and brake linings the material is often moulded with fillers and resins and may contain metallic particles. The resulting materials normally retain coefficients of friction in the range 0·3–0·4 up to temperatures ranging from 250°C to 450°C (482–842°F).

Sintered metals

In more recent times the severity of clutch and brake operating conditions in equipment like automobiles, earth-moving equipment, and electro-magnetic machinery has called for the introduction of alternative high-friction materials. Sintered metals, generally bronze, with additions of lead, graphite, iron or silica have proved to be useful for these heavy duties. The material has a somewhat lower coefficient of friction than the asbestos-based materials, but its better mechanical properties, low wear rate and vastly increased thermal conductivity permits it to operate at higher pressures and temperatures. Furthermore, it can function immersed in oil, the lower effective coefficient of friction being compensated by higher permissible contact pressures.

Ceramics

For very high temperature operation, friction materials with a high ceramic content have proved their worth in recent years. The cermet friction materials have very good wear characteristics, high thermal conductivity and density and can operate at temperatures of the order of 1000°C. They have been extremely useful for aircraft brakes and for brakes and clutches on earth-moving equipment.

Sections A46–A51 of the *Tribology Handbook* (Neale, ed., 1973) and the *Ferodo Design Manual on Friction Materials For Engineers* (Ferodo, 1961) provide useful background reading to these subjects, while a most valuable book on the wider issues of the braking of read vehicles has been written by Newcomb and Spurr (1967).

Rubber

When the early explorers brought back to Europe from South America samples of a remarkable, resilient, vegetable gum which '. . . rebounded so much as to be alive . . .' they treated it merely as a curiosity and a substance effective in rubbing out lead pencil marks. The English chemist Joseph Priestley proposed the name *rubber* for this novel substance in 1770.

Natural and synthetic rubbers have found wide application in tribology during the past fifty years. Many of these applications have resulted from the unique characteristics of rubber in deforming to high strains under small loads while retaining elasticity. In such cases the conformity with a mating component readily provided the basis for a good seal or a bearing.

In the case of belt drives, footwear, windscreen wipers and tyres on wheeled vehicles, it has been the dry friction and wear characteristics of the material which has attracted attention. Gough (1958a,b) described the general characteristics of the friction of rubber and pointed out that the force of friction initially rose rapidly with sliding velocity in a region of creep relative to the counterface, reached a maximum and then fell as the sliding velocity increased. In due course it was recognized that rubber was a viscoelastic material and Williams' et al. (1955) study of the relaxation of polymers proved particularly useful in representing friction data obtained at different temperatures and speeds on a single curve. At this stage the futility of quoting a coefficient of friction of rubber without specifying the conditions was readily appreciated; the range covering values of μ slightly above zero to 1, 2, 3 or even higher.

Two mechanisms of rubber friction have been studied: adhesion and deformation. Bartenev (1954), Bartenev and Lavrentjev (1961) and Bowden and Tabor (1964) proposed that adhesion between rubber and hard solids under dry conditions arose mainly from van der Waals forces. Some workers assumed that parts of the rubber adhered to the counterface and then extended until the interfacial bonds broke – each rupture and the snapping back of rubber strands being associated with a loss of energy. Ludema and Tabor (1966) used an alternative approach in which the interfacial bonds were assumed to remain intact while the rubber failed by tearing just beneath the interface, but it failed to achieve close quantitative agreement between predicted and measured coefficients of friction. It also implied a gradual transfer to rubber from the bulk material to the counterface. Ludema (1977) has recently reviewed the mechanism of rubber wear.

A phenomenological view of rubber friction has been presented by Savkoor (1965), while Kummer (1966) has attempted to reconcile molecular and macroscopic views of the process. Adhesion, the deformation of a viscoelastic material and electrostatic interactions have all been examined and a good introduction to this complex field has been provided by Roberts (1976). One of the most intriguing features of the subject came to light when Shallamach (1971) caused a rubber hemisphere to slide over transparent Perspex. The rubber was seen to buckle at the interface and *waves of detachment* swept across the contact region, suggesting a caterpillar-like movement and a lack of relative motion between the rubber and the Perspex in the regions of intimate contact. The observation of Shallamach waves has not only excited interest in rubber friction, it has done much to expose the physical nature of friction between the material and a smooth, dry surface.

Elastomeric friction has been the subject of books by Moore (1972, 1975b), who has also written extensively on the interesting topic of lubricated rubber friction and aquaplaning of tyres on wet roads. The latter topic, discussed in two substantial papers by Allbert et al. (1966a, 1966b) has been brought to the fore in recent years by the skidding of aircraft on runways and high-speed automobiles on wet roads. The problem involves both adhesion and elastohydrodynamics, with a film of water encroaching upon the interface between a tyre and the road as the speed increases. In due course a film of water penetrates through the complete nominal contact zone, adhesion is lost and the vehicle skids out of control on the wet surface.

Summary

The problems encountered in controlling and stopping high-speed trains, the present generation of large, wide-bodied passenger aircraft, heavy or fast road vehicles and the powerful range of earth-moving equipment has focused attention on the development of high-friction materials in the third quarter of the twentieth century. The tribological developments have been intimately linked with analytical and design studies of braking systems and vehicle dynamics. It is a challenging topic for engineers, with the need to balance the requirements for controlled deceleration against the thermal limitations of the materials. A simple calculation of the heat equivalent of the kinetic energy of a large fast-moving train or aircraft soon demonstrates the magnitude of the problem.

11.17   Surface contact

The phenomena of heat conduction and the passage of electricity between bodies, friction, wear and even lubrication under some conditions, are intimately related to the nature of contact between interacting solids. This observation is not new, since it formed the foundation of the early mechanistic views of friction and even lubrication, but detailed studies of its significance are of recent origin. It is convenient to review these important developments before proceeding to a discussion of wear in Section 11.18.

Plastic deformation

The view that plastic deformation of the asperities readily occurred when two solids were brought into contact dominated scientific approaches to friction, the conduction of heat and the passage of electricity between bodies until the middle of the twentieth century. Friction had been analyzed against this background by Bowden and Tabor (1950), heat conduction by Williams (1966) and the flow of electricity by Holm (1929). Interaction between the rounded summits of asperities was seen to be analogous to miniature Brinell hardness tests and this allowed the real contact area to be estimated with ease. Such calculations, supported by direct measurements based upon electrical and optical techniques, indicated that the real areas of contact were extremely small, being no more than a tiny fraction of the apparent contact areas under most conditions. Furthermore, since it was the magnitude of the real contact area rather than its distribution which was deemed to be important in the physical phenomena mentioned here, this approach made it unnecessary to involve the details of surface topography discussed in Section 11.2. It was necessary to assume only that the surfaces presented a geometrical form which would ensure that asperity deformation was plastic rather than elastic.

There is one simple yet substantial objection to the view that intimate contact between surfaces occurs only in regions which have deformed plastically. Since plastic flow involves permanent deformation of the material, it cannot be repeated indefinitely by successive interactions. There will be a gradual move towards elastic support and further consideration of contact between the interacting surfaces must consider the balance between elastic and plastic deformation and the relationship between load and real area of contact under these heterogeneous conditions. The main difficulty was that real contact area was seen to be proportional to the applied load to powers of either two-thirds or unity for fully elastic and fully plastic conditions, respectively.

Elastic deformation

Early in the 1950s Blok (1952) and Halliday (1955) initiated an interesting debate on the extent to which an asperity in the form of a spherical cap of radius (R) and height (h) above the surrounding material could be pressed into the surface without suffering plastic deformation. They found that for a material of hardness (H), elastic modulus (E) and Poisson's ratio (v) the criterion for the avoidance of plastic deformation was

$$\left[\frac{h}{2R}\right]^{1/2} < k\,\frac{H(1-v^2)}{E}, \qquad \qquad \ldots [11.13]$$

where k is a constant in the range 0·8–1·7.

In a later paper Halliday (1957) used reflection electron microscopy to study the topography of end faces of Duralumin cylinders before and after they were pressed against a highly polished, hard-chrome bearing steel. He not only found that the asperities could sustain high loads without suffering plastic deformation, but that the indications of equation [11.13] were reasonably correct.

Archard (1957, 1961) restored confidence in the linear relationship between real contact area and load when he neatly demonstrated that the power on load increased from two-thirds and approached unity, even for purely elastic conditions, when multiple rather than single-asperity interactions were considered. This finding might once again have deflected interest away from the detailed nature of surface contact, but in a number of fields, like the question of surface temperatures between rubbing solids discussed in a most valuable paper by Archard (1959), it was becoming increasingly evident that it was necessary to know the size and distribution of individual contact spots, in addition to their total area.

In the 1960s an important extension of profilometric studies of surfaces enabled further progress to be made in the study of surface contact. Greenwood and Williamson (1966) used digital analysis of surface profiles to demonstrate that the distribution of asperity heights on many machined surfaces was Gaussian. Even a worn surface, with a basically non-Gaussian height distribution, exhibited an essentially Gaussian form for those parts of the profile which were likely to be involved in contact.

An interesting discussion of the role played by surface energy in the mechanism of contact between elastic solids has been presented by Johnson et al. (1971).

Plasticity index

Greenwood and Williamson (1966) used these findings to establish a model of random surfaces consisting of asperities of identical radius (R) whose heights had a Gaussian distribution. Contact between this model surface and a smooth, rigid plane was governed by three parameters; the radius of curvature (R), the standard deviation of the asperity heights (σ^*) and the number of asperities per unit area (η). Once again it was found that this more realistic model of surface topography indicated that the real area of contact would be almost directly proportional to load, even when the latter was supported entirely by elastic contact. Greenwood and Williamson introduced a useful parameter known as the *plasticity index* (ψ) to indicate the extent of plastic deformation, where

$$\psi = \left[\frac{E'}{H}\right]\left[\frac{\sigma^*}{R}\right]^{1/2} \qquad \qquad \ldots [11.14]$$

E' being equal to $E/(1 - v^2)$.

Whitehouse and Archard (1970) further refined the model when they introduced a scale effect into the representation of surface features by treating the profile as a random signal having a Gaussian distribution of heights and an exponential function. Onions and Archard (1973) utilized this representation in their studies of surface contact. They found that the introduction of a distribution of asperity radii markedly increased the contact pressures and that the extent of plastic flow was determined by a two-parameter plasticity index (ψ^*), defined by the standard deviation of the height distribution (σ) and the correlation distance (β^*) as follows,

$$\psi^* = \left[\frac{E'}{H}\right]\left[\frac{\sigma}{\beta^*}\right]. \qquad \qquad \ldots [11.15]$$

It was found that, notwithstanding the greater pressures predicted by the Onions and Archard model, the parameters ψ and ψ^* had similar implications. This can be summarized as follows:

$$[\psi \text{ or } \psi^*] > 1 \qquad \text{significant plastic flow,}$$

$$1 > [\psi \text{ or } \psi^*] > 0\cdot 6 \qquad \text{some elastic and some plastic deformation,}$$

$$[\psi \text{ or } \psi^*] < 0\cdot 6 \qquad \text{plastic flow unlikely.}$$

It is thus clear that it is surface topography and material properties, rather than load, which determines the extent of plastic deformation between contacting surfaces, since load does not enter either of the relationships [11.14] or [11.15]. It is also evident that if initial surface topography introduces a plasticity index greater than unity, surface interactions might smooth the profile during the process familiarly known as running in, until the standard deviation decreases and the plasticity index itself moves into the range corresponding to elastic contact.

Recent studies of surface contact

In the 1970s great progress was made in the analysis of surfaces and their interactions. Greenwood and Tripp (1971) demonstrated theoretically that the essential features of earlier studies involving the contact between a rough surface and an idealized smooth, rigid plane, were unaffected by the more general assumption that both surfaces were rough. Nayak (1971, 1973a, 1973b) distinguished between statistical interpretations of a random surface and a profile of that surface, introduced random process theory in this field and drew attention to the merits of the power spectral density in surface quantification and studies of surface contact. This random process model was used to good effect by Bush et al. (1975) in an analysis of the static elastic contact of rough surfaces.

These more sophisticated approaches to studies of surface contact confirmed, rather than disturbed, the general findings based upon simpler models presented earlier. This provides a sound base for current studies of heat transfer between solids, electrical conduction, friction, wear and some aspects of lubrication. The subject is nevertheless more complicated than had previously been supposed and it was therefore refreshing for the casual student of the subject to find Tabor (1975) drawing attention to the physical significance of the two plasticity indices represented by equations [11.14] and [11.15]. In essence he demonstrated the importance of surface slopes in determining the contact behaviour of asperity-covered surfaces. Valuable introductions to surface topography and surface contact have been presented by Ling (1973), Archard (1974) and Archard et al. (1976).

Hisakado (1969, 1972, 1974) and Hisakado and Tsukizoe (1974) of Osaka University, Japan, have discussed the mechanics of surface contact in some detail. Tsukizoe and Hisakado (1972a, 1972b) have also applied the concepts of surface interactions to studies of heat transfer between metals. Williams (1966, 1968, 1971, 1972, 1974) has also reported on a number of interesting experimental studies of heat flow between metals. Valuable experimental studies of the real area of contact between metals have been reported by Uppal and Probert (1972) and Uppal et al. (1972). A bold attempt to incorporate studies of surface interactions between populations of spherical asperities into a theory of friction and wear for materials represented by a generalized stress–strain relationship has been presented by Halling (1976b). In the field of lubrication, particularly of counter-conformal surfaces, surface topography and contact studies have now been combined with the essential features of elasto-hydrodynamic lubrication by Johnson et al. (1972) and Tallian (1972).

It seems certain that further progress in tribology will lean heavily on the emerging knowledge of surface topography and surface contact. The latter represents one of the most exciting and important topics presently under investigation and one that can be expected to influence further progress in the fields of friction, wear and even lubrication.

11.18 Wear

Wear has long been recognized as a most important and usually detrimental process in mechanical devices, yet scientific studies of the phenomenon started only recently. The historical order of tribological studies has been friction, lubrication, then wear, as noted in the introduction to the present chapter. Indeed, the vast majority of work on wear has been packed into the last thirty years–a mere drop in the sea of time.

This chronology is all the more remarkable in view of the importance of wear in a technological society. Practically everything made by man *wears out*, yet the fundamental actions which govern the process remain elusive. Initially, attention was focused on friction, since this governed the ability of machines to function at all. The role of lubricants in controlling friction and wear was then explored, with little attention to the detailed mechanism of wear.

Machinery may cease to be useful as a result of fracture of one or more components, or by attrition and the impairment of interacting surfaces. The former process is often sudden, spectacular and sometimes catastrophic; the latter affects efficiency, reliability and can itself bring equipment to a standstill.

In former years, engineering education and training concentrated upon the strength of materials, the bulk properties of fluids and solids and continuum mechanics, to ensure that machines could be designed to function correctly without the embarrassment of failure associated with breakage of components. It is partly because these subjects are now understood in some depth, but also because of a growing appreciation of the importance of reliability, efficiency and life of equipment, that attention is now focused upon the properties of surfaces, the mechanism of surface interactions and the wear process itself.

An understanding of some of the fundamental aspects of wear is now emerging and valuable empirical guidance is available to designers, but it remains true that the engineer is better equipped to construct a machine to withstand known loads than he is to incorporate within it moving parts with a specified life.

Definition	In the Organization for Economic Cooperation and Development (OECD) *Glossary of Terms and Definitions in the Field of Friction, Wear and Lubrication (Tribology)* (OECD, 1969), Wear was defined as, '. . . The progressive loss of substance from the operating surface of a body occurring as a result of relative motion at the surface.' We usually think of the detrimental aspects of wear, but in some cases the process plays an important and useful role as in *grinding, polishing, writing with pencil or chalk* and, of course, in *the running-in process*.
Types of wear	There are two broad approaches to the classification of wear: the first being descriptive of the results of wear, with the second being based upon the physical nature of the underlying processes. In the former, such terms as *pitting, fretting, scuffing* and *scoring* are highly descriptive of the

appearance of the worn surfaces and in a sense indicative of the wear mechanism. It is, however, the second form of classification which is most useful. According to this approach five wear processes have been clearly recognized and are termed: *abrasion, adhesion, fatigue, erosion, corrosion.*

It is beyond the scope of the present text to discuss in detail each form of wear. Suffice it to say that *abrasion* is associated with the displacement of material from a relatively soft solid by the protuberances on a harder counterface, or by loose, hard particles between two surfaces. *Adhesive wear* is the process by which material is transferred from one surface to another during relative motion as a result of cold-welding at points of asperity interaction. The transferred material may be attached to the counterface, but in some cases part of the detached material returns to its original surface or it becomes loose wear debris. There is thus a difference between transfer and wear.

When a surface is subjected to repeated and high stresses, as in gears and rolling bearings, it might suffer *fatigue wear*. Particles are removed from the surface as a result of the formation of surface or sub-surface cracks. The disfigured surface exhibits pits or shallow craters from which flakes of material have been removed and is, of course, prone to further and often rapid deterioration.

If a stream of fluid containing solid particles impinges upon a surface *erosive wear* might occur. This is a problem encountered in either liquid–solid or gas–solid flow systems like the transport of crushed minerals in water-filled pipes or the flow of dust-laden air through ducts and valves in an internal combustion engine.

Corrosive wear requires both chemical or electrochemical action combined with relative motion. Corrosion can take place without wear, but if the reaction products are subsequently removed by relative surface motion, the phenomenon of corrosive wear prevails. Perhaps the best-known form of corrosive wear is that associated with oxidation of surfaces.

Relative importance of wear mechanisms

Since it has been agreed that wear cannot be totally prevented, it is of interest to consider which of the five wear mechanisms is most common. Most can be attenuated, and in some cases particular wear mechanisms can be avoided altogether, but as long as load-bearing, interacting surfaces are in relative motion, some wear can be anticipated.

It is found that abrasion, fatigue, erosion and corrosion can be controlled by careful attention to the selection of materials, the preparation of the surfaces, component design to minimize both the magnitude and perhaps the frequency of the stress cycles, filtration and environmental protection. However, as long as asperities interact and deform, cold-welding takes place most readily and adhesive wear emerges as the mechanism most widely encountered over the full spectrum of engineering situations. It can, of course, be most appropriately controlled by means of lubrication, but it still represents the most persistent mode of wear.

History of wear studies

It was Ragnar Holm, working first in the Siemens–Konzern laboratories in Berlin and later with the Stackpole Carbon Company of St Marys,

Pennsylvania, who made one of the earliest substantial contributions to the study of wear (Holm, 1946). His interest lay in the performance of electrical contacts, such as those in circuit-breakers, relays, terminals, microphones, current collectors and commutators, and it is interesting that such significant studies of friction and wear arose in a branch of electrical engineering.

At a later stage Ragnar Holm and his wife Else Holm published a further edition of their book, which contained a most valuable account of the fundamentals of surface interactions and wear mechanisms (Holm and Holm, 1958). This text still forms a good starting point for the student of wear.

It appears that Holm initially considered wear to be a uniform atomic transfer process taking place at asperity contacts, although it was clear from his later work that he introduced the concept of a uniform layer of transferred material having a thickness equal to an integral number of molecules mainly to establish a physical appreciation of the wear process. He established a relationship for the volume of material removed by wear (V), in a sliding distance (x) and related (V/x) to the true area of contact.

In pioneering studies involving the use of radioactive tracers to monitor the progress of wear, Rabinowicz and Tabor (1951) were able to distinguish between *transfer* and *wear*. In the former, material plucked from one surface adheres strongly to the counterface, while the production of loose wear debris calls for the subsequent detachment of the transferred particles.

A wide-ranging experimental study of wear involving a pin-on-disc machine was reported in two valuable papers by Burwell and Strang (1952*a*, 1952*b*). As they started their investigations, the laws of wear were ill-defined and by no means as well established as those of friction or lubrication. Their findings led them to question the atomic transfer hypothesis proposed by Holm and to propose instead that the wear process involved the detachment of material from a proportion of the asperity junctions formed between contacting surfaces. This view was well supported by the results of a study on an electron microscope which revealed wear particles of dimensions of the order of 100 Å, which clearly contained about 1 million atoms.

Adhesive wear

For adhesive wear, which is by far the most common form of attrition between rubbing metals, there was, in the 1950s, growing support for the view that the volume of material removed by wear (V) was directly proportional to the sliding distance (x) and the normal load (P), and inversely proportional to the hardness of the softer material (H). Burwell and Strang did much to confirm this understanding, although it is important to note that there are many exceptions to these rules. The latter reservation has been emphasized on several occasions by Professor Tabor, who, in his book with F. P. Bowden (Bowden and Tabor, 1973), still felt it necessary to draw our attention to the '. . . non-existent laws of wear'. Nevertheless, Burwell and Strang laid the foundations of an empirical relationship of appealing simplicity and great utility for those wishing to quantify the rate of wear of rubbing metals.

A further important finding by Burwell and Strang was the dramatic increase in wear rate when the nominal contact stress exceeded about one-third of the hardness of the softer material–a result which was attributed to plastic flow over a large proportion of the nominal contact area.

Archard (1953, 1961) provided theoretical support for the experimental indications of the factors governing adhesive wear in two most valuable papers. His analysis revealed a relationship of the form,

$$V = k_1 \left(\frac{Px}{3H} \right), \qquad \qquad \ldots [11.16]$$

where k_1 is a dimensionless wear coefficient which is a measure of the probability that any interaction between asperities will result in the production of a wear particle.

Experiments have shown that values of k_1 normally lie in the range 10^{-1}–10^{-8}, thus indicating that innumerable asperity encounters take place without yielding wear particles. At this stage it was possible to make two interesting and pertinent observations about friction and wear.

1. Coefficients of friction between dry solids normally cover a range of about one order of magnitude.
2. Whereas all asperity interactions contribute to the friction process, few lead to the formation of wear particles.

Archard and Hirst (1956a) carried out experiments on a wide range of material combinations and provided confirmation for the proportionality between the volume of material removed by wear (V) and load (P) in many cases. They demonstrated that the wear rate was independent of apparent contact area and drew attention to two major regimes of wear, distinguished by scale, which they classified as *mild* and *severe*. A detailed examination of the mild wear process was published in the same year by Archard and Hirst (1956b).

The mechanism by which material is removed from a surface during sliding is central to the understanding of adhesive wear. Rabinowicz (1965) proposed a strain energy criterion in which repeated rubbing was deemed to build up strain energy in the contact zones until such time that it equalled the interfacial energy associated with the removal of a fragment and the formation of a new surface. Kragelskii (1965), Kragelskii and Neponmyashchii (1965) and others preferred the concept of a fatigue mechanism. Rowe (1966) modified Archard's theory to take account of surface films and suggested that such films accounted for the relatively low values of the wear coefficient between similar metals.

Abrasive wear

In *abrasive wear*, hard asperities or particles form grooves in the softer material. The softer solid might flow to the side of the wear groove, resulting in little removal of material, or it might be removed in a series of chips, as in a micro-cutting process. It is somewhat surprising to learn that two physical processes like abrasive and adhesive wear, which appear to differ so much, both obey the general laws enunciated earlier. In each case there is generally a direct relationship between the volume of material displaced (V) and the product of load (P) and sliding distance (x), and an inverse relationship with the hardness of the softer material (H). The latter relationship appeared with great clarity for pure metals in the extensive studies of abrasive wear carried out in the USSR by Kruschov (1957) and his colleagues. These studies also revealed that initial cold-working of a metal had little effect upon its wear resistance, which prompted Kruschov to surmise that abrasive wear was accompanied by local work hardening much greater than that achieved by the initial working. He also found that heat treatment of steels had a pronounced effect upon their resistance to abrasive wear and that the latter was directly related to the difference in hardness between the heat-treated and annealed material.

The store of knowledge on abrasive wear was enhanced from diverse sources in the 1960s. Investigations related to the wear of tillage tools in agricultural soils were carried out by Richardson (1967, 1968a, 1968b) at the National Institute of Agricultural Engineering in the United Kingdom, while Wright (1969) studied the abrasive wear resistance of teeth at the National Engineering Laboratory. A review paper on the latter subject was prepared by Powers et al. (1973).

While the role of hardness in abrasive wear has attracted most attention, other factors such as the surface topography of the harder surface and the size and form of abrasive particles have also been investigated. Analytical studies of the abrasive wear process have revealed relationships of the form,

$$V = k_2 \left(\frac{Px}{H} \right), \qquad \qquad \dots [11.17]$$

where the wear coefficient (k_2), for hard asperities of conical shape, whose sides are inclined at an angle (θ) to the mean plane of the surface is given by $2 \tan \theta / \pi$.

A similarity of form is evident between equations [11.16] and [11.17] and it is clear that both are in accord with the general laws of wear. The equations reveal the role of the principal variables, but there are sufficient exceptions to the predicted behaviour to justify a measure of caution in their utilization. Furthermore, wear rates or the life of rubbing components, cannot be determined without some knowledge of the wear coefficients.

The 'zero wear' model

An interesting approach to the problem of predicting wear characteristics of a wide range of materials was initiated in the Endicott Development Laboratory of the International Business Machines Corporation in 1952. The study was described in a paper by Bayer et al. (1962) and in a book by Bayer, Ku et al. (1964). A distinction was drawn between 'zero wear' and 'measurable wear'; the former implying that changes in surfaces contours as a result of wear were smaller than, or of the same order of magnitude as, the surface finish, and the latter that wear exceeded the original roughness. In this sense the term 'zero wear' was misleading, but the effects of the process were nevertheless small and usually quite acceptable in many machine elements. The basic model relied upon the view that wear could be controlled by limiting the maximum shear stress in the contact region. This procedure attempted to overcome the limitations of conventional approaches to theoretical studies of wear by providing empirical data for designers. This wide range of experimental data, covering many of the materials used in tribological components, provided a valuable starting point for many engineering predictions of the life of machinery.

Delamination

A novel and stimulating contribution to the study of wear was reported by Suh (1973) of the Massachusetts Institute of Technology. It was founded on dislocation theory and the plastic deformation and fracture of metals near a surface, and envisaged wear to result from the development of sub-surface cracks and voids and the subsequent formation of flake-like wear particles. This 'delamination' theory of wear was supported by direct studies of sub-surface cracks and the size and shape of wear particles. It resulted in a relationship of the form,

$$V = \kappa P x, \qquad \qquad \dots [11.18]$$

where the dimensional constant (κ) is a wear factor determined by the shear modulus, friction stress, Burger's vector, Poisson's ratio and the sliding distance corresponding to the removal of a complete layer of wear flakes for the materials.

This expression differs from the earlier representations for adhesive and abrasive wear in the omission of a term directly representing the hardness of the materials. Nevertheless, the effect is incorporated in some of the terms defining κ, and Suh pointed out that the expression permitted a more sophisticated view of the role of hardness in the wear process. He further concluded that the wear processes identified as 'adhesive', 'fretting' and 'fatigue' were all caused by the same mechanisms.

These new ideas have opened up additional lines of approach to an old subject, and the concept is still being developed. It is, unfortunately, still necessary to introduce experimentally determined coefficients, as in the case of previous theories, before useful quantitative predictions can be made for specific situations. It appears that these theories of wear will have to be supported by laboratory tests for some time to come, although the expanding study of fracture mechanics is expected to enhance the approach based upon the delamination concept of wear.

Non-metallic materials

The increasing use of non-metallic materials in engineering has focused attention upon their wear characteristics. Polymers have attracted a great deal of attention and Lancaster (1973) has presented a wide-ranging review of the basic mechanisms of friction and wear of these materials. Lee (1974) edited a two-volume collection of papers presented at an

International Symposium on Polymer Friction and Wear organized by the American Chemical Society. Dowson et al. (1978) edited the *Proceedings of the 3rd Leeds–Lyon Symposium on Tribology* which was devoted to the wear of non-metallic materials, including polymers, polymer composites, elastomers, biological materials, textiles, lignum-vitae and graphite.

Summary

The theoretical basis of our understanding of most of the common forms of wear is still in its infancy. More detailed consideration of surface topography, surface interactions, the elastic and plastic deformation of real materials and the influence of micro-structure on the response of materials to repeated stress cycles will be necessary before the life of tribological components can be predicted with confidence.

There are now several textbooks and articles devoted wholly or partly to the subject of wear, including Bowden and Tabor (1950, 1956, 1964, 1973), Burwell (1950), Barwell (1956, 1978), Davies (1959*b*), Kruschov and Babichev (1960), Bayer and Ku (1964), Kragelskii (1965), Rabinowicz (1965), Tabor (1965), Lipson (1967), Hirst (1968), Richardson (1968*c*), Archard (1969), Summers-Smith (1970), Buckley (1971), Shaw (1971), Lancaster (1972), Wright (1973), Kruschov (1974), Halling (1975, 1976*a*), Engel (1976) and Sarkar (1976). A recent addition to the literature is the *Proceedings of an International Conference on Wear of Materials* (ASME, 1977) published by the American Society of Mechanical Engineers. The reader wishing to walk in the diffuse field of wear should find these articles and books useful gates of entry.

Wear is probably the most important aspect of tribology, yet it has remained largely unexplored until recent times. The recent flood of publications is difficult to appreciate in historical terms, partly because of the sheer volume of material, but also because we must wait awhile before some of the basic and intriguing concepts can be confirmed by experiment and practice as valid signposts on the road of truth.

Sometimes the steady progress towards a full understanding of a subject is disturbed by large-scale and immediate problems. Two recent examples in the field of wear are the severe damage experienced by some rotating shafts supported by journal and thrust bearings lined by relatively soft materials, and the scuffing of piston rings and cylinder liners in diesel engines. The former type of distress, widely and urgently studied in the late 1950s and 1960s, became known as *machining* or *wire-wool* failures, owing to the formation of long slivers of steel in and around the bearings on large turbines and generators. The phenomenon has been described by Dawson and Fidler (1966) and Karpe (1968). The former suggested that the basic mechanism was one of the spinning of the journal surface following embedment of a hard particle and the formation of a black scab in the white-metal bearing material.

The second example of a wear problem of immediate concern arose from the diesel engine field. The topic of piston-ring and liner scuffing attracted much research activity in the 1970s, as outlined in Section 11.11, and a valuable collection of papers on the subject was published by the

Institution of Mechanical Engineers (Institution of Mechanical Engineers, 1976*b*).

It is clear from this brief review that wear is one of the most challenging topics in tribology at the present time. If wear remained the Cinderella of tribology until the 1950s, it seems almost certain that it will dominate not only the latter half of the twentieth century, but also, perhaps, much of the twenty-first.

11.19   Professional bodies, meetings and the literature

The subject of tribology has been favoured by the formation of societies or divisions of existing professional bodies to serve and promote it. Similarly, specialist journals emerged to cater for the upsurge in interest in the subject. The role of the Institution of Mechanical Engineers, the American Society of Mechanical Engineers, the Japanese Society of Mechanical Engineers, the Royal Society and more recently the American and Japanese Societies of Lubrication Engineers in promoting meetings on friction, wear and lubrication in the central decades of the twentieth century is worthy of special mention.

Prior to the Second World War the learned societies and professional engineering bodies published a number of valuable papers on subjects which formed the elements of tribology, but the information remained largely dispersed, except for a very small number of books and the proceedings of a remarkable conference held in London in 1937.

The Institution of Mechanical Engineers

Table 11.2
Organising Committee for the Institution of Mechanical Engineers General Discussion on Lubrication and Lubricants–1937.

Dr H. G. Gough *Chairman*
Mr E. A. Evans Sir Nigel Gresley Mr H. L. Guy Mr Aubrey B. Smith Professor H. W. Swift

A General Discussion on Lubrication and Lubricants was held at the Central Hall, Westminster, from 13 to 15 October 1937. The discussion was organized by the Institution of Mechanical Engineers with the cooperation of some fifty-three scientific and technical bodies in the United Kingdom and overseas. The object was that of, '. . . reviewing the present state of the science and practice of lubrication in order to correlate theory and pure research with practice, to consider methods of bearing design, to obtain current views upon bearing metals, and to bring out the significance of laboratory tests, including wear and friction tests'.

The Organising Committee appointed by the Institution for the 1937 General Discussion on Lubrication and Lubricants is listed in Table 11.2.

In all, 136 papers were presented, and no less than 600 delegates registered for the meeting. An exhibition of lubricants, lubricators, bearings, bearing materials, and testing and research apparatus was held at the Science Museum, South Kensington, in association with the Discussion and by the end of October some 18,300 members of the public had visited the event.

The Discussion was opened by Major-General A. E. Davidson, Past President of the Institution, who said at a conversazione held at the

Table 11.3
Initial Lubrication Group
Committee of the Institution
of Mechanical Engineers – 1956

D. Clayton
Chairman

F. T. Barwell
Professor H. Ford
Eng. Rear-Admiral D. J.
Hoare
K. Holland
G. D. Jordan
H. Peter Jost
P. P. Love
E. V. Paterson
Professor O. A. Saunders
C. G. Williams

Table 11.4
Chairmen of the Lubrication
and Wear/Tribology Group of
the Institution of Mechanical
Engineers

1956 & 1958   D. Clayton
1959 & 60   F. T. Barwell
1961 & 62   P. P. Love
1963 & 64   L. F. Hall
1965 & 66   N. Balmforth
1967 & 68   D. Dowson
(The title *Lubrication and
Wear* Group was changed
to *Tribology* Group in
1967)

1969 & 70   M. J. Neale
1971   T. I. Fowle
1972   J. D. Summers-
       Smith
1973   W. H. Roberts
1974   D. Scott
1975   J. A. Robertson
1976   G. R. Higginson
1977   P. G. F. Selden

Science Museum on 13 October 1937: '. . . if the nations could be persuaded to discuss affairs in the spirit in which engineers of all nations were prepared to deal with engineering developments, great advance towards the solution of world problems, many of which were, in essence, due to new developments, would be achieved'.

Major-General Davidson's words proved to be timely, but unheeded, for hostilities leading to the Second World War broke out within two years of the remarkable General Discussion of 1937.

This was not only the first large professional meeting on lubrication, friction, wear and bearings, but one of the finest. The two volumes of *Proceedings of the General Discussion* (1937) are now classical and prized collections of papers. They provide an excellent account of the pre-war state of the art and science of tribology, and a good starting point for studies of many aspects of the subject.

In the postwar years, those who had recognized the importance of the subjects of the 1937 meeting began to consider how the initiative could be recaptured. These considerations led to the formation of the Lubrication and Wear Group of the Institution. In the *Annual Report* of the Institution of Mechanical Engineers for the year 1955 it is written: '. . . In December (1955) the Council approved the formation of a new Specialist Group to cover the activities in the field of lubrication. Dr D. Clayton was appointed to be the first Chairman of the new Lubrication Group. At the end of the year the Group was still in its formatory stage.'

The *Annual Report* for the following year reported briefly that, '. . . The Lubrication Group, formed in December 1955, arranged a full programme during 1956, as endorsed by the two General Meetings, and two Group Discussions with which the Group was associated.'

The composition of the first Committee of the Group is shown in Table 11.3.

The subsequent progression of chairmen, as recorded in December each year, is shown in Table 11.4.

Twenty years after the successful 1937 meeting, the Institution of Mechanical Engineers again held a Conference on Lubrication and Wear in London, this time in collaboration with the American Society of Mechanical Engineers. The Conference took place from 1–3 October 1957, 104 papers were presented and about 1,000 delegates representing 21 countries registered for the event. The American Society of Mechanical Engineers was represented by Dr Donald F. Wilcock, who was appointed an Honorary Vice-President of ASME for the occasion. The Organising Committee had the composition shown in Table 11.5.

This second Conference was a memorable event, both for the author who found that it provided the opportunity for his first publication on lubrication, and for everyone who participated in the stimulating sessions and discussions. The *Proceedings* (Institution of Mechanical Engineers, 1957) once again formed a most valuable reference volume. The subject-matter was spread over many aspects of tribology, but it was significant that wear attracted about one-third of the contributions.

The quickening pulse of tribological activity prompted the Institution to hold its next major international meeting in 1967. The Conference on

Table 11.5
Organizing Committee of the
Institution of Mechanical
Engineers Conference on
Lubrication and Wear – 1957

Dr D. Clayton
Chairman

Dr F. T. Barwell
Dr F. P. Bowden
Mr W. F. Cartwright
Professor D. G.
Christopherson
Mr E. S. Cox
Dr C. B. Davies
Dr D. F. Galloway
Mr L. F. Hall
Admiral D. J. Hoare
Mr P. P. Love
Dr S. Livingston Smith
Mr A. T. Wilford
Dr C. G. Williams

Table 11.6
Organizing Committee for the
Institution of Mechanical
Engineers' Conference on
Lubrication and Wear:
Fundamentals and Application
to Design – 1967

Professor F. T. Barwell
Chairman

Mr C. M. Allen
*(USA corresponding
member)*

Professor H. Blok
Dr D. F. Denny
Professor D. Dowson
Professor P. G. Forrester
Mr L. F. Hall
Mr A. A. Milne
Mr W. A. Pope
Dr W. Rizk
Dr D. Tabor
Mr H. J. Watson

Lubrication and Wear: Fundamentals and Application to Design (Institution of Mechanical Engineers, 1968) was held at Church House, Westminster, London from 25 to 29 September. The Organising Committee is listed in Table 11.6.

Government interest in the expanding concept of tribology was evident at the Conference and the Minister of Technology, The Rt Hon. Anthony Wedgwood Benn, MP, received overseas delegates at Lancaster House.

The 1960s and early 1970s encompassed the peak of professional activity, particularly in the form of meetings, promoted by the Institution of Mechanical Engineers. From 1963 to 1972 ten valuable annual tribology conventions were held for the presentation and discussion of any aspect of tribology, in Bournemouth (1963), Eastbourne (1964), London (1965), Scheveningen, Holland (1966), Plymouth (1967), Pitlochry, Scotland (1968), Göteborg, Sweden (1969), Brighton (1970), Douglas, Isle of Man (1971) and Keele (1972). The next convention was held in Durham (1976).

These tribology conventions in a hectic decade not only provided a forum for the bubbling output of research papers, they also nurtured the wider needs of a growing and energetic community of tribologists. The opportunity to meet others striving to resolve research problems, to unify design approaches, to introduce the essence of tribological concepts into industrial planning and to influence government, ensured that informal contacts and discussions were as valuable as the formal sessions.

In addition to the tribology conventions, the Institution also held special topic conferences and symposia on such subjects as fatigue in rolling contact (1964), elastohydrodynamic lubrication (1965, 1972), journal Bearings for reciprocating and turbo-machinery (1966), lubrication and wear in living and artificial human joints (1967), experimental methods in tribology (1968) and externally pressurized bearings (1971). The Lubrication and Wear Group also joined with the Iron and Steel Institute in arranging a conference held in Cardiff on Iron and Steelworks Lubrication (1964) which was to have a special significance as described in Section 11.20. Papers on tribology were also published in the *Proceedings* of the Institution and in the *Journal of Mechanical Engineering Science*, established in 1959. A newssheet entitled *Tribology News* was launched in 1966.

It is, perhaps, far too early to judge the full impact of the Institution of Mechanical Engineers upon the recognition and development of the concept of tribology, but there is little doubt that the enterprise of the 1870s and 1880s was matched by sound and energetic professional work during the forty years following 1937.

The American Society of Mechanical Engineers (ASME)

The oldest technical Committee of the Society is the Research Committee on Lubrication (RCL), established in 1915 with Mayo D. Hersey as its first Secretary. In the early 1940s[8] the Petroleum Committee of ASME

---

8. I am grateful to Dr D. F. Wilcock for kindly allowing me to have access to his notes on the *History of the ASME Lubrication Division*.

Table 11.7
Chairmen of the Lubrication
Division of the ASME

1955	D. F. Wilcock
1956	O. C. Bridgman
1957	J. C. Bunting
1958	S. Abramovitz
1959	C. C. Moore
1960–61	P. C. Werner
1962–63	W. E. Campbell
1964	V. S. Wagner
1965	R. A. Burton
1966	R. L. Wehe
1967	W. J. Anderson
1968	C. M. Allen
1969	R. P. Shevchenko
1970	J. M. Gruber
1971	R. L. Adamczak
1972	E. R. Maki
1973	V. Hopkins
1974–75	F. F. Ling
1976–77	W. O. Winer

Table 11.8
Mayo D. Hersey Awards

1965	Mayo D. Hersey
1966	Harmen Blok
1967	M. C. Shaw
1968	Ragnar Holm
1969	W. A. Zisman
1970	M. R. Fenske
1971	D. D. Fuller
1972	S. J. Needs
1973	D. F. Wilcock
1974	D. Tabor
1975	A. F. Underwood
1976	J. Boyd
1977	R. L. Johnson

was afforded the status of a Division, and since most of its members were primarily interested in problems of refining and transportation, others interested in lubrication formed a Lubrication Coordinating Committee in 1949. The Chairman was A. C. Stutson, who represented the Petroleum Division, while other representatives included D. F. Wilcock (RCL), M. E. Merchant (ASLE) and N. J. Gothard (ASTM).

In 1952 the Committee was renamed the Lubrication Activity and in 1955 it became a formal Division of ASME. The chairmen who have guided the Lubrication Division of the ASME since its formation in 1955 are listed in Table 11.7.

Initially, the Lubrication Division only participated in the annual winter meeting of the ASME, but it later launched its own spring symposium and joined with the American Society of Lubrication Engineers (ASLE) in presenting a joint conference each autumn. The latter event arose fortuitously in 1955, as a solution to the situation in which ASLE had too few papers for its own symposium while ASME had too many papers for its annual winter meeting. The Joint (ASME–ASLE) Conference has become the prestige meeting for the presentation and discussion of papers on lubrication in North America.

In 1967 the Lubrication Division was given its own section of the *Transactions of the ASME*, known officially as the *Journal of Lubrication Technology*, but colloquially and affectionately as 'JOLT'. The Division also worked closely with the RCL in the formation of specific research projects and the publishing of valuable volumes of literature. A good example of this initiative was the *Appraisal of World Literature on Boundary Lubrication* (Ling et al. eds., 1969). The ASME has published a number of valuable booklets based upon presentations on specific topics at its Winter Annual Meetings. The volume on *Surface Mechanics* promoted by the Applied Mechanics Division (Ling, 1969) and the more recent one on *Computer-Aided Design of Bearings and Seals* sponsored by the Lubrication Division (Kennedy and Cheng, 1976) are good examples of this activity.

In 1965 the Lubrication Division and the Research Committee on Lubrication established the Mayo D. Hersey Award to recognize and honour the leadership in lubrication science and engineering of Mayo D. Hersey, Professor Emeritus of Brown University. Ten years later the Burt L. Newkirk Award was established for achievements by young research workers in the field of tribology. The first recipient of the Burt L. Newkirk Award was Dr. F. E. Kennedy of the Thayer School of Engineering, Dartmouth College. The distinguished recipients of the Mayo D. Hersey Award are shown in Table 11.8.

The ASME Lubrication Division and the RCL have actively encouraged work on lubrication, friction and wear through their meetings, publications and awards. In 1977 ASME was actively participating with the Energy Research and Development Administration (ERDA) and the Office of Naval Research (ONR) in drawing up a tribology research and development plan aimed at achieving energy conservation. The seeds sown by ASME's first President, Robert Henry Thurston, are now sprouting vigorously.

The American Society of Lubrication Engineers (ASLE)

In 1942 a number of informal meetings were held in the United States to consider the art of lubrication and the need for a professional lubrication society. These meetings led to the formation of a Society devoted solely to the subject of lubrication which in due course became known as the American Society of Lubrication Engineers. The founding father and guiding influence in the Society was Walter D. Hodson and when the national organizational meeting was held in Pittsburgh, Pennsylvania, on 27 September 1944, about 100 members elected the first officers. Before the end of the 1960s the membership had exceeded 3,000.

The distinctive, energetic nature of ASLE soon attracted the interest and respect of those concerned with the art and science of lubrication. The Society promoted meetings, general educational activities and the publication of literature. It was always anxious to encourage the development of the highest standards of lubrication practice.

Past presidents of the ASLE since the formation of the Society are shown in Table 11.9.

The ASLE commenced publication of its journal *Lubrication Engineering* in 1945. In due course the need arose to publish papers presented at meetings, particularly those concerned with research on lubrication, and the *ASLE Transactions* was initiated in 1958. Some thirteen papers were published in 1945 and the annual figure had risen to eighty-seven in 1969. The Society has also produced a number of special publications of which Dudley D. Fuller's *A Survey of Journal Bearing Literature* (Fuller, 1958) and the *Standard Handbook of Lubrication Engineering* (O'Connor et al., 1968) will be well known to many readers. An interesting article on ASLE publications was written by Godfrey (1970).

The ASLE recognizes the merit of outstanding papers published in its journals through the Captain Alfred E. Hunt Memorial Award, the Walter D. Hodson Junior Award and the Wilbur Deutsch Memorial Award. Its highest award, which bestows life membership in recognition of outstanding contributions in the field of lubrication, is known as the

Table 11.9
Presidents of the ASLE

Years	President	Years	President
1944–46	C. E. Pritchard	1962–63	L. B. Sargent, Jr
1946–47	E. M. Kipp	1963–64	E. E. Bisson
1947–48	O. L. Maag	1964–65	S. R. Calish, Jr
1948–49	G. L. Sumner	1965–66	A. A. Raimondi
1949–50	C. L. Pope	1966–67	R. J. Torrens
1950–51	D. Hollingsworth	1967–68	M. M. Gurgo
1951–52	C. E. Schmitz	1968–69	R. L. Johnson
1952–53	M. E. Merchant	1969–70	P. M. Ku
1953–54	W. E. Campbell	1970–71	A. A. Manteuffel
1954–55	J. Boyd	1971–72	E. J. Gesdorf
1955–56	J. Hopkinson	1972–73	C. H. West
1956–57	W. Deutsch and E. R. Booser	1973–74	L. Robinette
1957–58	J. McLean	1974–75	W. H. Mann
1958–59	J. D. Lykins	1975–76	H. Tankus
1959–60	A. B. Wilder	1976–77	D. G. Flom
1960–61	L. O. Witzenburg	1977–78	W. K. Stair
1961–62	D. M. Cleaveland		

Table 11.10
ASLE National Awards

1948	W. D. Hodson	1963	A. E. Cichelli and R. Holm
1949	C. F. Kettering	1964	M. C. Shaw
1950	L. G. Benton	1965	D. Tabor
1951	M. D. Hersey	1966	M. R. Fenske
1952	J. W. Stack	1967	E. E. Bisson
1953	J. J. Simon	1968	A. A. Raimondi
1954	A. F. Brewer	1969	A. Cameron
1955	F. P. Bowden	1970	C. A. Bailey
1956	O. L. Maag	1971	R. L. Johnson
1957	D. D. Fuller	1972	E. A. Saibel
1958	E. A. Ryder	1973	P. M. Ku
1959	M. E. Merchant	1974	D. Dowson
1960	J. Boyd	1975	A. A. Manteuffel and T. A. Tallian
1961	W. A. Zisman	1976	E. E. Klaus
1962	C. L. Pope	1977	F. F. Ling

National Award. ASLE National Award winners are shown in Table 11.10.

The ASLE arose spontaneously in response to a growing recognition of the importance of lubrication. It has pursued its objectives with vigour and commands a respected place in the current range of professional activities in the field of tribology. A valuable series of articles dealing with various aspects of the history of the ASLE was published in the journal *Lubrication Engineering* throughout 1969–70 to celebrate the silver jubilee of the Society. These articles were later collected together and published as a separate volume under the title, *The 25 Years* (ASLE, 1970).

The Royal Society of London

The *Transactions* and *Proceedings* of this highly esteemed and long-established Society have included numerous papers dealing with the scientific basis of lubrication, friction and wear. The early studies of friction by Vince (1785) and Rennie (1829), Hatchett's (1803) study of wear and, of course, Reynolds' (1886) and Hardy and Doubleday's (1922*a,b*) investigations of fluid-film and boundary lubrication, were all reported in publications of the Royal Society of London.

On 19 April 1951 the Society held a Discussion on Friction, in London. The Discussion was introduced by Philip Bowden of the University of Cambridge and the collection of some eighteen papers, together with substantial contributions in the form of notes and comments was published in Volume 212 of the *Proceedings of the Royal Society* (1952).

The Japan Societies of Mechanical (JSME) and Lubrication (JSLE) Engineers

Formed in 1955 the JSLE has promoted meetings and encouraged the publication of papers of high quality on lubricants and lubrication. Papers published by the JSME and the JSLE were accounting for about 9 per cent of the world literature on tribology early in the 1970s.

On the occasion of the twentieth anniversary of the JSLE, an International Lubrication Conference was held at the Keidanren Kaikan, Tokyo, from 9 to 11 June 1975. The Conference was sponsored by the

JSLE with the cooperation of the ASLE and the event was supported by 575 delegates from 16 countries. Professor T. Sakurai of the Tokyo Institute of Technology chaired the Organizing Committee and guided the preparation of a fine volume of *Proceedings* (Sakurai, Ed., 1976).

The USSR Academy of Sciences

Work on tribology in the Soviet Union comes under the aegis of the Academy of Sciences. A Scientific Council of Friction of the Academy of Sciences was established in 1961. The stream of publications since the Second World War has been impressive in both quantity and quality. Work on friction, wear and elastohydrodynamic lubrication has been outstanding and about one-fifth of the annual output of publications on tribology in the early 1970s was attributable to the USSR.

The journal *Friction and Wear in Machinery*, which was published in English by the ASME from 1956, has made a notable contribution to the store of knowledge on tribology. International conferences in various countries have been supported by contributions from distinguished scientists and engineers from the Soviet Union. Several textbooks have been produced in recent years but it is regretted that few have been translated into English.

The Gordon conferences

These excellent informal meetings were established to stimulate research, and for some years one of the gatherings has been reserved for *friction, lubrication and wear*. Sessions are held in the mornings and evenings with the afternoon being free for leisure pursuits or informal discussion. No publications emanate directly from the Gordon Conferences, and indeed participants may not use the information presented in subsequent events without the agreement of the Gordon Conference speaker. In this way a free exchange of ideas is promoted and respected to the benefit of all participants.

Most of the Friction, Lubrication and Wear Gordon conferences have been held in New Hampshire, although on one occasion the venue was Wisconsin. Few people who have had the opportunity to attend the conferences have failed to find them stimulating and rewarding, with the high level of debate reflecting the wisdom and skill of the elected chairmen.

Limits of lubrication conferences

In 1969 Professor Cameron initiated a transatlantic version of the Gordon conferences. The meetings are held in alternate years at Imperial College, London, the aim being to discuss informally new concepts in lubrication. The format follows that of the Gordon conferences, with discussion proceeding freely without the constraints of subsequent publication. General topics are selected for each event and presentations are grouped to support particular themes.

The Leeds–Lyon symposia on tribology

A basic similarity of interests between the Institute of Tribology in the University of Leeds and the Laboratoire de Mécanique des Contacts of

Table 11.11
Leeds–Lyon symposia on
tribology

September 1974	Leeds	Cavitation and Related Phenomena in Lubrication
September 1975	Lyon	Superlaminar Flow in Bearings
September 1976	Leeds	The Wear of Non-Metallic Materials
September 1977	Lyon	Surface Roughness Effects in Lubrication
September 1978	Leeds	Elastohydrodynamic Lubrication and Related Topics

the Institut National des Sciences Appliquées de Lyon, prompted the creation of an annual symposium held alternately in mid-September in Leeds and Lyon. The idea emerged naturally during a convivial meeting between the author and Professor Maurice Godet of Lyon, and the first symposium was held in Leeds in September 1974. A pleasing feature of this Anglo-French venture has been the extent and international nature of the support.

The purpose of the meetings is to promote a discussion in depth of a single topic in tribology, normally reflecting the research interests of the two institutes, but also a subject of current concern. Papers and discussion are subsequently published and the topics selected for the first five events are listed in Table 11.11.

The National Aeronautics and Space Administration (NASA) interdisciplinary symposia

An excellent series of three NASA-sponsored symposia and one workshop meeting were organized by P. M. Ku of the Southwest Research Institute, San Antonio, Texas, between 1967 and 1972. The aim of the symposia was to foster interdisciplinary dialogue between professional groups active or vitally interested in the subject of lubrication and its theoretical and practical implications.

The volumes of papers and discussions which followed each of the three symposia were soon recognized as valuable accounts of the state of the art in various important fields of lubrication. The date, venue and subject of the symposia and workshop is noted in Table 11.12 and full details of the proceedings are listed in the References.

Table 11.12
NASA interdisciplinary
symposia and workshops
on lubrication (1967–72)

1967	San Antonio, Texas	Interdisciplinary Approach to Friction and Wear (Ku, Ed. (1968))
1968	Cleveland, Ohio	Workshop on Friction and Wear (Bisson and Ku, Eds. (1970))
1969	Troy, NY	Interdisciplinary Approach to the Lubrication of Concentrated Contacts (Ku, Ed. (1970))
1972	Cleveland, Ohio	Interdisciplinary Approach to Liquid Lubricant Technology (Ku, Ed. (1973))

The University of Southampton Gas Bearing Symposia

The specialist topic of gas bearings formed the subject of a number of international symposia initiated by the University of Southampton in the 1960s which continue to the present day. Strong transatlantic support has been a notable feature of the events.

The University Gas Bearing Advisory Service sponsored and organized the early events, but the sixth meeting held in Southampton in 1974 and the 1976 Symposium in Cambridge have been jointly promoted by the British Hydromechanics Research Association (BHRA) Fluid Engineering and the University of Southampton.

European Tribology Congress (Eurotrib)

The First European Congress on Tribology was held in London from 25 to 27 September 1973. Some 500 delegates from 23 countries provided the international basis for the event and 60 papers were presented. The Congress was organized by the Institution of Mechanical Engineers with cosponsorship from the Royal Institute of Chemistry, the Institute of Physics, the Institution of Metallurgists and the Institute of Petroleum in the United Kingdon, together with Gesellschaft für Tribologie und Schmierungstechnik (Federal Republic of Germany), the Scientific Council for Friction and Lubrication of the Academy of Sciences (USSR), Kring Tribologie-Bond voor Materialenkennis (Netherlands) and the Club Français de Tribologie (France).

The Congress was as much concerned with the formation of an international structure of tribological societies as it was with the science and technology of tribology, with the formation of an International Tribology Council being one of its major achievements. The Second Eurotrib Congress was held in Dusseldorf from 3 to 5 October 1977.

Individual conferences and symposia

Apart from the frequent meetings arranged by the professional bodies and the organizations mentioned earlier, a number of valuable individual conferences have been promoted throughout the world during the formative years of the subject and concept of tribology. A fine conference was held in Houston Texas in 1963 under the energetic leadership of Professor D. Muster and Dr B. Sternlicht. The *Proceedings* (Muster and Sternlicht, ed., 1965) of this event provided one of the foundation-stones for a number of subsequent research investigations.

Likewise, the *Proceedings of Research Symposia on Friction and Wear* (Davies, 1959) and *Rolling Contact Phenomena* (Bidwell, 1962) which were held at the General Motors Research Laboratories, Warren, Michigan, provided valuable contributions to the literature which have been appreciated by research workers in many countries.

There have been too many events of this nature to list them all in the present section, but it is worth noting one of them, since it demonstrated that the subject of tribology was of interest to emerging as well as established industrial nations. The First World Conference in Industrial Tribology (Malhotra and Sharma, 1972) was held in New Delhi and supported by a good number of Indian and overseas delegates.

The Literature

Papers and journals   Many papers on tribology have been published as individual items in the proceedings and transactions of the scientific and professional institutions mentioned earlier. In addition, conferences devoted entirely to various aspects of tribology have yielded valuable collections of papers. A few specialist journals have also been established to the great benefit of the subject.

The journal *Scientific Lubrication* was launched in October 1948, with a content reflecting a balance between science and practice. In January 1967 the change in emphasis towards industrial aspects of lubrication problems was reflected in a change in title to *Industrial Lubrication incorporating Scientific Lubrication*. As the concept of tribology became widely recognized, a further change introduced the present title *Industrial Lubrication and Tribology incorporating Scientific Lubrication*.

A journal devoted to the publication of research papers has been published under the title *Wear* since 1957. It is clear from the contents list of each issue that the subject extends well beyond the title on the cover. Papers on surfaces, contact, wear, friction and lubrication are all included and the journal provides a most important contribution to the developing science of tribology.

The first issue of the journal *Tribology* appeared in January 1968, but as the subject unfolded and interest widened, the title changed to *Tribology International*. Each issue carries a wide range of articles including reviews of important subjects, accounts of developments in design, research and practice, together with topical news, book reviews and a calendar of events in the field of tribology.

The tribology thesaurus known as *Tribos* prepared and published by the BHRA also appeared for the first time in January 1968. It is a journal of abstracts which provides a current awareness service of great value to those wishing to keep up to date with published work. The number of abstracts has now achieved quite large proportions, being 1,372 in the year 1971 and 1,694 in 1972.

An analysis of the number of published papers on tribology on a year-by-year basis emphasizes the literature explosion in the 1960s and 1970s. To illustrate this point and to draw attention to the evolution of tribology literature in recent times the last entry in this section will be devoted to books.

Books   It is not uncommon to find that many people believe that there is a dearth of books devoted to tribology. It is certainly true that texts bearing the title tribology, and structured round the new-found unity of the subject, are both recent and rare, yet many valuable books dealing with elements of the subject have been published in the fifty-year period which forms the setting for this chapter.

A list of books is presented in chronological order in Table 11.13. Most, but not all of the books, are English-language texts, and the list is in no sense comprehensive, although it is based largely upon the collections housed in the libraries of the University of Leeds, the Institution of Mechanical Engineers and the author's private collection. Texts resulting from a collection of conference papers and the many fine books produced by bearing manufacturers and the oil companies have generally been excluded, but the remaining list offers an interesting picture in historical

Table 11.13
Tribology Books

Year	Author	Title	Publisher

1920 **Thomsen, T. C.** *The Practice of Lubrication* McGraw-Hill
1921 **Evans, E. A.** *Lubricating and Allied Oils* Chapman and Hall
1923 **Stanton, T. E.** *Friction* Longmans
1925 **Gümbel, L. and Everling, E.** *Reibung und Schmierung in Maschinenbau* M. Krayn
1928 **Boswall, R. O.** *The Theory of Film Lubrication* Longmans

1936 **Hersey, M. D.** *Theory of Lubrication* Wiley

1942 **Norton, A. E.** *Lubrication* McGraw-Hill
1945 **Allan, R. K.** *Rolling Bearings* Pitman
1945 **Palmgren, A.** *Ball and Roller Bearing Engineering* SKF
1946 **Holm, R.** *Electrical Contacts* Almqvist and Wiksells
1949 **Purday, H. P. F.** *Streamline Flow* Constable
1949 **Shaw, M. C. and Macks, F.** *Analysis and Lubrication of Bearings* McGraw-Hill

1950 **Bouman, C. A.** *Properties of Lubricating Oil and Engine Deposits* Macmillan
1950 **Bowden, F. P. and Tabor, D.** *The Friction and Lubrication of Solids*, Part I Oxford U.P.
1950 **Burwell, J. T.** *Mechanical Wear* American Society for Metals
1950 **Gemant, A.** *Frictional Phenomena* Chemical Publishing Co.
1950 **Georgi, C. W.** *Motor Oils and Engine Lubrication* Reinhold
1950 **Michell, A. G. M.** *Lubrication–Its Principles and Practice* Blackie
1951 **Bastian, E. L. H.** *Metalworking Lubricants–Their Selection, Application and Maintenance* McGraw-Hill
1951 **Bondi, A.** *Physical Chemistry of Lubricating Oils* Reinhold
1952 **Kühnel, R.** *Werkstöffe für Gleitlager* Springer-Verlag
1953 **Ellis, E. G.** *Lubricant Testing–Introducing Recent Developments in Testing Techniques* Scientific
Publications (GB)
1953 **Schmid, E. and Weber, R.** *Gleitlager* Springer-Verlag
1954 **Boner, C. J.** *Manufacture and Application of Lubricating Greases* Reinhold
1954 **Forbes, W. G.** *Lubrication of Industrial and Marine Machinery* Wiley
1954 **Palmgren, A.** *Grundlagen der Wälzlagertechnik* Franckh'sche Verlagshandlung
1955 **Brewer, A. F.** *Basic Lubrication Practice* Reinhold
1955 **Cazaud, R.** *Le Frottement et l'Usure des Métaux les Anti-Frictions* Dunod
1955 **Hobson, P. D.** *Industrial Lubrication Practice* Industrial Press
1955 **Riddle, J.** *Ball Bearing Maintenance* Norman: Univ. of Oklahoma Press
1955 **Slaymaker, R. R.** *Bearing Lubrication Analysis* Wiley
1956 **Barwell, F. T.** *Lubrication of Bearings* Butterworths
1956 **Bowden, F. P. and Tabor, D.** *Friction and Lubrication* Methuen
1956 **Fuller, D. D.** *Theory and Practice of Lubrication for Engineers* Wiley
1956 **Kragelskii, I. V. and Shchedrov, V. S.** *Development of the Science and Friction* Academie Nauk SSR
1957 **Wilcock, D. F. and Booser, E. R.** *Bearing Design and Application* McGraw-Hill
1958 **Eschmann, P., Hasbargen, L., and Weigand, K.** *Ball and Roller Bearings–Their Theory, Design and Heyden
Application*
1958 **Fuller, D. D.** *A Survey of Journal Bearing Literature* ASLE
1958 **Vogelpohl, G.** *Betriebsichere Gleitlager. Berechnungsverfahen für Konstruction und Betrieb* Springer-Verlag
1958 **Holm, R. and Holm, E.** *Electric Contacts Handbook* Springer-Verlag
1959 **Clarke, G. H.** *Marine Lubrication* Scientific Lubrication (GB)
1959 **Howell, H. G., Mieskis, K. W., and Tabor, D.** *Friction in Textiles* Butterworths
1959 **Radzimovsky, E. I.** *Lubrication of Bearings–Theoretical Principles and Design* Ronald Press

1960 **Kruschov, M. M. and Babichev, M. A.** *Investigations into the Wear of Metals* USSR Academy of Sciences
1961 **Lipson, C. and Colwell, L. V. (eds.)** *Handbook of Mechanical Wear–Wear,* Univ. of Michigan Press
Frettage, Pitting, Cavitation, Corrosion
1961 **Pinkus, O. and Sternlicht, B.** *Theory of Hydrodynamic Lubrication* McGraw-Hill

Table 11.13 (*Continued*)

Year	Author	Title	Publisher
1961	**Tipei, N., Constantinescu, V. N., Nica, A. L., and Bitä, O.**	*Sliding Bearings – Computation – Design – Lubrication*	Academiei Republicii Populare Romine
1962	**Freeman, P.**	*Lubrication and Friction* Pitman	
1962	**Gross, W. A.**	*Gas Film Lubrication* Wiley	
1962	**Geary, P. J.**	*Fluid Film Bearings – A Survey of Their Design, Construction and Use*	British Scientific Research Association
1962	**Gunderson, R. C. and Hart, W. A. (eds.)**	*Synthetic Lubricants* Reinhold	
1962	**Kragelskii, I. V. and Vinogradova, I. A.**	*Coefficients of Friction* (in Russian) Mashgiz	
1962	**Tipei, N. (ed. W. A. Gross)**	*Theory of Lubrication* Stanford Univ. Press	
1963	**Constantinescu, V. N.**	*Lubrificatiacu Gaze* (see Eng. trans. 'Gas Lubrication', ed. R. L. Wehe ASME, 1969)	Academiei Republicii Populare Romine
1963	**Kodnir, D.**	*Contact-Hydrodynamic Theory of Lubrication* Kuibishevskoe knizkno Izdat, USSR	
1964	**Bayer, R. G., Ku, T. C. et al. (ed. C. W. MacGregor)**	*Handbook of Analytical Design for Wear*	Plenum Press
1964	**Bisson, E. E. and Anderson, W. J.**	*Advanced Bearing Technology* NASA	
1964	**Boner, C. J.**	*Gear and Transmission Lubricants* Reinhold	
1964	**Bowden, F. P. and Tabor, D.**	*The Friction and Lubrication of Solids*, Part II Oxford U.P.	
1964	**Braithwaite, E. R.**	*Solid Lubricants and Surfaces* Pergamon Press	
1964	**Caubet, J. J.**	*Theory and Industrial Practice of Friction* (in French) Technip and Dunod	
1964	**Eschmann, P.**	*Das Leitungsvermögen der Wälzlager* Springer-Verlag	
1964	**Grassam, N. S. and Powell, J. W. (eds.)**	*Gas Lubricated Bearings* Butterworths	
1965	**Kragelskii, I. V. (trans. by L. Ronson in collaboration with J. K. Lancaster)**	*Friction and Wear* Butterworths	
1965	**Rabinowicz, E.**	*Friction and Wear of Materials* Wiley	
1965	**Tamm, P. and Ulms, W.**	*Lubrication Practice – Introduction for Lubrication Personnel* VEB Verlag-Technik (in German)	
1966	**Akhmatov, A. S.**	*Molecular Physics of Boundary Lubrication* Israel Program for Scientific Translations	
1966	**Cameron, A.**	*Principles of Lubrication* Longmans	
1966	**Dowson, D. and Higginson, G. R.**	*Elasto-hydrodynamic Lubrication – The Fundamentals of Roller and Gear Lubrication* Pergamon Press	
1966	**Harris, T. A.**	*Rolling Bearing Analysis* Wiley	
1966	**Hersey, M. D.**	*Theory and Research in Lubrication* Wiley	
1966	**Muijderman, E. A.**	*Spiral Groove Bearings* Philips Technical Library	
1966	**Trumpler, P. R.**	*Design of Film Bearings* Macmillan	
1967	**Braithwaite, E. R. (ed.)**	*Lubrication and Lubricants* Elsevier	
1967	**Gruse, W. A.**	*Motor Oils – Performance and Evaluation* Reinhold	
1967	**Lipson, C.**	*Wear Considerations in Design* Prentice-Hall	
1967	**Stavoselsky, A. A. and Garkonov, D. N.**	*Duration Life of Working Machine Parts* Machinestrone, Moscow	
1968	**Christensen, H.**	*Smøring og Slitasjeteknik* Teknologisk Forlag, Oslo	
1968a	**Constantinescu, V. N.**	*Aplicatii Industriale Ale Lagärelor Cu Aer*	Academiei Republicii Socialiste Roumania
1968b	**Constantinescu, V. N.**	*Lubrication in Turbulent Regime* US Atomic Energy Commission	
1968	**Ellis, E. G.**	*Fundamentals of Lubrication* Scientific Publications (GB)	
1968	**O'Connor, J. J., Boyd, J., and Avallone, E. A. (eds.)**	*Standard Handbook of Lubrication Engineering* McGraw-Hill	
1968	**Scheel, W. S.**	*Technische Betriebsstoffe* VEB Deutscher Verlag für Grundstaffindustrie	
1968	**Schilling, A.**	*Motor Oils and Engine Lubrication* Scientific Publications (GB)	
1968	**Thomas, R. H.**	*Lubrication Mechanics* Carlton Press	
1969	**Ling, F. F., Klaus, E. E., and Fein, R. S. (eds.)**	*Boundary Lubrication – An Appraisal of World Literature* ASME	
1969	**Nica, A.**	*Theory and Practice of Lubrication Systems* (trans. of *Sisteme de Lubrificatie*)	Academiei Republicii Socialiste and Scientific Publications (GB)
1969	**Smith, D. M.**	*Journal Bearings in Turbomachinery* Chapman & Hall	
1960	**Summers-Smith, D.**	*An Introduction to Tribology in Industry* Machinery Publishing	
1969		*Glossary of Terms and Definitions in the Field of Friction, Wear and Lubrication – Tribology* OECD	

Table 11.13 (*Continued*)

Year	Author	Title	Publisher

1970   **Powell, J. W.**   *Design of Aerostatic Bearings*   Machinery Publishing
1970   **Pugh, B.**   *Practical Lubrication – An Introductory Text*   Newnes-Butterwoth
1970   **Schey, J. A.**   *Metal Deformation Processes – Friction and Lubrication*   Marcel Dekker
1970   **Stansfield, F. M.**   *Hydrostatic Bearings For Machine Tools and Similar Applications*   Machinery Publishing
1971   **Buckley, D. H.**   *Friction, Wear and Lubrication in Vacuum*   NASA
1971   **Cameron, A.**   *Basic Lubrication Theory*   Longmans
1971   **Floberg, L.**   *Maskinelement*   Almqvist & Wiksell Farlag
1971   **Hondros, E. D.**   *Tribology*   Mills & Boon
1971   **Matveevsky, R. M.**   *Effect of Heat on Lubricated Metal Surfaces* (in Russian)   USSR Academy of Sciences (Nauka)
1971   **Pavelescu, D.**   *New Conceptions on The Friction and Wear of Deformable Solids; Calculation and Applications* (in Roumanian)   Academiei Republicii Socialiste Roumania
1971   **Peck, H.**   *Ball and Parallel Roller Bearings: Design Application*   Pitman
1971   **Quinn, T. F. J.**   *The Application of Modern Physical Techniques to Tribology*   Newnes-Butterworth
1971   **Rowe, W. B. and O'Donoghue, J. P.**   *Design Procedures For Hydrostatic Bearings*   Machinery Publishing
1971   **Brewer, A. F.**   *Effective Lubrication – Management Responsibility including Basic Lubrication Practice*   Robert Krieger
1972   **Clauss, F. J.**   *Solid Lubricants and Self-Lubricating Solids*   Academic Press
1972   **Gunther, R. C.**   *Lubrication*   Bailey Bros & Swinfen
1972   **Moore, D. F.**   *The Friction and Lubrication of Elastomers*   Pergamon Press
1972   **Schilling, A.**   *Automobile Engine Lubrication*   Scientific Publications (GB)
1972   **Simons, E. N.**   *Metal Wear – A Brief Outline*   Frederick Muller
1972   **D.T.I.**   *Tribology Projects for Schools*   Department of Trade and Industry
1973   **Bowden, F. P. and Tabor, D.**   *Friction – An Introduction to Tribology*   Anchor Books
1973   **Kragelskii, Lubarsky, E. M., Tryovkia, G. E. and Odovniko, V. F.**   *Friction and Wear in Vacuum*   Machinestrone, Moscow
1973   **Ling, F. F.**   *Surface Mechanics*   Wiley
1973   **Neale, M. J. (ed.)**   *Tribology Handbook*   Butterworths
1973   **Pugh, B.**   *Friction and Wear*   Butterworths
1975   **Baumma, N. E.**   *Calculation Methods for The Degree of Friction and Wear*   Biryansic Institute Press, USSR
1975   **Halling, J. (ed.)**   *Principles of Tribology*   Macmillan
1975a   **Moore, D. F.**   *Principles and Applications of Tribology*   Pergamon
1975b   **Moore, D. F.**   *The Friction of Pneumatic Tyres*   Elsevier
1975   **Walowit, J. A. and Anno, J. N.**   *Modern Developments in Lubrication Mechanics*   Applied Science Publishers
1976   **Boner, C. J.**   *Modern Lubricating Greases*   Scientific Publications (GB)
1976   **Engel, P. A.**   *Impact Wear of Materials*   Elsevier
1976a   **Halling, J.**   *Introduction to Tribology*   Wykeham Publications
1976   **Houghton, P. S.**   *Ball and Roller Bearings*   Applied Science Publishers
1976   **Pinegin, S. V.**   *Vibrational Friction in Machines and Machine Elements*   Machinestrone, Moscow
1976   **Sarkar, A. D.**   *Wear of Metals*   Pergamon
1977   **Thomas, T. R. and King, M. (ed. N. G. Guy)**   *Surface Topography in Engineering – A State of the Art Review and Bibliography*   BHRA
1978   **Barwell, F. T.**   *Bearing Systems: Principles and Practice*   Oxford U.P.
1978   **Czichos, H.**   *Tribology – a systems approach to the science and technology of friction, lubrication and wear*   Elsevier

terms. Full references to the texts mentioned briefly in Table 11.13 are included in the References section of this volume.

Several of the books listed in the early sections of Table 11.13 are now considered classic texts. Gümbel's posthumous text (Gümbel and Everling, 1925) and Boswall's (1928) book on lubrication were remarkable representatives of the literature to emerge in the 1920s. Boswall's book grew in stature as the years rolled by, and even today it is a valuable text in both technical and historical terms. Boswall was born in 1884, the year that Osborne Reynolds presented his differential equation of fluid-film lubrication to the British Association Meeting in Montreal. He took an academic post in the Manchester College of Technology in 1920 and was stimulated into doing research on bearings by Professor Stoney, whose background was in steam turbines.

In many ways the book was decades ahead of general appreciation of the potential of lubrication theory in the bearing design process, and one marvels at its far-sighted approach to the subject. On a memorable occasion in 1976 the author was privileged to visit Dr Boswall, then aged ninety-two with one of his former students from the Manchester College of Technology, Dr K. L. Johnson. Boswall was then in the penultimate year of his life, but his alertness and interest in talking about developments in lubrication was salutary. His reminiscences of persons he had known like Goodman, Swift, Martin and Michell provided a direct link with the pioneering days of analytical and experimental studies of fluid-film lubrication.

Hersey's (1936) book has achieved a reputation which distinguishes the man and his work throughout the world. Early in the 1940s, Norton (1942) produced a book of great clarity based upon his lectures at Harvard, but published posthumously, which remained an essential text until well after the Second World War. In the immediate postwar years books on rolling-element bearings by Allan (1945) and Palmgren (1945) were eagerly recieved. Shaw and Macks (1949) combined lubrication theory with bearing analysis and indeed many aspects of tribology in a comprehensive test. It seems strange that the inventor of tilting-pad bearings should wait until 1950 before publishing a distinctive text on lubrication. Michell's (1950) book set the highest standards, included new work and became a prized publication. In the same year Bowden and Tabor (1950) produced the first volume of their highly respected texts on friction and lubrication.

If those entering the field of tribology feel that the subject is short of textbooks, perhaps Table 11.13 will allay their fears. A brief glance shows that students entering the field in 1950 were unlikely to be confused by a vast array of books.

Many of the texts which appeared in the 1950s and the 1960s were

Fig. 11.1
Publication pattern of some books on tribology topics (1920–77).

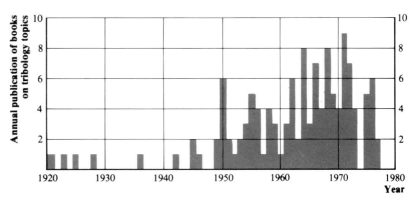

devoted to the practice of lubrication. This clearly reflected a need of the period, but at the same time a number of important books concerned with the underlying science of lubrication, friction and wear established themselves.

A final point of some historical interest emerges from an analysis of the entries listed in Table 11.13. The histogram shown in Fig. 11.1 indicates that the expansion of literature on tribological topics was already well advanced before the word *tribology* was introduced in 1966. If it is recalled that authors of texts put pen to paper some years before the publication of their books, it is quite clear that the events described in Section 11.20 represented a culmination, and not the initiation of, a widespread appreciation of the significance of the subjects subsequently and collectively known as *tribology'*.

11.20 Tribology

In this closing section the emergence and meaning of the concept of tribology will be described, but first the wider considerations which promoted the events of the 1960s and 1970s will be discussed.

The prime objective of any engineer is to produce, to an agreed specification, a machine or system which will perform some desired function. This aim so dominated the early history of engineering that friction, wear, reliability and overall efficiency were subservient considerations. Inventiveness, practical skills and in due course scientific principles were directed towards the construction of machines which worked, and the history of technology abounds with evidence of great achievements. As manufacturing processes yielded materials of construction of adequate and reliable strength, the engineer enjoyed the opportunity to apply with confidence the basic principles of thermodynamics, fluid mechanics and solid mechanics to the design of sound machinery. In due course he was able, and indeed found it necessary, to give more attention to the life, reliability and efficiency of machinery, considerations which inevitably focused attention upon friction, wear and lubrication.

The process started well before the twentieth century, as we have seen in earlier chapters, but it has impinged most urgently upon technological societies since about 1940. There are very sound reasons for seeking reliability and efficiency in plant as large as that in the chemical and steel-making industries and as complicated and sophisticated as that in land, sea, air and space transportation and conventional and nuclear electric power generating equipment. With the increase in size of much equipment, the enhancement of speeds and loads carried by machinery and the desire to provide communities with power, heat and transportation all taking place against a background of limited resources, it was inevitable that economic pressures alone would at some stage direct attention to the subject now known as tribology. That time arrived in the 1960s, but it is important that the developments described in the next sub-section should be seen not in isolation, but as a logical step in a process covering many centuries.

Formation of the Lubrication Engineering Working Group in the United Kingdom

Table 11.14
Lubrication Engineering (Education and Research) Working Group (1965)

Mr H. Peter Jost
Chairman

Dr D. Dowson
Dr J. E. Garside
Mr L. F. Hall
Mr R. A. Lake
Mr R. P. Langston
Dr J. G. Lavender
Mr A. A. Milne
Mr A. D. Newman
Mr J. Romney
Mr D. Scott
(replacing Mr Milne after July 1965)

Mr J. C. Veale
Mr J. G. Withers

Mr S. E. B. Solomons
Mr D. W. Tanner
Assessors

It is evident from the earlier sections of this chapter that research, professional meetings and publications in lubrication, friction and wear all experienced dramatic growth in the past half-century. However, a development of particular note took place in 1964.

At a meeting on Iron and Steel Works Lubrication (1965) held in Cardiff from 27 to 29 October 1964, two points emerged with great clarity. One was the schism between engineers operating large-scale equipment and those researching on bearings and lubrication, and the other was a clear recognition by professional engineers, several of whom were active members of the Institution of Mechanical Engineers' energetic Lubrication and Wear Group, that many of the serious plant-operating problems associated with lubricants or lubricated machine elements like bearings, gears and seals could be solved by the application of existing knowledge and sound engineering principles and practice. The reader might detect evidence of this conflict of viewpoints in the papers by Neale (1965), Barwell (1965), Chisholm (1965) and contributions to the discussion by Hall (1965) and Jost (1965). Another factor of some importance was dismay at the change in fortunes of important and highly regarded research laboratories concerned with these subjects a few years earlier.

After the Second World War Dr F. T. Barwell and a number of colleagues moved north from the National Physical Laboratory at Teddington to form the Mechanical Engineering Research Laboratory Lubrication and Wear Division at Thorntonhall, near Glasgow. The laboratory established an international reputation for its work under the wide guidance of Dr Barwell, but unfortunately lost its identity in a reorganization of the National Engineering Laboratory (NEL) in 1961. Much of the work continued in other divisions of NEL, but the advantages of a single multi-disciplinary group were lost at the dawn of the decade of recognition of the importance of work in its sphere of interest. Work on fundamental aspects of lubrication and wear carried out by a group of physicists at the Aldermaston Physical Science Centre of AEI was also terminated when the laboratory was disbanded for economy reasons. Once again, the excellent work carried out at Aldermaston came to an abrupt halt just as a widespread recognition of its significance was emerging.

Soon after the 1964 Cardiff meeting, the Rt Hon. the Lord Bowden of Chesterfield, Minister of State for Education and Science, addressed himself to some of the wider issues regarding lubrication which had then emerged. On 22 December 1964 he requested H. Peter Jost to '. . . consult with persons and bodies to establish the present position of lubrication education and research in this country, and to give an opinion on industry's needs thereof.'

The Lubrication Engineering Working Group formed to investigate this matter had the composition shown in Table 11.14.

The Working Group held ten meetings, received evidence from experts at a further eight sessions, sought the views of over 400 colleges and universities and a large number of companies, industrial research establishments, government laboratories, research associations, and individuals

Fig. 11.2
The Department of Education and
Science Lubrication Engineering
(Education and Research)
Working Group (1965).

in the United Kingdom and overseas. Meetings were held in Richmond Terrace, off Whitehall in London and a photograph of the Group at one of its sessions is shown in Fig. 11.2.

The Group examined general considerations, education, research, and undertook an assessment of industry's needs before presenting its findings and recommendations. It soon became evident that the economic significance of the subject featured large in the minds of those in industry. Early attempts to make this point by Rowe (1734), Thurston (1887), the DSIR (1920) Report, Parish (1935), Hersey (1936), and Vogelpohl (1951) were thus taken a stage further in emphatic manner in the 1960s. The Group concluded that, on a conservative estimate, the potential savings to British industry resulting from improvements in education and research, were about £515 million per annum, the biggest single item being savings in maintenance and replacement costs of £230 million per annum.

A further general point which emerged with some force was that the word 'lubrication' meant different things to different people. It was generally used in a narrow sense which totally belied its economic and technical significance. Furthermore, the essential unity of subjects separately labelled friction, lubrication and wear was clearly seen and for these reasons it was felt to be desirable to introduce a single word to embrace these interlinked, multi-disciplinary topics. The final choice, based upon advice from the English dictionary department of the Oxford University press, was *tribology*. The word is based upon the Greek *tribos* (rubbing) and is defined as follows: 'Tribology is the science and practice of interacting surfaces in relative motion and of the practices related thereto.'

The Report of the Working Group, entitled *Lubrication (Tribology) Education and Research–A Report on the Present Position and Industry's Needs* (HMSO, 1966) was forwarded to Mr R. E. Prentice, Minister of Education and Science on 23 November 1965. It included the following major recommendations.

Education

Courses of both general and specialist nature should be arranged for shop floor, technician, undergraduate and postgraduate levels.

It was envisaged that tribology would be introduced as a subject forming part of some existing or new courses in engineering and science, and that a limited number of M.Sc. courses in tribology would be established for the education of specialists in the subject.

Research

Some institutes of tribology should be established with the following principal functions.

1. To undertake both basic and applied research in tribology.
2. To give postgraduate instruction and to provide facilities for postgraduate research in tribology.
3. To form and maintain a two-way bridge with industry.

A coordinating and advisory Committee on Tribology should be established.

General Recommendations

1. An information centre on tribology should be established.
2. The Institution of Mechanical Engineers should be invited to extend the membership and scope of its Lubrication and Wear Group in order to provide for organized professional activities in tribology.
3. A handbook on triboengineering design and practice should be published.

The Report was published on 9 March 1966, the official birthday of *Tribology*.

The Committee on Tribology

The Minister of Technology, the Rt Hon. Anthony Wedgwood Benn, MP, announced in the House of Commons on 11 August 1966 the formation of a *Committee on Tribology* under the chairmanship of Mr H. Peter Jost, to advise on the implementation of measures designed to meet the objectives proposed in the DES Report. The Committee remained in existence until 1972, with Dr I. Maddock acting as Deputy Chairman until 1970 and Mr L. F. Hall from 1970 to 1972.

Three operating committees were established to deal with research and liaison, education and training and information and publicity, respectively. The Research and Liaison Committee was chaired by Dr D. Tabor

(1966–69) and then by Mr P. P. Love (1969–72); the Education and Training Committee by Professor H. Ford (1966–68) and then by Professor D. Dowson (1968–72); and the Information and Publicity Committee by Mr H. S. Winterbourne (1966–70) and Mr N. Shepherd (1970–72). A Tribology Handbook Exploratory Committee was established under the chairmanship of Mr M. J. Neale in 1967. By 1970, 128 contributors had been commissioned and the comprehensive volume edited by M. J. Neale soon established itself as a valuable work of reference (Neale, 1973).

Progress by the Committee on Tribology is charted in four annual reports published in (1968), (1969), (1970) and (1971). The final report of (1972) reviewed the whole exercise from 1966 to 1972. Most of the recommendations of the 1966 (Jost) Report had been introduced by 1972.

A vigorous approach was made to the problem of linking the research worker and persons with expertise in tribology with those requiring information in industry. Meetings for managers and designers were held in various centres and a directory of experts in tribology was established.

Three centres of tribology were established, two being based upon universities as envisaged in the 1966 Report and one in a government establishment. The Centres were located at the University of Leeds, The University College of Swansea and the Reactor Engineering Laboratory of the United Kingdom Atomic Energy Authority (UKAEA). The Leeds centre, known as The Industrial Unit of Tribology, was incorporated within an Institute of Tribology in the Department of Mechanical Engineering. The Institute fulfills the original aims of the 1966 Report through its undergraduate and postgraduate courses, research on various aspects of tribology and the work of its Industrial Unit. The Swansea Tribology Centre also enjoys proximity with the Engineering Department of the University College. The UKAEA centre, housed in the Reactor Engineering Laboratory at Risley to enable it to continue to provide a service to the Authority while operating a commercial service for industry, became known as the National Centre of Tribology. It was originally intended that each centre should offer a specialist service: Leeds on bearing design, Swansea on problems related to the metallurgical industries and Risley on hostile environments, in addition to general advice on tribology. This has happened naturally to some extent, but all three centres found themselves involved in work across the full spectrum of tribology.

The Research Sub-Committee also drew attention to specific topics in tribology requiring special attention. The problem of scuffing of internal combustion engine piston rings and cylinder liners being one example and an important review of the subject was published by Neale (1971).

In the field of education, the subject of tribology soon established itself in colleges, universities and to a lesser extent in schools. While new courses were being planned and introduced in colleges and universities, the urgent requirement to disseminate information on tribology to a large number of practising engineers was met by a substantial number of short courses in various centres. These short courses reached their peak in 1968–69, when nearly 950 persons attended full-time courses of up to two weeks' duration, or special evening lecture courses. Many universities and polytechnics now offer tribology as a subject within B.Sc. degree schemes, particularly in mechanical engineering. Interest in all aspects of teaching tribology was evident in the proceedings of a special conference (Dowson and Jones, 1970) at the Institute of Tribology, Leeds. In general the best courses reflect the special enthusiasm and interest of the lecturer, but guidelines for the essential content of undergraduate courses in tribology were contained within a *Basic Tribology Module* (Nightingale, 1975) and outlined at a Seminar on Teaching Tribology held at the Institution of Mechanical Engineers headquarters in London on 4 July 1972.

Postgraduate courses leading to the degree of M.Sc. in tribology were offered by the University of Leeds and, for a few years, by a combination of Reading and Swansea universities. The call for postgraduate courses for those wishing to specialize in tribology contained within the 1966 Report was anticipated by the presentation of an M.Sc. course in bearing design, lubrication, friction and wear in the Department of Mechanical Engineering in the University of Leeds in October 1964. In due course the title was reduced to tribology, and from 1964 to the present day specialist education has been offered to successful candidates from twenty-nine countries.

The tendency to encourage learning through projects provided a useful vehicle for the introduction of tribology in schools. Professor John Halling chaired a Committee which produced a booklet on *Tribology Projects for Schools* (Department of Trade and Industry, 1972) published by the Department of Trade and Industry.

This brief survey of the emergence of tribology in the 1960s and 1970s has been necessarily selective rather than comprehensive. The 1966 Report triggered off an immense activity in education, research and liaison. The subject is now widely recognized and appreciated, and although it first found expression in the United Kingdom, the basic concepts now enjoy international support.

International developments

An International Tribology Council was formed at the end of the First European Tribology Congress on 24 September 1973 with Dr H. Peter Jost as its first President. By 1974 some twenty-one countries were represented on the Council.

Several countries have formed societies to organize tribological activities. These include the Gesellschaft für Tribologie und Schmierungstechnik in the Federal Republic of Germany, Kring Tribologie–Bond voor Materialenkennis in The Netherlands, the Société Française de Tribologie in France, the Norwegian Tribological Society, The Tribology Group of the Royal Swedish Academy of Engineering Sciences, The Indian Society for Industrial Tribology and The Tribology Group of the Yugoslav Society for Fuels and Lubricants Application. Some of these societies produce their own journals, like *Schmiertechnik & Tribologie*, *Schmietungstechnik* and *Tribologia e Lubrificazione*, and several promote meetings.

In the United States, where the word 'lubrication' was well established in industry and in professional circles through the activities of ASLE and the Lubrication Division of ASME, the term 'tribology' has been received more cautiously, even though the basic philosophy is widely appreciated. Similar considerations to those which prompted the DES inquiry in the UK from 1964 to 1966 emerged in the twentieth meeting of the Mechanical Failures Prevention Group (1976) held at the National Bureau of Standards in Washington, DC from 8 to 10 May 1974. In 1977 the ASME announced that it was actively participating with the Energy Research and Development Administration (ERDA) and the Office of Naval Research (ONR) in drawing up a tribology research and development plan aimed at achieving energy conservation.

Tribology awards

In September 1969 the Institution of Mechanical Engineers announced the formation of a Tribology Trust. Awards by the Trust recognize outstanding achievement in tribology and the furthering of interest and study in the subject.

Gold, silver and bronze medals can be awarded annually, and the first five distinguished recipients of the gold medal, representing four countries, are listed in Table 11.15.

Summary

This final section of the text has been devoted to the organizational and wider aspects of tribology, since such considerations have impinged upon and to some extent dominated the field in recent times. We have seen that the word, and indeed the philosophy, of tribology emerged in the 1960s, initially in the United Kingdom, but then throughout the world.

Why did it happen in the 1960s? This question will no doubt be argued for many years to come. In the first place there was the vast increase in research and professional activities dating from the early 1950s, which was in turn a reflection of the increasing problems of a tribological nature encountered by the larger scale and increasingly complex machinery in industry. Then came the economic arguments, spurred on by an increasing concern for the efficient use of resources, an enhanced awareness of the importance of reliability and a desire to know more about the factors affecting the life of machinery. It is, however, important to keep a sense of historical perspective, and to realize that this was not the first time that the economic case had been advanced. History shows that scientists and engineers had drawn attention to the economic aspects of tribological subjects at least as early as the mid-eighteenth century, with the size and vigour of the bubble increasing on each occasion. It is probably more important to realize that boiling was inevitable, than to seek reasons for it happening in the 1960s. It was not merely the energy and enthusiasm of a few individuals that tipped the scales, for this alone was a necessary but not the sole requirement for the introduction of a new word and all that it implied in scientific, technological and economic terms. It was, indeed, an inevitable consequence of industrial development.

Table 11.15
Tribology gold medallists (1972–77)

1972	Professor David Tabor
1973	Professor Harmen Blok
1974	Mayo D. Hersey
1975	Professor Igor Kragel'skii
1976	Robert Lawrence Johnson
1977	Professor F. T. Barwell

An interesting feature of the 1960s and 1970s is the extent to which governments in various countries demonstrated interest in the issues raised by a recognition of the full significance of tribology. We have now emerged from the stage of vigorous inquiry to find much greater awareness of tribological matters in industry, educational establishments and government circles.

It is now time for taking stock, and several reviews of the recent developments have been published including those by Stone (1968), Dowson (1970b), Woodyard (1973), Stephenson (1976), Jost (1976a,b), Neale (1976) and Parker (1976). Specific references to the economic significance of tribology have also been made by Nevin (see SSRC Newsletter, 1974), Jost (1976a, 1976b) and Summers-Smith (1976).

It would, however, be unfortunate if the heavy emphasis on economic and organizational questions in recent years caused us to lose sight of the intellectual challenge of tribology. In the long run it is the teaching of tribology to successive generations of engineers and scientists which will provide the foundations for the remedy of ills reported in 1966. It is therefore fortunate that this multi-disciplinary subject provides great interest and intellectual stimulus to those who learn and those who teach it.

Writing this book has been rather like riding on a train. The immediate panorama is vivid, yet sometimes difficult to place in context with the overall geography. The early chapters were relatively easy to prepare, since major events from early history, like features from the initial part of a journey, stand out ever more clearly as the less significant characteristics fade from the memory and are excluded from the records.

Equally it has to be recognized that a journey across a plain does little to alert the traveller to his approach to a range of mountains, or vice versa, and perhaps the suddenness of the appearance and full expression of both the word and philosophy of tribology in the 1960s took many by surprise.

I can only hope that the reader will find interest in this account of the foundations and background to the subject, and some sense of direction from his journey through the book. Some day, perhaps, another author will tell us whether the experience helped to anticipate the terrain ahead.

Appendix

Men of tribology

A.1 Introduction

Throughout the text the contributions of individuals to the total store of knowledge of tribology have been noted. To have given brief backgrounds to the lives of all the contributors would have diverted attention away from their major scientific and technical achievements, whereas to have written fully about a few would have unbalanced the story. Yet the contributions of a few have been so dominant that no history of tribology would be complete without a more detailed discussion of their lives. An appreciation of their contributions to tribology usually benefits from a knowledge of their wider experiences and the factors which shaped their lives. In this Appendix to the main text such information is provided in the form of biographical sketches of a select group of outstanding men of tribology.

The selection of the short list of names for this special treatment was by no means easy. In general I have restricted consideration to those persons who exposed the fundamental nature of friction, lubrication or wear, or who provided practical solutions to tribological problems which were of great merit. Coulomb, Reynolds and Hardy fall into the former category, Michell and Kingsbury into the latter. Within these restrictions the basis of selection has been simply the outstanding nature and timing of the individual contribution. This necessarily excluded many outstanding men of science, like Newton, who never considered directly problems in tribology, but who contributed fundamental concepts which were embodied in the work of others. I have, however, included some like Heinrich Hertz, who never wrote on the subject of bearings, but whose fundamental studies of contact between elastic solids provided the scientific corner-stone of a great industry.

Some years ago Archibald (1955*a*,*b*,*c*,*d*,*e*, *f*, 1956a, 1957) wrote a number of articles published by the American Society of Lubrication Engineers under the title *Men of Lubrication*. These articles stimulated my writing of the present biographical sketches and I am grateful to Mr Archibald for his encouragement and generosity in providing access to much of the material which he collected together in the 1950s. I have endeavoured to include a reproduction of a portait or photograph of each subject and many people have assisted in seeking information for both the illustrations and the texts. I hope that all my friends will find the fruits of their encouragement and assistance enjoyable and rewarding.

A.2 Leonardo da Vinci (1452–1519)

c. 1499 (age forty-seven)
Portrait by Ambrogio de Predis
Royal Library – Windsor.

Leonardo was born near the small hill-top village of Vinci some 20 miles (32 km) west of Florence at 3 o'clock in the morning of Saturday, 15 April 1452. The year is known from an income tax return submitted by his grandfather Ser Antonio in 1457 in which the list of residents in his household includes a reference to '. . . Leonardo, aged five, the illegitimate son of Ser Piero, his mother being Caterina, now married to Chartabriga di Piero del Vaccha da Vinci'.

Little is known of the early life of Leonardo's mother Caterina, who has been variously discribed as a *peasant girl* and *of gentle birth*. His father Ser Piero da Vinci was a Florentine lawyer in a long line of notaries of some repute in both Vinci and Florence. Shortly after Leonardo was born, Caterina married a cattle herdsman, while Ser Piero embarked upon the first of four marriages.

While Leonardo was growing up in his father's homes in Vinci and Florence another young boy of similar age was living in a weaver's home about 160 miles (257 km) to the north in Genoa. His name was Christopher Columbus.

When Leonardo was seventeen the family moved permanently to Florence where his father showed some of his early drawings to an acquaintance, Andre del Verrocchio, who readily received him as a pupil. Between 1470 and 1477 Leonardo worked in his master's studios and it was during this period that he left his father's house for Verrochio's. A pleasant but disputed story claims that the master renounced his own career as a painter when he saw the angel painted by Leonardo as a contribution to Verrocchio's picture of Christ's baptism. Leonardo was admitted to membership of the Company of Painters in Florence in 1472 and his first dated drawing was signed in 1473.

In 1476 Leonardo was anonymously accused along with four other Florentine youths of having unnatural sexual relations with a seventeen-year-old boy. After two months the accusation was rejected as unproven, but the shame and humiliation had left its mark.

c. 1512 (age sixty)
Self-portrait
Royal Library – Turin.

From 1477 until about 1483 Leonardo worked independently but largely in the favour of the powerful Medici family and Lorenzo the Magnificent in particular. Other commissions of this period were received from the monks of San Donato in Scopeto, but there are few completed works of the time which can with certainty be attributed to him. He was already expressing his talents in various fields and it is known that he was concerned with a variety of projects in mechanics, hydraulics and architecture in this period. Towards the end of this era Leonardo moved to Milan and the prospect of a post at the court of Ludovico Sforza, but there is no firm evidence of his employment in the city until 1487. Suggestions have been made that some of the intervening years were spent in travel in the East. It was in 1494 that Leonardo started work on a joint commission from Ludovica, then Duke of Milan, and the local monks which was to become his greatest painting, the *Last Supper*.

In a brief but eventful period in 1502–3 Leonardo travelled as chief engineer over much of central Italy in the service of Duke Caesar. He then returned to Florence where he completed by 1505 a cartoon of an episode from a battle fought in 1440. In the period 1503–6 he completed his great portrait of Madonna Lisa, the Neapolitan wife of Zanobi del Giocondi, which is now one of the great glories of the Louvre.

In 1506 Leonardo returned to Milan on the invitation of Charles d'Amboise, Lieutenant of the French King in Lombardy and in 1507 he formally transferred his services to Louis XII with the title of painter and engineer. His father had died in 1504 and Ser Piero's seven surviving legitimate sons disputed his claim to a share of the estate in litigation which was not resolved until 1511. It was during this second period of residence in Milan from 1506 to 1513 that two events of note took place. In the first place he became attached to a new and devoted young friend and pupil of noble birth named Francesco Melzi who will appear later in our story. Secondly, he must have completed towards the end of this period the self-portrait in red chalk shown at the beginning of this section which is now to be found in Turin.

After a short unhappy period in Rome about 1512 Leonardo finally left Italy for good in 1516 to spend the last two and a half years of his life in France with a handsome pension in the Castle of Cloux near Amboise. He made his will on Easter Eve 1519 and died on 2 May 1519.

Bequests of money were made to his half-brothers, his servants and the local hospital in Amboise, and his vineyard near Milan was divided equally between two beneficiaries. However, it was his favourite pupil Melzi from the days of Leonardo's second period in Milan who was named as executor and the recipient of all his manuscripts and the contents of his studio.

It is far from easy to trace the fate of Leonardo's manuscripts on scientific matters, but the effort is worth while. It appears from several notes that his writing extended over a period of some forty years to the time when he was in Amboise. Francesco Melzi returned to Italy with the manuscripts shortly after Leonardo's death and he appears to have retained most of them in good order until his death in 1570. Melzi's heirs failed to show the same respect for the papers and in due course Dr Orazio Melzi permitted thirteen volumes, together with a number of paintings and other works, to be retained by a Milanese studying law at the University of Pisa named Giovanni Ambrogio Mazzenta. Giovanni joined the Barnabite Order in 1590 and in due course a considerable number of manuscripts were secured by a sculptor Pompeo Leoni who was employed in the service of Philip of Spain.

It was Pompeo Leoni who cut several of the manuscripts to assemble the single volume of 402 sheets and over 1,700 drawings now known as the *Codex Atlanticus*. The latter is a composite, not of Leonardo's making but based upon his writings between 1483 and 1518. It eventually reached the Ambrosian Library founded in Milan in 1603 along with a number of other manuscripts.

On 19 May 1796 Napoleon Bonaparte decreed that all men of genius were really French, irrespective of their country of birth, and many works were transported to France. The *Codex Atlanticus* was certainly in the Bibliothèque Nationale in Paris in August 1796 and twelve other manuscripts were deposited in the library of the Institut de France. The latter are still in France but the *Codex Atlanticus* was returned to the Ambrosian Library in Milan in 1815.

Numerous manuscripts including those at Windsor and the British Museum were acquired by individuals and subsequently changed hands a number of times. Some belonged to Lord Arundel, the Earl of Leicester and Mr Forster, and these manuscripts are now identified by their names.

In 1966 Professor Reti stated that there were probably about 75 per cent of Leonardo's writings still missing. In 1967 Jules Piccus, a specialist in early Spanish literature of the University of Massachusetts, was searching in the Biblioteca Nacional de Madrid for medieval ballads when he found the now famous manuscripts Codex Madrid I and II. The first manuscript of 382 pages and 1,600 sketches and drawings deals almost exclusively with theoretical and applied mechanics. It appears to have been written between 1493 and 1497 and one of its greatest merits is that it represents Leonardo's own systematic treatise in contrast to Leoni's scrapbook formulation of the *Codex Atlanticus*. The second volume in the Madrid Library appears to have been a notebook used by Leonardo for about eighteen years.

The story of the manuscripts is itself fascinating, but it is fitting to conclude this section by returning to a few personal details of the man himself. Leonardo wrote and painted with his left hand and further he wrote backwards, from right to left, like the Orientals. There are no known paintings of the young Leonardo, but he was apparently tall, remarkably handsome, an athlete with a mastery over horses and a lute player with a beautiful voice. The promise of youth flourished and there is every justification for the inheritance by our first tribological personality of the title . . . *the greatest genius of all time.*

Bibliography

Richie-Calder, Baron P. R. (1970) *Leonardo and the Age of the Eye* (Heinemann, London).

Reti, L. (1974) *The Unknown Leonardo*, Ladislao Reti (ed.) (Hutchinson, London).

MacCurdy, E. (1956) *The Notebooks of Leonardo da Vinci*, new edn, 2 vols (Jonathan Cape, London).

Taylor, R. A. (1927) *Leonardo The Florentine – A Study in Personality* (Richard Press, London).

A.3 Guillaume Amontons (1663–1705)

Amontons was born in Paris on 31 August 1663 and he died on 11 October 1705. His opportunity to pursue a career in a conventional profession was curtailed by deafness during adolescence. As a result of this disability, Amontons devoted himself to his studies and, like many others before and since his time, he indulged in vain attempts to develop a perpetual-motion machine. This frustration clearly influenced the direction of his early studies, for he decided to read mathematics and works on the physical sciences. It might also have been an important ingredient in his resolve to study the question of losses caused by friction in machines in later years.

Amontons' father was a lawyer who moved from Normandy to Paris before Guillaume was born. It is known that the family opposed Guillaume's idea of studying the physical sciences, but it appears that his resolve won the day. After studying drawing, surveying and architecture he obtained experience of the practical side of applied mechanics by working on various public works projects. At a later stage he studied celestial mechanics and also applied himself to the improvement of hygrometers, barometers and thermometers. Thermometry was to be one of Amontons' major areas of scientific investigation, but he also developed an optical telegraph system and championed the case for the clepsydra as a timing apparatus to improve navigation at sea.

The first known study of the question of losses attributable to friction was published by Amontons in 1699. The significance of this classical contribution to tribology has been discussed earlier, but it is worth recalling that it yielded the first two laws of dry friction which in due course became known as Amontons' laws. Published in the middle of the period 1695–1703, when Amontons was much concerned with thermometry, it represented his sole yet distinctive contribution to our subject.

Apart from some extracts from letters in the *Journal des savants* for 8 March and 10 May 1688, Amontons published some thirteen papers in the *Mémoires de l'Académie des Sciences* in the six years following 1699. His first substantial published work was a book on the clepsydra published in Paris in 1695.

Guillaume Amontons lived for a mere forty-two years, yet his early and major contribution to our understanding of the subject of friction ensures his honoured position in our history. I have been unable to trace a portrait for reproduction in this text, but I am grateful to Professors Courtel and Godet for their efforts and advice in this quest. It appears that we must remain ignorant of the appearance of the man whose name is invariably linked with the laws of dry friction.

Bibliography   Magie, W. F. (1963) *A Source Book in Physics* (Harvard U.P., Cambridge, Massachusetts, USA).

From a portrait in Nichols (1815), Vol. 9, pp. 640–1.

A.4 John Theophilus Desaguliers (1683–1744)

Desaguliers was a natural philosopher who was born on 13 March 1683 at La Rochelle. His father was pastor of a Protestant congregation at Aitré who fled to England with the boy John after the Edict of Nantes in 1685. It has been said[1] that John was concealed in a barrel while fleeing on board the refugee ship. After a short stay in Guernsey, Desaguliers' father moved to London, where he established a school at Islington and also became minister of the French chapel in Swallow Street.

Desaguliers entered Oxford shortly after the death of his father. He matriculated at Christ Church and took the degree of B.A. in 1710. He was appointed Lecturer on Experimental Philosophy in Hart Hill in the same year and proceeded to the degree of M.A. in 1712. The following year he moved to London where he took up residence in Channel-Row, Westminster. From this address Desaguliers continued to present lectures and he is said to have been the first person to deliver lectures on learned subjects to the general public. A notice at the end of Volume I of his work, *A Course of Experimental Philosophy* (Desaguliers, 1734) gives details of the arrangement.

> *The Course of Experimental Philosophy, of which the first Volume contains half, is performed by the Author at his house Channel-Row, Westminster (where Catalogues of the Experiments may be had) on such Days, and at such Hourse, as shall be agreed upon by the Majority of the Auditors. N.B. Every Auditor is to pay Three Guineas, when the number is not less than twelve Persons; but any Three or Four, nay any one Person, may have a Course to themselves paying the Price of Twelve.*
>
> *A Short but full Course of Astronomy, will also be performed by means of the PLANETARIUM, to any Number of Persons, not less than Ten, at a Guinea each; or to any less Number, who are willing to pay 10 Guineas, upon giving a Days Notice.*

It appears that Desaguliers was the inventor of the planetarium and his skill as an experimenter and lecturer was recognized by his election as a Fellow of the Royal Society and the subsequent invitation to become Demonstrator and Curator of the Society. He was concerned with studies of light, colours, the barometer and, of course, mechanics. He lost no opportunity to emphasize that observations by our senses could be explained in terms of the laws of physics without recourse to the occult. He was apparently held in high esteem by the current President of the Royal Society, Sir Isaac Newton.

Desaguliers completed the degrees of bachelor and doctor of laws at Oxford in 1718, and in 1742 he was awarded the Copley gold medal of the Royal Society in recognition of his successful experiments. His religious teaching continued throughout this time. He became chaplain to the Duke of Chandos in 1714 having taken deacon's orders in 1710. He lectured before King George I at Hampton Court in 1717 and he was later appointed chaplain to Frederick, Prince of Wales.

---

1. See *Dictionary of National Biography* (1888), Vol. 14, p. 400.

The rebuilding of Westminster Bridge, about which Desaguliers was consulted frequently by Parliament, led to the demolition of Channel-Row and his own home. He then moved to lodgings in the Bedford coffee-house overlooking Covent Garden, where he continued his lectures with great success. He also erected a ventilator, at the request of Parliament, in a room over the House of Commons.

Desaguliers' name merits recognition in this section of our History of Tribology primarily because he recognized the possible role of cohesion or adhesion in the friction process. His ideas were not fully developed, but an important new element had entered the unfolding story of tribological research. An equally impressive feature of his studies of friction which was totally in keeping with other contributions by great physical scientists of the day, was his concern that the work should be related to the functioning of machines. In discussing the role of *stickage or friction* upon the imperfections of engines he pleaded that '. . . we are to take care to allow enough to be deducted from the calculation made concerning an Engine suppos'd Mathematically true'.

Desaguliers was said to have an unattractive personal appearance[2] to be short and thickset and of irregular features; some but not all these characteristics being evident in the portrait from Nichols' *Literary Anecdotes* shown at the beginning of this section.

At the time of his death Desaguliers was apparently suffering from poverty. He died just short of his sixty-first birthday on 29 February 1744.

Bibliography	**Bowden, F. P.** and **Tabor, D.** (1964) *The Friction and Lubrication of Solids* (Oxford U.P., Oxford), Pt. II.

Bowden, F. P. and **Tabor, D.** (1964) *The Friction and Lubrication of Solids* (Oxford U.P., Oxford), Pt. II.
Desaguliers, J. T. (1734) *A Course of Experimental Philosophy* (2 vols, London).
Dictionary of National Biography (1888) Leslie Stephen (ed.), (London) Vol. 14.
Nichols, J. (1815) *Literary Anecdotes of the Eighteenth Century* (London), Vol. IX.

A.5 Charles Augustin Coulomb (1736–1806)

Charles Augustin Coulomb was born on 14 June 1736 in Angoulême in the Angoumois and he was baptized in the parish church of St André on 16 June. Professor C. Stewart Gillmor has prepared a full account of Coulomb's genealogy (Gillmor, 1968) and a most valuable appraisal of the man and his times (Gillmor, 1971). Coulomb's family had been prominent in law and administration in Languedoc and had lived for several generations in Montpellier. Charles Augustin's father, Henry Coulomb, first served in the military and then accepted a minor administrative post with the grand title of Inspecteur des Domaines du Roi. His mother, Catherine Bajet was related to the wealthy de Sénac family.

Early in Charles Augustin's childhood the family moved to Paris where Henry was involved in the tax-farm system. Catherine was most

---

2. See *Dictionary of National Biography* (1888), Vol. 14, p. 400.

anxious that her son should study medicine and she arranged for him to attend lectures at the Collège des quatre-nations. The college, established by the will of Cardinal Mazarin in 1661, normally received the sons of '. . . prominent residents who live like nobles . . .' between the ages of ten and fifteen years. It had a good reputation for the teaching of mathematics, d'Alembert and Lavoisier having studied there, and it is clear that Coulomb developed a liking for the subject about this time. He moved to the Collège royal de France, where Pierre Charles Le Monnier presented his lectures on mathematics, and soon produced a rift in the family by announcing that he wished to be a mathematician and not a doctor. In the meantime Henry Coulomb had engaged in financial speculations which had left him penniless and he returned to Montpellier while the family remained in Paris. Charles Augustin steadfastly defied his mother's desire that he should study medicine, and feelings were so strong that he was forced to return to his father in Montpellier after being temporarily disowned by his mother. His stay in Montpellier proved to be intellectually rewarding and formative. He became a student member or *membre adjoint* of the Société royale des sciences de Montpellier at the age of twenty-one and presented two mathematical and three astronomical memoirs during his subsequent sixteen-month membership. Mathematics was his chief interest during this brief period in Montpellier, but the need to earn a living caused him to enter the School of Military Engineering or Ecole du génie at Mézières. He passed the entrance examination administered by the abbé Charles Camus after studying in Paris for about nine months and prepared to enter the Ecole du génie early in 1760.

Coulomb graduated from the Ecole du génie with the rank of *lieutenant en premier* on 12 November 1761. The course was a mixture of theory and practice, with three days of each week being devoted to each activity. Coulomb finished high in his group of eight graduates, but not all the reports on his work engendered confidence in his future as a military engineer. The commandant, Chastillon, reported:[1]

> . . . *M. Coulomb is from the Academy of Montpellier. His conduct is good, he understands and executes drafting rather well. His siege memoir is worse than average, very badly portrayed, with erasures and jottings. It is carelessly done and employs incorrect nomenclature. . . . He has a certain intelligence, but not that which will make him advance in the Corps.*

Coulomb's first posting was to Brest, where he was charged with minor mapping tasks on the coast between Brest and La Rochelle. This was during the Seven Years War between England and France (1756–63), and the return of Martinique to France after the Treaty of Paris in 1763 caused Coulomb to be moved from Brest to Port Royal to assist with the reconstruction of the island's fortifications. He sailed from Brest to the West Indies in the *Brillant* in February 1764. In due course he was put in charge of the construction of Fort Bourbon on Mount Garnier, one of the two peaks dominating the entrance to Port Royal. He was twenty-seven

---

1. See Gillmor, C. S. (1971), p. 17.

years old when he arrived in Martinique, and during the next eight years he gained valuable and varied field experience. In due course much of this experience was to be distilled and presented in the form of memoirs on statics, architecture and the efficiency of labouring men to the Academy of Sciences in Paris. The corps was plagued with illness during the construction of Fort Bourbon and Coulomb became seriously ill on numerous occasions. When he left Martinique in the June of 1772 his health had deteriorated considerably.

On his return to France, Coulomb was posted to Bouchain, where he found time to write about his work and experiences in the West Indies. His 'Essay on an application of the rules of maxima and minima to some problems in statics relating to architecture' was read before the Academy in Paris in February and March of 1773. The wide-ranging essay was well received and Coulomb was appointed Bossut's correspondent to the Paris Academy of Sciences on 6 July 1774. In the same year he was posted to Cherbourg for a two-year tour of duty and it was while he was at La Hougue, near Cherbourg, that he completed his important essay on 'Investigations of the best method of making magnetic needles'. The Academy of Sciences had announced a competition for the best means of constructing magnetic compasses, with the prize being awarded in 1775. No winner was selected in 1775 and the prize was therefore doubled for 1777. Coulomb shared the doubled award with J. H. Van Swindon. Gillmor (1971:28) has drawn attention to the fact that this particular memoir, written by Coulomb the engineer, contained all the elements, including friction, of Coulomb's subsequent major contributions to physics.

Coulomb's next posting was to Besançon in 1777 and it was intended that he should move to Marseilles in 1779. His father had died during his service in the West Indies, but his mother survived until 1779. Charles-Augustin shared the inheritance with his sisters as a result of a reconciliation following the strained relationship which arose over the question of his early career. It is not clear whether he ever went to his post in Marseilles, but he certainly spent some time in Paris attending to his mother's estate. He also read a paper to the Academy in May 1779 on the subject of dredging machinery.

In May 1779 Coulomb served in the Brigade of Toulon at Marseilles, where he was involved in the construction of a fort near Rochefort. It was here that he undertook his extensive studies of friction. Once again the spur was provided by the Academy of Sciences in Paris through the announcement in 1777 of a prize to be awarded in 1779 for the solution of '. . . problems of friction of sliding and rolling surfaces, the resistance to bending in cords, and the application of these solutions to simple machines used in the navy'. With no winner selected in 1779, the prize was once again doubled for an award in 1781. Coulomb started work on the subject late in 1779, but there are suggestions that his interest in the topic had been aroused at an earlier stage. His experience of architecture and construction in Martinique and the preparation of a memoir on a '. . . Means of eliminating friction in arched doors and giving them the same mobility as a floating body . . .' provided some basis for his work. He enjoyed the facilities of a modern shipyard, the support of the port commander and the assistance of two men in the conduct of his experiments over several months. He considered the work of Amontons and Desaguliers and in due course submitted a lengthy essay which was judged to be the winner in the spring of 1781. The tribological content and impact of his work, Théorie des Machines Simples . . .', are discussed in detail in Chapter 9, but it should be noted that this was his second, consecutive success in the naval prize contests of the Academy of Sciences in Paris. Shortly afterwards he requested a posting in Paris to enable him to pursue his interests in the Academy. He moved in September, was awarded the Croix de Saint Louis on 30 September and elected to the Academy of Sciences on 12 December 1781. These hectic events late in 1781, when Coulomb was forty-five years old, marked a transition from his earlier career in engineering to his later career in physics. Few men have changed disciplines so effectively halfway through their professional lives.

The move to Paris in 1781 provided Coulomb with an established home for the first time in many years. Military service and transfers to new locations every two years or so prevented the establishment of a permanent laboratory facility. On the other hand it exposed Coulomb to a wide range of problems in engineering and physics and he clearly responded to the challenge. He investigated architecture and construction in Martinique, friction in Rochefort, coastal defences in Cherbourg and windmills in Lille. In Paris his studies were directed more towards physics and he became active in the Academy. Academy membership brought its share of administrative work and Coulomb was an active committee member. He became involved in committees dealing with hospital reform, weights and measures and water-pumps. He was a member of the Academy's Library Committee and on 19 March 1783 he became the first person in France to describe publicly the principle of the Boulton and Watt improved (condensing) steam engine developed across the English Channel. He nevertheless found time to read thirty-two scientific memoirs to the Academy and its successor the Institut de France between 1773 and 1806. His contemporaries in the Academy included Borda, Bossut, Le Roy, Laplace, Legendre, Lavoisier, Berthollet, Lagrange, d'Alembert and Lazare Carnot.

Coulomb's prize-winning studies of the magnetic compass in 1777 and friction in 1781 provided the foundation for his subsequent work in physics. The use of a torsion wire on his magnetic compass prompted him to study torsion in some depth. The outcome was his major memoir of 1784 on the subject, 'Theoretical and experimental investigations of the force of torsion and of elasticity in metal wires.' Both the title and the content of this memoir are of interest. Coulomb was not satisfied to determine the factors responsible for the response of tubes and rods to torsion, he needed to explain the phenomenon in greater detail. This caused him to re-examine concepts in elasticity, the limits of elastic behaviour and the role of cohesive bonds between molecules. In his earlier work on friction Coulomb had recognized that there was more to friction than the mechanistic view associated with surface irregularities. He appeared to accept Desaguliers' concept of cohesion, but since it contributed less than 5 per cent to the total resistance he did not pursue the matter further. The role of cohesive forces between molecules in problems of mechanics was developed in his 'Statics' memoir of 1773 and again in his famous 'Torsion' memoir of 1784.

The other aspect of physics with which Coulomb's name is intimately connected is electricity and magnetism. In the seven memoirs on these subjects presented between 1785 and 1791, he outlined his electric torsion balance, the inverse-square law for forces between charged bodies, the leakage of electric charge and the distribution of charge on conducting bodies of various shapes and sizes. The last of these memoirs was presented in 1791 when Coulomb was fifty-five years old. Even when allowance is made for his wide-ranging experience as a military engineer, Coulomb's second career as a physicist achieved outstanding recognition in the decade following his move to Paris in 1781.

In April 1783 the Minister of War, Ségur, nominated Coulomb as a member of a team of consultants charged with the task of examining a proposal from the Estates of Brittany Canal Commission. The proposal involved the construction of new canals and locks to provide a new internal supply route, and the fortification of St Malo to protect the river Rance from the British. The team of consultants included no less a person than Chézy, Inspecteur-Général des ponts at chaussées. The study of the canal scheme was completed by the early autumn of 1783 and a new group of consultants including Coulomb, the abbé Rochon, member of the Academy of Sciences and Director of the Marine Observatory at Brest, an engineer called Bavre and his old friend Jean Charles de Borda moved to St Malo to consider the second set of proposals. This scheme, prepared by Rosnyvinen, Comte de Piré, leader of the Estates of Brittany Canal Commission, involved the construction of two large ports at St Malo, one commercial and one military. The consultants rejected this grandiose part of the scheme in December 1783.

Before moving to St Malo Coulomb had applied for leave in Paris. His application to Ségur was forwarded to the Commandant of the province of Brittany where Coulomb was acting as a consultant and then to the Comte de Piré of the Canal Commission. The request for leave was refused, but by this time Coulomb had returned to Paris with the other members of the group of consultants from St Malo. His return to the capital without being granted leave caused him to be censured, although it should be noted that he had received no response at all to his application by the time his work at St Malo was completed. He was deeply offended by the official reaction and offered to resign. His case was reviewed by an engineering committee of the Corps du génie which agreed with Coulomb's position. It recommended that the application for leave be granted and that Ségur should refuse to accept Coulomb's resignation because it would represent '. . . a great loss for the King's Service'. The War Minister was under pressure from the Corps du génie on the one hand and the Estates of Brittany, still smarting under the rejection of its plans for the expansion of St Malo, on the other. He resolved his position by sentencing Coulomb to prison in the Abbaye de Saint-Germain for a period of one week. Coulomb was later to write of Ségur's decision[2] '. . . the reason that was given in the order for my imprisonment can only be the pretext that a poor devil, hard pressed by his mistakes and his conscience, employs to escape the pressure'.

Coulomb's social conscience was illustrated by his concern for public health. He had written about the deplorable conditions in the hospital at Cherbourg in 1775. In 1777 he joined a committee established by the Academy of Sciences to consider a plan for the reform of the Paris hospitals submitted by Jean-Baptiste Le Roy. The work of the Academy in this field was to continue for many years and it caused Coulomb to visit England with Jacques René Tenon in 1787. The purpose of this journey was to survey the latest developments in hospital design and operative procedures and the eight-week period encompassed visits to London, Birmingham and Plymouth. The visiting pair also had meetings with Sir Joseph Banks, who had accompanied Captain Cook on his voyage round the world in the *Endeavour* (1768–71), and James Watt, whose engine had been described in France by Coulomb some four years earlier.

The French Revolution of 1789 was to lead to changes in the organization of both scientific and military institutions, although the direct effect upon Coulomb's work appears to have been slight. In 1786 he was promoted from the rank of captain to major, but the latter rank was abolished in 1790 and all majors became lieutenant-colonels. He tendered his resignation from the Army on 18 December 1790 and this was formally accepted on 1 April 1791. He retired with a pension, but his work at the Academy continued until his last appearance at a meeting in July 1793. The Reign of Terror for science began the following month when the Academy of Sciences was abolished on 8 August. Coulomb remained close to Paris for a further two months, for he was still active in the work of the Commission of Weights and Measures. He was purged from the Commission on 23 December 1793, and he then retired from the centre of the Reign of Terror with his good friend Borda to some property near Blois. He had previously bought from Lavoisier some property at Thoisy, 8 miles (12·9 km) north of Blois, and in 1803 he purchased a farm 10 miles (16 km) south of Thoisy and 3 miles (4·8 km) to the west of Blois.[3]

Coulomb was elected to the Institut de France which replaced the Academy of Science on 9 December 1795. He had entered Paris only rarely during the Reign of Terror – once to attend Lavoisier's funeral in May 1794 and again when he was appointed to a reconstituted Commission on Weights and Measures in April 1795. Between 1795 and 1799 he presented memoirs to the new institut on such diverse topics as fluid resistance, the efficiency of labouring men and plant physiology. He had observed the working habits of men throughout his career and his detailed study of particular actions persuaded him that on heavy work maximum output could be achieved with seven or eight hours' effort each day, while ten hours or so were optimum for lighter tasks like shopkeeping. He also recognized that frequent rest periods in various occupations produced higher overall output.

2. See Gillmor, C. S. (1971), p. 58.
3. See ibid., (1971), p. 75.

It is fitting to conclude this survey of Coulomb's scientific work with reference to his work on plant physiology. The study illustrated his insatiable interest in his immediate environment and was related to his period in the country at Blois during the Reign of Terror. In the spring of 1796 he heard a noise like the bubbling of air in a liquid when some Italian poplars were being cut down. He ordered more trees to be felled and found by removing the outer layers of the trunk that the phenomenon was restricted to a central core having a diameter of only a few centimetres. He discovered that when the trunks of trees were reduced to this small section near their base they nevertheless retained great resistance to bending, that the bubbles issued with a clear tasteless liquid like water when the core was penetrated and that the noise ceased during the night and on cold and humid days. These observations later formed the basis for a delightful memoir to the Institut in 1799 under the title 'Experiments concerning the circulation of sap in trees'.

It is difficult to assess the effect of the Revolution upon Charles Augustin Coulomb. It clearly caused some slight interruption to his work and he withdrew to the countryside during the Reign of Terror, but in due course he resumed his work in the Institut which replaced the Academy of Sciences. He bought a house in Paris in 1791 and he certainly spent some time there in 1797. It is sometimes suggested that Coulomb's absence from the capital during the height of the Revolution indicated that he was regarded as being of the nobility, but scholars like Gillmor (1971) have dismissed this interpretation. It seems more likely that his departure was no more than could be expected of a prudent family man. In general the *génie militaire* supported the revolutionary movement.

Coulomb's family life came late. It was not until he established his home in Paris after his roving military career that he was joined by his young partner, the twenty-year-old Louise Françoise Le Proust Desormeaux. Their first son, Charles Augustin II was born in Paris on 26 February 1790 when Coulomb was fifty-four years old. His second son, Henry Louis was also born in Paris on 30 July 1797, but it was not until 8 November 1802 that the marriage was legitimized.

Charles Augustin Coulomb died in Paris at the age of seventy on the morning of 23 August 1806. He lived through an age of stirring scientific activity and political turmoil. France dominated much of the work in the physical sciences and Coulomb stood in the front rank of a galaxy of mathematicians and physicists. Coulomb completed an impressive career as a military engineer before he began in earnest his second career in the realm of physics. His great work on friction came at the divide of his career, embodying the fruits of a successful life as an engineer with the skill of the physicist. His name has been firmly linked with the laws of friction throughout the past two centuries.

In preparing this note I have enjoyed the generous help of Professor C. Stewart Gillmor of the Wesleyan University, Middletown, Connecticut, USA, Mr C. A. Coulomb Jr kindly allowed the portrait of his great-great-grandfather to grace the opening to this biographical text. It appears that the portrait was completed between 1803 and 1806. The marble bust which is also shown at the front of this note was probably produced after the loan to the Académie des Sciences in 1850 of the original portrait of Coulomb mentioned above. Permission to reproduce this picture of the marble bust of Coulomb was kindly granted by the Musée du Conservatoire de Arts et Métiers, Paris. I would also like to acknowledge the search for material which Professor Robert Courtel of the Laboratoire de Mécanique des Surfaces, CNRS, Bellevue, and Professor Maurice Godet of the Laboratoire de Mécanique des Contacts, INSA, Lyon, undertook on my behalf. I can only hope that my appreciation of this opportunity to discover something of the life of the man whose steps they now follow will be matched by their satisfaction with the views expressed on one of the greatest French tribologists of all times.

Bibliography

Bowden, F. P. and Tabor, D. (1964) *The Friction and Lubrication of Solids* (Clarendon Press, Oxford), Pt II, Ch. xxiv, pp. 502–16.

Coulomb, C. A. (1785) 'Theorie des machines simples, en Ayant égard au frottement de leurs parties, et a la roideur des cordages', *Mém. Math. Phys.* (Paris) **x**, 161–342.

Gillmor, C. S. (1968) 'Charles Augustin Coulomb: Physics and engineering in eighteenth-century France', Ph.D. dissertation (University of Princeton).

Gillmor, C. S. (1971) *Coulomb and the Evolution of Physics and Engineering in Eighteenth Century France* (Princeton U.P., Princeton, New Jersey, USA).

Kragelskii, I. V. and Shchedrov, V. S. (1956) *Razvite Nauki o Trenii* (*Development of the Science of Friction*), Ch. 4, 'Investigations by Coulomb' (Soviet Academy of Sciences, Moscow).

Magie, W. F. (1963) *A Source Book in Physics* (Harvard U.P., Cambridge, Massachusetts, USA), pp. 97–8.

The Engineer (1944) 'Historic researches; No. I Friction–Coulomb and Morin's experiments', *The Engineer*, 14 July, **178**, 22–3.

A.6   Arthur-Jules Morin   (1795–1880)

Arthur-Jules Morin belonged to that group of Frenchmen who successfully combined a military career with outstanding work in science and applied science. He achieved distinction in both science and military service, being a general, mathematician and scholar.

He was born in Paris on 17 October 1795. He entered the Ecole Polytechnique in 1813 and left in 1815. For the next two years he displayed some uncertainty about his career and he spent the time working in an iron forge. In 1817 he was appointed to the Ecole d'Application at Metz as an artillery officer and assistant to Poncelet. He taught at Metz until 1839 and during this time he was promoted from lieutenant (1817) to captain (1829).

On 26 September 1839 Morin took up the Chair of Applied Mechanics at the Conservatoire des Arts et Métiers in Paris. It appears that the Chair had been created with Morin in mind. Four years later, in December 1843, he was elected to the Académie des Sciences and he became Chairman of the Mechanics Section in succession to Coriolis. He became Vice-President of the Académie in 1863 and President in 1864. In 1862 he was appointed President of the Société des Ingénieurs Civils.

In 1848 the third French Revolution broke out in Paris. King Louis-Philippe was driven from his throne and fled to England. A republic was then established in France based upon the principle of universal suffrage. Following the events of 1848, Morin was appointed Director of the Conservatoire des Arts et Métiers in succession to Claude Pouillet. He held this post until his death in 1880. On the establishment of the Second Empire, Morin's advice and help were frequently sought by Napoleon III in carrying out his grandiose schemes for the reconstruction of Paris and other large French cities. In the latter years of his life he was concerned with problems of ventilation for theatres and other public buildings.

Morin played a full part on various public service committees. In 1850 he served as a member of the Commission set up to organize the National Institute of Agronomy, and in the following year he was a member of the French Commission to the Great Exhibition of 1851 in London. General Poncelet and Colonel Morin were among the prominent foreign engineers who spoke at a banquet to recognize the first London meeting of the Institution of Mechanical Engineers on the occasion of the Great Exhibition. He was President of the Imperial Commission for the Universal Exhibition in Paris in 1855. In 1867 the Institution of Mechanical Engineers held its first meeting in Paris. Morin and Tresca read papers and both were subsequently elected as honorary members of the Institution.[1] His zeal and patriotism found expression in the setting up of an International Commission to establish high-precision metric standard measures in 1869. For over ten years he was a member of the Council for Public Instruction and he took a deep interest in both technical and liberal education. He was associated with the reorganization of the Ecole Polytechnique and was strongly in favour of the widespread introduction of applied sciences at all levels in industrial schools.

During his time at the Conservatoire des Arts et Métiers, Morin's military standing was enhanced by steady promotion. He became a major in 1841, a lieutenant-colonel in 1846, a colonel in 1848, a brigadier-general in 1852 and a divisional general in 1855. In 1852 he commanded the artillery of the 'Camp du Nord'. He was made a Commander of the Legion d'Honneur in 1854.

Arthur-Jules Morin was a specialist in mechanics. His interests covered a wide range and he published on such topics as hydraulic machinery, projectiles, air resistance, the motion of falling bodies, practical mechanics, highway engineering and public works, steam engines and, of course, friction. He invented numerous instruments including a dynamometer and other devices to assist him with his investigations. He has been variously described as a mathematician and a physicist, but much of his work on engineering problems was characterized by carefully conducted and wide-ranging experiments.

In the field of tribology Morin is noted for his studies of friction. The work was carried out in Metz before he moved to the Chair in Applied Mechanics in Paris. The investigations must have played an important part in bringing his name to the attention of the authorities in the Conservatoire des Arts et Métiers. It was half a century since Coulomb had

---

1. See Parsons, R. H. (1947).

undertaken his well-known work, but Morin was to adopt a similar approach to the problem of friction. Both men responded to military interests in the subject and Morin's first work on 'New experiments in friction' (Morin, 1833) was undertaken between 1831 and 1833 on the orders of the Minister of War.

It was Morin who introduced the term and demonstrated the utility of the *coefficient of friction*. He devoted much effort to the determination and recording of friction coefficients for rolling and sliding conditions. The value of his coefficients to engineers in the period of rapid expansion of machinery and the steam engine was enormous. Indeed, the utility of his work was such that engineers generally referred to Morin's, rather than to Coulomb's, coefficients of friction, even though much of the data available came from the latter's investigations.

Morin's studies of rolling friction have been mentioned in Section 9.4. He was concerned with the relationship between tractive force and radius of the wheel and concluded that a simple inverse relationship applied. Dupuit disputed the result, in finding that the force of friction was inversely related to the square root of the radius of the wheel. Although Morin's findings were accepted by the Académie, he was quite willing to pursue the matter further. He subsequently repeated much of the work and could be seen working with convoys of heavy vehicles on the roads of France.

Arthur-Jules Morin was a liberal and generous man. In the discourse delivered at the funeral service to Morin, his colleague at the Académie des Sciences, M. Tresca gave a number of illustrations of this judgement (Tresca, 1880). At the age of eighty-four he visited his grandson in all weathers to help him with his preparations for the examinations for the Ecole de Saint-Cyr. He also intervened directly with the Minister when the rights of a distinguished colleague were in jeopardy in 1851. Tresca's funeral oration provides clear evidence of his affection and admiration of Morin. They had worked together for nearly thirty years in the Conservatoire des Arts et Métiers and Tresca described Morin as '. . . a great man whom I admired above all others'. Saint-Venant was another famous mechanician of the period counted as a friend and fellow student of Arthur-Jules Morin.

Morin died in his eighty-fifth year in the city of his birth on 7 February 1880. Perhaps the best way to close this review of his life and to indicate the high esteem in which he was held, is to quote from Tresca's closing sentences spoken at the graveside.

> As a scholar his career was arduous, useful and full; in this career he attained the first rank.
>
> As a soldier, he saw where his duty lay in all circumstances, on the field of battle, as in committee; he had to wait for the first promotions, but he never complained; in contrast, the highest ranks, as it were, were waiting for him to fill them.
>
> As an administrator, his progress was rich in serious undertakings, and the Conservatoire des Arts et Métiers, alone, would do ample honour to his memory.
>
> His previous life was crowned by these three haloes, combined with that of a man of conscience and of sincere faith, who could always conduct himself in perfect accord with his convictions.

In preparing this account of the life of a great French engineer and contributor to the subject of friction, I have been helped greatly in the collection of biographical information by Professor Maurice Godet of the Institut National des Sciences Appliquées de Lyon and Mr R. Butler of the South Library of the University of Leeds. The assistance of the Conservatoire National Des Arts and Métiers and L'Ecole Nationale Supérieure des Mines is also appreciated. I am also grateful to Mrs Juliet White for her invaluable translation of some of the documents.

Bibliography

Academy of Sciences (1666–1954) *Biographical Index of Members and Corresponding Members of the Academy of Sciences* – Morin, Arthur-Jules, p. 370.

Ecole Polytechnique (1995) *Book of the Centenary of the Ecole Polytechnique; 1794–1894* (Gauthier-Villars, Paris), Vol. 1, 10. pp. 176–81.

Hoefer, (1861) *Nouvelle Biographie Generale* (Firmin Didot, Paris), Vol. 35, pp. 599–602.

Morin, A. J. (1833) 'Memoire concernant de nouvelles expériences sur le frottement faites à Metz en 1831', *Mém. Savans Etrang.* (Paris), **iv**, 1–128; *Ann. Min.*, **iv**, 271–321.

Morin, A. J. (1834) 'Nouvelles expériences sur le frottement faites en 1832', *Mém. Savans Étrang.* (Paris), **iv** 591–696; *Ann. Min.*, **vi**, 73–96.

Morin, A. J. (1834) 'Lettre à M. Arago sur diverses expériences relatives au frottement et au choc des corps', *Ann. Chim.*, **lvi**, 194–8.

Morin, A. J. (1835) 'Nouvelles expériences faites à Metz en 1833 sur le frottement, sur la transmission due mouvement par le choc, sur la résistance des milieux imparfaits a la pénétration des projectiles, et sur le frottement pendant le choc', *Mém. Savans Etrang.* (Paris), **vi**, 1835, 641–785; *Ann. Min.*, **x**, 1836, 27–56.

Morin, A. J. (1840) 'Sur le tirage des voitures et sur les effects destructeurs qu'elles exercent sur les routes', *Comptes Rendus* (Paris), **x**, 101–4.

Morin, A. J. (1841) 'Note sur la résistance au roulement des corps les uns sur les autres, et sur la reaction élastique des corps qui se compriment réciproquement', *Comptes Rendus* (Paris), **xiii**, 1022–23.

Morin, A. J. (1845) 'Note sur le roideur des cordes', *Comptes Rendus* (Paris), **xx**, 228–31.

Parsons, R. H. (1947) *History of the Institution of Mechanical Engineers* (The Institution of Mechanical Engineers, London), pp. 18, 24.

Poggendorf, J. C. (1863) *Biographisch Literansch Handwörterbuch*, pp. 208–9 (see also 1898 ed.).

Private communication (1976) 'Short Biography of General Morin' (Private communication), Ministère de L'Education Nationale, Conservatoire National des Arts et Métiers, Bibliothéque.

The Engineer (1944) 'Historic researches: No. I – Friction – Coulomb and Morin's experiments', *The Engineer*, **178**, 14 July, 22.

Tresca, M. (1880) 'Discours Prononcés aux Funérailles De M. Morin, Membre de l'Académie des Sciences, Directeur du Conservatoire des Arts et Métiers', pp. 1–9.

A.7 Gustav Adolph Hirn (1815–1890)

With Clausius, Joule and Rankine, Hirn is recognized as one of the founders of the science of applied thermodynamics. It is no mere coincidence that a scientist noted for his work in thermodynamics was also outstanding in the field of friction and lubrication, for one of the most significant developments in science in the mid-nineteenth century was the recognition of an equivalence between work and heat. Hirn himself established that 365 kgf m (3579 Nm) of frictional work was equivalent to 1 kg cal (4187 J) of heat without knowing that Mayer and Joule had established similar relationships some five or six years earlier in 1842 and 1843.

Born at Logelbach near Colmar in Alsace on 21 August 1815 Gustav Adolph Hirn was a delicate boy who received no formal education. His father was a partner in the cotton manufacturing firm of Haussmann, Jordan and Hirn and the young Gustav was permitted to work in the chemical laboratory before he took charge of the mechanical department at a nominal salary until the year 1880. Throughout this time he took every opportunity to expand his knowledge of a wide range of subjects. In an appreciation of Hirn published in the journal *Engineering* on 31 January and 14 February 1890, Professor Dwelshauvers-Dery of Liège University argued that the range of his interests and achievements was probably aided by the non-conformist nature of self-education. His life was devoted to scientific research, but he became acquainted with German, English, Italian and Latin while developing his knowledge of chemistry, mathematics and the natural sciences. He was a talented musician and he no doubt inherited his deep appreciation of art from a father who was sufficiently distinguished in the field to have had a painting hung in the great gallery of the Louvre.

In philosophy he developed his own concept of the universe which culminated in his final essay and book entitled, *Constitution de l'Espace Celéste* (see *Engineering*, 1889). When presenting this work to the Académie des Sciences of the Institute of France, M. Flaye described it as '. . . one of the most interesting and original works of modern times'. It is an interesting essay which again reflects Hirn's view of an energy balance. He grappled with the apparent conflict between the observations of a near-constant solar temperature yet an enormous rate of solar radiation, and concluded that '. . . the sun has been receiving from a source as yet unknown a quantity of heat equal to the amount it loses by radiation'. He was, in fact, anticipating the twentieth-century discovery of nuclear reactions and the fusion process. The essay also provided a further glimpse of Hirn's understanding of the relationship between frictional force and heat and of his concern to explain physical phenomena in terms of the molecular theory of matter. He wrote: '. . . Friction continually disturbs the relative position of the atoms towards each other, and destroys the true equilibrium of the electric force; the presence of heat is merely a concomitant phenomenon, that is to say, necessary but consecutive.'

Hirn's important experiments on friction were completed by 1847, but this significant contribution to tribology and the culmination of much patient research was almost doomed to oblivion by the reluctance of scientific societies to publish the work. It appears that the paper was submitted to the Académie des Sciences and then the Royal Society and was rejected by both. A very full biography published in the *Bulletin of the Society of Natural History of the town of Colmar* (1889/90), where

Hirn lived from 1881 to 1890, mentions that the paper was submitted to the Académie des Sciences on 26 February 1846, but Cameron (1966) has pointed out that in *Comptes Rendus* the Académie dated the receipt of the manuscript as 26 February 1849. Dwelshauvers-Dery (1890) wrote that '. . . With the temerity of youth [he was then thirty-three], Hirn, who had arrived at positive conclusions from the study of nature, and had found them opposed in many instances to the notions generally received, applied, in 1848, with naive confidence to various scientific societies, but did not meet with the reception he expected.'

The Royal Society did not start their *Register of Papers Submitted* until 1853, but the evidence strongly suggests that Hirn completed his experiments by 1847 and first submitted his paper in 1848 or early 1849. This clarification of dates is important since it is clear that Hirn was the first person to undertake scientific studies of journal bearings. The work undertaken in the Nuremberg Railway Carriage Works under von Pauli started in 1847, was completed in 1848 and published in 1849. If Hirn's first manuscript had been accepted he might well have been recognized more widely as the originator of scientific studies of journal bearings through the timing of both his experiments and his publications.

The reasons for the rejection of Hirn's paper by two premier scientific societies are not known to us. Was it because the views expressed were contrary to established views on friction in the two countries which had contributed so much to the subject? Or, as mentioned by Professor Dwelshauvers-Dery, merely a consequence of the greater difficulties experienced by experimenters rather than theorists in having their findings published?

Emile Dollfus, President of the Société Industrielle de Mulhouse, fortunately persuaded Hirn to rewrite his manuscript and the shortened, revised paper was presented at a sitting of the Société on 28 June 1854. It was published in the *Bulletins* of the Society in the same year and republished later by the Société d'Encouragement, of Paris.

Whereas von Pauli was interested in the performance of various bearing metals lubricated by one lubricant, Hirn was solely concerned with lubricants and lubrication of a bronze bearing of fixed composition. Hirn constructed a fine friction balance and confirmed the laws of dry friction established by Amontons and Coulomb. His experiments on lubricated bearings revealed relationships between the coefficient of friction and load, speed, bearing area and fluidity (or viscosity). He tested some twelve oils, fat, water and air, and his recognition that viscosity of the lubricant greatly influenced friction in lightly loaded bearings was an early and significant finding. He noted that air could reduce the friction to one-hundredth of its normal value and he is widely acclaimed as the first man to demonstrate the feasibility of gas bearings. He also pioneered the use of mineral oil rather than the conventional animal and vegetable oils of the day by setting up a small oil-manufacturing plant in the cotton mill at Logelbach. It had previously been thought that mineral oil was unsuited as a machine lubricant, but Hirn established both its economy and its efficacy in reducing friction losses in machines.

Hirn was a kind and patient man who was held in high esteem throughout Alsace. He was generous with his time, particularly to the young and the weak. His unerring judgement and keen sense of justice often caused him to be chosen as arbitrator on delicate matters in the district. In due course his contributions to science and engineering were widely recognized, with honours and titles being conferred by societies in America, Belgium, Brazil, Denmark, France, England, Russia, Spain, Sweden and Switzerland.

Gustav Adolph Hirn was undoubtedly a man of great talent, industry and character. He was equally at ease in the company of men of science, philosophy, music and art. Professor R. H. Thurston, first President of the American Society of Mechanical Engineers and himself a highly esteemed figure in the history of tribology, had no doubt about the value of Hirn's contribution to the subject. The dedication of Thurston's (1885) book to Hirn provides a fitting closure to this biographical sketch.

To
The Memory of
The Engineer, Physicist and Mathematician
G. A. HIRN,
one of the earliest workers in this field,
This Little Work
is inscribed, in grateful appreciation of personal, as well
as of professional, aid and encouragement, and in
recognition of a most stimulating example
of noble work, inspired by nobler
thoughts and noblest aims

Bibliography

Academy of Science (1849) *Comptes Rendus*, **28**, 290.

Cameron, A. (1966) *Principles of Lubrication* (Longman, London), p. 264.

Colmar National History Society (1890) 'Biography of G. A. Hirn', in *Bull. Soc. d'Hist. Naturelk de Colmar* (1889/1890), **1** (New Series), 183–309.

Dwelshauvers-Dery (1890) 'Reminiscences of the life of G. A. Hirn', *Engineering*, **49**, 31 Jan., 120–1; 14 Feb., 174–5.

Engineering (1889) Review of G. A. Hirn's book, *'Constitution de l'Espace Celéste'* (published by Gauthier Villars et Fils), in *Engineering*, **48**, 8 Nov., 1889, 549–50.

A.8   Nikolai Pavlovitch Petrov (1836–1920)

Petrov's scientific work and his explanation of mediate friction in journal bearings is well known, but knowledge of his life and his attributes as a person is somewhat meagre. He was born of the gentry of Novgorod Province in 1836. He attended the Constantine Artillery Academy and the Nicholas Engineering Academy and completed his courses in these institutions in 1855 and 1857, respectively. Petrov's interest in mathematics was already evident at this stage, when he was twenty or twenty-one years old, for he remained at the Nicholas Engineering Academy for a while as an instructor in mathematics. He went on to study mechanics under the famous mathematician M. V. Ostrogradsky in the former

Pedagogical Institute and the Technological Institute of St Petersburg. He is said to have designed new machines for the Ochta Powder Factory during this period.

In 1865 Petrov was sent abroad to study applied mechanics and during the following year he became a lecturer in the Nicholas Engineering Academy and in the Technological Institute in St Petersburg. In 1867 his growing accomplishments were recognized and he was named as an adjunct-professor in mechanics in the Nicholas Engineering Academy. He held the Chair of Steam Engineering at the Technological Institute and from 1871 the Chair of Railway Vehicles. During this period his lectures covered such topics as steam engines and boilers.

The main Society of Russian Railways invited Petrov to become one of its engineering members in 1873. In 1876 the Society appointed him as an 'expert' or 'judge' for the World Exhibition in Philadelphia, USA. His standing as a railway engineer was thus recognized both in Russia and internationally for at least a decade before his famous work on mediate friction in journal bearings was published. It is in the reflective introduction to his paper of 1883 that the influence of the development of the Russian petroleum industry upon his studies can be detected.[1] There is no doubt that it was the twin pressures of problems with bearings on the Russian railways and the need to explore the merit of lubricants derived from the 70 per cent residue produced by refining the expanding production of Russian crude oil which provided the stimulus for Petrov's work in the 1870s and early 1880s.

Between 1882 and 1892 Petrov was President of the Temporary Council of the Directorate of State Railways, Director of the Department of Railways and Chairman of the Engineering Council of the Ministry of Communications. During the seven years leading up to 1900 he was Freight Minister in the Ministry of Communications. He was elected Chairman of the Russian Technical Society in 1897.

His well-known work on the friction of lubricated journal bearings was characterized by extensive and careful experimental work combined with skill in the interpretation and analysis of his results. The outcome, widely known as Petrov's law, is one of the best-known relationships in tribology. It expresses the friction torques on concentric journals and bearings in terms of the small radial clearance, the viscosity of the lubricant at the effective operating temperature, the journal speed and radius and the bearing length. This classical work was recognized by the award of the Lomonosov Prize of the Imperial Russian Academy of Sciences. His further work in this field (Petrov, 1886) merited the award of the Metropolitan Makaria Prize of the Academy of Sciences.

Petrov's outstanding contribution to the understanding of journal-bearing friction represented but one of his fields of study. He brought a fresh approach to the work of the Technological Institute in St Petersburg in the 1870s. His first paper appears to have been published when he was

---

1. See Ch. 10.

about thirty-four years old and was an account of his investigation of 'The outlining of the cogs of large cylindrical wheels by means of the arcs of circles'. It was published by and awarded the prize of the *Engineering Journal* in 1870. He also published other but less well-known papers on tribology, including 'Concerning the wearing-out of and the testing of steel tyres'; 'Concerning continual braking systems'; 'Friction in machinery and the effect of lubricating oil', and 'The practical results of experimentation with the hydrodynamic theory of friction and its application to railroads and cotton mills'. This list of publications demonstrates two things: Petrov's wider appreciation of the importance of tribology in engineering and his great involvement in the problems of the Russian railways. He was also acutely aware, like his contemporary in the United States, Robert H. Thurston, of the economic significance of tribology. He wrote on the subjects of railway economics and the need to balance the cost of improved lubrication against the savings resulting from lower fuel consumption. He also pointed out the necessity of the link between technology and life.

Nikolai Pavlovitch Petrov died in 1920. He published some eighty papers during his full life and received numerous high awards. He received the gold medal of the Lord Heir of the Tsarevich, the highest award of the Imperial Russian Technological Society and he was elected an honourable member of the Moscow Polytechnic Society. He was undoubtedly one of the most distinguished Russian engineers of the period. In tribology, it is not just the intrinsic value of his work, but his contributions in the 1880s which can in retrospect be recognized, alongside those of Beauchamp Tower in England, as the historic corner-stones of the foundations of understanding of fluid-film lubrication.

Bibliography

Archibald, F. R. (1956) 'Men of lubrication: Nikolay P. Petrov', *Lubr. Enging.*, **12**, Jan.–Feb. 1956, 15–16, 72.

Cameron, A. (1966) *Principles of Lubrication* (Longman, London).

Dowson, D. (1974) 'Lubricants and lubrication in the nineteenth century', paper presented at a Joint Meeting of The Institution of Mechanical Engineers and the Newcomen Society.

Hersey, M. D. (1933) 'Notes on the history of lubrication; Part I. General survey', *J. Am. Soc. Nav. Engrs.*, **xlv**, No. 4, Nov. 1933, 411–29.

Leibenson, L. D. (1948) '*N. P. Petrov, Hydrodynamic Theory of Friction. Selected Works*' (Academy of Sciences, USSR, Moscow).

Ostwald, W. (1927) *Abhandlungen über die hydrodynamische Theorie der Schmiermittebreibung* (Akad. Verlag, Leipzig) (Ostwald's *Klassiker der exakten Wissenschaft*), No. 218, pp. 1–38.

Petrov, N. P. (1883) 'Friction in machines and the effect of the lubricant', *Inzh. Zh.*, St Petersburg, **1**, 71–140; **2**, 227–79; **3**, 377–436; **4**, 535–64.

Petrov, N. P. (1886) 'Friction in machines', *Izv. St. Pet. Prakt. Tekh. Inst.*, **5**, 1–438.

Petrov, N. P. (1887) 'Friction in machines', *Inzh. Zh. St. Pet.*, **1**, 83–145; **2**, 229–88.

Velichko, K. I. (1915) *Vojennaja Enciklopedija* (I. D. Sytinid, St. Petersburg). Vol. XVIII, p. 394.

Wurzel, L. (1887) *Neue Theorie der Reibung* (Leopold Voss, Leipzig).

A.9   Robert Henry Thurston   (1839–1903)

Robert Henry Thurston was born in Providence, Rhode Island, in the United States of America on 25 October 1839. His father, Robert Lawton Thurston, manufactured steam engines in Providence and provided his son with a workshop training. The young Robert Henry Thurston went to Brown University, where he graduated as a civil engineer in 1859. He then spent two years working for the firm in which his father was a senior partner before joining the Navy as officer of engineers in 1861. He served on various vessels throughout the Civil War and was present at the Battle of Port Royal and the Seige of Charleston. At the end of 1865 he was transferred to the Department of Natural and Experimental Philosophy at the United States Naval Academy at Annapolis and was for a time acting head of the Department. He accepted the Chair of Mechanical Engineering at the new Stevens Institute of Technology in 1870 and was much involved in the planning of the curriculum before the Institute opened in the autumn of 1871.

An important feature of his work at the Institute was that it formed at that time an unusual combination of research, instruction and commercial work. Most of his work was concerned with the mechanical properties of materials, but he also initiated his extensive studies of the friction of lubricated surfaces and developed his famous pendulum lubricant tester at the Stevens Institute. His recognition of the importance of friction and lubrication, both technically and economically, and his individualistic and enthusiastic preaching of the essential features of the arguments, won him renown on both sides of the Atlantic. His lectures on 'Friction and lubrication', published by the Railroad Gazette Publication Company of New York (Thurston, 1879), were most valuable in drawing attention to the significance of tribological topics in a rapidly developing technological society. His pragmatic approach did much to stimulate interest in the friction of lubricated surfaces and his work was well known to Petrov and Tower who, in 1883, were to confirm the fluid-film nature of well-lubricated journal bearings.

Thurston represented the United States at the International Exposition in Vienna in 1873 and he edited four large volumes of reports on the exhibition for the Government. A notable feature of his life was his energetic involvement in writing, teaching, research and the work of government committees and professional institutions. He published some 300 papers in scientific and technical journals. Perhaps his best-known contribution to the literature on tribology was his book, written while he was at the Stevens Institute, on *Friction and Lost Work in Machinery and Mill Work* (Thurston, 1885). The book ran to seven editions, the last one being published in 1903.

Professor Thurston was an active figure in the American Society of Civil Engineers and he was accorded the distinction of nomination as the first President of the American Society of Mechanical Engineers at the age of forty-one. He had been instrumental, along with Professor J. E. Sweet and Mr A. L. Holley, in calling the initial organizing meeting on 16 February 1880, and his presidential inaugural address was delivered at the first annual meeting held in the Union League Theatre (NY) on 4 and 5 November of the same year. He spoke with '. . . much diffidence, although with pride and pleasure' and clearly thought that the honour should have been bestowed upon one of the '. . . veterans of the profession, many of whom are with us in person, and more of whom are with us in spirit today'. He referred to what he described as a good saying '. . . old men for council and young men for war'. The objects of the Society was '. . . the promotion of the arts and sciences connected with engineering and mechanical construction . . . to publish and circulate papers of sufficient value . . . and the enlightenment of our national legislators in regard to the needs, the wishes and the legal and moral rights of the industrial classes in our country'. The presidential address provided a valuable insight into both the aims of the Society and the individual and wide-ranging views of Robert H. Thurston. Reference was made to ethics, the importance of the Society for all branches of industry, the reputation of the United States as the home of all ingenious and effective labour-saving devices, and aspects of mechanical and scientific philosophy. It is interesting to note how energetically and successfully the American Society of Mechanical Engineers strives today to achieve the objectives outlined by Thurston.

Thurston remained at the Stevens Institute of Technology for fourteen years, but in 1885 he accepted an invitation from the trustees of Cornell University to organize a course in mechanical engineering for the Sibley College of Engineering and Mechanic Arts. In due course Albert Kingsbury became one of Thurston's students and Sibley College developed at an early stage in its history a long-standing connection with various aspects of tribology. According to Hersey (1966)[1] Thurston held a consultancy with the Pennyslvania Railroad, and one aspect of his agreement called for the testing of new shipments of babbitt for the rolling-stock bearings. This was achieved by scraping a bearing of the babbitt to fit one of Thurston's testing machines and then determining the coefficient of friction. Kingsbury, who was skilled in machine-shop practice, did such a professional scraping job that the bearings on which he worked acted in a hydrodynamic manner and produced a coefficient of friction lower than anything Thurston has seen previously. In any event it seems clear that Thurston introduced Kingsbury to the subject of bearings and lubrication, but the cost was high, since it resulted in the loss of his consultancy with the railroad when all babbitt bearings suitably scraped henceforth produced the same results!

Robert Henry Thurston's marriage to his first wife Susan T. Gladding of Providence, RI, lasted from 1865 until her death in 1878. Two years later he married Leonora Boughton of New York. He died suddenly on his birthday in Ithaca, NY, at the age of sixty-four. During his lifetime he became a father figure in mechanical engineering in the USA and the founder of studies of bearings, friction and lubrication in that country.

---

1. I am grateful to Mr F. R. Archibald for drawing this to my attention.

Bibliography

Durand, W. F. (1924) *Robert Henry Thurston, A Biography* (ASME), 297 pp.
Hersey, M. D. (1966) *Theory and Research in Lubrication: Foundations For Future Developments* (Wiley, New York).
Stevens Institute of Technology (1895) 'Robert Henry Thurston, Ph.B., C.E., U.D., Professor of Mechanical Engineering 1871–1885', Morton Memorial to celebrate the twenty-fifth anniversary of the Stevens Institute of Technology, pp. 210–17.
Thurston, R. H. (1878) 'Friction and its laws', *Proc. Am. Ass. Adv. Sci.*, **27**, 61–71.
Thurston, R. H. (1879) 'Friction and lubrication', *The Railroad Gazette* (New York), 212 pp.
Thurston, R. H. (1880) 'President's inaugural address', *Trans. Am. Soc. mech. Engrs.* **1**, 1–16.
Thurston, R. H. (1885) *Friction and Lost Work in Machinery and Millwork* seventh edn. (Wiley, New York), 1903.

Portrait of Osborne Reynolds (1904) by Hon. John Collier.

A.10 Osborne Reynolds (1842–1912)

Few would deny the pre-eminence of Osborne Reynolds in the field of lubrication, while many hold him in similar esteem in the wider field of tribology. Archibald (1955) has described him as the founder of the science of lubrication, Barwell (1970) proposed him as the founder of modern tribology and Allen (1970) suggested that he was the most distinguished man ever to occupy a chair of engineering in any British university.

Osborne Reynolds was born in Belfast on 23 August 1842. He came from a clerical family and both his grandfather and great-grandfather had been Rectors of Debach-with-Boulge in Suffolk. His father, the Rev. Osborne Reynolds, also had a distinguished academic background, being thirteenth wrangler in 1837, subsequently a Fellow of Queens' College, Cambridge, Principal of the Belfast Collegiate School, Headmaster of Dedham Grammar School, Essex, and finally, in his turn, Rector of Debach-with-Boulge.

Reynolds received his early education from his father, first at Dedham Grammar School and later privately. He developed a strong interest and ability in mathematics in general and mechanics in particular and the warmth of his appreciation for his father's guidance can be seen in the following extract from his letter of application for the Chair of Engineering at Owens College, Manchester, dated 1868.

> *. . . From my earliest recollection I have had an irresistible liking for mechanics; and the studies to which I have specially devoted my time have been mechanics, and the physical laws on which mechanics as a science are based. In my boyhood I had the advantage of the constant guidance of my father, also a lover of mechanics, and a man of no mean achievements in mathematics and their application to physics.*

In 1861, at the age of nineteen, he entered the workshop of Mr Edward Hayes, a mechanical engineer at Stony Stratford, to begin a short apprenticeship. Clearly, Reynolds had developed a strong interest in engineering by this time, for the purpose of his year with Mr Hayes was to obtain practical experience before entering university, or as Mr Hayes himself expressed it, '. . . to learn in the shortest time possible how work should be done, and, as far as time would permit, to be made a working mechanic before going to Cambridge to work for Honours'.

This workshop experience strengthened Reynolds' resolve to study mathematics and mechanics and the impact of his industrial training is reflected by the second extract from his letter of application for the Manchester Chair.

> *. . . Having now sufficiently mastered the details of the workshops, and my attention at the same time being drawn to various mechanical phenomena, for the explanation of which I discovered that a knowledge of mathematics was essential, I entered at Queens' College, Cambridge, for the purpose of going through the University course, previously to going into the office of a civil engineer.*

Mr Hayes clearly knew that Reynolds was destined for Cambridge, but Sir Horace Lamb has suggested that the decision to proceed to the University may have been taken rather suddenly (Lamb, 1913). Reynolds had not previously studied Greek, but he nevertheless succeeded '. . . by the obstinate labour of a few weeks . . .' in reaching the standard of the '. . . Previous Examination'.

Reynolds' career at Cambridge was highly successful, for he graduated in 1867 as seventh wrangler, thereby closely emulating his father, and was immediately afterwards elected to a Fellowship at Queens' College. On leaving Cambridge, Reynolds entered the office of Mr John Lawson, of the firm of Lawson and Mansergh, civil engineers, of London, but he was to spend only a short time in the capital before being appointed to the newly established Professorship of Engineering in Owens College, Manchester, at the age of twenty-six.

Most of the early chairs in engineering in Great Britain and Ireland were occupied, sometimes on a part-time basis, by well-established practising engineers. It is therefore particularly relevant to recognize the wisdom of the selectors for the Manchester Chair, many of whom were highly successful men in industry and commerce, in recommending the appointment of such a young and relatively inexperienced engineer to the post. It must surely rank as one of the most successful gambles or inspired choices ever made by an appointing committee.

The move to establish the Manchester Chair had its formal origin on 11 December 1866, when a number of engineers from the Manchester area held a meeting in the Town Hall and resolved that,

> *. . . It is expedient to establish a professorship for civil and mechanical engineering, together with a special library, a museum of models, a drawing class, etc. in connection with and under the management of the trustees for the time being of Owens College; and that a subscription be at once entered into with a view to raise a sum of £10,000 for this purpose, and for such adjustments as the fund may be adequate to.*

By 1867 a sum of £9,505 had been raised and the Chair was advertised at a salary of £250 per annum. None of the eighteen applications were deemed to be suitable, but an immigrant from Saxony by the name of Charles Frederick Beyer, who had arrived in Manchester in 1834, undertook to supplement the salary and the post was readvertised at a stipend not less than £500 per annum. Reynolds was appointed to the Chair he was to hold throughout his active life on 26 March 1868. Initially the facilities were poor, the Engineering Department being accommodated in the stables of a house in Quay Street. The ground-floor stable was converted into a lecture room and the hayloft above it into a drawing office entered by an outside, uncovered wooden staircase. In 1873 the College moved to Oxford Road, but it was some years later that the experimental facilities, including apparatus conceived by Reynolds, were recognized to be excellent by the standards of the time. A most valuable account of the life and works of Osborne Reynolds, which includes a full description of the origins and history of Owens College, has been presented by Allen (1970).

Reynolds was a man of ingenuity and great intellect. Sir Horace Lamb has described the course which he established for his students as being '... remarkable for the thoroughness and completeness of the theoretical groundwork' (Lamb, 1913). Lamb also remarks that many students felt the course to be too severe and that Reynolds' lectures were not always easy to follow. Some support for this latter point can be found in the revealing words of Reynolds' most eminent student, J. J. Thomson, later Sir Joseph J. Thomson O.M., Nobel Laureate, President of the Royal Society and Master of Trinity College, Cambridge (Thomson, 1936).

> *... As I was taking the engineering course, the Professor I had most to do with in my first three years at Owens was Osborne Reynolds, the Professor of Engineering. He was one of the most original and independent of men and never did anything or expressed himself like anybody else. The result was that it was very difficult to take notes at his lectures so that we had to trust mainly to Rankine's text books. Occasionally in the higher classes he would forget all about having a lecture and after waiting for ten minutes or so, we sent the janitor to tell him that the class was waiting. He would come rushing into the room pulling on his gown as he came through the door, take a volume of Rankine from the table, open it apparently at random, see some formula or other and say it was wrong. He then went up to the blackboard to prove this. He wrote on the board with his back to us, talking to himself, and every now and then rubbed it all out and said that was wrong. He would then start afresh on a new line, and so on. Generally, towards the end of the lecture he would finish one which he did not rub out and say that this proved that Rankine was right after all.*

To some eminent practical engineers and other friends of the Owen College who had worked for the creation of the professorship, the individual and scientific approach adopted by the first holder of the Chair induced at first some shade of disappointment. Lamb (1913) mentioned these sentiments in his Royal Society obituary notice and wrote:

> *... Few could have forseen at that time how splendidly the appointment was destined to be justified, not only by the distinguished scientific career for which it served as a base, but also by the succession of students who derived stimulus and inspiration from the genius of their teacher, and who came afterwards to occupy important positions in professional as well as in the academical world.*

Osborne Reynolds developed firm views on the form of education and training suitable for engineers. While he believed in the need to relate subjects to the work which engineering students would ultimately be called upon to undertake, he took great pains to establish a course in which an understanding of the laws and principles of mechanics and an opportunity to apply these principles to practical problems were essential features. The course was systematic, notwithstanding the observations on lecturing techniques mentioned earlier, covering the fundamentals of civil and mechanical engineering in three years. His emphasis on the fundamentals and unity of the subject was noted by Lamb (1913):

> *. . . On one point he was uncompromising. In his mind all engineering was one, so far as the student is concerned, and the same fundamental training was required whatever the nature of the specialisation which was to come afterwards in practice. As an ideal principle this can hardly be gainsaid, although the varied ramifications of mechanical science, and the increasing multiplicity of 'subjects', have in more recent times compelled a deviation from it.*

The debate on this issue continues and it is interesting to speculate on the attitude which Reynolds might have adopted in the current situation, for there are still many advocates of essentially the same views.

To the right kind of student his lectures were stimulating if sometimes rather bewildering. He became increasingly convinced that laboratory work should be an essential feature of engineering training, and in due course the Manchester laboratories were noted for their excellence. The Whitworth Engineering Laboratory was opened in 1888 with several items of apparatus specially designed by Osborne Reynolds.

Many who studied engineering at Manchester found the course demanding yet stimulating. They found Osborne Reynolds uncompromising as a teacher yet friendly and kind to his students. Sir J. J. Thomson was to write, '. . . My personal relations with him when I was a student are a very pleasant recollection; he was always very kind to me, had a winning way with him and a charming smile.' He was described by Fiddes (1937),[1] as '. . . the kindly and eccentric Professor of Engineering'.

An interesting insight into Reynolds' personal qualities is provided by the penultimate paragraph of Sir Horace Lamb's obituary notice.

> *. . . The character of Reynolds was, like his writings, strongly individual. He was conscious of the value of his work, but was content to leave it to the mature judgment of the scientific world. For advertisement he had no taste; and undue pretensions on the part of others only elicited a tolerant smile. To his pupils he was most generous in the opportunities for valuable work which he put in their way, and in the share of credit which he assigned to them in cases of cooperation. Somewhat reserved in serious or personal matters, and occasionally combative and tenacious in debate, he was in the ordinary relations of life the most kindly and genial of companions. He had a keen sense of humour, and delighted in startling paradoxes, which he would maintain, half seriously and half playfully, with astonishing ingenuity and resource. The illness which at length compelled his retirement was felt as a grievous personal calamity by his pupils, his colleagues, and by other friends throughout the country.*

Reynolds established a distinguished reputation for his original research during the thirty-seven years that he held the Manchester Chair.

---

1. See Allen, J. (1970), p. 6.

Between 1901 and 1903 three volumes of his collected *Papers on Mechanical and Physical Subjects*, which contained most, but by no means all of his writings, were published by the Cambridge University Press. These volumes have been of great value to subsequent research workers, but the compliment extended by the Syndics of the University Press was also greatly appreciated by Reynolds. Of the 68 collected works more than one-third had been communicated to the Manchester Literary and Philosophical Society, 13 had been published in the *Proceedings of the Philosophical Transactions of the Royal Society*, 15 had been presented to the British Association for the Advancement of Science and a further 9 had been published in engineering journals including the *Proceedings of the Institution of Civil Engineers* and the *Transactions of the Institution of Naval Architects*.

The papers were concerned with a wide range of physical and engineering problems, but there is a consistency of approach and an underlying affinity between many of the topics considered. In many of his writings Reynolds demonstrated his great skill in unravelling and explaining a mass of detail in terms of simple, well-established mechanical principles. He did not hesitate to apply his considerable mathematical skill to this end, although Lamb (1913) has noted that, '. . . he sometimes affected, not yet seriously, to despise mathematics.'

Nevertheless, Reynolds sought to isolate and demonstrate the essential physical principles underlying a practical problem, and the careful balance between experimentation and necessary mathematics is evident in much of his work. He always sought a simple explanation of a phenomenon, although his style of writing has not escaped criticism.[2] Sir J. J. Thomson explained that,

> . . . *The novelty of his method of approach made his papers very hard reading—in fact I think it is probable that some of them have never been read through by anyone.*
>   . . . *This paper (On Certain Dimensional Properties of Matter in the Gaseous State) is very difficult reading, so much so that a severe criticism of certain parts of it by Professor G. F. Fitzgerald . . . was shown by Reynolds to be based on a wrong interpretation of his meaning.*

Reynolds himself was not unaware of the difficulty and on at least one occasion[3] sought the advice of his highly respected friend Sir George G. Stokes. However, Lamb (1913) has pointed out that when Reynolds took up a topic on which he had written previously for a second time, with a view to explaining it to a popular audience, he was both lucid and forcible.

It is customary today, as indeed it was in the nineteenth century, to advise research workers to commence their study of a problem with a thorough review of the literature. There is ample evidence to show that Reynolds kept abreast of the publications of others in the field of mechanics, as the excitement which he experienced on reading Beauchamp Tower's account of his experiments on journal bearings fortunately

---

2. See Lamb, Sir H. (1913), p. xix, Kingsbury, A. (1932), and Cameron (1966), p. 271.
3. See Allen, J. (1970), p. 33.

demonstrates, but it is also clear that he preferred to keep his mind un-fettered by the views of others once he had decided to apply himself to a particular problem. The approach was well described by Sir J. J. Thomson,

> ... *When he took up a problem, he did not begin by making a bibliography and reading the literature about the subject, but thought it out for himself from the beginning before reading what others had written about it. There is, I think, a good deal to be said for this method. Many people's minds are more alert when they are thinking than when they are reading, and less liable to accept a plausible hypothesis which will not bear criticism.*

Cameron (1966) has expressed some surprise that Reynolds missed the direct integration of the trigonometric function $d\theta/(1 + \varepsilon \cos \theta)^3$ which occurred in the analysis of journal-bearing lubrication, particularly since the integral had already been given in Todhunter's standard textbook on integral calculus published some ten to twelve years previously. Perhaps on this occasion Reynolds' individual approach to research worked to his disadvantage and caused him much unnecessary labour.

Most of Reynolds' publications were in the general field of mechanics with the emphasis being on hydrodynamics. In the field of tribology his name is linked primarily with his classical formulation of the theory of fluid-film lubrication (Reynolds, 1886) and his earlier study of rolling friction (Reynolds, 1875). The former is a wonderful example of his ability to unravel a mass of experimental data and to provide a physical under-standing of an important practical problem. It is worth noting that this outstanding contribution to tribology followed closely upon his famous experiments which revealed the transition from laminar to turbulent flow in pipes in 1883.

His interest in friction had been aroused long before he undertook his famous work on lubrication. As early as 1873 he discussed the influence of friction upon the work done in giving rotation to the shot in rifled guns, and the following year he wrote on the efficiency of belts as communicators of work. The latter study appears to have provided the essential ingredients for his subsequent work on rolling friction. His most comprehensive experimental study was concerned with the determination of the mechanical equivalent of heat.

A paper worthy of special mention is his delightful explanation of the slipperiness of ice. It reflects the excitement of research, the ability of a researcher to utilize observations in one field in the explanation of phenomena in others, the frankness and modesty of Reynolds and the final frustration, when all the concepts have fallen into place, that the beautiful and simple truth had not been recognized much earlier. In this case it was the chance sliding of a hot iron on solder that caused Reynolds to wonder whether the low friction encountered by skaters could be attributed to the melting of ice under pressure and the subsequent lubrication of the skate by water. It was in 1886, when Beauchamp Tower's experiments on friction and his own theory of lubrication were very much in his mind, that Reynolds took note of the behaviour of hot soldering-irons. He convinced himself of the explanation of the slipperiness of ice, evaluated the area of skate required to depress the melting point of ice under the weight of a man, and noted with obvious satisfaction after waiting in vain for a winter of sufficient severity in England, '. . . a casual but emphatic statement by Dr Nansen, in his book on Greenland, that at the low temperatures he there encountered the ice completely lost its slipperiness.'

Reynolds took a lively interest in the affairs of the Manchester Literary and Philosophical Society. Many of his papers were first presented to the Society and he was the Dalton Medallist in 1903. He acted as Secretary to the Society from 1874 to 1883 and he was elected President for the Session 1888–89. He wrote a fine memorial volume for James Prescott Joule, a previous President of the Society for whom he had the highest admiration, and he was the leading spirit in the movement for the public monument by Gilbert which was placed in Manchester Town Hall.

He was also an active participant in the British Association for the Advancement of Science and he served as President of Section G (Mechanical Sciences) for the Manchester meeting of 1887. His delightful explanation, which apparently gave great delight to Lord Kelvin, of the drying of the space around a foot placed upon moist sand on the seashore and the wetness of the footprint immediately afterwards, was presented at the 1885 meeting in Aberdeen in a paper entitled 'On the dilatancy of media composed of rigid particles in contact'.

Reynolds is reported to have presented two papers to the 1884 meeting of the British Association in Montreal: one in Section G (Mechanical Sciences) entitled 'On the friction of journals' and the other in Section A (Mathematical and Physical Societies), 'On the action of lubricant'. It is a matter of great regret and considerable frustration that no record of these papers has been found. A brief and tantalizing report of these presentations appeared in the Montreal *Daily Witness* of 3 September 1884: '. . . Professor Osborne Reynolds gave a paper on "The Friction of Journals" which was entirely theoretical and only of interest to the initiated'.

Professor Osborne Reynolds was elected a Fellow of the Royal Society in 1877 and he was awarded a Royal medal in 1888. He received his Honorary Fellowship of Queens' College, Cambridge in 1882 and was elected to Membership of the Institution of Civil Engineers in 1883. Two years later the Institution awarded him a Telford Premium for his paper on steam-engine indicators. The University of Glasgow conferred the degree of LL.D. *honoris causa* on him in 1884.

By the time Reynolds completed his long memoir on the 'Submechanics of the universe', in 1902, illness had already begun to impair his powers of expression. Failing health finally caused him to withdraw from active work and to retire from his Chair in 1905. His last years were spent in retirement at Watchet, Somerset, where he died on 21 February 1912 at the age of sixty-nine. He was married twice: first to a daughter of Dr Chadwick of Leeds in June of the year in which he was appointed to the Owens Chair (1868) and then to a daughter of the Rev. H. Wilkinson, Rector of Otley, Suffolk in December 1881. His first wife died in July 1869 and a son by this marriage died in 1879. Reynolds was survived by three sons and a daughter of his second marriage. One of his sons graduated at

Manchester in 1908 and later held the Vulcan and Osborne Reynolds Fellowships.

In 1919, Albert Kingsbury of Pittsburgh in the United States of America founded the Osborne Reynolds Research Fellowship in the University of Manchester '. . . as some recognition of the debt which he owed to Reynolds' researches in lubrication.'

In this biographical sketch I have endeavoured to convey something of the genius, personality and humility of Osborne Reynolds. I do not disagree with the assessment by others of his pre-eminent position and this alone justifies the space devoted to the life of this great man in this limited text. The illustration shown at the beginning of the section is based upon the portrait in oils by the Hon. John Collier, painted in 1904 on a commission by a number of Reynolds' scientific friends. This admirable portrait now hangs in the Simon Engineering Laboratories and I am grateful to Professor J. Diamond for making the print available to me.

Bibliography

Allen, J. (1970) 'The life and work of Osborne Reynolds', in *Osborne Reynolds and Engineering Science Today* (Manchester U.P.), 1, pp. 1–82.

Archibald, F. R. (1955) Men of lubrication–Osborne Reynolds, *Lubr. Enging*, **11**, Mar.–Apr. 1955, 84–5, 128–9.

Barwell, F. T. (1970) 'The founder of modern tribology', in *Osborne Reynolds and Engineering Science Today* (Manchester U.P.), ch. 10, pp. 240–63.

Cameron, A. (1966) *The Principles of Lubrication* (Longman, London).

Centenary Symposium, 1968 (1970) 'Osborne Reynolds and Engineering Science Today' (Manchester University Press). (Papers presented at the Osborne Reynolds Centenary Symposium, University of Manchester, September, 1968).

Fiddes, E. (1937) 'Some teachers at Owens College', *J. Univ. Manchr*, No. 1, **1**.

Inst. Civil Engineers (1913) 'Obituary notice (on O. Reynolds)', *Proc. Inst. Civ. Engrs*, **cxci**, 314.

Kingsbury, A. (1932) 'Optimum conditions in journal bearings', *Trans. Am. Soc. mech. Engrs*, **54**, 123–48.

Lamb, Sir H. (1913) 'Osborne Reynolds, 1842–1912' (Obituary Notices of Fellows Deceased), *Proc. Roy. Soc.*, A **lxxxviii**, xv–xxi.

Reynolds, O. (1900–3) *Papers on Mechanical and Physical Subjects* (*collected works*), Vol. I (1900); Vol. II (1901); Vol. III (1903) (Clay, London).

The Engineer (1944) 'Historic researches: No. III–Friction; Reynolds analysis', *The Engineer*, **178**, 28 July, 60–2.

Thomson, Sir J. J. (1936) *Recollections and Reflections*, (Bell, London).

A.11 John William Strutt, third Baron Rayleigh

(1842–1919)

John William Strutt, third Baron Rayleigh, was born at Langford, near Maldon, Essex on 12 November 1842. His father, John James, the second Baron Rayleigh, of Terling Place, Essex, was educated at Winchester and Oriel College, Oxford. The second Baron, who had retired from the Eastern Battalion of the Essex Militia with the rank of major, was a country squire of strong religious inclinations.

It is certain that the Strutt family had lived in Essex since 1660, and they are known to have been located in Terling since 1720. The family owned several water-mills in the vicinity of Maldon and had acquired its wealth by milling corn. In 1761 John Strutt bought the manor of Terling Place. John William Strutt's grandfather, Joseph Holden Strutt (1758–1845) and great-grandfather, John Strutt (1727–1816) were both Members of Parliament. The Rayleigh peerage dated from 1821, when it was bestowed upon the wife of Colonel Joseph Holden Strutt as Baroness Rayleigh in recognition of her husband's public services. The territorial title was taken from Rayleigh, a small market town in Essex.

John William Strutt's mother was Clara Elizabeth La Touche Vicars. She lived with her widowed mother at Shenfield, about 20 miles (32 km) from Terling, and she married Lord Rayleigh at the age of seventeen in 1842. William Strutt was born in November of the same year, being a seven months' child. Clare Elizabeth's family were of Irish origin and there is greater evidence of scientific and engineering background on the maternal than on the paternal side of John William Strutt's ancestry.

Lord Rayleigh's early education was greatly affected by ill-health. He went to Eton at the age of ten and was left in the care of the Rev. J. W. Hawtrey on 12 April 1853. Illness, including a mild attack of smallpox, caused him to return to his home and a private tutor after a few months. He later went to Mr George Murray's school at Wimbledon Common and then, at the age of fourteen, to Harrow. During his second term at Harrow, anxiety arose about his chest trouble and he was once again removed from school on the grounds of serious ill-health. In the autumn of 1857 he was placed under the care of the Rev. G. T. Warner, who took pupils at Highstead, Torquay. It was here that he received his final preparation for entry to university, although he also studied mathematics under a scholar of Trinity College by the name of Frederick Thompson during the summer of 1861.

In October 1861 Lord Rayleigh entered Trinity College, Cambridge, as a fellow commoner. Three years later he was successful in the competition open to all undergraduates for the Sheepshanks Astronomical Exhibition. He took the degree of Bachelor of Arts as senior wrangler in 1865, and the degree of Master of Arts in 1868. Four years elapsed before the publication of his first paper, but in the next decade he made contributions to almost every branch of physics. His major work culminated in the publication in two volumes of his *Theory of Sound* (Lord Rayleigh, 1877, 1878).

At the age of thirty-seven Lord Rayleigh succeeded Clerk Maxwell as head of the Cavendish Laboratory. He was elected on 12 December 1879 and he took up his post at the beginning of the following Lent term. The Laboratory expanded considerably under his guidance and by the time he resigned in 1884 the reputation of the Cavendish as a research institution was firmly established. Lord and Lady Rayleigh returned to Terling in Essex on 13 December 1884. The family home became the head-quarters of his scientific activity for the rest of his life, a converted stable loft being the laboratory in which many famous experiments were to be conducted.

In November 1885 Lord Rayleigh succeeded Sir George Stokes as Secretary of the Royal Society. Two years later he was appointed to the

Professorship of Natural Philosophy at the Royal Institution. Between 1871 and 1919 he published at a steady and impressive rate. The move from Cambridge to Terling and the inevitable accumulation of outside duties and responsibilities failed to disturb his output. In all he was to publish almost 450 papers, now contained in six volumes of his *Collected Papers*, and Sir J. J. Thomson's commentary on the feat in a memorial address in Westminster Abbey is of interest. '. . . Among the 446 papers which fill these volumes, there is not one that is trivial, there is not one which does not advance the subject with which it deals, there is not one which does not clear away difficulties, and among that great number there are scarcely any which time has shown to require correction.'

Lord Rayleigh's knowledge of the literature of science was extensive and thorough and his own papers contained frequent references to historical matters. His personal, colloquial, yet direct style of writing adds greatly to the pleasure of reading his work. How refreshing it is to read a scientific paper which opens in the manner of the introduction to Lord Rayleigh's (1918) paper on the effect of surface contaminants upon the friction between solids.

> . . . *A cup of tea, standing in a dry saucer, is apt to slip about in an awkward manner, for which a remedy is found in the introduction of a few drops of water, or tea, wetting the parts in contact. The explanation is not obvious, and I remember discussing the question with Kelvin many years ago, with but little progress.*

A dominant feature of Lord Rayleigh's research and writing was the directness of his approach. This is so well reflected in the clarity of his papers and in a remark he made to John Aitken in 1917: '. . . A good instinct and a little mathematics is often better than a lot of calculations.'

In an appreciation of his work on hydrodynamics Professor Horace Lamb[1] wrote: '. . . His mathematical analysis never lost touch with realities, so that the lucid presentation of the nature of the problem, and the simple but precise reasoning along which the argument proceeds, allows even the non-mathematical reader to appreciate the bearing of the results, and the limitations to which the problem is subject.'

Rayleigh's great skill and thoroughness as an experimental physicist was probably best illustrated by his discovery of the rare gas, argon, in the atmosphere. The painstaking work was reported to the Royal Society by Lord Rayleigh and Sir William Ramsay in January 1895. In 1904 he was awarded the Nobel Prize for Physics in recognition of his work on the densities of the permanent gases and his discovery of argon. He characteristically presented the proceeds of the prize to the University of Cambridge for an extension of the Cavendish Laboratory.

Lord Rayleigh's contribution to contemporary science was not restricted to his vast, personal contributions to physical research. He was extremely active in the learned societies and on committees for government and various professional bodies. He was a member, Professor of Natural Philosophy and later Honorary Professor of the Royal Institution; Fellow of the Royal Astronomical Society of London; Fellow,

---

1. See Schuster, Sir A. (1921), pp. i–l.

Royal medallist, Secretary, Copley medallist, President and Rumford medallist of the Royal Society; Fellow of the Royal Society of Edinburgh; honorary member of the Institutions of Mechanical and Civil Engineers; member of the Cambridge Philosophical Society and associate, honorary or foreign member of innumerable scientific societies overseas. He became Chancellor of the University of Cambridge in 1908 and received honorary doctorates from the universities of Cambridge, Oxford, McGill, Edinburgh, Heidelberg, Glasgow, Dublin, Erlangen, Victoria (Manchester), Christiania, Birmingham, Leeds and Durham. In 1902 he was among the first recipients of the Order of Merit. In 1900 he declined a pressing invitation to become President of the Royal Society on the grounds that he would be as good a president ten years later when he did not then expect to be equally productive as a man of science. He succumbed, however, five years later.

At the turn of the century Rayleigh was much involved in the creation of the National Physical Laboratory, and during the First World War his advice was frequently sought and respected. He accepted the position of Scientific Advisor to Trinity House in 1896, and became Chief Gas Examiner in 1901. His scientific writings on the soaring of birds made him a natural choice for the presidency of a committee set up by the Prime Minister, Mr Asquith, in 1909 to advise on aeronautical matters. The first report from the Advisory Committee on Aeronautics drew attention to the important conclusions that would be drawn from a careful application of the principle of 'dynamical similarity', thus opening the way for wind-tunnel tests on models of flying machines.

Lord Rayleigh's work on lubrication was quite insignificant compared with his great and wide-ranging contributions to physics as a whole. Yet the researches which he undertook and the papers which he wrote on such subjects as the optimum profile of fluid-film thrust bearings, hydrostatic lubrication, surface actions and boundary lubrication are now recognized as early and classical contributions to tribology. Furthermore, he soon recognized the importance of studying friction and lubrication and actively promoted work on the subject by the Institution of Mechanical Engineers and the Department of Scientific and Industrial Research (DSIR). He was a member of the Institution of Mechanical Engineers' Committee on Friction at High Velocities which promoted Beauchamp Tower's famous discovery of high pressures in the oil-films of journal bearings and a consultative member, along with Sir Charles Parsons, of the DSIR Lubricants and Lubrication Inquiry Committee established in 1917. He also participated in that scene of tribological history enacted in Montreal in 1884, when Osborne Reynolds first presented his equation of fluid-film lubrication and Rayleigh referred to the interest of Stokes in the matter. His contributions to lubrication theory and his friendship with Beauchamp Tower are considered in Chapter 10. It appears that Lord Rayleigh first met Tower at Weald Hall, Brentwood in the summer of 1875. They took to each other at once[2] and Rayleigh invited Tower to

---

2.   See Lord Rayleigh (Robert John Strutt) (1924) p. 71.

come and assist him in his laboratory at Terling. Tower spent some months at Terling in the winter of 1875/76, living initially in the house and later at the Rayleigh Arms inn. It appears that Rayleigh acquired most of his knowledge of practical metal-work from Tower. Tower was appointed by the Institution of Mechanical Engineers for the study of journal-bearing friction in 1882 and it seems quite possible that the confidence Lord Rayleigh developed in Tower's practical skills could have influenced the selection.

Lord Rayleigh appears to have been a kindly man with a quiet sense of humour. He was always anxious to hear about the views and work of others and it was not in his nature to make personal criticism. He was interested in political questions of the day and, interestingly, in psychical research. He first experimented in the latter field in 1874 and late in life became President of the Society for Psychical Research. For the reader who is interested in learning more about this great man, good starting points are provided by the obituary notice by Sir Arthur Schuster (Schuster, 1921) and the full biography by his son, Robert John Strutt (Lord Rayleigh, 1924). These references make delightful reading and generate a suitable respect for his scientific studies and a warm appreciation of the human qualities of John William Strutt, Lord Rayleigh.

John William Strutt married Evelyn, daughter of James Maitland Balfour, and sister of the Rt. Hon. Arthur Balfour, on 19 July 1871. He assumed the title Lord Rayleigh on the death of his father in 1873. Their first son, Robert John Strutt, was born on 28 August 1875. His youngest son, Williams Maitland, was struck down by a spinal disease while preparing for the bar and he died on 22 November 1912.

Lord Rayleigh died at Terling on 30 June 1919 at the age of seventy-six. The grave is in a family plot in the churchyard adjoining the garden of Terling Place. A memorial tablet was later erected in Westminster Abbey and unveiled by Sir J. J. Thomson on 30 November 1921. The inscription reads:

John William Strutt, O.M., P.C., 3rd Baron Rayleigh, Chancellor of the University of Cambridge 1908–1919, President of the Royal Society, 1905–1908. An unerring Leader in the Advancement of Natural Knowledge.

Bibliography

Archibald, F. R. (1955*f*) 'Men of lubrication–John William Strutt', *Lubr. Enging,* **11**, Nov.–Dec., 375, 420, 422.

Lord Rayleigh (John William Strutt) (1877, 1878) *The Theory of Sound,* Vol. 1 (1877); Vol. 2 (1878) (Macmillan, London and New York).

Lord Rayleigh (John William Strutt) (1918) 'On the lubricating and other properties of thin oily films', *Phil. Mag. and J. Sci.,* **xxxv**, 157–63.

Lord Rayleigh (John William Strutt) (1918*a*) 'Notes on the theory of lubrication', *Phil. Mag. J. Sci.,* **35**, No. 205, 1–12.

Lord Rayleigh (Robert John Strutt) (1924) *John William Strutt; Third Baron Rayleigh* (E. Arnold, London).

Schuster, Sir A. (1921) 'John William Strutt, Baron Rayleigh' (Obituary Notices of Fellows Deceased), *Proc. R. Soc.,* *A***98**, i–l.

A.12 Beauchamp Tower (1845–1904)

Few tribologists are unaware of the catalytic effect upon the science of fluid-film lubrication created by the direct and memorable observation of pressures in a half-bearing by an engineer by the name of Beauchamp Tower (Tower, 1883).

Born on 13 January 1845 at Moreton near Epping in Essex, he was christened Beauchamp on 16 February[1] after his father, Robert Beauchamp Tower, Rector of the parish. The lectern bible in use in the church in the early 1960s carried the name of Beauchamp's father and the date 1868, but it appears that it has now been removed and its whereabouts is unknown. His grandfather, Christopher Thomas Tower (1775–1867) had married Harriet, daughter of Sir Thomas Beauchamp-Proctor in 1803 and this appears to explain how the somewhat unusual Christian name of Beauchamp entered the Tower family tree.

Beauchamp, who had one brother and sister, was educated at Uppingham School under Dr Thring. He was resident in West Deyne, one of the boarding houses of the school, from 1859 to 1862.[2] Little is known about the factors which shaped his early life, but by the age of sixteen he had decided to become an engineer. He received his training in the Armstrong Engine Works, Elswick, Newcastle upon Tyne, and stayed for a few months as a draughtsman after completing his four-year pupilage. He left the company in April 1866 to take charge of the construction of a number of iron steamers at the Tyne Iron Works.

In 1869 Tower became assistant to that distinguished investigator, Mr William Froude, F.R.S., who is remembered primarily for his pioneering studies of the nature and magnitude of ship resistance. Froude was engaged in the preparation of plant for the Admiralty Experimental Works which was then being set up at Torquay, and Tower was responsible for the design of a large part of the apparatus. During this period Tower invented a speed indicator which was fitted to several ships in the Royal Navy. Froude imparted to Tower great knowledge and enthusiasm for studies of hydraulic problems in the three-year period from 1869 to 1872. In addition to his official employment during the day, Tower worked extremely hard at schemes of his own during his leisure time. The strain became so great that his health gave way and he had to leave Mr Froude in June 1872 to take a year's trip in a sailing vessel to the South Sea Islands.

On his return to England, Tower undertook an extensive series of experiments on torpedoes for Sir William Armstrong and Co. during 1874 and 1875. During the summer of 1875 he first met John William Strutt, Baron Rayleigh, at Weald Hall, Brentwood, home of Mrs Christopher Tower,[3] Lord Rayleigh and Beauchamp Tower appear to have taken to each other at once and Lord Rayleigh soon recognized

1. See Cameron, A. (1966), p. 271.
2. I am grateful to Mr Macdonald, Headmaster of Uppingham School, for this background information.
3. See Sect. A.11 and Lord Rayleigh (Robert John Strutt) (1924).

Tower as the ideal person to assist him in the development of his private laboratory at his home at Terling in Essex. He wrote to Tower about the matter and the latter agreed on the understanding that '... his professional prospects would not allow of more than a short stay'. He appears to have stayed for a few months during the winter of 1875/76, spending some of the time living in Lord Rayleigh's home and some in the village at the Rayleigh Arms.

During his stay at Terling, Tower assisted Lord Rayleigh with his classical experiments on sound, installed for the first time a foot-lathe and taught Lord Rayleigh the essentials of practical metal-work, built a *hydraulics laboratory* in the grounds in close proximity to the Swan Pond, participated in some of the fluid-flow experiments and continued model experiments on his scheme for obtaining motive power for shipborne machinery from the motion of a ship upon the waves. The possibility of utilizing wave power had originated with Mr Spencer Deverell of Portland, Australia, and his brother, Mr W. Deverell, had shown such confidence in the scheme that he returned home from the Antipodes to promulgate the idea. It clearly caught Tower's imagination and he worked on the idea with the assistance and interest of Mr Froude. The general concept was to mount a heavy mass on springs on board ship and to utilize the relative movements between the mass and the hull to cause irregular or continuous circular motion of a drive shaft attached to the propeller screw. A 6 ft (1·83 m) long, 16 in (0·41 m) broad model boat of displacement 80 lbf (356 N) was filled with a 7 lb (3·2 kg) mass on a long leaf spring running fore and aft and tested with some success in a 30 ft (9·1 m) tank. Mr Froude was present when the same model was tested at sea in Torbay and when waves having a period similar to the natural frequency of the spring arrangement hit the model the machine '... worked with great vigour, causing the screw to go at the rate of about 150 revolutions per minute, but at other times it almost stopped.'

At a later stage the flexing beam connected to double-acting ratchet pawls was replaced by a large mass on a rotating arm. Tower (1875) presented a paper on his wave-power machine to the Institution of Naval Architects on 20 March 1875, and while the idea created interest, several of the speakers in the discussion expressed doubts about the practicalities of the system. Lord Rayleigh never had any confidence in the scheme as a practical arrangement and events appeared to prove him right. The device never developed to the stage of a full-scale trial, but the story is interesting since it illustrates through Tower's passion and enthusiasm for the scheme all the characteristics of his inventive mind. It is interesting to note that 100 years after Beauchamp Tower's energetic approach to the subject, wave power is once again a subject of considerable interest to engineers.

In 1877 Tower returned to Mr Froude and was involved in the design of a turbine dynamometer for marine engines which was described in a paper read at the Bristol meeting of the Institution of Mechanical Engineers later that year. Tower was also intimately involved in many of the beautiful experiments conducted by Froude for the Admiralty on the models of ships and on full-size ships and engines for the Navy. Froude's health began to fail soon after Tower rejoined him and he asked his good friend and associate to join him on a voyage to the Cape of Good

Hope. Froude died at the Cape and Tower returned to England in 1878 to enter practice on his own account. It was, of course, at the end of this year that the Council of the Institution of Mechanical Engineers established its Research Committee for Experimental Research, which in turn created its Committee for the Investigation of Friction at High Velocities, although it was not until 1882 that Mr Beauchamp Tower entered this particular scene.

In his private practice in Westminster, Tower invented or improved several interesting devices. The ones which dominated much of the second half of his life were his well known *spherical engine* and his *gyroscopically controlled steady platform* for guns and searchlights at sea. Dynamos for the provision of electric light were being introduced rapidly in the 1880s and engineers were faced with the problem of building steam engines suitable for driving these high-speed machines. American and continental European engineers chose to use belt drives from conventional engines, but English engineers correctly anticipated that the long-term solution would be found in the direct coupling of engine and dynamo. This called for engine speeds much in advance of current practice, and various experiments were conducted with ingenious forms of rotary engine. Tower designed and developed a spherical steam engine of great merit which was later promoted by a syndicate including Sir Frederick Bramwell. Before the agreement was completed Tower had to demonstrate that the engine could run satisfactorily for 3 million revolutions, and once this had been achieved a considerable number of engines were manufactured. The engines found application at sea where they were used by the Admiralty for lighting ships and, together with the dynamos to which they were coupled, they apparently formed the lightest units then available. They were also used as dynamo engines on railway locomotives and, being built in sizes up to 12 in (0·3 m) diameter, they developed up to 34·9 hp (26 kW) at speeds up to 1,100 rpm (18 Hz) and mechanical efficiencies of about 80 per cent.

In 1882 Tower was appointed to carry out the study of journal-bearing friction on behalf of the Institution of Mechanical Engineers which was to be responsible for his outstanding qualifications as a contributor to the science of tribology. He continued this work from 1887 to 1891. It might be thought that his all-important observation of high pressures in the clearance space of lubricated journal bearings could be attributed to chance, but his very thorough approach to the problem in hand, his inventiveness and skill as an experimenter and the soundness of his training and experience with men of the calibre of Froude and Lord Rayleigh make it far more likely to be the outcome of careful observation by an engineer skilled in experimental research. The details of this period in Tower's life and his contributions to tribology are presented in Section 10.4, but it is worth while linking the story to the discussion in the previous paragraph. Tower used one of his spherical engines to drive his journal-bearing testing machine at the Edgware Road Station, Chapel Street Works of the Metropolitan Railway.[4]

---

4. See Fowle, T. I. (1975), pp. 4–5.

Furthermore, it appears that the engine was one of the first to use a mechanical lubricator. The bearing pedestal on which the drive shaft was mounted contained an oil reservoir and a plunger pump. A ratchet mechanism drove the pump intermittently and oil was forced through drilled passageways to reach the working parts. The engine was described by Heenan (1885) in the *Proceedings of the Institution of Mechanical Engineers* and in the discussion Beauchamp Tower himself used the term *forced lubrication* to describe the mechanical lubricator.

In 1889 Tower presented a paper to the Institution of Naval Architects in which he described his apparatus for providing a steady platform for guns, etc. at sea. The paper attracted considerable attention, particularly when contributors to the discussion like H. M. Brunel and R. E. Froude spoke of a tilt of only $\frac{1}{2}°$ (0·005 rad) on the platform when the ship experienced a total roll of 30° (0·52 rad). Tower himself described the efficacy of the device in a persuasive manner in the last sentence of his paper, '. . . The sensation of calmly sitting on the machine at one's ease, and seeing the ship roll about under one and everybody else holding on is a very novel one.'

The $\frac{1}{2}°$ error was attributed by Tower to an error in a spherical bearing in which he had,

> . . . *unfortunately, tried a bell-metal ball in a phosphor bronze cup, which turned out to be a most unfortunate combination of materials. The bearing was continually heating, and I had to keep on continually oiling it. I believe it seized three or four times, and the machine actually stopped from the fact that the bell metal was continually abraiding the phosphor bronze cup. I found that the oil which came away from it was completely charged with brass.*

Tower had great hopes for this particular invention and after thorough testing it was installed by the Admiralty in two gunboats. The platform worked well and Tower was encouraged to devote his time to further experiments. However, in due course the invention was rejected by the Admiralty, apparently because it was felt that the sacrifice in weight which could otherwise be used for extra guns or ammunition was too great. The labour and ingenuity which Tower had devoted with characteristic determination and enthusiasm brought him no reward, and his treatment by the Admiralty, which he regarded as a breach of faith, cast a cloud over the closing years of his life. He nevertheless proceeded in such discouraging times to adapt the platform for passenger seats on cross-Channel steamers. He was still engaged on this work at the time of his death in 1904, and many uneasy travellers could have wished that he had lived to bring his invention into service.

Beauchamp Tower was a member of the Institutions of Mechanical and Civil Engineers (1884) and the Institution of Naval Architects (1884). He married late in life in the year 1902, his wife Mary Alice being the daughter of Edward Christopher Egerton, sometime Member of Parliament for Macclesfield and Cheshire East.

Tower died suddenly of a brain haemorrhage on 31 December 1904, when in his sixtieth year.[5] He lived at Hillstead, Cornsland, Brentwood,

---

5. See Cameron, A. (1966), p. 271.

in Essex, and was cycling to Wealdcote, South Weald, to visit his mother when he was struck down at the bottom of the hill in Weald Lane. He was taken to a nearby cottage and then to his home but died within an hour. The funeral took place on Thursday, 5 January 1905 and he is buried at South Weald Church, Brentwood.[6]

According to the *Essex* (now *Romford*) *Times* of 4 January 1905, Beauchamp Tower was extremely quiet and retiring in his nature, yet of a very affable spirit and highly esteemed by all who knew him. Another curious and interesting point is that he was said to have been closely connected with the designing and erection of the famous Tower Bridge across the Thames in London, as one of the architects.

Beauchamp Tower will forever be remembered as the man who provided the stimulus for analytical studies of fluid-film lubrication through his inspired experimental work. He was an outstanding engineer, an experimenter of great repute and an inventor of notable flair.

Bibliography

Archibald, F. R. (1955) 'Men of Lubrication–Beauchamp Tower', *J. Am. Soc. lubr. Engrs*, **13**, 13 and 63.
Cameron, A. (1966) *Principles of Lubrication* (Longman, London).
Engineering (1905) 'The Late Mr Beauchamp Tower' (obituary) *Engineering*, 13 Jan. 1905.
Essex Times (1905) 'Sudden Death of Mr Beauchamp Tower' (obituary notice), *Essex Times*, 4.1.1905.
Fowle, T. I. (1975) 'Beauchamp Tower and Joseph Tomlinson', *Tribology News*, Issue 28, Dec. 1975, pp. 4–5.
Heenan, R. H. (1885) 'Tower spherical engine', *Proc. Inst. mech. Engrs*, 96–120.
Inst. Civil Engineers (1905) 'Beauchamp Tower' (obituary notice), *Minutes of Proceedings Inst. Civil Engrs*, **162**, 420.
Inst. Mechanical Engineers (1905) 'Beauchamp Tower' (obituary notice), *Proc. Inst. mech. Engrs*, **1**, 163.
Lord Rayleigh (Robert John Strutt) (1924) *John William Strutt; Third Baron Rayleigh* (E. Arnold, London).
The Engineer (1944) 'Historic researchers: No. 11–Friction–Tower's experiments', *The Engineer*, **178**, 21 July, 40.
The Times (1905) 'Beauchamp Tower' (obituary notice) *The Times*, 7.1.1905.
Tower, B. (1875) *On a Method of Obtaining Motive Power From Wave Motion* (Institution of Naval Architects), pp. 1–12.
Tower, B. (1883) 'First report on friction experiments', *Proc. Inst. mech. Engrs*, 632–59.
Tower, B. (1889) *An Apparatus for Providing a Steady Platform for Guns, etc., at Sea* (Institution of Naval Architects), pp. 1–14.

A.13 Heinrich Rudolph Hertz (1857–1894)

Heinrich Hertz was born in Hamburg, Germany, on 22 February 1857. In October 1877, at the age of twenty, he went to Munich to continue his studies in engineering after having completed a year of practical work and preliminary studies in mathematics and other science subjects. Indeed,

6. I am grateful to Professor A. Cameron for making his correspondence with the *Romford Times* available to me.

it was these early studies of science which caused him to have doubts about a career in engineering at the time that he moved to Munich. He enjoyed the natural sciences so much that he began to fear that his chosen profession of engineering would be less rewarding. He was tortured by the question of his future studies and finally sought a decision from his parents in a letter written on 1 November 1877. He left his parents in little doubt about the drift of his feelings, for he reminded them in the letter that he had always said, '. . . I would rather be a great scientific investigator than a great engineer, but would rather be a second-rate engineer than a second-rate investigator'. He then went on to say, '. . . But of this I feel positive, that if the decision is in favour of natural science, I shall never look back with regret towards engineering science, whereas if I became an engineer I shall always be longing for the other. . . .'

He declared that business capacity, experience, and knowledge of data and formulae which appeared to be so important in practice did not interest him, even though he knew he would be more certain to earn a living as an engineer.

His parents made the appropriate decision and Heinrich completed a year in Munich studying mathematics, mechanics and practical physics. In October 1878, at the age of twenty-one, he moved to Berlin to become a pupil of von Helmholtz and Kirchhoff. He noticed at an early stage that the Philosophical Faculty of the University had offered a prize for the solution of a problem in physics dealing with *electric inertia*. He discussed the matter with Professor Helmholtz and by early November had started his experiments. During the progress of his investigation he experienced the frustration and doubts associated with research. He wondered if it was right for him to devote so much time to the particular subject of the prize when there was so much to learn, but defended his position in a further letter to his parents dated 24 November 1878.

> . . . And yet I feel that it is right; to get information for myself and for others direct from nature gives me so much more satisfaction than to be always learning it from others and for myself alone. . . .
> . . . When I am only studying books I am never free of the feeling that I am a perfectly useless member of society.

In due course his experimental findings started to reveal a similarity to his theoretical predictions, but the ultimate accord was achieved only when he found a missing '2' in his calculations. '. . . At first my calculations gave a value which was much greater than the observed value. Then I happened to notice that it was just twice as great. After a long search amongst the calculations I came upon a 2 which had been forgotten, and then both agreed better than I could have expected.'

His paper was written during a period of military service at Freiburg. It won the prize – a fine gold medal.

The above extracts from letters written to his parents not only provide an interesting insight into Hertz's youthful enthusiasm and attitude towards his physical investigations, they show the closeness of the family bond.

In the following year Hertz prepared papers on electricity and induction together with a dissertation for the doctorate awarded in 1880 by the

Berlin Philosophical Faculty. He became an assistant to Helmholtz in October 1880 and it was then that he undertook his theoretical investigation on the contact of elastic solids. It was not the development of the ball-bearing industry, which was at that time mainly restricted to the manufacture of lightly loaded bearings for bicycles, but discussions of Newton's rings in the Physical Society of Berlin which attracted the attention of Hertz to the problem. He recognized that little was known about the effects of changes in form of the glasses which were pressed together in demonstrations of optical phenomena and then proceeded to solve the problem during the Christmas vacation of 1880. His classical paper, which was to provide the basis for contact stress and deformation calculations in many tribological situations, was first read to the Physical Society of Berlin on 31 January 1881. Hertz was still one month short of his twenty-fourth birthday when the lecture was presented. The paper attracted much attention from scientists and engineers alike, and in due course it was published not only in *Borchardt's Journal* (see Hertz, 1881), but also in a technical journal with a supplement on hardness (see Hertz, 1882).

The analysis, developed in terms of an electrical analogy, yielded expressions for the elliptical distribution of normal stress (potential), together with relationships for the dimensions of the contact region and the compression of the spheres. Examples were presented for the contact of glass lenses and steel spheres and the paper concluded with a discussion of the impact of elastic bodies. The final example, occupying but two and a half lines, possessed a magnitude to match the content of the paper. '. . . For two steel spheres as large as the earth impinging, with an initial velocity of 10 mm/sec, the duration of contact would be nearly 27 hours.'

The year 1883 saw Hertz's induction to the position of *Privatdozent* at Kiel and in the following year he became Professor of Physics in the Technische Hochschule at Karlsrule. It was during this period that he developed his views on electromagnetic radiation. In subsequent years he completed his great electrical investigations for which he is most famous. The work on contact stresses and deformation upon which so many tribological concepts rest was thus completed at a very early stage in his career.

Hertz made his last move in 1889 when he became Professor of Physics at the University of Bonn. His health soon failed and he died at the age of thirty-six on 1 January 1894. Few men of science have achieved so much in so short a time.

Bibliography

Hertz, H. (1881) 'On the contact of elastic solids', *J. reine und angew. Math.*, **92**, 156–71.

Hertz, H. (1882) 'On the contact of rigid elastic solids and on hardness', *Verh. Ver. Beförderung des Gewerlefleisses*, Nov.

Lenard, P. (1896) *Miscellaneous Papers by Heinrich Hertz – With an Introduction by Professor Phillip Lenard*, authorized English translation by D. E. Jones and G. A. Schott (Macmillan, London).

Magie, W. F. (1935) 'Heinrich Rudolph Hertz', in *Source Book in Physics*, (McGraw-Hill, New York and London), p. 549.

A.14   Richard Stribeck   (1861–1950)

Heinrich Hertz,[1] Richard Stribeck and John Goodman[2] were undoubtedly the outstanding contributors to the scientific studies upon which the rolling-element bearing industry was established at the end of the nineteenth century. Hertz provided the well-known and brilliantly executed general theoretical analysis of contact between elastic bodies without reference to bearings, but Stribeck and Goodman provided a firm experimental basis for the development of rolling-element bearings. The contributions of Stribeck and Goodman were not only similar in timing, but also in many respects in nature. Both men brought to the subject a direct experimental approach, built upon a firm foundation in mechanical engineering and steam-engine development. In this and the following biographical sketch we take the opportunity to examine the similarities and differences in the lives of these great contributors to the ball-bearing industry.

Richard Stribeck was born in Stuttgart, Germany on 7 July 1861. He studied at the Technische Hochschule in Stuttgart, then gained practical experience in engineering in Königsberg and Esslingen. In 1888, at the age of twenty-seven, he was appointed Professor of Machine Construction at the Building School in Stuttgart. Two years later he moved to the Technische Hochschule in Darmstadt and then on 1 April 1893 he was appointed to the Technische Hochschule in Dresden. His inaugural address as Professor für Maschineningenieurwesen in the latter institution was delivered on 6 May 1893 with the title 'Progress in the production of steam and in the exploitation of its energy in steam engines during the last 20 years'. On 1 July 1893 he set out on an educational visit to the United States of America. In the academic year 1896/97 he was appointed to the governing body of the newly constructed Machines Laboratory I (Strengths), but he left Dresden at the end of the winter semester 1897/98 and moved to Berlin in March 1898.

During the period 1889–93, when Stribeck was in Stuttgart and Darmstadt, he published some ten papers. This prolific early output was addressed mainly to problems of boilers and steam turbines for marine applications. It is only in the last paper in this group, which dealt with dynamic loading on the main and crosshead bearings of steam engines, that we find evidence of direct interest in tribological problems. In Dresden, Stribeck directed his attention to gears, with particular emphasis on experimental studies of worm drives. He determined experimentally the load (F) and speed (n) for the limit of correct functioning of worm drives and established empirically the hyperbolic relationship $Fn = a$ constant. In 1897 he published a paper on the wear of gears and its consequences. Stribeck was the first of a line of professors who established the reputation of Dresden as a centre for research on machine elements.

In 1898 Stribeck was appointed as one of the directors of the newly created Centralstelle für wissenschaftlich-technische Untersuchungen in

1.   See Sect. A.13.
2.   See Sect. A.15.

Neubabelsberg, Berlin. As head of the Metallurgical Division of the Central Authority, Stribeck was to undertake and publish the findings of his most basic tribological studies between 1898 and 1902. Professor Czichos[3] has pointed out that the institute in which Stribeck worked was one of the predecessors of the present Bundesanstalt für Material-prüfung (BAM), the Federal Institute for Testing of Materials. Stribeck's major contributions to tribology during this period were in two distinct areas: the load-carrying capacity of ball-bearings and the frictional characteristics of plain and rolling-element journal bearings. In both cases his approach was experimental, with the application of sound physical principles and the interpretation of theoretical studies of the problems being much in evidence.

In his classical study of the load-carrying capacity of ball-bearings, Stribeck (1901) provided a scientific basis for the development of a large industry. He studied the Hertzian theory of elastic contact and the onset of plastic deformation of contacts between three hardened steel balls, a steel plate sandwiched between two hardened steel balls and more conformal contacts between the balls and a grooved plate. Stribeck's studies, which were commissioned by the Deutsche Waffen- und Munitionsfabriken, Berlin, have been described by Allan (1945) as epoch-making in their effect on the ball-bearing industry. Details of the work are discussed in Chapter 10, but it is worth noting that its major impact arose from the analysis of carefully conducted experiments and the sound appreciation of Hertzian contact theory, together with the presentation of practical formulae in a form eminently suitable for the bearing designer. Professor Blok[4] has drawn my attention to the fact that Stribeck's approach to the question of load-carrying capacity of ball-bearings was some seven years in advance of similar considerations in gear design.

Stribeck's (1902) celebrated paper on the friction characteristics of both plain and rolling-element journal bearings was published only one year after his masterly attack on the problem of ball-bearing load capacity. There had been much confusion about the influence of load and speed upon lubricated plain bearings in Germany and elsewhere, but Stribeck's experiments clarified the situation with great authority. A discussion of the development of the well-known friction coefficient relationship indicating boundary, mixed and hydrodynamic lubrication, now known by many as the Stribeck curve, is presented in Section 10.4. Stribeck's experiments were to provide the guidelines for future theoretical studies by men like Sommerfeld and Gümbel. It is quite remarkable, and a measure of Stribeck's contributions to tribology, that he should have concluded such impressive studies of both rolling-element and plain bearings in successive years. This nevertheless represented only part of his work during this period. He continued to work on, and publish about, gear performance and by 1909 he had written some twenty-three

3. Czichos, H. (1976*b*).
4. Blok, H. (1976*b*).

papers, all were published in the journal of the Society of German Engineers (*Zeitschrift des Vereines deutscher Ingenieure*). He also collaborated on the development of Duralumin during this period.

In the autumn of 1908 Richard Stribeck moved from Berlin to Essen, where he became a director of the Friedrick Krupp Company. He remained with the company from 1 October 1908 to 30 June 1919 and his published work during this period shows that he was concerned with the mechanical properties of materials. In particular, he wrote about the fatigue strength of iron and steel and the notch impact test. Again, there is an interesting parallel with his English counterpart, John Goodman.

After the First World War, Stribeck returned to his home town of Stuttgart to become technical–scientific adviser to a company owned by his friend Robert Bosch. He joined the board of directors in 1924 and the technical contributions during the latter days of his active professional life were devoted to the further development of diesel engines. He worked on problems of materials research, particularly in relation to spark-plug manufacture and magnetic steels. Some of his work on diesel engines was published as late as 1927.

Richard Stribeck died in the city of his birth on 29 March 1950. He lived to see the development of a vast ball-bearing industry and the recognition by later generations of research workers of his contributions to the science and technology of plain and rolling-element bearings.

In preparing this brief biographical note I have been greatly assisted in my search for information by Professor Blok of Delft, who kindly contributed several pages of notes from his own files; Professor Czichos of BAM, Berlin, who also prepared for me a valuable review of Stribeck's work; Dr Ingemar Fernlund of SKF, Göteborg, who provided copies of some of Stribeck's papers; the Archivist, Karl-Heinz Adolph, at the Technische Universität, Dresden, who kindly supplied background information on Stribeck's formative years in Dresden and the Friedrich Krupp Company with confirmation of Stribeck's appointment during the period 1908 to 1919. For some months I feared that these notes would not be supported by a portrait of Richard Stribeck, but the Archivist in Dresden kindly drew my attention to the picture reproduced here from the comprehensive text by Seherr-Thoss (1965) dealing with the development of gear technology.

Bibliography

Adolph, K. H. (1976) 'Richard Hermann Stribeck', private communication from the Archivist, Technische Universität Dresden.

Allan, R. K. (1945) *Rolling Bearings* (Pitman, London).

Blok, H. (1976) 'Notes about Professor Richard Stribeck', private communication.

Czichos, H. (1976) 'Professor Dr. Richard Stribeck', private communication.

Seherr-Thoss, (1965) *Die Entwicklung Der Zahnrad-Technik* (Springer-Verlag, Berlin), p. 410.

Stribeck, R. (1901) 'Kugellager für beliebige Belastungen', *Z. Verein. Deut. Ing.*, **45**, No. 3, 73–125.

Stribeck, R. (1902) 'Die Wesentlichen Eigenschaften der Gleit- und Rollenlager', *Z. Verein. Deut. Ing.*, **46**, No. 38, 1341–8, 1432–8; No. 39, 1463–70.

A.15 John Goodman (1862–1935)

John Goodman, being the youngest son of Thomas Goodman, was born in Royston, Hertfordshire on 1 May 1862. He went to school at Gravesend and Cambridge before serving a five-and-a-half-year apprenticeship as an engineer. He became an assistant to William Stroudley in the Brighton Works of the London, Brighton and South Coast Railway and in 1883, the year in which Beauchamp Tower's first report was presented to the Institution of Mechanical Engineers' Research Committee on Friction, he designed a dead-weight testing machine for measuring the friction of railway bearings. The results of his studies of friction on Mr Stroudley's machine formed the basis of his first paper, read at a meeting of the students of the Institution of Civil Engineers on 20 November 1885. Thus began John Goodman's interest in tribological subjects which was to remain with him throughout his life. He served Stroudley for two years and in 1885 was awarded a Whitworth scholarship. He studied in the Engineering Department of University College, London, and then became an assistant to Sir Alex B. W. Kennedy in 1886. In 1887 he was appointed chief assistant to Mr W. H. Stranger at the Broadway Testing Works, Westminster.

On 30 April 1874 a constitution of the Yorkshire College of Science was approved in Leeds. The object was '. . . to promote the education of persons of both sexes and, in particular, to provide instruction in such sciences and arts as are applicable or ancillary to the manufacturing, mining, engineering, and agricultural industries of the county of Yorkshire.'

The title was reduced to 'Yorkshire College' in 1878 when the teaching of arts was begun, and on 3 November 1887 the College joined Owens College, Manchester and Liverpool College by becoming established in the Federal Victoria University. The first Professor of Civil and Mechanical Engineering was G. F. Armstrong, a former member of the chief engineer's staff of the Great Northern Railway. He came to Leeds from McGill University in 1876 and held the post until 1884 when he became Regius Professor at Edinburgh. Archibald Barr of the University of Glasgow replaced Armstrong. Described as '. . . a Scott whose accent was sometimes incomprehensible' he did much to strengthen the curriculum and to extend the period of study to three years. He was perhaps best known for his cooperation with Stroud, Professor of Physics, for work on rangefinders.

In 1890 Goodman became the third professor to hold the Chair of Civil and Mechanical Engineering in the Yorkshire College. His essentially practical approach and his industrial background satisfied the shrewd practical engineers of Leeds who managed the affairs of the department and who '. . . would not tolerate for one moment an unpractical man at the head of the teaching staff'.

The challenge of developing an academic department was approached with vigour and thoroughness. Goodman first familiarized himself with teaching systems in engineering in the United States and Canada and then visited European schools in the following summer. He developed the theoretical content of the course, established an honours degree and promoted research. In due course he was to develop firm ideas on engineering education and professional training (Goodman, 1906*b*). The lecture courses were soundly based and practical in nature, the laboratory work was extensive and realistic and the teaching methods novel for their times. Goodman provided examples classes for his students and introduced open book examinations. He viewed with contempt the *toy experiments* which constituted laboratory exercises in some departments. His view was, '... that the practical experience so essential to every engineer can only be obtained in workshops carried on on commercial lines, and that no ideas of cost, time, and, above all, knowledge of men, can possibly be obtained in a college toy workshop'.

Goodman firmly established research in engineering at Leeds. His investigations over the years were wide ranging but always relevant to the problems of the day. He is best known for his studies of tribological problems, but he also wrote on such matters as drifting of washers, air compressors and the transmission of power by compressed air in the mining industry, the efficiency of steam engines, stresses in crane hooks, governing of impulse water-wheels, feed-water heaters on boilers, determination of the inclination and direction of bore holes and the flooding of culverts. His book *Mechanics Applied to Engineering* (Goodman, 1906*a*), first published in 1906 and extended in 1927 after his retirement, became an established and respected text.

His studies in tribology reflected some of the major developments of bearings at the close of the nineteenth and the dawn of the twentieth centuries. His reports on friction (Goodman, 1886, 1887) were a direct consequence of his experience on the railways and the problems being faced in connection with bearings and brakes for locomotives and trains. He considered the friction of both dry and lubricated surfaces and was at pains to interpret his observations in the light of current understanding of the physical situation. In his very first paper (Goodman, 1886), written at the age of twenty-three, he recognized that the frictional resistance in well-lubricated journal bearings could be attributed to viscous shearing. The work was carried out after Tower and Petrov had published their findings on the subject, but before Osborne Reynolds had presented his classical analysis of fluid-film lubrication. It was Beauchamp Tower who first measured the pressure in films in bath-lubricated bearings, but the young John Goodman made the first measurements of film thickness. At the end of the section dealing with lubricated surfaces he mentioned that an insulated micrometer attached to the brass on Mr Stroudley's machine had revealed a lift which increased steadily from 0·0013 to 0·0029 in (0·033–0·074 mm). These early papers also described experiments on a Rudge bicycle ball-bearing and the wear of brasses. They also carried a condemnation of current practice of cutting oil-grooves in the crowns of bearings.

A comprehensive review of work, including his own, on the friction and lubrication of cylindrical journals was read before the Manchester Association of Engineers at the Grand Hotel, Manchester on 22 March of the year in which he was appointed to the Chair at Leeds (1890). This paper reaffirmed his interest in ball-bearings, even though they '. . . are but little used for engineering purposes'. He felt, however, that it was '. . . only a matter of time before they will be extensively used for heavy bearings'.

Goodman not only witnessed the truth of his prediction, he contributed in an impressive and authoritative way to the development of rolling-element bearing science and technology. His papers on roller- and ball-bearings and the testing of antifriction bearing materials (Goodman, 1912), based upon fifteen years of investigation, are recognized as landmarks in the history of rolling-element bearings. Allan (1945) in his account of the development of rolling-element bearings has written that, '. . . While the contributions of Heinrich Hertz, Professor Stribeck, and Professor Goodman were doubtless the most outstanding in the early days of the industry. . .'.

Goodman was concerned for the life of ball- and roller-bearings and he developed empirical equations to take account of fatigue; the most common mode of failure of the rolling elements. He developed a procedure of optical inspection to reveal signs of '. . . pitting and flaking, and occasionally of scratching and peeling'. His interest in the subject of fatigue was to grow, even in retirement, and the 'Goodman diagram' will not be unknown to many in this field today.

In 1887 Goodman became an associate member of the Institution of Civil Engineers and in 1890 he was transferred to full membership. He was awarded the Miller Scholarship and Prize of that Institution, together with a Telford Premium. He became a full member of the Institution of Mechanical Engineers in 1890.

John Goodman retired from the Chair at Leeds at the age of sixty in 1922. In his thirty-two years of service to the University he developed a flourishing and respected Department of Engineering. The *Yorkshire College* had been transformed into the *University of Leeds* following the granting of a charter on 25 April 1904. Goodman was thus the first Professor of Engineering in the independent University. His work at Leeds was interrupted only by the First World War. During this time he was a major attached to the Royal Air Force and was for a long time engaged at a large aeroplane factory in Scotland.

The University bestowed upon him the title of Emeritus Professor in 1922 and the Council assigned to him special accommodation so that he might continue his independent research work. After two years this well-meaning but awkward arrangement proved too difficult for both Goodman and his successor, and access to the facilities of the University was discontinued. Goodman's journal-bearing testing machine, on which so much of his work had been based, was then moved to the workshop of Bradley's engineering garage at Addingham, near his Beamsley home. He continued working on the machine until his death in 1935. The test bearings which he required for his retirement studies were provided by his former colleague Dr W. H. Swift, who had moved from Leeds in 1926 to an appointment as Head of the Mechanical Engineering Department at Bradford Technical College. Swift, who was later to become a professor at Sheffield, had his interest in hydrodynamic lubrication engendered by his contact with John Goodman at Leeds during the period 1922–24. Swift had joined the lecturing staff at Leeds in 1922 and he developed an admiration for Goodman's careful experimental approach. The admiration was mutual, for Goodman greatly respected Swift's analytical skill. Goodman also appreciated the way in which H. M. Martin brought analytical skills to the subject of lubrication. It was Martin who had solved the Reynolds equation for involute spur gears in 1916 and there was some correspondence between him and Goodman about Petrov's work in 1920.[1]

Goodman indicated his own preference for experimental work and his respect for the analytical ability of H. W. Swift and H. M. Martin with a nice reference in the discussion of Swift's (1931) paper on the stability of lubricating films in journal bearings to '. . . the mathematical flights of the Swifts and the Martins'.

Goodman continued to write about his work after retirement. He lived in the country to the north and west of Leeds at Beamsley, near Skipton, and was fond of cycling, swimming and gardening. Shimmin wrote that, '. . . John Goodman, who looked every inch a farmer, trained engineers whose skill was in demand the world over.'

A bachelor and man of presence and authority, John Goodman won the respect and affection of students and colleagues alike. On his death a colleague said, '. . . He was not only a deeply respected tutor, but one who was deeply loved by all who knew him.'

He died in a Leeds nursing home on 28 October 1935.

Bibliography

Allan, R. K. (1945) *Rolling Bearings* (Pitman, London).

Cameron, A. (1966) *Principles of Lubrication* (Longman, London), p. 266.

Cole, B. N. (1975) 'Engineering department a part of industry–and never apart from it; Leeds University's philosophy for first 100 years', *Mech. Eng. News*, April 1975.

Engineering (1896) 'The Engineering Department of the Yorkshire College, Leeds', *Engineering*, London, **62**, 28 Aug., pp. 262–4; 11 Sept., pp. 327–30; 25 Sept., pp. 390–1; 5 Oct., pp. 449–52.

Engineering (1935) 'The late Professor John Goodman', *Engineering*, **140**, 1 Nov., 1935, 483.

Goodman, J. (1886) 'Recent researches in friction', *Proc. Inst. civ. Engrs*, **ixxxx**, Session 1885–86, Pt iii, 3–19.

Goodman, J. (1887) 'Recent researches in friction–Part II', *Proc. Inst. civ. Engrs*, **ixxxix**, Session 1886–87, Pt iii, 3–36.

Goodman, J. (1890) *The Friction and Lubrication of Cylindrical Journals* (The Manchester Association of Engineers), pp. 87–135.

Goodman, J. (1906a) *Mechanics Applied to Engineering* (Longman, London).

Goodman, J. (1906b) *The Results of Technical Education in Engineering* (The Manchester Association of Engineers).

Goodman, J. (1912) '(1) Roller and ball bearings; (2) The testing of anti-friction bearing materials', *Proc. Inst. civ. Engrs*, **clxxxix**, Session 1911–12, Pt iii, 4–88.

Shimmin, A. N. (1954) *The University of Leeds–the First Half Century* (published for the University of Leeds at the University Press, Cambridge).

The Engineer (1935) 'The late Professor John Goodman', *The Engineer*, **60**, 1 Nov., 1935, 445.

---

1. Private communication. I am grateful to Dr P. B. Neal of the University of Sheffield for bringing this to my attention.

The Times (1935) 'Professor Goodman; engineering at Leeds University' (obituary notice), *The Times*, 30.10.1935.

Yorkshire Post (1935) 'Professor John Goodman' (obituary notice), *Yorkshire Post*, 29.10.1935, p. 5.

A.16 Albert Kingsbury (1863–1943)

Albert Kingsbury was born at Morris, Illinois in 1863. His mother and father both came from established New England families and he appears to have inherited from them his strength of character and concern for the welfare of everyone with whom he became associated. His mother's family were Quakers, but Kingsbury followed his father as a member of the Presbyterian Church. His first paternal American ancestor was Joseph Kingsbury who came from England and settled with his wife, Millicent Ames, in Dedham, Massachusetts, in 1628. On the maternal side of the family he was twice descended from Elder William Brewster of the Plymouth Colony. His father, a superintendent of the Stoneware Manufacturing Company of Akron, Ohio, and an expert musician, was Lester Wayne Kingsbury, while his mother's maiden name was Eliza Emeline Fosdick. He attended school in the village, now a city, of Cuyahoga Falls, Ohio, where he was a member of the first graduation class, consisting of three boys and five girls, in June 1880. He then spent a year in a Latin–scientific course in Buchtel College, Akron, Ohio, the foundation College of the University of Akron.

Between 1881 and 1884 Albert Kingsbury served his time as an apprentice machinist with the Turner, Vaughn and Taylor Company in Cuyahoga Falls. He was involved mainly with clay-working and wire-drawing machines, the work being rather rough and heavy. In his auto-biographical notes, entitled 'Recollections', which were later published in the American Society of Mechanical Engineers journal *Mechanical Engineering* (ASME, 1950), Kingsbury stated that the experience gained during this period was most valuable to him.

In September 1884 Kingsbury entered the Ohio State University in Columbus, Ohio, as a freshman in the mechanical engineering course. He left the course at the end of his sophomore year, when he was '. . . low in mind and funds . . .', to work on a project on a wire-grip fastening machine offered by Professor S. W. Robinson of the Ohio State University. Professor Robinson had invented the machine and Kingsbury was involved in drafting work, inspection of the machines being made in the Carver Cotton Gin Company in East Bridgewater, Mass., and visits to various shoe factories in New England where the machines were in need of repair of modification. After less than a year Kingsbury became discouraged in this situation and he returned to Ohio to join the Warner and Swasey Company in Cleveland as a machinist. He worked mainly on 16 in (0·41 m) engine lathes and was clearly proud of the fact that he was involved in the manufacture of some parts for the famous 36 in (0·91 m) objective lens Lick telescope.

In the autumn of 1887 Albert Kingsbury entered Sibley College, Cornell University in the junior class in mechanical engineering. He graduated in 1889 at the age of twenty-six. It was during these two years at Cornell that Kingsbury came under the influence of Dr Robert H. Thurston.[1]

Thurston assigned to Kingsbury the problem of testing various bearing metals in the form of half-journal bearings for the Pennsylvania Railroad Company. The bearings were tested on the Thurston lubricant tester mentioned in Chapter 10 and previous reports by students and others at Sibley indicated a wide variation in the results from wear tests. It was at this stage that Kingsbury's training and skill in the mechanical arts played a role in the history of tribology. He carefully scraped and refitted all the test bearings, and then found on the basis of careful initial and final weighing that they all exhibited identical characteristics with no detectable wear. In some cases, coefficients of friction as low as 0·0005 were recorded. Neither Kingsbury nor Dr Thurston, who were both unfamiliar with Osborne Reynolds' work at that time, were able to offer an explanation of the observed behaviour, and this left Kingsbury with a desire to learn more about the mysterious and impressive performance of lubricated journals.

Kingsbury accepted a teaching post, for which he had been recommended by Dr Thurston, in the New Hampshire College of Agriculture and the Mechanic Arts in Hanover, New Hampshire, in 1889. After years of study and work to pay the expenses of college life, the compensation of $100 per month seemed huge. There were about eighteen students in all during the 1889–90 session and Kingsbury taught courses in mechanical engineering and physics. At the end of the college year Kingsbury moved back to Cuyahoga Falls, Ohio, to work for an older cousin, Horace B. Camp, for whom he had a great regard. Kingsbury was superintendent of a new machine shop erected to manufacture a brickmaking machine invented by his cousin, but business did not develop as anticipated and in the autumn of 1891 the Dean, Professor Pettee, persuaded him to return to the New Hampshire College in Hanover, as Professor of Mechanical Engineering.

In 1893 the College moved across the State to new buildings in Durham, and Kingsbury was much involved, along with Professor Pettee, with planning the move during the period 1891–92. The return to an academic post in 1891 enabled Kingsbury to reopen investigations of lubrication and this was to remain the central theme throughout his lifetime. He worked closely with a highly respected mechanic in the machine shop named John Brown and proceeded to his studies of a torsion–compression testing machine and an air-lubricated journal bearing as described in Section 10.4. He first came to the conclusion that air was acting as a lubricant when examining the free movement of the piston in a cylinder on the former machine, but it was only when Osborne Reynolds' famous paper of 1886 was brought to his attention by John H. MacAlpine after a demonstration of the air-lubricated piston before the Bureau of Steam

---

1. See Sects. A.9 and 10.4.

Engineering at the Navy Department, that the jigsaw of fluid-film lubricated bearings fell into place. The concept of a pivoting-pad bearing with the pivot located at the theoretical centre of pressure developed as he read Reynolds' paper, but he was concerned about the need for unidirectional operation and the lopsided arrangement which such a proposal necessitated. However, in 1898 Kingsbury and Brown built and successfully operated a centrally pivoted thrust bearing with spherical bosses projecting from the blocks.

An 'Osborne Reynolds Fellowship' was initiated by a contribution from Kingsbury to the University of Manchester in 1919. This form of appreciation of the work of Osborne Reynolds and the impact which it had upon Kingsbury's own work was typical of the man. A rather nice touch, which clearly pleased Kingsbury, was added to the story when F. D. Reynolds, son of Osborne, became the first incumbent of the fellowship to enable him to work on problems of the screw propeller.

Kingsbury's next move was to the Worcester Polytechnic Institute as Professor of Applied Mechanics in July 1899. He continued his work on bearings, but in June 1903 he was appointed *general engineer* in the East Pittsburgh works of the Westinghouse Electric and Manufacturing Company. In 1904 he spent some months in Canada preparing a report on electrical equipment for the Canadian Pacific Railway and in due course he became a consulting engineer for both the Westinghouse Electric and Manufacturing Company and the nearby Westinghouse Machine Company. He remained with the Westinghouse Electric Company until 1914 and it was during this period that he grasped the opportunity to demonstrate the merits of his pivoted-pad bearings.

The early attempts were not entirely successful, and the full chronicle of events leading up to the impressive and successful application of the principle to a vertical hydroelectric generator on the Susquehanna river in 1912 can be found in Section 10.7. This 48 in (1·22 m) bearing carried a load of 405,000 lbf (1·80 MN) with such facility that it ensured innumerable further applications. However, before this hurdle was overcome, Kingsbury suffered setbacks in 1904, when workshop tests indicated overheating of the bearing, and in 1909, when another test bearing ran perfectly but the company lacked conviction and installed a roller-bearing. Kingsbury's disappointment in the latter case was relieved fourteen years later when the engineers operating the motor in Idaho requested him to replace the roller-bearing with one of his tilting-pad bearings.

The date of Kingsbury's American patent is of interest in relation to the parallel development of tilting pad bearings by A. G. M. Michell.[2]

Kingsbury's application was filed in 1907, but rejected on the grounds of Michell's patent granted in England in 1904. However, he was able to show that he had built tilting-pad bearings prior to 1904 and Patent No. 947242 was awarded in 1910. It thus appears that Kingsbury's work on tilting-pad bearings commenced earlier than Michell's, although the

---

2. See Sects. 10.7 and A.19.

latter was several years in advance of the former in carrying out the fundamental analysis, writing about the subject and obtaining patents.

The Westinghouse Machine Company had manufactured bearings to Kingsbury's design for various applications, but by 1921 it could no longer keep up with demand. At that time Kingsbury had a small workshop on Cherry Street in Philadelphia, but he soon decided to form his own company and in 1924 the Kingsbury Machine Works was founded on Tackawanna Street, Frankford, in the northeastern part of the city. The company grew in size and strength and in 1966 moved to a new site in the Philadelphia Industrial Park.

Albert Kingsbury married Alison Mason, daughter of Erskine Mason, a prominent surgeon in New York City, on 25 July 1893. They had five daughters named Margaretta Mason, Alison Mason, Elizabeth Brewster, Katherine Knox and Theodora. Kingsbury was a man of character and great intellect, with a keen sense of humour, He was equally at ease in industry, teaching engineering or in a cultural setting. As a young man he played the flute and sang in an episcopal church choir. In later years he studied languages in his leisure time and read extensively in Italian, Spanish, Greek, French, German and Danish. He deplored the restricted conversational skills and common illiteracy of educated people in his own and other professions. He was courteous, calm, yet firm in his convictions. He avoided advertising, apparently on the grounds of professional ethics, but also because he believed it to be a waste of money.

Many honours were bestowed upon Kingsbury for his great contributions to engineering in general and bearing science and technology in particular. He became a member of the American Society of Mechanical Engineers in 1892 and an honorary member in 1940. He was also a member (1893), then Fellow (1898) and ultimately Emeritus life member (1939) of the American Association for the Advancement of Science, and a Member of the American Society for the Promotion of Engineering Education (1893). Associate membership of the American Institute of Electrical Engineers followed in 1908. Awards included the Elliott Cresson medal of the Franklin Institute of 1923, the gold medal of the American Society of Mechanical Engineers in 1931 and the John Scott medal of the city of Philadelphia in the same year. He was awarded honorary doctorates by Worcester Polytechnic Institute and the University of New Hampshire in 1933 and 1935, respectively. On 14 October 1950, the University of New Hampshire opened its newest and largest building to engineering subjects and dedicated it to the memory of Dr Albert Kingsbury by naming it 'Kingsbury Hall'. A booklet published at the time by the University of New Hampshire (1950) is rich in information about the subject of these notes.

Kingsbury was more than a successful engineer, businessman and teacher. He was a true inventor, with flair, coupled with a sound appreciation of the science and art of engineering. As the first man to construct a successful pivoted-pad bearing, a device of beautiful simplicity and performance, he ranks as a great engineer and an outstanding tribologist.

Kingsbury spent his last years in Greenwich, Connecticut. He died on 28 July 1943. His contributions to tribology are well recorded in the following extract from the citation prepared on the occasion of his election to honorary membership of the ASME in 1940. '. . . Engineer, inventor, manufacturer and student of the mechanics of lubrication, who has spent a lifetime preventing waste and destruction from friction, by the supporting of huge masses of machinery on thin films of oil, and has provided for hydraulic turbines and other weighty mechanisms the remarkable bearing known by his name.'

In preparing this biographical sketch I have been greatly assisted by Mr F. R. Archibald, who supplied notes, correspondence and a copy of his articles on Albert Kingsbury, and Mr R. S. Gregory, Manager of Research and Development of the company, Kingsbury, Inc., for background information.

Bibliography

Archibald, F. R. (1955) 'Men of Lubrication–Albert Kingsbury', *Lubr. Enging*, (ASME), **11**, May–June 1955, 162–3, 197–8.

Kingsbury, A. (1950) 'Development of the Kingsbury thrust bearing', *Mech. Enging*, **72**, No. 12, pp. 957–62.

University of New Hampshire (1950) 'Kingsbury Hall–named for Professor Albert Kingsbury 1863–1943', University of New Hampshire, dedicated 14 Oct., 1950.

A.17 William Bate Hardy (1864–1934)

Sir William Bate Hardy was the first person to use the term 'boundary lubrication'. He applied the techniques of physical chemistry to expose the existence and nature of the phenomenon with great skill and equal success to that achieved by Osborne Reynolds in his mathematical-physical approach to fluid-film lubrication. Yet Hardy was a biologist, and the path which led him to a study of friction and lubrication is most interesting.

Hardy, the only child of William Hardy of Llangollen, North Wales and Sarah, eldest daughter of William Bate, was born on 6 April 1864 at Erdington, Warwickshire. He was educated at Framlingham College and at Gonville and Caius College, Cambridge, which he entered in 1884. He was elected scholar in 1885 and in 1888 he obtained a first class in Part II of the Natural Sciences Tripos, his subject being zoology. He was elected to a Shuttleworth Research Scholarship after graduation, he became a Fellow of his college in 1892 and a tutor from 1900 to 1918. In 1900 he was Thurston Prizeman and in the University he was appointed Demonstrator and then, in 1913, Lecturer in Physiology. Successive generations of medical students benefited from his teaching. In an address forming part of the Centennial Celebration held in Cambridge on 16 and 17 June 1964 Professor A. V. Hill recorded his earliest recollection of Hardy. '. . . He always seemed to be dashing from one job to another; turning up at a lecture with a great bundle of papers he could not really use, at a practical class with no very clear idea of how the apparatus worked and leaving us to find out; but always unperturbed and over-flowing with endearing human qualities.'

In the six years prior to 1898 his main research interest was histology.

He was sceptical of much that was seen, after fixing and staining, under the microscope. During this period he published eleven papers on histology and his studies focused attention on the stability and behaviour of colloidal systems. Although he had little previous acquaintance with physical chemistry he clearly distinguished suspensoid from emulsoid systems, and his studies of the role of ionization upon the coagulative power of electrolytes was to form a useful background to later studies of surface activity in friction and boundary lubrication. Immediately before the First World War Hardy studied the effect of radium upon living tissues and the equilibrium of proteins in the blood. He had by this time become so involved in molecular physics that he resolved to refresh and extend his knowledge of mathematics. The zoologist had thus expanded his interests to encompass those aspects of physical chemistry and mathematics relevant to his major field of study. His ability to discuss quantitatively subjects which had previously been treated descriptively was demonstrated in the publication by the Royal Society of his papers, 'A general theory of colloid solutions' and 'The tension of composite fluid surfaces'.

Hardy was elected a Fellow of the Royal Society in 1902 and from 1915 to 1925 he was the biological secretary. At the beginning of the First World War the Board of Trade invited the Society, through Hardy, to set up a physiology sub-committee to inquire into the food supply of the United Kingdom. The outcome of this request was the formation of the Food (War) Committee, which advised government from 1916 to 1919. Following the formation of the Department of Scientific and Industrial Research, Hardy was appointed the first Chairman of the Food Investigation Board (1917–28) and Director of Food Investigation (1917–34). He became responsible for and helped to establish important food research establishments at Cambridge (Low Temperature Station for Research in Biochemistry and Biophysics), Ditton Laboratory (fruit and vegetable) and the Torry Research Station (fish preservation).

Hardy's contributions to tribology were undertaken after the First World War when he studied the static friction of dry and lubricated surfaces. This work was a natural extension of his work on colloids and surface tension. By the time his studies of the latter subject were published (Hardy, 1912) his views on molecular orientation at interfaces were well developed. He wrote:

> ... If the stray fields of a molecule, that is, of a complex of these atomic systems be unsymmetrical the surface layer of fluids and solids which are close packed states of matter must differ from the interior in the orientation of the axis of the fields with respect to the normal to the surface and so form a skin on the surface of a pure substance having all the molecules oriented in the same way instead of purely in random degrees.

There seems to be little doubt that Hardy's friction experiments would have been started earlier if it had not been for the intervention of the First World War. His two papers on static friction were published in the *Philosophical Magazine* in 1919 (Hardy and Hardy, 1919) and 1920 (Hardy, 1920). In the latter paper he drew attention to the relationship between the molecular structure of substances and their effectiveness as lubricants. The work was summarized in an evening address presented to the Royal Institution in February, 1920. The concept of fluid-film lubrication dominated by the physical (viscous) properties of matter had been established some thirty-four years earlier by Osborne Reynolds, and Hardy's demonstration of a boundary state governed by chemical properties of lubricants exposed a completely new field of study. The mechanism was later studied extensively and termed 'boundary lubrication' by Hardy. He was aware of the work of Langmuir and of Harkins reported in the *Journal of the American Chemical Society* in 1917 which emphasized the orientated and monomolecular nature of surface films, but his own experiments led him to the view that orientation effects extended well beyond the range of a single molecular layer. The debate on mono- and multi-molecular layers in boundary lubrication continues to the present day. It was Hardy's joint paper with Miss Ida Doubleday which firmly established the basic concepts in boundary lubrication (Hardy and Doubleday, 1922). An impressive feature of Hardy's studies of friction and boundary lubrication was the simplicity of the apparatus. For example, for his work on glass he used watchglasses loaded by different weights of lead.

Hardy achieved the unique distinction of presenting both the Croonian and the Bakerian Lecture to the Royal Society. The latter, presented jointly with Dr I. Bircumshaw in 1925, was devoted to the subject of boundary lubrication.

Hardy received many honours, including degrees from Aberdeen, Birmingham, Edinburgh and Oxford. He was a Royal Medallist and, as previously mentioned, Croonian and Bakerian Lecturer of the Royal Society. He was knighted in 1925 and in 1931 he delivered the Abraham Flexner Lectures at Vanderbilt University in the United States. He was a member of the Economic Advisory Council, President of the British Association of Refrigeration, a member of the governing body of Charterhouse and of the Leverhulme Trust Committee and, from 1922, a trustee of the National Portrait Gallery. At the time of his death he was President of the British Association.

In his leisure pursuits Hardy displayed zest and enthusiasm. He loved nature, good literature and greatly valued close friendships. He was an adventurer with a great love of the sea, having been described by Professor A. V. Hill as one of the finest yachtsmen of his time and a sailor who combined perfect seamanship with a capacity for instant decision in an emergency. He married Alice Mary Finch, the daughter of a barrister, in 1898 and upon his death in Cambridge on 23 January 1934 he was survived by his wife, son and two daughters.

It is fitting to close this biographical sketch of the man who opened up such an important aspect of tribology with an extract from the 1926 Anniversary Address of no less a person than Rutherford. Referring to the Royal Medal Rutherford said:

> *... For the first time in the history of science the dependence of friction and lubrication on the structure and molecular orientation of surface films and the force fields of molecules in relation to their structure and polarity have been elucidated in*

a series of beautiful and highly important researches. . . . In all these various fields researches . . . have been characterised by an originality of outlook and an imaginative insight which bears the impress of genius.

Bibliography

Bate-Smith, E. C. (ed.) (1964) *Sir W. B. Hardy–Biologist, Physicist and Food Scientist; Centenary Tributes* (editor: Director of the Low Temperature Research Station, Cambridge).

Dictionary of National Biography (1940) 'Hardy, Sir William Bate (1864–1934)', *Dictionary of National Biography (1931–1940)*, pp. 397–8.

Hardy, W. B. (1936) *Collected Scientific Papers of Sir William Bate Hardy*, Sir Eric K. Rideal (ed.). Published under the auspices of the Colloid Committee of the Faraday Society, Cambridge U. P.

Hardy, W. B. and **Doubleday, I.** (1922) 'Boundary lubrication–the paraffin series', *Proc. R. Soc.*, **A100**, 550–74.

Royal Society (1934) 'William Bate Hardy 1864–1933', *Obituary Notices of Fellows of the Royal Society*, No. 3, Dec., pp. 327–33.

The Times (1934) Obituary notices, *The Times*, 24–25.1.1934.

A.18   Arnold Johannes Wilhelm Sommerfeld

(1868–1951)

Many of the notable early contributors to the science of tribology ventured briefly into the field, made a single and now classical contribution, and then moved on. They were generally philosophers, scientists or engineers of great distinction, unencumbered by present-day pressures of specialization. Such a man was the great theoretical physicist Arnold Sommerfeld, whose name is normally associated with atomic physics and the quantum theory.

Arnold Sommerfeld was born in Königsberg, Prussia, the son of a general practitioner, Dr Franz Sommerfeld and Cäcilie (Mathias) Sommerfeld, on 5 December 1858. He attended the Altstädtisches Gymnasium (High School) in Königsberg where he developed an interest in all the subjects which he studied. If anything he was more interested in literature and history than in science. He received his baccalaureate in 1886, the year in which Osborne Reynolds published his classical theory of lubrication in England, and after some wavering registered in mathematics in the distinguished University of Königsberg. He was taught by Lindemann, Hurwitz and Hilbert at a time when there was much excitement about electrodynamics following the experiments of Hertz. He was awarded his doctorate in 1891 for work on 'The arbitrary functions in mathematical physics'. In an autobiographical sketch prepared for the records of the Vienna Academy, Sommerfeld wrote of his thesis, '. . . I had it roughed out and written down in a few weeks'.

In the same year he had contributed to, but later withdrew from, a prize competition concerned with the evaluation of earth temperature observations at a location in the Botanical Gardens. He wrote that, '. . . My work did not reach the numerical evaluation stage, but in typical fashion, remained stuck in mathematical generalities.'

In 1892 he took his teacher's diploma and then completed his military service before moving to the University of Göttingen as an assistant at the

Mineralogical Institute in October 1893. In the following year he became an assistant to Felix Klein in the Mathematical Institute. Klein had a great influence upon Sommerfeld, and it was during this period that Sommerfeld completed his famous work on the theory of diffraction.

Arnold Sommerfeld was appointed Professor of Mathematics at the Mining Academy in Clausthal in the Harz Mountains in 1897 at the age of twenty-eight. In the Bergakademie his lecturing was concerned mainly with elementary mathematics, but he still managed to extend his important work on diffraction to Röntgen rays, the work being published in 1900. From 1900 to 1906 he held the Chair of Technical Mechanics in the Technological Academy (Technische Hochschule) in Aachen and it was here that he completed his solution of the Reynolds equation for journal bearings. He was exposed to technical problems of the day at both Clausthal and Aachen and he took some satisfaction from the contributions which a 'pure mathematician' was able to make to engineering science. During his few years at Aachen he completed papers on vibrations and electric generators, but his most fruitful endeavours were concerned with the *Braking Effect of Railroad Cars* and *Hydrodynamic Lubrication*. Sommerfeld considered the latter work to be his most important contribution to the technical field, for he wrote;

> *My most important work in this [technical] field was 'Zur hydrodynamischen Theorie der Schmiermittebreibung' (Z. Math. Phys., 50, 1904, pp. 97–155), in which older statements by Petroff and Osborne Reynolds were compared with newer experiences of Stribeck. I was pleased that even in this seemingly inaccessible field an exact solution helped to demonstrate the power of the mathematical-physical thought process.*

This work, carried out during Sommerfeld's six-year stay in Aachen and published in 1904, was the scientist's major contribution to tribology. He provided an analytical solution to the problem of fluid-film lubricated journal bearings of infinite width which still forms the basis of mathematical procedures in lubrication. The full impact of his work is discussed in Chapter 10, but he is best known for his trigonometric substitution which enabled him to solve the Reynolds equation for journal bearings. The curious thing is that he appears not to have used the relationship which is affectionately known as the *Sommerfeld substitution*, but to have employed reduction relationships and a well-known integral formula. Boswall (1928) provided a general form of the procedure which he attributed to Sommerfeld. Whether or not Sommerfeld used the mathematical substitution which now carries his name in the lubrication literature is perhaps of minor importance scientifically, even though the history of events holds great fascination. The important thing is that Sommerfeld succeeded in presenting a neat analytical solution to a problem which had remained only partially solved since Osborne Reynolds introduced it in 1886. Sommerfeld (1921) wrote one more paper on lubrication, but this was more of a review of the concept of fluid-film lubrication and a restatement of earlier work.

Sommerfeld's direct involvement with matters technical was not to last for long. Röntgen, the discoverer of X-rays, initiated a process which culminated in Sommerfeld's move to the Chair of Theoretical Physics at Munich University in the autumn of 1906. He succeeded Ludwig Boltzmann who had worked so fruitfully on the kinetic theory of gases and radiation from dark bodies. He confirmed his great reputation as a teacher and established an outstanding international name for his Institute of Theoretical Physics in Munich. He retired in 1935, but continued to give lectures until he was seventy. His work at Munich was concerned mainly with spectral analysis, quantum theory, relativity and atomic structure. He published some 276 papers together with 13 books and, as Heisenberg said, '. . . not only did Sommerfeld have students–he developed practically a whole generation of theoretical physicists, who are now widely distributed over the world.' Among his students and associates were Hans Berthe, Peter Debye, Carl Eckart, Werner Heisenberg, Heitler, Herzfeld, Landé, London, von Laue, W. Pauli Jr and Linus Pauling. The full list reads like a 'Who's-Who' of theoretical physicists for the period.

Honours were bestowed upon Sommerfeld by institutions in many nations. He was a member or Fellow of the academies of Berlin, Budapest, Calcutta, Göttingen, Madrid, Munich, Rome, Uppsala and Vienna, together with the Royal Society of London, the National Academy of Sciences in Washington, the Academy of the USSR in Moscow and the Indian Academy of Science in Bangalore. He received many honorary degrees and medals and visited several nations. He journeyed round the world in 1928–29, visiting India, Japan and America.

Arnold Sommerfeld enjoyed the warmth of friendship with his students. He frequently joined them on skiing expeditions and he also enjoyed mountain climbing and classical music. His marriage to Johanna Höpfner daughter of the Curator of Göttingen University in December 1897 led to a happy family life. They had three sons and a daughter of whom all but one son survived him. He lived in and contributed to an exciting period in the history of mathematical physics, but he also endured the upheavals of political strife and war. He was a Protestant who described himself as a national liberal. He felt deeply about the impact of Adolf Hitler upon German science and found solace in his writing and the preparation of books on physics during the Second World War. He once wrote that, '. . . The next to the last volume of this series is now coming into print. Without this work I would hardly have been able to live through the political upheavals of war time.'

Sommerfeld died on 26 April 1951 as a result of injuries received in a traffic accident. His mental agility was retained to the end and he had been editing the last volume of his lectures prior to the accident. It is most fortunate that this great scientist found time during his creative life to make a single but most penetrating study of fluid-film lubrication in journal bearings.

Bibliography

Archibald, F. R. (1955*d*) 'Arnold Sommerfeld', *Lub. Enging*, July–Aug., 228–9 and 283.

Born, M. (1952) 'Arnold Johannes Wilhelm Sommerfeld; 1868–1951', *Obituary Notices of Fellows of the Royal Society*, **8**, No. 21, 274–96.

Boswall, R. O. (1928) *The Theory of Film Lubrication* (Longmans, London), see p. 196.

Condon, E. U. (1951) 'Arnold Sommerfeld', *J. Opt. Soc. Am.*, **41**, No. 2, Feb., 63.

Current Biography (1950) 'Sommerfeld, Arnold Johannes Wilhelm', *Current Biography*, pp. 537–8.

Heisenberg, Von W. (1951) 'Arnold Sommerfeld', *Die Naturwiss.*, Jakrgang 38, Heft 15, Aug.

Nature (1951) 'Arnold Sommerfeld', *Nature*, **168**, 364.

Pauling, L. (1951) 'Arnold Sommerfeld: 1868–1951', *Science*, **114**, 12 Oct., 383–4.

Sommerfeld, A. (1904) 'Zur hydrodynamischen Theorie der Schmiermittehreibung', *Z. Math. Phys.*, **50**, 97–155.

Sommerfeld, A. (1919 and 1950) 'Autobiographical sketch written by Arnold Sommerfeld for the records of the Vienna Academy'.

Sommerfeld, A. (1921) 'Zur Theorie der Schmiermittebreibung', *Z. Tech. Phys.*, **2**, No. 3, 58–62; No. 4, 89–93.

The Times (1951) 'Arnold Sommerfeld', *The Times*, 28.4.1951.

A.19   Anthony George Maldon Michell   (1870–1959)

Anthony George Maldon Michell, inventor of the tilting-pad bearing, was a first-generation Australian even though he was born in Islington, London on 21 June 1870 while his parents were visiting friends, parents and relatives from 1870 to 1873. His parents John Michell (1826–91) and Grace Michell, née Rowse (1828–1921) both came from Devon and were married in Tavistock. The menfolk of both families had been engaged in mining in Cornwall and Devon for many generations before John and Grace set out to seek their fortunes in the newly discovered Australian goldfields. They left England in 1854 and landed in Port Philip, Victoria, in 1855.

A. G. M. Michell[1] was the fifth and last child in the family. He lived in Maldon, Victoria, until he was seven years old and then in a suburb of Melbourne until his parents again returned to England in 1884 when he was fourteen years old.

His sisters and his only brother provided early and effective instruction in reading, writing and number. When he entered one of the newly established Victorian State Primary Schools he was placed in a class of children considerably older than himself. His primary school education continued in Melbourne and was supplemented by out-of-hours lessons from some of the teachers in more advanced subjects which included Latin and geometry.

The Michells formed a closely knit family. George described his parents as, '. . . energetic and adventurous, but serious minded people; unscholastic, but very respectful to scholarship and very quick to recognize intellectual superiority.'

---

1.   Michell himself has explained that the family name is pronounced with emphasis on the first syllable and that the latter should rhyme with 'rich' or 'which'. His first name, which was that of his maternal grandfather, was traditional in the family and thought to be connected with French ancestry, although the family had lived in the West Country for many years. His somewhat unusual third name was the same as that of the small rural mining field in Victoria where his parents had settled.

In a remarkable demonstration of these qualities the whole family moved from Melbourne to Cambridge in 1884 to launch the elder son John Henry on his mathematical career. Subsequent events were to justify this return to England, for John became senior wrangler, Smith's Prizeman, a Fellow of Trinity College, Professor of Mathematics at Melbourne and, in 1902, a Fellow of the Royal Society. John Henry Michell made significant contributions to the theory of elasticity and hydrodynamics at the turn of the century. He died in 1940 and a fine account of his life was written in the *Obituary Notices of Fellows of the Royal Society* by his brother.

While his brother was establishing himself in the realms of mathematics at Cambridge, George Michell attended the Perse School for three years with some distinction. He studied Latin, Greek and mathematics and cleared the University's local examinations at both junior and senior level with distinctions in both classical and mathematical subjects. On leaving the Perse School he was awarded most of the prizes for the senior class. Organized sport and formal games held no attraction for him, but while at school he became, in his own words, '. . . a strong and swift pedestrian'.

During one summer he walked over 1,000 miles (1,600 km) on the roads of eastern England visiting and sketching medieval cathedrals and churches, and in later years he turned his skill to practical advantage when visiting sites of engineering works in rough terrain in Victoria and Tasmania.

An explanation of the next and critical stage in his career which appeared in autobiographical form provides an interesting insight into the personality and modesty of A. G. M. Michell.

> *. . . Leaving 'Perse', he, knowing well that he had neither the special mathematical talent nor the industry necessary to follow his brother through the University course with like success, made up his mind to be trained in Australia for some occupation giving more scope for his constructive faculties–either architecture, or some branch of engineering or agriculture.'*

Before returning to Australia to pursue this aim and for reasons described as, '. . . Circumstances making return to Australia impracticable . . .', Michell spent a year at the University of Cambridge as a non-collegiate student. He attended lectures on mechanics in the new School of Engineering as well as lectures on physics, chemistry and Greek art. In this brief period there occurred one of those interesting cross-links which occur in the fabric of history, for his lectures on physics were presented by Osborne Reynolds' most distinguished student, J. J. Thomson. The family returned to Melbourne in 1890, but the size of the household was sadly reduced by the death of the father, John Michell, in 1891.

George Michell entered the University of Melbourne in 1890 to take courses in architecture, civil and mining engineering. The former was soon dropped, but Michell achieved first and second places respectively in the list of honours for the engineering subjects. Practical experience was obtained in the office of Bernhard Alexander Smith, one of his university teachers and a hydraulic engineer, and in the workshops of Johns and Waggood, hydraulic and structural engineers. Michell graduated in civil engineering in 1895 and was awarded the final scholarship for the course. The degree of Master of Science was awarded two years later. He joined Bernhard Smith's consulting firm as a pupil-assistant and later as a partner before establishing his own professional business centred on hydraulic engineering in 1903. This independent venture was started after Michell served as Examiner of Patents in the Victoria Patent Office from 1902 to 1903.

Soon after graduation Michell completed an analytical and experimental study of the instability of elastic beams which was published in the *Philosophical Magazine* (Michell, 1899) as his first paper. In 1904 he published a paper on the limits of economy in framed structures which clearly demonstrated his ingenuity. The paper was well in advance of wider interest in the optimum design of structures and it was not until the middle of the twentieth century that aeronautical engineers rediscovered and extended the concept.

Since the motivation for Michell's work on bearings originated in hydraulics it is of interest to examine his experience in this field. The firm of Bernhard Alexander Smith was concerned with the design and manufacture of pumps and turbines and the Smith–Michell regenerative centrifugal pump was invented and patented by the firm in 1901. In this machine the fluid leaving the impeller was directed on to a turbine wheel connected through gears to the pump shaft. Up to three-quarters of the kinetic energy imparted to the stream was thus recovered and many large-capacity machines were constructed in this design for irrigation schemes in Australia.

After forming his own company Michell took over the turbine business of the former partnership and specialized as a consultant on power plants with particular emphasis on hydroelectric applications. He was a consultant to the Melbourne firm which made and installed the pumping machinery for the Murray Valley Irrigation Scheme and to the Tasmanian company responsible for one of the first Australian hydroelectric schemes. As hydroelectric schemes advanced and the size of turbines increased, the problem of accommodating large thrust loads loomed large and soon presented a major obstacle to further development in this field. It was this problem which taxed the inventiveness and scientific skill of Anthony George Maldon Michell and which promoted the development of the famous tilting-pad bearing.

Michell's fundamental investigations of methods of supporting large thrust loads in hydraulic machinery were apparently made in the short period 1902–4 and his now classical solution was described in a paper published in English in the *Zeitschrift für Mathematik und Physik* (Michell, 1905*b*) and covered by patents in England and Australia on 16 January of the same year. This must have been an exhilarating period in Michell's career, for he was still in his early thirties when he solved a most difficult analytical problem and immediately invented an outstanding mechanical device to execute the required function. Osborne Reynolds had considered the lubrication of plane surfaces in 1886, but Michell was the first to obtain solutions which took account of side-leakage some nineteen years later. This was a truly remarkable achievement. On the occasion of the James Watt International Presentation in 1943 Professor Andrew Robertson, Vice-President of the Institution of Mechanical Engineers noted that the feat was even then regarded by some as the only really important extension of Reynolds' theory yet effected.

If the mathematical solution of the problem tackled by Michell was remarkable, the practical device which resulted was both beautiful in its simplicity and spectacular in its achievement. The tilting-pad bearing, which adjusts its inclination to the runner to operate in an optimum fashion under all operating conditions, is a wonderful example of engineering at its best. To quote Professor Robertson again, '. . . Few inventions have provided so complete a solution to an engineering problem, and it belongs to that small class in which there is little scope left for further practical development.'

In the United States Albert Kingsbury also conceived, about the turn of the century, the idea of the tilting-pad bearing thus providing a further example in the field of tribology of simultaneous and independent developments in different parts of the world. Kingsbury started his work in this field in 1898, but his first publication and award of an American patent were dated 1910. His early work appears to have been more empirical than Michell's and the latter's solution of the Reynolds equation for three-dimensional flow was undoubtedly unique. Kingsbury's early bearings had a central pivot, but it is worth noting the following authoritative and generous commentary on the question of priority written by Michell in 1929.

> . . . *In this connection, the author avails himself of the opportunity to explain that the pivoted thrust bearings known in America and Europe respectively as the* 'Kingsbury' *and* 'Michell' *bearings are in principle the same; and to state that the development of these bearings was affected in its early stages by Professor Kingsbury and himself independently in their respective countries without knowledge of each other's work. Professor Kingsbury's work was commenced a few years earlier, though his first publication on the subject was later than the author's.*

It is interesting to note that the bearing was initially developed by each man for use in hydraulic machinery, with the important marine applications following soon after. The first tilting-pad bearing in service was probably that built by George Weymouth (Proprietary) Ltd of Melbourne under Michell's guidance for a centrifugal pump at Cohuna on the Murray river, Victoria. Michell and Mr H. T. Newbiggin became partners in a company founded in England and based upon Newcastle upon Tyne for the manufacture of the bearings. By 1913 the great merits of the tilting-pad bearing had been recognized for marine applications. Ironically, it is said that it was the use by Krupps, who had declined to take licences from Michell, which stimulated interest elsewhere. The first English ship to be fitted with the bearing was the cross-Channel steamer the *Paris*, but many naval vessels were similarly equipped during the First World War. The practical results were spectacular. The troublesome thrust block became smaller, more efficient and remarkably free from maintenance troubles. Specific loadings rose by a factor of ten and typical coefficients of friction were reduced by a factor of twenty from

0·03 to 0·0015. It was estimated that the Royal Navy saved coal to a value of £500,000 in 1918 alone as a result of fitting Michell's tilting-pad bearings.

Michell later developed tilting-pad journal bearings of improved performance. He used an accurately manufactured air-lubricated demonstration model of a tilting-pad thrust bearing which was the fore-runner of many similar items of equipment to be found in present-day teaching laboratories. He was undoubtedly inventive, for in addition to the device for which he is best known, he also developed an impulse turbine, a telegraph cipher system, a cipher decoding machine (jointly with Mr H. C. Newton), a crankless engine and a workshop viscometer. In the Michell crankless engine developed in 1917, four or more cylinders were arranged with their axes parallel to the driving shaft, and reciprocating motion was converted into rotary action via a swash-plate inclined at $22\frac{1}{2}^{0}$ (0·39 rad) to the normal. Tilting-pad slippers were naturally used on the swash-plate. By 1925 the crankless engine had been used quite widely in Australia in air and gas compressors, diesel engines and motor cars. Michell's own Buick was fitted with one about 1921. He travelled extensively in Europe and the USA between 1925 and 1933 to promote the crankless engine and it has been said that this activity probably absorbed most of his royalties from tilting-pad bearings. This venture failed to match his success in the bearing field, notwithstanding many attractive features of the engine, and he returned to Australia in 1932 having left the future of the device in the hands of the Michell Crankless Engine Corporation (1928) of New York.

Michell's workshop viscometer consisted of a 1 in (25·4 mm) diameter steel ball sitting in a metal cup of slightly greater diameter. A small sample of the test lubricant was applied to the cup and the ball was then placed in position on three small projections arranged to provide some initial clearance. The time taken for the ball to separate from the cup was a measure of the viscosity of the lubricant. This simple device is now rarely used, but it demonstrated once again Michell's talent in relating theory to practice. He always strove to relate the science to the art of mechanical engineering. In his book on lubrication he quotes from Leonardo da Vinci the following motto: '. . . theory is the captain, practice the soldiers'.

During his visits to Europe and the USA between 1925 and 1932 Michell took the opportunity to meet some men famous for their contributions to science and engineering. Noteworthy in the present context are the names of A. Sommerfeld, P. Kapitza and, of course, A. Kingsbury.

In 1934 Michell became a Fellow of the Royal Society in the first year of his nomination. At this stage the family held the rare distinction of having two living Fellows and it is known that George was deeply sensitive of the honour. He wrote: '. . . On account of his having been resident in Australia during (and for some years before) his election (while his nominators and supporters were all in Britain), he was, and still is, quite ignorant of the circumstances of his election.'

He received from the University of Melbourne in 1938 the Kernot Memorial Medal which was awarded for distinguished engineering achievement in Australia. He was the recipient of the James Watt International Medal for 1942 on the nomination of the Institution of Engineers, Australia, the Engineering Institute of Canada and `the South African Institution of Engineers. The Institution published a fine account of the presentation ceremony which was held in London on 22 January 1943. The circumstances of the Second World War prevented Michell from being present in person, but he was able to listen to a direct broadcast of the proceedings through the suggestion and generous courtesy of the British Broadcasting Corporation. The Rt Hon. S. M. Bruce, then High Commissioner for the Commonwealth of Australia, received the medal on behalf of Michell before 172 members and 40 visitors and arranged for its transference to Australia. In typical fashion Michell gave the premium associated with the James Watt International Medal to Melbourne University for the funding of a prize in mechanical engineering.

In 1950, at the age of eighty, Michell published his fine book on lubrication. Of distinctive style, the book is wide-ranging yet concise, original and thorough, thus providing a good starting point for students of the subject.

Of his human characteristics it can be said that Anthony George Maldon Michell was a man of firm principle, precise thought and expression and considerable reticence. In his scientific and professional life he adhered to the most rigorous standards. In the 1920s he appeared as an expert witness in a number of law cases involving patents and their underlying physical principles. His modest demeanour and patience in presenting simple, clear and convincing arguments, even under cross-examination, caused him to be respected and much in demand. That he soon gave up this activity suggests that its semi-public nature and certain distasteful aspects of special pleading were discordant with his personality. He apparently charged only one-quarter of the minimum professional fees and responded to suggestions that he should charge more realistic rates with a slow smile and a shake of the head. He was averse to publicity and avoided it whenever possible. He appears to have been a genuinely modest man of great sincerity. He was considerate and helpful to others and, according to the Royal Society obituary notice written by T. M. Cherry in 1962, '. . . possessed of the sort of courtesy which is nowadays called old fashioned'. He had little time for small talk, but was every ready to discuss real questions with a keen intelligence sprinkled with a dry humour.

Michell found much pleasure and interest in horticulture and arboriculture. In 1911 he acquired and later extended his forest farm 'Ruramihi', some 50 miles (80 km) from Melbourne. He was quite a conservationist, for he protected the native flora, forest landscape and animal life against the ravages of forest burning which was, and still is, prevalent in parts of Australia. His depth of feeling and emotional concern for the environment of Ruramihi and its farm-homestead surfaced at one point in his autobiographical note for the Royal Society:

> . . . *The possession of such a home, in some like environment, has always been for him, as it was for his father, a condition essential for his mental health and comfort;*

and it has been his lasting conviction that no other dwelling places can ever be compatible with durable happiness for mankind, or with the permanent stability of any human society.

At the time of writing this biographical sketch I was privileged to spend some six months in Australia where I took the opportunity to visit the State of Victoria, Michell's home city of Melbourne and his university. Such visits greatly enhanced my appreciation of the personality, life and achievements of Anthony George Maldon Michell. He was a bachelor and lived at 413 Collins Street, Melbourne, until his death on 17 February 1959 at the age of eighty-eight. The fine photograph of this gentle genius of the art and science of tribology which is reproduced here was included in the Institution of Mechanical Engineers' account of the James Watt International Medal presentation proceedings. I would like to record my appreciation of the opportunity to include this autographed version, which was supplied by Mr F. R. Archibald, and the permission to publish afforded by the Institution of Mechanical Engineers.

It seems fitting to close with a quotation from William Wordsworth on James Watt which was repeated by Sir Henry Barraclough on one of the few occasions when Michell was persuaded to lecture on his work.

*. . . I look upon him, co~ ⋅   ⋅tude and universality of his genius,
as perhaps the m⋅   ~ntry has ever produced: he
never sought disp   ⋅~ness and humility both
of spirit and of   ⋅s truly great and
good was ever*

Bibliography

Archibald, F.
    Sept.–O⋅
Cherry, T
    of Fel⋅
Instn. o⋅
    to ⋅
    E
Mi⋅

N

work on engine vibration problems at Harland & Wolf, Belfast, during the summer vacation of 1897.

It was in 1897, when Gümbel was twenty-three years old, that he commenced his first scientific investigation. This early study of the stability of ships, which was published in the *Transactions of the Institution of Naval Architects of London* in 1898 and as a book, launched Ludwig Gümbel on a distinguished career. Upon graduation he first joined the firm of F. Schichau in Elbing as a *kongstruktur für Schiffsmaschinenbau* on 1 January 1899. One year later he moved to the shipping company of Hapag-Hamburg, where he became head of the mechanical engineering division by 1905. The next move, again on the first day of a new year, took him to Bremen on 1 January 1906, where he became Deputy-Director of the company then known as Norddeutsche Maschinenund Armaturenfabrik, but which subsequently became Atlas Werke.

During his eleven and a half years of activity in industry, Gümbel worked on a wide range of vibration problems in marine engineering, including the transverse vibrations of bars and ships' hulls and the torsional vibration of shafts. He developed graphical methods for solving the differential equation of vibration theory and evolved 'Gümbel's' method for the determination of torsion eigenfrequencies of systems of rotating masses. His ability to analyse complex problems associated with large-scale systems was amply illustrated by the publication of respected papers during this period. His contributions to engineering science were recognized by the award of a doctorate, Dr -Ing., in 1909.

Gümbel returned to the Technische Hochschule, Berlin-Charlottenburg, to assume the title Professor of Schiffsmaschinenbau on 1 October 1910. In his new academic career he offered introductory lectures on mechanical engineering, together with courses on ships and their machinery, based upon his wide experience of marine engineering. In the next few years his publications reflected a shift in emphasis in his interests from solid to fluid mechanics. In 1913 he published two extensive papers dealing with the skin friction of moving ships and the theory of screw propellers which clarified contemporary views on these subjects and were soon recognized as outstanding contributions to knowledge.

It was in the field of fluid-film lubrication that Gümbel achieved his greatest success as a hydrodynamicist. In 1914, when he was forty years old, he applied himself to a study of self-acting journal bearings. He analysed Stribeck's[1] experimental findings on the friction of journal ~ings and showed for the first time that the various curves to the the minimum could be condensed into a single graph if the friction was plotted against the parameter [viscosity × load per unit area]. He also determined theoretically entre within the clearance between bearing and ~iation of external operating conditions. A ~strated by a summary paper published ~ionships which could be used by bearing designers unfamiliar with the intricacies of the Reynolds equation. He recognized that detrimental grooving patterns were introduced into various journal-bearing design procedures and endeavoured to correct the practice, but it was to be many decades before designers learnt this particular lesson.

In the First World War Gümbel served his country and was involved for over three years in heavy fighting. He rose to the rank of company commander before being recalled by the Navy to help to build up the U-boat fleet. He was decorated for his wartime services as can be seen in the photograph at the beginning of this section.

It should be recalled that in the two opening decades of the twentieth century, the nature of boundary lubrication was not appreciated, even though several experiments from Thurston to Stribeck had demonstrated the existence of distinct fluid-film, boundary and even mixed modes of lubrication. The great conflict between those who championed the role of either fluid-film or boundary lubrication in the functioning of machine elements was nowhere more hotly disputed than in Germany. Gümbel became embroiled in this dispute and was moved to write on the subject (Gümbel, 1920) with a contribution entitled 'Wer ist der wirklich Blinde?'– who is actually the blind one?

The question of cavitation and the appropriate boundary conditions to be applied to the Reynolds equation in cases where the lubricating film exhibited a divergent form was considered in 1921. It was an early and significant contribution to a difficult subject which did much to establish for designers a more realistic set of solutions to the Reynolds equation than that proposed by Sommerfeld in 1904.

In all, Ludwig Gümbel published some seventy-five papers, of which sixteen were devoted to lubrication topics. It was, however, a posthumous publication which confirmed his position in the history of our subject. Early in the 1920s he worked on the manuscript of a book and much of it was complete on the occurrence of his untimely death in 1923. In his will he requested his friend and colleague Professor E. Everling to complete the manuscript and to edit the text. The work was carried out expeditiously and the Gümbel–Everling (1925) book was published only two years after Gümbel's death.

The book had a great impact in Germany, although it appears that only a limited number of copies were printed. It was published in German and not widely read in other countries. The major part of the text was devoted to applications of the Reynolds equation and much of it had not been published previously in papers. For example, it included what was probably the first systematic approach to externally pressurized bearings; a solution of the Reynolds equation for meshing gear teeth similar to, but apparently independent of, that by Martin (1916);[2] and, perhaps most surprising of all, an exercise for the student reader of the text on a type of self-acting spiral-groove type of bearing. Blok (1976*a*) has mentioned that this exercise was to prevent himself and others who

---

2. See Sect. 10.4.

independently developed the idea many years later, from safeguarding the principle of their invention by means of patents.

The text included not only derivations of the major theoretical results, but also exercises and design drawings. It was clearly intended for students and machine designers alike. When full solutions to the governing equations were not available, as in the case of journal bearings of finite width, Gümbel was willing to introduce sensible approximations which provided valuable data for the designer. Sound physical and engineering foundations supported novel analysis to yield a book which contained valuable design rules for bearings and which had a long-lasting influence on German designers and tribologists. This was seen by many to be his greatest achievement. Gümbel turned to the problem of lubrication and bearings relatively late in his restricted life-span, yet he achieved so much in the spell of a single decade.

Gümbel's work on lubrication and bearings yielded such a fine statement of basic principles and a penetrating analysis of fluid-film bearings, crowned by the compilation of design guides, that it was a truly remarkable contribution to the subject of tribology early in the twentieth century. The work was well known in Europe, and Germany in particular, but it remained in obscurity in many English-speaking countries. In a touching acknowledgement of the impact of Gümbel's work upon 'a budding tribologist' in the 1930s Professor H. Blok has described Gümbel's posthumous book as '. . . one of his major sources of both inspiration and basic knowledge'.

Gümbel married Olga Catharine Dietz, daughter of a New York wholesale merchant, in 1902. They had three sons and one daughter. He died early, about one month before his forty-ninth birthday, after suffering for many years from an illness which was a legacy of the First World War. His untimely death took place in Berlin-Charlottenburg on 8 February 1923.

In preparing this account of the life of Ludwig Karl Friedrich Gümbel I have been greatly helped by a number of people. Professor H. Blok of the University of Technology, Delft, Holland, kindly prepared a set of notes based upon his personal records; Professor Dr-Ing. H. Czichos of the Bundenstalt für Materialprüfung (BAM), Berlin, sought background information, prepared biographical notes and unearthed the photograph of Gümbel; Professor Dr-Ing. Jörn Holland of the Institut für Reibungstechnik and Maschinenkinetik, Technische Universitat, Clausthal, kindly drew my attention to a biography on Gümbel prepared by Georg Schnadel; and my colleague John Schwarzenbach translated a number of German documents. I acknowledge with pleasure this assistance which made it possible for me to prepare these notes on a truly remarkable German engineer and tribologist.

Bibliography

Blok, H. (1976) 'Notes on Professor Dr-Ing. Ludwig Gümbel, private communication.

Czichos, H. (1976) 'Professor Dr Ludwig Gümbel–biographical Notes', private communication.

Gümbel, L. (1941*a*) 'Das Problem der Lagerreibung', *Monatsblätter, Berl. Bez. Verein. deut. Ing.* (V.D.I.), No. 5, 1 Apr. and No. 5 May, 6 June, 97–104, and 109–120 (also July 1916).

Gümbel, L. (1914*b*) 'Die Schubkraftmaschine', *Z. Ges. Turbinenwesen*, **11**, Nos. 23, 24 and 25, pp. 357–60; 372–6 and 381–4.

Gümbel, L. (1914*c*) 'Über geschmierte Arbeitsräder', *Z. Ges. Turbinenwesen*, **11**, 22–26.

Gümbel, L. (1920) 'Wer ist der wirklich Blinde?', *Eine Frage im Interesse von Wissenschaft und Technik; offener Brief an die Herren A. Reidler und St. Löffler; mit einen Beitrag: Due unmittelbare Reibung fester Körper* (Springer, Berlin).

Gümbel, L. (1921) 'Verleich der Ergebnisse der rechnerischen Behandlung des Lagerschmierungsproblem mit neueren Versuchsergebnissen', *Monatsblätter Berl. Bez.–V.D.I.*, Sept., 125–8.

Gümbel, L. and **Everling, E.** (1925) *Reibung und Schmierung im Maschinenbau* (M. Krayn, Berlin), pp. 1–240.

Jahrbuch der schiffsbautechnischen Gesellschaft (1924) *Jahrbuch der schiffsbautechnischen Gesellschaft*, 25 Jakrgang, pp. 37–8.

Schnadel, G. (1966) 'Ludwig Karl Friedrich Gümbel', *Neue Deutsche Biographie*, Band 7, Seite 258 (Verlag Duncker und Humblot, Berlin).

Z. Verein. deut. Ing. (1923) *Z. Verein. deut. Ing.*, **67**, 762.

A.21 Herbert Walker Swift (1894–1960)

Herbert Walker Swift was born on 15 December 1894. He was educated first at elementary school, but then won a place at Christ's Hospital on the basis of '. . . a few sums and an essay one Saturday morning'. His undergraduate days at Cambridge, where he was a scholar in St John's College, were interrupted by active service during the First World War. After being first rejected on medical grounds, he was accepted by the Army in 1915. He served in France, attained the rank of captain, was wounded and mentioned in dispatches before returning to Cambridge in 1919. He had initially intended to graduate first in mathematics and then in engineering, but the intrusion of the war caused him to proceed directly to the Mechanical Sciences Tripos, in which he gained first-class honours and was a prizeman in 1920. He also achieved a half-blue for swimming during his days at Cambridge.

After leaving Cambridge, Swift obtained industrial experience with Hollins Bros, a firm of hydraulic turbine manufacturers in Kendal. He was to retain an interest in hydraulics throughout his life. He rapidly achieved the position of chief engineer to the company, but in 1922 he turned away from industry to enter the academic world. His move to the University of Leeds as a demonstrator and assistant lecturer brought him into contact with Professor John Goodman,[1] and the two men soon became good friends and collaborators. John Goodman was a talented experimenter, but he found in Herbert Walker Swift a skilful theoretician who was able to analyse the intriguing observations of journal-bearing performance. It was during this period that Swift developed his penetrating researches on lubrication which greatly advanced the subject. His stay at Leeds was short, but in a fruitful four years in which he carried a heavy teaching

---

1. See Sect. A.15.

load, he also published papers on stress analysis and flow through orifices. He also met and married in 1924 a lecturer in biology in the University by the name of Maisie Hobbins.

In 1926 Swift was appointed Head of the Department of Mechanical Engineering at Bradford Technical College. He accepted a heavy teaching and administrative responsibility, but once again he somehow found time to apply his agile mind to research. He developed his researches on lubrication, but also considered other aspects of the mechanical sciences. His study of mechanical power transmission in general, and belt drives in particular, enabled the latter subject to be placed on a sound scientific basis. It also attracted the attention of industry and confirmed his reputation as an engineer capable of combining sound theory with careful experimentation. Further recognition of his scholarship came in 1928, when he was awarded the degree of D.Sc. of the University of London on the basis of work completed with a single grant to the value of £5. He introduced and developed several courses at Bradford, including one which enabled students to read for a London external degree entirely by evening work. Swift remained in Bradford for ten years and during this time he published important papers on the lubrication of steadily and dynamically loaded journal bearings. These included his well-known minimum potential energy approach to the specification of the cavitation boundary condition (Swift, 1931) and the development of bearing design procedures (Swift, 1935) on the basis of hydrodynamic theory. He retained his contact with John Goodman throughout this period and even provided test bearings for his former professor's retirement studies carried out in a garage.

Swift was appointed to the Chair of Engineering in the University of Sheffield in 1936 with responsibility for activities in civil, mechanical and electrical engineering. The move at once enabled him to enlarge his range of research topics, while preserving and extending his interest in lubrication and bearings. His analytical skill was directed to problems encountered in the pressing of sheet-steel panels for motor cars, and in due course his work on metal forming and the basic theory of plasticity achieved international recognition. Able research students anxious to work with Swift soon formed part of an excellent postgraduate school, while his work was encouraged by grants from industry, the British Iron and Steel Research Association, professional institutions and the Royal Society. During this period Swift (1937a) also completed his pioneering work on dynamically loaded bearings.

Baildon and Boulton (1960) have testified to Swift's quality as a teacher. He combined lucidity with inspiration and wit with logic. Always willing to give guidance and assistance to his students, he was particularly sensitive and sympathetic to their personal problems. He clearly generated affection and respect and many informal technical discussions were accompanied by beer and cakes on the lawns of his home. In short, Swift had time for his students.

A combination of shrewdness, mature judgement, enthusiasm and ceaseless effort inevitably drew Swift into wider activities within and outside the University. He was Dean of the Faculty of Engineering for nine years, which included the difficult period of the Second World War. During this time a fourth term was added to the academic year in the summer months to enable degree courses to be completed in two years and three months. Swift's organizational skill smoothed the introduction of such a scheme in the Engineering School and he took direct control of the workshops when they were turned over to production for war.

Swift was for many years an active member of the Institution of Mechanical Engineers, being a founder member of the Yorkshire branch and its first honorary secretary. He became Chairman of the branch in 1936 and he served on the Council of the Institution from 1946 to 1952. He also became a member of the Council of the Sheffield Society of Engineers and Metallurgists in 1936 and was President in the 1950–51 Session. He devoted time and interest to the Regional Academic Board (Yorkshire) and the National Advisory Council for Higher Technological Training. He became a member of the newly created Mechanical Engineering Research Board in 1948 and he was the first Chairman of the British Standards Institution Committee on Industrial Instrumentation.

For his outstanding contributions to research and engineering education Swift was awarded the Thomas Hawksley Medal, the Joseph Whitworth Prize and the Clayton Prize (1953) of the Institution of Mechanical Engineers, together with the Diploma of the Institution of Foundrymen. His work on lubrication led to the award of the Manby Premium by the Institution of Civil Engineers.

Swift's active professional life was complemented by leisure interests which included football, gardening and caravanning. Ill-health caused him to retire at the early age of sixty in 1955. He retired with his wife to a cottage in Suffolk, where he died suddenly on 14 October 1960. He was survived by his wife Maisie and their two daughters.

Herbert Walker Swift made significant contributions to many aspects of mechanical engineering science, but his contributions to the subject of lubrication alone justify this recognition of his stature.

In the preparation of this note I had access to personal notes prepared by Dr Philip Neal and obituary notices written by his colleagues in the University of Sheffield. I am most grateful to Dr Neal for providing this background information and the photograph of Herbert Walker Swift.

Bibliography

Baildon, E. and Boulton, N. S. (1960) 'Professor H. W. Swift–obituary notice', *University of Sheffield Gazette*.

Boulton, N. S. (1955) 'Appreciation–Professor H. W. Swift', *University of Sheffield Gazette*, pp. 3–4.

Swift, H. W. (1931) 'The stability of lubricating films in journal bearings', *Proc. Inst. civ. Engrs*, **233**, 1931–32, 267–288; Discussion 289–322.

Swift, H. W. (1935) 'Hydrodynamic principles of journal bearing design', *Proc. Inst. mech. Engrs*, **129**, 399–433.

Swift, H. W. (1937) 'Fluctuating loads in sleeve bearings', *J. Inst. civ. Engrs*, **5**, 161–95.

W. J. (1961) 'Professor H. W. Swift, M.A., D.Sc. (Eng.)' *The Chartered Mechanical Engineer*, Inst. Mech. Engrs, Personal Notes, January, pp. 48–9.

Who Was Who (1960) 'Swift, Herbert Walker', *Who Was Who*, 1951/60, p. 1061.

A.22 Georg Vogelpohl (1900–75)

Georg Vogelpohl was for forty years an outstanding personality in the field of lubrication research in Germany in the middle of the twentieth century. He was born in Osnabrück on 15 July 1900 and he attended school in the town until the Easter of 1915. He then obtained practical experience in mechanical engineering in various workshops in the steelworks before joining the Osnabrück firm of Brück, Kretschel & Co. on 1 December 1917. During the period in which he worked in the design office he developed an appreciation of the value of education and he attended evening classes to study for his school-leaving certificate.

In September 1918 he was called up for a brief period of military service, being released again on 31 December of the same year. He passed his examination for upper secondary schooling in Hameln in January 1919 and then attended a special course for ex-servicemen at the Scientific Secondary School in Osnabrück. He was awarded his school-leaving certificate in September 1919.

By the age of nineteen Vogelpohl had already experienced formal education at school, workshop training, night school, military service and a special course for ex-servicemen. He was determined to proceed further with his studies and after a short period as a designer with his former firm, he entered the Technical High School in Hanover in September 1920. He soon experienced financial hardship and since his parents could not support him fully when there were other brothers and sisters to be cared for, he again interrupted his studies in order to earn some money.

For a period of about eighteen months between 1922 and 1924 Vogelpohl worked as a designer and works manager and helped to build up a factory at Aranjuez in Spain. By the time he returned to Germany inflation in his country was reaching its peak, and he found it necessary to work in industry until 30 September. It was during this time that he encountered design problems related to hydrodynamics and he became so absorbed by the subject that he resolved to study further the subjects of mechanics, mathematics and physics. He spent the next three years studying in Münster and Berlin before entering the Technical High School in Berlin in the autumn of 1927. He completed his studies for the Diploma Engineer in the Faculty of Mathematics in 1929, having become a temporary assistant to Professor H. Föttinger in the Institute of Fluid Mechanics in 1928. Vogelpohl's reputation in research was to be established in the next few years, first in the field of fluid mechanics and then in lubrication and friction.

A major piece of work undertaken by Vogelpohl in the early 1930s was related to high-speed trains. In 1932 the Borsig Locomotive Works of the German Federal Railway in Berlin suggested that the speed of express trains could be increased on existing routes if the locomotives were constructed to have a low aerodynamic drag. Vogelpohl carried out wind-tunnel tests in the Technical High School in Berlin and demonstrated the potential for reducing power requirements by about 70 per cent by careful attention to the external profile of the locomotive. His data provided the basis for the construction of the first steam locomotives in the world to reach speeds in excess of 124 mph (200 km/h); the large, streamlined Series No. 5 engines of the German Federal Railway. These studies of the resistance to motion inevitably caused Vogelpohl to consider the nature of friction. He recognized the great importance of the phenomenon, yet found the current level of scientific and technological understanding quite inadequate. At the same time he became quite fascinated by the Reynolds equation of fluid-film lubrication, and for forty years from 1935 these basic tribological subjects were to form his major field of investigation.

In 1936 Vogelpohl completed his dissertation entitled 'Contributions to the understanding of slider bearing friction' and in 1937 he was promoted from the post of assistant to Chief Engineer in Professor Föttinger's Institute. In his dissertation he provided a valuable solution for the circumferential and longitudinal pressure distribution in a journal bearing based upon trigonometric functions and the procedures of Ritz. There followed within a space of about fifteen years a number of significant publications which established Vogelpohl as the leading authority on fluid-film bearings in Germany. Considerable interest was generated by his reference to the work of Helmholtz and the proposition that the pressure distribution in plain bearings adopted the form which minimized the rate of energy dissipation. He wrote about the integration of the Reynolds equation and preferred the Reynolds cavitation boundary condition to the full- or half-Sommerfeld conditions discussed by Gümbel. Perhaps his major contributions arose from a recognition of the importance of thermal considerations in bearings and the need to generate reliable design guides.

Vogelpohl's work became known to Professor Ludwig Prandtl and the two men developed a fruitful collaboration in Göttingen. It appears that Prandtl sought a position for Vogelpohl in the Kaiser-Wilhelm Company from about 1940; his efforts being rewarded by the creation of a special Department of Friction Research at the Kaiser-Wilhelm Institute for Fluid Mechanics Research in Göttingen. The declared aim of his new department was to investigate the nature of friction and to apply the results of research to engineering problems, but the plans were thwarted by air raids and the buildings were heavily and repeatedly damaged.

In April 1943 Vogelpohl began his work as a teacher in the Technical High School, Hanover. He lectured on the physics of oils and hydro-dynamics and in August 1944 was appointed academic lecturer. An inaugural address by Vogelpohl (1949), delivered in 1944, was devoted to the question of heat transfer between lubricating films and the bearing surfaces. At the end of the Second World War, Vogelpohl found himself in the Eastern Zone of Berlin, and with the links with Hanover broken he accepted an invitation to become Professor of Mechanics and Thermo-dynamics in the University of Berlin. In 1946 he was invited by the British authorities to return to Göttingen, where the Max-Planck Company was developing from the former Kaiser-Wilhelm Company. Vogelpohl was at last in a position to develop the Friction Research Department after the long interruption caused by the Second World War.

In 1950 Vogelpohl became a scientific member of the Max-Planck

Institute for Fluid Mechanics Research and this also brought him a place on the Chemical–Physical Section of the Scientific Council of the Max-Planck Company. He was awarded the *venia legendi* for machine elements and fluid mechanics by the Technical High School, Braunschweig in 1952. In 1958 he was appointed apl. Professor, and in 1963 the Braunschweig Scientific Society elected him to corresponding membership. It was in 1958 that he became an honorary professor at the Technical University, Clausthal.

Throughout the 1950s and 1960s the stature of Vogelpohl and his institute in the field of friction and lubrication research became recognized throughout the world. A well-known characteristic of his work was a firm commitment to convert research findings into results of practical value. He was particularly concerned to utilize the theory of hydrodynamic lubrication in the development of bearing-design procedures–an approach exemplified by the book *Reliable Slider Bearings* (Vogelpohl, 1958) which formed his major work in the 1950s. The development of his *Ubergangsformal* (transition formula or volume rule) in which the volume to be enveloped by a bearing was related directly to the load and inversely to the product of speed and viscosity, provided valuable initial guidance for bearing designers and was the epitome of his approach.[1] He developed the concept of a *wear spectrum* based upon measured linear wear rates of a wide range of machine elements operating under various conditions which was to form the basis of many future design guides for tribological components. He was also concerned about the misuse of results from experiments which he believed to be totally unrealistic of engineering situations, particularly where lubricant testing was concerned. He was critical of the use of terms which were so vague that they were of little value to the engineer, *oiliness* and even *boundary lubrication* being included in this category. He found that various authors had used no less than twelve definitions of boundary lubrication and he vigorously contested the relevance of Hardy's[2] watch-glass experiments for engineering situations. He never became reconciled with, and was antagonistic to, the view developed so convincingly in Cambridge by Bowden[3] and Tabor of the role of adhesion in the friction process.

In 1971, when the Department of Friction Research was separated from the Max-Planck Company in Göttingen, Vogelpohl retired and many former colleagues moved to the new Institute for Friction Technology and Machine Technology at the Technical University at Clausthal. Formal retirement caused little interruption to the work of Georg Vogelpohl. He gave lectures, read papers at conferences in Germany and abroad and even continued various research projects. The first volume of the second edition of his book was published in 1967 and he was working towards the completion of the second volume when he died in

---

1. Professor H. Blok who first met and interviewed Vogelpohl in the early summer of 1946 as part of a Dutch postwar intelligence team, spoke of this approach on various occasions and has expressed his appreciation of it in a private communication.
2. See Sect. A.17.
3. See Sect. A.23.

1975. He published over eighty papers and gave support to various scientific societies. He was an active member of the Braunschweig Scientific Society, an Honorary President of the Lubricants Division of the DGMK and an Honorary President of the German Society for Tribology. In 1975 the latter Society not only endowed a prize for outstanding achievement in tribology, but also recognized the contributions of one of Germany's leading tribologists, by naming it the Georg Vogelpohl Prize.[4]

Those who had the privilege to know and work with Vogelpohl developed a high regard and affection for him. He was a man of learning whose interests eveloped philosophy, literature, music and history. He did much to stimulate interest in the history of friction and lubrication, particularly in two papers (Vogelpohl, 1940, 1969), while he had a life-long interest in steam engines and railways. A fine appreciation of Vogelpohl the man emerges from the article by his co-worker Dr -Ing. G. Noack (1976). His curiosity and critical mind, combined with an insistence on the highest standards, established a unique and respected personality in tribological circles. He was both courteous and generous with his time for others, as I had occasion to experience when I met him in Houston, Texas in 1963.

Georg Vogelpohl died on 9 March 1975, just a few months short of his seventy-fifth birthday. He was buried at the city cemetery in Göttingen, the town where he had accomplished so much, on 18 March 1975. Perhaps the best lesson to be extracted from his life and contributions to tribology are contained in the words which, according to Dr-Ing. Noack, would have introduced his final book: 'New thoughts are not numerous, just tell us the old ones fluently.'

In preparing these notes I have been greatly assisted by Professor H. Blok of the University of Technology, Delft, Professor Dr G. H. Göttner of the Institut für Erdölforschung, Abteilung Materialprüfung und Tribologie, Hanover, Dr G. Graue, Gessellschaft für Tribologie, Duisberg (Homberg) and Dr H. Peter Jost. My colleague Mr J. Schwarzenbach translated several German articles and I gratefully acknowledge all this support for the preparation of a biographical sketch of Georg Vogelpohl.

Bibliography

Graue, G. (1967) 'Jubilee in the Max-Planck-Institute for Friction Research, Göttingen; Prof. Vogelpohl 25 Years with the Max-Planck Company', personal communication.

Gülker, E. (1976) 'Tribologie in Deutschland preiswürdig?', *Schmiertechnik & Tribologie*, **23**, No. 2, 26.

Holland, J. (1975) 'Prof. G. Vogelpohl', *MTZ Motortech. Z.*, **36**, Nos. 7/8, 229–230.

Noack, G. (1976) 'Die Epoche Vogelpohl in der Reibungsforschung', *Schmiertechnik & Tribologie*, **23**, No. 2, 27–32.

Rodermund, H. (1975) 'Georg Vogelpohl zum Gedenken', *Antriebstechnik*, **14**, No. 4, 190.

4. See Gülker, E. (1976).

Vogelpohl, G. (1937) 'Bietrage zur Kenntris der Gleitlagerreibung', *V.D.I. For-schungsheft*, No. 386.

Vogelpohl, G. (1940) 'Die Geschientliche Entwicklung unseres Wissens ueber Reibung und Schmierung', *Oel u Kohle*, **36**, Nos. 9 and 13, 1 Mar., 89–93; 1 Apr., 129–34.

Vogelpohl, G. (1949) 'Der Uebergang der Reibungswaerme von Lagern aus der Schmierschicht in die Gleitflaechen', *V.D.I. Forschungsheft*, No. 425 (supplement to *Forschung auf dem Gebiete des Ingenieurwesens*), **16**, July–Aug.

Vogelpohl, G. (1958) *Betriebsichere Gleitlager. Berechnungsverfahren für Konstruction und Betrieb* (Springer-Verlag, Berlin).

Vogelpohl, G. (1969) Uber die Ursachen der unzureichenden Bewertung von Schmierungsfragen im vorigen Jahrhundert', *Schmiertechnik & Tribologie*, **16**, No. 5, 191–200.

A.23 Frank Philip Bowden (1903–68)

Frank Philip Bowden is the second of our select *Men of Tribology* to be linked with Australia. He was born fifth of a family of six children in Hobart, Tasmania, on 2 May 1903. Both his parents had been born in Tasmania, his father Frank Prosser Bowden and his mother Grace Elizabeth Hill being the only children in their respective families. His father was a telegraphist, while Grace assisted her country postmistress mother, and it is said that his parents first met over a telegraph key. Grace was of Irish descent and she died when Philip was fourteen years old. Philip appears to have developed a strong bond with his father, particularly during adolescence.

Philip Bowden first went to school at the age of seven. He attended a dame's school for about two years before entering Hutchins School in Hobart. Progress through school was typified more by a conscientious and hard-working approach than by outstanding success. His results were satisfactory but by no means remarkable, and a failure in mathematics at the matriculation stage denied him a place in a university. Tabor (1969) has noted that Bowden's weakness in mathematics had far-reaching consequences upon his scientific work. It caused him to acquire, initially through laboratory work with the Electrolytic Zinc Company in Tasmania, an appreciation of the experimental method, and it heightened his ability to seek out solutions to physical problems by means of a direct approach, uncluttered by mathematics, which was to form the hallmark of his scientific work.

Bowden continued his studies with a private tutor, matriculated and entered the University of Tasmania as a science student in 1921. Illness caused him to withdraw in his second year and he was advised to rest for at least six months in a warmer climate. He found the rest and the warmth at an Australian inland station in New South Wales, where he recovered his health while acting as a jackaroo, hunting kangaroos, riding horses and reading. He returned to the University and in his final B.Sc. degree examinations in 1924 gained high distinctions in three subjects and distinctions in another four. In 1925 he worked with Dr A. L. McAulay in the Physics Department, submitted a series of papers on topics in electrochemistry and was awarded the degree of M.Sc. with first-class honours. He worked for a further year in the Physics Department as the first recipient of a scholarship offered by the Electrolytic Zinc Company and published, jointly with McAulay, his first three papers. In 1926 he was awarded an 1851 scholarship to work under Eric Rideal in Cambridge. The move from Australia to England came just a few years after Hardy had established the firm foundations of boundary lubrication studies and Bowden joined the colloid chemist who in due course was to edit Sir William Bate Hardy's *Collected Scientific Papers*.[1]

It is interesting to note that Bowden had little interest in competitive or team sports, but a great love of sailing, skiing and walking in the Tasmanian highlands. In the latter pursuit he paralleled our earlier Australian, A. G. M. Michell.[2] Skiing remained a lifelong interest, both as a hobby and the subject of one of his many research investigations.

On 14 January 1927 Philip Bowden entered Gonville and Caius College as a research student. He settled easily into the Cambridge environment and was to remain there throughout his scientific career, apart from the period of wartime separation. He worked initially with Eric Rideal, then a lecturer in physical chemistry, and from 1928 forged a close friendship with a fellow research student in the same department by the name of C. P. Snow. Tabor (1969) has confirmed that in novels of later years C. P. Snow was to base the character, Getliffe, a gifted, wise and sensitive scientist, upon Philip Bowden. In 1930 the University of Cambridge set up a new Department of Colloid Science under the professorship of Eric Rideal. Bowden, who had been made an unofficial Drosier Fellow in 1929, became Director of Studies in Natural Sciences in 1933 and an official Fellow in 1934. In the meantime Margot Hutchison had left Hobart, Tasmania, to become his wife.

Tabor (1969) had noted that Bowden's college teaching of under-graduates was not an unqualified success, but his reputation in research went from strength to strength. The early years in Cambridge were devoted mainly to electrochemistry, but in 1931 he published with Stewart Bastow a paper dealing with *contact* between smooth surfaces which marked the beginning of his tribological studies. His attention was drawn to this subject by the unusual assertion by Hardy and Miss Nottage in 1928 that a polished steel cylinder remained suspended above a steel plate at a height of about 157 μ in (4,000 nm) by long-range intermolecular forces. Bowden and Bastow could not reconcile this with their under-standing of the effective range of orientated layers of molecules on solid surfaces, and after discussions with Hardy they investigated the matter and concluded that no floatation occurred when the surfaces were scrupulously clean. They attributed Hardy's findings to dust and Sir William Bate Hardy himself communicated their paper to the Royal Society. Once again we find firm evidence of the handing down of the tribological torch.

Bowden followed this initial work on surface contact with studies of

---

1. See Sect. A.17.
2. See Sect. A.19.

kinetic friction, frictional heating at asperity interactions, the formation of Beilby layers, adhesion of clean metals, the real area of contact between solids, and the friction of ice and snow. Osborne Reynolds[3] had concluded that the low friction between skates and ice could be attributed to lubrication by water resulting from pressure melting of the ice, but Bowden took the view that the most likely mechanism for the creation of the lubricant was frictional heating of the sliding surfaces. He later (1953) recommended that ski and sledge runners should be coated with polytetrafluoroethylene (PTFE) to reduce friction and included an acknowledgement to his wife Margot for her assistance with the model ski experiments at the Jungfraujoch Experimental Station.

By the late 1930s Bowden's work had attracted the interest and support of the Lubrication Committee of DSIR, the oil companies, industrial concerns in England, Europe and the United States, the Air Ministry, the Fuel Research Board and the War Office. A Shell research unit for the study of lubrication and wear was established under his supervision in 1937. In 1939 Bowden was joined by David Tabor, who was to become a close friend, collaborator and co-author of so many major contributions to tribology, that their names became inextricably linked.

In the summer of 1939 Bowden made his first visit to America, where he gave a number of lectures. He decided to return via Australia, where he was joined by his wife and their first child. The outbreak of the Second World War caused him to spend the next four years in Melbourne, for he decided that he could make the greatest contribution to the war effort in his native land. This called for the creation of a new group with laboratory facilities and Bowden discussed the possibility with a number of officials including Sir David Rivett, Chief Executive Officer of the Council for Scientific and Industrial Research (CSIR) and the Minister, the Rt Hon. R. G. Casey. In the meantime the Cambridge Group disbanded, but not everyone in Australia felt that Bowden's proposed work for industry was a better outlet for his talent than a continuation of his fundamental work in Cambridge. In due course Bowden's proposals were accepted, and this was probably as much a reflection of his conviction of the lack of dichotomy between fundamental and applied research as it was of his genuine belief that he could contribute to the war effort most effectively in the Australian situation.

On 1 November 1939 Bowden was appointed to the staff of CSIR as officer in charge of a new section entitled 'Lubricants and Bearings'. For a few months early in 1940 the section was housed in the Engineering School of the University of Melbourne, but it soon transferred to the new chemistry building. It was initially a small section, with David Tabor joining the team in 1940, but by the end of hostilities the research staff numbered almost twenty. Others who joined the wartime team were J. S. Courtney-Pratt and, in December 1941, A. J. W. Moore.

The wartime studies included the development of lubricants for

---

3. See Sect. A.10.

machine tools and aircraft, the performance and manufacture of aircraft bearings, the penetration of metal sheet by bullets and the initiation of explosions and the detonation of explosive materials by adiabatic heating of trapped air bubbles, frictional heating at asperity contacts and viscous dissipation within viscous liquids. The two-element model of metallic friction, combining both adhesion at asperity interactions and the ploughing of hard asperities through a softer material, which is linked primarily with the names of Bowden and Tabor (Bowden and Tabor, 1942; Bowden, Moore and Tabor, 1943), developed at this time, while the role of fatty acids in boundary lubrication (Bowden, Gregory and Tabor, 1945) and the breakdown of lubricating films between piston rings and cylinder walls were also investigated.

During an official visit to England in 1944 Bowden explored the possibilities of re-establishing his Cambridge Group after the war. The idea was well received and he resigned from his post in Melbourne in 1945 to return to Cambridge. He was accommodated within the Physical Chemistry Department where he was made a Reader in 1946 and he began to rebuild his Group with financial support from the Ministry of Supply (Air). Bowden named the laboratory the Research Group on the Physics and Chemistry of Rubbing Solids (PCRS) and he insisted that the work of the Group should be balanced between fundamental studies and their application to practical problems. The main areas of work were to be friction, lubrication and the initiation and detonation of explosives.

Early in 1946 David Tabor followed Bowden back to England to join the reformed Cambridge Group. In the meantime a development of some historical interest had taken place in Melbourne. Before leaving Australia Bowden had suggested that it would be to the advantage of the Lubricants and Bearings section if a new, more scientific and romantic title could be found. It was David Tabor who proposed the word *tribophysics* and the designation Section of Tribophysics was adopted in 1946. In 1948 the section was afforded the status of a Division of Tribophysics, with Dr Stewart Bastow as its first chief. In 1953 the Division moved into new buildings and an account of its activities was written by the chief, Dr W. Boas (Boas, 1953). A second article by Boas (1954) included a photograph of the three-storey brick building in the heart of the University of Melbourne. By this time the staff totalled fifty and about half held academic qualifications. It is this building, with the now weathered title 'CSIRO Tribophysics Laboratory' sitting comfortably above the door, which houses the Division of Tribophysics today, and I found much interest in its early history when I visited Melbourne in 1975. It is a remarkable tribute to the energy and inspiration of Philip Bowden. Diffusion of the word *tribo-physics* from the Antipodes to the United Kingdom was quite slow, but by 1966 the wider significance of the subject was recognized with the publication of the Jost Report.[4] Philip

---

4. See Ch. 11.

Bowden was one of the experts interviewed by the Jost Committee during its deliberations.

Meanwhile the work in Cambridge on friction and boundary lubrication was exerting a dominant influence on postwar thinking on these subjects, both in England and overseas. Many talented research workers joined Bowden's Group in this period and one of the most significant events was the publication of impressive monographs by Bowden and Tabor in 1950 and 1964. A single yet noteworthy illustration of Bowden's work came with his observation of the low friction of PTFE. It was soon suggested that incorporation of the material in a porous metal would provide a dry bearing of adequate strength and low friction and a new and most valuable form of bearing was initiated.

Bowden's work was not restricted to the development of the Research Group on the Physics and Chemistry of Rubbing Solids in Cambridge. He was invited by the Chairman of Tube Investments Ltd, Ivan Stedeford, to advise on the creation of new central research facilities for the company. The outcome was the Tube Investments Research Laboratories at Hinxton Hall, the unique development of this laboratory owing much to Philip Bowden. In 1958 Bowden became a director of the English Electric Company, with special responsibility for research. He recommended the formation of a Company Research Council and became Chairman of the Council in 1962. He gave much of his time to government committees, particularly the Science Research Council, Ministry research boards and the National Physical Laboratory.

In 1956 the Physical Chemistry Department started a move to new accommodation and Bowden's Research Group transferred to the Cavendish Laboratory. Bowden himself became a reader in physics in 1957, the name of the Group was shortened to Physics and Chemistry of Solids (PCS) and it became a sub-department of the Cavendish Laboratory in Free School Lane. In 1966 Bowden was appointed to a personal Chair in Surface Physics. He had been awarded the Tasmanian degree of D.Sc. in 1933 and the Cambridge Sc.D. in 1938. By 1948 he was a Fellow of the Royal Society and his career was interspersed with awards and medals which included the Redwood Medal of the Institute of Petroleum in 1953, the Elliott Cresson Medal of the Franklin Institute in 1955, the Rumford Medal of the Royal Society in 1956, the Medal of the Société Française de Metallurgie in 1957, and the Glazebrook Medal and Prize of the Institute of Physics and the Physical Society together with the Bernard Lewis Gold Medal of the Combustion Institute in 1968. He received the C.B.E. in 1956. His impressive list of publications ran to no less than 185 papers and 6 books.

Although I met Bowden I regret that I did not have the opportunity to know him well. Tabor's (1969) comprehensive biography presents a picture of a man who was shy and reserved yet tough and ambitious, a man entirely at home in the University yet responsive to the problems and stimulus of industry. He was truly a twentieth-century giant in tribology whose researches have greatly influenced development of the subject. He had little time for theoretical work and his experimental approach was direct rather than derivative.

The Bowdens had three sons and a daughter, and it is once again clear from Tabor's (1969) writing that family life was both happy and harmonious.

Frank Philip Bowden died in Cambridge after a long illness on 3 September 1968.

My task in preparing this biographical sketch was greatly relieved by access to Professor David Tabor's full and fascinating biographical memoir prepared for the Royal Society (Tabor, 1969) and by the opportunity to talk with both Professor Tabor in Cambridge and Dr Alan J. W. Moore in Melbourne. Both were closely associated with Philip Bowden, and in offering my appreciation of their help I trust that we can all benefit from their intimate knowledge of one of the great men of tribology.

Bibliography

Boas, W. (1953) 'Metal physics work of the Division of Tribophysics, C.S.I.R.O.', *Nature*, **171**, No. 4360, 23 May, 908–10.

Boas, W. (1954) 'New tribophysics laboratory of the Commonwealth Scientific and Industrial Research Organization, Australia', *Nature*, **171**, No. 4412, 22 May, 969–70.

Bowden, F. P., Gregory, J. N. and Tabor, D. (1945) 'Lubrication of metal surfaces by fatty acids', *Nature, Lond.*, **156**, 97.

Bowden, F. P., Moore, A. J. W., and Tabor, D. (1943) 'The ploughing and adhesion of sliding metals', *J. Appl. Phys.*, **14**, 141–51.

Bowden, F. P. and Tabor, D. (1953) 'Friction on snow and ice', *Proc. R. Soc.*, **A217**, 462–78.

Bowden, F. P. and Tabor, D. (1942) 'The mechanism of metallic friction', *Nature, Lond.*, **150**, 197.

Bowden, F. P. and Tabor, D. (1950) *The Friction and Lubrication of Solids – Part I* (Clarendon Press, Oxford).

Bowden, F. P. and Tabor, D. (1964) *The Friction and Lubrication of Solids – Part II* (Clarendon Press, Oxford).

Tabor, D. (1969) 'Frank Philip Bowden', *Biographical Memoirs of Fellows of the Royal Society*, **15**, Nov. 1969, 1–38.

A.24 Fred William Ocvirk (1913–67)

Fred William Ocvirk was born in Chicago, Illinois on 28 December 1913. His father, Joseph, who was a tailor, had emigrated from Yugoslavia to the United States in 1907 and his mother's maiden name was Louise Eckl.

Fred Ocvirk attended the Cass Technical High School in Detroit, Michigan, and then entered Wayne State University to study civil engineering. He was awarded his B.Sc. degree in 1938 and an M.S. degree from the University of Illinois, where he held a research scholarship and assistantship, in 1940. He accumulated industrial experience during six summers between 1935 and 1940, working as a layout draftsman on conveyor systems for the International Conveyor and Washer Corporation, Detroit, Michigan.

In 1940 he began a lifelong association with Cornell University when he became an instructor in the engineering science management war training programme carried out in Buffalo, N.Y., before and during the Second World War. He taught courses on mechanics, aerodynamics, aircraft and civil engineering structures and in 1944 he was appointed Assistant Professor of Aeronautical Engineering. Later in that year he was afforded leave of absence to accept a position as a senior engineer concerned with the design of structural components for gun directors and guided missiles in the Johns Hopkins University Applied Physics Laboratory. In November 1945 he returned to Cornell as Assistant Professor in the Graduate School of Aeronautical Engineering, where he taught aircraft and civil engineering structures at the main campus in Ithaca, NY. Two years later he transferred to the Sibley School of Mechanical Engineering and he soon established a permanent place at Cornell as a member of the Department of Machine Design.

Two events of great importance in the life of Fred William Ocvirk occurred in the late 1940s. On 1 August 1948 he married in Fabius, NY, Milacent Grimes, daughter of a merchant John Dudley Grimes of Groton, NY, and in 1948 he became involved in lubrication research. His interest in the application of engineering mechanics to problems of machine design led him to join Professor DuBois and others in a project sponsored by the National Advisory Committee for Aeronautics (NACA) and by its successor, the National Aeronautics and Space Administration (NASA). He became a full-time research associate in 1950, and in 1951 he was appointed Associate Professor of Mechanical Engineering in the Department of Machine Design.

It was the lack of agreement between conventional journal-bearing theory and experimental observations by Fred W. Ocvirk and Professor George B. DuBois that caused Ocvirk to reconsider the question of hydrodynamic theory for short bearings in the early 1950s. It was becoming increasingly popular for reasons of efficiency and economy to use journal bearings whose axial lengths were less than the shaft diameter in the postwar years and this provided the spur for Ocvirk's important simplification of the governing Reynolds equation. He argued that as the axial length of the bearing was reduced in relation to the diameter, the importance of Poiseuille or pressure flow in the circumferential direction relative to that in the axial direction would diminish and in due course become negligible. This was the basis of Ocvirk's impressive *short journal-bearing theory* which was first published in NACA reports in 1952 and 1953. This novel approach rejuvenated hydrodynamic lubrication analysis, shocked the purists, yet produced analytical solutions of great utility. *Short bearing theory* or the *Ocvirk solution* had the great merit that it provided designers with data of adequate accuracy for the bearings most commonly used and it provided a ready indication of the changes required to improve bearing performance without recourse to numerical solutions of the full Reynolds equation. In trying to place this development in valid historical perspective it should be recalled that computing equipment in the late 1940s and early 1950s was relatively primitive compared with present-day systems. The dimensionless group which proved to be most valuable and all-embracing in the short bearing theory was initially known as the *load number*, but it is now universally referred to as the *Ocvirk number*. Michell[1] (1929) had previously developed the general concepts of short bearing theory, but it appears that Ocvirk not only rediscovered the idea but also extended it most completely and convincingly to journal bearings.

Ocvirk's work on bearings also included a valuable and early investigation of the effect of misalignment upon pressure distribution and restoring torques. He was an authority on high-speed, rotating machinery and in later years was to be joined by J. F. Booker in a study of dynamically loaded bearings. He published some fifteen papers and was co-author with H. H. Mabie of a textbook on mechanisms and dynamics (Mabie and Ocvirk, 1957).

In 1959 Ocvirk was promoted to a full professorship in mechanical engineering at Cornell. It is worth noting the strong and early connection between Cornell and tribology through Thurston and Kingsbury. His outstanding research on bearings and lubrication was but one of his highly respected talents. He taught both undergraduate and graduate courses and was an enthusiastic contributor to the development of new courses. He was an outstanding classroom teacher and he spent much of his time helping individual students with their problems. Much of his work on research and teaching reflected his philosophy on engineering in general and design in particular.

Fred Ocvirk's talents and sound engineering approach were recognized though consultancy arrangements with the Bendix Corporation, the Boeing Airplane Company, the Cornell Aeronautical Laboratory, the Glacier Metal Company of England and the Atomic Energy Commission at the University of California. He spent his sabbatical leave of 1955–56 as an analytical engineer in the axial compressor division of the Carrier Corporation in Syracuse, NY, and in 1962–63 he was a Fulbright Lecturer in the University of New South Wales, Sydney, Australia.

Professional societies benefited from Ocvirk's enthusiastic support. He was a member of the American Society for Engineering Education and the American Societies of Civil and Mechanical Engineers and he acted as a member and chairman of several important committess in the Lubrication Division of the latter Society. He was the 1955 recipient of the Captain Edward E. Hunt Memorial Medal of the American Society of Lubrication Engineers for his paper, 'Measured oil film pressure distribution in misaligned plain bearings' which was judged to be the best paper of the year on the subject of lubrication.

In his personal life Fred Ocvirk had a passionate interest in travel, particularly by ship. He enjoyed several overseas visits with his wife, including a journey to Hawaii through the Panama Canal, the West Indies, a number of visits to Europe and a journey round the world. His wife Milacent was also connected with education, being a supervisor of English in the Ithaca public schools. They had no children.

Fred William Ocvirk died suddenly in Ithaca, NY, on 21 May 1967, at the age of fifty-three.

---

1. See Sect. A.19.

I am grateful to Professor J. F. Booker of Cornell University and his wife Barbara for providing the photograph of Fred William Ocvirk, personal notes and references to much of the background information upon which this biographical sketch has been constructed. I would also like to thank Mr D. F. Hays of the General Motors Technical Center, Warren, Michigan for providing additional background information.

Bibliography

Cornell University (1967) 'Fred William Ocvirk', *Necrology of the Faculty of Cornell University, 1966–67*, Ithaca, New York, pp. 46–8.

DuBois, G. B. and **Ocvirk, F. W.** (1952) 'Experimental investigation of eccentricity ratio, friction, and oil flow of short journal bearings', *NACA, Tech. Note 2809*.

DuBois, G. B. and **Ocvirk, F. W.** (1953) 'Analytical derivation and experimental evaluation of short bearing approximation for full journal bearings', *NACA, Report 1157*.

Ithaca Journal (1967) 'Prof. Ocvirk, Engineer, Dies at 53', *Ithaca Journal*, 22.5.1967, p. 5.

Mabie, H. H. and **Ocvirk, F. W.** (1957) *Mechanisms and Dynamics of Machinery* (Wiley, New York).

Michell, A. G. M. (1929) 'Progress in fluid film lubrication', *Trans. Am. Soc. mech. Engrs*, **51**, MSP 51–2, 153–63.

National Cyclopedia of American Biography (1971) 'Ocvirk, Fred William', *National Cyclopedia of American Biography*, Vol. 53 (James T. White, New York).

Ocvirk, F. W. (1952) 'Short bearing approximation for full journal bearings', *NACA, Tech. Note 2808*.

Wehe, R. L. (1967) 'Memorial talk on Fred W. Ocvirk at ASME Lubrication Luncheon'.

A.25 Summary

The initial reason for this appendix of biographical sketches was simply to provide more ample accounts of notable contributors to the art and science of tribology, without disturbing the main historical account of the subject contained within the body of the text. It seems to be both appropriate and tidy, but it also led to unexpected benefits. In the first place it afforded the opportunity to delve more deeply into the backgrounds of the authors of tribological works and by so doing the writer developed a fuller appreciation of their scientific and technical contributions. It also focused attention on a number of historical links which might otherwise have been overlooked.

The choice of these twenty-three names was inevitably personal and I cannot expect all my readers to agree with the selection. I suspect that there would be many names common to similar lists prepared by fellow tribologists throughout the world and I did take the precaution of asking a number of them for their views. To those who responded to requests for the names of the ten or twenty most influential workers in tribology at conferences, in trains and in their homes I offer my appreciation.

The final list includes the names of workers from many disciplines and some seven countries of origin. Their contributions range over the fields of friction, wear, lubrication and bearings, and in this sense at least the breadth of tribology is evident. The dates of birth of the selected men span 461 years, but the majority are contained within the nineteenth century. This is partly a reflection of the growing importance of tribology at a time of industrialization, particularly in relation to steam power and the railways, and partly because I declined the temptation to write about any of the numerous, outstanding contributors still alive today.

Steam engines and the railways did much to promote basic studies of lubrication and bearings, as evident in the writings of Petrov (Sect. A.8), Thurston (Sect. A.9), Tower (Sect. A.12), Kingsbury (Sect A.16) and Goodman (Sect. A.15). Three of our subjects, Hertz (Sect. A.13), Stribeck (Sect. A.14) and Goodman (Sect. A.15) are accredited with the scientific foundations of the rolling-element bearing industry, although the classical studies of elastic contact by Hertz were not performed with such pragmatic matters in mind.

One of the most interesting results of the present exercise has been the emergence of evidence of links between a number of our subjects. Tower (Sect. A.12) not only knew but worked for Lord Rayleigh before he undertook his exemplary experimental studies of axle-box bearings on behalf of the Institution of Mechanical Engineers' Committee on Friction at High Velocities. Michell (Sect. A.19) attended a source of lectures on physics given in Cambridge by J. J. Thomson, Osborne Reynolds' (Sect. A.10) most famous student, before returning to Australia to develop both the concept and analysis of the tilting-pad bearing. Kingsbury (Sect. A.16) was Thurston's (Sect. A.9) student at Cornell, a university which also claims the name of our last subject, Ocvirk (Sect. A.24). Swift (Sect. A.21) worked under Goodman (Sect. A.15) in the University of Leeds and had his interest in lubrication awakened at that time. Bowden (Sect. A.23) became interested in friction and boundary lubrication as a result of working in Hardy's (Sect. A.17) Department of Physical Chemistry at Cambridge.

I have known but three of the subjects listed in this appendix, but discussions with many persons who reminisced and talked so freely of the past with anecdotes of such persons as Reynolds, Goodman, Michell, Sommerfeld and Kingsbury almost convinced me that I knew many more. I wish that time and space had permitted me to do more justice to the small band of selected *Men of Tribology* and to have included many more. As the study progressed I found myself wondering what the outcome might have been if all our subjects had enjoyed simultaneous lives in an Institute of Tribology, but I doubt if their resulting and no doubt stimulating interactions could have yielded greater contributions than the sum of their individual findings recorded in the text. Many showed such heightened individuality that their close association might well have been disastrous!

In preparing my biographical sketches I have found so much interest and such willing assistance from colleagues, fellow tribologists, historians and librarians that I cannot conclude without expressing to all my sincere appreciation. My investigations have proved to be so stimulating and rewarding that I have benefited greatly by the exercise and I shall welcome any further biographical information on workers in tribology that readers care to communicate to me.

References and Selected Bibliography

Abbott, E. J. and **Firestone, F. A.** (1933) 'Specifying surface quality', *Mech. Enging*, **55**, 569–72.

Abramovitz, S. (1956) 'Turbulence in a tilting pad thrust bearing', *Trans. Am. Soc. mech. Engrs*, **78**, 7–11.

Academy of Sciences (1945) *Biographical Index of Members and Corresponding members of the Academy of Sciences 1666–1954* (Morin, Arthur Jules, p. 370).

Adams, W. B. (1853) 'On railway axle lubrication', *Proc. Instn mech. Engrs*, 57–65.

Adolph, K. H. (1976) 'Richard Hermann Stribeck', Private communication from the Archivist, Technische Universität Dresden.

Agricola, G. (1556) *De Re Metallica* (Basel).

Aked, C. K. (1968) 'Oil for chronometers', *Antiq. Horol.*, **5**, No. 12, 454–5.

Akhmatov, A. S. (1966) *Molecular Physics of Boundary Lubrication*, (*Molekulyarnaya Fizika Granichnogo Treniya*– 1963). Translated from the Russian by N. Kaner (Israel Program for Scientific Translations, Jerusalem).

Allan, R. K. (1945) *Rolling Bearings*, (Pitman, London).

Allan, T. (1972) 'The application of finite element analysis to hydrodynamic and externally pressurized pocket bearings', *Wear*, **19**, Feb., 169–206.

Allbert, B. J., Walker, J. C. and **Maycock, G.** (1966*a*) 'Tyre to wet road friction', *Proc. Instn mech. Engrs*, **180**, Pt 2A, No. 4, 105–21.

Allbert, B. J., Walker, J. C. and **Maycock, G.** (1966*b*) 'Studies of the skidding resistance of passenger-car tyres on wet surfaces', *Proc. Instn mech. Engrs*, **180**, Pt 2A, No. 4, 122–57.

Allen, C. M. and **Drauglis, E.** (1969) 'Boundary layer lubrication: monolayer or multilayer', *Wear*, **14**, 363–84.

Allen, J. (1970) 'The life and work of Osborne Reynolds', in *Osborne Reynolds and Engineering Science Today*, Manchester U.P., **1**, 1–82.

Al'shits, I. Ya. (1959), 'The use of plastic materials for plain bearings', *Vest Mashin.*, No. 7, 13.

Amontons, G. (1699), 'De la resistance caus'ee dans les machines', *Mémoires de l'Académie Royale*, **A**, (Chez Gerard Kuyper, Amsterdam, 1706), 257–82.

Anderson, W. J. (1964), 'Lubrication of bearings with liquid metals', in *Advanced Bearing Technology* (National Aeronautics and Space Administration, Washington, D.C.) Ch. 14, 469–96.

Anderson, W. J. and **Glenn, D. C.** (1968), 'Effect of the space environment on lubricants and rolling element bearings'. Conference on Lubrication and wear: Fundamentals and Application to Design, *Instn of mech. Engrs, Proceedings 1967–1968*, **182**, Pt 3A, 505–19.

Archard, J. F. (1953), 'Contact and rubbing of flat surfaces', *J. Appl. Phys.* **24**, No. 8, Aug. 981–8.

Archard, J. F. (1957), 'Elastic deformation and the laws of friction', *Proc. R. Soc.*, **A243**, 190–205.

Archard, J. F. (1959) 'The temperature of rubbing surfaces', *Wear*, **2**, No. 6, 438–55.

Archard, J. F. (1961) 'Single contacts and multiple encounters', *J. app. Phys.*, **32**, No. 8, Aug., 1420–5.

Archard, J. F. (1969) 'Wear', in *Interdisciplinary Approach to Friction and Wear*. Ku, P. M. (ed.). Proceedings of the NASA-sponsored Symposium held in San Antonio, Texas, NASA SP-181, 267–333.

Archard, J. F. (1974) 'Surface topography and tribology', *Tribology Int.*, Oct., 213–20.

Archard, J. F. and **Cowking, E. W.** (1966) 'A simplified treatment of elastohydro-dynamic lubrication theory for a point contact', *Proc. Instn mech. Engrs*, **180**, Pt 3B, 47–56.

Archard, J. F. and **Hirst, W.** (1956*a*) 'The wear of metals under unlubricated conditions', *Proc. R. Soc.*, **A236**, 397–410.

Archard, J. F. and **Hirst, W.** (1956*b*) 'An examination of the mild wear process', *Proc. R. Soc.*, **A238**, 515–28.

Archard, J. F., Hunt, R. T. and **Onions, R. A.** (1975) 'Stylus profilometry and the analysis of the contact of rough surfaces', *Proceedings of the Symposium on the Mechanics of the Contact of Deformable Bodies*. Pater and Calker (eds.), IUTAM, Delft (Delft U.P.), 282–303.

Archbutt, L. and **Deeley, R. M.** (1900) *Lubrication and Lubricants; a Treatise on the Theory and Practice of Lubrication, and on the Nature, Properties, and Testing of Lubricants* (Charles Griffin, London).

Archibald, F. R. (1955*a*) 'Men of lubrication – Beauchamp Tower', *Lubr. Engng*, **13**, 13, 63.

Archibald, F. R. (1955*b*) 'Men of lubrication – Osborne Reynolds', *Lubr. Engng*, Mar.–Apr., **11**, 84–5, 128–9.

Archibald, F. R. (1955*c*) 'Men of lubrication – Albert Kingsbury', *Lubr. Engng*, May–June, **11**, 162–3, 197–8.

Archibald, F. R. (1955*d*) 'Arnold Sommerfeld', *Lubr. Engng*, **11**, Aug., 228–9, 283.

Archibald, F. R. (1955*e*) 'Men of lubrication – Anthony G. M. Michell', *Lubr. Engng*, Sept.–Oct., **11**, 304–5, 346.

Archibald, F. R. (1955*f*) 'Men of lubrication – Lord Rayleigh (John William Strutt)', *Lubr. Engng*, Nov–Dec., **11**, 375, 420, 422.

Archibald, F. R. (1956*a*) 'Men of lubrication – Nikolay P. Petrov', *Lubr. Engng*, Jan.–Feb., **12**, 15–6, 72.

Archibald, F. R. (1956*b*) 'Load capacity and time relations for squeeze-films', *Trans. Soc. mech. Engrs*, **A78**, 231–45.

Archibald, F. R. (1957) *Men of Lubrication* (ASLE, Chicago).

Argyris, J. H. and **Scharpf, D. W.** (1969) 'The incompressible lubrication problem' (12th Lanchester Memorial Lecture, Appendix IV), *Aeronaut. J.* (Royal Aeronautical Society) **73**, 1044–6.

Armitage, W. H. G. (1961) *A Social History of Engineering* (Faber and Faber, London).

ASLE (1970) *The 25 Years*, A Special Publication Commemorating ASLE's 25 Years of Service, The American Society of Lubrication Engineers (ASLE), Chicago.

ASME (1971) *Life Adjustment Factors for Ball and Roller Bearings*, An engineering design guide sponsored by the Rolling Elements Committee of the Lubrication Division of American Society of Mechanical Engineers (ASME), New York.

ASME (1974*b*) 'Power from the wind', *Mech. Engng*, **96**, No. 4, April 1974, 55.

ASME (1974*a*) '"Comeback" the wind-mill', *Mech. Engng*, **96**, No. 7, July 1974, 29.

ASME (1974*c*) 'Film lubrication: turbulence and related phenomena', *Trans. Am. Soc. Mech. Engrs; J. lubri. Technol.*, **96**, F, No. 1, Jan.

ASME (1977) *Wear of Materials*, papers presented at an International Conference on Wear of Materials, St Louis, Missouri, 25–28 Apr., American Society of Mechanical Engineers.

Ausherman, V. K., Nagaraj, H. S., Sanborn, D. M. and **Winer, W. O.** (1976) 'Infrared temperature mapping in elastohydrodynamic lubrication', *Trans. Am. Soc. mech. Engrs*; *J. lubr. Technol.*, **F98**, No. 2, 236–43.

Ausman, J. S. (1965) 'Gas-lubricated bearings', in *Proceedings, International Symposium on Lubrication and Wear*, Muster, D. and Sternlicht, B. (eds), (McCutchan Pub. Corp., Calif.), 825–53.

Baildon, E. and **Boulton, N. S.** (1960) 'Professor H. W. Swift' (obituary notice), *University of Sheffield Gazette*.

Baker, A. J. S., Dowson, D. and **Strachan, P.** (1973), 'Dynamic operating factors in piston rings', *International Symposium of Marine Engineers*, Tokyo, Dec., paper 256, 2.5.59–2.5.70.

Baker, A. J. S., Dowson, D. and **Strachan, P.** (1976), 'Dynamic factors related to piston ring scuffing', in *Piston Ring Scuffing*, (Mechanical Engineering Publs, London), 25–34.

Ball Bearing Journal (1977) 'Sven Wingquist–founder of SKF', *Ball Bearing J.*, No. 190, Feb. (supplement).

Barlow, A. J., Lamb, J., Matheson, A. J., Padinini, P. R. K. L. and **Richter, J.** (1967) 'Viscoelastic relaxation for supercooled liquids', *Proc. R. Soc. Lond.*, **A298**, No. 1455, 467–80 (Pt I), 481–94 (Pt II).

Barnard, D. P. (1925) *S.A.E. Trans.*, **20**, Pt 2, 66 (see also *Ind. Engng Chem.* (1926), **18**, 460).

Barnett, C. H. and **Cobbold, A. F.** (1962) 'Lubrication within living joints', *J. Bone and Joint Surg.*, **B44**, 662.

Barr, A. (1888) 'Contribution to the discussion on the third report of the Research Committee on Friction', *Proc. Instn mech. Engrs*, May 1888, 187–9.

Barr, G. (1931) *A Monograph of Viscometry* (Oxford U.P., London).

Barrans, J. (1851) 'On an improved axle box for railway engines and carriages', *Proc. Inst. mech. Engrs*, 22 Jan., 30–5; 23 Apr., 3–8.

Bartenev, G. M. (1954) 'Theory of dry friction of rubber', *Dokl. Akad. Nauk. SSSR*, **96**, 1161–4.

Bartenev, G. M. and **Lavrentjev, V. V.** (1961) 'The law of vulcanized rubber friction', *Wear*, **4**, 154–60.

Barwell, F. T. (1956) *Lubrication of Bearings* (Butterworths Scientific Publs, London).

Barwell, F. T. (1965) 'Bearing data, can they be made more useful to designers', *Proc. Instn mech. Engrs*, **179**, Pt 3D, 28–38.

Barwell, F. T. (1970*a*) *Lubrication of Bearings* (Butterworths Sci. Publs, London).

Barwell, F. T. (1970*b*) 'The founder of modern tribology', in *Osborne Reynolds and Engineering Science Today*, Manchester U.P., 240–63.

Barwell, F. T. (1978) *Bearing Systems: Principles and Practice* (Oxford U.P., Oxford).

Bastian, E. L. H. (1951) *Metalworking Lubricants–Their Selection, Application and Maintenance* (McGraw-Hill, New York).

Bate, J. (1654) *The Mysteries of Nature and Art in Four Severall Parts*, printed by R. Bishop for Andrew Crook at the Green Dragon in Paul's Churchyard, London.

Bate-Smith, E. C. (ed.) (1964) *Sir W. B. Hardy–Biologist, Physicist and Food Scientist; Centenary Tributes* (Editor: Director, of the Low Temperature Research Station, Cambridge).

Baumma, N. E. (1975) *Calculation Methods for the Degree of Friction and Wear*, Biryansic Institute Press, USSR.

Bayer, R. G. and **Ku, T. C.** (1964) *Handbook of Analytical Design for Wear*, MacGregor, C. W. (ed.) with contributions by Clinton, W. C., Schumacher, R. A., Sirico, J. L., Wayson, A. R. and Nelson, C. W. (Plenum Press, New York).

Bayer, R. G., Clinton, W. C., Nelson, C. W. and **Schumacher, R. A.** (1962) 'Engineering models for wear', *Wear*, **5**, 378–91.

Beazley, M. (1975) *The Mitchell Beazley Illustrated Biographical Dictionary, Who Did What* (Mitchell Beazley, London).

Bélidor, B. F. (1737) *Architecture hydraulique, ou l'art de conduire d'élever et de ménager les eaux* (C. A. Jombert, Paris).

Benjamin, C. H. (1898) 'The design and construction of ball bearings', *The Practical Engineer*, 9 and 16 Dec.

Bentall, R. H. and **Johnson, K. L.** (1967) 'Slip in the rolling contact of two dissimilar elastic rollers', *Int. J. Mech. Sci.*, **9**, 389–404.

Bernal, J. D. (1954) *Science in History* (Watts, London).

Bernoulli, D. (1738) *Hydrodynamica, sive viribus et motibus fluidorum commentarii*, Argentorati.

Bernoulli, D. (1751) See *P.H. Full Correspondence Mathematique et Physique*, **2**, (1843).

Besson, J. (1569) *Theatre des Instruments et Machines* (Lyon, France).

Bidwell, J. B. (ed.) (1962) *Rolling Contact Phenomena*. Proceedings of a symposium held at the General Motors Research Laboratories, Warren, Michigan, Oct. 1960 (Elsevier, Amsterdam).

Biringuccio, V. (1540) *De La Pirotechnia* (Venice).

Birkhoff, G. and **Hays, D. F.** (1963) 'Free boundaries in partial lubrication', *J. Math. Phys.*, **42**, No. 2, **126** (MIT).

Bishop, R. E. D. (1959) 'The vibration of rotating shafts', *J. mech. Engng Sci.*, **1**, No. 1, 50–64.

Bishop, R. E. D. and **Johnson, D. C.** (1960) '*The Mechanics of Vibration* (Cambridge U.P.).

Bisson, E. E. and **Anderson, W. J.** (1964) '*Advanced Bearing Technology*' (National Aeronautics and Space Administration, Washington, DC).

Bisson, E. E., Johnson, R. L. and **Anderson, W. J.** (1958) 'Friction and lubrication with solid lubricants at temperatures up to 1000°F with particular reference to graphite'. Instn Mech. Engrs *Proceedings of the Conference on Lubrication and Wear* (1957), 348–54.

Bisson, E. E. and **Ku, P. M.** (eds) (1970) *Friction and wear interdisciplinary workshop*. Proceedings of a NASA Workshop on Friction and Wear, 19–21 Nov. 1968, Cleveland, Ohio, NASA TMX-52748.

Black, H. F. (1973) 'Calculation of forced whirling and stability of centrifugal pump rotor systems', ASME Paper 73-DET-131.

Black, H. F. and **Loch, N. E.** (1976) 'Computation of lateral vibration and stability of pump rotors', *Proc. Instn mech. Engrs*.

Blanchard, P. M. (1968) 'Lubricants for hot environments with special reference to aero gas turbines operating under severe conditions', Conference on Lubrication and Wear: Fundamentals and Application to Design. Instn Mech. Engrs, *Proceedings 1967–1968*, **182**, Pt 3A, 472–8.

Blok, H. (1937a) 'Les températures de surface dans des conditions de graissage sous pression extrème', *Proceedings of Second World Petroleum Congress* (Paris), Sect. 4, 151–82.

Blok, H. (1937b) 'Measurement of temperature flashes on gear teeth under extreme pressure conditions'. Instn Mech. Engrs, *Proceedings of the General Discussion on Lubrication and Lubricants*, **2**, 14–20.

Blok, H. (1937c) 'Theoretical study of temperature rise at surfaces of actual contact under oiliness lubricating conditions'. Instn Mech. Engrs, *Proceedings of the General Discussion on Lubrication and Lubricants*, **2**, 222–35.

Blok, H. (1952) 'Comments on a paper by R. W. Wilson', *Proc. R. Soc. Lond.*, **A212**, 480–2.

Blok, H. (1973) 'The impulse capacity as a design criterion for full journal bearings under severe dynamic duty, *WTHD 45* (Technische Hogeschool, Delft).

Blok, H. (1976a) 'Notes on Professor Dr.-Ing Ludwig Gümbel, Private communication.

Blok, H. (1976b) 'Notes about Professor Richard Stribeck', Private communication.

Blok, H. and **Van Rossum, J. J.** (1948) 'Lifting of liquids by viscous forces', *Proceedings of the 7th International Conference on Applied Mechanics*, **2**, I, 31.

Boas, W. (1953) 'Metal physics work of the Division of Tribophysics, C.S.I.R.O.', *Nature*, 23 May, 908–10.

Boas, W. (1954) 'New Tribophysics Laboratory of the Commonwealth Scientific and Industrial Research Organization, Australia', *Nature*, 22 May, 969–70.

Bochet, M. (1861) 'Nouvelles recherches expérimentales sur le frottement de glissement', *Ann. Mines*, **xix**, 27–120.

Bochvar, A. M. (1918) *Issledovanie Belykh Antifriktsionnykeh Metallov'* (*An Investigation of Anti-Friction White Metals*), (Izd. ob-va im. Kh. S. Ledentsova, Moscow).

Bochvar, A. M. (1921) 'Anti-friction white-metal alloys', Coll. *Publications on Scientific and Technical Matters* **5** (Odessa politekh, Inst. (OPTI)).

Bondi, A. (1951) *Physical Chemistry of Lubricating Oils* (Reinhold, New York).

Bone, J. H. A. (1865) *Petroleum and Petroleum Wells* (J. B. Lippincott, Philadelphia).

Boner, C. J. (1954) *Manufacture and Application of Lubricating Greases* (Reinhold, New York).

Boner, C. J. (1964) *Gear and Transmission Lubricants* (Reinhold, New York).

Boner, C. J. (1976) *Modern Lubricating Greases* (Scientific Publs. (GB), Broseley, Salop).

Booker, J. F. (1965*a*) 'A table of the journal-bearing integral', *Trans. Am. Soc. mech. Engrs; J. Basic Engrng*, June 1965, D, **87**, 2, 533.

Booker, J. F. (1965*b*) 'Dynamically loaded journal bearings – mobility method of solution', *Trans. Am. Soc. mech. Engrs; J. Basic Engng*, **D187**, 537.

Booker, J. F. (1971) 'Dynamically-loaded journal bearings: Numerical application of the mobility method', *Trans. Am. Soc. mech. Engrs; J. lubr. Technol.*, **F93**, 1, 168, 2, 315.

Booker, J. F. and **Heubner, K. H.** (1972) 'Application of finite-element methods to lubrication: An engineering approach', *Trans. Am. Soc. mech. Engrs; J. lubr. Technol.*, Oct., 313–23.

Booth, H. (1835) 'Composition for greasing axles of wheels and parts of machinery', British Patent No. 6814.

Born, M. (1952) 'Arnold Johannes Wilhelm Sommerfeld; 1868–1951', *Obituary Notices of Fellows of the Royal Society*, **8**, No. 21, 274–96.

Boswall, R. O. (1928) *The Theory of Film Lubrication* (Longmans, London).

Boswall, R. O. and **Brierley, J. C.** (1932) 'The film lubrication of the journal bearing', *Proc. Instn mech. Engrs*, **122**, 423–517; discussion, 518–69.

Boulton, N. S. (1955) 'Appreciation – Professor H. W. Swift', *University of Sheffield Gazette*, 3–4.

Bouman, C. A. (1937) 'The lubrication of piston rings'. Instn Mech. Engrs, *Proceedings of the General Discussion on Lubrication and Lubricants*, **1**, 426–31.

Bouman, C. A. (1950) *Properties of Lubricating Oil and Engine Deposits* (Macmillan, London).

Bourne, J. C. (1846) '*The History and Description of the Great Western Railway*' (David Bogue, Fleet Street, London). See also the reprints of *Bourne's Great Western Railway* (David and Charles, Newton Abbot, Devon).

Bowden, F. P. (1950) 'BBC broadcast'.

Bowden, F. P. and **Tabor, D.** (1939) 'The area of contact between stationary and between moving surfaces', *Proc. R. Soc., Lond.*, **A169**, 391.

Bowden, F. P. and **Tabor, D.** (1942) 'The mechanism of metallic friction', *Nature, Lond.*, **150**, 197.

Bowden, F. P. and **Tabor, D.** (1950) *The Friction and Lubrication of Solids – Part I*, (Oxford U.P., Oxford).

Bowden, F. P. and **Tabor, D.** (1953) 'Friction on snow and ice', *Proc. R. Soc.*, **A217**, 462–78.

Bowden, F. P. and **Tabor, D.** (1956) *Friction and Lubrication* (Methuen, London).

Bowden, F. P. and **Tabor, D.** (1964) *The Friction and Lubrication of Solids – Part II*, (Oxford U.P., Oxford).

Bowden, F. P. and **Tabor, D.** (1973) *Friction – An Introduction to Tribology* (Anchor Books, Anchor Press/Doubleday, New York).

Bowden, F. P., Gregory, J. N. and **Tabor, D.** (1945) 'Lubrication of metal surfaces by fatty acids', *Nature, Lond.*, **156**, 97.

Bowden, F. P., Moore, A. J. W. and **Tabor, D.** (1943) 'The ploughing and adhesion of sliding metals', *J. appl. Phys.*, **14**, 141–51.

Boyd, J. (ed.) (1971) *Proceedings of International Conference on Solid Lubrication* (American Society of Lubrication Engineers, Park Ridge, Illinois).

Bracegirdle, B. (1973) *The Archaeology of the Industrial Revolution*, (Heinemann, London).

Brackner, A. (1974) 'Taking power off the wind', *New Scientist*, 28 Mar. 1974.

Bradley, D. (1966) 'Thermodynamics–a daughter of steam', *Engineering Heritage*, **2**, 26–31.

Braithwaite, E. R. (1964) *Solid Lubricants and Surfaces* (Pergamon Press, Oxford).

Braithwaite, E. R. (ed.) (1967) *Lubrication and Lubricants* (Elsevier, Amsterdam).

Brand, R. S. (1955) 'Inertia forces in lubricating films', *J. appl. Mech.*, **22**, **T77**, 363–4.

Brazier, S. A. and **Holland-Bowyer, W.** (1937) 'Rubber as a material for bearings'. Instn Mech. Engrs, *Proceedings of the General Discussion on Lubrication and Lubricants*, **1**, 30–7.

Brewer, A. F. (1955) *Basic Lubrication Practice* (Reinhold, New York).

Brewer, A. F. (1972) *Effective Lubrication–Management Responsibility including Basic Lubrication Practice* (Robert Krieger, Huntington, New York).

Brillié, H. (1928) 'Théorie du graissage rationelle', *Bull. Tech. Bur. Veritas*, **10**, 105–10, 130–3, 151–3, 170–3, 198–201, 215–7, 237–42.

Brillié, H. (1937) 'The influence of increase of temperature, of oiliness and of surface conditions on viscosity'. Instn Mech. Engrs, *Proceedings of the General Discussion on Lubrication and Lubricants*, II, 241–53.

Broster, M., Pritchard, C. and **Smith, D. A.** (1974) 'Wheel/rail adhesion: Its relation to rail contamination on British railways', *Wear*, **29**, 309–21.

Brown, S. R. and **Hamilton, G. M.** (1976) 'Pressure measurements between the rings and cylinder liner of an engine', in *Piston Ring Scuffing*, (Mechanical Engineering Publs, London), 99–106.

Buckingham, E. (1914*a*) 'On physically similar systems; Illustrations of the use of dimensional equations', *Phys. Rev.*, **4**, 345–76.

Buckingham, E. (1914*b*) 'Physically similar systems', *J. Wash. Acad. Sci.*, **4**, 347–53.

Buckley, D. H. (1971) 'Friction, wear and lubrication in vacuum', *NASA Sp-277* (National Aeronautics and Space Administration, Washington, DC).

Burstall, A. F. (1963) *A History of Mechanical Engineering* (Faber and Faber, London).

Burton, R. A. (1963) 'Effect of two-dimensional, sinusoidal roughness on the load support characteristics of a lubricant film', *Trans. Am. Soc. mech. Engrs; J. Basic Engng*, **D85**, No. 2, June, 258.

Burwell, J. T. (1947, 1949, 1951*a*) 'The calculated performance of dynamically loaded sleeve bearings', *J. appl. Mech.* (1947) **A69**, 231–45; (1949) Pt II, **71**, 358–60 and 1951(*a*), Pt III, **73**, 393–404.

Burwell, J. T. (ed.) (1950) *Mechanical Wear* (American Society for Metals).

Burwell, J. T. (1951*b*) 'Recent developments in full fluid lubrication', *Ann. N.Y. Acad. Sci.*, **53**, Art. 4, 27 June, 759–78.

Burwell, J. T. and **Strang, C. D.** (1952*a*) 'On the empirical law of adhesive wear', *J. appl. Phys.*, **23**, No. 1, Jan. 18–28.

Burwell, J. T. and **Strang, C. D.** (1952*b*) 'Metallic wear', *Proc. R. Soc., Lond.*, **A212**, 470–7.

Bush, A. W., Gibson, R. D. and **Thomas, T. R.** (1975) 'The elastic contact of a rough surface', *Wear*, **35**, 87–111.

Butler, J. F. and **Henshall, S. H.** (1976) 'Piston ring performance in a highly rated two-cycle engine', in *Piston Ring Scuffing*, (Mechanical Engineering Publs, London), 141–56.

Ritchie-Calder, Baron P. R. (1970) *Leonardo and The Age of the Eye* (Heinemann, London).

Cameron, A. (1951) 'Hydrodynamic lubrication in rotating discs in pure sliding', *J. Inst. Petrol.*, **37**, 471–86.

Cameron, A. (1955) 'Oil whirl in bearings. Theoretical deduction of a further criterion', *Engineering*, **179**, 237–9.

Cameron, A. (1958) 'The viscosity wedge', *Trans. Am. Soc. lubr. Engrs*, **1**, 248–53.

Cameron, A. (1960) 'New theory for parallel surface thrust bearing', *Engineering* (London), **190**, 194.

Cameron, A. (1964) 'Discussion on Dowson and Hudson's papers on thermo-hydrodynamic analysis of the infinite slider bearing'. Instn Mech. Engrs *Lubrication and Wear Group Convention* (1963), 149.

Cameron, A. (1966) *The Principles of Lubrication* (Longman, London).

Cameron, A. (1971) *Basic Lubrication Theory* (Longman, London).

Cameron, A. and **Solomon, P. J. B.** (1957) 'Vibrations in journal bearings: Preliminary observations'. Instn Mech. Engrs, *Proceedings of the Conference on Lubrication and Wear* (1957), 191–7.

Cameron, A. and **Wood, W. L.** (1946), *International Congress on Applied Mechanics*, Paris. Extended and reprinted in *Trans. Am. Soc. Lubr. Engrs* (1958), **1**, 254–8.

Cameron, A. and **Wood, W. L.** (1949) 'The full journal bearing', *Proc. Instn mech. Engrs*, **161**, 59–72.

Campbell, J., Love, P. P., Martin, F. A. and **Rafique, S. O.** (1968) 'Bearings for reciprocating machinery: A review of the present state of theoretical, experimental and service knowledge'. Conference on Lubrication and Wear: Fundamentals and Application to Design, Instn Mech. Engrs, *Proceedings, 1967–68*, **182**, Pt 3A, 51–74.

Campbell, M. E. (1970) 'Solid lubricants–A technical survey' *NASA Report SP-50 59(01)*.

Campbell, R. B. (1957) 'Sulphur as an extreme pressure lubricant'. Instn Mech. Engrs, *Proceedings of the Conference on Lubrication and Wear*, (1957), 534–8.

Campbell, R. B. (1958) 'Liquid sodium as a lubricant'. Instn Mech. Engrs, *Proceedings of Conference on Lubrication and Wear* (1957), 529–33.

Campbell, W. E. (1969a) 'Boundary lubrication' in *Boundary Lubrication: An Appraisal of World Literature*, Ling, F. F., Klaus, E. E. and Fein, R. S. (eds), (ASME, New York), Ch. 6, 87–117.

Campbell, W. E. (1969b) 'Solid lubricants', in *Boundary Lubrication: An Appraisal of World Literature*, Ling, F. F., Klaus, E. E. and Fein, R. S. (eds) (American Society of Mechanical Engineers, New York), Ch. 10, 197–227.

Carburi, M. (1777) 'Monument Élevé a la Gloire De Pierre-le-Grand, ou Relation Des Travaux et des Moyens Mechaniques Qui ont été employés pour transporter à Pétersbourg un Rocher de trois millions pesant, destiné à servir de base à la Statue équestre de cet Empereur; avec un Examen Physique et Chymique du meme Rocher'. Paris. (Bookseller: Nyon âiné, Libraire, rue Saint-Jean-de-Beauvais; Printer: Imprimeur-Librairé, rue de la Harpe, vis-à-vis la rue S. Severin.)

Carnot, S. (1824) *Reflections on the Motive Power of Fire*. See Mendoza, E. (ed.) (Dover, New York, 1960).

Castle, P. and **Mobbs, F. R.** (1968) 'Hydrodynamic stability of the flow between eccentric rotating cylinders: Visual observations and torque measurements', *Proc. Instn mech. Engrs*, **182**, Pt 3N, 41–52.

Castleman, R. A. (1937) 'A hydrodynamic theory of piston lubrication', *Phys. Rev.*, **49**, 410, 886.

Caubet, J. J. (1964) *Theory and Industrial Practice of Friction* (in French) (Technip and Dunod, Paris).

Cazaud, R. (1955) *Le Frottement et l'Usure des Métaux les Anti-Frictions* (Dunod, Paris).

Charnes, A., Osterle, F. and **Saibel, E.** (1953) 'Parallel surface slider bearing without side leakage', *Trans. Am. Soc. mech. Engrs*, **75**, 1133–6.

Charnley, J. (1959) 'The lubrication of animal joints', *Biomechanics, Proceedings of the Symposium 1959* (Institution of Mechanical Engineers, London), 12–9.

Charnley, J. (1960) 'The lubrication of animal joints in relation to surgical reconstruction by arthroplasty', *Ann. rheum. Dis.*, **19**, 10.

Charron, M. E. (1910) 'Rôle lubrifiant de l'air dans le frottement des solids. Frottement dans le vide', *Comptes Rendus*, **150**, 960–8.

Cheng, H. S. (1970) 'A numerical solution to the elastohydrodynamic film thickness in an elliptical contact', *Trans. Am. Soc. mech. Engrs; J. lubr. Technol.*, **F92**, Jan., 155–62.

Cheng, H. S. and **Orcutt, F. K.** (1966), 'A correlation between the theoretical and experimental results on the elastohydrodynamic lubrication of rolling and sliding contacts, in elastohydrodynamic lubrication', *Proc. Instn mech. Engrs*, **180**, Pt 3B, 158–168.

Cherry, T. M. (1962) 'Anthony George Maldon Michell, 1870–1959', *Biographical Memoirs of Fellows of the Royal Society*, **8**, 91–103.

Childe, V. G. (1951) 'The first waggons and carts–from the Tigris to the Severn', *Proc. prehist. Soc.*, **xvii**, Pt 2, No. 8, 177–94.

Chisholm, S. F. (1965) 'Some factors affecting the choice of rolling lubricants', *Proc. Instn mech. Engrs*, **179**, Pt 3D, 56–64.

Christensen, H. (1968) *Smøring og Slitasjeteknikk*, Hovedredaktør (ed.) (Teknologisk Forlag, Oslo).

Christensen, H. (1970) 'Some aspects of the functional influence of surface roughness in lubrication', *Wear*, **17**, 149–62.

Christensen, H. and **Tonder, K.** (1971) 'The hydrodynamic lubrication of rough bearing surfaces of finite width', *Trans. Am. Soc. mech. Engrs; J. lubr. Technol.*, 324–30.

Christensen, H. and **Tonder, K.** (1972) 'The hydrodynamic lubrication of rough journal bearings', ASME paper 72-Lub-3.

Christopherson, D. G. (1941) 'A new mathematical method for the solution of film lubrication problems', *Proc. Instn mech. Engrs*, **146**, 126–35.

Christopherson, D. G. (1957) 'A review of hydrodynamic lubrication with particular reference to the conference papers'. Instn Mech. Engrs, *Proceedings of the Conference on Lubrication and Wear*, 9–15.

Clark, R. H. (1938) 'Earliest known ball thrust bearing used in windmill', *Engl. Mech.*, 30 Dec., 223.

Clark, R. H. (1963) 'The steam engine in industry and road transport', in *Engineering Heritage* (Heinemann, on behalf of the Institution of Mechanical Engineers, London), **1**, 62–9.

Clarke, G. H. (1959) *Marine Lubrication* (Scientific Lubrication (GB), Broseley, Shropshire).

Clarke, S. and **Engelbach, R.** (1930) *Ancient Egyptian Masonry–The Building Craft* (Oxford U.P., London).

Clauss, F. J. (1972) *Solid Lubricants and Self-Lubricating Solids* (Academic Press, New York).

Clayton, D. and **Wilkie, M. J.** (1948) 'Temperature distribution in the bush of a journal bearing', *Engineering*, **166**, July, 49–52.

Cohen, M. R. and **Drabkin, I. E.** (1948) *A Source Book in Greek Science* (McGraw-Hill, New York).

Cole, B. N. (1975) 'Engineering department a part of industry–and never apart from it: Leeds University's philosophy for first 100 years', *Mech. Engng News*, Apr. (Institution of Mechanical Engineers, London).

Cole, J. A. and **Hughes, C. J.** (1956) 'Oil flow and film extent in complete journal bearings', *The Engineer*, **201**, 255–63.

Cole, J. A. and **Hughes, C. J.** (1957) 'Visual studies of film extent in dynamically loaded complete journal bearings'. Instn Mech. Engrs, *Proceedings of the Conference on Lubrication and Wear*, 147–50.

Coleman, T. (1965) *The Railway Navvies*, (Hutchinson, 1965, and Penguin Books, 1968).

Collins, I. F. (1972) 'A simplified analysis of the rolling of a cylinder on a rigid/perfectly plastic half-space', *Int. J. mech. Sci.*, **14**, 1–14.

Colmar Natural History Society (1890) 'Biography of G. A. Hirn', in *Bull. Soc. d'Hist. Nat. de Colmar* (1889–90), **1**, (NS), 183–309.

Comptes Rendus (1849) *Comptes Rendus*, **28**, 290.

Condon, E. U. (1951) 'Arnold Sommerfeld', *J. opt. Soc. Am.* **41**, No. 2, Feb., 63.

Constantinescu, V. N. (1959) 'On turbulent lubrication', *Proc. Instn mech. Engrs*, **173**, No. 38, 881–900.

Constantinescu, V. N. (1963) *Lubrificatia cu Gaze* (Editura Academiei Republicii Populare Romine, Bucuresti). English trans., *Gas Lubrication* (1969), Wehe, R. L. (ed.), American Society of Mechanical Engineers.

Constantinescu, V. N. (1968*a*) *Aplicatii Industriale Ale Lagărelor Cu Aer* (Editura Academiei Republicii Socialiste România, Bucharest).

Constantinescu, V. N. (1968*b*) *Lubrication in Turbulent Regime*, US Atomic Energy Commission, Burton, R. A. technical editor of translated version. Original title *Teoria Lubrificatiei in Regim Turbulent* (Editura Academiei Republicii Populare Romane, Bucharest, 1965).

Conti, P. (1875) 'Sulla resistenza di attrito', *Atti R. Accad. Lincei*, **11**, 16.

Conway-Jones, J. M. (1975) 'Discussion on Dr Wilson's paper on "Cavitation damage in plain bearings"', *Proceedings of the 1st Leeds–Lyon Symposium on Tribology on Cavitation and Related Phenomena in Lubrication* (Mechanical Engineering Publs., London), 201–17.

Cornell University (1967) 'Fred William Ocvirk', *Necrology of the Faculty of Cornell University, 1966–67*, Ithaca, New York, 46–8.

Corrie, J. M. (1914) 'Notes on a collection of polishers and other objects found on the site of the Roman Fort at Newstead, Melrose', *Proc. Soc. Antiq. Scotl.*, **xlviii** (1913–14), 338–41.

Couette, M. (1890) 'Études sur le frottement des liquides', *Ann. Chim. Phys.* **21**, 433–510.

Coulomb, C. A. (1780) 'Recherches sur la meilleure manière de fabriquer les aiguilles aimantées', *Mém. Math. Phys.* presantés à l'Académie Royale des Sciences (magnetism memoir submitted for the 1777 Paris Academy of Sciences prize), **ix**, Paris, 166–264.

Coulomb, C. A. (1785) 'Théorie des machines simples, en ayant égard au frottement de leurs parties, et a la roideur des cordages', *Mém. Math. Phys.*, **x**, Paris, 161–342.

Courtel, R. and **Tichvinsky, L. M.** (1963) 'A brief history of friction', *Mech. Engng*, Pt I, Sept., 55–9, Pt II, Oct., 33–7.

Courtney-Pratt, J. S. and **Eisner, E.** (1957) 'The effect of a tangential force on the contact of metallic bodies', *Proc. R. Soc.*, **A238**, 529–50.

Courtney-Pratt, J. S. and **Tudor, G. K.** (1946) 'Analysis of the lubrication between the piston rings and cylinder wall of a running engine', *Proc. Instn mech. Engrs*, **155**, 293.

Coyne, J. C. and **Elrod, H. G.** (1970) 'Conditions for the rupture of a lubricating film, Pt I, Theoretical model', *Trans. Am. Soc. mech. Engrs; J. lubr. Technol.*, **92**, F3, 451.

Craig, W. G. (1855) 'On an improved axle box and spring fittings for railway carriages', *Proc. Instn mech. Engrs*, 182–91.

Crook, A. W. (1958) 'The lubrication of rollers, I', *Phil. Trans. R. Soc.*, **A250**, 387–409.

Crook, A. W. (1961*a*) 'The lubrication of rollers, II–film thickness with relation to viscosity and speed', *Phil. Trans. R. Soc.*, **A254**, 223–36.

Crook, A. W. (1961*b*) 'The lubrication of rollers III–a theoretical discussion of friction and the temperatures in the oil film', *Phil. Trans. R. Soc.*, **A254**, 237–58.

Crook, A. W. (1963) 'The lubrication of rollers, IV–measurements of friction and effective viscosity', *Phil. Trans. R. Soc.*, **A255**, 281–312.

Current Biography (1950) 'Sommerfield, Arnold Johannes Wilhelm', *Current Biography*, 537–8.

Czichos, H. (1976*a*) 'Professor Dr Ludwig Gümbel–biographical notes', Private communication.

Czichos, H. (1976*b*) 'Professor Dr Richard Stribeck', private communication.

Czichos, H. (1978) *Tribology–a systems approach to the science and technology of friction, lubrication and wear* (Elsevier, Amsterdam).

Daniel, G. (1971) *The First Civilizations: The Archaeology of Their Origins* (Penguin, Harmondsworth).

Davenport, T. C. (ed.) (1973) *The Rheology of Lubricants*, Applied Science Publishers, Barking, Essex.

Davies, R. (1959*a*) 'A tentative model for the mechanical wear process', *Proceedings of the Symposium on Friction and Wear* held at the General Motors Research Laboratories (Detroit, 1957), Davies, R. (ed.) (Elsevier, Amsterdam).

Davies, R. (ed.) (1959*b*) 'Friction and wear', *Proceedings of the Symposium on Friction and Wear* (Detroit 1957) (Elsevier, Amsterdam).

Davies, R. (ed.) (1964) *Cavitation in Real Liquids*, Proceedings of a symposium at General Motors Research Laboratories (Elsevier, Amsterdam).

Davison, C. St. C. B. (1957*a*) 'Wear prevention in early history', *Wear*, **1**, No. 2, 155–9.

Davison, C. St. C. B. (1957*b*) 'Bearings since the Stone Age; A short history of their development', *Engineering*, 4 Jan., 2–5.

Davison, C. St. C. B. (1958) 'Wear prevention between 25 B.C. and 1700 A.D.', *Wear*, **2**, No. 1, 59–63.

Davison, C. St. C. B. (1961) 'Transporting sixty-ton statues in early Assyria and Egypt', *Technol. and Cult.*, **2**, No. 1, 11–6.

Dawson, P. H. (1962) 'Effect of metallic contact on the pitting of lubricated rolling surfaces', *J. mech. Engng Sci.*, **4**, No. 1, 16.

Dawson, P. H. and **Fidler, F.** (1966) 'Wire-wool type bearing failures: the formation of the wire wool', *Proc. Instn mech. Engrs*, **180**, Pt I, No. 1, 513–30.

De Baader, J. (1815) 'Railroads and carriages', British Patent No. 3939, 1–8.

De Camus, F. J. (1724) *Traité des forces mouvantes; pour la pratique des arts et métiers, avec une explication de vingt machines nouvelles et utiles* (Paris), 304–25.

De Caus, S. (1615) *Les Raisons des Forces Mouvantes*. Paris (See also later edn. with additional figures. Charles Serestne, Paris 1624).

De Hart, A. O. (1959) 'Which bearing and why', Paper 59-MD-12, ASME Design Engineering Conference, *General Motors Corporation Research Laboratories Report GMR-213*, 1–12.

De Monconys, B. (1666) *Journal des Voyages de Monsieur de Monconys*, II, 29.

De Vries, L. (1971) *Victorian Inventions*, compiled in collaboration with Ilonka van Amstel (John Murray, London).

Deacon, R. F. and **Goodman, J. F.** (1958) 'Orientation and frictional behaviour of lamellar solids on metals'. Instn Mech. Engrs, *Proceedings of the Conference on Lubrication and Wear* (1957), 344–7.

Dean, E. W. and **Davis, G. H. B.** (1929) 'Viscosity variation of oils with temperature', *Chem. Metall. Engng*, **36**, 618–9.

Dean, E. W. and **Davis, G. H. B.** (1940) 'Viscosity index of lubricating oils', *Ind. Engng Chem.*, **32**, 102.

Deeley, R. M. and **Parr, P. H.** (1913) 'On the viscosity of glacier ice', *Phil. Mag.* **26**, No. 6, 85–111.

Denne, T. (1847) 'Compositions for atmospheric pipes, and for lubricating machinery', British Patent No. 11674.

Den Hartog, J. P. (1934) *Mechanical Vibrations* (McGraw-Hill, New York).

Department of Education and Science (1966) *Lubrication (Tribology) Education and Research. A Report on the Present Position and Industry's Needs*, Department of Education and Science (HMSO, London).

Department of Trade and Industry (1971) *Committee on Tribology Report 1969–70*, Department of Trade and Industry (HMSO, London).

Department of Trade and Industry (1973) *The Introduction of a New Technology–Committee on Tribology Report 1966–72*, Department of Trade and Industry (HMSO, London).

Department of Trade and Industry (1972) *Tribology Projects for Schools*, Department of Trade and Industry.

Derry, T. K. and **Williams, T. I.** (1960) *A Short History of Technology: From the Earliest Times to A.D. 1900* (Oxford U.P., Oxford).

Deryagin, B. V. (1934) *Zh. Fiz. Khim.* **5**, No. 9.

Desaguliers, J. T. (1725*a*) 'Some experiments concerning the cohesion of lead', *Phil. Trans. R. Soc. Lond.*, **33**, 345.

Desaguliers, J. T. (1725*b*) *Phil. Trans. R. Soc. Lond.*, abridged, **vii**, 1724–34, 100.

Desaguliers, J. T. (1734) *A Course of Experimental Philosophy*, 2 vols, (London), vol. I, with thirty-two copper plates.

Dintenfass, L. (1963) 'Lubrication in synovial joints: a theoretical analysis: a rheological approach to the problems of joint movements and joint lubrication', *J. Bone and Joint Surg.*, **A45**, 1241–56.

Di Prima, R. C. (1959) 'The stability of viscous flow between rotating concentric cylinders with a pressure gradient acting round the cylinders', *J. fluid Mech.*, **6**, 462.

Di Prima, R. C. (1963) 'A note on the stability of flow in loaded journal bearings', *Trans. Am. Soc. lubr. Engrs*, **6**, 249.

DNB (1940) 'Hardy, Sir William Bate, 1864–1934', *Dictionary of National Biography* (1931–40), 397–8.

Donlan, M. J. J. (1848) 'Compounds for lubricating machinery', British Patent No. 12109.

Dowson, D. (1957) 'Cavitation in lubricating films supporting small loads'. Instn Mech. Engrs, *Proceedings of the Conference on Lubrication and Wear*, 93–9.

Dowson, D. (1961) 'Inertia effects in hydrostatic thrust bearings', *Trans. Am. Soc. mech. Engrs; J. Basic Engng*, **D83**, No. 2, 227–34.

Dowson, D. (1967) 'Modes of lubrication in human joints', in 'Lubrication and wear in living and artificial human joints', *Proc. Instn mech. Engrs*, **181**, Pt 3J, 45–54 (London).

Dowson, D. (1969) *Tribology*, inaugural lecture delivered in the University of Leeds on 5 Feb. 1968 (Leeds U.P.).

Dowson, D. (1970*a*) 'Monitoring, 1–introduction to tribological prognosis', *Tribology*, Aug., 138–9.

Dowson, D. (1970*b*), 'Whither tribology', *Proc. Instn mech. Engrs*, **184**, Pt 3L, 181–5.

Dowson, D. (1973) 'Tribology before Columbus', (ASME) *Mech. Engng*, Apr., 12–20 and (ASLE) *J. Am. Soc. lubr. Engrs* (*Lubr. Engng*), **29**, No. 6, 245–53.

Dowson, D. (1974*a*) *Lubricants and Lubrication in The Nineteenth Century*, Joint Institution of Mechanical Engineers–Newcomen Society Lecture, 1–8.

Dowson, D. (1974*b*) 'Stranger than fiction', the 6th Isaac Newton Lecture of the Institution of Mechanical Engineers, London, 1–31.

Dowson, D. (1975) 'The early history of tribology', *Proceedings of the First European Tribology Congress*, held in London, Sep. 1973 (Mechanical Engineering Publs), 1–14.

Dowson, D. (1976) 'The origins of rolling contact bearings', *Proceedings of the JSLE–ASLE International Lubrication Conference*, Sakuri, T. (ed.) (Elsevier, Amsterdam), 20–38.

Dowson, D. and **Ashton, J. N.** (1976) 'Optimum computerised design of hydrodynamic journal bearings', *Int. J. mech. Sci.*, **18**, No. 5, 215–22.

Dowson, D. and **Blount, G. N.** (1976) 'The design of optimum tilting-pad thrust bearings', *Proceedings of the JSLE–ASLE International Lubrication Conference*, Tokyo, 1975, Sakurai, T. (ed.) (Elsevier, Amsterdam), 247–71.

Dowson, D. and **Economou, P. N.** (1978) 'The wear of lignum vitae', *Proceedings of the 3rd Leeds–Lyons Symposium on Tribology* (Mechanical Engineering Publications, Bury St Edmunds).

Dowson, D. and **Higginson, G. R.** (1959) 'A numerical solution to the elastohydrodynamic problem', *J. mech. Engng Sci.*, **1**, No. 1, 6–15.

Dowson, D. and **Higginson, G. R.** (1966) *Elasto-Hydrodynamic Lubrication–The Fundamentals of Roller and Gear Lubrication* (Pergamon Press, Oxford).

Dowson, D. and **Hudson, J. D.** (1964*a*) 'Thermo-hydrodynamic analysis of the infinite slider-bearing, I. The plane-inclined slider bearing'. Instn Mech. Engrs, *Proceedings of the Lubrication and Wear Group Convention* (1963), 34–44.

Dowson, D. and **Hudson, J. D.** (1964*b*) 'Thermo-hydrodynamic analysis of the infinite slider-bearing, II. The parallel-surface bearing'. Instn Mech. Engrs, *Proceedings of the Lubrication and Wear Group Convention*, 45–51.

Dowson, D. and **Jones, D. A.** (1970) *Teaching Tribology*. Proceedings of a conference arranged by the Institute of Tribology, the University of Leeds, 14–16 Apr. (Iliffe Science and Technology Publ., Guildford).

Dowson, D. and **Taylor, C. M.** (1967) 'Fluid inertia effects in spherical hydrostatic thrust bearings', *Trans. Am. Soc. lubr. Engrs*, **10**, 316–24.

Dowson, D. and **Taylor, C. M.** (1975) 'Fundamental aspects of cavitation in bearings', *Proceedings of the 1st Leeds–Lyon Symposium on Tribology on Cavitation and Related Phenomena in Lubrication* (Mechanical Engineering Publ. London), 15–26.

Dowson, D. and **Whomes, T. L.** (1967) 'Side-leakage factors for a rigid cylinder lubricated by an isoviscous fluid', *Proc. Instn mech. Engrs* (1966–67), **181**, Pt 30, 165–76.

Dowson, D. and **Whomes, T. L.** (1971) 'The effect of surface roughness upon the lubrication of rigid cylindrical rollers', *Wear*, **18**, Pt I: Theoretical, 129–40; Pt II: Experimental, 141–51.

Dowson, D. and **Wright, V.** (1973) 'Bio-tribology', in *The Rheology of Lubricants*, Davenport, T. C. (ed.) (Applied Science Publishers, Barking), 81–8.

Dowson, D. and **Wright, V.** (eds) (1978) *An Introduction to the Bio-Mechanics of Joints and Joint Replacements* (Mechanical Engineering Publs, Bury St Edmunds).

Dowson, D., Hudson, J. D., Hunter, B. and **March, C. N.** (1967), 'An experimental investigation of the thermal equilibrium of steadily loaded journal bearings', *Proc. Instn mech Engrs*, **181**, Pt 3B, 70–80.

Dowson, D., Unsworth, A. and **Wright, V.** (1970) 'Analysis of "boosted lubrication" in human joints', *J. mech. Engng Sci.*, **12**, No. 5, 364–9.

Dowson, D., Godet, M. and **Taylor, C. M.** (1975) 'Cavitation and related phenomena in lubrication', *Proceedings of the 1st Leeds–Lyon Symposium on Tribology*, (Leeds) Sep. 1974 (Mechanical Engineering Publs, Bury St Edmunds).

Dowson, D., Blount, G. N. and **Ashton, J. N.** (1977) 'Optimisation methods applied to hydrodynamic bearing design', *Int. J. for Numerical Methods in Engineering*, **11**, 1005–27.

Dowson, D., Godet, M. and **Taylor, C. M.** (1977*a*) 'Superlaminar flow in bearings', *Proceedings of the 2nd Leeds–Lyon Symposium on Tribology*, (Leeds) Sep. 1975 (Mechanical Engineering Publs, Bury St Edmunds).

Dowson, D., Godet, M. and **Taylor, C. M.** (1978) 'The wear of non-metallic materials', *Proceedings of the 3rd Leeds–Lyon Symposium on Tribology*, (Leeds) Sep. 1976 (Mechanical Engineering Publs, Bury St Edmunds).

Dowson, J. (1937) 'Thrust bearings for steam turbine machinery'. Instn Mech Engrs, *Proceedings of the General Discussion on Lubrication and Lubricants*, **1**, 72–83.

Drachmann, A. G. (1932) 'Ancient oil mills and presses', *Kgl Danske Videnskabernes Selskab* (*Archaeol. Kunsthist Medd.*), **1**, No. 1, Sep. publ. (Levin and Munksgaard, Copenhagen).

Drachmann, A. G. (1948) 'Ktesibios, Philon and Heron – a study in ancient pneumatics', *Acta Hist. Sci. Nat. Med.*, **iv**, 74–7.

Drachmann, A. G. (1963) *The Mechanical Technology of Greek and Roman Antiquity* (Munksgaard, Copenhagen).

DSIR (1920) *Report of the Lubricants and Lubrication Inquiry Committee*, Advisory Council of the Department of Scientific and Industrial Research (HMSO, London).

DuBois, G. B. and **Ocvirk, F. W.** (1952) 'Experimental investigation of eccentricity ratio, friction, and oil flow of short journal bearings', *NACA, Tech. Note 2809*.

DuBois, G. B. and **Ocvirk, F. W.** (1953) 'Analytical derivation and experimental evaluation of short bearing approximation for full journal bearings', *NACA Report 1157*

Dudley, D. W. (1969) *The Evolution of the Gear Art* (American Gear Manufacturers Association, Washington, DC).

Duhem, P. M. M. (1906), *Études sur Léonard de Vinci. Ceux qu'il a lus et ceux qui l'ont lu*, Paris.

Duley, A. J. (1969) 'The Liechtis of Winterthur – clockmakers for three centuries', *Antiq. Horol.*, **6**, No. 2, 83–90.

Dunkerley, S. (1894) 'On the whirling and vibration of shafts', *Trans. R. Soc.*, **A185**, 279–360.

Dupuit, A. J. E. J. (1839) 'Résumé du Mémoire sur le tirage des voitures et sur le frottement de seconde espèce', *Comptes rendus hebdomadaires des séances de l'Académie des sciences* (Paris), **ix**, 689–700, 775.

Dupuit, A. J. E. J. (1840) *Comptes rendus*, **x**, 194.

Dupuit, A. J. E. J. (1841) *Comptes rendus*, **xii**, 482.

Dupuit, A. J. E. J. (1842) 'Sur le tirage des coitures et sur le frottement de roulement', *Ann. des Ponts et Chaussées*, **3**, 261.

Durand, W. F. (1924) *Robert Henry Thurston, A Biography* (American Society of Mechanical Engineers), 297 pp.

Dwelshauvers-Dery, (1890) 'Reminiscences of the life of G. A. Hirn', *Engineering*, 31 Jan. 120–1; 14 Feb. 174–5.

Dykes, P. de K. (1947) 'Piston ring movement during blow-by in high speed petrol engines', *Proc. Instn mech. Engrs* (1947–48), 71.

Dyson, A. (1970) 'Flow properties of mineral oils in elastohydrodynamic lubrication', *Phil. Trans. R. Soc. Lond.*, No. 1093, **A258**, 529–64.

Dyson, A., Naylor, H. and **Wilson, A. R.** (1966) 'The measurement of oil-film thickness in elastohydrodynamic contacts', *Proc. Instn mech. Engrs*, **180**, Pt 3B, 119–34.

Eaton, J. T. H. (ed.) (1969) 'A trip down memory lane', *The Dragon*, **xliv**, No. 5, 5–7.

Ecole Polytechnique (1895) *Book of the Centenary of the Ecole Polytechnique; 1794–1894*, **1**, 10, 176–81 (Gauthier-Villars, Paris).

Economou, P. N. (1973) 'A survey of lignum-vitae', Project Report, M.Sc. Course in Tribology, the University of Leeds.

Economou, P. N. (1976) 'The lubrication and dynamics of piston rings', Ph.D. Thesis, the University of Leeds.

Edgell, J. (1795) 'Axles for carriages', British Patent No. 2057 of A.D. 1795, 1–3.

Eilson, S. and **Saunders, O. A.** (1957) 'A study of piston-ring lubrication', *Proc. Instn mech. Engrs*, **171**, No. 11, 427–43.

Elco, R. A. and **Hughes, W. F.** (1962) 'Magnetohydrodynamic pressurization of liquid metal bearings', *Wear*, **5**, 198–212.

Eldredge, K. R. and **Tabor, D.** (1954) 'The mechanism of rolling friction 1. The plastic range', *Proc. R. Soc.*, **A229**, 181–98.

Elliott, J. S. and **Edwards, E. D.** (1957) 'Load carrying additives for steam turbine oils'. Instn Mech. Engrs, *Proceedings of the Conference on Lubrication and Wear*, 472–91.

Ellis, E. G. (1968) *Fundamentals of Lubrication* (Scientific Publs (GB), Broseley, Salop).

Ellis, E. G. (1953) *Lubricant Testing – Introducing Recent Developments in Testing Technique* (Scientific Publs (GB) Broseley, Salop).

Ellis, H. (1759) *Voyage de la Baye de Hudson*, Tome 2, Paris, 22.

Elrod, H. G. and **Ng, C. W.** (1967) 'A theory of turbulent fluid films and its application to bearings', *Trans. Am. Soc. mech. Engrs; J. lubr. Technol.*, **F89**, No. 3, 346.

Emmerton, J. T. (1958) 'Which should turn – wheel or axle', *Engineering*, 11 Apr. 1958, 474–7.

Engel, P. A. (1976) *Impact Wear of Materials* (Elsevier, Amsterdam).

Engelbach, R. (1923) *The Problem of the Obelisks* (T. Fisher Unwin, London).

Engineering (1889) Review of G. A. Hirn's book, *Construction de l'Espace Céléste* (published by Gauthier Villars et Fil), *Engineering*, 8 Nov., 549–50.

Engineering (1896) 'The Engineering Department of the Yorkshire College, Leeds', *Engineering*, London, 1–12.

Engineering (1905) 'The late Mr Beauchamp Tower', obituary notice in *Engineering*, 13 Jan.

Engineering (1935) 'The late Professor John Goodman', *Engineering*, 1 Nov., 483.

Engler, C. (1885) 'Ein Apparat zur Bestimmung der sogenannten Viscosität der Schmieröle', *Chem. Z.*, **9**, 189–90.

Ernst, H. and **Merchant, M. E.** (1940) 'Surface friction of clean metals–a basic factor in the metal cutting process', *Proceedings of a Special Summer Conference on Friction and Surface Finish*, (Tech. Press, M.I.T.), 76. (Also, *Friction–Selected Reprints*, Am. Inst. Phys. (1964), 34.)

Eschmann, P. (1964) *Das Leistungsvermögen der Wälzlager* (Springer-Verlag, Berlin).

Eschmann, P., Hasbargen, L. and **Weigand, K.** (1958) *Ball and Roller Bearings– Their Theory, Design, and Application* (K. G. Heyden, London).

ESDU (1965*a*) 'General guide to the choice of journal bearing type', *Engineering Sciences Data Unit, Item 65007* (Institution of Mechanical Engineers, London).

ESDU (1965*b*) 'Calculation methods for steadily loaded pressure fed hydrodynamic journal bearings', *Engineering Sciences Data Unit, Item 66023* (Institution of Mechanical Engineers, London).

ESDU (1967) 'General guide to the choice of thrust bearing type', *Engineering Sciences Data Unit, Item 67033* (Institution of Mechanical Engineers, London).

ESDU (1969) 'Computer service for prediction of performance of steadily loaded pressure fed hydrodynamic journal bearings', *Engineering Sciences Data Unit, Item 69002* (Institution of Mechanical Engineers, London).

ESDU (1973) 'Calculation methods for steadily loaded fixed-inclined-pad thrust bearings', *Engineering Sciences Data Unit, Item 73030* (Institution of Mechanical Engineers, London).

ESDU (1975) 'Calculation methods for steadily loaded tilting-pad thrust bearings', *Engineering Sciences Data Unit, Item 75023* (Institution of Mechanical Engineers, London).

ESDU (1976) 'A guide on the design and selection of dry rubbing bearings', *Engineering Sciences Data Unit, Item 76029,* (Institution of Mechanical Engineers, London).

Eshleman, R. L. (1972) *Flexible Rotor-Bearing System Dynamics–I. Critical Speeds and Response of Flexible Rotor Systems* (ASME, New York).

Essex Times (1905) 'Sudden death of Mr Beauchamp Tower', obituary notice, *Essex Times*, 4 Jan. 1905.

Ettles, C., Wells, D. E., Stokes, M. and **Matthews, J. C.** (1974) 'Investigation of bearing misalignment problems in a 500 MW turbo-generator set', *Proc. Instn mech. Engrs*, **188**, No. 35, 403–14.

Euler, L. (1744) 'De Curvis elasticic', *Acta Acad. Pretropolitanae* (1778).

Euler, L. (1750*a*) 'Sur le frottement des corps solides', *Mém. Acad. Sci. Berl.*, No. 4 (1748), 122–32.

Euler, L. (1750*b*) 'Sur la diminution de la resistance du frottement', *Mém. Acad. Sci. Berl.*, No. 4 (1748), 133–48.

Euler, L. (1762) 'Remarques sur l'effect du frottement dans l'equilibre', *Mém. Acad. Sci. Berl.*, No. 18 (1762), 265–78.

Evans, E. A. (1921) *Lubricating and Allied Oils* (Chapman & Hall, London).

Evans, E. A. (1937) 'Extreme-pressure lubricants'. Instn Mech. Engrs, *Proceedings of the General Discussion on Lubrication and Lubricants*, **2**, 52–9.

Eyssen, G. R. (1937) 'Properties and performance of bearing materials bonded with synthetic resin'. Instn Mech. Engrs, *Proceedings of the General Discussion on Lubrication and Lubricants*, **1**, 84–92.

Faro Barros, A. de, and **Dyson, A.** (1960) 'Piston ring friction–rig measurements with low viscosity oils', *J. Inst. Petrol.*, **46**, No. 433, Jan., 1–18.

Fein, R. S. (1967) 'Are synovial joints squeeze-film lubricated', in 'Lubrication and wear in living and artificial human joints', *Proc. Instn mech. Engrs*, **181**, Pt 3J, 125–8.

Felton, W. (1794) 'Treatise on carriages' (2 vols).

Fenton, A. (1965) 'Early and traditional cultivating implements in Scotland', *Proc. Soc. Antiq. Scot.*, Session 1962–63, **xcvi**, 264–317.

Ferodo, (1961) *Friction Materials for Engineers* (Ferodo, Chapel-en-le-Frith, Derbyshire).

Ferranti, F. J. de (1909) 'Air bearing for high speeds', US Patent No. 930851.

Ferranti, G. R. de (1934) 'Improvements in and relating to spinning, doubling, and twisting machinery', British Patents Nos 416717, 416718 and 416719.

Ferranti, S. Z. de (1904) 'Improvements in, and relating to methods of and apparatus for, spinning, doubling and the like', British Patent No. 115583.

Fiddes, E. (1937) 'Some teachers at Owens College', *J. Univ. Manch.*, **1**, No. 1.

Fisher, H. A. L. (1935) *A History of Europe–Volume II, From The Early Eighteenth Century to 1935* (Eyre and Spottiswoode).

Fitz-Gerald, J. M. (1969) 'Mechanics of red-cell motion through very narrow capillaries', *Proc. R. Soc. Lond.*, **B174**, 193–227.

Floberg, L. (1957) 'The infinite journal bearing considering vaporization', *Trans Chalmers Univ. Technol.*, Gothenburg, Sweden, No. 189.

Floberg, L. (1961) 'Attitude–eccentricity curves and stability conditions for the infinite journal bearing', *Trans. Chalmers Univ. Technol.*, Gothenburg, Sweden, No. 235.

Floberg, L. (1968) 'Sub-cavity pressures and number of oil steamers in cavitation regions with special reference to the infinite journal bearing', *Acta Polytech. Scand.*; Mechanical Engineering Series, No. 37.

Floberg, L. (1969) 'On the optimum design of sector-shaped tilting-pad thrust bearings', *Acta Polytech. Scand.*, Mechanical Engineering Series, No. 45.

Floberg, L. (1971) *Maskinelement* (Almqvist & Wiksell Förlag AB, Stockholm).

Flowers, A. E. and **Hersey, M. D.** (1932) 'Lubrication research of the American Society of Mechanical Engineers', *Mech. Engng*, **54**, 269–70.

Fogg, A. (1946) 'Fluid film lubrication of parallel surface thrust surfaces', *Proc. Instn mech. Engrs*, **155**, 49–67.

Fogg, A. and **Hunwicks, S. A.** (1937) 'Some experiments with water-lubricated rubber bearings'. Instn Mech. Engrs, *Proceedings of the General Discussion on Lubrication and Lubricants*, **1**, 101–6.

Foord, C. A., Wedeven, L. D., Westlake, F. J. and **Cameron, A.** (1970) 'Optical elastohydrodynamics', *Proc. Instn mech. Engrs*, **184**, Pt I, 487–503.

Forbes, R. J. (1958) 'Petroleum', in *A History of Technology*, Volume V, *The Late Nineteenth Century c. 1850–1900*, 102–23, Singer, C., Holmyard, E. J., Hall, A. R. and Williams, T. I. (eds) (Clarendon Press, Oxford).

Forbes, W. G. (1954) *Lubrication of Industrial and Marine Machinery*, (Wiley, New York).

Ford, G. W. K., Harris, D. M. and **Pantall, D.** (1957) 'Principles and applications of hydrodynamic-type gas bearings', *Proc. Instn mech. Engrs*, **171**, 93–128.

Forrester, P. G. (1968) 'Soft-metal bearings'. Conference on Lubrication and Wear: Fundamentals and Application to Design, Instn Mech. Engrs, *Proceedings, 1967–68*, **182**, Pt 3A, 321–4.

Fowle, T. I. (1968) 'Lubricants for fluid film and Hertzian contact conditions'. Conference on Lubrication and Wear: Fundamentals and Application to Design, Instn Mech. Engrs, *Proceedings, 1967–68*, **182**, Pt. 3A, 568–84.

Fowle, T. I. (1974) 'For the record: the reciprocating steam engine', *Tribology News*, Issue 24, July 1974, Institution of Mechanical Engineers.

Fowle, T. I. (1975) 'Beauchamp Tower and Joseph Tomlinson', *Tribology News*, Issue 28, Dec., 4–5.

Fränkel, A. (1944) 'Berechung von Zylindrischen Gleitlasern', *Mitt. Inst. Thermodyn. VerbrennMotBau* (Federal Institute of Technology, Zurich), No. 4, 134 pp.

Francis, Sir F. (ed.) (1971) *Treasures of the British Museum* (Thames and Hudson, London).

Frazer, Sir J. G. (1922) *The Golden Bough – a Study in Magic and Religion* (Macmillan, London).

Freeman, P. (1962) *Lubrication and Friction* (Pitman, London).

Freeman, P. (1962) *Lubrication and Friction*, (Pitman, London).

Freese, S. (1971) *Windmills and Millwrighting* (David & Charles, Newton Abbot, Devon).

Froessel, W. Von (1938) 'Nachprufungeder Hydrodynamischen Schmierungs theorie durch Versuche', *Forsch. Geb. IngWes.*, **9**, Nov./Dec. 261–78.

Fujino, T. (1969) 'Analysis of hydrodynamic problems by the finite element method', Paper No. J5-4, Japan–USA Seminar on Matrix Methods of Structural Analysis and Design, Tokyo, Japan, 25–30 Aug.

Fuller, D. D. (1947) 'Hydrostatic lubrication', *Mach. Des.*, **19**: I, 'Oil pad bearings', June, 110–6; II, 'Oil lifts', July, 117–22; III, 'Step bearings', Aug., 115–20; IV, 'Oil cushions', Sep., 127–31, 188, 190.

Fuller, D. D. (1956) *Theory and Practice of Lubrication for Engineers*, (Wiley, New York).

Fuller, D. D. (1958) *A Survey of Journal Bearing Literature*, prepared for the Journal Bearing Research Committee, the American Society of Lubrication Engineers (Chicago).

Furey, M. J. (1969) 'Friction wear and lubrication', *Ind. Engng Chem.* (American Chemical Society) **61**, No. 3, Mar., 12–29.

Furuhama, S. (1959–61) 'A dynamic theory of piston-ring lubrication' (1959); (i) First Report–Calculation, *Bull. Jap. Soc. mech. Engrs*, **2**, 423 (1960); (ii) Second Report–Experiment, *Bull. Jap. Soc. mech. Engrs*, **3**, 291 (1961); (iii) Third Report–Measurement of oil film thickness, *Bull. Jap. Soc. mech. Engrs*, **4**, 744.

Galloway, E. (1831) *History and Progress of the Steam Engine* (Thomas Kelly, London).

Galton, D. (1878*a*) 'The action of brakes: on the effect of brakes upon railway trains', *Engineering*, 14 June, 469–72.

Galton, D. (1878*b*) 'Railway brakes: on the coefficient of friction from experiments on railway brakes', *Engineering*, 23 Aug., 153–4.

Gaman, I. D. C., Higginson, G. R. and **Norman, R.** (1974) 'Fluid entrapment by a soft surface layer', *Wear*, **28**, 345–52.

Garnett, J. (1787) 'Method of reducing friction in axles, etc.', British Patent No. 1580, A.D. 1787, 1–4, accompanied by one figure.

Geary, P. J. (1962) *Fluid Film Bearings–A Survey of Their Design, Construction and Use* (British Scientific Research Association).

Gemant, A. (1950) *Frictional Phenomena* (Chemical Publ. Co., New York).

George, W. (1787) 'Destroying friction in all kinds of axles and shafts', British Patent No. 1602, A.D. 1787, 1–3, accompanied by two figures.

Georgi, C. W. (1950) *Motor Oils and Engine Lubrication* (Reinhold, New York).

Gille, B. (1966) *The Renaissance Engineers* (Lund, Humphries, London).

Gillmor, C. S. (1968) *Charles Augustin Coulomb: Physics and Engineering in Eighteenth Century France* (Princeton U.P., Princeton, NJ).

Gillmor, C. S. (1971) *Coulomb and the Evolution of Physics and Engineering in Eighteenth-Century France* (Princeton U.P., Princeton, NJ).

Girard, L. D. (1865) *Hydraulique appliquée. Note sur le chemin de fer glissant* (Gauthier-Villars, Paris) 8p.

Glienicke, J. (1967) 'Experimental investigation of the stiffness and damping coefficients of turbine bearings and their application to instability prediction', *Proc. Instn mech. Engrs*, **181**, Pt 3B, 116–29.

Goddard, D. R. (1967) 'Synthetic lubricants', in *Lubrication and Lubricants*, Braithwaite, E. R. (ed.) (Elsevier, Amsterdam), Ch. 4, 166–96.

Godfrey, D. (1965) 'Boundary lubrication', in *Proceedings of International Symposium on Lubrication and Wear*, Muster, D. and Sternlicht, B. (eds) McCutchan Publ. Corp., Berkeley, Calif.), 283–306.

Godfrey, D. (1968) 'Boundary lubrication', in *Interdisciplinary Approach to Friction and Wear*, Ku, P. M. (ed.), Proceedings of a National Aeronautics and Space Administration Symposium held in San Antonio, Texas, in 1967.

Godfrey, D. (1970) 'A.S.L.E. publications, the growth of A.S.L.E.–Part 8', *J. Am. Soc. lubr. Engrs*, 100a.

Goldblatt, J. (1973) 'The life and times of Paraffin Young', *Esso Mag.* (London), Spring 1973, 21–5.

Goodman, J. (1886) 'Recent researches in friction', *Proc. Instn civ. Engrs*, **ixxxv**, Session 1885–6, Pt III, 1–19.

Goodman, J. (1887) 'Recent researches in friction–Part II', *Proc. Instn civ. Engrs*, **lxxxix**, Session 1886–7, Pt III, 3–36.

Goodman, J. (1890) *The Friction and Lubrication of Cylindrical Journals* (Manchester Association of Engineers, Manchester), 87–135.

Goodman, J. (1906*a*) *Mechanics Applied to Engineering* (Longmans, London).

Goodman, J. (1906*b*) *The Results of Technical Education in Engineering* (Manchester Association of Engineers, Manchester), 401–24.

Goodman, J. (1912) '(1) Roller and ball bearings'; '(2) The testing of antifriction bearing materials', *Proc. Instn civ. Engrs*, **clxxxix**, Session 1911–2, Pt III, 4–88.

Gough, V. E. (1958*a*) 'Friction of rubber, No. 1', *The Engineer*, 31 Oct., 701–4.

Gough, V. E. (1958*b*) 'Friction of rubber, No. 2', *The Engineer*, 7 Nov., 741–3.

Grassam, N. S. and **Powell, J. W.** (eds) (1964) *Gas Lubricated Bearings* (Butterworths, London).

Graue, G. (1967) 'Jubilee in the Max-Planck institute for friction research, Göttingen; Prof. Vogelpohl 25 years with the Max-Planck Company', Personal communication.

Greene, A. B. (1967) 'Automotive lubrication', in *Lubrication and Lubricants*, Braithwaite, E. R. (ed.) (Elsevier, Amsterdam) Ch. 9, 427–71.

Greene, A. B. (1969) 'Photographic recording of dynamic oil-film characteristics in a piston cylinder simulation rig'. Instn Mech. Engrs, *Proceedings of the Conference on Photography in Engineering*, 37.

Greenway, C. (1840) 'Reducing friction in Carriage wheels and machinery', British Patent No. 8333, 1–4, accompanied by ten figures.

Greenwood, J. A. and **Tripp, J. H.** (1971) 'The contact of two nominally flat rough surfaces', *Proc. Instn mech. Engrs*, **185**, 625–33.

Greenwood, J. A. and **Williamson, J. B. P.** (1966) 'Contact of nominally flat surfaces', *Proc. R. Soc., Lond.*, **A295**, 300–19.

Grieve, W. H. (1944) 'Historic researches', letter to *The Engineer*, 25 Aug., 150.

Groenveld, P. (1970) *Dip-Coating by Withdrawal of Liquid Films* (Delftsche Vitegevers Maatshappÿ, N.V., Delft).

Gross, W. A. (1962) *Gas Film Lubrication* (Wiley, New York).

Grubin, A. N. and **Vinogradova, I. E.** (1949) 'Investigation of the contact of machine components', Kh, F. Ketova (ed.) *Central Scientific Research Institute for Technology and Mechanical Engineering* (Moscow), Book No. 30, (DSIR translation No. 337).

Gruse, W. A. (1967) *Motor Oils – Performance and Evaluation* (Reinhold, New York).

Gülker, E. (1976) 'Tribologie in Deutschland preiswürdig?', *Schmiertechnik + Tribologie*, **23**, No. 2, 26.

Gümbel, L. (1914*a*) Das Problem der Lagerreibung', *Mbl. berl. BezVer. dt. Ing.* (VDI), 1 Apr. and No. 5, May, 6 June, 87–104 and 109–20 (also July 1916).

Gümbel, L. (1914*b*) 'Die Schubkraftmaschine', *Z. Ges. Turbinenwesen*.

Gümbel, L. (1914*c*) 'Uber geschmierte Arbeitsräder', *Z. Ges. Turbinenwesen*, 22–6.

Gümbel, L. (1920) 'Wer ist der wirklich Blinde?' *Eine Frage im Interesse von Wissenschaft und Technik; offener Brief an die Herren A. Riedler und St. Löffler; mit einen Beirtag: Die unmittelbare Reibung fester Körper* (Springer, Berlin).

Gümbel, L. (1921) 'Verleich der Ergebnisse der rechnerischen Behandlung des Lagerschmierungsproblem mit neueren Versuchsergebnissen', *Mbl. berl. Bez.* (VD 1), Sep., 125–8.

Gümbel, L. and **Everling, E.** (1925) *Reibung und Schmierung im Maschinenbau* (M. Krayn, Berlin), 1–240.

Gunderson, R. C. and **Hart, W. A.** (eds), (1962) *Synthetic Lubricants* (Reinhold, New York).

Gunn, B. (1975) *The Timetables of History – A Chronology of World Events from 5000 B.C. to the Present Day* (Thames and Hudson, London).

Gunter, E. J. (1966) 'Dynamic stability of rotor-bearing systems', *NASA SP-111* (*N67-11942*) (National Technical Information Service, Washington).

Gunter, E. J. (1970) 'Influence of flexibly mounted rolling element bearings on rotor response, Part I – Linear analysis, *Trans. Am. Soc. mech. Engrs; J. lubr. Technol.* **92**, No. 1, Jan., 59–75.

Gunther, R. C. (1972) *Lubrication* (Bailey Bros & Swinfen, Folkestone).

Gunther, R. T. (1930) *Early Science in Oxford*, Vols VI and VII, *The Life and Work of Robert Hooke*, printed for the author at the Oxford University Press by John Johnson (Oxford).

Hagg, A. C. (1946) 'The influence of oil-film journal bearings on the stability of rotating machines', *J. appl. Mech.*, **A13**, No. 68, 211–20; **A14**, No. 69, 77–8 (1947).

Hagg, A. C. (1948) 'Some vibration aspects of lubrication', *Lubr. Engng*, **4**, No. 4, 166–9.

Hagg, A. C. and **Warner, P. C.** (1953) 'Oil-whirl of flexible rotors', *Trans. Am. Soc. mech. Engrs*, **75**, 1339–44.

Hahn, E. J. (1975) 'The excitability of flexible rotors in short sleeve bearings', *Am. Soc. mech. Engrs; J. lubr. Technol.*, **97**, No. 1, 105–15.

Hahn, E. J. and **Simandiri, S.** (1974) 'Squeeze film mounts for vibration attenuation in rigid rotors', *Proceedings, Noise, Shock and Vibration Conference*, Monash University, Melbourne, 435–44.

Hahn, H. W. and **Radermacher, K.** (1967) 'Hydrodynamic theory of dynamically loaded bearings and its application to engine bearing design', *Proc. Instn mech. Engrs*, (1966–67), **181**, Pt 3B, 35–44.

Håkansson, B. (1964) 'The journal bearing considering variable viscosity', *Trans. Chalmers Univ. Technol.*, Sweden, No. 298.

Håkansson, B. (1967) 'The infinite partial fitted journal bearing', *Acta Polytech. Scand.*, Mechanical Engineering Series, No. 30.

Hale, J. R. (1971) *Renaissance Europe 1480–1520* (Collins, London).

Hall, L. F. (1957) 'A review of the papers on the lubrication of rotating bearings and gears'. Instn Mech. Engrs, *Proceedings of the Conference on Lubrication and Wear* (1957), 425–9.

Hall, L. F. (1965) 'Discussion during Session 4', *Proc. Instn mech. Engrs*, **179**, Pt 3D, 281–2, 284.

Halliday, J. S. (1955) 'Surface examination by reflection electron microscopy', *Proc. Instn mech. Engrs*, **169**, No. 38, 177.

Halliday, J. S. (1957) 'Application of reflection electron microscopy to the study of wear'. Instn Mech. Engrs, *Proceedings of the Conference on Lubrication and Wear* (1957), 647–51.

Halling, J. (ed.) (1975) *Principles of Tribology* (Macmillan Press, London and Basingstoke).

Halling, J. (1976*a*) *Introduction to Tribology* (Wykeham Publ. (London), London).

Halling, J. (1976*b*) 'A contribution to the theory of friction and wear and the relationship between them', *Proc. Instn mech. Engrs*, **190**, 477–88.

Hamilton, G. M. and **Moore, S. L.** (1971) 'Deformation and pressure in an elasto-hydrodynamic contact', *Proc. R. Soc., Lond.*, **A322**, 313–30.

Hamilton, G. M. and **Moore, S. L.** (1974*a*) 'Measurement of the oil-film thickness between the piston rings and liner of a small diesel engine', *Proc. Instn mech. Engrs*, **180**, 253.

Hamilton, G. M. and **Moore, S. L.** (1974*b*) 'Comparison between measured and calculated thicknesses of the oil-film lubricating piston rings', *Proc. Instn mech. Engrs*, **180**, 262.

Hamilton, G. M. and **Moore, S. L.** (1976) 'Measurement of piston ring profile during running-in', *Piston Ring Scuffing*, (Mechanical Engineering Publs, London), 61–70.

Hamrock, B. J. and **Dowson, D.** (1976*a*) 'Isothermal elastohydrodynamic lubrication of point contacts. Pt I–Theoretical formulation', *Trans. Am. mech. Engrs; J. lubr. Technol.*, **F98**, No. 2, 223–9.

Hamrock, B. J. and **Dowson, D.** (1976*b*) 'Isothermal elastohydrodynamic lubrication of point contacts: Part II–Ellipticity parameter results', *Trans. Am. Soc. mech. Engrs; J. lubr. Technol.*, **F98**, No. 3, 375–83.

Hamrock, B. J. and **Dowson, D.** (1977*a*) 'Isothermal elastohydrodynamic lubrication of point contacts–Pt III–Fully flooded results', *Trans. Am. Soc. mech. Engrs*, **F99**, No. 2, 264–76.

Hamrock, B. J. and **Dowson, D.** (1977*b*) 'Isothermal elastohydrodynamic lubrication of point contacts: Part IV–Starvation results', *Trans. Am. Soc. mech. Engrs*, **F99**, No. 1, 15–23.

Hansen, G. (1975) *Steinöl und Brunnenfeuer* (Wintershall Aktiengesellschaft, Kassel).

Hardacre, H. T. (1812) 'Composition to prevent friction', British Patent No. 3573.

Hardy, W. B. (1912) 'The general theory of colloidal solutions', *Proc. Roy. Soc.*, **A86**, 601–10.

Hardy, W. B. (1920) 'Problems of lubrication', Address to the Royal Institution of Great Britain (see *Collected Scientific Papers of Sir William Bate Hardy*, Cambridge U.P. (1936), 639–44).

Hardy, W. B. (1936) 'Collected scientific papers of Sir William Bate Hardy', Rideal, Sir Eric K. (ed.). Published under the auspices of the Colloid Committee of the Faraday Society, Cambridge U.P.

Hardy, W. B. and **Doubleday, I.** (1922*a*) 'Boundary lubrication – the temperature coefficient', *Proc. R. Soc.*, **A101**, 487–92.

Hardy, W. B. and **Doubleday, I.** (1922*b*) 'Boundary lubrication – the paraffin series', *Proc. R. Soc.*, **A100**, 550–74.

Hardy, W. B. and **Doubleday, I.** (1923) 'Boundary lubrication – the latent period and mixtures of two lubricants', *Proc. R. Soc.*, **A104**, 25–39.

Hardy, W. B. and **Hardy, J. K.** (1919) 'Note on static friction and on the lubricating properties of certain chemical substances', *Phil. Mag.*, S6, **38**, 32–40.

Harkins, W. D., Brown, F. E. and **Davies, E. C. H.** (1917) 'The structure of the surfaces of liquids, and solubility as related to the work done by the attraction of two liquid surfaces as they approach each other', *J. Am. Chem. Soc.*, **39**, March., 354–64.

Harkins, W. D., Davies, E. C. H. and **Clark, G. L.** (1917) 'The orientation of molecules in the surfaces of liquids, the energy relations at the surfaces, solubility, adsorption, emulsification, molecular association and the effect of acids and bases on interfacial tension', *J. Am. chem. Soc.*, **39**, Apr., 541–96.

Harris, T. A. (1966) *Rolling Bearing Analysis* (Wiley, New York).

Harrison, H. C. (1949) *The Story of Sprowston Mill* (Phoenix House, London).

Harrison, W. J. (1913) 'The hydrodynamical theory of lubrication with special reference to air as a lubricant', *Trans. Camb. Phil. Soc.*, **xxii**, (1912–25), 6–54.

Harrison, W. J. (1919) 'The hydrodynamical theory of the lubrication of a cylindrical bearing under variable load, and of a pivot bearing', *Trans. Camb. Phil. Soc.*, **xxii**, (1912–23), 373–88.

Hart, I. B. (1963) *The Mechanical Investigations of Leonardo da Vinci* (University of California Press, Berkeley and Los Angeles, Calif.).

Hatchett, C. (1803) 'Experiments and observations on the various alloys, on the specific gravity, and on the comparative wear of gold. Being the substance of a report made to the Right Honourable the Lord of the Committee of Privy Council, appointed to take into consideration the state of the coins of the Kingdom, and the present establishment and Constitution of his Majesty's Mint', *Phil. Trans. R. Soc. Lond.*, for the year MDCCCIII, Part I, 43–194.

Haudricourt, A. G. and **Jean-Brunhes, M.** (1955) *L'Homme et la Charrue a Travers le Monde* (Gallimard (Geographic Humaine No. 25), Paris), 229.

Hawkes, C. J. and **Hardy, G. F.** (1936) 'Friction of piston rings', *Trans. N.E. Coast Inst. Civil Engrs. and Shipbuilders*, **52**, 143–78, D, 49–58.

Hays, T. J. (1962) 'A steam turbine driven, water lubricated, high speed centrifugal compressor for oxygen service', *Mech. Engng*, **84**, Oct., 72.

Hays, D. F. (1959) 'A variational approach to lubrication problems and the solution of the finite journal bearings', *Trans. Am. Soc. mech. Engrs; J. Basic Engng*, Mar., 13–23.

Hays, D. F. (1960) 'Design curves for journal bearings', ASME Paper 59-MD-11.

Hays, D. F. (1961) 'Squeeze films: a finite journal bearing with a fluctuating load', *Trans. Am. Soc. mech. Engrs; J. Basic Engng*, **D83**, 579–88.

Hays, D. F. (1962) 'Squeeze films for rectangular plates', ASME Paper 62-Lub S-9.

Heathcote, H. L. (1921) 'The ball bearing: in the making, under test, and on service', *Proc. Instn automot. Engrs*, **15**, 569–702.

Heenan, R. H. (1885) 'Tower spherical engine', *Proc. Instn mech. Engrs*, 96–120.

Heisenberg, Von W. (1951) 'Arnold Sommerfeld', *Die Naturw.*, Jahrgang 38, Heft 15, Aug.

Helsham, R. (1767) *A Course of Lectures in Natural Philosophy*, by the late Richard Helsham, M.D., published by Bryan Robinson, M.D., printed for J. Nourse, opposite Katherine Street in the Strand, Bookseller in Ordinary to his Majesty.

Herschel, W. H. (1918) 'Standardisation of the Saybolt universal viscometer', *Technological Papers of the Bureau of Standards* (Washington), No. 112.

Hersey, M. D. (1914) 'The laws of lubrication of horizontal journal bearings', *J. Wash. Acad. Sci.*, **4**, 542–52.

Hersey, M. D. (1933) 'Notes on the history of lubrication, Part I. General survey', *J. Am. Soc. nav. Engrs*, **xlv**, No. 4, Nov. 1933, 411–29.

Hersey, M. D. (1934) 'Notes on the history of lubrication, Part II. Journal bearing experiments', *J. Am. Soc. nav. Engrs*, **xlvi**, No. 3, Aug., 369–85.

Hersey, M. D. (1936) *Theory of Lubrication* (Wiley, New York).

Hersey, M. D. (1966) *Theory and Research in Lubrication–Foundations for Future Developments* (Wiley, New York).

Hertz, H. (1881) 'On the contact of elastic solids', *J. reine und angew. Math.*, **92**, 156–71.

Hertz, H. (1882) 'On the contact of rigid elastic solids and on hardness', *Verh. Ver. Beförd. GewFleiss.*, Nov.

Heubner, K. H. (1975) *The Finite Element Method for Engineers* (Wiley, New York).

Hill, E. C. (1969) 'Microbes and lubricants', *Tribology*, **2**, No. 1, Feb., 5–10.

Hire, P. de la (1699) 'Sur les frottements des machines', *Histoire de l'Academie Royale*, **A**, (Chez Gerard Kuyper, Amsterdam, 1706), 128–34.

Hirn, G. (1854) 'Sur les principaux phénomènes qui présentent les frottements médiats', *Bull. Soc. ind. Mulhouse*, **26**, 188–277.

Hirs, G. G. (1972) 'A bulk-flow theory for turbulence in lubricant films', Paper 72-Lub 12, ASME–ASLE Conference, New York, Oct.

Hirst, W. (1968) 'Basic mechanisms of wear'. Conference on Lubrication and Wear: Fundamentals and Application to Design, Inst Mech. Engrs, *Proceedings, 1967–68*, **182**, Pt 3A, 281–92.

Hisakado, T. (1969) 'On the mechanism of contact between solid surfaces' (1st–3rd Reports), *Bull. Jap. Soc. mech. Engrs*, **12**, No. 54, 1519–45.

Hisakado, T. (1972) 'On the mechanism of contact between solid surfaces, Part 5', *Trans. Jap. Soc. mech. Engrs*, **38**, 2657–65.

Hisakado, T. (1974) 'Effect of surface roughness on contact between solid surfaces', *Wear*, **28**, 217–34.

Hisakado, T. and **Tsukizoe, T.** (1974) 'Effects of distribution of surface slopes and flow pressures of contact asperities on contact between solid surfaces', *Wear*, **30**, 213–27.

Hobson, P. D. (1955) *Industrial Lubrication Practice* (Industrial Press, New York).

Hodge, P. R. (1852) 'On a new self-lubricating axlebox for railway engines and carriages, and a self-acting spring crossing point', *Proc. Instn mech. Engrs*, 213–22.

Hoefer, F. (1861) *Nouvelle Biographic Générale* (Firmin Didot, Paris), **35**, 599–602.

Holland, J. (1975) 'Prof. G. Vogelpohl', *MTZ Motortech. Z.*, **36**, Nos. 7/8, 229–30.

Holligan, P. T. (1966) 'Plain bearings', in *The Principles of Lubrication*, Cameron, A., (Longmans, London), 561–74.

Holm, R. (1929) 'On metallic contact resistance', *Wiss. Veröff. Siemens-Werk*, Nos 7/2, 217.

Holm, R. (1938) 'The friction force over the real area of contact' (in German), *Wiss. Veröff. Siemens-Werk*, **17**, No. 4, 38–42.

Holm, R. (1946) *Electrical Contacts* (Almqvist and Wiksells, Stockholm).

Holm, R., assisted by **Holm, E.** (1958) *Electric Contacts Handbook*, 3rd edn, (Springer-Verlag, Berlin).

Holmes, R. (1960) 'The vibration of a rigid shaft in short sleeve bearings', *J. mech. Engng Sci.*, **2**, 337–41.

Holzer, H. (1921) *Die Berechnung der Drehschwingungen* (Springer-Verlag, Berlin).

Hompesch, T. A. W. Count de (1841) 'Obtaining oils, etc. from bituminous matters', British Patent No. 9060.

Hondros, E. D. (1971) *Tribology* (Mills & Boon, London).

Hooke, R. (1684) See Gunther, R. T. (1930), *Early Science in Oxford*, Vols VI and VII, *The Life and Work of Robert Hooke* (Oxford).

Hooke, R. (1705) 'The posthumous works of Robert Hooke' (Waller, and Seer, London).

Hoover, H. C. and **Hoover, L. H.** (1950) *Georgius Agricola–De Re Metallica*, translated from the first Latin edition of 1556 (Dover, New York).

Hori, Y. (1956) 'A theory of oil whip', *Proc. Fifth Jap. Natn. Congr. appl. Mech.*, 395–8; *J. appl. Mech.* (1959), **26**, **T81**, 189–98.

Hother-Lushington, S. (1971) 'Monitoring, 4–capacitive and inductive techniques', *Tribology*, Feb., 33–7.

Hother-Lushington, S. (1976) 'Water-lubricated bearings', *Tribology Int.*, Dec., 257–60.

Hother-Lushington, S. and **Sellors, P.** (1964) 'Water-lubricated bearings: initial studies and future prospects in the power generation industry'. Instn Mech. Engrs, *Proceedings of the Lubrication and Wear Convention* (1963), 139–46.

Houghton, P. S. (1976) *Ball and Roller Bearings* (Applied Science Publishers, London).

Howarth, H. A. S. (1919) 'Slow speed and other tests of Kingsbury thrust bearings', *Trans. Am. Soc. mech. Engrs.*, **41**, 685–707.

Howarth, H. A. S. (1929) 'Journal running positions', *Trans. Am. Soc. mech. Engrs*, **51**, No. 1, Paper APM-51-3, 21–35.

Howarth, H. A. S. (1934) 'Current practice in pressures, speeds, clearances and lubrication of oil-film bearings', *Trans. Am. Soc. mech. Engrs*, **56**, 891–902.

Howell, H. G., Mieskis, K. W. and **Tabor, D.** (1959) *Friction in Textiles*, (Butterworths, London).

Hulse, W. W. (1882) 'Contribution to the discussion at the spring meeting of the Institution of Mechanical Engineers', held at the Institution of Civil Engineers, London, on 20 Apr., 1883. *Proc. Instn mech. Engrs*, 149–50.

Hummel, C. (1926) 'Kritische Drehzahlen als Folge der Nachgiebigkeit des Schmiermittels im Lager', *Verein. Deut. Ing. Forschungs*, Pt No. 287.

Hunter, W. B. and **Zienkiewicz, O. C.** (1960) 'Effect of temperature variations across the lubricant films in the theory of hydrodynamic lubrication', *J. mech. Engng Sci.*, **2**, 52–8.

Illmer, L. (1937) 'Piston ring friction in high speed engines', *Trans. Am. Soc. mech. Engrs*, **59**, 1–6.

Instn of Civil Engineers (1905) 'Beauchamp Tower', obituary notice in minutes of *Proc. Instn civ. Engrs*, **162**, 420.

Instn of Civil Engineers (1913) 'Obituary notice (on O. Reynolds)', *Proc. Instn Civ. Engrs*, **cxci**, 314.

Institute of Petroleum (1959*a*) 'The Golden Jubilee of B.P.', *Inst. Petrol. Rev.*, May 1959, 135–7.

Institute of Petroleum (1959*b*) 'The Drake centenary', *Inst. Petrol. Rev.*, **13**, No. 152, Aug., 233–5.

Instn of Mechanical Engineers (1888) 'Third report of the research committee on friction (experiments on the friction of a collar bearing)', *Proc. Instn mech. Engrs*, May 1888, 173–205.

Instn of Mechanical Engineers (1891) 'Fourth report of the research committee on friction (experiments on the friction of a pivot bearing)', *Proc. Instn mech. Engrs*, March 1891, 111–40.

Instn of Mechanical Engineers (1905) 'Beauchamp Tower', obituary notice in *Proc. Instn mech. Engrs*, **1**, 163.

Instn of Mechanical Engineers (1937*a*) *Proceedings of the General Discussion on Lubrication and Lubricants* (13–15 Oct.) Vol. 1 (Institution of Mechanical Engineers, London).

Instn of Mechanical Engineers (1937*b*) *Proceedings of the General Discussion on Lubrication and Lubricants* (13–15 Oct.) Vol. 2 (Institution of Mechanical Engineers, London).

Instn of Mechanical Engineers (1943) Award of the James Watt International Medal to Mr A. G. M. Michell, F.R.S., 22 Jan. 1943 (Institution of Mechanical Engineers, London).

Instn of Mechanical Engineers (1957) *Proceedings of the Conference on Lubrication and Wear*, Oct. 1957 (Institution of Mechanical Engineers, London).

Instn of Mechanical Engineers (1963*a*) *Engineering Heritage*, Vol. 1 (Heinemann, on behalf of the Institution of Mechanical Engineers, London).

Instn of Mechanical Engineers (1963*b*) *Fatigue in rolling contact*. Instn Mech. Engrs,

Proceedings of a Symposium arranged by the Applied Mechanics and Lubrication and Wear Groups, (London) 28 Mar. 1963.

Instn of Mechanical Engineers (1964) *Proceedings of the Lubrication and Wear Convention* (Bournemouth) 23–25 May 1963 (Institution of Mechanical Engineers, London).

Instn of Mechanical Engineers (1964) *Proceedings of the 2nd Convention on Lubrication and Wear* (Eastbourne) 28–30 May 1964 (Institution of Mechanical Engineers, London), **178**, Pt 3N.

Instn of Mechanical Engineers (1965*a*) 'Iron and steel works lubrication'. Proceedings of the Third Annual Meeting of the Lubrication and Wear Group in Association with the Iron and Steel Institute, (Cardiff) 27–29 Oct. 1964. *Proc. Instn mech. Engrs*, **179**, Pt 3D.

Instn of Mechanical Engineers (1965) *Proceedings of the 3rd Convention on Lubrication and Wear* (London) 27–29 May 1965 (Institution of Mechanical Engineers, London) **178**, Pt 3N.

Instn of Mechanical Engineers (1966*a*) *Elastohydrodynamic Lubrication*. Proceedings of a Symposium held in Leeds (Institution of Mechanical Engineers, London), **180**, Pt 3B.

Instn of Mechanical Engineers (1966*b*) *Proceedings of the 4th Convention on Lubrication and Wear* (Scheveningen) 12–14 May 1966 (Institution of Mechanical Engineers, London), **180**, Pt 3K.

Instn of Mechanical Engineering (1966*c*) *Engineering Heritage*, Vol. 2 (Heinemann Educational Books, on behalf of the Institution of Mechanical Engineers, London).

Instn of Mechanical Engineers (1967*a*) 'Journal bearings for reciprocating and turbo machinery'. Symposium (Nottingham) 20–22 Sep. 1966, *Proc. Instn mech. Engrs*, **181**, Pt 3B.

Instn of Mechanical Engineers (1967*b*) *Proceedings of the 5th Convention on Lubrication and Wear* (Plymouth) 3–6 May 1967 (Institution of Mechanical Engineers), **181**, Pt 30.

Instn of Mechanical Engineers (1967*c*) 'Lubrication and wear in living and artificial human joints', *Proc. Instn mech. Engrs*, **181**, Pt 3J (London).

Instn of Mechanical Engineers (1968*a*) 'Experimental methods in tribology', *Proc. Instn mech. Engrs*, **182**, Pt 3G, London.

Instn of Mechanical Engineers (1968*b*) 'Lubrication and wear: fundamentals and application to design', *Proc. Instn mech. Engrs*, **182**, Pt 3A (London).

Instn of Mechanical Engineers (1968*c*) *Tribology Convention 1968*. Proceedings of a Convention (Pitlochry) 15–17 May 1968 (Institution of Mechanical Engineers, London), **182**, Pt 3N.

Instn of Mechanical Engineers (1969) *Tribology Convention 1969*. Proceedings of a Convention (Gothenburg) 28–30 May 1969 (Institution of Mechanical Engineers, London), **183**, Pt 3P.

Instn of Mechanical Engineers (1970) *Tribology Convention 1970*. Proceedings of a Convention (Brighton) 27–29 May 1970 (Institution of Mechanical Engineers, London), **184**, Pt 3L.

Instn of Mechanical Engineers (1971) *Tribology Convention 1971*. Proceedings of a Convention (Douglas) 12–15 May 1971 (Institution of Mechanical Engineers, London).

Instn of Mechanical Engineers (1972*a*) *Elastohydrodynamic Lubrication 1972 Convention*. Proceedings of a Symposium (Leeds) 11–13 Aug. 1972 (Institution of Mechanical Engineers, London).

Instn of Mechanical Engineers (1972*b*) *Externally Pressurized Bearings*. Proceedings of a joint Conference of the Institution of Mechanical and Production Engineers, (London) 17–18 Nov. 1971 (Institution of Mechanical Engineers, London).

Instn of Mechanical Engineers (1972*c*) *Tribology Convention 1972*. Proceedings of a Convention (Keele) 27–28 Sep. 1972 (Institution of Mechanical Engineers, London).

Instn of Mechanical Engineers, et al. (1976*a*) *Piston Ring Scuffing*. Proceedings of a Conference (London) 13–14 May 1975, sponsored by the Institution of Mechanical Engineers, the Department of Industry, the Diesel Engineers and

Users' Association and the British Compressed Air Society (Mechanical Engineering Publs., London).

Instn of Mechanical Engineers (1976*b*) 'Total knee replacement', Proceedings of a Conference (London) 16–18 Sep. 1974 (Institution of Mechanical Engineers, London).

Instn of Mechanical Engineers (1978) *Tribology Convention 1976*. Proceedings of a Convention (Durham) 30 Mar. to 1 Apr. 1976 (Institution of Mechanical Engineers, London).

Instn of Mechanical Engineers (1978) *Joint Replacement in the Upper Limb*. Instn Mech. Engrs, Proceedings of a Conference, (London) 18–20 April 1977.

Ithaca Journal (1067) 'Prof. Ocvirk, engineer, dies at 53', *Ithaca Journal*, 22 May 1967.

Jahrbuch der schiffbautechnischen Gesellschaft (1924) *Jb. schiffbautech. Ges.*, 25 Jahrgang, 37–8.

Jakob, M. (1928) 'Dynamisch und Kinematisch Zähligkeit Zahl', *Z. tech. Phys.*, **9**, 21–2.

Jakobsson, B. and **Floberg, L.** (1957) 'The finite journal bearing, considering vaporization', *Trans. Chalmers Univ. Technol.*, Gothenburg, Sweden, No. 190.

Jakobsson, B. and **Floberg, L.** (1958*a*) 'The partial journal bearing', *Trans. Chalmers Univ. Technol.*, Gothenburg, Sweden, No. 200.

Jakobsson, B. and **Floberg, L.** (1958*b*), 'The rectangular plane pad bearing', *Trans. Chalmers Univ. Technol.*, Gothenburg, Sweden, No. 203.

Jakobsson, B. and **Floberg, L.** (1959) 'The centrally loaded partial journal bearing', *Trans. Chalmers Univ. Technol.*, No. 214.

J. C. H. (1884) 'Reader's letter', *The Engineer*, lvii, 29 Feb., 164.

Jeffcott, H. H. (1919) 'The lateral vibration on loaded shafts in the neighbourhood of a whirling speed – the effect of want of balance', *Phil. Mag.*, **37**, 304–14.

Jenkin, F. and **Ewing, J. A.** (1877) 'On friction between surfaces moving at low speeds', *Phil. Trans. R. Soc., Lond.*, **167**, 509–28.

Jenkins, R. (1930) 'A chapter in the history of the water supply of London: a Thames-side pumping installation and Sir Edward Ford's patent from Cromwell', *Trans. Newcomen Soc.*, ix, 1928–29, 43–51.

Johnson, K. L. (1962) 'Tangential tractions and micro-slip in rolling contact', in *Rolling Contact Phenomena*, Bidwell, J. B. (ed.) (Elsevier, Amsterdam), 6–28.

Johnson, K. L. (1972) 'Rolling resistance of a rigid cylinder on an elastic-plastic surface', *Int. J. mech. Sci.*, **14**, 145–8.

Johnson, K. L. and **Roberts, A. D.** (1974) 'Observations of viscoelastic behaviour of an elastohydrodynamic lubricant film', *Proc. R. Soc. Lond.*, **A324**, 301–13.

Johnson, K. L., Kendall, K. and **Roberts, A. D.** (1971) 'Surface energy and the contact of elastic solids', *Proc. R. Soc., Lond.*, **A324**, 301–13.

Johnson, K. L., Greenwood, J. A. and **Poon, S. Y.** (1972) 'A simple theory of asperity contact in elastohydrodynamic lubrication', *Wear*, **19**, 91–108.

Jones, A. B. (1946) *Analysis of Stresses and Deflections* – 2 vols (New Departure Division, General Motors Corp., Bristol, Connecticut, USA).

Jones, E. S. (1936) 'Joint lubrication', *The Lancet*, 1043.

Jost, H. P. (1965) 'Authors' replies to discussion', *Proc. Instn mech. Engrs*, **179**, Pt 3D, 404–5.

Jost, H. P. (1976*a*) 'Economic impact of tribology, *Proceedings of the 20th Meeting of the Mechanical Failures Prevention Group, National Bureau of Standards*, Special publ. 423, 117–39.

Jost, H. P. (1976*b*) 'Ten years of tribology – the story of the birth and growth of a new and vital science in just one decade', *Ind. Lubr. Tribol.*, May–June.

Joule, J. P. (1884) *Scientific Papers* (Taylor & Francis, London).

Joy, D. (1855) 'Description of a spiral coil packing', *Proc. Instn mech. Engrs*, 171–6.

Kahlert, W. (1948) 'Der Einfluss der Trägheitskräfte, bei der Hydrodynamischen Schmiermittel Theorie', *Ing.-Arch.*, **16**, 341–2.

Kannel, J. W., Bell, J. C. and **Allen, C. M.** (1964) 'Methods for determining pressure distribution in lubricated rolling contact', ASLE Paper No. 64 LC-24, presented to the joint ASME–ASLE International Lubrication Conference, (Washington DC) 13–16 Oct. 1964.

Kapitza, P. L. (1939) 'Influence of friction forces on the stability of high-speed rotors', *J. Phys. USSR*, **1**, No. 29, 551–80.

Kapitza, P. L. (1955) Hydrodynamic theory of lubrication during rolling', *Zh. Tek. Fiz.*, **25**, 747–62.

Karelitz, M. B. (1938) 'Oil pad bearings and driving gears of 200 inch telescope', *Mech. Engng*, **60**, 541.

Karpe, S. A. (1968) 'Influence of rotor metal on bearing failures generally classified as the machining type', *Proc. Instn mech. Engrs*, **182**, Pt 1, 1–18.

Keller, A. G. (1964) *A Theatre of Machines* (Chapman & Hall, London).

Kenedi, R. M. (ed.) (1973) *Perspectives in Biomedical Engineering* (Macmillan, London and Basingstoke).

Kennedy, A. B. W. (1883) 'First report to the Council of the Committee on Friction at High Velocities', *Proc. Instn mech. Engrs*, 660–7.

Kennedy, F. E. and **Cheng, H. S.** (eds) (1976) *Computer-Aided Design of Bearings and Seals* (ASME, New York).

Kettleborough, C. F. (1955) 'An electrolytic tank investigation into stepped thrust bearings', *Proc. Instn mech. Engrs*, **169**, 679–88.

Kettleborough, C. F. (1956) 'The hydrodynamic pocket thrust-bearing', *Proc. Instn mech. Engrs*, **70**, No. 17, 535–44.

Kimball, A. L. (1877) 'A new investigation on the laws of friction', *Am. J. Sci. Arts*, **xiii** (*Silliman's Journal*), 353.

Kimball, A. L. (1924) 'Internal friction theory of shaft whipping', *Gen. Elect. Rev.*, **27**, 244–51; (1925) **28**, 554–8.

Kingsbury, A. (1897) 'Experiments with an air-lubricated journal', *J. Am. Soc. nav. Engrs*, **9**, 267–92.

Kingsbury, A. (1910) 'Thrust bearings', US Patent No. 947242.

Kingsbury, A. (1931) 'On problems in the theory of fluid-film lubrication with an experimental method of solution', *Trans. Am. Soc. mech. Engrs*, **53**, PME-53-5, 59–75.

Kingsbury, A. (1932) 'Optimum conditions in journal bearings', *Trans. Am. Soc. mech. Engrs*, **54**, 123–48.

Kingsbury, A. (1950) 'Development of the Kingsbury thrust bearing', *Mech. Engng*, Dec., 957–62.

Klindt-Jensen, O. (1949) 'Foreign influences in Denmark's early Iron Age', *Acta Archaeol.*, **xx**, 86–108.

Kodnir, D. (1963) *Contact-Hydrodynamic Theory of Lubrication*, (Kuibishervskoe Knizhno Izdat, USSR).

Korovchinskii, M. V. (1953) 'Journal bearing stability', *Trenie i Iznos v Mashinak*, Institut Mashinovedentiia Akademiia Nauk, USSR, **7**, 223–37.

Korovchinskii, M. V. (1956) 'Stability of position of equilibrium of a journal on an oil film', *Trenie i Iznos v Mashinak*, Institut Mashinovedentiia Akademiia Nauk, USSR, **11**, 264–323.

Korovchinskii, M. V. (1960) 'Non-steady motion of the journal in a bearing', *Trenie i Iznos v Mashinak*, Institut Mashinovedentiia, Akademiia Nauk, USSR, **14**, 267–83.

Klein, F. and **Sommerfeld, A.** (1903) *Über die Theorie des Kreisels* (Teubner, Leipzig), **2**, 546.

Kotel'nikov, S. K. (1774) *A Book containing Instruction on the Equilibrium and Movement of Bodies* (St Petersburg).

Kragelskii, I. V. (1965) *Friction and Wear*, translated from the Russian text of the same title (1962), by Ronson, L., in collaboration with Lancaster, J. K. (Butterworths, London).

Kragelskii, I. V., Lubarsky, E. M., et al. (1973) *Friction and Wear in Vacuum*, Machinestrone, Moscow.

Kragelskii, I. V. and **Nepomnyashchii, E. F.** (1965) 'Fatigue wear under elastic contact conditions', *Wear*, **8**, 303–19.

Kragelskii, I. V. and **Shchedrov, V. S.** (1956) *Razvite Nauki o Trenii (Development of the science of Friction)* (Academie Nauk SSR, Moscow).

Kragelskii, I. V. and **Vinogradova, I. A.** (1962) *Coefficients of Friction* (in Russian), (Mashgiz, Moscow).

Kranzberg, M. and **Pursell, Jr., C. W.** (1967) *Technology in Western Civilization*, Vol. I, *The Emergence of Modern Industrial Society. Earliest Times to 1900* (Oxford U.P., New York).

Krieger, R. J., Day, H. J. and **Hughes, W. F.** (1966) 'The MHD hydrostatic thrust bearing–theory and experiments', ASME Paper 66-LUB S-8.

Kruschov, M. M. (1957) 'Resistance of metals to wear by abrasion, as related to hardness'. Instn Mech. Engrs, *Proceedings of the Conference on Lubrication and Wear*, 655–9.

Kruschov, M. M. (1974) 'Principles of abrasive wear', *Wear*, **28**, 69–88.

Kruschov, M. M. and **Babichev, M. A.** (1960) *Investigations Into The Wear of Metals*, (in Russian) (USSR Academy of Sciences Publishing House, Moscow).

Ku, P. M. (ed.) (1968) *Interdisciplinary Approach to Friction and Wear*. Proceedings of a NASA-sponsored Symposium (San Antonio, Texas), 28–30 Nov. 1967, NASA SP-181, (National Aeronautics and Space Administration, Washington, DC).

Ku, P. M. (ed.) (1970) *Interdisciplinary Approach to the Lubrication of Concentrated Contacts*. Proceedings of a NASA-sponsored Symposium, (Troy, NY) 15–17 July 1969, NASA SP-237 (National Aeronautics and Space Administration, Washington, DC).

Ku, P. M. (ed) (1973) *Interdisciplinary Approach to Liquid Lubricant Technology*. Proceedings of a NASA-sponsored Symposium (Cleveland, Ohio) 11–13 Jan. 1972, NASA SP-318 (National Aeronautics and Space Administration, Washington, DC).

Kühnel, R. (1952) *Werkstoffe für Gleitlager* (Springer-Verlag, Berlin).

Kummer, H. W. (1966) 'Unified theory of rubber tire friction', *Engng Res. Bull.*, **B94**, Penn. State University, USA.

Kuzma, D. C. (1963) 'The magnetohydrodynamic journal bearing', *Trans. Am. Soc. mech. Engrs; J. Basic Engng*, **D85**, 424–8.

Kuzma, D. C. (1964) 'Magnetohydrodynamic squeeze films', *Trans. Am. Soc. mech. Engrs; J. Basic Engng*, **D86**, No. 3, 441–4.

Kuzma, D. C. (1965) 'The magnetohydrodynamic parallel plate slider bearing', *Trans. Am. Soc. mech. Engrs; J. Basic Engng*, **D87**, 778–80.

Kuzma, D. C., Maki, E. R. and **Donnelly, R. J.** (1964) 'The magnetohydrodynamic squeeze film', *J. fluid Mech.*, **19**, Pt 3, 395–400.

Ladd, J. (1757) 'Wheel carriages, land rollers, etc.', British Patent No. 714, A.D. 1757, 1–6, accompanied by thirteen figures.

Lamb, Sir H. (1913) 'Osborne Reynolds, 1842–1912', Obituary Notices of Fellows Deceased), *Proc. R. Soc.*, **A lxxxviii**, xv–xxi.

Lancaster, J. K. (1966) 'Solid lubrication', in *Principles of Lubrication*, Cameron, A., (Longmans, London), Ch. 21, 468–97.

Lancaster, J. K. (1972) 'Friction and wear', in *Polymer Science*, Jenkins, A. D. (ed.) (North-Holland, Amsterdam), Ch. 14.

Lancaster, J. K. (1973) 'Basic mechanisms of friction and wear of polymers', *Plast. and Polym.*, Dec., 297–306.

Langmuir, I. (1917) 'The constitution and fundamental properties of solids and liquids: II. Liquids, *J. Am. Chem. Soc.*, **39**, 1848–1906.

Langmuir, I. (1920) 'The mechanism of the surface phenomena of flotation', *Trans. Faraday Soc.*, **15**, Pt 3, 62–74.

Lansdown, A. R. (1973) 'Liquid lubricants–functions and requirements', *Interdisciplinary Approach to Liquid Lubricant Technology* (National Aeronautics and Space Administration, Washington, DC), 1–55.

Larmor, J. (1907) *Sir George Gabriel Stokes. Memoir and Scientific Correspondence*, selected and edited by Larmor, J. (Cambridge U.P., Cambridge), 246–8.

Lasche, O. and **Kieser, W.** (1927) *Materials and Design in Turbo-Generator Plant*, translation by Mellanby and Cooper (Oliver and Boyd, London).

Layard, A. H. (1849) *Nineveh and Its Remains*, Vols I and II (John Murray, London).

Layard, A. H. (1853) *Discoveries in the Ruins of Nineveh and Babylon*, Vols I and II (John Murray, London).

Lea, J. (1853) 'On a new lubricating material', *Proc. Instn mech. Engrs*, 65–9.

Leach, B. (1940) *A Potter's Book* (Faber and Faber, London).

Lecount, P. (1839) *The History of the Railway Connecting London and Birmingham* (London).

Lee, L. H. (1974) 'Advances in polymer friction and wear', in *Polymer Science and Technology* (Plenum Press, New York and London), Vols 5A, 5B.

Leibenson, L. D. (1948) 'N. P. Petrov, hydrodynamic theory of friction', *Selected Works* (Academy of Sciences, USSR, Moscow).

Leibnitz, G. W. (1706) 'Tentamen de natura et remedlie resistenziarum in machines', *Miscellanea Berolinensia. Class. mathem, 1710*, (Jean Boudot, Paris), **1**, 307.

Lenard, P. (1896) *Miscellaneous Papers by Heinrich Hertz – With an Introduction by Professor Philipp Lenard*, authorized English translation by Jones, D. E. and Schott, G. A. (Macmillan, London).

Lerche, G. (1970) 'The ploughs of medieval Denmark', *Tools and Tillage*, **1**, No. 3, 131–49.

Leslie, John (1804) *An Experimental Inquiry Into The Nature and Propagation of Heat*, printed for J. Newman, No. 22. Poultry; also sold by Bell and Bradfute, Edinburgh.

Leupold, J. (1735) *Theatrum Machinarium* (Wolfgang Deer, Leipzig).

Lighthill, M. J. (1968) 'Pressure-forcing of tightly fitting pellets along fluid-filled elastic tubes', *J. fluid Mech.*, **34**, Pt 1, 113–43.

Lilley, S. (1965) *Men, Machines and History* (Lawrence & Wishart, London).

Ling, F. F. (1969) *Surface Mechanics* (ASME, New York).

Ling, F. F. (1973) *Surface Mechanics* (Wiley, New York).

Ling, F. F., Klaus, E. E. and **Fein, R. S.** (eds) (1969) *Boundary Lubrication – An Appraisal of World Literature* (ASME, New York).

Ling, F. F., Whitely, R. L., Ku, P. M. and **Peterson, M. B.** (eds), (1966) *Friction and Lubrication in Metal Processing* (ASME, New York).

Lipson, C. (1967) *Wear Considerations in Design* (Prentice-Hall, Englewood Cliffs, NJ).

Lipson, C. and **Colwell, L. V.** (eds) (1961) *Handbook of Mechanical Wear – Wear, Frettage, Pitting, Cavitation, Corrosion* (University of Michigan Press, Ann Arbor).

Little, W. (1849) 'Materials for lubricating machinery', British Patent No. 12571.

Little, T., Freeman, M. A. R. and **Swanson, S. A. V.** (1969) 'Experience on friction in the human joint', in *Lubrication and Wear in Joints*, Wright, V. (ed.) (Sector Publishing, London), 110–4.

Littler, E. H. and **Purkiss, B. E.** (1970) 'The control of biodegradation of cutting fluid', *Ind. Lubr. Tribol.*, Apr., 108–11.

Liveing, E. (1959) *Pioneers of Petrol. A Centenary History of Carless, Capel & Leonard 1859–1959* (H. F. & G. Witherby).

Lloyd, H. A. (1953) *Suisse Horlog.*, Oct.

Lloyd, S. (1860) 'Description of Aerts' water axlebox', *Proc. Instn mech. Engrs*, 170–87.

Lloyd, T. (1969), 'The hydrodynamic lubrication of piston rings', *Proc. Instn mech. Engrs*, **183**, Pt 3P, 28.

Lloyd, T., Horsnell, R. and **McCallion, H.** (1967a), 'An investigation into the performance of dynamically loaded journal bearings: theory'. Instn Mech. Engrs, *Proceedings, 1966–67*, **181**, Pt 3B, 1–8.

Lloyd, T., Horsnell, R. and **McCallion, H.** (1967b), 'An investigation into the performance of dynamically loaded journal bearings: design study'. Instn Mech. Engrs, *Proceedings, 1966–67*, **181**, Pt 3B, 28–34.

Love, P. P. (1957), 'A review of the papers on bearing materials, solid lubricants, surface treatments, and glands and seals'. Instn Mech. Engrs, *Proceedings of the Conference on Lubrication and Wear* (Oct. 1957), 361–5.

Ludema, K. C. (1977) 'The wear of rubber'. *Proceedings of the 3rd Leeds–Lyon*

Symposium on Tribology, Paper VI(1) (Mechanical Engineering Publs, Bury St Edmunds).

Ludema, K. C. and **Tabor, D.** (1966) 'The friction and visco-elastic properties of polymeric solids', *Wear*, **9**, 329–48.

Lund, J. W. (1965) 'The stability of an elastic rotor in journal bearings with flexible, damped supports', *Trans. Am. Soc. mech. Engrs; J. appl. Mech.*, 1–10.

Lund, J. W. (1975) 'Some unstable whirl phenomena in rotating machinery', *The Shock and Vibration Digest*, **7**, No. 6, June, 5–12.

Lykoudis, P. S. and **Roos, R.** (1970) 'The fluid mechanics of the ureter from a lubrication theory point of view', *J. fluid Mech.*, **43**, Pt 4, 661–74.

MacAdam, R. (1856) 'Ancient water-mills', *Ulster J. Archaeol.*, 1st Series, **iv**, 6–15.

MacConaill, M. A. (1932) 'The function of intra-articular fibrocartilages; with special reference to the knee and radio-ulnar joints, *J. Anat.*, **66**, 210–27.

MacGregor, C. W. (1964) *Handbook of Analytical Design For Wear* (Plenum Press, New York).

McCallion, H. (1973) *Vibration of Linear Mechanical Systems* (Longman, London).

McCallion, H., Lloyd, T. and **Large, J.** (1972) 'The formation of piston profiles', Tribology Convention 1972, *Proc. Instn mech. Engrs*, 55.

MacCurdy, E. (1938) *The Notebooks of Leonardo da Vinci*, 2 vols, (Jonathan Cape, London). See also new edition, (2 vols), 1956.

McCutchen, C. W. (1959) 'Sponge-hydrostatic and weeping bearings', *Nature, Lond.*, **184**, 1284–5.

McKee, S. A. and **McKee, T. R.** (1932) 'Pressure distribution in oil films of journal bearings', *Trans. Am. Soc. mech. Engrs*, **54**, 149–65.

McLaren, K. G. and **Tabor, D.** (1961) 'The frictional properties of lignum vitae', *Br. J. appl. Phys.*

Mabie, H. H. and **Ocvirk, F. W.** (1957) *Mechanisms and Dynamics of Machinery* (Wiley, New York).

Magie, W. F. (1935) 'Heinrich Rudolph Hertz', in *Source Book in Physics* (McGraw-Hill, New York and London), 549.

Magie, W. F. (1963) *A Source Book in Physics* (Harvard U.P., Cambridge, Mass., USA).

Malhotra, R. C. and **Sharma, J. P.** (eds) (1972) *Proceedings of the First World Conference in Industrial Tribology, New Delhi* (Indian Society for Industrial Tribology, New Delhi).

March, C. N. (1970) 'Monitoring, 2–periodic checks', *Tribology*, Aug., 140–4.

Margules, M. (1881) 'Über die Bestimmung der Reibung und Gleitungs Coefficienten aus ebenen Bewegungen einer Flüssigkeit', *Sber. Akad. Wiss. Wein* (1881), **83**, 558–605.

Maroudas, A. (1967) 'Hyaluronic acid films'. Symposium on Lubrication and Wear in Living and Artificial Human Joints, *Proc. Instn mech. Engrs*, **181**, Pt 3J, 122–4.

Marsh, H. (1965) 'The stability of aerodynamic gas bearings', *Mechanical Engineering Science Monograph* No. 2, (Institution of Mechanical Engineers, London), 1–44.

Martin, H. M. (1916) 'Lubrication of gear teeth', *Engineering* (London), **102**, 199.

Martin, F. A. (1970) 'Tilting pad thrust bearings: rapid design aids'. Tribology Convention, *Proc. Instn mech. Engrs*, **184**, Pt 3L, 120–38.

Martin, F. A. (1973) 'Ring and disc fed journal bearings', in *Tribology Handbook*, Neale, M. J. (ed.) (Butterworths, London), Sect. A6.

Martin, F. A. and **Booker, J. F.** (1967) 'Influence of engine inertia forces on minimum film thickness in con-rod big-end bearings', *Proc. Instn mech. Engrs*, (1966–67), **181**, Pt 1, 30–6.

Martin, F. A. and **Garner, D. R.** (1976) 'Plain journal bearings under steady loads: design guidance for safe operation', *Proceedings of the First European Tribology Congress*, '*Tribology and Reliability*', London 1973, 449–63.

Maxwell, C. (1860) 'Illustrations of the dynamical theory of gases', *Phil. Mag.*, Series 4, **14**, 19.

Merchant, M. E. (1968) 'Friction and adhesion', in *Interdisciplinary Approach to Friction and Wear*. Proceedings of a NASA-sponsored Symposium (held in San Antonio, Texas) Washington, DC, 181–265.

Merwin, J. E. and **Johnson, K. L.** (1963) 'An analysis of plastic deformation in rolling contact', *Proc. Instn mech. Engrs*, **177**, 676–90.

The Metallurgist (1932) 'Bearings for heavy service', *The Metallurgist*-supplement to *The Engineer*, 25 Mar., 37–9.

Michell, A. G. M. (1899) 'Elastic stability of long beams under transverse forces', *Phil. Mag.*, Series 5, **48**, 298.

Michell, A. G. M. (1905*a*) 'Improvements in thrust and like bearings', British Patent No. 875.

Michell, A. G. M. (1905*b*) 'The lubrication of plane surfaces', *Z. Math. Phys.*, **52**, Pt 2, 123–37.

Michell, A. G. M. (1929) 'Progress in fluid-film lubrication', *Trans. Am. Soc. mech. Engrs*, **51**, No. 2, Paper MSP-51-21, 153–63.

Michell, A. G. M. (1950) *Lubrication–Its Principles and Practice* (Blackie, London and Glasgow).

Middleton, V., Dudley, B. A. and **McCallion, H.** (1967) 'An investigation into the performance of dynamically loaded journal bearings: experiment', *Proc. Instn mech. Engrs*, (1966–67), **181**, Pt 3B, 9–27.

Miller, F. L. (1937) 'Extreme-pressure lubricants and lubrication'. Instn Mech. Engrs, *Proceedings of the General Discussion on Lubrication and Lubricants*, **2**, 107–17.

Miller, G. M. (1862) 'On a packing for pistons of steam engines and pumps', *Proc. Instn mech. Engrs*.

Milne, A. A. (1965) 'Inertia effects in self-acting bearing lubrication theory', in *Proceedings of the International Symposium on Lubrication and Wear*, Muster, D. and Sternlicht, B. (eds) (McCutchan Publ. Corp., Berkeley, Calif.), 429–527.

Ministry of Technology (1968) *Committee on Tribology Report 1966–67*, Ministry of Technology (HMSO, London).

Ministry of Technology (1969) *Committee on Tribology Report 1967–68*, Ministry of Technology (HMSO, London).

Ministry of Technology (1970) *Committee on Tribology Report 1968–69*, Ministry of Technology (HMSO, London).

Moes, H. and **Bosma, A.** (1971) 'Design charts for optimum bearing configuration: I. The full journal bearing', ASME Paper No. 70-Lub S-I.

Mohan, S. and **Hahn, E. J.** (1974) 'Design of squeeze film damper supports for rigid rotors', *Trans. Am. Soc. mech. Engrs*, 976–82.

Molyneux, P. H. (1967) 'Chemical additives', in *Lubrication and Lubricants*, Braithwaite, E. R. (ed.) (Elsevier, Amsterdam), Ch. 3, 119–65.

Moore, D. F. (1965) 'A review of squeeze-films', *Wear*, **8**, 245–63.

Moore, D. F. (1972) *The Friction and Lubrication of Elastomers*, International Series of Monographs on Materials Science and Technology (Pergamon Press, Oxford), **9**.

Moore, D. F. (1975*a*) *Principles and Applications of Tribology*, International Series in Materials Science and Technology (Pergamon Press, Oxford), **14**.

Moore, D. F. (1975*b*) *The Friction of Pneumatic Tyres* (Elsevier, Amsterdam).

Moreton, D. H. (1964) 'Liquid lubricants', in *Advanced Bearing Technology* (National Aeronautics and Space Administration, Washington, DC), Ch. 7, 175–201.

Morgan, F., Muskat, M. and **Reed, D. W.** (1940) 'Studies in lubrication VI. Electrolytic models of full journal bearings', *J. Appl. Phys.*, **ii** (Feb.), 141–52.

Morgan, M. H. and **Warren, H. L.** (1960) *Translation of 'Vitruvius: The Ten Books on Architecture'* (Dover, New York).

Morgan, V. T. (1958) 'Study of the design criteria for porous metal bearings'. Instn Mech. Engrs, *Proceedings of the Conference on Lubrication and Wear* (1957), 405–8.

Morgan, V. T. (1966) 'Porous metal bearings', in *Principles of Lubrication*, Cameron, A. (Longmans, London), 543–59.

Morgan, V. T. (1969) 'Porous metal bearings', *Powder Metall.*, **12**, No. 24, 426–51.

Mori, H. (1961) 'Theoretical investigation of pressure depression in externally pressurised gas lubricated circular thrust bearings', *Trans. Am. Soc. mech. Engrs*, **83**, No. 2 (June), 201–8.

Morin, A. J. (1833) 'Memoire concernant de nouvelles expériences sur le frottement faites à Metz en 1831', *Mém. Savans Etrang.* (Paris), **iv**, 1–128; *Ann. Min.*, **iv**, 271–321.

Morin, A. J. (1834*a*) 'Nouvelles expériences sur le frottement faites en 1832', *Mém. Savans Etrang.* (Paris), **iv**, 591–696; *Ann. Min.*, **vi**, 73–96.

Morin, A. J. (1834*b*) 'Lettre à M. Arago sur diverses expériences relatives au frottement et au choc des corps', *Ann. Chim.*, **lvi**, 194–8.

Morin, A. J. (1835) 'Nouvelles expériences faites à Metz en 1833 sur le frottement, sur la transmission due movement par le choc, sur le résistance des milieux imparfaits a le pénétration des projectiles, et sur le frottement pendant le choc', *Mém. Savans Etrang.* (Paris), **vi**, (1835), 641–785; *Ann. Min.*, **x**, (1836), 27–56.

Morin, A. J. (1840) 'Sur le tirage des voitures et sur les effects destructeurs qu'elles exercent sur les routes', *Comptes Rendus* (Paris), **x**, 101–4.

Morin, A. J. (1841*a*) *Comptes Rendus* (Paris), **xii**, 211.

Morin, A. J. (1841*b*) 'Note sur la résistance au roulement des corps les uns sur les autres, et sur la reaction élastique des corps qui se compriment réciproquement', *Comptes rendus* (Paris), **xiii**, 1022–3.

Morin, A. J. (1845) 'Note sur le roideur des cordes' *Comptes Rendus* (Paris), **xx**, 228–31.

Morris, J. A. (1967) 'Metallic bearing materials' in *Lubrication and Lubricants*, Braithwaite, E. R. (ed.) (Elsevier, Amsterdam), Ch. 7, 310–75.

Morton, P. G. (1966) 'On the dynamics of large turbo-generator rotors', *Proc. Instn mech. Engrs*, **180**, Pt 1, 295–313.

Morton, P. G. (1968) 'The influence of coupled asymmetric bearings on the motion of a massive flexible rotor', *Proc. Instn mech. Engrs*, **182**, Pt 1, 255.

Morton, P. G. (1970) 'Measurement of the dynamic characteristics of a large sleeve bearing', *ASME Paper* 70-Lub-14.

Morton, P. G. (1972) 'Analysis of rotors supported upon many bearings', *J. mech. Engng Sci.*, **14**, No. 1, 25–33.

Muijderman, E. A. (1966*a*) *Spiral Groove Bearings* (Philips Technical Library, Eindhoven).

Muijderman, E. A. (1966*b*), 'Bearings', *Sci. Am.*, Mar., 60–9.

Mumford, L. (1934) *Technics and Civilization* (Routledge, London).

Munday, A. J. (1974) 'Development of gas bearing technology', *Chart. mech. Engr*, Feb. 78–83.

Musschenbrock, P. (1762) *Introductio ad Philosophiam Natwalem* (Luchtmans), cap **ix**, 145–59.

Muster, D. and **Sternlicht, B.** (eds) (1965) *Proceedings of the International Symposium on Lubrication and Wear* (McCutchan Publ. Corp., Berkeley, Calif.).

Nagaraj, H. S., Sanborn, D. M. and **Winer, W. O.** (1977) 'Effects of load, speed, and surface roughness on sliding EHD contact temperatures', *Trans. Am. Soc. mech. Engrs; J. lubr. Technol.*, **F99**, No. 2, 254–63.

National Bureau of Standards (1976) 'Mechanical failure: definition of the problem', *Proceedings of the 20th Meeting of the Mechanical Failures Prevention Group* (Washington, DC) 8–10 May 1974, NBS Special Publication 423.

National Cyclopedia of American Biography (1971) 'Ocvirk, Fred William', *National Cyclopedia of American Biography*, **53**, (James T. White, New York).

Nature (1951) 'Arnold Sommerfield', obituary notice in *Nature*, **168**, 364.

Navier, C. L. M. H. (1823) 'Mémoire sur les lois du mouvement des fluides', *Mém. Acad. R. Sci.*, **6**, No. 2, 389–440.

Nayak, P. R. (1971) 'Random process model of rough surfaces', *Trans. Am. Soc. mech. Engrs; J. lubr. Technol.*, **93**, 398–407.

Nayak, P. R. (1973*a*) 'Some aspects of surface roughness measurement', *Wear*, **26**, 165–74.

Nayak, P. R. (1973*b*) 'Random process model of rough surfaces in plastic contact', *Wear*, **26**, 305–33.

Naylor, H. (1965) 'Bearings and lubrication', *Chart. mech. Engr, Dec.* 1965 (see also *Engineering Heritage* (1966), Semler, E. G. (ed.), **2**, 86–92.

Neal, P. B. (1970) 'Influence of film inlet conditions on the performance of fluid-film bearings', *J. mech. Engng Sci.*, **12**, No. 2, 153.

Neal, P. B., Wallis, J. F. and **Duncan, J. P.** (1961) 'A university's research for industry', *Engineering*, London, **191**, 434.

Neale, M. J. (1964) 'Selection of bearings', *Engng Mater. Des.*, Jan., 12–5.

Neale, M. J. (1965) 'The design of plain bearings', *Proc. Instn mech. Engrs*, **179**, Pt 3D, 1–11.

Neale, M. J. (1968) 'Selection of bearings', in 'Lubrication and wear: fundamentals and application to design', *Proc. Instn mech. Engrs*, **182**, Pt 3A, 546–56.

Neale, M. J. (1970) *The Problems of Piston Ring and Cylinder Scuffing in Internal Combustion Engines* (Ministry of Technology), Sep.

Neale, M. J. (1971) 'Piston ring scuffing–a broad survey of problems and practice', *Proc. Instn mech. Engrs*, **185**, 21–32.

Neale, M. J. (ed.) (1973) *Tribology Handbook* (Butterworths, London).

Neale, M. J. (1976) 'Tribology past and future', *Tribology Convention 1976*, University of Durham, 30 Mar.–1 Apr. 1976 (Institution of Mechanical Engineers, London), 25–8.

Neave, D. P. C. (1937) 'Copper alloys as bearing materials'. Instn Mech. Engrs, *Proceedings of the General Discussion on Lubrication and Lubricants*, **1**, 208–15.

Needham, J. (1945) *Chinese Science* (Pilot Press, London).

Needham, J. (1965) *Science and Civilization in China*, Vol. 4. *Physics and Physical Technology*, Pt II. 'Mechanical engineering' (Cambridge U.P., Cambridge).

Needs, S. J. (1934) 'Effects of side leakage in 120 degree centrally supported journal bearings', *Trans. Am. Soc. mech. Engrs*, **56**, 721–32. (Also, Discussion, **57**, (1935), 135–8.)

Neugebauer, O. (1938) 'Über eine Methode zur Distanzbestimmung Alexandria–Rom bei Heron', *Vidensk. Selsk. Hist.-filol. Meddn.*, **26**, No. 2, Ch. 35.

New Scientist (1973) 'Aero-engineers build better windmills', *New Scientist*, 4 Jan. 1973.

Newberry, P. E. (1893) 'El-Bersheh–Part I, The tomb of Tehuti-Hetep', *Archaeological Survey of Egypt*, Griffith, F. L. (ed.). Special publ. of the Egypt Exploration Fund, 37, Great Russell Street, London.

Newberry, P. E. (1900) *The Life of Rekhmara–Vezir of Upper Egypt Under Thothmes III and Amenhetop II (circa 1471–1448 B.C.)*, (Archibald Constable, Westminster, London).

Newcomb, T. P. and **Spurr, R. T.** (1967) *Braking of Road Vehicles* (Chapman & Hall, London).

Newkirk, B. L. (1924) 'Shaft whipping', *Gen. Elect. Rev.*, **27**, 169–78.

Newkirk, B. L. (1931) 'Whirling balanced shafts', *Proceedings of the Third International Congress on Applied Mechanics* (Stockholm), **3**, 105–10.

Newkirk, B. L. (1937) 'Instability of oil films and more stable bearings'. Instn Mech. Engrs, *Proceedings of the General Discussion on Lubrication and Lubricants*, **1**, 223–6.

Newkirk, B. L. (1957) 'Journal bearing instability'. Instn Mech. Engrs, *Proceedings of the Conference on Lubrication and Wear* (1957), 179–85.

Newkirk, B. L. and **Taylor, H. D.** (1925) 'Shaft whipping due to oil action in journal bearings', *Gen. Elect. Rev.*, **28**, 559–68.

Newkirk, B. L. and **Grobel, L. P.** (1934) 'Oil-film whirl–a non-whirling bearing', *Trans. Am. Soc. mech. Engrs*, **56**, APM-56-10, 607.

Newman, A. D. (1958) 'Water-lubricated bearings for marine use', *Trans N.E. Coast Inst. Engrs and Shipbuilders*, **74**, Pt 7., 357–70.

Newton, I. (1687) *Philosophiae Naturales Principia Mathematica*, Imprimatur S. Pepys, Reg. Soc. Praeses, 5 Julii, 1686. (See A. Motte's trans and 1729). Revised and supplied with an historical and explanatory appendix by F. Cajori, edited by R. T. Crawford (1934) and published by the University of California Press, Berkeley and Los Angeles (1966).

Newton, I. (1709) *Opticks*, See reprint from 4th edn (G. Bell, London).

Ng, C. W. (1964) 'Fluid dynamic foundation of turbulent lubrication theory *Trans. Am. Soc. Lubr. Engrs*, **11**, 311.

Ng, C. W. and **Pan, C. H. T.** (1965) 'A linearized turbulent lubrication theory', *Trans. Am. Soc. mech. Engrs ; J. Basic Engng*, **87**, No. 4 (Dec.), 675.

Nica, A. (1969) *Theory and Practice of Lubrication Systems*. Translation of *Sisteme de lubrificatie* published by Editura Academiei Republicii Socialiste, Bucharest (1967), by A. Nica and P. A. J. Scott (Publishing House of the Rumanian Academy and Scientific Publ. (GB), Broseley, Salop.).

Nichols, J. (1815) *Literary Anecdotes of the Eighteenth Century* (London), **ix**.

Nightingale, A. F. (ed.) (1975) *Basic Tribology Module : Teaching Tribology to Undergraduate Engineers* (Department of Trade and Industry).

Noack, G. (1976) 'Die Epoche Vogelpohl in der Reibungsforschung', *Schmiertechnik + Tribologie*, **23**, No. 2, 27–32.

North, J. D. (1967) *Isaac Newton* (Oxford U.P., London).

Norton, A. E. (1942) *Lubrication*, Muerger, J. R. (ed.) (McGraw-Hill, New York).

O'Connor, J. J., Boyd, J. and **Avallone, E. A.** (eds) (1968) *Standard Handbook of Lubrication Engineering* (McGraw-Hill, New York).

Ocvirk, F. W. (1952) 'Short bearing approximation for full journal bearings', *NACA, Tech. Note 2808*.

OECD (1969 *Glossary of Terms and Definitions in the Field of Friction, Wear and Lubrication – Tribology* (Research Group on Wear of Engineering Materials, Organisation for Economic Cooperation and Development (OECD), Paris).

Olsson, K. O. (1965) 'Cavitation in dynamically loaded bearings', *Chalmers Tek. Högskolas Handl.*, No. 308, 1–60.

Onions, R. A. and **Archard, J. F.** (1973) 'The contact of surfaces having a random structure', *J. Phys. D: Appl. Phys.*, **6**, 289–304.

Opitz, H. (1968) 'Pressure pad bearings', in 'Lubrication and wear: fundamentals and application to design', *Proc. Instn mech. Engrs*, **182**, Pt 3A, 100–15.

Orloff, P. I. (1947) 'Coefficient of friction, oil flow, and heat balance in a complete cylindrical bearing', *Aeronaut. Engng* (Moscow) 9th year, No. 1, Jan. (1935), 25–56 (trans. S. Reiss, *NACA Tech. Mem.*, 1165, Washington, Oct. 1947).

Ormondroyd, J. (1936) 'Oil pad bearings prove effective for massive equipment', *Mach. Des.*, **8**, 37.

Osterle, J. F. and **Hughes, W. F.** (1958) 'The effect of lubricant inertia in hydrostatic thrust bearing lubrication', *Wear*, **i**, (1957–58), 465–71.

Osterle, F. and **Saibel, E.** (1955) 'On the effect of lubricant inertia in hydrodynamic lubrication', *Z. angew. Math. Phys.*, **6**, No. 4, 334–9.

Osterle, J. F. and **Young, F. J.** (1962) 'On the load capacity of the hydromagnetically lubricated slider bearing', *Wear*, **5**, 227.

Osterle, F., Chou, Y. T. and **Saibel, E. A.** (1958) 'The effect of lubricant inertia on journal bearing lubrication', *J. appl. Mech.*, **24**, **T79**, 494–6; **80**, 420.

Ostwald, W. (1927) *Abhandlungen über die hydrodynamische Theorie der Schmiermittebreibung* (Akad. Verlag, Leipzig) (Ostwald's *Klassiker der exakten Wissenschaft*), No. 218, 1–38.

Palmgren, A. (1945) *Ball and Roller Bearing Engineering* (SKF, Philadelphia).

Palmgren, A. (1954) *Grundlagen der Wälzlagertechnik* (Franckh'sche Verlagshandlung, Stuttgart).

Parent, A. (1704) 'Nouvelle Statique Avec Frotemens et Sons Frotemens ou Regles pour calculer les Frotemens des machines dans l'étal de l'equilibre' *Mémoires de l'Académie Royale*, **A**, (Chez Gerard Kuyper, Amsterdam, 1708). Premier Mémoire, 25 juin 1704, 235–54; Second Mémoire, 2 juillet 1704, 255–69.

Parish, W. F. (1929) 'Lubricants', *Encyclopaedia Britannica*, **14**, edn 14, 453.

Parish, W. F. (1935) 'Three thousand years of progress in the development of machinery and lubricants for the hand crafts', *Mill and Factory*, **16–17**, (Mar.), 27–30, 86–7. See also: Apr., 33–6, 104; May, 81–6, 180, 182, 184, 186, 188;

June, 53–6, 106, 109, 111; July, 41–3, 122; Aug., 45–7; Sep., 75–6, 154–5; Oct., 43–4, 102, 104, 107, 108; Nov., 48–9, 107, 108, 111.

Parker, R. C. (1976) 'The contributions of research in tribology'. *Tribology Convention* 1976, University of Durham, 30 Mar.–1 Apr. 1976 (Institution of Mechanical Engineers, London), 29–34.

Parr, P. H. (1912) *A Theoretical and Experimental Study of Mediate Friction*, by Prof. N. Petroff. Translated from the French by P. H. Parr, *The Engineer*, No. I, 8 Mar., 244–5; No. II, 22 Mar., 294–5; No. III, 5 Apr., 351–2; No. IV, 12 Apr., 376–7.

Parsons, R. H. (1947) *A History of The Institution of Mechanical Engineers, 1847–1947*, (Institution of Mechanical Engineers, London).

Partridge, N. (1835) 'Composition for lubricating bearings of wheels, etc.,' British Patent No. 6945.

Von Pauli, F. A. (1849) 'Uber den Widerstand der Zapfenreibung', *Kunst Gewerb. Polytech. Verein. Königreich Bayern*, **8/9**, 452–69.

Pauling, L. (1951) 'Arnold Sommerfeld: 1868–1951', *Science*, **114**, 12 Oct., 383–4.

Pavelescu, D. (1971) *New Conceptions on the Friction and Wear of Deformable Solids; Calculation and Applications, (Conceptii Noi, Calcul Si Aplicatii in Frecarea Si Uzarea Solidelor Deformabile)* (Editura Academiei Republicii Socialiste Romania, Bucharest, Romania).

Peace, J. B. (1967) 'Solid lubricants', in *Lubrication and Lubricants*, Braithwaite, E. R. (ed.) (Elsevier, Amsterdam), Ch. 2, 67–118.

Pearson, J. (1807) *Observations on the Effects of Various Articles of the Materia Medica in the Cure of the Lues Venerea*, Medical Catalogue, 2nd edn (J. Callow, London).

Peck, H. (1971) 'Ball and parallel roller bearings: design application', in *Topics in Engineering Design* (Pitman, London).

Penn, J. (1856) 'On wood bearings for screw propeller shafts', *Proc. Instn mech. Engrs*, 24–34.

Penn, J. (1858) 'On wood bearings as applied to the shafts of screw steam vessels', *Proc. Instn mech. Engrs*, 81–91.

Petersen, H. (1888) *Vognfundene i Dejbjerg Praestegaardsmose ved Ringkjøping* (Copenhagen).

Petrie, W. M. F. (1883) *The Pyramids and Temples of Gizeh* (Field and Tuer, London).

Petrov, N. P. (1883) 'Friction in machines and the effect of the lubricant' *Inzh. Zh., St-Peterb.*, **1**, 71–140; **2**, 227–79; **3**, 377–436; **4**, 535–64.

Petrov, N. P. (1886) 'Friction in machines', *Izv. St-Peterb. Prakt. Tekh. Inst.*, **5**, 1–438.

Petrov, N. P. (1887) 'Friction in machines', *Inzh. Zh. St-Peterb.*, **1**, 83–145, **2**, 229–88.

Petrov, N. P. (1900) 'Frottement Dans Les Machines', *Mém. l'Acad. Imp. Sci. St-Peterb. Classe Phys.-Math.*, **x**, No. 4, 1–84.

Petrusevich, A. I. (1951) 'Fundamental conclusions from the contact-hydrodynamic theory of lubrication', *Izv. Akad. Nauk. SSSR (OTN)*, **2**, 209.

Phillips, C. W. (1938*a*) 'Some early mould-board ploughs from Jutland', *Proc. Prehist. Soc.*, **4**, 230.

Phillips, C. W. (1938*b*) 'Pebbles from early ploughs in England', *Proc. Prehist. Soc.*, **4**, 338–9.

Piccolpasso, C. (*c.* 1550) *Three Books of the Potters Art*, Italy.

Piggott, S. (ed.) (1961) *The Dawn of Civilization: The First World Survey of Human Cultures in Early Times* (Thames and Hudson, London).

Piggott, S. (1968*a*) 'The earliest wheeled vehicles and the Caucasian evidence', *Proc. Prehist. Soc.*, **xxxiv**, No. 8, 266–318.

Piggott, S. (1968*b*) 'The beginnings of wheeled transport', *Sci. Am.*, **129**, No. 1, 82–90.

Pinegin, S. V. (1976) *Vibrational Friction in Machines and Machine Elements*, Machinestrone, Moscow.

Pinkus, O. (1958) 'Solution of Reynolds equation for finite journal bearings', *Trans. Am. Soc. mech. Engrs*, **80**, 858–64.

Pinkus, O. and **Sternlicht, B.** (1961) *Theory of Hydrodynamic Lubrication* (McGraw-Hill, New York).

Plumier, C. (1701) *L'art de tourner en perfection* (Certe, Lyon).

Pockels, A. (1891) 'Surface tension' (with a translation and introduction by Lord Rayleigh), *Nature*, **xliii**, 12 Mar., 437–9.

Poggendorf, J. C. (1863) *Biographisch Literansch Handwörterbuch*, 208–9 (see also 1893 edn).

Poiseuille, J. L. M. (1846) 'Recherches expérimentales sur la mouvement des liquides dans les tubes de très petits diametres', (1840), **11**, 961–7; 1041–8, (1841), **12**, 112–5. Mémoires presentées par divers savants à l'Académie Royale des Sciences de l'Institut de France. *Sci. Math. Phys.*, (1846), **9**, 433–545.

Poritsky, H. (1953) 'Contribution to the theory of oil whip', *Trans. Am. Soc. mech. Engrs*, **75**, 1153–61.

Porta, G. B. della, (1606) *I tre Libri de Spiritali* (G. I. Carbine, Naples, Italy).

Powell, J. W. (1970) *Design of Aerostatic Bearings* (Machinery Publishing, Brighton).

Powers, J. M., Craig, R. G. and **Ludema, K. C.** (1973) 'Review paper: wear of dental enamel', *Wear*, **23**, 141–52.

Prandtl, L. (1928) 'Ein Gedankenmodell zur kinetischen Theorie der festen Körper', *Z. angew. Math. Mech.*, **8**, 85.

Pratt, G. C. (1967) 'Plastic-based bearings', in *Lubrication and Lubricants*, Braithwaite, E. R. (ed.) (Elsevier, Amsterdam), Ch. 8, 377–426.

Coservatoire National des Arts et Métiers (Private communication 1976) 'Short biography of General Morin', Private communication, Ministère de l'Education Nationale, Conservatoire National des Arts et Métiers, Bibliothéque.

Pugh, B. (1970) *Practical Lubrication – An Introductory Text* (Newnes-Butterworths, London).

Pugh, B. (1973) *Friction and Wear* (Butterworths, London).

Purday, H. F. P. (1949) *Streamline Flow: An Introduction to the Mechanics of Viscous Flow, Film Lubrication, the Flow of Heat by Conduction, and Heat Transfer by Convection* (Constable, London).

Quinn, T. F. J. (1971) *The Application of Modern Physical Techniques to Tribology* (Newnes-Butterworths, London).

Rabinowicz, E. (1965) *Friction and Wear of Materials*, (Wiley, New York).

Rabinowicz, E. and **Tabor, D.** (1951) 'Metallic transfer between sliding metals; an autoradiographic study', *Proc. R. Soc. Lond.*, **A208**, 455–75.

Robinowitz, M. D. and **Hahn, E. J.** (1975) 'Squeeze film bearing supports for flexible rotors', *Proc. Instn mech. Engrs*, 575–80.

Rackham, B. and **Van de Put, A.** (1934) Fascimile edition of Piccolpasso's *Three Books of the Potters Art*, with translation and introduction (Victoria and Albert Museum, London).

Radzimovsky, E. I. (1959) *Lubrication of Bearings – Theoretical Principles and Design* (Ronald Press, New York).

Rahtz, P. and **Sheridan, K.** (1972) 'Fifth report of excavations at Tamworth, Staffs, 1971 – a Saxon water-mill in Bolebridge Street. An Interim note', *Trans. Lichfield and S. Staffs. Archaeol. and Hist. Soc.*, 1971–72, **xiii**, 9–16.

Raimondi, A. A. and **Boyd, J.** (1957) 'An analysis of orifice and capillary-compensated hydrostatic journal bearings', *Lubr. Engng*, **13**, Jan., 28–37.

Raimondi, A. A. and **Boyd, J.** (1958) 'A solution for the finite journal bearing and its application to analysis and design', *Trans. Am. Soc. lubr. Engrs*, **1**, Pt 1, 159–74; Pt 2, 175–94; Pt 3, 194–209.

Ramelli, A. (1588) *Le Diverse Et Artificiose Machine* (Paris, France).

Ramsbottom, J. (1854) 'On an improved piston for steam engines', *Proc. Instn mech. Engrs*, 70–4.

Ramsbottom, J. (1855) 'On the construction of packing rings for pistons', *Proc. Instn mech. Engrs*, 206–8.

Rankine, W. M. M. (1869) 'On the centrifugal force of rotating shafts', *The Engineer*, **27**, 249.

Rankine, W. J. M. (1881) *Miscellaneous Scientific Papers* (Charles Griffin, London).

Lord Rayleigh (John William Strutt) (1877, 1878) *The Theory of Sound*, **1**, (1877); **2**, (1878) (Macmillan, London and New York).

Lord Rayleigh (1884) 'Presidential address to the British Association for the Advancement of Science', Montreal, *Report on the Fifty-Fourth Meeting of the British Association for the Advancement of Science*, 1–23 (John Murray, London).

Lord Rayleigh (1894) *Theory of Sound* (Dover, New York, 1945).

Lord Rayleigh (1899) 'Investigations in capillarity: the size of drops – the liberation of gas from super-saturated solutions – colliding jets – the tension of contaminated water surfaces', *Phil. Mag. J. Sci.*, fifth series, **48**, No. 293 (Oct.), 321–37.

Lord Rayleigh (1917a) 'A simple problem in forced lubrication', *Engineering*, 14 Dec., 617.

Lord Rayleigh (1917b) 'On the pressure developed in a liquid during the collapse of a spherical cavity, *Phil. Mag.*, sixth series, **34**, 94.

Lord Rayleigh (1918a) 'Notes on the theory of lubrication', *Phil. Mag. J. Sci.*, 1–12.

Lord Rayleigh (1918b) 'On the lubricating and other properties of thin oily films', *Phil. Mag. J. Sci.*, sixth series, **35**, No. 206 (Feb.), 157–63.

Lord Rayleigh (Robert John Strutt) (1924) *John William Strutt; Third Baron Rayleigh* (Edward Arnold, London).

Record, S. J. and **Mell, C. D.** (1924) *Timbers of Tropical America* (Yale U.P., New Haven).

Record, S. J. and **Hess, R. W.** (1943) *Timbers of The New World* (Yale U.P., New Haven).

Reddi, M. M. (1969) 'Finite-element solution of the incompressible lubrication problem', *Trans. Am. Soc. mech. Engrs; J. lubr. Technol.*, **F91**, No. 3, (July), 524.

Reddi, M. M. and **Chu, T. Y.** (1970) 'Finite-element solution of the steady-state compressible lubrication problem', *Trans. Am. Soc. mech. Engrs; J. lubri. Technol.*, **F22**, No. 3 (July), 495.

Redwood, B. (1885) 'The Russian petroleum industry', *J. Soc. Chem. Ind.*, **4**, 70.

Redwood, B. (1886) 'On viscosimetry or viscometry', *J. Soc. Chem. Ind.*, **5**, 121–9.

Rennie, G. (1829) 'Experiments on the friction and abrasion of the surfaces of solids', *Phil. Trans. R. Soc. Lond.*, **34**, Pt I, 143–70.

Reti, L. (1967a) 'The Leonardo de Vinci codices in the Biblioteca Nacional of Madrid', *Technol. and Cult.*, **8**, No. 4, 437–45.

Reti, L. (1967b) 'Die wiedergefundenen Leonardo-Manuskripte der Biblioteca Nacional in Madrid', *Technikgeschichte*, **24**, No. 3, 193–225.

Reti, L. (1967c) 'The codex of Juanelo Turriano (1500–1585)', *Technol. and Cult.*, **8**, No. 1, 53–66.

Reti, L. (1967d) 'El artificio de Juanelo en Toledo: su historia y su teenica', *Conferencia Pronunciada* el dia 15 de junio de 1967 en la Casa de la Culture de Toledo (Imprenta de la Diputacion Provincial, Plaza de la Merced, Toledo).

Reti, L. (1968) 'The horizontal water wheels of Juanelo Turriano (ca. 1565): a prelude to Basacle', *XIIᵉ Congrès International d'Histoire des Sciences*, Paris 1968, (Libraire Scientifique et Technique, Paris).

Reti, L. (1971) 'Leonardo on bearings and gears', *Sci. Am.*, **224**, No. 2, 101–10.

Reti, L. (1974) *The Unknown Leonardo*, Reti, L. (ed.) (Hutchinson, London).

Reynolds, J. (1970) *Windmills and Watermills*, Excursion into Architecture Series (Hugh Evelyn, London).

Reynolds, O. (1874) 'On the efficiency of belts or straps as communicators of work', *The Engineer*, 27 Nov.

Reynolds, O. (1875) 'On rolling friction', *Phil. Trans. R. Soc.*, **166**, Pt 1, 155.

Reynolds, O. (1882) 'On the internal cohesion of liquids and the suspension of a column of mercury to a height more than double that of the barometer', *Mem. Manchr Lit. Phil. Soc.*, third series, read 16 Apr. 1878, **7**, 1–19.

Reynolds, O. (1884) 'On the action of lubricants' and 'On the friction of journals', *Report of the Fifty-Fourth Meeting of the British Association for the Advancement of Science*, Montreal, 622, 895 (John Murray, London).

Reynolds, O. (1886) 'On the theory of lubrication and its application to Mr Beauchamp Tower's experiments, including an experimental determination of the viscosity of olive oil', *Phil. Trans. R. Soc.*, **177**, 157–234.

Reynolds, O. (1900–03) *Papers on Mechanical and Physical Subjects*, (*Collected Works*), **I**, (1900); **II** (1901); **III** (1903).

Ricardo, H. R. (1922) 'Some recent research work on the i.c. engine'. *Auto. Engr*, **12**, (Sep.), 265; (Oct.), 299; (Nov.), 329.

Richardson, H. M. (1937) 'Heavy-duty bearings of phenolic plastics'. Instn Mech. Engrs, *Proceedings of the General Discussion on Lubrication and Lubricants*, **1**, 249–59.

Richardson, R. C. D. (1967) 'The wear of metals by hard abrasives', *Wear*, **10**, 291–309.

Richardson, R. C. D. (1968*a*) 'The wear of metals by relatively soft abrasives', *Wear*, **11**, 245–75.

Richardson, R. C. D. (1968*b*) 'Laboratory simulation of abrasive wear, such as that imposed by soil'. Conference on Lubrication and Wear: Fundamentals and Application to Design, Instn Mech. Engrs, *Proceedings, 1967–68*, **182**, Pt 3A, 29–32.

Richardson, R. C. D. (1968*c*) 'The abrasive wear of metals and alloys'. Conference on Lubrication and Wear: Fundamentals and Application to Design, Instn Mech. Engrs, *Proceedings, 1967–68*, **182**, Pt 3A, 410–4.

Richter, I. A. (1952) *Selections from the Notebooks of Leonardo da Vinci* (Oxford U.P., London).

Richter, J. P. (1883) *The Literary Works of Leonardo da Vinci* (S. Low, Marstan, Searle and Rivington, London).

Riddle, J. (1955) *Ball Bearing Maintenance* (University of Oklahoma Press, Norman, Okla.).

Rieger, N. R. (1973) *Flexible Rotor-Bearing System Dynamics–III. Unbalance Response and Balancing of Flexible Rotors in Bearings* (ASME, New York).

Ripple, H. C. (1963) *Cast Bronze Hydrostatic Bearing Design Manual* (Cast Bronze Bearing Institute, Cleveland, Ohio).

Roberts, A. D. (1976) 'Theories of dry rubber friction', *Tribology Int.*, Apr., 75–81.

Robertson, D. (1933) 'Whirling of a journal in a sleeve bearing', *Phil. Mag.* 7, No. 15, 113–30.

Robinson, E. and **Musson, A. E.** (1969) *James Watt and the Steam Revolution* (Adams and Dart, London).

Rockham, B. and **van de Put, A.** (1934) *Three Books of the Potter's Art*, by Cavaliere Cipriano Piccolpasso, The Victoria and Albert Museum, London.

Rodermund, H. (1975) 'Georg Vogelpohl zum Gedenken', *Antriebstechnik*, **14**, No. 4, 190.

Rolt, L. T. C. (1957) *Isambard Kingdom Brunel* (Longman, London).

Rolt, L. T. C. (1967) *The Mechanicals–Progress of a Profession* (Heinemann, London).

Roscoe, T. (1922) *Memoirs of Benvenuto Cellini*, with the notes and observations of G. P. Carpani, translated by Thomas Roscoe, 2 vols., (H. Colbern and Co., London).

Routh, E. Y. (1877) *A Treatise on the Stability of a Given State of Motion; Particularly Steady Motion* (Macmillan, London), 1–108.

Rowe, C. N. (1966) 'Some aspects of the heat of absorption in the function of a boundary lubricant', *Trans. Am. Soc. lubr. Engrs*, **9**, 100–11.

Rowe, G. W. (1966) 'Boundary lubrication', in *The Principles of Lubrication*, Cameron, A., (Longman, London), 451–67.

Rowe, J. (1734) *All Sorts of Wheel-Carriage Improved*, printed for Alexander Lyon under Tom's Coffee House in Russell Street, Covent Garden, London.

Rowe, W. B. and **O'Donoghue, J. P.** (1971) *Design Procedures for Hydrostatic Bearings* (Machinery Publishing, Brighton).

Royal Society (1934) 'William Bate Hardy 1864–1933', *Obituary Notices of Fellows of The Royal Society*, No. 3, Dec., 327–33.

Royal Society (1952) 'A discussion on friction', *Proc. R. Soc. Lond.*, **A212**, 439–519.

Royle, J. K., Howarth, R. B., and **Casely-Hayford, A. L.** (1962) 'Applications of automatic control to pressurized oil-film bearings', *Proc. Instn mech. Engrs*, **176**, No. 22, 532.

Rumpf, A. (1938) 'Reibung und Temperatur Verlauf im Gleitlager', *Forschung*, **9**, 149–58.

Russell, C. A. and **Goodman, D. C.** (eds) (1922) *Science and the Rise of Technology Since 1800* (Open University; John Wright, Bristol).

Sakurai, T. (ed.) (1976) *Proceedings of the JSLE–ASLE International Lubrication Conference*, Tokyo, Japan, 9–11 June 1975 (Elsevier, Amsterdam).

Salomon, G. (1968) 'Plastics and rubbers in machine design', 'Lubrication and wear: fundamentals and application to design', *Proc. Instn mech. Engrs*, **182**, Pt 3A, 341–8.

Sampson, A. (1975) *The Seven Sisters: The Great Oil Companies and the World They Made* (Hodder and Stoughton, London).

Sarkar, A. D. (1976) *Wear of Metals*, International Series in Materials Science and Technology (Pergamon Press, Oxford), **18**.

Sassenfeld, H. and **Walther, A.** (1954) 'Gleitlagerberechnungen', *Verein. Deut. Ing. Forschungsheft*, 441, 28 pp. (a) *Forschung*, **20**, 36.

Savkoor, A. R. (1965) 'On the friction of rubber', *Wear*, **8**, 222–37.

Sawtelle, W. H. (1909) 'Some points regarding the traverse spindle grinders', *Am. Mach.*, **32**, Pt 2, 5 Aug., 248–9.

Scheel, W. S. (1968) *Technical Energy Materials*, in German (*Technische Betriebsstoffe*) (VEB Deutscher Verlag für Grundstoffindustrie, Leipzig).

Seherr-Thoss, (1965) *Die Entwicklung Der Zahnrad-Technik* (Springer-Verlag, Berlin), 410.

Schey, J. A. (1970) *Metal Deformation Processes–Friction and Lubrication* (Marcel Dekker, New York).

Schilling, A. (1968) *Motor Oils and Engine Lubrication* (Scientific Publs (GB), Broseley, Salop).

Schilling, A. (1972) *Automobile Engine Lubrication* (Scientific Publs (GB), Broseley, Salop).

Schmid, E. and **Weber, R.** (1953) *Gleitlager* (Springer-Verlag, Berlin).

Schnadel, G. (1966) 'Ludwig Karl Friedrich Gümbel', *Neue Deutsche Biographie* (Verlag Duncker und Humblot, Berlin), **7**, 258.

Schuster, Sir A. (1921) 'John William Strutt, Baron Rayleigh' (Obituary Notices of Fellows Deceased), *Proc. R. Soc.*, **A98**, i–l.

Scott, D. (1975) 'Debris examination–a prognostic approach to failure prevention', *Wear*, **34**, 15–22.

Segner, J. A. (1758) *De fricto corporum solidorum in motu constitutorum*, Diss, Halle.

Seifert, W. W. and **Westcott, V. C.** (1972) 'A method for the study of wear particles in lubricating oil', *Wear*, **21**, 27–42.

Seireg, A. and **Ezzat, H.** (1969) 'Optimum design of hydrodynamic journal bearings', *Trans. Am. Soc. mech. Engrs; J. lubr. Technol.*, **F91**, 516–23.

Shallamach, A. (1971) 'How does rubber slide?', *Wear*, **17**, 301–12.

Shapiro, W. and **Rumbarger, J. H.** (1972) *Flexible Rotor-Bearing System Dynamics– II. Bearing Influence and Representation in Rotor Dynamics Analysis* (ASME, New York).

Sharp, G. R. (1934) 'The physical work of Charles August de Coulomb, and its influence upon the development of natural philosophy', M.Sc. Thesis, University of London.

Shaw, M. C. (1947) 'Analysis of the parallel surface thrust bearing', *Trans. Am. Soc. mech. Engrs*, **69**, 381–7.

Shaw, M. C. (1971) 'Fundamentals of wear', *Ann. C.I.R.P.*, **xviv**, 533–43.

Shaw, M. C. and **Macks, F.** (1949) *Analysis and Lubrication of Bearings* (McGraw-Hill, New York).

Shimmin, A. N. (1954) *The University of Leeds–The First Half Century* (University Press, Cambridge, for the University of Leeds).

Shukla, J. B. (1963) 'Principles of hydromagnetic lubrication', *J. phys. Soc. Japan*, **18**, No. 7, 1086–8.

Shukla, J. B. (1970) 'The optimum one-dimensional magneto-hydrodynamic slider bearing', *Trans. Am. Soc. mech. Engrs; J. lubr. Technol.* (July), 530–4.

Shukla, J. B. and **Prasad, R.** (1966) 'Squeezing effects in a magnetohydrostatic circular step bearing', *Bull. Nation. Inst. Sci. India*, No. 3, 62–9.

Sibley, L. B., Bell, J. C., Orcutt, F. K. and **Allen, C. M.** (1960) 'A study of the influence of lubricant properties on the performance of aircraft gas turbine engine rolling contact bearings', *WADD Technical Report*, 60–189.

Simons, E. N. (1972) *Metal Wear – A Brief Outline* (Frederick Muller, London).

Singer, C., Holmyard, E. J. and **Hall, A. R.** (eds) (1954) *A History of Technology*. Vol. I: *From Ancient Times to Fall of Ancient Empires* (Oxford U.P., Oxford).

Singer, C., Holmyard, E. J., Hall, A. R. and **Williams, T. I.** (eds) (1956) *A History of Technology*. Vol. II: *The Mediterranean Civilizations and the Middle Ages c. 700 B.C. to c. A.D. 1500* (Oxford U.P., Oxford).

Singer, C., Holmyard, E. J., Hall, A. R. and **Williams, T. I.** (eds) (1957) *A History of Technology*. Vol. III: *From The Renaissance to the Industrial Revolution c. 1500 – c. 1750* (Oxford U.P., Oxford).

Singer, C., Holmyard, E. J., Hall, A. R. and **Williams, T. I.** (1958) *A History of Technology*. Vol. IV: *The Industrial Revolution c. 1750–c. 1850* (Oxford U.P., London).

Singh, S. (1969) 'Conversational techniques in computer-aided bearing design', *NEL Report 429*, presented at the International Conference on Computer-aided Design, Southampton.

Skeat, W. O. (1973) *George Stephenson – The Engineer and his Letters* (Institution of Mechanical Engineers, London).

Slaymaker, R. R. (1955) *Bearing Lubrication Analysis*, (Wiley, New York).

Slotte, K. F. (1881) *Wied. Ann.*, **14**, 13.

Smalley, A. J. and **Malanoski, S. B.** (1976) 'The use of the computer in the design of rotor-bearing systems', in *Computer-aided Design of Bearings and Seals* (ASME, New York), 1–17.

Smeaton, J. (1759) 'Experimental enquiry concerning the natural powers of wind and water to turn mills', *Phil. Trans.*, **51**, 100–32.

Smith, C. S. and **Forbes, R. J.** (1957) 'Metallurgy and assaying', in *A History of Technology*, Vol. III, Singer, C., Holmyard, E. J., Hall, A. R. and Williams, T. I. (eds) (Oxford U.P., London).

Smith, D. M. (1933) 'The motion of a rotor carried by a flexible shaft in flexible bearings', *Proc. R. Soc.*, **A142**, 92–118.

Smith, D. M. (1969) *Journal Bearings in Turbomachinery* (Chapman & Hall, London).

Smith, M. I. and **Fuller, D. D.** (1956) 'Journal bearing operation at super-laminar speeds', *Trans. Am. Soc. mech. Engrs*, **78**, 469.

Snyder, W. T. (1962) 'The magnetohydrodynamic slider bearing', *Trans. Am. Soc. mech. Engrs*, **D84**, 197–204.

Sommerfeld, A. (1904) 'Zur hydrodynamischen Theorie der Schmiermittehreibung', *Z. Math. Phys.*, **50**, 97–155.

Sommerfeld, A. (1919 and 1950) *Autobiographical Sketch Written by Arnold Sommerfeld for the Records of the Vienna Academy.*

Sommerfeld, A. (1921) 'Zur Theorie der Schmiermittebreibung', *Z. Tech. Phys.*, **2**, No. 3, 58–62; No. 4, 89–93.

Southcombe, J. E., Wells, J. H. and **Waters, J. H.** (1937) 'An experimental study of lubrication under conditions of extreme pressure'. Instn Mech. Engrs, *Proceedings of the General Discussion on Lubrication and Lubricants*, **2**, 400–11.

Spurr, R. T. (1965) 'The wear rate of coins', *Wear*, **8**, 487–9.

SSRC (1974) 'The economics of tribology', report on a research project by Nevin, E. T., Egermayer, P. and Evans, J., *SSRC Newsletter*, 23 May, 24–6.

Stansfield, F. M. (1970) *Hydrostatic Bearings For Machine Tools – and Similar Applications*, (Machinery Publishing, Brighton).

Stanton, T. E. (1923a) *Friction* (Longmans, London).

Stanton, T. E. (1923*b*) 'On the characteristics of cylindrical journal lubrication at high values of the eccentricity', *Proc. R. Soc.* **A** cii, 241–55.

Stanton, T. E. (1925) 'The friction of pistons and piston rings', *The Engineer*, **139**, 70, 72.

Steckzen, B. (1957) *Svenska Kullager–Fabrikens Historia* (SKF, Göteborg, Sweden).

Stefan, J. (1874) 'Versuche Über die scheinbare Adhäsion', *Sber. Acad. Wiss. (Math-Naturw.) Wien*, **69**, Pt 2, 713–35.

Steensberg, A. (1936) 'North West European plough types of prehistoric times and the Middle Ages', *Acta Archaeol.*, **vii**, 244–80.

Steensberg, A. (1963) *Indborede Sten Og Traepløkke Som Erstatning For Beslag*, särtryck ur Varbergs Museum–årsbok 1963, 69–76.

Steindorff, G. (1913) *Das Grab des Ti* (Hinrichs, Leipzig).

Stephen, L. (ed.) (1888) *Dictionary of National Biography* (London).

Stephenson, H. (1976) 'Tribology: a moral tale about government aid', *The Times*, Monday, 8.3.1976, London.

Sternlicht, B. (1959) 'Elastic and damping properties of cylindrical journal bearings', *Trans. Am. Soc. mech. Engrs; Basic Engng*, **81**, 101–8; **82**, 249–50 (1960).

Sternlicht, B. (1965) 'Influence of bearings on rotor behaviour', in *Proceedings of International Symposium on Lubrication and Wear*, Muster, D. and Sternlicht, B. (eds) (McCutchan Publ. Corp., Berkeley, Calif.), 529–699.

Sternlicht, B. and **Rieger, N. R.** (1968) 'Rotor stability'. Conference on Lubrication and Wear: Fundamentals and Application to Design, Instn Mech. Engrs, *Proceedings, 1967–68*, **182**, Pt 3A, 82–99.

Stevens Inst. of Technology (1895) 'Robert Henry Thurston, Ph.D., C.E., U.D., Professor of Mechanical Engineering 1871–1885', Morton Memorial to celebrate the 25th anniversary of the Stevens Institute of Technology, 210–7.

Stieber, W. (1933) *Das Schwimmlager* Krayn (Berlin).

Stodola, A. (1925) 'Kritische Wellenstörung infolge der Nachgiebigkeit des Olpolsters im Lager', *Schweiz. Bauzeitung*, **85**, 265–6.

Stodola, A. (1927) *Steam and Gas Turbines*, trans. L. C. Leowenstein (McGraw-Hill, New York).

Stokes, G. G. (1845) 'On the theory of internal friction of fluids in motion', *Trans. Camb. Phil. Soc.*, **8**, 287–319.

Stokhuyzen, F. (1962) *The Dutch Windmill* (Bussum).

Stone, J. M. and **Underwood, A. F.** (1947) 'Load-carrying capacity of journal bearings', *Soc. Auto. Engrs* q. *Trans.* **1**, 56–70.

Stone, M. D. (1956) 'Rolling of thin strip–Part II', *Iron and Steel Engineer*, Dec., **33**, 55–69.

Stone, P. (1968) 'Tribology–new word for an old problem', British Industry Week, 16 Feb.

Stone, W. (1915) *Engineering*, **100**, 26 Nov., 554.

Stone, W. (1921) 'A proposed method for solving some problems in lubrication', *Common. Engr*, **9**, 114–22, 139–49.

Stowers, A. (1963) 'Pumps and water-raising machinery', in *Engineering Heritage* (Heinemann on behalf of the Institution of Mechanical Engineers, London), **I**, 18–25.

Stribeck, R. (1901) 'Kugellager für beliebige Belastungen', *Z. Ver. dt. Ing'.* **45**, No. 3, 73–125.

Stribeck, R. (1902) 'Die Wesentlichen Eigenschaften der Gleit und Rollenlager', *Z. Ver. dt. Ing.*, **46**, No. 38, 1341–8; 1432–8; No. 39, 1463–70.

Stuart, R. (1829) *Historical and Descriptive Anecdotes on Steam Engines* (John Knight and Henry Lacey, London, 1922).

Suh, N. P. (1973) 'The delamination theory of wear', *Wear*, **25**, 111–24.

Summers-Smith, D. (1969) *An Introduction to Tribology in Industry* (Machinery Publishing, Brighton).

Summers-Smith, D. (1970) 'Fundamentals of wear–details of the basic forms of wear and an outline of their effects', *Mach. Prod. Engng*, **116**, No. 2986, Feb., 169–74.

Summers-Smith, D. (1976) '10 years after Jost: the effect on industry', *Tribology*

Convention 1976, University of Durham, 30 Mar.–1 Apr. 1976 (Institution of Mechanical Engineers, London), 20–4.

Sunday Times (1971) 'A nose for lignum vitae', *Sunday Times*, 21 Feb. 1971, 42.

Sung Ying-Hsing (1637) *T'ien-kung k'ai-wu*. See the translation, *Chinese Technology in the Seventeenth Century*, by E-tu Zen Sun and Shiou-Chuan Sun (1966) (Pennsylvania State U.P., University Park and London).

Swift, H. W. (1932) 'The stability of lubricating films in journal bearings', *Proc. Instn civ. Engrs*, **233**, 1931–32, 267–88; Discussion, 289–322.

Swift, H. W. (1935) 'Hydrodynamic principles of journal bearing design', *Proc. Instn mech. Engrs*, **129**, 399–433.

Swift, H. W. (1937a) 'Fluctuating loads in sleeve bearings', *J. Instn civ. Engrs*, **5**, 161–95.

Swift, H. W. (1937b) 'Theory and experiment applied to journal bearing design', Instn Mech. Engrs, *Proceedings of the General Discussion on Lubrication and Lubricants*, **1**, 309–16.

Swift, H. W. (1946) 'Discussion on Fogg's paper on fluid-film lubrication of parallel surface thrust surfaces', *Proc. Instn mech. Engrs*, **155**, 58.

Tabor, D. (1951) *The Hardness of Metals* (Oxford U.P., Oxford).

Tabor, D. (1955) 'The mechanism of rolling friction II. The elastic range', *Proc. R. Soc.*, **A229**, 198–220.

Tabor, D. (1962) 'Introductory remarks', in *Rolling Contact Phenomena*. Proceedings of a Symposium held at the General Motors Research Laboratories, Warren, Michigan, Oct., 1960 (Elsevier, Amsterdam), 1–5.

Tabor, D. (1965) 'Friction and wear', *Proceedings of an International Symposium on Lubrication and Wear*, Muster, D. and Sternlicht, B. (eds) (McCutchan Publ. Corp., Berkeley, Calif.), 701–61.

Tabor, D. (1969) 'Frank Philip Bowden, *Biographical Memoirs of Fellows of the Royal Society*, **15**, Nov., 1–38.

Tabor, D. (1972) 'Friction, lubrication and wear', in *Surface and Colloid Science*, Matijevic, E. (ed.) (Wiley, New York), **5**, 245–312.

Tabor, D. (1975) 'A simplified account of surface topography and the contact between solids', *Wear*, **32**, 269–71.

Tallian, T. E. (1969) 'Progress in rolling contact technology', *Report AL 690007* (SKF) Industries, King of Prussia, Pa., USA).

Tallian, T. E. (1972) 'Theory of partial elastohydrodynamic contacts', *Wear*, **21**, 49–101.

Tallian, T. E. (1976a) 'Prediction of rolling contact fatigue life in contaminated lubricant: Part I–Mathematical model', *Trans. Am. Soc. mech. Engrs; J. lubri. Technol.*, **F98**, No. 2, 251–7.

Tallian, T. E. (1976b) 'Prediction of rolling contact fatigue life in contaminated lubricant: Part II–Experimental', *Trans. Am. Soc. mech. Engrs; J. lubr. Technol.*, **F98**, No. 3, 384–92.

Tamm, P. and **Ulms, W.** (1965) *Lubrication Practice–Introduction for Lubrication Personnel*, in German, (*Schmierpraxis*) (VEB Verlag Technik, Berlin).

Tanner, R. I. (1961) 'The estimation of bearing loads using Galerkin's method', *Int. J. mech. Sci.*, **3**, 13–27.

Tanner, R. I. (1966) 'An alternative mechanism for the lubrication of synovial joints', *Phys. Med. Biol.*, **11**, 119–28.

Tao, L. N. (1959) 'On journal bearings of finite length with variable viscosity', *Trans. Am. Soc. mech. Engrs; J. appl. Mech.*, **26**, 179–83.

Tarbell, I. M. (1904) *The History of The Standard Oil Company* (McClure, Phillips, New York).

Taton, R. (1963) *Ancient and Medieval Science: From Prehistory to* A.D. *1450*, trans. A. J. Pomerans (Thames and Hudson, London).

Taub, A. (1939) 'Cylinder bore wear', *Proc. Instn mech. Engrs*, **141**, 87.

Taylor, C. M. (1972) 'Bearings–which, why and how', paper presented at a conference sponsored by *Engineering*, Cheltenham, 8.1–8.12.

Taylor, C. M. (1973) 'Designing for turbulent lubrication', *Int. J. mech. Sci.*, **15**, 895–904.

Taylor, C. M. (1975) 'The use of the cylinder-plane geometry in the study of steady-state film rupture boundary conditions', *Proceedings of the 1st Leeds–Lyon Symposium on Tribology on Cavitation and Related Phenomena in Lubrication* (Mechanical Engineering Publs, London), 59–62.

Taylor, C. M. and **Dowson, D.** (1974) 'Turbulent lubrication theory–application to design', *Trans. Am. Soc. mech. Engrs; J. lubr. Technol.*, **F96**, No. 1 (Jan.), 36–47.

Taylor, G. I. (1923) 'Stability of a viscous liquid contained between two rotating cylinders', *Phil. Trans.*, **A223**, 289–343.

Taylor, R. A. (1927) *Leonardo The Florentine–A Study in Personality–with a note on the authors work by Gilbert Murrary* (Harpers and Brothers, New York and London).

Teetor, M. O. (1938) 'The reduction of piston ring and cylinder wear', *Trans. S.A.E.*, **33**, 137.

Temperley, H. N. V. (1975) 'The tensile strength of liquids', in 'Cavitation and related phenomena in lubrication', *Proceedings of the 1st Leeds–Lyon Symposium on Tribology* (Mechanical Engineering Publications, London), Paper 1, **ii**, 11–3.

The Engineer (1884) 'What is friction', *The Engineer*, **lvii**, 22 Feb., 149–50.

The Engineer (1932) 'Bearings for heavy service', *The Metallurgist*. Supplement to *The Engineer*, 25 Mar., 8.

The Engineer (1935) 'The late Professor John Goodman', *The Engineer*, 1 Nov., 445.

The Engineer (1944*a*) 'Historic researches: No. I–Friction–Coulomb and Morin's experiments', *The Engineer*, 14 July, 22–3.

The Engineer (1944*b*) 'Historic researches: No. II–Friction–Tower's experiments', *The Engineer*, 21 July, 40–2.

The Engineer (1944*c*) 'Historic researches: No. III–Friction–Reynolds' analysis', *The Engineer*, 28 July, 60–2.

The Times (1905) 'Beauchamp Tower', obituary notice in *The Times*, 7 Jan. 1905.

The Times (1934) Sir William Bate Hardy', obituary notices in *The Times*, 24 and 25 Jan., 1934.

The Times (1935) 'Professor Goodman: Engineering at Leeds University', obituary notice in *The Times*, 30 Oct. 1935.

The Times (1951) 'Arnold Sommerfield', obituary notice in *The Times*, 28 Apr. 1951.

Thomas, R. H. (1968) *Lubrication Mechanics*, a Hearthstone book (Carlton Press, New York).

Thomas, T. R. and **King, M.** (1977) *Surface Topography in Engineering–A State of the Art Review and Bibliography*, Guy, N. G. (ed.) (BHRA Fluid Engineering, UK).

Thomsen, T. C. (1920) *The Practice of Lubrication* (McGraw-Hill, New York).

Thomson, Sir J. J. (1936) *Recollections and Reflections* (Bell, London).

Thurston, R. H. (1878) 'Friction and its laws', *Proc. Am. Ass. Adv. Sci.*, **27**, 61–71.

Thurston, R. H. (1879) 'Friction and lubrication', *The Railroad Gazette*, New York, 212 pp.

Thurston, R. H. (1880) 'President's inaugural address', *Trans. Am. Soc. mech. Engrs*, **1**, 1–16.

Thurston, R. H. (1885) *A Treatise on Friction and Lost Work in Machinery and Millwork* (Wiley, New York), 7th edn, 1903.

Ting, L. L. and **Mayer, (Jr) J. E.** (1973) 'Piston ring lubrication and cylinder bore wear analysis', Pt I–Theory, ASME Paper No. 73-Lub-25; Pt II–Theory verification, ASME Paper No. 73-Lub-27.

Tipei, N. (1956) 'La stabilité du mouvement dans les paliers à charge dynamique', *Rev. Méc. Appl., Acad. RPR*, **1**, 115–22.

Tipei, N. (1957) *Hidro-Aerodinamica Lubrificatiei*. English trans. *Theory of Lubrication–With Applications to Liquid- and Gas-Film Lubrication* (1962) Gross, W. A. (ed.) (Stanford U.P., Stanford, Calif.).

Tipei, N., Constantinescu, V. N., Nica, A. L. and **Bitä, O.** (1961) *Lagare cu Alunecare*

(*Sliding Bearings–Computation–Design–Lubrication*) (Editura Academiei Republicii Populare Romine, Bucharest).

Tipei, N. and **Pascal, H.** (1966) 'Micro- and macrogeometry effects on pressure in three-dimensional lubrication', *Rev. Ro. Sci. Technol. mech. Appl.*, **ii**, No. 3, 699.

Tokaty, G. A. (1971) *A History and Philosophy of Fluid Mechanics* (G. T. Foulis, Henley-on-Thames, Oxon.).

Tomlinson, C. (1863) 'On the action of oils in arresting the motions of camphor on water', *Phil. Mag.*, **xxvi**, 187–90.

Tomlinson, C. (1867) 'On the so-called "inactive" conditions of solids', *Phil. Mag.*, **xlix**, 305.

Tomlinson, C. (1875) 'On the action of solids in liberating gas from solution', *Phil. Mag.*, **xlix**, 302–7.

Tomlinson, G. A. (1929) 'A molecular theory of friction', *Phil. Mag.*, **7**, 905–39.

Tondl, A. (1965) *Some Problems of Rotor Dynamics* (Publishing House of the Czechoslovak Academy of Sciences, Prague).

Tower, B. (1875) 'On a method of obtaining motive power from wave motion', *Instn nav. Archit.*, 1–12.

Tower, B. (1883) 'First report on friction experiments (friction of lubricated bearings)', *Proc. Instn mech. Engrs*, Nov. 1883, 632–59. (see also 'Adjourned discussion', *Proc. Instn mech. Engrs*, Jan. 1884, 29–35).

Tower, B. (1885) 'Second report on friction experiments (experiments on the oil pressure in a bearing), *Proc. Instn mech. Engrs*, Jan. 1885, 58–70.

Tower, B. (1889) 'An apparatus for providing a steady platform for guns, etc. at sea', *Instn nav. Archit.*, 1–14.

Town, H. C. (1973) 'A history of the boring machine', *Chart. mech. Engr*, July, 73–8.

Treadgold, T. (1838) *The Steam Engine* (John Weale, London).

Tresca, M. (1880) 'Discours prononcés aux funérailles de M. Morin, membre de l'Académie des Sciences, Directeur du Conservatoire des Arts et Metiers', *Institute de France, Académie de Science*, 1–9.

Trumpler, P. R. (1966) *Design of Film Bearings* (Macmillan, New York).

Tsukizoe, T. and **Hisakado, T.** (1972*a*) 'On the mechanism of heat transfer between metal surfaces in contact: Part 1–Heat transfer' *Jap. Res.*, **1**, 104–112.

Tsukizoe, T. and **Hisakado, T.** (1972*b*) 'On the mechanism of heat transfer between metal surfaces in contact: Part 2–Heat transfer' *Jap. Res.*, **1**, 23–31.

Tugendhat, C. and **Hamilton, A.** (1968) *Oil–The Biggest Business* (Eyre Methuen, London).

Turchina, V., Sanborn, D. M. and **Winer, W. O.** (1974) 'Temperature measurements in sliding elastohydrodynamic point contacts', *Trans. Am. Soc. mech. Engrs; J. lubr. Technol.*, **F96**, No. 3, 464–71.

Turner, E. (1832) 'Sources of caloric', in *Elements of Chemistry*, 4th American edn (Grigg and Elliot, Philadelphia), 68.

Tzeng, S. T. and **Saibel, E.** (1967) 'On the effects of surface roughness in the hydrodynamic lubrication theory of a short journal bearing', *Wear*, **10**, 179–84.

Ucelli, G. (1950) *Le Navi Di Nemi* (La Liberia Dello Stato, Roma).

University of Manchester (1970) *Osborne Reynolds and Engineering Science Today.* (Papers presented at the Osbourne Reynolds Centenary Symposium, University of Manchester, Sep. 1968 (Manchester U.P.).

University of New Hampshire (1950) 'Kingsbury Hall–named for Professor Albert Kingsbury 1863–1943', University of New Hampshire, dedicated 14 Oct.

Unsworth, A., Dowson, D. and **Wright, V.** (1975) 'The frictional behaviour of human synovial joints. Part I–Natural joints', *Trans. Am. Soc. mech. Engrs; J. lubr. Technol.*, **F97**, No. 3, 369–76.

Uppal, A. H. and **Probert, S. D.** (1972) 'Deformation of single and multiple asperities on metal surfaces, *Wear*, **20**, 381–400.

Uppal, A. H., Probert, S. D. and **Thomas, T. R.** (1972) 'The real area of contact between a rough and a flat surface', *Wear*, **22**, 163–83.

Van Musschenbroek, P. (1762) 'Cours de physique experimentale et mathematique', in Vol. I, Ch. ix. Du Frottement des Machines (Leyden).

Van Natrus, L., Polly, J. and **Van Vuuren, C.** (1734 and 1736) *Groot Volkomen Moolenbock*, 2 vols (Amsterdam).

Van Rossum, J. J. (1958) 'Viscous lifting and drainage of liquids', *Appl. Sci. Res.*, **A7**, 121.

Varlo, C. (1772) *Reflections Upon Friction With a Plan of the new Machine for taking it off in Wheel-carriages, Windlasses of Ships etc., Together With Metal proper for the machine, the full Directions for making it. To which is annexed, Stonehenge, one of the wonders of the world, unriddled* (printed for the author MDCCLXXII, London), 29 pp.

Vaughan, P. (1794) 'Axle trees, arms, and boxes', British Patent No. 2006 of A.D. 1794, 1–2, accompanied by eleven diagrams on one sheet.

Velichko, K. I. (1915) *Vojennaja Enciklopedija*, **xviii**, 394 (I.D. Sytinid, St Petersburg).

Vince, S. (1785) 'On the motion of bodies affected by friction', *Phil. Trans. R. Soc. Lond.*, **75**, Pt I, 165–89.

Vogelpohl, G. (1937) 'Bietrage zur Kenntris der Gelitlagerreibung', *Verein. Deut. Ing.*, Forschungsheft, No. 386.

Vogelpohl, G. (1940) 'Die Geschientliche Entwicklung unseres Wissens ueber Reibung und Schmierung', *Oel u Kohle*, **36**, Nos. 9 and 13, 1 Mar., 89–93; 1 Apr., 129–34.

Vogelpohl, G. (1949) 'Der Uebergang der Reibungswaerme von Lagern aus der Schmierschicht in die Gleitflaechen', *Verein. Deut. Ing.*, Forschungsheft No. 425 (supplement to *Forschung auf dem Gebiete des Ingenieurwesens*), **16**, July–Aug.

Vogelpohl, G. (1951) 'Lubrication problems', in *Proceedings of the Third World Petroleum Congress* (E. J. Brill, The Hague), Sect. 7, 298–303 (see also *Sci. Lubr.*, **3**, 9, Sep.).

Vogelpohl, G. (1958) *Betriebsichere Gleitlager. Berechnungsverfahren für Konstruction und Betrieb* (Springer-Verlag, Berlin).

Vogelpohl, G. (1965) 'Thermal effects and elasto-kinetics in self-acting bearing lubrication', in *Proceedings of International Symposium on Lubrication and Wear*, Muster, D. and Sternlicht, B. (eds) (McCutchan Publ. Corp., Berkeley, Calif.), 766–822.

Vogelpohl, G. (1969) 'Uber die Ursachen der unzureichenden Bewertung von Schmierungsfragen im vorigen Jahrhundert', *Schmiertechnik + Tribologie*, **16**, No. 5, 191–200.

Voice, E. H. (1950) 'The history of the manufacture of pencils', the Newcomen Society for the study of the History of Engineering and Technology, *Trans., Newcomen Soc.*, **xxvii** (1949–50, 1950–51), 131–41.

Wada, S., Hayashi, H. and **Migita, M.** (1971) 'Application of finite-element method to hydrodynamic lubrication problems (Part I: Infinite width bearings)', *Bull. Japan Soc. mech. Engrs*, **14**, No. 77, Nov. 1222.

Wada, S. and **Hayashi, H.** (1971) 'Application of finite-element method to hydrodynamic lubrication problems (Part 2: Finite width bearings)', *Bull. Japan Soc. mech. Engrs*, **14**, No. 77, Nov., 1234.

Wailes, R. (1947) 'Windshafts', The Newcomen Society for the Study of the History of Engineering and Technology, *Trans. Newcomen Soc.*, **xxvi** (1947–48 and 1949–50), 1–11.

Wailes, R. (1954) *The English Windmill* (Routledge & Kegan Paul, London).

Wailes, R. (1957) 'Windmills', in *A History of Technology*, Volume III (1957), by Singer, C., Holmyard, E. J., Hall, A. R. and Williams, T. I. (Oxford U.P., London), 89–109.

Wailes, R. (1963) 'Windmills–their rise and decline', in *Engineering Heritage* (Heinemann on behalf of the Institution of Mechanical Engineers, Norwich, UK), **1**, 10–6.

Wakelin, R. J. (1974) 'Tribology: the friction, lubrication and wear of moving parts', *Ann. Rev. Mater. Sci.*, **4**, 221–53.

Wakuri, Y., Tsuge, M., Yamashita, M. and **Hase, M.** (1970) 'A study of the oil loss past a series of piston rings', *Bull. Japan Soc. mech. Engrs*, **13**, No. 55, 150.

Walker, P. S., Dowson, D., Longfield, M. D. and **Wright, V.** (1968) 'Boosted lubrication in synovial joints by fluid entrapment and enrichment', *Ann. rheum. Dis.*, **27**, No. 6, 512–20.

Wallace, J. F. (1960) 'Death of Professor Swift', personal notes.

Walowit, J. A. and **Anno, J. N.** (1975) *Modern Developments in Lubrication Mechanics* (Applied Science Publishers, London).

Warner, P. C. (1963) 'Static and dynamic properties of partial journal bearings', *Trans. Am. Soc. mech. Engrs; J. Basic Engng*, **D85**, 247.

Wehe, R. L. (1967) 'Memorial talk on Fred W. Ocvirk at ASME Lubrication Luncheon'.

Wellauer, E. J. (1967) 'AGMA experience in establishing coordinated gear rating standards', invited paper presented to the Semi-International Symposium, Japan Society of Mechanical Engineers.

Wellington, A. M. (1884) 'Experiments with new apparatus on journal friction at low velocities', *Trans. Am. Soc. civ. Engrs*, **ccxcv**, 409–75.

Welter, G. and **Brasch, W.** (1937) 'Tests on plain bearings with a new method of lubrication under very high pressure'. Instn Mech. Engrs, *Proceedings of the General Discussion on Lubrication and Lubricants*, **1**, 330–5.

Wepfer, W. M. and **Cattabiani, E. J.** (1955) 'Water-lubricated-bearing-development programme', *Mech. Engng*, **77**, 413–8, 923.

West, G. H. and **Rathie, G. J.** (1972) 'Some important past and present developments in tribology', *Wear*, **20**, 227–33.

Wheeler, M. (1961) 'Ancient India – the civilization of a sub-continent', in *The Dawn of Civilization: The First World Survey of Human Cultures in Early Times*, Piggott, S. (ed.) (Thames and Hudson, London), Ch. viii, 229–52.

Wheeler, M. (1966) *Civilizations of the Indus Valley and Beyond*, (Thames and Hudson, London).

Whipple, R. T. P. (1949) 'Herringbone pattern thrust bearings, *AERE Tech. Mem. 29*.

Whipple, R. T. P. (1951) 'Theory of the spiral grooved thrust bearing with liquid or gas lubricant', **AERE Tech. Rep.,** *622*.

Whitcher, J., Pickford, M. and **Whitbourn, J.** (1822) 'Wheels for carriages etc.', British Patent No. 4709, 1–7, accompanied by six figures.

Whitehouse, D. J. and **Archard, J. F.** (1970) 'The properties of random surfaces of significance in their contact', *Proc. R. Soc. Lond.*, **A316**, 97–121.

Whitt, F. R. (1969) 'Bicycle bearings–1869–1969'.

Whitt, F. R. (1970) 'Bicycle bearings', *Bicycling*, Sep., 22–3.

Who Was Who (1960) 'Swift, Herbert Walker', *Who Was Who*, 1951–60, 1061.

Wilcock, D. F. (1949) 'Turbulence in high-speed journal bearings', *Trans. Am. Soc. mech. Engrs*, **72**, 825–34.

Wilcock, D. F. (1955) 'The hydrodynamic pocket thrust bearing', *Trans. Am. Soc. mech. Engrs*, **77**, 311–9.

Wilcock, D. F. (1957) 'Predicting sleeve bearing performance'. Instn Mech. Engrs, *Proceedings of the Conference on Lubrication and Wear* (1957), 82–92.

Wilcock, D. F. (1974a) 'Tribology–then, now and when', *Mech. Engng*, Mar., 22–7.

Wilcock, D. F. (1974b) 'Turbulent lubrication–its genesis and role in modern design', *Trans. Am. Soc. mech. Engrs; J. lubr. Technol.*, **F96**, No. 1, Jan., 2–6.

Wilcock, D. F. and **Booser, E. R.** (1957) *Bearing Design and Application* (McGraw-Hill, New York).

Williams, A. (1966) 'Heat transfer through metal to metal joints', *Proceedings of the Third International Heat Transfer Conference*, Chicago, Illinois, 7–12 Aug., paper No. 125, American Institute of Chemical Engineers, **iv**, 109–17.

Williams, A. (1968) 'Heat transfer across metallic joints', *Mech. chem. Engng Trans. Instn Engrs, Aust.*, Nov., 247–54.

Williams, A. (1971) 'Heat flow across stacks of steel laminations', *J. mech. Engng Sci.*, **13**, No. 3, 217–23.

Williams, A. (1972) 'Experiments on the flow of heat across stacks of steel laminations', *J. mech. Engng Sci.*, **14**, No. 2, 151–4.

Williams, A. (1974) 'Heat flow through single points of metallic contacts of simple shapes', in *Thermophysics and Heat Transfer Conference* (Boston, Massachusetts) 1974, American Institute of Aeronautics and Astronautics/American Society of Mechanical Engineers, AIAA Paper No. 74-692, 1–7.

Williams, C. G. (1936) 'Cylinder wear in gasoline engines', *Trans. S.A.E.*, **38**, No. 5, May, 191.

Williams, C. G. (1937) 'Temperature rise in bearings of automobile engines and its influence on durability'. Instn Mech. Engrs, *Proceedings of the General Discussion on Lubrication and Lubricants*, **1**, 336–42.

Williams, C. G. and **Young, H. A.** (1939) 'Piston ring blow-by in high speed petrol engines', *Engineering*, **147**: 9 June, 693; 16 June, 723.

Williams, M. L., Landel, R. F. and **Ferry, J. D.** (1955) 'Temperature dependence on relaxation mechanisms in amorphous polymers', *J. Am. Chem. Soc.*, **77**, July, 3701–7.

Williamson, J. B. P. (1967) 'Topography of solid surfaces', *Interdisciplinary Approach to Friction and Wear*. Proceedings of a NASA-sponsored Symposium, San Antonio, Texas, 85–179.

Wilson, R. E. and **Barnard, D. P. 4th** (1922) 'The mechanism of lubrication', *Trans. S.A.E.*, **17**, Pt I, 203–38.

Wilson, R. W. (1972) 'Metallic bearing materials', paper presented to the Materials Science Club Frictional Materials Conference, Brunel University, Apr. 1972.

Wilson, R. W. (1975) 'Cavitation damage in plain bearings', *Proceedings of the 1st Leeds–Lyon Symposium on Tribology on Cavitation and Related Phenomenon in Lubrication* (Mechanical Engineering Publs, London), 177–84.

Wolf, A. (1935) *A History of Science, Technology and Philosophy in the 16th and 17th Centuries* (Allen & Unwin, London).

Wood, N. (1838) *A Practical Treatise on Rail-Roads, and Interior Communications in General* (Longman, Orme, Brown, Green, and Longmans, London).

Woodbridge, E. (1959) 'The United Kingdom motor vehicle industry', *Inst. Petrol. Rev.*, Jan., 1–3.

Woodyard, D. F. (1973) 'Tribology comes to town – the First European Tribology Congress', *Chart. mech. Enger*, Nov., 1–7.

Woolley, C. L. (1930) 'Excavations at Ur: 1929–30', *Antiq. J.*, **x**, No. 4, Oct.

Wright, K. H. R. (1969) 'The abrasive wear resistance of human dental tissues', *Wear*, **14**, 263–84.

Wright, K. H. R. (1973) 'Mechanisms of wear', in *Tribology Handbook*, Neale, M. J. (ed.) (Butterworths, London), Sect. F6.

Wright, V. (ed.) (1969) *Lubrication and Wear in Joints* (Sector Publishing, London).

Wurzel, (1887) *Neue Theorie der Reibung* (Verlag Leopold Voss, Leipzig), 187.

Yonge, J. G. and **Barclay, A.** (1766) 'Sugar mills', British Patent No. 862, A.D. 1766, 1–2.

Yorkshire Post (1935) 'Professor John Goodman', obituary notice in the *Yorkshire Post*, 29 Oct. 1935.

Young, J. (1850) 'Treating bituminous coals to obtain paraffine and oil containing paraffine therefrom', British Patent No. 13292 (enrolled on 17 Apr. 1851).

Young, T. (1805) 'An essay on the cohesion of fluids', *Phil. Trans. R. Soc.*, Pt I, 65–87.

Zeitschrift des Vereins deutscher Ingenieure (1923) *Z. Ver. dt. Ing.*, **67**, 762.

Zienkiewicz, O. C. (1957) 'Temperature distribution within lubricating films between parallel bearing surfaces and its effect upon the pressure developed'. Instn Mech. Engrs, *Proceedings of the Conference on Lubrication and Wear* (1957), 135–41.

Zienkiewicz, O. C. (1970) *The Finite Element Method in Engineering* (McGraw-Hill, London).

Zonca, V. (1607) *Teatro Nuovo di Machine et Edificii* (Bertelli, Padua).

Subject Index

Name Index